# CADMIUM TOXICITY AND TOLERANCE IN PLANTS

# CADMIUM TOXICITY AND TOLERANCE IN PLANTS

## From Physiology to Remediation

Edited by

MIRZA HASANUZZAMAN

MAJETI NARASIMHA VARA PRASAD

MASAYUKI FUJITA

ELSEVIER

ACADEMIC PRESS

An imprint of Elsevier

Academic Press is an imprint of Elsevier
125 London Wall, London EC2Y 5AS, United Kingdom
525 B Street, Suite 1650, San Diego, CA 92101, United States
50 Hampshire Street, 5th Floor, Cambridge, MA 02139, United States
The Boulevard, Langford Lane, Kidlington, Oxford OX5 1GB, United Kingdom



**Library of Congress Cataloging-in-Publication Data**
A catalog record for this book is available from the Library of Congress

**British Library Cataloguing-in-Publication Data**
A catalogue record for this book is available from the British Library

ISBN 978-0-12-814864-8

For information on all Academic Press publications visit our website at https://www.elsevier.com/books-and-journals



*Publisher:* John Fedor
*Acquisition Editor:* Nancy Maragioglio
*Editorial Project Manager:* Ruby Smith
*Production Project Manager:* Swapna Srinivasan
*Cover Designer:* Mark Rogers

Typeset by SPi Global, India

# CONTENTS

# CONTRIBUTORS

**Zaigham Abbas**
Ministry of Climate Change, Islamabad, Pakistan

**Ghassen Abid**
Laboratory of Legumes, Centre Biotechnology Borj Cedria, University Tunis El Manar, Hammam Lif, Tunisia

**Javed Ahmad**
Department of Biotechnology, Jamia Millia Islamia, New Delhi, India

**Zahoor Ahmad**
Cholistan Institute of Desert Studies (CIDS), The Islamia University of Bahawalpur (IUB), Bahawalpur, Pakistan

**Temoor Ahmed**
Department of Bioinformatics and Biotechnology, Government College University, Faisalabad, Pakistan

**Toqeer Ahmed**
Centre for Climate Research and Development (CCRD), COMSATS, Islamabad, Pakistan

**Muhammad Sohail Akram**
Department of Botany, Faculty of Life Sciences, Government College University, Faisalabad, Pakistan

**Shafaqat Ali**
Department of Environmental Sciences and Engineering, Government College University, Faisalabad, Pakistan

**Qasim Ali**
Department of Botany, Government College University, Faisalabad, Pakistan

**Arlene Asthana Ali**
Department of Biotechnology, Jamia Millia Islamia, New Delhi, India

**Emilia L. Apostolova**
Institute of Biophysics and Biomedical Engineering, Bulgarian Academy of Sciences, Sofia, Bulgaria

**Mahnoor Asif**
Department of Environmental Sciences, COMSATS, Abbottabad, Pakistan

**Souhir Abdelkrim Ayed**
Laboratory of Legumes, Centre Biotechnology Borj Cedria, University Tunis El Manar, Hammam Lif, Tunisia

**Muhammad Ashar Ayub**
Institute of Soil and Environmental Sciences, University of Agriculture, Faisalabad, Pakistan

**Muhammad Azeem**
Department of Botany, Faculty of Life Sciences, Government College University, Faisalabad, Pakistan

**Mohd Affan Baig**
Department of Biotechnology, Jamia Millia Islamia, New Delhi, India

**Aditi Shreeya Bali**
Department of Botany, M.C.M. DAV College for Women, Chandigarh, India

**Nazneen Bangash**
Department of Bio Sciences, COMSATS, Islamabad, Pakistan

**María Patricia Benavides**
Universidad de Buenos Aires—Consejo Nacional de Investigaciones Cientificas y Tecnicas, Instituto de Química y Fisicoquímica Biológicas "Profesor Alejandro C. Paladini" (IQUIFIB), Facultad de Farmacia y Bioquímica, Buenos Aires, Argentina

**Renu Bhardwaj**
Department of Botanical and Environmental Sciences, Guru Nanak Dev University, Amritsar, India

**Vyomendra Chaturvedi**
Institute of Environment and Sustainable Development (IESD), Banaras Hindu University, Varanasi, India

**Manel Chiboub**
Laboratory of Legumes, Centre Biotechnology Borj Cedria, University Tunis El Manar, Hammam Lif, Tunisia

**Anelia G. Dobrikova**
Institute of Biophysics and Biomedical Engineering, Bulgarian Academy of Sciences, Sofia, Bulgaria

**M. Farooq Qayyum**
Department of Soil Science, Faculty of Agricultural Sciences & Technology, Bahaudin Zakariya University Multan, Multan, Pakistan

**Imen Challougui Fatnassi**
Laboratory of Legumes, Centre Biotechnology Borj Cedria, University Tunis El Manar, Hammam Lif, Tunisia

**Susana Mabel Gallego**
Universidad de Buenos Aires—Consejo Nacional de Investigaciones Cientificas y Tecnicas, Instituto de Química y Fisicoquímica Biológicas "Profesor Alejandro C. Paladini" (IQUIFIB), Facultad de Farmacia y Bioquímica, Buenos Aires, Argentina

**Sílvia C. Gonçalves**
MARE—Marine and Environmental Sciences Centre, ESTM, Polytechnic Institute of Leiria, Peniche; MARE—Marine and Environmental Sciences Centre, Department of Life Sciences, Faculty of Sciences and Technology, University of Coimbra, Coimbra, Portugal

**Anna Grobelak**
Department of Environmental Engineering, Faculty of Infrastructure and Environment, Czestochowa University of Technology, Czestochowa, Poland

**Bin Guo**
Institute of Environment, Resource, Soil and Fertilizer, Zhejiang Academy of Agricultural Sciences, Hangzhou, People's Republic of China

**Inamul Haq**
Department of Biotechnology, Jamia Millia Islamia, New Delhi, India

**Malik Tahir Hayat**
Department of Environmental Sciences, COMSATS, Abbottabad, Pakistan

**Afzal Hussain**
Department of Environmental Sciences and Engineering, Government College University, Faisalabad, Pakistan

**Iqbal Hussain**
Department of Botany, Government College University, Faisalabad, Pakistan

**Syed Makhdoom Hussain**
Department of Zoology, Government College University, Faisalabad, Pakistan

**Muhammad Iqbal**
Department of Botany, Jamia Hamdard, Hamdard Nagar, New Delhi, India

**Nadeem Iqbal**
Department of Environmental Sciences, COMSATS, Abbottabad, Pakistan

**M. Irfan Qureshi**
Department of Biotechnology, Jamia Millia Islamia, New Delhi, India

**Muhammad Tariq Javed**
Department of Botany, Faculty of Life Sciences, Government College University, Faisalabad, Pakistan

**Marta Jaskulak**
Department of Environmental Engineering, Faculty of Infrastructure and Environment, Czestochowa University of Technology, Czestochowa, Poland

**Salwa Harzalli Jebara**
Laboratory of Legumes, Centre Biotechnology Borj Cedria, University Tunis El Manar, Hammam Lif, Tunisia

**Moez Jebara**
Laboratory of Legumes, Centre Biotechnology Borj Cedria, University Tunis El Manar, Hammam Lif, Tunisia

**Hinnan Khalid**
Punjab Agriculture Department, Shahkot; Faculty of Agriculture, Institute of Soil and Environmental Sciences, University of Agriculture, Faisalabad, Pakistan

**Muhammad Usman Khalid**
Department of Fisheries and Aquaculture, University of Veterinary and Animal Sciences, Lahore, Pakistan

**Vinod Kumar**
Department of Botany, DAV University, Jalandhar, India

**Inês Leitão**
LEAF, School of Agriculture, University of Lisbon, Lisbon, Portugal

**Sylvia Lindberg**
Department of Ecology, Environment and Plant Sciences, Stockholm University, Stockholm, Sweden

**Qaisar Mahmood**
Department of Environmental Sciences, COMSATS, Abbottabad, Pakistan

**Faisal Mahmood**
Department of Environmental Sciences and Engineering, Government College University, Faisalabad, Pakistan

**Ashwini Malla**
Plant Genetic Engineering Laboratory, Department of Biotechnology, Bharathiar University, Coimbatore; Bharat Biotech International Limited, Genome Valley, Hyderabad, India

**Luisa Louro Martins**
LEAF, School of Agriculture, University of Lisbon, Lisbon, Portugal

**Orapan Meesungnoen**
Department of Biology, Faculty of Science, Mahasarakharm University, Khamriang, Thailand

**Bodin Mongkhonsin**
Department of General Science, Faculty of Education, Kalasin University, Kalasin, Thailand

**Inês Moreira**
LEAF, School of Agriculture, University of Lisbon, Lisbon, Portugal

**Miguel Mourato**
LEAF, School of Agriculture, University of Lisbon, Lisbon, Portugal

**Aneta Murtaś**
Department of Environmental Engineering, Faculty of Infrastructure and Environment, Czestochowa University of Technology, Czestochowa, Poland

**Aqsa Mushtaq**
Department of Bioinformatics and Biotechnology, Government College University, Faisalabad, Pakistan

**Asif Naeem**
Soil and Environmental Sciences Division, Nuclear Institute for Agriculture and Biology (NIAB); Institute of Soil and Environmental Sciences, University of Agriculture, Faisalabad, Pakistan

**Sareeta Nahakpam**
Department of Biochemistry, Institute of Science, Banaras Hindu University, Varanasi; Department of Biochemistry and Crop Physiology, Bihar Agricultural University, Bhagalpur, India

**Woranan Nakbanpote**
Department of Biology, Faculty of Science, Mahasarakharm University, Khamriang, Thailand

**Ashwini Nangia**
School of Chemistry, University of Hyderabad, Central University P.O., Hyderabad, India

**Muhammad Nauman**
Department of Environmental Sciences, COMSATS, Abbottabad, Pakistan

**Nida Nazir**
Department of Environmental Sciences, COMSATS, Abbottabad, Pakistan

**Nabeel Khan Niazi**
Institute of Soil and Environmental Sciences, University of Agriculture Faisalabad, Faisalabad, Pakistan

**Muhammad Noman**
Department of Bioinformatics and Biotechnology, Government College University, Faisalabad, Pakistan

**Carmen A. Pedro**
MARE—Marine and Environmental Sciences Centre, ESTM, Polytechnic Institute of Leiria, Peniche, Portugal

**Filipa Pinto**
LEAF, School of Agriculture, University of Lisbon, Lisbon, Portugal

**Majeti Narasimha Vara Prasad**
Department of Plant Sciences, University of Hyderabad, Hyderabad, India

**Sathishkumar Ramalingam**
Plant Genetic Engineering Laboratory, Department of Biotechnology, Bharathiar University, Coimbatore, India

**Ladawan Rattanapolsan**
Department of Biology, Faculty of Science, Mahasarakharm University, Khamriang, Thailand

**Muhammad Rizwan**
Department of Environmental Sciences and Engineering, Government College University, Faisalabad, Pakistan

**Omar Saadani**
Laboratory of Legumes, Centre Biotechnology Borj Cedria, University Tunis El Manar, Hammam Lif, Tunisia

**Joana Sales**
LEAF, School of Agriculture, University of Lisbon, Lisbon, Portugal

**Abin Sebastian**
School of Chemistry, University of Hyderabad, Central University P.O., Hyderabad, India

**Kavita Shah**
Institute of Environment and Sustainable Development (IESD), Banaras Hindu University, Varanasi, India

**Shahida Shaheen**
Department of Environmental Sciences, COMSATS, Abbottabad, Pakistan

**Muhammad Shahid**
Department of Bioinformatics and Biotechnology, Government College University, Faisalabad, Pakistan

**Bala Murugan Shanmugaraj**
Plant Genetic Engineering Laboratory, Department of Biotechnology, Bharathiar University, Coimbatore; Bharat Biotech International Limited, Genome Valley, Hyderabad, India

**Gagan Preet Singh Sidhu**
Department of Environment Education, Government College of Commerce and Business Administration, Chandigarh, India

**Prerna Singh**
Department of Biochemistry and Crop Physiology, Bihar Agricultural University, Bhagalpur, India

**Muhammad Irfan Sohail**
Faculty of Agriculture, Institute of Soil and Environmental Sciences, University of Agriculture, Faisalabad, Pakistan; Indian River Research and Education Center, Institute of Food and Agricultural Sciences, University of Florida, Fort Pierce, FL, United States

**Jakub Świątek**
Department of Environmental Engineering, Faculty of Infrastructure and Environment, Czestochowa University of Technology, Czestochowa, Poland

**Kashif Tanwir**
Department of Botany, Faculty of Life Sciences, Government College University, Faisalabad, Pakistan

**Muhammad Umair**
Faculty of Agriculture, Institute of Soil and Environmental Sciences, University of Agriculture, Faisalabad, Pakistan

**Tahira Yasmeen**
Department of Environmental Sciences and Engineering, Government College University, Faisalabad, Pakistan

**Mehwish Zafar**
Institute of Soil and Environmental Sciences, University of Agriculture, Faisalabad, Pakistan

**Bibi Saima Zeb**
Department of Environmental Sciences, COMSATS, Abbottabad, Pakistan

**Muhammad Zia-ur-Rehman**
Faculty of Agriculture, Institute of Soil and Environmental Sciences, University of Agriculture, Faisalabad, Pakistan

# ABOUT THE EDITORS

**Dr. Mirza Hasanuzzaman** is professor of agronomy at Sher-e-Bangla Agricultural University in Dhaka, Bangladesh. He received his Ph.D. on "Plant Stress Physiology and Antioxidant Metabolism" from Ehime University, Japan, with a scholarship from the Japanese government. Later, he completed his postdoctoral research at the Center of Molecular Biosciences at University of the Ryukyus, Japan, as a recipient of the Japan Society for the Promotion of Science (JSPS) postdoctoral fellowship. He was also the recipient of the Australian Government's Endeavour Research Fellowship for postdoctoral research as an adjunct senior researcher at the University of Tasmania, Australia. Dr. Hasanuzzaman's current work is focused on the physiological and molecular mechanisms of environmental stress tolerance (salinity, drought, flood, and heavy metals/metalloids). Dr. Hasanuzzaman has published over 80 articles in peer-reviewed journals. He has edited 6 books and written 30 book chapters on important aspects of plant physiology, plant stress tolerance, and crop production. According to Scopus, Dr. Hasanuzzaman's publications have received over 2700 citations with an h-index of 26. He is an editor and reviewer for more than 50 peer-reviewed international journals and was a recipient of the "Publons Peer Review Award 2017 and 2018." He has been honored by different authorities for his outstanding performance in different fields such as research and education, and he has received the World Academy of Science Young Scientist Award (2014). He has presented 25 papers, abstracts, and posters at international conferences in several countries (United States, United Kingdom, Germany, Australia, Japan, Austria, Sweden, Russia, etc.). Dr. Hasanuzzaman is a member of 40 professional societies and is the acting publication secretary of the Bangladesh Society of Agronomy and the Bangladesh JSPS Alumni Association. He is a fellow of The Linnean Society of London.

**Dr. Majeti Narasimha Vara Prasad** currently Emeritus Professor in School of Life Sciences, formerly professor of environmental biology at the University of Hyderabad, India. He made significant contributions to the fields of plant-metal interactions, bioremediation, and bioeconomy. Dr. Prasad is the author, coauthor, editor, or coeditor of six books and more than 200 research papers on environmental botany and heavy metal stress in plants. He also published 127 book chapters and edited over 20 books with Elsevier, Wiley, Kluwer Academic, Springer, Taylor & Francis, Fizmatlit Russia, Marcel Dekker, Narosa, Russian Academy of Sciences, etc. According to Google Scholar, Dr. Prasad's publications have received over 14,000 citations with an h-index of 59 and an i10-index of 156. Dr. Prasad is an elected fellow of the Linnean Society of London, England, and the National Institute of Ecology, New Delhi, India; a life member of the National Institute of Ecology and the Bioenergy Society of India; and a member of the International Allelopathy Society and the Indian Network for Soil Contamination Research. He has visited many countries as a visiting professor and research fellow.

**Dr. Masayuki Fujita** is professor in the Laboratory of Plant Stress Responses, Faculty of Agriculture, Kagawa University, Kagawa, Japan. He received his B.Sc. in chemistry from Shizuoka University, Shizuoka, Japan, and his M.Agr. and Ph.D. in plant biochemistry from Nagoya University, Nagoya, Japan. His research interests include physiological, biochemical, and molecular biological responses based on secondary metabolism in plants under various abiotic and biotic stresses; phytoalexin, cytochrome P450, glutathione *S*-transferase, and phytochelatin; and redox reaction and antioxidants. In the last decade, his work has focused on oxidative stress and antioxidant defense in plants under environmental stress. His group investigates the role of different exogenous protectants in enhancing antioxidant defense and methylglyoxal detoxification systems in plants. He has supervised 4 M.S. students and 13 Ph.D. students as their main supervisor. He has about 150 publications in journals and books and has edited 4 books.

# ACKNOWLEDGMENTS

The editors would like to thank all the authors of this volume for their cogent and comprehensive contributory chapters.

The editors would also like to record their appreciation and thanks to Nancy Maragioglio (ELS-CMA), senior acquisitions editor, and Ruby Smith (ELS), editorial assistant, for their excellent coordination of this fascinating project, suggestions, and help in many ways that resulted in a record-setting publication time for this volume. Thanks are also due to the production project manager Swapna Srinivasan.

Last, but not least, we wish to thank numerous colleagues for sharing knowledge, ideas, and assistance which helped in shaping this book.

**Mirza Hasanuzzaman, Majeti Narasimha Vara Prasad, and Masayuki Fujita**

# PREFACE

Cadmium (Cd) ranks the highest in terms of damage to plant growth and productivity. It accumulates in the soil naturally or through anthropogenic activities, such as mining, industrial waste disposal, use and disposal of batteries and sludges, and the application of pesticides and phosphate fertilizers.

In the last couple of decades, considerable research work has been done to understand the physiological and molecular mechanisms of cadmium uptake and transport in plants, and its negative effects on plant productivity raised interest in researchers. However, the information on plants' responses and tolerance to cadmium toxicity and the underlying mechanisms as well as appropriate approaches for improvement of crop plants against cadmium stresses lies unorganized in different articles of journals and edited books. Therefore, this book, the first comprehensive one deals with all aspects of cadmium toxicity and tolerance in plants in a single platform.

The Agency for Toxic Substances and Disease Registry [ASTDR] (United States), in its toxicological profile for cadmium, reported that approximately 3 kg of cadmium are produced for each ton of zinc produced. Cadmium is found in the earth's crust, associated with zinc, lead, and copper ores. Pure cadmium is a soft, silver-white metal. Cadmium chloride and cadmium sulfate are soluble in water. Cadmium has a number of uses for manufacturing industrial and consumer products, including batteries (83%), pigments (8%), coatings and electroplatings (7%), stabilizers for plastics (1.2%) nonferrous alloys, photovoltaic devices, and other uses (0.8%). Cadmium is emitted into the soil, water, and air by nonferrous metal mining and refining, manufacture, and application of phosphate fertilizers, fossil fuel combustion, and waste incineration and disposal. Cadmium can accumulate in aquatic organisms and agricultural crops. Agroecosystems receiving large amounts of phosphate fertilizer are the main source of cadmium contamination in edible crops. Millions of tons of cadmium-contaminated rice were discarded from 2010 to 2013 in Asian countries such as China and Thailand. A daily intake of 20–40 μg of cadmium is reported through rice in regions where rice is used as a food staple, leading to health implications such as bone mineralization. As a result, cadmium minimization is a topic of priority investigation in agriculture.

The remediation potential of cadmium-tolerant plants, and risk assessment and remediation of soil, coupled with various amendments such as biosolids' applications toward the phytoavailability of soil cadmium, are priority areas of research. Cadmium minimization strategies can be either field remediation or manifestations in plant functions. Ecophysiological factors that influence cadmium uptake and distribution in plants assure checkpoints in the process of cadmium contamination of rice by controlling mineral nutrient supply and field management. Organic farming using ecofriendly amendments exhibits correlation with drop-in cadmium uptake and allows immediate sustainable field remediation.

The scope of cadmium minimization research in rice cultivation is very high. The development of biosensors that indicate field cadmium levels is feasible for screening cadmium-contaminated fields on a large scale. The selection of low-cost and widely applicable methods for cadmium reduction in rice is preferred. The efficiency of field-based techniques (such as the application of soil amendments and phytoremediation) needs further improvement through a technological approach. Establishment of a database of amendments will be helpful for immediate remediation programs in contaminated fields. The scope of biofortification of rice with micronutrients that reduce cadmium accumulation remains unexplored.

Mineral nutrient management is gaining attention as a cadmium minimization strategy because of its direct field applications. Studies on the interaction between cadmium accumulation and plant nutrients deserve attention because of micronutrient deficiency in soils of arable lands due to the repeated use of nitrogen fertilizers. It must be taken into account that NPK fertilization is inevitable in rice cultivation. As micronutrients compete with cadmium uptake, deficiency of these elements leads to cadmium accumulation in plants. Comprehensive studies on the relationship between cadmium accumulation in rice and plant nutrients availability are hardly ever performed. Nitrogen fertilization strategies need to be revised with sustainable agriculture practices, such as the application of farmyard manures for preserving soil structure. Cadmium accumulation in rice plants during the above sustainable agronomical practices needs to be characterized. A study on cadmium-iron interaction in rice is also essential for disclosing the influence of iron fortification on cadmium uptake and alleviation of cadmium stress in rice. The ability of cadmium to participate in redox reactions points to the regulatory role of this metal in iron-dependent plant growth and regulation. This is because many of the iron-mediated metabolic processes are redox reactions. Apart from these, iron fortification of staple foods such as rice is a practical solution for iron-deficiency mediated anemia, which affects about 30% of the world's population.

Cadmium Toxicity and Tolerance in Plants: From Physiology to Remediation broadly deals in different sections with topics such as (i) toxicity and tolerance, (ii) cadmium contamination and hazards, (iii) detoxification and oxidative stress, (iv) stress physiology, (v) growth regulation, (vi) myco-microbe mitigation, and (vii) phytoremediation.

In preparing this book, we aimed to provide the readers with a background for understanding cadmium toxicity, its effect on plants, and remediation mechanisms. We are thankful to all the authors for their valuable contributions. The editors and authors hope that this book will serve as a practical update on our knowledge about cadmium toxicity in plants and will lead to new discussions and efforts in the use of various techniques for the mitigation of cadmium-induced toxicity.

**Mirza Hasanuzzaman**
Dhaka, Bangladesh

**Majeti Narasimha Vara Prasad**
Hyderabad, India

**Masayuki Fujita**
Kagawa, Japan

# CHAPTER 1

# Cadmium Stress and Toxicity in Plants: An Overview

**Bala Murugan Shanmugaraj**[*,†,a], **Ashwini Malla**[*,†,a], **Sathishkumar Ramalingam**[*]

[*]Plant Genetic Engineering Laboratory, Department of Biotechnology, Bharathiar University, Coimbatore, India

[†]Bharat Biotech International Limited, Genome Valley, Hyderabad, India

## Abstract

Heavy metal pollution has long been a major environmental problem, and threatens all living forms globally. Cadmium (Cd) is one of the nonessential, highly toxic environmental pollutants worldwide that causes deleterious effects and serious problems in agriculture. Plants growing in Cd-contaminated soil uptake the heavy metal through their roots, which accumulates in different organs, eventually reducing plant growth and productivity. Bioaccumulation of high concentrations of Cd in plants enters the food chain and affects both animals and humans. In plants, several metabolic processes are associated with Cd toxicity and its tolerance. This chapter summarizes the effects of Cd toxicity in plant growth and other related physiological and metabolic processes, including an overview of phytoremediation technologies to clean up Cd contamination in the environment.

**Keywords:** Abiotic stress, Accumulation, Cadmium, Heavy metals, Phytoremediation, Toxicity

## 1 INTRODUCTION

Heavy metal pollution in soil has dramatically increased in recent decades, causing a potential threat to all living forms (Ghosh and Singh, 2005). Heavy metals may be essential or nonessential. Essential heavy metals include Mn, Fe, Ni, and Zn, which are necessary for the growth of living organisms, and for their development and physiological functions (Gohre and Paszkowski, 2006). Nonessential heavy metals, such as Cd, Pb, Hg, and As, are not required for any known physiological functions of living organisms (Peng et al., 2009). Nonessential heavy metal contamination from industrial

[a] These authors contributed equally.

*Cadmium Toxicity and Tolerance in Plants*
https://doi.org/10.1016/B978-0-12-814864-8.00001-2

effluents, mining, pesticides, and disposal of municipal waste interfere with physiological, biochemical, and molecular processes of all biological systems, posing a severe risk to the environment and other living organisms, causing ecological imbalances affecting the food chain and ecosystem (Subhashini and Swamy, 2013).

Cd is one of the nonessential, highly toxic heavy metals with a long biological half-life that is dangerous for most living organisms. Cd is mainly found in the earth's crust, occurring in combination with zinc. Cd is released into the environment by the zinc, lead, and copper industries as a byproduct, used in the application of phosphate-based fertilizers, urban composts, wastewater irrigation, metal working industries, and metal-based pesticides, and accumulated in many parts of the world (di Toppi and Gabbrielli, 1999; Benavides et al., 2005; Gratao et al., 2005; Zhou et al., 2017). It is also released into rivers through the weathering of rocks, and into the air through forest fires and volcanoes. The heavy metals like Cd poses a significant threat to natural resources, such as soil and water, due to its nondegradable and persistent nature in the environment. Humans are exposed to heavy metals by contaminated food intake, drinking water from contaminated areas, and inhalation of polluted air (Alissa and Ferns, 2011). Cd is teratogenic and mutagenic, and chronic Cd exposure is associated with respiratory and bone disorders, hypertension, and diabetes in humans, and also different types of cancers, including lung, kidney, and prostate cancer (Das et al., 1997; Nawrot et al., 2006; Edwards and Prozialeck, 2009; Gallagher and Meliker, 2010; Ueno et al., 2010). Itai-itai disease is caused by the intake of Cd through contaminated rice consumption, which is still a threat for human health (Ueno et al., 2010).

Both biotic and abiotic stress cause adverse effects on plant growth, resulting in significant crop loss worldwide (Rodriguez et al., 2005). Heavy metals are phytotoxic, even at low concentrations that can be easily transferred from contaminated soil to different plant organs, and its accumulation causes toxic effects to plants, and affects animals as well as humans via the food chain (Clemens et al., 2013; Paunov et al., 2018). Cd accumulation in plants could reduce the chlorophyll content, inhibiting photosynthesis and suppressing plant growth. Cadmium alters the redox potential of the cell by inducing increased reactive oxygen species (ROS) formation, causing oxidative damage to cell membranes and other biomolecules (Stohs et al., 2000; Schutzendubel et al., 2001; Yu et al., 2017). Many reports have shown the effects of Cd on morphological, physiological, and biochemical alterations in different plant species, including *Brassica*, *Lactuca sativa*

(Shanmugaraj et al., 2013; Jibril et al., 2017). This chapter provides a brief overview of the physiological and metabolic response of plants to heavy metal stress, and also phytoremediation strategies to clean up environmental pollutants in the soil.

# 2 PLANT RESPONSES TO CADMIUM TOXICITY

## 2.1 Morphological and Structural Changes

Cd is one of the most toxic heavy metals causing deleterious effects in crops (Sethy and Ghosh, 2013; Shanmugaraj et al., 2013). The uptake of Cd by plants is determined by the biological availability of Cd in the contaminated soil (Clemens, 2006). Cd is a readily absorbed and rapidly translocated heavy metal, which is transported through the root cortex to the shoots symplastically (Tudoreanu and Phillips, 2004). The toxic symptoms of Cd in plants are growth retardation, alternations in photosynthetic activity, changes in stomatal movement, enzymatic activities, protein metabolism, and membrane functioning.

Cd taken up by plants is the major source of contamination in food (Chunhabundit, 2016). Cd is toxic to most living organisms, even at low concentrations. Cd is among the elements that can be accumulated in plants to levels above 0.01% of shoot dry weight, without causing toxic symptoms (Reeves and Baker, 2000; Verbruggen et al., 2009). Cd gathers in the topsoil, in close relationship with the organic fraction, which is highly accessible for plants growing in acidic soils (Tudoreanu and Phillips, 2004; Kirkham, 2006), thereby enhancing its solubility in root exudates (Luo et al., 2014). It occurs predominantly as $Cd^{2+}$ in soil (Tudoreanu and Phillips, 2004; Verbruggen et al., 2009). The cell wall is identified as a major site of heavy metal storage in plants, and its deposition in the cell wall is considered a vital mechanism for heavy metal tolerance (Vazquez et al., 2006). Roots are the first ones to be affected by heavy metals, as metal ions are accumulated more in roots than shoots (Singh et al., 2015). Cd localization in plants, when observed under an electron microscope, showed that the root cell walls contain the majority of the metal, as compared with cytoplasm, because the heavy metal binds to the cell wall due to its negative charge (Polle and Schutzendubel, 2003; Dal Corso et al., 2010).

Cd can be easily absorbed by plant roots and transported to shoots, resulting in cellular, molecular, biochemical, and physiological alterations that affect plant growth and morphology (Sgherri et al., 2002; Shanmugaraj et al., 2013; Song et al., 2017). Cd toxicity apparently inhibits plant root growth, affecting the root morphology (Daud et al., 2009). Prolonged exposure of plants to Cd may cause the roots to become mucilaginous, brown, and decomposed, with a reduction of shoots and root elongation, rolling of leaves, and chlorosis. Cd accumulation inhibits lateral root formation, while the main root becomes brown, rigid, and twisted (Krantev et al., 2008; Yadav, 2010; Rascio and Navari-Izzo, 2011). It is observed by the disordered division and abnormal enlargement of epidermal and cortical cell layers in the apical region. Cd stress also causes an unusual number of nucleus populations in the differentiated roots (Fusconi et al., 2006), inhibits the mitotic index, induces chromosomal aberrations and mitotic aberrations, and delays the micronucleus formation. It also damages the DNA in root-cap cells (Seth et al., 2008). Cd damages the nucleoli in root tip cells, inducing chromosomal and mitotic aberrations in onions (Liu et al., 1995; Seth et al., 2008), and it alters the synthesis of RNA by inhibiting the ribonuclease activity in rice (Shah and Dubey, 1995).

## 2.2 Effect on Physiological Process

Cd intake alters phytochemicals, the nutritional content of plants, and hinders stomatal opening by intermingling with the water balance in plants (Feng et al., 2010); altering Calvin cycle enzymes, photosynthesis, and carbohydrate metabolism (Mobin and Khan, 2007; Shi et al., 2010), which changes the antioxidant metabolism (Khan et al., 2009), affecting plant growth. Plants have a well-developed defense system to protect them from heavy metal toxicity (Shanmugaraj et al., 2013).

When plants are exposed to a high concentration of Cd, it causes leaf roll, reduces the chlorophyll content, and inhibits leaf photosynthesis by suppressing the biosynthesis of chlorophyll (He et al., 2008; Rascio et al., 2008; Lee et al., 2010; Liu et al., 2010; Miyadate et al., 2011). It was reported that Cd exposure inhibits photosynthesis in *Brassica napus*, *Helianthus annus*, *Thlaspi caerulescens*, *Zea mays*, *Pisum sativum*, *Hordeum vulgare*, *Vigna radiata*, and *Triticum* (Baryla et al., 2001; Di Cagno et al., 2001; Kupper et al., 2007; Popova et al., 2008; Wahid et al., 2008; Moussa and El-Gamal, 2010). Photosystem II (PSII) is affected by Cd toxicity (Baker, 1991), and its damage or failure under stress can be easily identified by the changes of the chlorophyll in fluorescence structures (Maxwell and Johnson, 2000). The two enzymes involved in $CO_2$ fixation that is, Ribulose-1, 5-bisphosphate carboxylase (RuBisCO) and phosphoenol pyruvate carboxylase (PEPCase), are the main targets for Cd damage. The activity of RuBisCO is lowered by altering its structure, replacing the Mg ions, which are the vital cofactors of carboxylation reactions, and swing toward oxygenation reactions (Siedlecka et al., 1998). The water oxidizing complex (OEC) of PSII is also affected by replacing the $Ca^{2+}$ in Ca/Mn clusters that constitute the oxygen evolving centers (Sigfridsson et al., 2004), and also modifying the Qb binding site (Geiken et al., 1998).

In *B. napus*, Cd lowers the total chlorophyll and carotenoid content, and increases the nonphotochemical quenching (Larsson et al., 1998). Cd significantly reduces the normal $H^+$-$K^+$ exchange, activity of the plasma membrane ATPase, and several other enzymes, such as glucose-6-phosphate dehydrogenase, glutamate dehydrogenase, malic enzymes, isocitrate dehydrogenase (Van Assche and Clijsters, 1990; Mattioni et al., 1997), and Rubisco and carbonic anhydrase (Siedlecka et al., 1997). It was reported that there was a significant increase in the phosphoenolpyruvate carboxylase polypeptide, with no further synthesis of glutamate dehydrogenase and glutamate synthase polypeptides in *Z. mays* seedlings when exposed to 20 mM Cd (Ju et al., 1997); whereas in pea plants, chromatin alterations were observed (Hadwiger et al., 1973). Cd ions are chemically similar to Zn ions, and can hamper the activity of Zn-finger transcription factors, substituting the Zn ions, thereby interfering with transcription mechanisms (di Toppi and Gabbrielli, 1999). It also replaced $Ca^{2+}$ ions in calmodulin proteins, causing the perturbation of intracellular calcium levels and altering calcium-dependent signaling pathways (Perfus-Barbeoch et al., 2002) through a similar mechanism to the one observed with Zn transcription factors. $Ca^{2+}$ concentration is significantly increased in cells during Cd stress (Dal Corso et al., 2008), stimulating the calmodulin-like proteins to interact with $Ca^{2+}$ ions by changing their conformation to regulate a variety of mechanisms, including ion transport, gene regulation, metabolism, and stress tolerance (Yang and Poovaiah, 2003).

Cd reduces the absorption of micro- and macronutrients, thereby affecting the transportation activities in plants (Hernandez et al., 1996). Cd shows adverse effects by interfering with the uptake, transport, and subsequent distribution of nutrient elements, such as Ca, Mg, P, K, S, Fe, Mo, Zn, Mn, B, and Cu in plant species such as sugar beets (Chang et al., 2003), peas (Metwally et al., 2005), and barley (Guo et al., 2007). It has an effect on the uptake of nitrate, and its transportation from roots to shoots, by inhibiting the nitrate reductase activity, which was observed in plants of *Silene cucubalus* (Mathys, 1975). Cd also induces changes in the lipid and fatty acid composition, altering the functionality of the membranes (Popova et al.,

2009). High concentrations of Cd also lead to abnormal nutrient metabolism, such as imbalances between proteins and sugars in plants (Costa and Spitz, 1997). It also has an adverse effect on sugars and amino acid contents in certain species of plants (Moya et al., 1993; Costa and Spitz, 1997; Wu et al., 2004) by increasing their concentrations, indicating an inhibition of starch hydrolysis (Bishnoi et al., 1993).

## 2.3 Oxidative Stress

Heavy metals also cause oxidative damage to plants by the direct or indirect formation of ROS that can cause damage to cell membrane lipids and protein directly or indirectly (Lesser, 2006). The ROS are strong oxidizing agents, causing oxidative damage to biomolecules such as lipids and proteins, eventually leading to cell death (Molassiotis et al., 2006).

Cd toxicity is related to oxidative stress in plant cells, promoting the generation of ROS (Olmos et al., 2003), and inhibiting or activating the antioxidant enzymes (Iannelli et al., 2002); thereby resulting in cellular oxidative damage and lipid peroxidation (Chien et al., 2002). Cd, being a nonredox metal, cannot perform single electron transfers, and does not generate ROS such as superoxide anion ($O_2^-$), singlet oxygen ($^1O_2$), hydrogen peroxide ($H_2O_2$), or hydroxyl radicals ($OH^-$); however, it induces oxidative stress by its interference with the antioxidant defense systems (Benavides et al., 2005; Cho and Seo, 2005; Gratao et al., 2005). Cd slows down the activity of antioxidative enzymes (Maksymiec and Krupa, 2006; Lin et al., 2007), resulting in foliar damage and plant growth inhibition (Liu et al., 2003; Rai et al., 2005; Wu et al., 2007; Dell'Amico et al., 2008). Heyno et al. (2008) reported that Cd stimulates the production of ROS in the mitochondrial electron transfer chain. It also stimulates the NADPH oxidase in peroxisomes by accumulation of hydrogen peroxide, followed by cellular oxygen and fatty acid hydroperoxide accumulation (Gill and Tuteja, 2010). Of the alterations in the free radical scavengers such as superoxide dismutase (SOD) and catalase (CAT), ascorbate peroxidase (APOX), mono-dehydroascorbate reductase (MDHAR), dehydroascorbate reductase (DHAR), peroxidases (POD), glutathione reducatse (GR) and some nonenzymatic scavengers glutathione (GSH), ascorbic acid (AsA) was observed in different species, including *P. sativum*, *Glycine max*, *Oryza sativa*, and *Arabidopsis*, depending on the Cd concentrations (Benavides et al., 2005).

During Cd intoxication, higher activities of antioxidant enzymes and nonenzymatic constituents were observed in the plants to tolerate the stress.

ROS can be quenched by the induction of enzymatic defense systems, such as SOD and CAT, to reduce the oxidative damage in Cd-stressed plants (Wu et al., 2003). The biochemical changes occurring in plants subject to Cd stress include the production of ROS such as $O_2$, $H_2O_2$, and $OH^-$ (Cho and Park, 2000). When plants are subjected to stress due to Cd accumulation, oxidative damage may result due to the imbalance between the production of ROS, and their detoxification by the altered antioxidative system (Gomez et al., 1999). Tolerance to Cd stress is correlated with an increased capacity to scavenge or detoxify activated oxygen species (Hashem et al., 2016). The enzymatic antioxidant defense systems in the plant cell are SOD, CAT, APOX, and nonenzymatic antioxidants such as GSH, AsA, and α-tocopherol. SOD, a major scavenger, is a metalloprotein catalyzing the dismutation of super oxide to $H_2O_2$ and molecular oxygen (Allen, 1995). $H_2O_2$ is also toxic to cells that need to be additionally detoxified to water and oxygen by CAT and/or POD (Zhu et al., 2004).

Plants also respond to Cd metal toxicity by synthesizing phytochelatins (PCs) that act as heavy metal chelators. Sulphur, which is a component of phytochelatin, has an important role in their synthesis, and ultimately, in detoxification of Cd by the formation of Cd-binding peptides (CdBP) (Cobbett and Goldsbrough, 2002). Phytochelatins [(γ-Glu-Cys)n-Gly] are the Cys-rich peptides that are synthesized enzymatically using GSH as a substrate by phytochelatin synthase (PCS), which is found to be activated upon exposure to heavy metal Cd (Thangavel et al., 2007). PCs are transported to the vacuole by binding to Cd ions, and play an important function in Cd stress-tolerant plants (Mendoza Cozatl et al., 2005). Transgenic approaches have also revealed that PCs play an important role in Cd tolerance in many plants, as observed by the overexpression of the γ-glutamyl cysteine synthetase gene in *Brassica juncea*, resulting in increased biosynthesis of GSH and PCs, and an increased tolerance to Cd (Zhu et al., 1999). In a study by Shanmugaraj et al. (2013), the transcripts of the *PCS* gene increased in the cultivars of *Brassica* during Cd treatment, indicating the enhanced phytochelatin biosynthesis in order to protect the plant from heavy metal stress.

## 3 PHYTOREMEDIATION OF Cd CONTAMINATED SOIL

Environmental cleanup of Cd-contaminated soil by mechanical and chemical remediation is expensive, which may take years to decades (McGrath and Zhao, 2003), and it cannot be implemented for the remediation of large areas and cultivated lands, as it may affect the soil structure and fertility (Zhu et al., 2004). Earlier reports have demonstrated the bioremediation of Cd using plants or microbes to restore polluted soil and water (Gan et al., 2009; Haritash and Kaushik, 2009; Tak et al., 2013; Elekwachi et al., 2014).

The term "phytoremediation" is derived from the Greek and Latin Words "Phyto" meaning plants, and "remedium," meaning restoring balance (Sood et al., 2012). Phytoremediation is an eco-friendly, low cost, in situ remediation technology utilizing naturally occurring or genetically engineered plants to mitigate environmental pollutant levels in contaminated lands (Meagher et al., 2000). Phytoremediation can be carried out in different ways, such as rhizofiltration, phytostabilization, phytovolatilization, phytodegradation, and phytoextraction (Long et al., 2002; Jadia and Fulekar, 2009). Some plants are adapted to live in challenging environments to tackle the accumulated pollutants that occur in the environment. Growing plants on pollutant sites is a passive technique that can be used to clean up places with low to moderate levels of contamination. Plants act as filters that can stabilize the metal contaminants, either by breakdown, or by degrading the pollutants.

Some plants are hyperaccumulators that are able to grow in highly contaminated soil, and accumulate high concentrations of heavy metal, that is, 10–500 times higher than the normal plant (Chaney et al., 1999). Hyperaccumulators absorb toxic heavy metals in the soil by their roots, and transport/translocate them to shoots where the heavy metals are detoxified and stored (Doty et al., 2000; Greenberg et al., 2006). The plants of the Brassicaceae or cruciferae families are reported as hyper accumulators of Cd (Clemens et al., 2002; Pilon-Smits and Pilon, 2002). The species of the *Brassica* genus (*B. napus*, *B. juncea*, and *B. rapa*) are fast growing, with high biomass hyperaccumulators (Shanmugaraj et al., 2013). Plants are capable of growing in soils with very high concentrations of Cd by absorbing the metal through roots and building up in the various tissues. Based on the accumulation capacity, they are categorized as accumulators and hyperaccumulators. Many species of plants have been identified as hyper-accumulators/accumulators of Cd from soil and water such as *Athrium yokoscense*, *Bacopa monnieri*, *Salix viminalis*, *Lemna minor*, *H. annus*, *B. juncea*, *Spirodela polyrhiza*, *Vallisneria spiralis*, and so forth (Prasad, 2005; Shanmugaraj et al., 2013; Shimpei et al., 2006). The use of different hyper-accumulators in phytoremediation can be effective for cleaning heavily Cd-contaminated soils

(Chaney et al., 1999; Zhu et al., 2004). There are also certain plant species such as *Tagetes erecta* and *T. caerulescens*, which can tolerate the Cd concentrations in soil up to a certain extent, and are very useful for phytoextraction process (Lombi et al., 2000, 2001).

Despite Cd toxicity, high tolerance to this metal element has evolved in several plant species. These plants have developed a number of different mechanisms to survive Cd stress through metal exclusion, active excretion, restricted distribution of the metal in sensitive tissues, metal binding to the cell wall, chelation by organic molecules, and compartmentalization in vacuoles (Benavides et al., 2005; Gratao et al., 2005). The uptake of Cd in plants from soil mainly occurs via $Ca^{2+}$, $Fe^{2+}$, $Mn^{2+}$, and $Zn^{2+}$ transporters (Rascio and Navari-Izzo, 2011). In nonhyper accumulating plants, the uptake of Cd is not specific. The transportation of heavy metals from root to shoot in hyper-accumulating plants differs from nonhyper accumulators. The mechanism is that Cd, which is taken by root cells, is detoxified by chelation in the cytoplasm, stored in vacuoles, and further translocated to the shoot through xylem. There are many small organic molecules acting as metal binding ligands in hyper-accumulator roots (Tran and Popova, 2013), and free amino acids such as histidine and nicotinamide form stable complexes with bivalent cations (Hassan and Aarts, 2011). The key step for increased metal flow toward the shoot is enhanced metal xylem loading and its translocation. The Cd is detoxified or sequestered by hyper-accumulators in the epidermis, trichomes, and cuticle, so the photosynthetic machinery damage is minimized. In certain cases, the subsidiary and guard cells of the stomata exclude the Cd, thereby protecting the functional stomatal cells from phytotoxic effects of the metal (Tran and Popova, 2013). The heavy metal complexation with ligands is responsible for its removal from the cytoplasm by moving into inactive compartments such as vacuoles and cell walls in the aerial regions of hyper-accumulators (Rascio and Navari-Izzo, 2011). Another approach for detoxification of nonessential metals is the synthesis of low molecular weight chelators that avoid binding to physiologically important proteins, thereby facilitating their movement into the vacuoles. The thiols present in glutathione and phytochelatins favor the ligands of $Cd^{2+}$. The tripeptide glutathione Glu-Cys-Gly (GSH) involved in redox defense can bind to many metals and metalloids. The presence of small ligands such as organic acids act as detoxifying factors, preventing the persistence of heavy metals as free ions in the cytoplasm by sequestering them into vacuoles (Rascio and Navari-Izzo, 2011).

## 4 CONCLUSION

Cd concentration in the soil has increased in recent years due to anthropogenic activities. Cd alters physiological and biochemical activities in plants, which eventually hampers plant growth. The damaging effects of Cd in agricultural land can be mitigated by implementing feasible strategies to counteract the deleterious effects of Cd. Although regulatory measures have been implemented in a number of countries to reduce the heavy metal release from metal industries and metal-based pesticides, it is still a major health problem for high-risk communities in many countries. Bioremediation of environmental contaminants is carried out by combining plants and microorganisms with, or as a substitute for, conventional clean-up technologies, such as mechanical and chemical remediation, that often require high capital inputs and consume a great deal of energy. Although heavy metals in the soil can be removed by phytoremediation, it is a slow and time-consuming process. Through genetic engineering approaches, plants can be engineered to enhance their tolerance and accumulation of heavy metals by targeting multiple genes in crop plants (Clemens et al., 2002). Furthermore, many stress-resistant genes and transcription factors have been identified, which can be used to develop a suitable plant prototype ideal for heavy metal remediation. Furthermore, by implementing multifaceted Cd-management, such as reducing the anthropogenic inputs, minimizing Cd accumulation in plants through agronomic practices may ensure a safer food supply, and better sustainability of natural resources.

## ACKNOWLEDGMENTS

The authors are very thankful to The Department of Biotechnology, Bharathiar University, Tamil Nadu, India for supporting this research through DST-FIST and UGC-SAP.

## CONFLICT OF INTEREST STATEMENT

The authors declare that they have no conflict of interest.

## REFERENCES

Alissa, E.M., Ferns, G.A., 2011. Heavy metal poisoning and cardiovascular disease. J. Toxicol. 2011870125.

Allen, R.D., 1995. Dissection of oxidative stress tolerance using transgenic plants. Plant Physiol. 107, 1049–1054.

Baker, N.R., 1991. A possible role for photosystem II in environmental perturbations of photosynthesis. Physiol. Plant. 81, 563–570.

Baryla, A., Carrier, P., Franck, F., Coulomb, C., Sahut, C., Havaux, M., 2001. Leaf chlorosis in oilseed rape plants (*Brassica napus*) grown on cadmium-polluted soil: causes and consequences for photosynthesis and growth. Planta 212, 696–709.

Benavides, M.P., Gallego, S.M., Tomaro, M.L., 2005. Cadmium toxicity in plants. Braz. J. Plant Physiol. 17, 21–34.

Bishnoi, N., Sheoran, I., Singh, R., 1993. Effect of cadmium and nickel on mobilization of food reserves and activities of hydrolytic enzymes in germinating pigeon pea seeds. Biol. Plant. 35, 583–589.

Chaney, R.L., Ryan, J.A., Li, Y.M., Brown, S.L., 1999. Soil cadmium as a threat to human health. In: McLaughlin, M.J., Singh, B.R. (Eds.), Cadmium in Soil and Plants. Kluwer Academic Publisher, Dordrecht, pp. 219–256.

Chang, Y., Zouari, M., Gogorcena, Y., Lucena, J.J., Abadía, J., 2003. Effects of cadmium and lead on ferric chelate reductase activities in sugar beet roots. Plant Physiol. Biochem. 41, 999–1005.

Chien, H.F., Lin, C.C., Wang, J.W., Chen, C.T., Kao, C.H., 2002. Changes in ammonium ion content and glutamine synthetase activity in rice leaves caused by excess cadmium are a consequence of oxidative damage. Plant Growth Regul. 36, 41–47.

Cho, U.H., Park, J.O., 2000. Mercury-induced oxidative stress in tomato seedlings. Plant Sci. 156, 1–9.

Cho, U., Seo, N.H., 2005. Oxidative stress in *Arabidopsis thaliana* exposed to cadmium is due to hydrogen peroxide accumulation. Plant Sci. 168, 113–120.

Chunhabundit, R., 2016. Cadmium exposure and potential health risk from foods in contaminated area, Thailand. Toxicol. Res. 32, 65–72.

Clemens, S., 2006. Toxic metal accumulation, responses to exposure and mechanisms of tolerance in plants. Biochimie 88, 1707–1719.

Clemens, S., Palmgreen, M.G., Kramer, U., 2002. A long way ahead: understanding and engineering plant metal accumulation. Trends Plant Sci. 7, 309–315.

Clemens, S., Aarts, M.G., Thomine, S., Verbruggen, N., 2013. Plant science: the key to preventing slow cadmium poisoning. Trends Plant Sci. 18, 92–99.

Cobbett, C., Goldsbrough, P., 2002. Phytochelatins and Metallothioneins: roles in heavy metal detoxification and homeostasis. Annu. Rev. Plant Biol. 53, 159–182.

Costa, G., Spitz, E., 1997. Influence of cadmium on soluble carbohydrates, free amino acids, protein content of in vitro cultured *Lupinus albus*. Plant Sci. 128, 131–140.

Dal Corso, G., Farinati, S., Maistri, S., Furini, A., 2008. How plants cope with cadmium: staking all on metabolism and gene expression. J. Integr. Plant Biol. 50, 1268–1280.

Dal Corso, G., Farinati, S., Furini, A., 2010. Regulatory networks of cadmium stress in plants. Plant Signal. Behav. 5, 663–667.

Das, P., Samantaray, S., Rout, G.R., 1997. Studies on cadmium toxicity in plants: a review. Environ. Pollut. 98, 29–36.

Daud, M.K., Sun, Y., Dawood, M., Hayat, Y., Variath, M.T., Wu, Y.X., Raziuddin, Mishkat, U., Salahuddin, Najeeb, U., Zhu, S., 2009. Cadmium-induced functional and ultrastructural alterations in roots of two transgenic cotton cultivars. J. Hazard. Mater. 161, 463–473.

Dell'Amico, E., Cavalca, L., Andreoni, V., 2008. Improvement of *Brassica napus* growth under cadmium stress by cadmium-resistant rhizobacteria. Soil Biol. Biochem. 40, 74–84.

Di Cagno, R., Guidi, L., De Gara, L., Soldatini, G.F., 2001. Combined cadmium and ozone treatments affect photosynthesis and ascorbate-dependent defences in sunflower. New Phytol. 151, 627–636.

di Toppi, L.S., Gabbrielli, R., 1999. Response to cadmium in higher plants. Environ. Exp. Bot. 41, 105–130.

Doty, S.L., Shang, Q.T., Wilson, A.M., Tangen, J., Westergreen, A.D., Newman, L.A., Strand, S.E., Gordon, M.P., 2000. Enhanced metabolism of halogenated hydrocarbons in transgenic plants containing mammalian cytochrome P450 2E1. Proc. Natl. Acad. Sci. U. S. A. 97, 6287–6291.

Edwards, J.R., Prozialeck, W.C., 2009. Cadmium, diabetes and chronic kidney disease. Toxicol. Appl. Pharmacol. 238, 289–293.

Elekwachi, C.O., Andresen, J., Hodgman, T.C., 2014. Global use of bioremediation technologies for decontamination of ecosystems. J. Bioremed. Biodeg. 5, 1–9.

Feng, J., Shi, Q., Wang, X., Wei, M., Yang, F., Xu, H., 2010. Silicon supplementation ameliorated the inhibition of photosynthesis and nitrate metabolism by cadmium (Cd) toxicity in *Cucumis sativus* L. Sci. Hortic. 123, 521–530.

Fusconi, A., Repetto, O., Bona, E., Massa, N., Gallo, C., Dumas-Gaudot, E., Berta, G., 2006. Effects of cadmium on meristem activity and nucleus ploidy in roots of *Pisum sativum* L. cv. Frisson seedlings. Environ. Exp. Bot. 58, 253–260.

Gallagher, C.M., Meliker, J.R., 2010. Blood and urine cadmium, blood pressure, and hypertension: a systematic review and meta-analysis. Environ. Health Perspect. 118, 1676–1684.

Gan, S., Lau, E.V., Ng, H.K., 2009. Remediation of soils contaminated with polycyclic aromatic hydrocarbons (PAHs). J. Hazard. Mater. 172, 532–549.

Geiken, B., Masojidek, J., Rizzuto, M., Pompili, M.L., Giardi, M.T., 1998. Incorporation of [$^{35}$S]methionine in higher plants reveals that stimulation of the D1 reaction Centre II protein turnover accompanies tolerance to heavy metal stress. Plant Cell Environ. 21, 1265–1273.

Ghosh, M., Singh, S.P., 2005. A comparative study of cadmium phyto extraction by accumulator and weed species. Environ. Pollut. 133, 365–371.

Gill, S.S., Tuteja, N., 2010. Reactive oxygen species and antioxidant machinery in abiotic stress tolerance in crop plants. Plant Physiol. Biochem. 48, 909–930.

Gohre, V., Paszkowski, U., 2006. Contribution of the arbuscular mycorrhizal symbiosis to heavy metal phytoremediation. Planta 223, 1115–1122.

Gomez, J.M., Hernández, J.A., Jimenez, A., del Rio, L.A., Sevilla, F., 1999. Differential response of antioxidative enzymes of chloroplasts and mitochondria to long-term NaCl stress of pea plants. Free Radic. Res. 31, S11–S18.

Gratao, P.L., Polle, A., Lea, P.J., Azevedo, R.A., 2005. Making the life of heavy metal stressed plants a little easier. Funct. Plant Biol. 32, 481–494.

Greenberg, B.M., Hunag, X.D., Gurska, Y., Gerhardt, K.E., Wang, W., et al., 2006. Successful field tests of a multi process phytoremediation system for decontamination of persistent petroleum and organic contaminants. Proceedings of the Twenty Ninth Artic and Marine Oil Spill Program (AMOP), Canada, pp. 389–400.

Guo, T.R., Zhang, G.P., Zhou, M.X., Wu, F.B., Chen, J.X., 2007. Influence of aluminum and cadmium stresses on mineral nutrition and root exudates in two barley cultivars. Pedosphere 17, 505–512.

Hadwiger, L.A., Broembsen, S., Eddy, R., 1973. Increased template activity in chromatin from cadmium treated pea tissues. Biochem. Biophys. Res. Commun. 50, 1120–1128.

Haritash, A.K., Kaushik, C.P., 2009. Biodegradation aspects of polycyclic aromatic hydrocarbons (PAHs): a review. J. Hazard. Mater. 169, 1–15.

Hashem, A., Abd Allah, E.F., Alqarawi, A.A., Al Huqail, A.A., Egamberdieva, D., Wirth, S., 2016. Alleviation of cadmium stress in *Solanum lycopersicum* L. by arbuscular mycorrhizal fungi via induction of acquired systemic tolerance. Saudi J. Biol. Sci. 23, 272–281.

Hassan, Z., Aarts, M.G.M., 2011. Opportunities and feasibilities for biotechnological improvement of Zn, Cd or Ni tolerance accumulation in plants. Environ. Exp. Bot. 72, 53–63.

He, J.Y., Ren, Y.F., Zhu, C., Jiang, D.A., 2008. Effects of cadmium stress on seed germination, seedling growth and seed amylase activities in rice (*Oryza sativa*). Rice Sci. 15, 319–325.

Hernandez, L.E., Carpena-Ruiz, R., Garate, A., 1996. Alterations in the mineral nutrition of pea seedlings exposed to cadmium. J. Plant Nutr. 19, 1581–1598.

Heyno, E., Klose, C., Krieger-Liszkay, A., 2008. Origin of cadmium-induced reactive oxygen species production: mitochondrial electron transfer versus plasma membrane NADPH oxidase. New Phytol. 179, 687–699.

Iannelli, M.A., Pietrini, F., Fiore, L., Petrilli, L., Massacci, A., 2002. Antioxidant response to cadmium in *Phragmites australis* plants. Plant Physiol. Biochem. 40, 977–982.

Jadia, C.D., Fulekar, M.H., 2009. Phytoremediation of heavy metals: recent techniques. Afr. J. Biotechnol. 8, 921–928.

Jibril, S.A., Hassan, S.A., Ishak, C.F., Wahab, P.E.M., 2017. Cadmium toxicity affects phytochemicals and nutrient elements composition of Lettuce (*Lactuca sativa* L.). Adv. Agric. 2017, 1–3.

Ju, G.C., Li, X.Z., Rauser, W.E., Oaks, A., 1997. Influence of cadmium on the production of g-glutamylcysteine peptides and enzymes of nitrogen assimilation in *Zea mays* seedlings. Physiol. Plant. 101, 793–799.

Khan, N.A., Anjum, N.A., Nazar, R., Iqbal, N., 2009. Increased activity of ATP-sulfurylase and increased contents of cysteine and glutathione reduce high cadmium-induced oxidative stress in mustard cultivar with high photosynthetic potential. Russ. J. Plant Physiol. 56, 670–677.

Kirkham, M.B., 2006. Cadmium in plants on polluted soils: effects of soil factors, hyper accumulation and amendments. Geoderma 137, 19–32.

Krantev, A., Yordanova, R., Janda, T., Szalai, G., Popova, L., 2008. Treatment with salicylic acid decreases the effect of cadmium on photosynthesis in maize plants. J. Plant Physiol. 165, 920–931.

Kupper, H., Parameswaran, A., Leitenmaier, B., Trtílek, M., Setlik, I., 2007. Cadmium-induced inhibition of photosynthesis and long-term acclimation to cadmium stress in the hyperaccumulator *Thlaspi caerulescens*. New Phytol. 175, 655–674.

Larsson, E.H., Bornman, J.F., Asp, H., 1998. Influence of UV-B radiation and $Cd^{2+}$ on chlorophyll fluorescence, growth and nutrient content in *Brassica napus*. J. Exp. Bot. 49, 1031–1039.

Lee, K., Bae, D.W., Kim, S.H., Han, H.J., Liu, X., Park, H.C., Lim, C.O., Lee, S.Y., Chung, W.S., 2010. Comparative proteomic analysis of the short-term responses of rice roots and leaves to cadmium. J. Plant Physiol. 167, 161–168.

Lesser, M.P., 2006. Oxidative stress in marine environments: biochemistry and physiological ecology. Annu. Rev. Physiol. 68, 253–278.

Lin, R., Wang, X., Luo, Y., Du, W., Guo, H., Yin, D., 2007. Effects of soil cadmium on growth, oxidative stress and antioxidant system in wheat seedlings (*Triticum aestivum* L.). Chemosphere 69, 89–98.

Liu, D., Jiang, W., Wang, W., Zhai, L., 1995. Evaluation of metal ion toxicity on root tip cells by the *Allium* test. Isr. J. Plant Sci. 43, 125–133.

Liu, J., Li, K., Xu, J., Liang, J., Lu, X., Yang, J., Zhu, Q., 2003. Interaction of Cd and five mineral nutrients for uptake and accumulation in different rice cultivars and genotypes. Field Crop Res. 83, 271–281.

Liu, J., Cao, C., Wong, M., Zhang, Z., Chai, Y., 2010. Variations between rice cultivars in iron and manganese plaque on roots and the relation with plant cadmium uptake. J. Environ. Sci. 22, 1067–1072.

Lombi, E., Zhao, F.J., Dunham, S.J., McGrath, S.P., 2000. Cadmium accumulation in populations of *Thlaspi caerulescens* and *Thlaspi goesingense*. New Phytol. 145, 11–20.
Lombi, E., Zhao, F.J., McGrath, S.P., Young, S.D., Sacchi, G.A., 2001. Physiological evidence for a high-affinity cadmium transporter highly expressed in a *Thlaspi caerulescens* ecotype. New Phytol. 149, 53–60.
Long, X., Yang, X., Ni, W., 2002. Current situation and prospect on the remediation of soils contaminated by heavy metals. J. Appl. Ecol. 13, 757–762.
Luo, Q., Sun, L., Hu, X., Zhou, R., 2014. The variation of root exudates from the hyperaccumulator *Sedum alfredii* under cadmium stress: metabonomics analysis. PLoS One 9, e115581.
Maksymiec, W., Krupa, Z., 2006. The effects of short-term exposition to Cd, excess cu ions and jasmonate on oxidative stress appearing in *Arabidopsis thaliana*. Environ. Exp. Bot. 57, 187–194.
Mathys, W., 1975. Enzymes of heavy-metal–resistant and nonresistant populations of *Silene cucubalus* and their interaction with some heavy metals *in vitro* and *in vivo*. Physiol. Plant. 33, 161–165.
Mattioni, C., Gabbrielli, R., Vangronsveld, J., Clijsters, H., 1997. Nickel and cadmium toxicity and enzymatic activity in Ni-tolerant and non-tolerant populations of *Silene italica* Pers. J. Plant Physiol. 150, 173–177.
Maxwell, K., Johnson, G.N., 2000. Chlorophyll fluorescence-a practical guide. J. Exp. Bot. 51, 659–668.
McGrath, S.P., Zhao, F.J., 2003. Phytoextraction of metals and metalloids from contaminated soils. Curr. Opin. Biotechnol. 14, 277–282.
Meagher, R.B., Rugh, C.L., Kandasamy, M.K., Gragson, G., Wang, N.J., 2000. Engineered phytoremediation of mercury pollution in soil and water using bacterial genes. In: - Terry, N., Banuelos, G. (Eds.), Phytoremediation of Contaminated Soil and Water. Lewis Publishers, Boca Raton, USA, pp. 201–221.
Mendoza Cozatl, D., Loza Tavera, H., Hernandez Navarro, A., Moreno-Sanchez, R., 2005. Sulfur assimilation and glutathione metabolism under cadmium stress in yeast, protists and plants. FEMS Microbiol. Rev. 29, 653–671.
Metwally, A., Safronova, V.I., Belimov, A.A., Dietz, K.J., 2005. Genotypic variation of the response to cadmium toxicity in *Pisum sativum* L. J. Exp. Bot. 56, 167–178.
Miyadate, H., Adachi, S., Hiraizumi, A., Tezuka, K., Nakazawa, N., Kawamoto, T., Katou, K., Kodama, I., Sakurai, K., Takahashi, H., Satoh-Nagasawa, N., Watanabe, A., Fujimura, T., Akagi, H., 2011. OsHMA3, a P18-type of ATPase affects root-to-shoot cadmium translocation in rice by mediating efflux into vacuoles. New Phytol. 189, 190–199.
Mobin, M., Khan, N.A., 2007. Photosynthetic activity, pigment composition and antioxidative response of two mustard (*Brassica juncea*) cultivars differing in photo-synthetic capacity subjected to cadmium stress. J. Plant Physiol. 164, 601–610.
Molassiotis, A., Sotiropoulos, T., Tanou, G., Diamantidis, G., Therios, I., 2006. Boron-induced oxidative damage and antioxidant and nucleolytic responses in shoot tips culture of the apple rootstock EM9 (*Malus domestica* Borkh). Environ. Exp. Bot. 56, 54–62.
Moussa, H., El-Gamal, S., 2010. Effect of salicylic acid pretreatment on cadmium toxicity in wheat. Biol. Plant. 54, 315–320.
Moya, J.L., Ros, R., Picazo, I., 1993. Influence of cadmium and nickel on growth, net photosynthesis and carbohydrate distribution in rice plants. Photosynth. Res. 36, 75–80.
Nawrot, T., Plusquin, M., Hogervorst, J., Roels, H.A., Celis, H., Thijs, L., Vangronsveld, J., Van Hecke, E., Staessen, J.A., 2006. Environmental exposure to cadmium and risk of cancer: a prospective population-based study. Lancet Oncol. 7, 119–126.
Olmos, E., Martínez-Solano, J.R., Piqueras, A., Hellin, E., 2003. Early steps in the oxidative burst induced by cadmium in cultured tobacco cell (BY-2 line). J. Exp. Bot. 54, 291–301.

Paunov, M., Koleva, L., Vassilev, A., Vangronsveld, J., Goltsev, V., 2018. Effects of different metals on photosynthesis: cadmium and zinc affect chlorophyll fluorescence in Durum Wheat. Int. J. Mol. Sci. 9, 19.

Peng, K.J., Luo, C.L., Chen, Y.H., Wang, G.P., Li, X.D., Shen, Z.G., 2009. Cadmium and other metal uptake by *Lobelia chinensis* and *Solanum nigrum* from contaminated soils. Bull. Environ. Contam. Toxicol. 83, 260–264.

Perfus-Barbeoch, L., Leonhardt, N., Vavasseur, A., Forestier, C., 2002. Heavy metal toxicity: cadmium permeates through calcium channels and disturbs the plant water status. Plant J. 32, 539–548.

Pilon-Smits, E., Pilon, M., 2002. Phytoremediation of metals using transgenic plants. Crit. Rev. Plant Sci. 21, 439–456.

Polle, A., Schutzendubel, A., 2003. Heavy metal signalling in plants: linking cellular and organismic responses. In: Hirt, H., Shinozaki, K. (Eds.), Plant Responses to Abiotic Stress. Springer-Verlag, Berlin, Heidelberg, Germany, pp. 187–215, ISBN 3540200371.

Popova, L., Maslenkova, L., Yordanova, R., Krantev, A., Szalai, G., Janda, T., 2008. Salicylic acid protects photosynthesis against cadmium toxicity in pea plants. Gen. Appl. Plant Physiol. 34, 133–144 (Special issue).

Popova, L.P., Maslenkova, L.T., Yordanova, R.Y., Ivanova, A.P., Krantev, A.P., Szalai, G., Janda, T., 2009. Exogenous treatment with salicylic acid attenuates cadmium toxicity in pea seedlings. Plant Physiol. Biochem. 47, 224–231.

Prasad, M.N.V., 2005. Nickelophilous plants and their significance in phytotechnologies. Braz. J. Plant Physiol. 17, 113–128.

Rai, V., Khatoon, S., Bisht, S.S., Mehrotra, S., 2005. Effect of cadmium on growth, ultramorphology of leaf and secondary metabolites of *Phyllanthus amarus* Schum. and Thonn. Chemosphere 61, 1644–1650.

Rascio, N., Navari-Izzo, F., 2011. Heavy metal hyperaccumulating plants: how and why do they do it? And what makes them so interesting? Plant Sci. 180, 169–181.

Rascio, N., DallaVecchia, F., La Rocca, N., Barbato, R., Pagliano, C., Raviolo, M., Gonnelli, C., Gabbrielli, R., 2008. Metal accumulation and damage in rice (cv. *Vialone nano*) seedlings exposed to cadmium. Environ. Exp. Bot. 62, 267–278.

Reeves, R.D., Baker, A.J.M., 2000. Metal-accumulating plants. In: Raskin, I., Ensley, B.D. (Eds.), Phytoremediation of Toxic Metals: Using Plants to Clean Up the Environment. John Wiley and Sons Inc, New York, USA, pp. 193–229, ISBN 9780471192541.

Rodriguez, M., Canales, E., Borras-Hidalgo, O., 2005. Molecular aspects of abiotic stress in plants. Biotecnol. Apl. 22, 1–10.

Schutzendubel, A., Schwanz, P., Teichmann, T., Gross, K., Heyser, R.L., Godbold, D.L., Polle, A., 2001. Cadmium induced changes in antioxidative systems, hydrogen peroxide content, and differentiation in Scots pine roots. Plant Physiol. 127, 887–898.

Seth, C.S., Misra, V., Chauhan, L.K., Singh, R.R., 2008. Genotoxicity of cadmium on root meristem cells of *Allium cepa*: cytogenetic and Comet assay approach. Ecotoxicol. Environ. Saf. 71, 711–716.

Sethy, S.K., Ghosh, S., 2013. Effect of heavy metals on germination of seeds. J. Nat. Sci. Biol. Med. 4, 272–275.

Sgherri, C., Quartacci, M.F., Izzo, R., Navari-Izzo, F., 2002. Relation between lipoic acid and cell redox status in wheat grown in excess copper. Plant Physiol. Biochem. 40, 591–597.

Shah, K., Dubey, R.S., 1995. Effect of cadmium on RNA level as well as activity and molecular forms of ribonuclease in growing rice seedlings. Plant Physiol. Biochem. 33, 577–584.

Shanmugaraj, B.M., Harish, M.C., Balamurugan, S., Sathishkumar, R., 2013. Cadmium induced physio-biochemical and molecular response in *Brassica juncea*. Int. J. Phytoremediation 15, 206–218.

Shi, G., Liu, C., Cai, Q., Liu, Q., Hou, C., 2010. Cadmium accumulation and tolerance of two safflower cultivars in relation to photosynthesis and antioxidative enzymes. Bull. Environ. Contam. Toxicol. 85, 256–263.

Shimpei, U., Watanabe, I., Yoshitomi, A., Kiyono, M., Katsuji, K., 2006. Characteristics of cadmium accumulation and tolerance in novel Cd-accumulating crops, *Avena strigosa* and *Crotalaria juncea*. J. Exp. Bot. 57, 2955–2965.

Siedlecka, A., Krupa, Z., Samuelsson, G., Oquist, G., Gardestrom, P., 1997. Primary carbon metabolism in *Phaseolus vulgaris* plants under Cd:Fe interaction. Plant Physiol. Biochem. 35, 951–957.

Siedlecka, A., Samuelsson, G., Gardenstrom, P., Kleczkowski, L.A., Krupa, Z., 1998. The "activatory model" of plant response to moderate cadmium stress-relationship between carbonic anhydrase and Rubisco. In: Garab, G. (Ed.), Photosynthesis: Mechanisms and Effects. Kluwer Academic, Dordrecht, Boston, London, pp. 2677–2680.

Sigfridsson, K.G.V., Bernát, G., Mamedov, F., Styring, S., 2004. Molecular interference of $Cd^{2+}$ with Photosystem II. Biochim. Biophys. Acta 1659, 19–31.

Singh, S., Parihar, P., Singh, R., Singh, V.P., Prasad, S.M., 2015. Heavy metal tolerance in plants: role of transcriptomics, proteomics, metabolomics, and ionomics. Front. Plant Sci. 6, 1143.

Song, Y., Jin, L., Wang, X., 2017. Cadmium absorption and transportation pathways in plants. Int. J. Phytoremediation 19, 133–141.

Sood, A., Uniyal, P.L., Prasanna, R., Ahluwalia, A.S., 2012. Phytoremediation potential of aquatic macrophyte Azolla. Ambio 41, 122–137.

Stohs, S.J., Bagchi, D., Hassoun, E., Bagchi, M., 2000. Oxidative mechanisms in the toxicity of chromium and cadmium ions. J. Environ. Pathol. Toxicol. Oncol. 19, 201–213.

Subhashini, V., Swamy, A.V.V.S., 2013. Phytoremediation of cadmium and chromium from contaminated soils using *Physalis minima* Linn. Am. Int. J. Res. Form. Appl. Nat. Sci. 3, 119–122.

Tak, H.I., Ahmad, F., Babalola, O.O., 2013. Advances in the application of plant growth-promoting rhizobacteria in phytoremediation of heavy metals. In: Whitacre, D.M. (Ed.), Reviews of Environmental Contamination and Toxicology. In: vol. 223. Springer, New York, USA, pp. 33–52.

Thangavel, P., Long, S., Minocha, R., 2007. Changes in phytochelatins and their biosynthetic intermediates in red spruce (*Picea rubens* Sarg.) cell suspension culture under cadmium and zinc stress. Plant Cell Tissue Organ Cult. 88, 201–216.

Tran, T.A., Popova, L.P., 2013. Functions and toxicity of cadmium in plants: recent advances and future prospects. Turk. J. Bot. 37, 1–13.

Tudoreanu, L., Phillips, C.J.C., 2004. Modelling cadmium uptake and accumulation in plants. Adv. Agron. 84, 121–157.

Ueno, D., Yamaji, N., Kono, I., Huang, C.F., Ando, T., Yano, M., Ma, J.F., 2010. Gene limiting cadmium accumulation in rice. Proc. Natl. Acad. Sci. U. S. A. 107, 16500–16505.

Van Assche, F., Clijsters, H., 1990. Effects of metals on enzyme activity in plants. Plant Cell Environ. 13, 195–206.

Vazquez, S., Goldsbrough, P., Carpena, R.O., 2006. Assessing the relative contributions of phytochelatins and the cell wall to cadmium resistance in white lupin. Physiol. Plant. 128, 487–495.

Verbruggen, N., Hermans, C., Schat, H., 2009. Mechanisms to cope with arsenic or cadmium excess in plants. Curr. Opin. Plant Biol. 12, 364–372.

Wahid, A., Ghani, A., Javed, F., 2008. Effect of cadmium on photosynthesis, nutrition and growth of mungbean. Agron. Sustain. Dev. 28, 273–280.

Wu, F.B., Zhang, G.P., Dominy, P., 2003. Four barley genotypes respond differently to cadmium: lipid peroxidation and activities of antioxidant capacity. Environ. Exp. Bot. 50, 67–78.

Wu, F.B., Chen, F., Wei, K., Zhang, G.P., 2004. Effect of cadmium on free amino acid, glutathione and ascorbic acid concentrations in two barley genotypes (*Hordeum vulgare* L.) differing in cadmium tolerance. Chemosphere 57, 447–454.

Wu, F., Zhang, G., Dominy, P., Wu, H., Bachir, D.M.L., 2007. Differences in yield components and kernel Cd accumulation in response to Cd toxicity in four barley genotypes. Chemosphere 70, 83–92.

Yadav, S.K., 2010. Heavy metals toxicity in plants: an overview on the role of glutathione and phytochelatins in heavy metal stress tolerance of plants. S. Afr. J. Bot. 76, 167–179.

Yang, T., Poovaiah, B.W., 2003. Calcium/calmodulin-mediated signal network in plants. Trends Plant Sci. 8, 505–512.

Yu, R., Tang, Y., Liu, C., Du, X., Miao, C., Shi, G., 2017. Comparative transcriptomic analysis reveals the roles of ROS scavenging genes in response to cadmium in two pak choi cultivars. Sci. Rep. 7, 9217.

Zhou, J., Wan, H., He, J., Lyu, D., Li, H., 2017. Integration of cadmium accumulation, subcellular distribution, and physiological responses to understand cadmium tolerance in apple rootstocks. Front. Plant Sci. 8, 966.

Zhu, Y.L., Pilon-Smits, E.A., Tarun, A.S., Weber, S.U., Jouanin, L., Terry, N., 1999. Cadmium tolerance and accumulation in Indian mustard is enhanced by overexpressing γ-glutamylcysteine synthetase. Plant Physiol. 121, 1169–1178.

Zhu, Y.G., Chen, S.B., Yang, J.C., 2004. Effects of soil amendments on lead uptake by two vegetable crops from a lead-contaminated soil from Anhui, China. Environ. Int. 30, 351–356.

## FURTHER READING

Chugh, L.K., Sawhney, S.K., 1999. Photosynthetic activities of *Pisum sativum* seedlings grown in presence of cadmium. Plant Physiol. Biochem. 37, 297–303.

Gustav, R., 1974. Hazardous heavy metals: cadmium, mercury, lead and arsenic. WHO International Reference Centre for Waste Disposal, IRCWD News, Switzerland, pp. 1–11.

Krupa, Z., 1988. Cadmium-induced changes in the composition and structure of the light-harvesting complex II in radish cotyledons. Physiol. Plant. 73, 518–524.

Prasad, M.N.V., 1995. Cadmium toxicity and tolerance in vascular plants. Environ. Exp. Bot. 35, 525–545.

# CHAPTER 2

# Cadmium Toxicity and Tolerance in Micro- and Phytobiomes

**Abin Sebastian*, Ashwini Nangia*, Majeti Narasimha Vara Prasad†, Ladawan Rattanapolsan‡, Woranan Nakbanpote‡**
*School of Chemistry, University of Hyderabad, Central University P.O., Hyderabad, India
†Department of Plant Sciences, University of Hyderabad, Hyderabad, India
‡Department of Biology, Faculty of Science, Mahasarakharm University, Khamriang, Thailand

## Abstract

Heavy metals, such as Cd, are contaminating soil and water bodies. Toxicity of Cd retards the growth of microbes and plants. The entry of Cd into the cells occurs through transition metal transporters. Diminished cellular micronutrient stores, and a decrease in the activity of enzymes such as dehydrogenases, alkaline phosphatases, and so forth, are the immediate toxic effect of Cd on microbes. Green plants experience oxidative stress upon Cd exposure. The deleterious effects of Cd retard photosynthesis and biomass productivity in plants. However, there are Cd defense mechanisms in plants. These mechanisms are root exudation, Cd exclusion, and speciation of Cd with carboxylic acids, or metal binding peptides such as phytochelatins. Cadmium stress on bacteria and plants has ecological impacts, such as inhibition of organic matter decomposition and loss of biodiversity, respectively. Ecological assessment of Cd toxicity also helps determine Cd-tolerant plant species that can be used in the ecofriendly cleanup of Cd in the environment.

**Keywords:** Cadmium, Environment, Metal transporters, Oxidative stress, Metal tolerance, Ecology

## 1 INTRODUCTION

Cadmium, a group IIB transition metal in the periodic table of elements, is categorized as a hazardous metal. There are 8 isotopic forms of Cd in nature, and two of them are radioactive (Rosman and De Laeter, 1978). Cadmium has many industrial uses (Tchounwou et al., 2012) (Fig. 1). For example, cadmium hydroxide is one of the two principal electrode materials in Ni-Cd batteries. About 70.0% of refined Cd is used for the manufacture of Ni-Cd batteries. These batteries perform well under low and high temperatures, relative to other battery chemistries. Cadmium sulfide and

*Cadmium Toxicity and Tolerance in Plants*
https://doi.org/10.1016/B978-0-12-814864-8.00002-4

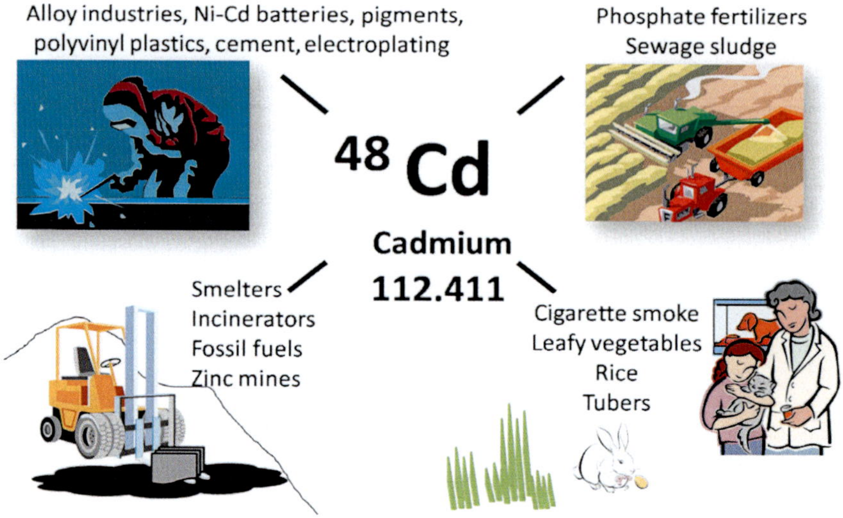


**Fig. 1** Cadmium—industrial uses and environmental exposure.

cadmium sulfoselenide are used for coloring bright yellow to deep red in plastics, ceramics, glasses, and enamels. Polyvinyl chloride degradation is retarded in the presence of Cd; hence, the metal is used in plastic stabilizers. A coating made with Cd is stable under saline and high-temperature conditions. The high ductility of Cd makes it suitable for anticorrosive alloys. The addition of Cd to metals results in alloys with a low melting point. These alloys are used for the manufacture of aircraft, solar cells, and so forth. Electroplating of steel is another important industrial application that has ceased because of environmental concerns.

Chemical properties of Cd show similarities with that of Zn and Hg (Traina, 1999). The melting and boiling point of Cd are 321.0 and 765.0°C, respectively. Cadmium has a density of $8.65\,g\,cm^{-3}$. This metal mainly occurs in a +2 oxidation state, although a +1 oxidation state has been reported in some diatomic compounds. The electronic configuration of Cd is [Kr] $4d^{10}5s^2$. The metal reacts with air, halogens, and acids. Compounds such as CdO and CdS are being used as precursors for the preparation of Cd salts. Cadmium oxide is commonly used as a catalyst in redox reactions, hydrogenation reactions, polymerization, and cleavage. Greenockite is the most important ore of Cd (CdS). But the metal is being refined from Zn ores.

Cadmium causes health disorders such as bone mineralization, kidney damage, and cancer (Godt et al., 2006). The hazardous nature of Cd was revealed after the outbreak of Itai-Itai disease in Toyama Prefecture, Japan, in 1912 (Yoshida, 1999). The disease is characterized by osteomalacia, osteoporosis, renal tubular dysfunction, malabsorption, and anemia. The reason for the outbreak of this disease was the environmental pollution of Cd from a Zn mine leachate. The leachate mixed with the river, and the usage of water for the irrigation of rice paddies resulted in Cd contamination of rice paddies. The consumption of Cd-contaminated rice resulted in health disorders in humans. Recently, Cd-related health problems have been reported in China, Thailand, and Belgium (Degryse et al., 2003; Sebastian and Prasad, 2014a). The technology improvement in the production of Cd has resulted in a decrease of Cd emissions into the environment. But studies have indicated that consumption of Cd has increased from 8400 tons in 2003 to 20,400 tons in 2007. This has resulted in occupational exposure of Cd from industries dealing with heavy metal mining, metallurgy, manufacturing of nickel-cadmium batteries, pigments, plastic stabilizers, and anticorrosive products. Cigarette smoke is also a potential route of Cd exposure in humans (Piadé et al., 2015). Agriculture fields have also been contaminated with Cd because of the usage of rock phosphate fertilizer containing Cd (Roberts, 2014). The application of phosphate fertilizer-contaminated topsoil, and the leafy vegetables that rely on topsoil for nutrition accumulate Cd. All these reports indicate that the prevention of Cd exposure is an important matter in relation to human health. It is important to understand the toxic effects of Cd accumulation in plants, because the major route of Cd intake in humans is through plant-based food products.

The daily dietary intake of Cd should be limited to 60–70.0 μg (Satarug et al., 2010; Chunhabundit, 2016). The permissible level of Cd in cereal grains is 100.0 $\mu g\,kg^{-1}$. Cadmium accumulation in edible plants has been reported. The contamination of Cd in agriculture fields through the application of phosphate fertilizer has resulted in a more-than-permissible level of Cd accumulation (0.7–1.17 $mg\,kg^{-1}$) in the seeds of durum wheat, flax, sunflower, and potatoes (Grant et al., 1998). The accumulation of Cd in plants depends on factors associated with plants and soil (Sebastian and Prasad, 2014a). Cadmium-accumulating factors associated with soil include physicochemical properties of the soil and microbial action. Field experiments conducted with edible crops such as red beets, field pumpkins, chicory, common beans, barley, white cabbage, maize, alfalfa, and common parsnips indicated that the metal accumulates more in leaves, unlike cereals, where more Cd accumulates in roots (Sekara et al., 2005). Green manure crops such as *Avena strigosa* Schreb. and *Crotalaria juncea* L. have been found to accumulate Cd of more than 100.0 $mg\,kg^{-1}$ dry weight in leaves (Uraguchi et al., 2006). This kind of information is important to identify potential plant species for the usage of green technologies such as phytoremediation and phytoextraction. These technologies utilize the Cd accumulation capacity of the plants for removal of toxic metals in the environment. The environments in which these technologies are being applied include aqueous, submerged, or aerated soils. Also, the presence of microbes alters plant availability of Cd. Hence, understanding the mechanism of Cd accumulation and the factors associated with Cd tolerance in plants are important for formulating Cd-safe food products and improving green technologies for the removal of Cd from the environment.

## 2 CADMIUM UPTAKE AND TRANSPORT

The direct contact of microbes and plants with extracellular Cd occurs through the cell wall, which is made of polymers such as peptidoglycan and cellulose, respectively. These structural components form a matrix over the cells and act as the metal-adsorbing surface. Metal ions such as Cd bind with the cell wall, and the process is controlled by factors such as concentration of metal ions, cation exchange capacity of the soil, redox potential, pH, temperature, and so forth (Das et al., 2008; Parrotta et al., 2015). It is well known that a higher concentration of a metal ion increases the chance of metal accumulation. But soil with a higher cation exchange capacity tends to hold more Cd in the −ve sites, and results in decreased Cd accumulation (Sebastian and Prasad, 2014b). The presence of organic matter and clay increases cation exchange capacity of the soil, and hence decreases metal exposure to microbes and plants (Sebastian and Prasad, 2014b). Higher redox potential reduces Cd, and results in precipitation of the Cd ion (Sebastian and Prasad, 2014a). Similarly, higher pH also supports precipitation reactions of Cd, and minimizes Cd uptake in plants and microbes (Sebastian and Prasad, 2014a). It has been reported that a rise in temperature from 25.0 to 37.0°C enhances Cd accumulation in safflower and *Salmonella enterica* 43C (Pourghasemian et al., 2013; Khan et al., 2016). So it is clear that the initial steps in Cd uptake are under the control of environmental factors. The presence of extracellular mucilage-containing substances, such as root exudates and bacterial siderophores, also influences Cd uptake (Hill et al., 2002; Sinha and Mukherjee, 2008).

The uptake of Cd from the cell wall matrix to the protoplast of microbes and plant cells occurs through transition metal transporters (Table 1). The

**Table 1** Cadmium transporters in microbes and plants

| Transporter | Metal ion | Cellular compartment | Expression | Reference |
|---|---|---|---|---|
| *Nramp family* | | | | |
| AtNramp1 | Cd, Fe | Plasma membrane, Tonoplast | Root, leaf | Thomine et al. (2000) |
| AtNramp3 | Fe, Cd, Mn | Vacuole | Shoot, root | Thomine et al. (2003) |
| OsNramp1 | Cd, Mn, and Fe | Plasma membrane | Leaf, root | Takahashi et al. (2011) |
| OsNramp 5 | Cd, Mn, and Fe | Plasma membrane | Root | Sasaki et al. (2012) |
| Bacterial/ Yeast Nramp | Cd | Cytoplasmic membrane | Bacteria, yeast | Thomine et al. (2000) |
| *CDF family* | | | | |
| TgMIP1 | Cd, Co, Ni, Zn | Vacuole | Leaf | Persans et al. (2001) |
| OsMTP1 | Cd | Plasma membrane | Root | Yuan et al. (2012) |
| AtMTP1 | Cd, Zn, Ni | Tonoplast | Root, leaf, flower | Desbrosses-Fonrouge et al. (2005) |
| *ZIP family* | | | | |
| IRT1 | Fe, Zn, Mn, Cd | Plasma membrane | Root | Connolly et al. (2002) |
| TcZNT1 | Zn, Cd | Plasma membrane | Root, shoot | Pence et al. (2000) |
| OsZIP1 | Cd,Zn | Plasma membrane | Root, leaf | Ramesh et al. (2003) |
| Bacterial/ Yeast ZIP | Cd | Cytoplasmic membrane | Bacteria, yeast | Jeong and Eide (2013) |
| *Cation/H+ antiporters family* | | | | |
| At CAX 2 | Ca, Mn, Cd | Vacuole | Root | Hirschi et al. (2000) |
| AtCAX5 | Cd, Mn, Cu | Plasma membrane | Root, leaf | Hossain and Komatsu (2012) |

*Continued*

**Table 1** Cadmium transporters in microbes and plants—cont'd

| Transporter | Metal ion | Cellular compartment | Expression | Reference |
|---|---|---|---|---|
| *ABC transporters family* | | | | |
| ID17 | PC-Cd, GS-Cd | Vacuole | Roots | Yamaguchi et al. (2002) |
| OsPDR9 | Cd, Zn | Plasma membrane | Roots | Moons (2003) |
| AtPDR8 | Cd | Plasma membrane | Root, leaf | Kim et al. (2007) |
| AtMRP3 | Cd | Pericycle | Root, leaf | Gaillard et al. (2008) |
| *Heavymetal ATPase family* | | | | |
| OsHMA2 | Cd,Zn | Xylem | Root | Takahashi et al. (2012) |
| OsHMA3 | Cd | Vacuole | Root | Ueno et al. (2010) |
| AtHMA3 | Cu, Zn, Co, Cd | Tonoplast | Root, leaf | Morel et al. (2009) |
| OsYSL2 | Cd, Fe | Phloem | Root, leaf | Koike et al. (2004) |
| *LCT1* | Cd, K, Mg, Ca | Phloem | Leaf, panicle | Uraguchi et al. (2011) |
| *cadA* | Cd | Membrane vesicle | Bacteria | Tsai et al. (1992) |

active uptake of Cd in bacteria occurs through MntH (member of Natural Resistance-Associated Macrophage Proteins-Nramp) transporters (Nevo and Nelson, 2006). This family of metal transporters affects the uptake of Cd, Fe, Co, and Mn. The energy requirement of this transporter for Cd uptake in bacteria has been confirmed with the help of inhibition studies with protonophore carbonyl cyanide m-chlorophenyl hydrazone. It was found that the addition of this Nramp-specific inhibitor decreases Cd accumulation in *Staphylococcus aureus* (Tynecka et al., 1998). The uptake of Cd through this transporter is associated with lowering cytosolic pH; hence, it concluded that uptake of Cd through MntH occurs in symport with $H^+$ ions. Zrt and Irt-like protein (ZIP) transporters, such as ZupT and ZRT1, are another class of Cd transporters found in the plasma membrane of bacteria and yeast (Grass et al., 2002; Gitan et al., 2003). These transporters usually transport Zn ions. The glutamate and histidine residue in the transmembrane region helps Cd uptake through this transporter. MntA and CdtB (P-type ATPase transporter) are membrane transporters that have been reported to be involved in the uptake of both Mn and Cd (Hao et al., 1999). The generation of mutants lacking the preceding membrane transporters often results in Cd insensitivity in bacteria.

Plant uptake of Cd also occurs through transition metal transporters (Hall and Williams, 2003). There are 6 families of transition metal transporters involved in Cd uptake in plants. These families are NRAMP, ZIP, P-type ATPase, ABC (ATP-binding cassette transporter), CDF (cation diffusion facilitator family), and CAX (Cation/$H^+$ transporters), respectively. The involvement of these transporters in Cd uptake has been reported mainly in rice and Arabidopsis (Table 1). The function and activities of NRAMP, ZIP, and P-type ATPases (Heavy metal ATPases) are similar to that found in bacteria. But the ZIP transporters, such as IRT1, in plants, are well known for Fe transport (Connolly et al., 2002). The ABC transporter allows entry of Cd into the root with the help of ATP hydrolysis (Yamaguchi et al., 2002). This group of transporters acts as both Cd importer and exporter. ABC transporters contain at least one transmembrane subunit and a nucleotide binding domain, which helps in ATP hydrolysis for the uptake of Cd. It has been reported that these membrane proteins are critical for mineral nutrition, plant development, and biotic and abiotic stress tolerance (Wilkens, 2015). The role of ABC transporter has also been reported in the detoxification and transport of conjugated compounds (Yamaguchi et al., 2002). So it is clear that this family of proteins helps with both Cd uptake and intracellular trafficking of Cd. The cation diffusion facilitator (CDF) family is involved in the transport of Mn, Zn, Co, Pb, and Cd (Yuan et al., 2012). The CDF family of transporters is grouped into four lineages. These transporters contain six transmembrane domains, and the C-terminal of this protein involved in Cd binding. Cadmium uptake through the plasma membrane-located cation/$H^+$ transporters has also been reported in *Arabidospis* (Hirschi et al., 2000). This transporter generally mediates the transport of Ca in plants for regulation of the cytosolic Ca pool. Thus, the preceding report points out that Cd uptake in plants occurs through transition metal transporters meant for nutrient uptake in plants.

Cadmium in the cell is transported with the help of phytochelatins, metallothionein amino acids, organic acids, metal chaperones, secondary metabolites, and membrane transporters (Hall and Williams, 2003). The transport process may be a short distance or long distance. The intracellular transport of Cd mainly occurs in the plant vacuole. This mode of transport requires the help of Cd chelators such as organic acids and phytochelatin (Yamaguchi et al., 2002; Hall and Williams, 2003). The carboxylic and the sulphhydryl groups present in these chelators bind with Cd prior to transport into the vacuole. The entry of these metal complexes into the vacuole is achieved with vacuolar transporters that belong to families such as NRAMP, P-type ATPase, ABC, LCT, and CE (Hall and Williams, 2003). The intracellular transport of Cd helps in Cd detoxification in the root, and the process is critical for Cd tolerance.

The short-distance transport of the Cd occurs through both apoplastic and symplastic pathways. The apoplastic pathways mainly contribute to Cd binding in the cell wall, and the carboxyl carbon present in the cell wall helps in the immobilization of Cd (Parrotta et al., 2015). The chance of movement of Cd via this pathway depends on the plasticity of the cell wall. The suberin layer acts as a barrier for Cd transport in the apoplastic route (Tao et al., 2017). The transport of Cd from the epidermis to conductive elements takes place via the symplastic route. Xylem loading of Cd has been reported with the help of a membrane transporter, namely OsHMA3a, in rice plants (Uraguchi et al., 2009). But transporters such as OsHMA3 play an important role in the decrease of Cd transport to the shoot (Ueno et al., 2010). This transporter enhanced Cd sequestration in the vacuole of roots cells, and decreased the availability of Cd for aerial transport. P1B type ATPases such as AtHMA2 and AtHMA4 also reported participating in xylem loading of Cd and Zn in Arabidopsis (Morel et al., 2009). The aerial transport of Cd occurs with the transpiration stream, and the transport of Cd within the leaf takes place with the help of transition metal transporters. The transport of Cd in the xylem was either in the form of free ionic form or Cd complexes. The Cd complexes formed with organic acids, phytochelatins, sugars, phosphates, and amino acids such as cysteine (Sigel et al., 2013; Álvarez-Fernández et al., 2014).

The remobilization of Cd via phloem was also found to associate with metal complexes, and the process occurs with the activity of metal transporters such as those in the LCT1 family (Uraguchi et al., 2011). This transporter functions as a Cd exporter, and occurs in nodes and leaf blades. Phytochelatins and glutathione are the predominant Cd binding agents in the phloem (Mendoza-Cózatl et al., 2008). These peptides are also reported to take part in the transport of plant nutrients such as Fe. The amount of cysteine in the phloem increased after Cd exposure, and this indicates the role of this amino acid in the long-distance transport of Cd (Jozefczak et al., 2012). The production of the thiol chelators in the phloem depend on the availability of Cd, rather than changes in the expression of enzymes involved in the formation of these chelators. These studies also pointed to the occurrence of competitive binding of Cd to metal chelators present in phloem, which ultimately decreases the availability of essential micronutrients for storage in seeds.

## 3 CADMIUM TOXICITY

Cadmium stress affects the growth of microbes and plants. The uptake of trace metals, enzyme activities, protein synthesis, DNA replication, and cell division is altered in the course of Cd exposure in microbes (Trevors et al., 1986; Ron et al., 1992) (Table 2). Cyanobacteria and algae are more prone to Cd toxicity, compared with bacteria and fungi (Mohammadian Fazli et al., 2015). Plasmid-encoded resistance is found to restrict the accumulation of Cd in bacteria (Nies, 1992). One of the immediate outcomes of Cd exposure in bacteria could be depletion of cellular reserves of Zn and Mn (Palmer and Skaar, 2016). This effect occurs as an outcompeting action of Cd on Zn and Mn uptake. The ZupT transporters, which transport Zn in bacteria, have also been found to have more affinity to Cd compared with Zn (Grass et al., 2002). This affinity creates more bacterial uptake of Cd, compared with Zn during Cd exposure. Studies of endophytic bacteria isolated from *Murdannia spectabilis* growing in the vicinity of Zn mine sites indicates a relationship between Zn and Cd toxicity (Fig. 2) (Rattanapolsan, 2018). A cobalt–zinc-cadmium transporter (Czc) also allows passage of Cd, and assists in Cd exclusion (Nies, 1995). Thus, the disturbance in the active trace metal uptake system could be the reason for the hindering of micronutrient uptake during Cd stress in bacteria.

Microbial enzymes such as dehydrogenases and alkaline phosphatases show an immediate response to Cd toxicity (Wyszkowska and Wyszkowski, 2003). The activity of these enzymes is retarded because of the interaction of Cd with the enzyme-substrate complex, or active site of the enzymes. The binding of Cd with the active site of alkaline phosphatases has been found to completely diminish the enzyme activity (McGrath, 1999). The affinity of Cd toward cysteinyl and histidyl groups of proteins retards enzyme activity (Sóvágó and Várnagy, 2013). The ability of Cd to bind with phosphates also creates a block in the binding of substrates to alkaline phosphatases (Sigel et al., 2013). Protein synthesis in bacteria is inhibited during Cd exposure (Wang and Crowley, 2005). The decrease in the production of tRNA and aminoacyl tRNA synthetase activities were the immediate reason for

**Table 2** Cadmium toxicity in microbes and plants

| Target | Organism | Reference |
|---|---|---|
| Cell division | *Hymenomonas carterae*, *Lactobacillus plantarum* | Stillwell and Corum (1982)<br>Zhai et al. (2017) |
| Rhizotoxicity | *Sedum alfredii* | Tian et al. (2011) |
| Nuclear damage | *Allium sativum*, *Vicia faba* | Celik et al. (2006) |
| Mineral nutrition | *Pfaffia glomerata* | Gomes et al. (2013) |
| Hormonal changes | *Kosteletzkya virginica* | Han et al. (2013) |
| Signaling pathway | *Arabidopsis thaliana* | Smeets et al. (2013) |
| Flowering | *Tagets erecta* | Eid et al. (2016) |
| Water balance | *Urtica pilulifera* | Özyiğit and Akinci (2009) |
| Plant pigments | *Cajanus cajan* | Khudsar et al. (2001) |
| Xylem and Phloem | *Pisum sativum* | Rodríguez-serrano et al. (2006) |
| Stomatal conductance | *Nicotiana tabacum* | Liu et al. (2015) |
| Cell wall alterations | *Medicago truncatula* | Rahoui et al. (2017) |
| Photo synthesis | *Phragmites australis* | Pietrini et al. (2003) |
| Respiration | *Hordeum distichum* | Garmash and Golovko (2009) |
| Antioxidant enzymes | *Nannochloropsis oculata* | Lee and Shin (2003) |
| Antioxidants | *Glycyrrhiza uralensis* | Zheng et al. (2010) |
| DNA pattern | *Jatropha curcas* | Maravi et al. (2017) |
| Proteome alteration | *Oryza sativa* | Sebastian and Prasad (2015a, 2015b) |
| Osmolytes | *Cicer arietinum* | Ahmad et al. (2016) |
| Membrane damage | *Triticum aestivum* | Zhao et al. (2005) |
| Seed germination | *Vigna radiata* | Ashraf et al. (2016) |
| Ferrittin content | *Glycine max* | Balestrasse et al. (2004) |

inhibition of protein synthesis during Cd stress. Morphological aberrations were the major change found in bacteria after Cd exposure (Higham et al., 1986). This effect arose because of the interference of Cd with cell division. Bacterial DNA has undergone deleterious changes during Cd stress.

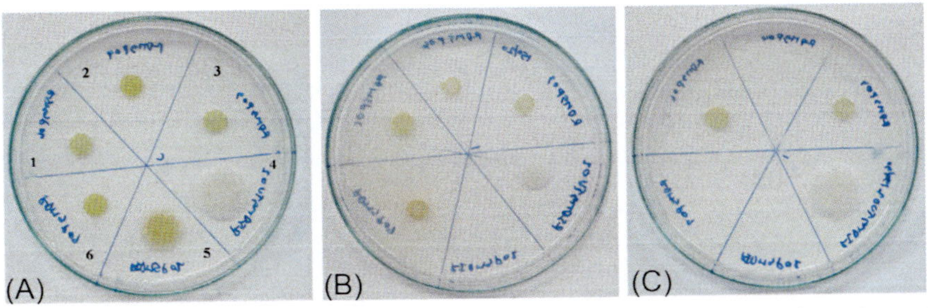


**Fig. 2** Relationship between Zn stress tolerance and Cd toxicity in endophytic bacteria isolated from *Murdannia spectabilis* growing in the vicinity of zinc mines, Mae Sot, Tak Province, Thailand. (A) Endophytic bacteria growing on ½ TSA medium. (B) Supplement with Zn 150.0 mg $L^{-1}$ plus Cd 20.0 mg $L^{-1}$ 24.0 h. (C) Supplement with Zn 150 mg $L^{-1}$ plus Cd 30.0 mg $L^{-1}$ for 48.0 h. Abbreviations: 1—*Microbacterium neimengense*, 2—*Curtobacterium* sp., 3—*Microbacterium neimengense*, 4—*Bacillus* sp., 5—*Pseudomonas stutzeri*, 6—*Curtobacterium* sp. The result indicated that the increase of Cd concentration decrease growth of Zn tolerant bacterial strains.

Malfunctioning of genes encoding ribosomal and zinc-binding proteins and oxidative stress has been reported to cause the DNA damage in bacterial cells containing Cd (Wang and Crowley, 2005). Single strand breaks occurred in DNA during oxidative stress, and this disrupted the DNA replication process. It has also been reported that Cd inhibits DNA mismatch repair, and affects cell division in bacteria (Lützen et al., 2004).

Green plants subjected to Cd stress also experience oxidative stress (Petrov et al., 2015). Reactive oxygen species formation from plasma membrane NADPH oxidase is very common during the uptake of Cd (Heyno et al., 2008). This enzyme converts molecular oxygen to superoxide radicals, which is then dismutated to hydrogen peroxide by superoxide dismutase. The increase in NADPH oxidase was the outcome of enhancement in $H^{+}$- ATPase activity in response to Cd accumulation. So it can be concluded that production of ROS begins at the plasma membrane during Cd stress. Cellular organelles, such as mitochondria and chloroplast development, are affected during Cd stress (Djebali et al., 2005; Keunen et al., 2011). The deleterious action of Cd on these organelles resulted in the malfunction of energy metabolism, and retarded housekeeping processes in plants.

Metalloproteins in photosynthetic electron transport and mitochondrial electron transport also malfunction with Cd stress (Belyaeva et al., 2012; Parmar et al., 2013). This creates a blockage in the electron transport system that ultimately produces ROS. A low concentration of Cd stimulates the activity of antioxidant enzyme systems, and increases proline content in plants (Chaia et al., 2013). These changes help prevent deleterious effect of ROS under mild Cd stress. But the higher-level accumulation of Cd in the cell retarded activity of antioxidant systems, and generated free radicals in the cell (Sebastian and Prasad, 2013). Lipid peroxidation in membrane systems and ionic leakage were the result of the generation of free radicals in the cells during Cd stress (Gonçalves et al., 2007). The oxidative stress that arises during Cd accumulation also retards photosynthesis (Parmar et al., 2013). The low photosynthetic efficiency lowers biomass productivity of plants during Cd stress. It has been found that Cd binds with Ca binding sites in PSII, decreasing the activity of enzymes involved in chlorophyll and antioxidants synthesis, and induces changes in chloroplast's ultrastructure (Djebali et al., 2005; Parmar et al., 2013). Thus, apart from oxidative stress, Cd-inducible structural changes in chloroplast, as well as a decrease of phytochemicals, is important for photosynthetic activity, retarding photosynthesis during Cd stress.

Cadmium-induced cell death can be found in root cells (Zhang et al., 2015). Chromosomal aberrations are very often noticed in plant roots exposed to Cd (Nefic et al., 2013). Disturbances in mitosis during Cd treatment have resulted in the formation of more than one nucleus in *Allium cepa* (Nefic et al., 2013). These changes in the root are reflected as an abnormal growth pattern in the epidermal and cortical layers of the apical meristem. Therefore, it has been concluded that the long-term exposure to Cd decreases root elongation, and promotes decomposition of roots. This drastically alters mineral nutrient uptake in plants. The alteration of the Ca and K level in plants upon Cd stress results in abnormal signal transduction and stomatal gas exchange, respectively (Thévenod and Lee, 2013). Cadmium-inducible Fe deficiency also decreases plant growth. The decrease in Fe accumulation during Cd stress not only retards photosynthesis, but also affects the fixation of nitrogen (Sebastian and Prasad, 2015a,b). These changes ultimately decreased both carbohydrate and protein turnover in plants, which has a devastating effect on growth and development.

## 4 MECHANISMS OF CADMIUM TOLERANCE

Accumulation of Cd in the cell initiates defensive responses against Cd toxicity (Shahid et al., 2017). Tolerance mechanisms operating in response to Cd can be extracellular or intracellular (Fig. 3). Both microbes and plants responded to Cd with the secretion of mucilage substances called siderophores and phytosiderophores, respectively (Rajkumar et al., 2010;

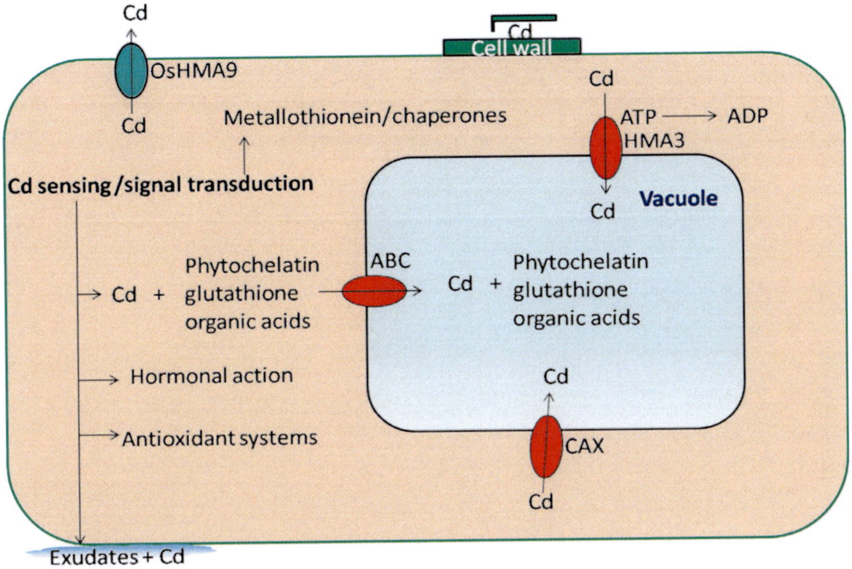


**Fig. 3** Mechanisms of cadmium stress tolerance.

Römheld, 1991). These compounds form hexadentate or octadentate complexes with Fe, and help the uptake of Fe. Siderophores can be catecholates (phenolates), hydroxamates, or carboxylates. Studies with *Pseudomonas aeruginosa* strain KUCd1 indicated that Cd tolerance in this bacterium is closely related to the production of siderophore (Sinha and Mukherjee, 2009). Similarly, *Streptomyces* spp. responds to Cd by increasing the production of hydroxamate siderophores, such as desferrioxamine and coelichelin (Dimkpa et al., 2008). It is assumed that the stimulation in the production of siderophore under Cd stress could be either the result of the stimulation of signaling cascades related to siderophore synthesis, or Cd-inducible Fe deficiency. The production of siderophores by plant growth promoting bacteria also helps to extract metals from the soil in the course of phytoextraction (Rajkumar et al., 2010). Also, the presence of microbial organic acids in the rhizosphere helps to solubilize plant nutrients such as phosphate, and results in Cd tolerance in plants because of phosphate-mediated Cd precipitation reactions. Phytosiderophores are very often released from plant roots during Fe and Zn deficiency in graminaceous plants (Römheld, 1991). The uptake of the metal phytosiderophore complexes occurs with the help of specific transporters belonging to the YSL family. Phytosiderophores are carboxylates containing mugineic acid, epi-mugineic acid, and deoxymugineic acid. The carboxylic group of these organic acids binds with Cd. The synthesis of phytosiderophores depends on amino acid metabolism, because methionine is the precursor molecule of mugineic acid. Phenolics, amino acids, and small peptides are also implicated in extracellular detoxification of Cd in microbes and plants (Ma et al., 2016). Similarly, pectin, callose, and histydyl sites in the cell wall also help to bind with Cd in the extracellular space, and thereby prevent Cd toxicity (Parrotta et al., 2015).

Metallothionein confers intracellular Cd tolerance in both bacteria and plants (Russo et al., 2003). Metallothionein is a cysteine-rich, low molecular weight (3.4–140.0 kDa) protein. These proteins confer Cd tolerance via Cd complexation and antioxidant activity. The sulfhydryl groups in these proteins bind with Cd, as well as electrophiles, which in turn prevents metal toxicity. Studies with *Pseudomonas aeruginosa* and *Pseudomonas putida* showed that metallothionein production correlates with Cd accumulation (Enshaei et al., 2010). Similar results were observed in rice and *Arabidopsis* (DalCorso et al., 2010; Gautam et al., 2012). It has also been observed that Cd-induced $H_2O_2$ production plays an important role in the induction of metallothionein synthesis in plants (Zhang et al., 2017).

Plants produce small Cd binding peptides called phytochelatins in response to Cd stress (Brunetti et al., 2011). These peptides are formed by the activity of the enzyme called phytochelatin synthase, which converts glutathione into a small peptide containing a glutamic acid-cysteine repeating unit with a terminal glycine (Cobbett, 2000). The presence of Cd triggers phytochelatin formation, and the Cd binds with the sulphhydril group of the phytochelatin. Cadmium bound to the glutathione or phytochelatin is trafficked into the vacuole for storage, and thereby protects it from Cd cytotoxicity (DalCorso et al., 2010; Shahid et al., 2017). Alleviation of Cd toxicity in the cell is also found to be associated with chaperones. These chaperones play an important role in the repair of protein misfolding under Cd stress. It has been reported that an endoplasmic reticulum chaperone binding protein called LcBiP confers Cd tolerance in tobacco plants (Guan et al., 2015). The activity of LcBiP can be attributed to a sensor that detects and prevents alteration in the endoplasmic reticulum (ER) under Cd stress. This prevents alteration in protein folding in ER under Cd stress. LcBiP accumulation in tobacco plants is triggered with ROS. So, it can be concluded that Cd-induced oxidative stress plays a key role in activation of LcBiP. Chaperones also bind with $Cd^{2+}$ ions in the cell, and thereby prevent toxicity due to the occurrence of free Cd ions in the cell.

Organic acids chelate Cd in the cell (Rauser, 1999). Organic acids such as citrate, malate, and oxalic acids bind with Cd, and the complex is trafficked into the plant vacuole (Rauser, 1999). This process is one of the key mechanism that allows metal hyperaccumulating plants to accumulate more Cd in the plant tissue. The presence of Cd retards the tricarboxylic acid cycle (Morgan et al., 2008). This change creates a disturbance in organic acid production during Cd stress. But the upregulation of antioxidant activity mediated through the thioredoxin system, with antioxidant enzymes such as superoxide dismutase, catalase, peroxidases, and so forth, and cellular antioxidants such as phenolics, flavonoids, polyamines, and so forth, helps to avoid Cd stress, which in turn enhances production of organic acids under Cd stress, which is important for the chelation of Cd (Xie et al., 2014). Apart from this secretion of organic acids, such as oxalic acid, reported as a Cd exclusion mechanism, operates in plant roots (Zhu et al., 2011). An increase of phenolics and flavonoids in response to Cd stress help in the chelation of Cd, as well as detoxification of reactive oxygen species (Sebastian and Prasad, 2014c). It has also been noted that polyamines such as putrescine, spermidine, and spermine increase after Cd treatment in sunflower plants (Groppa et al., 2007). The putative role of polyamine could be the binding of free Cd ions accumulated in the cytosol.

Signal transduction plays an important role in the detoxification of Cd stress. The sensing of Cd in the cell is achieved with the help of $Ca^{2+}$ ions, nitrous oxide, ROS, glutathione, and microRNAs (DalCorso et al., 2010). The outcome of Cd sensing is the upregulation of redox-dependent signal transduction pathways mediated by MAPK (mitogen-activated protein kinase) and CDPKs (calcium-dependent protein kinases), respectively (DalCorso et al., 2010). Transcription factors such as MYB, HSF, bZIP, WRKY, and DREB responded in a Cd signaling pathway that ultimately enhances Cd tolerance (DalCorso et al., 2010). An increase in lipoxygenase activity was also observed during Cd stress (Liptáková et al., 2013). This resulted in the generation of short chain fatty acids from membrane lipids that are able to trigger signal transduction pathways. Apart from this, signaling molecules, such as salicylic acid and jasmonate, have also been reported to initiate signaling pathways critical for Cd stress tolerance (Gondor et al., 2016; Ahmad et al., 2017). The presence of bacterial strains such as *Bacillus amyloliquefaciens* SAY09 have been found to stimulate auxin-dependent signaling pathways in Arabidospsis, leading to Cd tolerance (Zhou et al., 2017). Plant stress-responsive hormones such as abscisic acid, brassinosteroid, and ethylene also play an important role in Cd stress tolerance in plants (Bücker-Neto et al., 2017). These hormones play a role in signaling pathways and gene regulation critical for the alleviation of Cd toxicity.

## 5 ECOLOGICAL ASPECTS OF CADMIUM STRESS TOLERANCE

Ecological risk assessment of Cd is very important for the management of Cd pollution, and rehabilitation of mine sites (Yang et al., 2017). It is well known that Cd toxicity affects biological processes in an ecosystem. The presence of Cd alters colonization of microbes, plants, and animals, especially soil-dwelling arthropods and crustaceans. Among the microbial flora, bacterial colonization is more prone to the presence of Cd compared with fungi (An et al., 2015). Bacteria play an important role in the cycling of nutrients and decomposition of organic matter in the soil. These biogeochemical processes affect physicochemical properties of the soil, such as pH, conductivity, temperature, and cation exchange capacity. But the inhibitory action of Cd on bacterial respiratory metabolism inhibits bacterial growth in the soil. Thus retardation of bacterial activity under Cd stress retards decomposition of organic matter in the soil and blocks nutrient release (Laskowski et al., 1994; Vig et al., 2003). Hence, availability of nutrients required for plant growth is very low in Cd-contaminated soil, and this accounts for the loss of biodiversity in Cd-contaminated sites such as abandoned Zn mine sites. But the presence of Cd tolerant bacteria, such as *Variovorax paradoxus*, *Rhodococcus*, and *Flavobacterium* were found to stimulate plant growth in Cd-contaminated soil (Ahemad and Kibret, 2014). Similarly, mycorrhizal colonization was also found to promote plant growth in Cd-contaminated sites (Kumar et al., 2015). So the approaches based on plant growth promoting bacteria and fungi are feasible for restoring natural vegetation in Cd-polluted sites.

The microbial degradation of organic matter is important for the release of humic and fulvic acids in the soil (Trevisan et al., 2010). These organic acids have a critical role in seed germination, the vitality of plant roots, and nutrient uptake (Nikbakht et al., 2008). The release of these acids during decomposition of plant residues slightly lowers pH of the soil, which in turn enhances the plant uptake of minerals and Cd. Thus, humic acids released from organic matter enhance bioavailability of micronutrients and Cd because of metal chelating activity. The presence of organic matter also significantly increases −ve charges, and hence enhances Cd binding capacity of the soil (Olaniran et al., 2013). So it is clear that Cd-dependent hindering of microbial activity does not support plant growth, but helps immobilization of Cd in the soil. The affinity of the metal binding functional groups and the stability of transition metals vary, as per the Irving–Williams series (Irving and Williams, 1948). Therefore, the presence of a metal in excess may lead to leaching of Cd from organic matter too.

Metal hyperaccumulating plants are able to accumulate a high concentration of Cd with the help of internal Cd detoxification mechanisms, and these plants grow well in Cd-polluted sites (Milner and Kochian, 2008) (Fig. 4). *Chromolaena odoratum*, *Gynura pseudochina*, *Impatiens violaeflora*, and *Justicia procumbens* are Cd hyperaccumulators found in the vicinity of the Padaeng zinc mine area (Phaenark et al., 2009). These plants accumulated more than 100 mg Cd $kg^{-1}$ dry mass, and possessed a translocation factor of Cd >1. Similarly, *Solanum nigrum*, *Lobelia chinensis*, *Arabidopsis halleri*, and *Noccaea caerulescens* are well known to thrive in Cd enriched soils (Prasad and Freitas, 2002). But graminaceous plants such as *Vetiveria zizanioides* are found to accumulate more Cd in the root (2232.0 mg $kg^{-1}$ dry biomass), compared with the leaves (93 mg $kg^{-1}$ dry biomass) (Danh et al., 2009). Ferns are one of the dominant plant species found in Cd-polluted areas because of lower growth requirements (Kachenko et al., 2006). For example, *Athyrium yokoscence* has been found to accumulate Cd 851.0 mg $kg^{-1}$ dry mass (Zonghui et al., 2009). *Pteris vittata* is another fern that is able to survive under higher Cd concentrations (301.0 mg $kg^{-1}$ soil), and the plant accumulated Cd 186.0 mg $kg^{-1}$ dry font biomass (Xiyuan et al., 2008).

**Fig. 4** Cadmium tolerant plants growing in metalliferous substrates. (A) *Chrysopogon zizanioides*, (B) *Chromolaena odoratum*, (C) *Gynura pseudochina*, (D) *Pistia stratiotes*, (E) *Typha latifolia*, (F) *Cyperus rotundus*.

Bryophytes are sensitive to Cd stress, and these plants are used for environmental monitoring of Cd (Mazzoni et al., 2012). Accumulation of Cd in bryophytes is often noticed in Zn-rich blanket bogs (Lee et al., 1984). Bryophyte, namely *Hygrohypnum ochraceum*, was able to accumulate a significant amount of Cd when the Cd concentration of the media was $0.15\,\mu g\,L^{-1}$. This plant also shows seasonal variation in Cd accumulation; more Cd accumulation is observed during fall, compared with spring (Rehe Jr and Nimmo, 2001). *Marchantia polymorpha*, *Funaria hygrometrica*, and *Pseudoscleropodium purum* show Cd tolerance until 500.0 μM Cd in the growth media (Tyler, 1990). Algae such as *Cladophora glomerata* have been found to accumulate Cd more than $5.0\,g\,Cd\,kg^{-1}$ dry biomass under laboratory conditions (Akin and Unlu, 2014). The physical environment is also found to influence Cd accumulation in algae, such as *Chlamydomonas* CPCC121 (Samadani et al., 2017). At pH 4.0, the CPCC121 strain tends to exclude Cd from the cell, whereas at physiological pH, this alga accumulates more Cd. Thus, the study indicates the ability of this algae to withstand Cd toxicity in acid mine leachate. Lichen, namely *Xanthoria parietina* and *Peltigera membranacea*, have been found to exhibit Cd tolerance because of the maintenance of photosynthetic turnover, rather than restricting the uptake of Cd (Beckett and Brown, 1984; Dzubaj et al., 2008). So it is assumed that higher assimilation capacity enables Cd tolerance in Plantae. Aquatic macrophytes, such as *Chara australis*, grow rapidly in Cd-contaminated water bodies (Clabeaux et al., 2011). This plant is also found to be ideal the growth in Cd-containing sediments. *Ceratophyllum demersum*, *Bacopa monnieri*, *Rotala rotundifolia*, and *Pistia stratiotes* also grow well in Cd-contaminated water bodies (Das et al., 2014; Dogan et al., 2015). So these plants can be used for abatement and monitoring of Cd pollution.

Agroecosystems can become contaminated with Cd from mine site runoff water, and the application of rock phosphate and sewage-based fertilizers (Sebastian and Prasad, 2014a). The entry of Cd in rice fields has resulted in the accumulation of Cd in rice plants, and the consumption of Cd-contaminated rice has resulted in health disorders in humans. The Cd output from agriculture manifests not only in humans, but also in organisms in the soil and grazing animals. Long-term Cd exposure in agricultural land also contaminates groundwater and aquifers via the leaching of Cd with irrigation and rainwater. Long-term persistence of Cd in the field results in precipitation-dissolution, adsorption-desorption, and complexation of Cd, which plays an important role in bioavailability and mobility of Cd in agriculture fields (Wuana and Okieimen, 2011). Micronutrient deficiency is another noticeable impact of Cd stress in crop plants (Gill et al., 2011). Paddy fields are subjected to intermittent flooding periods. This results in oxidation and reduction reactions that affect Cd bioavailability and toxicity. The induction of Fe plaque in rice plants is common in submerged rice paddies, which are subject to decreased Cd exposure to rice roots by acting as a barrier to Cd exposure (Sebastian and Prasad, 2016). But the formation of Fe pan in the soil decreases leaching of Cd to more depth, and hence, crops with fibrous roots tend to accumulate more Cd in these soils (Sebastian and Prasad, 2015a,b). Accumulation of Cd in beneficial arthropods and earthworms has also been reported (van Straalen et al., 1989; Kumar et al., 2008). Toxicity of Cd on these organisms affects the longevity of soil structure, which is important for crop growth. Thus, it is clear that the environmental impact of Cd is complex, and the impact of Cd on microbes, plants, and animals needs to be monitored to prevent entry of Cd into the food chain.

## 6 OUTLOOK

Cadmium-safe agriculture food products are important for health. Monitoring the biogeochemical cycle of Cd helps to control the mobility of Cd in the environment. Accumulation of Cd in food products should not exceed the permissible limit. Strict policies to monitor Cd in food products and fertilizer are a matter of interest for ensuring public health. The transfer of Cd from soil to plants or microbes depends on physical factors in the soil. Studies of these factors help to formulate strategies to block entry of Cd in the food chain. It is well understood that Cd uptake and toxicity evoke molecular responses in plants and microbes. These studies are important for technogenic improvement of Cd tolerance in plant and microbes that are important in eco-friendly Cd cleanup methods such as phytoremediation and bioremediation, respectively. Ecological assessment of Cd toxicity helps to realize the mobility of Cd in the soil, Cd tolerant taxa, and efforts required for rehabilitation of abandoned mine sites.

## ACKNOWLEDGMENTS

Abin Sebastian gratefully acknowledges Dr. DS Kothari Post doctoral fellowship (No. BL/14-15/0162), UGC, India for financial support, and Crystalin research Pvt. Ltd, Technology business incubator, University of Hyderabad for academic support.

## REFERENCES


Ahemad, M., Kibret, M., 2014. Mechanisms and applications of plant growth promoting rhizobacteria: current perspective. J. King Saud Univ. Sci. 26, 1–20.

Ahmad, P., Arafat, A., Latef, A., Elsayed, F., Hashem, A., Sarwat, M., Anjum, N.A., Gucel, S., 2016. Calcium and potassium supplementation enhanced growth, osmolyte secondary metabolite production, and enzymatic antioxidant machinery in cadmium-exposed chickpea (*Cicer arietinum* L.). Front. Plant Sci. 7, 513. https://doi.org/10.3389/fpls.2016.00513.

Ahmad, P., Alyemeni, M.N., Wijaya, L., Alam, P., Ahanger, M.A., Alamri, S.A., 2017. Jasmonic acid alleviates negative impacts of cadmium stress by modifying osmolytes and antioxidants in faba bean (*Vicia faba* L.). Arch. Agron. Soil Sci. 63, 1889–1899.

Akin, H.K., Unlu, E., 2014. Cadmium accumulation by green algae *Cladophora glomerata* (L.) Kutz. (Chlorophyta) in presence of Nile tilapia *Oreochromis niloticus* (L.). Toxicol. Environ. Chem. 9 (5), 1565–1571.

Álvarez-Fernández, A., Díaz-Benito, P., Abadía, A., López-Millán, A., Abadía, J., 2014. Metal species involved in long distance metal transport in plants. Front. Plant Sci. 5 (105), 1–20.

An, H., Liu, Y., Zhao, X., Huang, Q., Yuan, S., Yang, X., Dong, J., 2015. Characterization of cadmium-resistant endophytic fungi from *Salix variegata* Franch. in Three Gorges Reservoir Region, China. Microbiol. Res. 176, 29–37.

Ashraf, M.Y., Roohi, M., Iqbal, Z., Ashraf, M., Öztürk, M., Gücel, S., 2016. Cadmium (Cd) and lead (Pb) induced changes in growth, some biochemical attributes, and mineral accumulation in two cultivars of mung bean [*Vigna radiata* (L.) Wilczek]. Commun. Soil Sci. Plant Anal. 47, 405–413.

Balestrasse, K.B., Gallego, S.M., Tomaro, M.L., 2004. Cadmium-induced senescence in nodules of soybean (*Glycine max* L.) plants. Plant Soil 262, 373–381.

Beckett, R.P., Brown, D.H., 1984. The relationship between cadmium uptake and heavy metal tolerance in the lichen genus *peltigera*. New Phytol. 97, 301–311.

Belyaeva, E.A., Sokolova, T.V., Emelyanova, L.V., Zakharova, I.O., 2012. Mitochondrial electron transport chain in heavy metal-induced neurotoxicity: effects of cadmium, mercury, copper. Sci. World J. https://doi.org/10.1100/2012/136063, Article ID 136063.

Brunetti, P., Zanella, L., Proia, A., et al., 2011. Cadmium tolerance and phytochelatin content of *Arabidopsis* seedlings over-expressing the phytochelatin synthase gene *AtPCS1*. J. Exp. Bot. 62 (15), 5509–5519.

Bücker-Neto, L., Paiva, A.L.S., Machado, R.D., Arenhart, R.A., Margis-Pinheiro, M., 2017. Interactions between plant hormones and heavy metals responses. Genet. Mol. Biol. 40, 373–386.

Celik, S.U.A., Çekiç, F.O., Gözel, A., 2006. Cadmium-induced genotoxicity, cytotoxicity and lipid peroxidation in *Allium sativum* and *Vicia faba*. Mutagenesis 21, 77–81.

Chaia, M.W., Shia, F.C., Lib, R.L., Liu, F.C., Qiu, G.Y., Liu, L.M., 2013. Effect of NaCl on growth and Cd accumulation of halophyte *Spartina alterniflora* under $CdCl_2$ stress. S. Afr. J. Bot. 85, 63–69.

Chunhabundit, R., 2016. Cadmium exposure and potential health risk from foods in contaminated area, Thailand. Toxicol. Res. 32 (1), 65–72.

Clabeaux, B.L., Navarro, D.A.G., Aga, D.S., Bisson, M.A., 2011. Cd tolerance and accumulation in the aquatic macrophyte, *Chara australis*: potential use for charophytes in phytoremediation. Environ. Sci. Technol. 45 (12), 5332–5338.

Cobbett, C.S., 2000. Phytochelatins and their roles in heavy metal detoxification. Plant Physiol. 123, 825–832.

Connolly, E.L., Fett, J.P., Guerinot, M.L., 2002. Expression of the IRT1 metal transporter is controlled by metals at the levels of transcript and protein accumulation. Plant Cell 14, 1347–1357.

DalCorso, G., Farinati, S., Furini, A., 2010. Regulatory networks of cadmium stress in plants. Plant Signal. Behav. 5 (6), 663–667.

Danh, L.T., Truong, P., Mammucari, R., Tran, T., Foster, N., 2009. Vetiver grass, *Vetiveria zizanioides*: a choice plant for phytoremediation of heavy metals and organic wastes. Int. J. Phytoremediation 11 (8), 664–691.

Das, N., Vimala, R., Karthika, P., 2008. Biosorption heavy metals. An overview. Indian J. Pathol. Bacteriol. 7, 159–169.

Das, S., Goswami, S., Talukdar, A.D., 2014. A study on cadmium phytoremediation potential of water lettuce, *Pistia stratiotes* L. Bull. Environ. Contam. Toxicol. 92 (2), 169–174.

Degryse, F., Broos, K., Smolders, E., Merckx, R., 2003. Soil solution concentration of Cd and Zn can be predicted with a $CaCl_2$ soil extract. Eur. J. Soil Sci. 54, 149–158.

Desbrosses-Fonrouge, A.G., Voigt, K., Schrder, A., Stephanie, A., Sebastien, T., 2005. Arabidopsis thaliana MTP1 is a Zn transporter in the vacuolar membrane which mediates Zn detoxification and drives leaf Zn accumulation. FEBS Lett. 579, 4165–4174.

Dimkpa, C., Svatos, A., Merten, D., Büchel, G., Kothe, E., 2008. Hydroxamate siderophores produced by *Streptomyces acidiscabies* E13 bind nickel and promote growth in cowpea (*Vigna unguiculata* L.) under nickel stress. Can. J. Microbiol. 54 (3), 163–172.

Djebali, W., Zarrouk, M., Brouquisse, R., El Kahoui, S., Limam, F., Ghorbel, M.H., Chaïbi, W., 2005. Ultrastructure and lipid alterations induced by cadmium in tomato (*Lycopersicon esculentum*) chloroplast membranes. Plant Biol. (Stuttg.) 7 (4), 358–368.

Dogan, M., Akgul, H., Inan, O.G., Zeren, H., 2015. Determination of cadmium accumulation capabilities of aquatic macrophytes *Ceratophyllum demersum*, *Bacopa monnieri* and *Rotala rotundifolia*. Iran. J. Fish. Sci. 14, 1010–1017.

Dzubaj, A., Backor, M., Tomko, J., Peli, E., Tuba, Z., 2008. Tolerance of the lichen *Xanthoria parietina* (L.) Th. Fr. to metal stress. Ecotoxicol. Environ. Saf. 70 (2), 319–326.

Eid, R.A., Mazher, A.A.M., Khalifa, R.K.M., Shaaban, S.H.A., 2016. Effect of cadmium concentrations on the vegetative growth, flowering and chemical constituents of T*agets erecta* L. plant. Int. J. PharmTech Res. 9 (4), 18–24.

Enshaei, M., Khanafari, A., Sepahey, A.A., 2010. Metallothionein induction in two species of *Pseudomonas* exposed to cadmium and copper contamination. Iran. J. Environ. Health Sci. Eng. 7, 287–298.

Gaillard, S., Jacquet, H., Vavasseur, A., Leonhardt, N., Forestier, C., 2008. AtMRP6/AtABCC6, an ATP-binding cassette transporter gene expressed during early steps of seedling development and up-regulated by cadmium in *Arabidopsis thaliana*. BMC Plant Biol. 8 (1), 1–11.

Garmash, E.V., Golovko, T.K., 2009. Effect of cadmium on growth and respiration of barley plants grown under two temperature regimes. Russ. J. Plant Physiol. 56, 343–347.

Gautam, N., Verma, P., Verma, S., Chakrabarty, D., 2012. Genome-wide identification of rice class I metallothionein gene: tissue expression patterns and induction in response to heavy metal stress. Funct. Integr. Genomics 12 (4), 635–647.

Gill, S.S., Khan, N.A., Anjum, N.A., Tuteja, N., 2011. Amelioration of cadmium stress in crop plants by nutrients management: morphological, physiological and biochemical aspects. Plant Stress 5, 1–23.

Gitan, R.S., Shababi, M., Kramer, M., Eide, D.J.A., 2003. Cytosolic domain of the yeast Zrt1 zinc transporter is required for its post-translational inactivation in response to zinc and cadmium. J. Biol. Chem. 278 (41), 39558–39564.

Godt, J., Scheidig, F., Grosse-Siestrup, C., et al., 2006. The toxicity of cadmium and resulting hazards for human health. J. Occup. Med. Toxicol. 1, 22.

Gomes, M.P., Marques, T.C., Soares, A.M., 2013. Cadmium effects on mineral nutrition of the Cd-hyperaccumulator *Pfaffia glomerata*. Biologia 68 (2), 223–230.

Gonçalves, J.F., Becker, A.G., Cargnelutti, D., Tabaldi, L.A., Pereira, L.B., Battisti, V., Spanevello, R.M., Morsch, V.M., Nicoloso, F.T., Schetinger, M.R.C., 2007. Cadmium

toxicity causes oxidative stress and induces response of the antioxidant system in cucumber seedlings. Braz. J. Plant Physiol. 19 (3), 223–232.
Gondor, O.K., Pál, M., Darkó, É., Janda, T., Szalai, G., 2016. Salicylic acid and sodium salicylate alleviate cadmium toxicity to different extents in maize (*Zea mays* L.). PLoS One 11 (8), e0160157.
Grant, C.A., Buckley, W.T., Bailey, L.D., Selles, F., 1998. Cadmium accumulation in crops. Can. J. Plant Sci. 78, 1–17.
Grass, G., Wong, M.D., Rosen, B.P., Smith, R.L., Rensing, C., 2002. ZupT Is a Zn(II) uptake system in *Escherichia coli*. J. Bacteriol. 184, 864–866.
Groppa, M.D., Ianuzzo, M.P., Tomaro, M.L., Benavides, M.P., 2007. Polyamine metabolism in sunflower plants under long-term cadmium or copper stress. Amino Acids 32, 265–275.
Guan, C., Jin, C., Ji, J., Wang, G., Li, X., 2015. LcBiP, a endoplasmic reticulum chaperone binding protein gene from *Lycium chinense*, confers cadmium tolerance in transgenic tobacco. Biotechnol. Prog. 31 (2), 358–368.
Hall, J.L., Williams, L.E., 2003. Transition metal transporters in plants. J. Exp. Bot. 54 (393), 2601–2613.
Han, R.M., Lefèvre, I., Albacete, A., Pérez-Alfocea, F., Barba-Espín, G., Díaz-Vivancos, P., Quinet, M., Ruan, C.J., Hernández, J.A., Cantero-Navarro, E., Lutts, S., 2013. Antioxidant enzyme activities and hormonal status in response to Cd stress in the wetland halophyte *Kosteletzkya virginica* under saline conditions. Physiol. Plant. 147 (3), 352–368.
Hao, Z., Chen, S., Wilson, D.B., 1999. Cloning, expression, and characterization of cadmium and manganese uptake genes from *Lactobacillus plantarum*. Appl. Environ. Microbiol. 65 (11), 4746–4752.
Heyno, E., Klose, C., Krieger-Liszkay, A., 2008. Origin of cadmium-induced reactive oxygen species production: mitochondrial electron transfer versus plasma membrane NADPH oxidase. New Phytol. 179 (3), 687–699.
Higham, D.P., Sadler, P.J., Scawen, M.D., 1986. Effect of cadmium on the morphology, membrane integrity and permeability of *Pseudomonas putida*. J. Gen. Microbiol. 132, 1475–1482.
Hill, K.A., Lion, L.W., Ahner, B.A., 2002. Reduced Cd accumulation in *Zea mays*: a protective role for phytosiderophores? Environ. Sci. Technol. 36 (24), 5363–5368.
Hirschi, K.D., Korenkov, V.D., Wilganowski, N.L., Wagner, G.J., 2000. Expression of Arabidopsis CAX2 in tobacco. Altered metal accumulation and increased manganese tolerance. Plant Physiol. 124 (1), 125–133.
Hossain, Z., Komatsu, S., 2012. Contribution of proteomic studies towards understanding plant heavy metal stress response. Front. Plant Sci. 3, 310.
Irving, H., Williams, R.J.P., 1948. Order of stability of metal complexes. Nature 162, 746–747.
Jeong, J., Eide, D.J., 2013. The SLC39 family of zinc transporters. Mol. Asp. Med. 34 (2-3), 612–619.
Jozefczak, M., Remans, T., Vangronsveld, J., Cuypers, A., 2012. Glutathione is a key player in metal-induced oxidative stress defenses. Int. J. Mol. Sci. 13 (3), 3145–3175.
Kachenko, A.G., Singh, B., Bhatia, N.P., 2006. Heavy metal tolerance in common fern species. Aust. J. Bot. 55 (1), 63–73.
Keunen, E., Remans, T., Bohler, S., Vangronsveld, J., Cuypers, A., 2011. Metal-induced oxidative stress and plant mitochondria. Int. J. Mol. Sci. 12 (10), 6894–6918.
Khan, Z., Rehman, A., Hussain, S.Z., Nisar, M.A., Zulfiqar, S., Shakoori, A.R., 2016. Cadmium resistance and uptake by bacterium, *Salmonella enterica* 43C, isolated from industrial effluent. AMB Express 6, 54.
Khudsar, T., Mahmooduzzafar, M., Iqbal, M., 2001. Cadmium-induced changes in leaf epidermes, photosynthetic rate and pigment concentrations in *Cajanus cajan*. Biol. Plant. 44, 59–64.

Kim, D.Y., Bovet, L., Maeshima, M., Martinoia, E., Lee, Y., 2007. The ABC transporter AtPDR8 is a cadmium extrusion pump conferring heavy metal resistance. Plant J. 50 (2), 207–218.

Koike, S., Inoue, H., Mizuno, D., Takahashi, M., Nakanishi, H., Mori, S., Nishizawa, N.K., 2004. OsYSL2 is a rice metal-nicotianamine transporter that is regulated by iron and expressed in the phloem. Plant J. 39, 415–424.

Kumar, S., Sharma, V., Bhoyar, R.V., Bhattacharyya, J.K., Chakrabarti, T., 2008. Effect of heavy metals on earthworm activities during vermicomposting of municipal solid waste. Water Environ. Res. 80 (2), 154–161.

Kumar, P., Lucini, L., Rouphael, Y., Cardarelli, M., Kalunke, R.M., Colla, G., 2015. Insight into the role of grafting and arbuscular mycorrhiza on cadmium stress tolerance in tomato. Front. Plant Sci. 26, 477.

Laskowski, R., Maryahski, M., Niklihska, M., 1994. Effect of heavy metals and mineral nutrients on forest litter respiration rate. Environ. Pollut. 84, 97–102.

Lee, M., Shin, H.W., 2003. Cadmium-induced changes in antioxidant enzymes from the marine alga *Nannochloropsis oculata*. J. Appl. Phycol. 15, 13–19.

Lee, J., Jonasson, I.R., Goodfellow, W.D., 1984. Metal accumulation by bryophytes in some zinc-rich blanket bogs, Selwyn Mountains, Yukon Territory. Can. J. Bot. 62 (4), 722–728.

Liptáková, Ľ., Huttová, J., Mistrík, I., Tamás, L., 2013. Enhanced lipoxygenase activity is involved in the stress response but not in the harmful lipid peroxidation and cell death of short-term cadmium-treated barley root tip. J. Plant Physiol. 170 (7), 646–652.

Liu, W., Shang, S., Feng, X., Zhang, G., Wu, F., 2015. Modulation of exogenous selenium in cadmium-induced changes in antioxidative metabolism, cadmium uptake, and photosynthetic performance in the 2 tobacco genotypes differing in cadmium tolerance. Environ. Toxicol. Chem. 34, 92–99.

Lützen, A., Liberti, S.E., Rasmussen, L.J., 2004. Cadmium inhibits human DNA mismatch repair in vivo. Biochem. Biophys. Res. Commun. 321 (1), 21–25.

Ma, Y., Oliveira, R.S., Freitas, H., Zhang, C., 2016. Biochemical and molecular mechanisms of plant-microbe-metal interactions: relevance for phytoremediation. Front. Plant Sci. 7, 918.

Maravi, D.K., Adreeja, B., Goud, V.V., Sahoo, L., 2017. Cadmium induced antioxidant activity and alteration in DNA pattern in biodiesel plant *Jatropha curcas* L. Int. J. Sci. Eng. Res. 8, 40–47.

Mazzoni, A.C., Lanzer, R., Bordin, J., Schäfer, A., Wasum, R., 2012. Mosses as indicators of atmospheric metal deposition in an industrial area of southern Brazil. Acta Bot. Bras. 26 (3), 553–558.

McGrath, S.P., 1999. Adverse effects of cadmium on soil microflora and fauna. In: - McLaughlin, M.J., Singh, B.R. (Eds.), Cadmium in Soils and Plant. Kluwer Academic, Netherlands, pp. 199–218.

Mendoza-Cózatl, D.G., Butko, E., Springer, F., et al., 2008. Identification of high levels of phytochelatins, glutathione and cadmium in the phloem sap of *Brassica napus*. A role for thiol-peptides in the long-distance transport of cadmium and the effect of cadmium on iron translocation. Plant J. 54 (2), 249–259.

Milner, M.J., Kochian, L.V., 2008. Investigating heavy-metal hyperaccumulation using *Thlaspi caerulescens* as a model system. Ann. Bot. 102 (1), 3–13.

Mohammadian Fazli, M., Soleimani, N., Mehrasbi, M., Darabian, S., Mohammadi, J., Ramazani, A., 2015. Highly cadmium tolerant fungi: their tolerance and removal potential. J. Environ. Health Sci. Eng. 13, 19.

Moons, A., 2003. OsPDR9, which encodes a PDR-type ABC transporter, is induced by heavy metals, hypoxic stress and redox perturbations in rice roots. FEBS Lett. 553, 370–376.

Morel, M., Crouzet, J., Gravot, A., Auroy, P., Leonhardt, N., Vavasseur, A., Richaud, P., 2009. AtHMA3 a P1B-ATPase allowing Cd/Zn/Co/Pb vacuolar storage in Arabidopsis. Plant Physiol. 149 (2), 894–904.

Morgan, M.J., Lehmann, M., Schwarzländer, M., et al., 2008. Decrease in manganese superoxide dismutase leads to reduced root growth and affects tricarboxylic acid cycle flux and mitochondrial redox homeostasis. Plant Physiol. 147 (1), 101–114.

Nefic, H., Musanovic, J., Metovic, A., Kurteshi, K., 2013. Chromosomal and nuclear alterations in root tip cells of *Allium cepa* L. induced by Alprazolam. Mediev. Archaeol. 67 (6), 388–392.

Nevo, Y., Nelson, N., 2006. The NRAMP family of metal-ion transporter. Biochim. Biophys. Acta 1763, 609–620.

Nies, D.H., 1992. Resistance to cadmium, cobalt, zinc, and nickel in microbes. Plasmid 27 (1), 17–28.

Nies, D.H., 1995. The cobalt, zinc, and cadmium efflux system CzcABC from *Alcaligenes eutrophus* functions as a cation-proton antiporter in *Escherichia coli*. J. Bacteriol. 177 (10), 2707–2712.

Nikbakht, A., Kafi, M., Babalar, M., Xia, Y.P., Luo, A., Etemadi, N., 2008. Effect of humic acid on plant growth, nutrient uptake, and postharvest life of gerbera. J. Plant Nutr. 31, 2155–21562.

Olaniran, A.O., Balgobind, A., Pillay, B., 2013. Bioavailability of heavy metals in soil: impact on microbial biodegradation of organic compounds and possible improvement strategies. Int. J. Mol. Sci. 14 (5), 10197–10228.

Özyiğit, İ.İ., Akinci, S., 2009. Effects of some stress factors (aluminum, cadmium and drought) on stomata of roman nettle (*Urtica pilulifera* L.). Not. Bot. Horti Agrobot. 37 (1), 108–115.

Palmer, L.D., Skaar, E.P., 2016. Transition metals and virulence in bacteria. Annu. Rev. Genet. 50, 67–91.

Parmar, P., Kumari, N., Sharma, V., 2013. Structural and functional alterations in photosynthetic apparatus of plants under cadmium stress. Bot. Stud. 54 (1), 45.

Parrotta, L., Guerriero, G., Sergeant, K., Cai, G., Hausman, J.F., 2015. Target or barrier? The cell wall of early- and later-diverging plants vs cadmium toxicity: differences in the response mechanisms. Front. Plant Sci. 6, 133.

Pence, N.S., Larsen, P.B., Ebbs, S.D., Letham, D.L.D., Lasat, M.M., Garvin, D.F., Eide, D., Kochian, L.V., 2000. The molecular physiology of heavy metal transport in the Zn/Cd hyperaccumulator *Thlaspi caerulescens*. Proc. Natl. Acad. Sci. U. S. A. 97, 4956–4960.

Persans, M.W., Nieman, K., Salt, D.E., 2001. Functional activity and role of cation-efflux family members in Ni hyperaccumulation in *Thlaspi goesingense*. Proc. Natl. Acad. Sci. U. S. A. 98, 9995–10000.

Petrov, V., Hille, J., Mueller-Roeber, B., Gechev, T.S., 2015. ROS-mediated abiotic stress-induced programmed cell death in plants. Front. Plant Sci. 6, 69.

Phaenark, C., Pokethitiyook, P., Kruatrachue, M., Ngernsansaruay, C., 2009. Cd and Zn accumulation in plants from the Padaeng zinc mine area. Int. J. Phytoremediation 11 (5), 479–495.

Piadé, J.J., Jaccard, G., Dolka, C., Belushkin, M., Wajrock, S., 2015. Differences in cadmium transfer from tobacco to cigarette smoke, compared to arsenic or lead. Toxicol. Rep. 2, 12–26.

Pietrini, F., Lannelli, M.A., Pasqualini, S., Massacci, A., 2003. Interaction of cadmium with glutathione and photosynthesis in developing leaves and chloroplasts of *Phragmites australis* (Cav.) Trin. ex steudel. Plant Physiol. 133, 829–837.

Pourghasemian, N., Ehsanzadeh, P., Greger, M., 2013. Genotypic variation in safflower (*Carthamus* spp.) cadmium accumulation and tolerance affected by temperature and cadmium levels. Environ. Exp. Bot. 87, 218–226.

Prasad, M.N.V., Freitas, H., 2002. Metal hyper accumulation in plants—biodiversity prospecting for phytoremediation technology. Electron. J. Bot. 6 (3), 285–321.

Rahoui, S., Martinez, Y., Sakouhi, L., Ben, C., Rickauer, M., El Ferjani, E., Gentzbittel, L., Chaoui, A., 2017. Cadmium-induced changes in antioxidative systems and differentiation in roots of contrasted *Medicago truncatula* lines. Protoplasma 254 (1), 473–489.

Rajkumar, M., Ae, N., Prasad, M.N.V., Freitas, H., 2010. Potential of siderophore-producing bacteria for improving heavy metal phytoextraction. Trends Biotechnol. 28 (3), 142–149.

Ramesh, S.A., Shin, R., Eide, D.J., Schachtman, D.P., 2003. Differential metal selectivity and gene expression of two zinc transporters from rice. Plant Physiol. 133, 126–134.

Rattanapolsan, L., 2018. Isolation and Characterization of Endophytic Bacteria from *Murdannia spectabilis* (Kurz) Faden from Zn/Cd Contaminated Area and a Potential to Promote Plant Growth Under the Metal Stress, (Ph.D. thesis). The Graduate School, Mahasarakham University, Maha Sarakham.

Rauser, W.E., 1999. Structure and function of metal chelators produced by plants: the case for organic acids, amino acids, phytin, and metallothioneins. Cell Biochem. Biophys. 31, 19–48.

Rehe Jr., W.G., Nimmo, D.W.R., 2001. Culturing the bryohphyte *Hygrohypnum ochraceum* for use as an instream monitor of metals. J. Freshw. Ecol. 16 (3), 375–379.

Roberts, T.L., 2014. Cadmium and phosphorous fertilizers: the issues and the science. Procedia Eng. 83, 52–59.

Rodríguez-serrano, M., Romero-puertas, M.C., Zabalza, A., Corpas, F.J., Gómez, M., Delrío, L.A., Sandalio, L.M., 2006. Cadmium effect on oxidative metabolism of pea (*Pisum sativum* L.) roots. Imaging of reactive oxygen species and nitric oxide accumulation in vivo. Plant Cell Environ. 29, 1532–1544.

Römheld, V., 1991. The role of phytosiderophores in acquisition of iron and other micronutrients in graminaceous species: an ecological approach. Plant Soil 130, 127–134.

Ron, E.Z., Minz, D., Finkelstein, N.P., Rosenberg, E., 1992. Interactions of bacteria with cadmium. Biodegradation 3, 161.

Rosman, K.J.R., De Laeter, J.R., 1978. A survey of cadmium isotopic abundances. J. Geophys. Res. 83 (B3), 1279–1287.

Russo, R., Bonaventura, R., Zito, F., Schröder, H.C., Müller, I., Müller, W.E., Matranga, V., 2003. Stress to cadmium monitored by metallothionein gene induction in *Paracentrotus lividus* embryos. Cell Stress Chaperones 8 (3), 232–241.

Samadani, M., Perreault, F., Oukarroum, A., Dewez, D., 2017. Effect of cadmium accumulation on green algae *Chlamydomonas reinhardtii* and acid-tolerant *Chlamydomonas* CPCC 121. Chemosphere 191, 174–182.

Sasaki, A., Yamaji, N., Yokosho, K., Ma, J.F., 2012. Nramp5 is a major transporter responsible for manganese and cadmium uptake in rice. Plant Cell 24 (5), 2155–2167.

Satarug, S., Garrett, S.H., Sens, M.A., Sens, D.A., 2010. Cadmium, environmental exposure, and health outcomes. Environ. Health Perspect. 118 (2), 182–190.

Sebastian, A., Prasad, M.N.V., 2013. Cadmium accumulation retard activity of functional components of photo assimilation and growth of rice cultivars amended with vermicompost. Int. J. Phytoremediation 15, 965–978.

Sebastian, A., Prasad, M.N.V., 2014a. Cadmium minimization in rice. A review. Agron. Sustain. Dev. 34, 155–173.

Sebastian, A., Prasad, M.N.V., 2014b. Vertisol prevents cadmium accumulation in rice: analysis by ecophysiological toxicity markers. Chemosphere 108, 85–92.

Sebastian, A., Prasad, M.N.V., 2014c. Red and blue lights induced oxidative stress tolerance promote cadmium rhizocomplexation in *Oryza sativa* L. J. Photochem. Photobiol. B Biol. 137, 135–143.

Sebastian, A., Prasad, M.N.V., 2015a. Trace element management in rice. Agron. J. 5, 374–404.

Sebastian, A., Prasad, M.N.V., 2015b. Operative photo assimilation associated proteome modulations are critical for iron-dependent cadmium tolerance in *Oryza sativa* L. Protoplasma 252 (5), 1375–1386.

Sebastian, A., Prasad, M.N.V., 2016. Iron plaque decreases cadmium accumulation in *Oryza sativa* L. and serves as a source of iron. J. Plant Biol. 18, 1008–1015.

Sekara, A., Poniedzialeek, M., Ciura, J., Jedrszczyk, E., 2005. Cadmium and lead accumulation and distribution in the organs of nine crops: implications for phytoremediation. Pol. J. Environ. Stud. 14, 509–516.

Shahid, M., Dumat, C., Khalid, S., Niazi, N.K., Antunes, P.M.C., 2017. Cadmium bioavailability, uptake, toxicity and detoxification in soil-plant system. Rev. Environ. Contam. Toxicol. 241, 73–137.

Sigel, R.K., Skilandat, M., Sigel, A., Operschall, B.P., Sigel, H., 2013. Complex formation of cadmium with sugar residues, nucleobases, phosphates, nucleotides, and nucleic acids. Met. Ions Life Sci. 11, 191–274.

Sinha, S., Mukherjee, S.K., 2008. Cadmium-induced siderophore production by a high Cd-resistant bacterial strain relieved Cd toxicity in plants through root colonization. Curr. Microbiol. 56, 55.

Sinha, S., Mukherjee, S.K., 2009. *Pseudomonas aeruginosa* KUCD1, a possible candidate for cadmium bioremediation. Braz. J. Microbiol. 40 (3), 655–662.

Smeets, K., Opdenakker, K., Remans, T., Forzani, C., Hirt, H., Vangronsveld, J., Cuyper, J., 2013. The role of the kinase OXI1 in cadmium- and copper-induced molecular responses in *Arabidopsis thaliana*. Plant Cell Environ. 36, 1228–1238.

Sóvágó, I., Várnagy, K., 2013. Cadmium(II) complexes of amino acids and peptides. Met. Ions Life Sci. 11, 275–302.

Stillwell, E.F., Corum, K.E., 1982. Effects of cadmium and low magnesium on cell division and calcification in the marine coccolithophorid *Cricosphaera* (*hymenomonas*) *carterae*. Mar. Biol. 66, 227–230.

Takahashi, R., Ishimaru, Y., Senoura, T., Shimo, H., Ishikawa, S., Arao, T., Nakanishi, H., Nishizawa, N.K., 2011. The OsNRAMP1 iron transporter is involved in Cd accumulation in rice. J. Exp. Bot. 62 (14), 4843–4850.

Takahashi, R., Ishimaru, Y., Shimo, H., Ogo, Y., Senoura, T., Nishizawa, N.K., Nakanishi, H., 2012. The OsHMA2 transporter is involved in root-to-shoot translocation of Zn and Cd in rice. Plant Cell Environ. 35 (11), 1948–1957.

Tao, Q., Jupa, R., Luo, J., Lux, A., Kováč, J., Wen, Y., Zhou, Y., Jan, J., Liang, Y., Li, T., 2017. The apoplasmic pathway via the root apex and lateral roots contributes to Cd hyperaccumulation in the hyperaccumulator *Sedum alfredii*. J. Exp. Bot. 68, 739–751.

Tchounwou, P.B., Yedjou, C.G., Patlolla, A.K., Sutton, D.J., 2012. Heavy metals toxicity and the environment. Experientia Suppl. 101, 133–164.

Thévenod, F., Lee, W.K., 2013. Cadmium and cellular signaling cascades: interactions between cell death and survival pathways. Arch. Toxicol. 87 (10), 1743–1786.

Thomine, S., Wang, R., Ward, J.M., Crawford, N.M., Schroeder, J.I., 2000. Cadmium and iron transport by members of a plant metal transporter family in Arabidopsis with homology to Nramp genes. Proc. Natl. Acad. Sci. U. S. A. 97 (9), 4991–4996.

Thomine, S., Lelievre, F., Debarbieux, E., Schroeder, J.I., Barbier-Brygoo, H., 2003. AtNRAMP3, a multispecific vacuolar metal transporter involved in plant responses to iron deficiency. Plant J. 34 (5), 685–695.

Tian, S., Lu, L., Zhang, J., Wang, K., Brown, P., Zhenli, H., Liang, J., Yang, X., 2011. Calcium protects roots of *Sedum alfredii* H. against cadmium-induced oxidative stress. Chemosphere 84, 63–69.

Traina, S.J., 1999. The environmental chemistry of cadmium. In: McLaughlin, M.J., Singh, B.R. (Eds.), Cadmium in Soils and Plants. Developments in Plant and Soil Sciences. In: vol. 85. Springer, Dordrecht, pp. 11–37.

Trevisan, S., Francioso, O., Quaggiotti, S., Nardi, S., 2010. Humic substances biological activity at the plant-soil interface: from environmental aspects to molecular factors. Plant Signal. Behav. 5 (6), 635–643.

Trevors, J.T., Stratton, G.W., Gadd, G.M., 1986. Cadmium transport, resistance, and toxicity in bacteria, algae, and fungi. Can. J. Microbiol. 32 (6), 447–464.

Tsai, K., Yoon, K.P., Lynn, A.R., 1992. ATP-dependent cadmium transport by the cadA cadmium resistance determinant in everted membrane vesicles of *Bacillus subtilis*. J. Bacteriol. 174, 116–121.

Tyler, G., 1990. Bryophytes and heavy metals: a literature review. Bot. J. Linn. Soc. 104, 231–253.

Tynecka, Z., Malm, A., Kosikowska, U., Kot, A., 1998. Substrate-dependent cadmium toxicity affecting energy-linked K+/86Rb transport in *Staphylococcus aureus*. Folia Microbiol. (Praha) 43 (6), 617–622.

Ueno, D., Yamaji, N., Kono, I., Huang, C.F.T., Yano, M., Ma, J.F., 2010. Gene limiting cadmium accumulation in rice. Proc. Natl. Acad. Sci. U. S. A. 107, 16500–16505.

Uraguchi, S., Watanabe, I., Yoshitomi, A., Kiyono, M., Kuno, K., 2006. Characteristics of cadmium accumulation and tolerance in novel Cd-accumulating crops, *Avena strigosa* and *Crotalaria juncea*. J. Exp. Bot. 57, 2955–2965.

Uraguchi, S., Mori, S., Kuramata, M., Kawasaki, A., Arao, T., Ishikawa, S., 2009. Root-to-shoot Cd translocation via the xylem is the major process determining shoot and grain cadmium accumulation in rice. J. Exp. Bot. 60 (9), 2677–2688.

Uraguchi, S., Kamiya, T., Sakamoto, T., Kasai, K., Sato, Y., Nagamura, Y., Yoshida, A., Kyozuka, J., Ishikawa, S., Fujiwara, T., 2011. Low affinity cation transporter (OsLCT1) regulates cadmium transport into rice grains. Proc. Natl. Acad. Sci. U. S. A. 108 (52), 20959–20964.

van Straalen, N.M., Schobben, J.H., de Goede, R.G., 1989. Population consequences of cadmium toxicity in soil microarthropods. Ecotoxicol. Environ. Saf. 17 (2), 190–204.

Vig, K., Megharaj, M., Sethunathan, N., Naidu, R., 2003. Bioavailability and toxicity of cadmium to microorganisms and their activities in soil: a review. Adv. Environ. Res. 8, 121–135.

Wang, A., Crowley, D.E., 2005. Global gene expression responses to cadmium toxicity in *Escherichia coli*. J. Bacteriol. 187, 3259–3266.

Wilkens, S., 2015. Structure and mechanism of ABC transporters. F1000Prime Rep. 7, 14.

Wuana, R.A., Okieimen, F.E., 2011. Heavy metals in contaminated soils: a review of sources, chemistry, risks and best available strategies for remediation. Int. Sch. Res. Notices Ecol.402647.

Wyszkowska, J., Wyszkowski, M., 2003. Effect of Cd and Mg on enzymatic activity in soil. Pol. J. Environ. Stud. 12, 473–479.

Xie, Y., Hu, L., Du, Z., Sun, X., Amombo, E., Fan, J., et al., 2014. Effects of cadmium exposure on growth and metabolic profile of bermudagrass [*Cynodon dactylon* (L.) Pers.]. PLoS One 9(12), e115279.

Xiyuan, X., Tongbin, C., Zhizhuang, A., Mei, L., Zechun, H., Xiaoyong, L., Yingru, L., 2008. Potential of *Pteris vittata* L. for phytoremediation of sites co-contaminated with cadmium and arsenic: the tolerance and accumulation. J. Environ. Sci. 20, 62–67.

Yamaguchi, H., Nishizawa, N.K., Nakanishi, H., Mori, S., 2002. IDI7, a new iron-regulated ABC transporter from barley roots, localizes to the tonoplast. J. Exp. Bot. 53, 727–735.

Yang, B., He, J.L., Wen, X.M., Tan, H., 2017. Cadmium pollution and ecological risk assessment for mining activity zone in Karst Area. IOP Conference Series: Earth and Environmental Science, 82, Conference 1.

Yoshida, F., 1999. Itai-Itai disease and the countermeasures against cadmium pollution by the Kamioka mine. Environ. Econ. Policy Stud. 2 (3), 215–229.

Yuan, L., Yang, S., Liu, B., Zhang, M., Wu, K., 2012. Molecular characterization of a rice metal tolerance protein, OsMTP1. Plant Cell Rep. 31 (1), 67–79.

Zhai, Q., Xiao, Y., Zhao, J., Tian, F., Zhang, H., Narbad, A., Che, W., 2017. Identification of key proteins and pathways in cadmium tolerance of *Lactobacillus plantarum* strains by proteomic analysis. Sci. Rep. **7**, 1182. https://doi.org/10.1038/s41598-017-01180-x.

Zhang, L., Pei, Y., Wang, H., et al., 2015. Hydrogen sulfide alleviates cadmium-induced cell death through restraining ROS accumulation in roots of *Brassica rapa* L. ssp. *pekinensis*. Oxidative Med. Cell. Longev. https://doi.org/10.1155/2015/804603, Article ID 804603.

Zhang, H., Lv, S., Xu, H., Hou, D., Li, Y., Wang, F., 2017. $H_2O_2$ is involved in the metallothionein-mediated rice tolerance to copper and cadmium toxicity. Int. J. Mol. Sci. 18, 1–12.

Zhao, Z., Cai, Y., Zhu, Y., Kneer, R., 2005. Cadmium-induced oxidative stress and protection by L-Galactono-1,4-lactone in winter wheat (*Triticum aestivum* L.). J. Plant Nutr. Soil Sci. 168, 759–763.

Zheng, G., Lv, H.P., Gao, S., Wang, S.R., 2010. Effects of cadmium on growth and antioxidant responses in *Glycyrrhiza uralensis* seedlings. Plant Soil Environ. 56 (11), 508–515.

Zhou, C., Zhu, L., Ma, Z., Wang, J., 2017. *Bacillus amyloliquefaciens* SAY09 increases cadmium resistance in plants by activation of auxin-mediated signaling pathways. (In Singh Gill, S., Tuteja, N., Gil, l R., Jimenez Bremont, J.F., Macovei, A., Anjum N.A., Eds.) Gene 8 (7), 173.

Zhu, X.F., Zheng, C., Hu, Y.T., Jiang, T., Liu, Y., Dong, N.Y., Yang, J.L., Zheng, S.J., 2011. Cadmium-induced oxalate secretion from root apex is associated with cadmium exclusion and resistance in *Lycopersicon esulentum*. Plant Cell Environ. 34 (7), 1055–1064.

Zonghui, C., Setagawa, M., Kang, Y., Iwasaki, K., 2009. Zinc and cadmium uptake from a metalliferous soil by a mixed culture of *Athyrium yokoscense* and *Arabis flagellosa*. Soil Sci. Plant Nutr. 55 (2), 315–324.

## FURTHER READING

Chaney, R.L., Malik, M., Li, Y.M., Brown, S.L., Brewer, E.P., Angle, J.S., Baker, A.J.M., 1997. Phytoremediation of soil metals. Curr. Opin. Biotechnol. 8, 279–284.

Matsuda, T., Kuramata, M., Takahashi, Y., Kitagawa, E., Youssefian, S., Kusano, T., 2009. A novel plant cysteine-rich peptide family conferring cadmium tolerance to yeast and plants. Plant Signal. Behav. (5), 419–421.

# CHAPTER 3

# Morphological and Physiological Responses of Plants to Cadmium Toxicity

**Afzal Hussain*, Shafaqat Ali*, Muhammad Rizwan*, Muhammad Zia-ur-Rehman†, Tahira Yasmeen*, Malik Tahir Hayat‡, Iqbal Hussain§, Qasim Ali§, Syed Makhdoom Hussain¶**

*Department of Environmental Sciences and Engineering, Government College University, Faisalabad, Pakistan
†Faculty of Agriculture, Institute of Soil and Environmental Sciences, University of Agriculture, Faisalabad, Pakistan
‡Department of Environmental Sciences, COMSATS, Abbottabad, Pakistan
§Department of Botany, Government College University, Faisalabad, Pakistan
¶Department of Zoology, Government College University, Faisalabad, Pakistan

## Abstract

Plants are exposed to various biotic and abiotic stresses, including heavy metals stress. Soil contamination due to heavy metals is a big environmental issue that impacts the deterioration of the environment and food quality. Phytoremediation is an environmentally friendly and cost-effective technique to address the problem of contaminated soil due to heavy metals. Among various heavy metals, cadmium (Cd) is very toxic to plants. Cadmium uptake reduces plant growth, root and shoot lengths, fresh and dry masses, chlorophyll contents, and gas exchange attributes. Also, Cd lowers seed germination and availability of essential nutrients, which results in lower yield and quality of the grains in cereals. Antioxidant enzymes play an important role in the defense mechanism of plants against reactive oxygen species (ROS) under heavy metal stress. Cadmium toxicity inhibits plant growth by stimulating ROS, which triggers certain alterations to antioxidant enzymes, and disturbs the normal functionality of physical and biochemical processes.

**Keywords:** Heavy metals, Cadmium, Toxicity, Antioxidants, Reactive oxygen species, Metal uptake, Physiological, Morphological

## 1 INTRODUCTION

Heavy metals accumulation in our environment occurs because of natural, as well as anthropogenic activities, including mining, sewage sludge application, wastewater use in agriculture, various industrial activities, and rock


https://doi.org/10.1016/B978-0-12-814864-8.00003-6

phosphate fertilizers (Gallego et al., 2012; Yousaf et al., 2016). Several biochemical, morpho-physiological, and ultra-structural alterations in plants are induced by heavy metals toxicity (Gill et al., 2015; Ali et al., 2015). Cadmium, among various heavy metals, is mobile in soil and toxic for animals and plants (Chaney, 2015; Rizwan et al., 2016b). By traversing the food chain, Cd accumulates in humans (Rizwan et al., 2016a), which can give rise to various disorders, such as nephrotoxicity and cancer (Khan et al., 2016). Through the use of sewage sludge and wastewater for irrigation purposes, Cd accumulates in food crops, and enters into our food chain (Abbasi et al., 2013; Anwar et al., 2016). Similarly, many other heavy metals, as well as Cd, accumulate in food crops grown in soils adjacent to industrial areas and mining (Bui et al., 2016; Rehman et al., 2017). Plant growth, photosynthesis, biomass, quality, and yield is diminished by Cd toxicity (Gharaibeh et al., 2015; Mombo et al., 2016). Alterations of mineral nutrients in many plants has been observed due to Cd toxicity (Bertoli et al., 2012). Growth and root morphology is affected by an excess of Cd (Finger-Teixeira et al., 2010), which causes the overproduction of ROS (Gratao et al., 2015; Hossain et al., 2010). Plants with broader leaves accumulate more concentration of Cd (Wang et al., 2014). Cadmium enters into the food chain by accumulating in various consumable parts of plants (Baldantoni et al., 2016).

When cadmium (Cd) concentration exceeds its threshold level, that is, $8\,mg\,kg^{-1}$ in soil, and $3–30\,mg\,kg^{-1}$ in plant tissues (Solis-Dominguez et al., 2007; Chen et al., 2008), visible signs of cadmium toxicity, such as growth retardation, chlorosis, necrosis, brown roots, and even death were observed in many plants (Gratao et al., 2009; Chang et al., 2013). Accumulation of cadmium in plants can profoundly affect a series of physiological processes, such as the interference of respiration and photosynthesis (Smiri et al., 2009; Feng et al., 2010; Andosch et al., 2012; Volland et al., 2014), transport and absorption of mineral nutrients (Wang et al., 2007; Nedjimi and Daoud, 2009), and water absorption (Polle et al., 2013). Stress-altering Cd gene (Qian et al., 2010; Irfan et al., 2013) and protein expression causes inhibition of enzymes, an increase in accumulation of lipid peroxidation, and alteration of metabolism and reactive oxygen species (ROS) (Lopez-Millan et al., 2009; Semane et al., 2010).

## 2 EFFECTS OF Cd ON BIOMASS AND PLANT GROWTH

Cd negatively effects plant growth and biomass. The Cd stress decreases the roots tips and length and surface area in pepper cultivars under hydroponic conditions (Huang et al., 2015). Also, 2 and 10 μM of applied Cd reduce the fresh weight of two pepper cultivars in a dose-dependent manner under hydroponic conditions (Xin et al., 2014). The application of $60\,\mathrm{mg\,kg^{-1}}$ of Cd reduces the dry weight and root and shoot lengths of the potato seedlings, as compared with a control set in a pot study (Hassan et al., 2016). In a nutrient solution of Cd (100 μM) for 90 days, the tomato exhibited no fruit production (Hediji et al., 2015).

Cd stress causes a significant reduction in dry weights and total leaf area of roots and shoots of cabbage (Jinadasa et al., 2016). A similar reduction was observed in the biomass and growth of soybeans (Wang et al., 2016), radishes (Varalakshmi and Ganeshamurthy, 2013), and lettuce (Monteiro et al., 2009).

## 3 Cd EFFECT ON ROOTS

Root browning is seen in plants when they exposed to Cd (Lux et al., 2010; Chang et al., 2013). Cd toxicity is described as reducing root length and dry matter, as well as increasing root diameter (Gratao et al., 2009). Due to an increase in the root length of Cd inhibition, it can be attributed to depolymerization of the microtubule cellular structure and the formation of chromosomal abnormalities cells in meristem cells less than that and consequently (Seth et al., 2008). Cd stress results in increased root diameter due to the increase in the size of parenchymal cells and cortical tissue expansion, which has a functional role in increasing the resistance of plants to the radial flow of water (Maksimovic et al., 2007). *Salix alba* and *Populus* × *euramericana* grow in $10\,\mathrm{\mu mol\,L^{-1}}$ $Cd(NO_3)_2$ (Lunackova et al., 2003), Miscanthus tea extract in $6.6\,\mathrm{mol\,L^{-1}}$ $Cd(NO_3)_2$ (Scebba et al., 2006), and corn $1\,\mathrm{mM\,L^{-1}}$ Cd (Maksimovic et al., 2007) showed higher diameters due to cadmium toxicity. Both Boot and Mensink (1990) and Lima et al. (2006) observed elongation of the roots, but the biomass in the grass and chickpea (*Pisum sativum* L.) was more influenced by Cd. Cd stress impacts the total and specific root length (Guerrero-Campo et al., 2006), the decrease in root length, surface area, specific root length and root number, and the increase in root diameter capacity (water, food, etc.), as well as the root system architecture in different species of plants (Stanova et al., 2012; Wei et al., 2012). Therefore, the morphological parameters of roots as indicators for evaluating the toxicity of cadmium (Lu et al., 2013a, 2013b) are presented (Tables 1 and 2).

Among herbal varieties and cultivars, the effects of Cd on root morphology may vary (Wei et al., 2012). Treatment with Cd and Cd + Zn reduced the entire root area of *Salix caprea* F20 (Vaculík et al., 2012). Gratao et al.

**Table 1** Cadmium accumulation potentials of various plant species

| Species | Cd level in growth medium (growth condition) | Cd concentration in plant parts ($mg\,Cd\,kg^{-1}$ DW) | Reference |
|---|---|---|---|
| *Populus tremula × tremuloides* | 5.1 $mg\,kg^{-1}$ DW soil (field, 2 years) | 35 (leaves), 14 (branches) | Hassinen et al., 2009 |
| *Salix matsudana, Salix alba* | 60.6 $mg\,kg^{-1}$ DW soil (pot) | 167 (leaves) | Robinson et al., 2000 |
| *Salix caprea* genotype KH21 | 0.5 $mg\,L^{-1}$ (pure perlite culture) | 400 (roots), 340 (leaves) | Vaculík et al., 2012 |
| *Salix dasycla-Dos* | 5.46 $mg\,kg^{-1}$ DW soil (pot) | 41.1 (shoots) | Fischerova et al., 2006 |
| *S. caprea* | 6.25–51.5 $mg\,kg^{-1}$ DW soil (field) | 116 (leaves) | Claire-Lise and Nathalie, 2012 |
| *S. caprea* | 0.5 $mg\,L^{-1}$ (perlite culture) | 115 (leaves) | Puschenreiter et al., 2010 |
| *Cynara cardunculus* cv. Peralta | 2–100 $mg\,kg^{-1}$ DW soil (pot, 90 d) | 10–280 (roots), 10–260 (shoots) | Hernandez-Allica et al., 2008 |
| *S. caprea* genotype F20 | 0.5 $mg\,L^{-1}$ (pure perlite culture) | 400 (roots), 200 (leaves) | Vaculík et al., 2012 |
| *Cichorium intibus* | 40–100 $mg\,kg^{-1}$ DW soil (pot, 35–50 d) | 96–207 (roots), 3680 (shoots) | Fotiadis and Lolas, 2011 |
| *Bromus sterilis* | 40–120 $mg\,kg^{-1}$ DW soil (pot, 35–50 d) | 114–463 (roots), 65–75 (shoots) | Fotiadis and Lolas, 2011 |
| *Vicia sativa* | 40–160 $mg\,kg^{-1}$ DW soil (pot, 35–50 d) | 189–495 (roots), 14–29 (shoots) | Fotiadis and Lolas, 2011 |
| *Zea mays* cv. Ranchero | 2–100 $mg\,kg^{-1}$ DW soil (pot, 90 d) | 10–500 (roots), 5–130 (shoots) | Hernandez-Allica et al., 2008 |
| *Z. mays* | 40–160 $mg\,kg^{-1}$ DW soil (pot, 35–50 d) | 199–243 (roots), 57–60 (shoots) | Fotiadis and Lolas, 2011 |

**Table 1** Cadmium accumulation potentials of various plant species—cont'd

| Species | Cd level in growth medium (growth condition) | Cd concentration in plant parts ($mg\,Cd\,kg^{-1}$ DW) | Reference |
|---|---|---|---|
| *Apium graveolens* | 40–120 $mg\,kg^{-1}$ DW soil (pot, 35–50 d) | 220–300 (roots), 16–40 (shoots) | Fotiadis and Lolas, 2011 |
| *Paspalum atratum* cv. Reyan No. 11 | 8 $mg\,kg^{-1}$ DW soil (pot) | 349 (roots), 46 (shoots) | Zhang et al., 2010 |
| Poplar clone (*Populus* × *canadensis*) | 54.3 $mg\,kg^{-1}$ DW soil (pot, 35 d), 163 $mg\,kg^{-1}$ DW soil (pot, 35 d) | 7.6 (leaves), 4.6 (stems), 57 (roots) 12.7 (leaves), 15 (stems), 80 (roots) | Di Baccio et al., 2014 |
| *Arundo donax* L. | 750 $g\,L^{-1}$ (hydroponics, 21 d) | 263 (stems) | Sabeen et al., 2013 |

**Table 2** Effect of cadmium on plants

| Crop | Cd levels | Effects | Reference |
|---|---|---|---|
| Wheat | 0, 5, 20, 50 and 80 $mg\,L^{-1}$ | Reduced seed germination | Ahmad et al., 2012 |
| Wheat | 0, 5, 20 $mg\,L^{-1}$ | Decreased seed germination | Ahmad et al., 2013 |
| Maize | Cd (0, 100 μM) | Inhibited growth and photosynthetic pigments | Anjum et al., 2017 |
| Wheat | 0, 42 $mg\,kg^{-1}$ | Reduced transfer of cadmium from mine tailings to wheat | Boussen et al., 2013 |
| Cucumber | 0, 10, 25, 50, 100, 200 μM | Decreased nutrients uptake and photosynthetic performance | Sun et al., 2017 |
| Wheat | 0, 10, 20, 30, 40 or 50 μM | Negative impact on seedlings, growth and photosynthesis | Ci et al., 2010 |
| Wheat | 0, 50, 100, 200 and 400 μM) | Negatively induced oxidative stress | Riaz et al., 2014 |

*Continued*

**Table 2** Effect of cadmium on plants—cont'd

| Crop | Cd levels | Effects | Reference |
|---|---|---|---|
| Wheat | 0, 10 mg L$^{-1}$ | Reduced photosynthesis and chlorophyll contents | Li et al., 2015 |
| Maize | Cd (0, 1, 3 and 5 ppm) | Decreased the antioxidants enzymes activities | Akinyemi et al., 2017 |
| Wheat | 0, 0.5mg kg$^{-1}$ | Detrimental effects on growth and physiological process | Lu et al., 2013a, 2013b |
| Wheat | 0, 20, 50, and 100 μM | Reduction in growth and antioxidant system activity of seedlings | Poghosyan et al., 2014 |
| Wheat | 0, 2, 4 μM | Negative effect on photosynthesis and chlorophyll fluorescence | Shafi et al., 2011 |
| Maize | (0, 75, 150, 225, 300, 375 μM) | Reduced antioxidants enzymes activities | Anjum et al., 2015 |
| Wheat | 0, 15 and 30 mg kg$^{-1}$ | Negative effect on growth, cadmium accumulation and macronutrient content of durum wheat | Yourtchi and Bayat, 2013 |
| Wheat | 10, 100 and 200 μM | Caused disturbance in roots proteins | Wang et al., 2011 |
| Maize | (0, 5, 10, and 15 μM) | Reduced growth and photosynthesis | Akhtar et al., 2017 |
| Wheat | 0–33 mg kg$^{-1}$ | Negative effected the growth, oxidative stress and antioxidant system in wheat seedlings | Lin et al., 2007 |
| Wheat | 0, 100 μM | Reduced glutathione reductase activity and isoforms in leaves and roots | Yannarelli et al., 2007 |

(2009) reported reduced root diameter after the use of Cd in the tomato (*Lycopersicon esculentum*) Micro-Tom. In addition, other studies that did not show the effect of *Phaseolus vulgaris* root physiology stress (Vazquez et al., 1992) included two different lines of coronary embryos (Florijn et al., 1993) and *S. caprea* KH21 (Vaculík et al., 2012). Vaculík et al. (2012) observed that the lack of response of KH21 tissues to high cadmium concentrations is likely to be related to their storage capacity.

## 4 Cd EFFECT ON LEAVES AND SHOOTS

Chlorosis, necrosis, and drying are toxic symptoms of Cd stress in the foliage of the plant. The normal concentration of Cd in the leaves is from 0.05 to 0.2 mg kg$^{-1}$ for many plant species (Solis-Dominguez et al., 2007). The concentration of Cd in leaves of many plants is >5–10 mg kg$^{-1}$ of dry weight of plants (White and Brown, 2010). Pietrini et al. (2010). Cd is used to reduce node formation and node dysfunction in olive plants, *Glycine max* (Balestrasse et al., 2001), peas (*P. sativum* L.) (Ausili et al., 2002), and white lupine (*Lupinus albus*) (Carpena et al., 2003).

## 5 Cd EFFECT ON PLANTS YIELD

The yield of grain was decreased for Mung beans (*Vigna radiata*) (Wani et al., 2007a) and chickpeas (*Cicer arietinum*) (Wani et al., 2007b) in Cd-polluted soil. Cadmium accumulates in the roots, which causes water stress in certain plants that are exposed to Cd, such as Holm cranberries (*Quercus ilex* subsp. *Ballota*) (Dominguez et al., 2011).

In Arqval (*Eruca sativa*), 81–810 μmol L$^{-1}$ Cd (as Cd $(NO_3)_2 \cdot 4H_2O$) reduced root length, but did not affect the germination rate. During the filming of 650 seeds of *E. sativa*, the germination μmol L$^{-1}$ Cd decreased at 240 μmol L$^{-1}$ Cd (Ozdener and Kutbay, 2009). Interestingly, with low concentrations of Cd, quantitative stimulation of germination was observed (Lefevre et al., 2009). This assumption is in accordance with the proposed concept of an oxidative window. For example, a high concentration of the ROS regime for germination is essential (Bailly et al., 2008). Similarly, the low concentration of Cd (Lefevre et al., 2009) results in little growth of the seedlings, probably due to the stimulation of cell cycle activity by increasing the ROS signal (Gapper and Dolan, 2006).

### 5.1 Effect of Cd on Seed Germination

Cadmium inhibits the absorption of seeds and reduces the amount of water in the seedlings, which affects germination. Water is very important for germination of seeds. The inhibition of seed germination, and water contents of seedlings is induced by Cd. Therefore, it seems that when seeds are exposed to Cd, they may not be able to absorb enough water, because of the toxicity of Cd. Lee et al. (2010) observed that the exposure of Cd to plants prevents seed germination, seedling growth, and so forth. However, exposure to 3–11 $\mu mol\,L^{-1}$ $CdCl_2 \cdot H_2O$ in sea full-color onion shoots (*Bowiea volubilis*) and Mountain Lilies, which belong to the family Hyacinthaceae, prevents full-red germination of the iris of the same family.

## 6 EFFECT OF Cd ON PHOTOSYNTHESIS AND RESPIRATION

Cd toxicity negatively affects plants' photosynthesis. Water use efficiency and the net photosynthetic rate decreased in pepper seedlings under cadmium stress (León et al., 2002). The Cd toxicity reduced the carotenoids and chlorophyll contents of tomato leaves (Hediji et al., 2010). Under Cd toxicity, the response of various crops varied toward chlorophyll contents and the photosynthetic rate in soybean genotypes (Shamsi et al., 2014). A significant decrease in stomatal conductance, the photosynthetic rate, and chlorophyll contents was reported in younger soybean leaves, as compared with mature leaves exposed to cadmium of less than 50 μM, and 100 μM of Cd, for 10 days under hydroponic conditions (Xue et al., 2014). In addition, Cd toxicity reduced the chlorophyll content in potatoes (Xu et al., 2013), lettuce (Monteiro et al., 2009), tomatoes (Hassan et al., 2016), soybeans (Xue et al., 2014), and peas (Agrawal and Mishra, 2009).

The efficiency of the reaction, the chlorophyll contents, and the intensity of photosynthesis are known critical indices for the growth state of plants with Cd toxicity (Chen et al., 2008). Several studies have shown that Cd is an effective inhibitor of photosynthesis. Cadmium, in fact, inhibits chlorophyll synthesis, and is stabled by binding proteins, resulting in damage to the photosynthetic apparatus, especially the reduction of the light collection complex II, photosystems I (PSI) and II (PSII) (Cooper et al., 2007), and the inhibition of certain Calvin cycle enzymes (Assche and Clijsters, 1990). In addition, aggregate starch and modified starch granules are used to bind heavy metals such as cadmium (Hao et al., 2009).

Also, Cd significantly affects the concentration of chlorophyll in the star cluster (Chang et al., 2013). Chen et al. (2008) reported that Cd interferes with repeat cutting, which indicates a reduction of the properties of the chlorophyll meter (CMR). The chlorophyll degradation, or biosynthesis inhibition, are responsible for the reduction of stress due to Cd toxicity (Wan et al., 2012). It has been reported that indirect effects of Cd on chlorophyll include inferior nutrients. Qureshi et al. (2010) found that the boundary state of Cd-iron, under a devastating multiprotein complex such as thylakoid pigment, can damage many of its parts, resulting in coordination loss between them, and a negative effect on chloroplast homeostasis in *Brassica juncea*. Nevertheless, the phosphorus plants produce for the prevention of oxidative stress also help maintain chlorophyll and chloroplasts under cadmium stress. In addition, Durrett et al. (2007) showed that a decrease in the chlorophyll content of plants under Cd stress also exhibit the decreased the expression of FRD3 transport citrate in the parenchyma of the vascular components of the plant.

Cadmium causes serious disturbances in growth activities, morphogenesis, photosynthesis, and respiration in microtacteric denticulata algae (Volland et al., 2011, 2012, 2014; Andosch et al., 2012). Alteration in the metabolism of respiratory system is a main cause that contributes to the destructive effects of the Cd on the germination of plants (Chugh and Sawhney, 1996; Smiri et al., 2009). In addition, Cd affects the activity of the enzyme in the Kreb cycle (Bansal et al., 2002), and also the transport chain of electrons (Bansal and Sharma, 2000) in pea seed cotyledon germination. Smiri et al. (2009) stated that the activities of respiratory enzymes are changed due to Cd stress.

## 7 Cd EFFECT ON NITROGEN METABOLISM

The assimilation of nitrogen is strongly affected by the stress of Cd in the *P. sativum* L. (Dhingra and Priefer, 2006), *P. vulgaris* (Gouia et al., 2003), and the *C. arietinum* (Hasan et al., 2008). Cadmium stress adversely affects N metabolism via inhibition of the nitrate uptake, and by reducing enzyme activity related to the uptake of nitrates (Chang et al., 2013). Cd presence reduces nitrogenase and nodulation activity in *V. radiata* (Wahid et al., 2007), *P. sativum L.* (Dhingra and Priefer, 2006), and *C. arietinum* (Hasan et al., 2008). A positive correlation has been noted between the content of leghemoglobin and nitrogenase activity in legume roots (Dakora, 1995); also, both of the parameters showed a reduction in *L. albus* Cd

(Fernandez-Pascual et al., 1996). Sanchez-Pardo et al. (2013) confirmed that some of the enzyme activity involved in the assimilation of ammonium into *L. albus* decreases under Cd stress, which promotes a decrease in the content of N in nodules, and the other organs of plants. Deterioration of the metabolism of sucrose, and sucrose synthase downregulation in pellets could be responsible for the decrease of binding in N2 Cd, due to the flow limitation of carbon for respiratory bacteroides. Thus, metabolism of carbon may play an important role in the altered function of root nodules of *L. albus* because of Cd, while metabolism of N creates secondary implications (Sanchez-Pardo et al., 2013).

Enhancement of endogenous NO content was exhibited in *Triticum aestivum* L. roots (Groppa et al., 2008) and *Hordeum vulgare*, upon exposure to Cd (Valentovicova et al., 2010). In contrast, Cd stress caused a reduction in NO content in plants (Rodriguez-Serrano et al., 2009). In addition, NO production under Cd stress shows a dose- and time-dependent pattern. Bartha et al. (2005) showed that the treatment of 50 mmol $L^{-1}$ of $CdCl_2$ caused enhancement of the contents of NO in arabidopsis cells after 24 h. Rodriguez-Serrano et al. (2009) reported that higher Cd doses and longer treatment time (>7 d) decreased the production of NO, which could be due to suppression of the activity of enzyme-type NO synthase caused by $Ca^{2+}$ deficiency, which is caused by Cd accumulation in the leaves of plants. Furthermore, Mahmood et al. (2009) found that in the roots of *T. aestivum* L., $NO_2$-dependent NO production improved under Cd stress, and the NO production stimulation was an indirect impact of the Cd on the inhibition of Cd in root respiration. In addition, the decrease in adenosine triphosphate values in roots induced the NO production of $NO_2$.

## 8 Cd EFFECT ON UPTAKE AND ASSIMILATION OF MINERAL ELEMENTS

Cd toxicity greatly changes the uptake of various mineral nutrients (Li et al., 2016). Cd (5 mg $kg^{-1}$) in soil reduces the uptake of many mineral nutrients, such as sulfur, zinc, calcium, potassium, manganese, and the phosphorus in pea seedlings (Metwally et al., 2004). It was found that 1.12 mg of Cd $kg^{-1}$ in soil reduced the Mn content, while enhancing the Zn and Fe content in soybean cultivars (Zhi et al., 2015). The application of Cd to a nutrient solution reduces K, Ca, Mn, and Zn content in tomato aerial parts (Bertoli et al., 2012). As Cd reduces the Mn contents, it shows the antagonistic effect between Mn and Cd (Dong et al., 2006). Moreover, the concentrations of mineral nutrients such as Ca, Fe, Cu, Zn, and Mn decreased both in roots and cabbage shoots cabbage under stress of Cd (Jinadasa et al., 2016). A recent study showed that Cd toxicity reduced mineral nutrients in tomatoes, lettuce, and potatoes (Khan et al., 2016). A similar reduction in the mineral nutrients was observed in lettuce under Cd stress (Monteiro et al., 2009). Alternately, it has been observed that application of Cd (1, 2.5 and 5 $mg kg^{-1}$) enhanced the contents of mineral nutrients such as Zn, Fe, Mg, K, Ca, and P in 25 welsh onion leaves and pseudostems, which might be ascribed to the synergistic effect between Cd and the mineral nutrients (Li et al., 2016). Nevertheless, sometimes the opposite result may be observed upon Cd higher concentrations, which needs to be further investigated.

Fourteen mineral elements are necessary for plant growth and development. These essential elements are taken up by plants roots, and translocate to shoots and other organs, depending on their requirements (Marschner, 2012). Due to anthropogenic activities, toxic chemicals became the part of food chain (Kong et al., 2017). These toxic compounds, such as Cd and Cr As, are also transported from the soil to arial parts (Sarwar et al., 2017). Uptake, transport, and assimilation of mineral elements require specific membrane proteins (Mitani-Ueno et al., 2018). Rice plants develop efficient root nodes and vascular systems, such as enlarged and diffuse vascular bundles (Yamaji and Ma, 2017). For healthy growth, six macronutrients (N, P, K, Mg, S, Ca) are required. N is an important building block for rice productivity, and is found mostly in two forms, that is, nitrate ions ($NO_3$), or ammonium ions ($NH_4^+$) and organic molecules, such as amino acids. Rice mostly grows in paddy conditions, where $NH_4^+$ is the main inorganic nitrogen form. According to Wang et al. (1993), rice plants develop two systems for uptake; a high-affinity transport system and a low-affinity transport system. Plants have efficient mechanisms for the uptake of mineral elements. Toxic chemicals compete with minerals during their uptake from soil and become part of metabolic activities (Ertani et al., 2017). Soil cadmium comes into contact with cations, such as $Mg^{2+}$, $Si^{2+}$, $Fe^{2+}$, $Ca^{2+}$, $Mn^{2+}$, and $Zn^{2+}$ (Degryse et al., 2004). In the roots and shoots of *Atriplex halimus*, schweinfurthii affected Ca and K nutrition when was Cd present in the growth medium (Nedjimi and Daoud, 2009). Calcium and cadmium compete with each other for the same Ca channels in plants. In guard cells Cd soaks into the plasma membrane through root cells (White, 2000) and Ca channels (Perfus-Barbeoch et al., 2002). After entering the plasma membrane, Cd interacts with Ca metabolic activities, disturbing signal processes. Cd presence in guard cells leads to stomatal closing (Perfus-Barbeoch et al., 2002). A reduction in the transpiration rate and the inhibition of growth disturbs the translocation of Mn and Cu (Sipos et al., 2013).

## 9 Cd EFFECT ON WATER TRANSPORT AND UPTAKE

Through reducing the transpiration rate, stomatal conductance, and the leaf water contents, the presence of cadmium in the growth medium induces water stress in plants (Dominguez et al., 2011). These adverse impacts are basically the result of physiological damage, the chloroplast amount, and cell enlargement and reduced intracellular space (Sandalio et al., 2001). Cd interacts with plasma membranes and disturbs their permeability, which results in a decrease in the water content of plants (Fernandez et al., 2013). Under Cd stress within the leaves of *Populus euphratica*, wilting under and water loss was observed (Polle et al. (2013). In this study, Cd-stressed seeds of sorghum (*Sorghum bicolor* L.) experience water stress conditions (Kuriakose and Prasad, 2008). Cells displayed a clear shrinkage of cytoplasm of *P. euphratica* when treated with Cd euphratica; consequently disturbing the water balance (Sun et al. (2013). The Cd presence in soil by inhibiting the growth of root hairs reduces the surface area for water absorption (Gouia et al., 2003).

## 10 Cd EFFECT ON CELL DEATH AND GENETIC ALTERATIONS

The worst effect of cadmium toxicity includes alteration of the structural features and anatomy of cells (Kupper et al., 2000). Under Cd stress, cell division, cell proliferation, a low mitotic index, and chromosomal aberration were found in various plant species (Shah and Dubey, 1995). According to Liu and Kottke (2004), increased vacuole numbers and nucleolus reduced the mitochondria number, condensed cytoplasm, stimulated plasmolysis, ruptured the nuclear envelope, enlarged vacuoles, and disturbed chloroplasts in leaf and root cells. *Brassica napus* grown under Cd concentration exhibited cracked cell walls, undeveloped mitochondria, plasmolysis, and the absence of endoplasmic reticulum in cells root tips at ($500\,\text{mmol}\,\text{L}^{-1}$) (Ali et al., 2013). Different cadmium levels ruptured the cell structure in *Sedum alfredii* (Jin et al., 2008). In various plants, cell death has been documented when systems were exposed to cadmium (Balestrazzi et al., 2009). Under Cd stress, different physiological effects serve as good biomarkers for observing significant genetic alterations in plants (Liu et al., 2012). Cd genotoxicity is linked to the function and structure of the DNA, such as the point mutations, base substitutions, single- and double-strand breaks, oxidized bases, small deletions and insertions, ploidy changes, rearrangements, and even the bulky adducts at specific DNA loci (Cambier et al., 2010; Liu et al., 2012). Cd toxicity damages $OH^-$ radials; the number of radials can increase up to 10,000, with 100,000 DNA base modifications in a single cell each day (Kranner and Colville, 2011).

## 11 Cd EFFECT ON REACTIVE OXYGEN SPECIES

Various mechanisms are responsible for the reduction of plant growth and nutrient uptake in plants. The overproduction of ROS is caused by higher Cd concentration in many plants, which enhances the malondialdehyde (MDA) contents in peas (Agrawal and Mishra, 2009) and soybean cultivars (Shamsi et al., 2014). A similar enhancement in MDA contents was found in tomatoes (Zhao et al., 2016), lettuce (Monteiro et al., 2009), potatoes (Xu et al., 2013), and mung bean plants (Hossain et al., 2010) under Cd stress. Hydrogen peroxide production is observed in mung beans because of Cd stress. Only an extra 0.3 mM of Cd enhanced the MDA concentration, as compared with a control group of seedlings of mung beans; however, the response differed in different cultivars (Hassan and Mansoor, 2014). It shows that the response of plants to cadmium stress varies within cultivars of the same species. In soybean roots, the Cd may increase the lignification, which decreases the root growth (Finger-Teixeira et al., 2010). Cadmium in excess amounts causes genotoxicity in various plant species. In potato seedlings, Cd damages the DNA in the root nuclei (Gichner et al., 2008).

Many research studies describe the toxic impacts of cadmium on plants (Benavides et al., 2005; Semane et al., 2010). Reactive oxygen species production increased when plants were exposed to a high Cd concentration (Sharma et al., 2010). MDA content is a common indictor that of an increase in oxidative damage. (Moller et al., 2007). Membrane lipid peroxidation initiates when ROS production increases, consequently creating a free radical reaction in tissues. Cd toxicity can result in production of reactive oxygen species, membrane distortion, toxic metabolite and photosynthesis inhibition (Shamsi et al., 2008). Cd has complex mechanisms for the oxidase induction, interaction with antioxidative defenses, electron transport chain interruption, and so forth (Schutzendubel and Polle, 2002; Gratao et al., 2005). Overproduction of ROS generates oxidative processes such as protein oxidation, enzyme inhibition, membrane lipid peroxidation, and DNA and RNA destruction, which consequently leads to cell damage, and finally results in cell death (Gratao et al., 2005).

## 12 Cd EFFECT ON ANTIOXIDANTS ENZYME ACTIVITIES

When reacting to stresses, including heavy metals stress, plants evolve their immune system through the enhancement of antioxidant enzymes. A number of studies have revealed that antioxidant enzymes activities are altered because of Cd stress (Dong et al., 2006). Also, the contents of catalase (CAT) and super oxidase (SOD) were shown to be enhanced in pepper cultivars to cope with the stress of Cd (León et al., 2002). A similar increase in CAT and SOD contents was found in pea seedlings under Cd stress (Agrawal and Mishra, 2009). The response of antioxidants can vary toward the different genotypes of the same species. As an example, the GSH contents increased while APX contents decreased for genotypes of peas under the stress of Cd (Metwally et al., 2004). A similar increase in POD and SOD contents in different genotypes of soybean seedlings (Shamsi et al., 2014) was shown. Also, the CAT, ascorbate peroxidase (APX), and glutathione peroxidase (GPX) contents decreased in genotypes of the mung bean under Cd toxicity (Hassan and Mansoor, 2014). Cadmium sequestration in leaf vacuoles and root cell walls may be ascribed to the detoxification of Cd in plants (Wang et al., 2015). Plants' cell wall play a key role in Cd immobilization in pepper cultivars (Xin and Huang, 2014). But in various metabolic parts, the Cd sequestration varied among genotypes or cultivars of the same plant species (Xin et al., 2014). Various plant species tolerate Cd stress through phytochelatin production (Cabala et al., 2011). It has been reported that phytochelatin production increased in shoots and the roots of cabbage under Cd stress (500 $\mu g L^{-1}$) as compared with the control group (Jinadasa et al., 2016).

With the passage of time, plants developed defensive systems that are also known as enzyme-catalyzed clean up systems that protect the cellular function of plants (Horvath et al., 2007). The equilibrium among activities of ROS production and antioxidative enzymes determines the possibility of oxidative signaling that causes cell damage (Moller et al., 2007). Defensive systems commonly are two types; enzymatic and nonenzymatic antioxidants, which reduce the adverse impacts of ROS such as peroxidase, catalase, ascorbate peroxidase, low molecular mass antioxidants reductase, superoxide dismutase, glutathione, ROS, and carotenoid scavenging enzymes and ascorbate (Apel and Hirt, 2004).

## 13 CONCLUSION

Heavy metal stress is one of the major concerns for researchers. As heavy metals are toxic in nature and make their way into the food chain, they cause harmful impacts on human health and plants. Cadmium is considered a dangerous heavy metal due to its toxic properties and consequences of exposure. From various studies, it has been observed that Cd affected the plants by reducing their growth, biomass, and yield. Cadmium lowered the seed germination, respiration, and photosynthesis, and disturbed the nitrogen metabolism. In addition, Cd reduced mineral uptake and their transport from roots to shoots by inducing genetic alterations and stimulating the reactive oxygen species, which also reduced the antioxidants' enzyme activities. However, further investigations are still needed to explore the detailed mechanisms of Cd at the cellular, as well as genetic levels.

## ACKNOWLEDGMENTS

The authors thank the Higher Education Commission (HEC) of Pakistan for financial support. This study was also financially assisted by HEC, Pakistan, under HEC Project No. 20-3653/NRPU/R&D/HEC/14/437 and NRPU project No. 5634/Punjab/NRPU/R&D/HEC/2016.

## CONFLICTS OF INTEREST

The authors declare that there are no conflicts of interest regarding the publication of this chapter.

## REFERENCES

Abbasi, A.M., Iqbal, J., Khan, M.A., Shah, M.H., 2013. Health risk assessment and multivariate apportionment of trace metals in wild leafy vegetables from Lesser Himalayas, Pakistan. Ecotoxicol. Environ. Saf. 92, 237–244.

Agrawal, S.B., Mishra, S., 2009. Effects of supplemental ultraviolet-B and cadmium on growth, antioxidants and yield of *Pisum sativum* L. Ecotoxicol. Environ. Saf. 72 (2), 610–618.

Ahmad, I., Akhtar, M.J., Zahir, Z.A., Jamil, A., 2012. Effect of cadmium on seed germination and seedling growth of four wheat (*Triticum aestivum* L.) cultivars. Pak. J. Bot. 44 (5), 1569–1574.

Ahmad, I., Javed Akhtar, M., Naeem Asghar, H., Ahmad, Z.Z., 2013. Comparative efficacy of growth media in causing cadmium toxicity to wheat at seed germination stage. Int. J. Agric. Biol. 15 (3), 134–139.

Akhtar, T., Zia-ur-Rehman, M., Naeem, A., Nawaz, R., Ali, S., Murtaza, G., Maqsood, M.A., Azhar, M., Khalid, H., Rizwan, M., 2017. Photosynthesis and growth response of maize (*Zea mays* L.) hybrids exposed to cadmium stress. Environ. Sci. Pollut. Res. 24 (6), 5521–5529.

Akinyemi, A.J., Faboya, O.L., Olayide, I., Faboya, O.A., Ijabadeniyi, T., 2017. Effect of cadmium stress on non-enzymatic antioxidant and nitric oxide levels in two varieties of maize (*Zea mays*). Bull. Environ. Contam. Toxicol. 98 (6), 845–849.

Ali, B., Tao, Q., Zhou, Y., Gill, R.A., Ali, S., Rafiq, M.T., Xu, L., Zhou, W., 2013. 5-Aminolevolinic acid mitigates the cadmium-induced changes in *Brassica napus* as revealed by the biochemical and ultra-structural evaluation of roots. Ecotoxicol. Environ. Saf. 92, 271–280.

Ali, B., Gill, R.A., Yang, S., Gill, M.B., Farooq, M.A., Liu, D., Daud, M.K., Ali, S., Zhou, W., 2015. Regulation of cadmium-induced proteomic and metabolic changes by 5-aminolevulinic acid in leaves of *Brassica napus* L. PLoS One 10 (4), 328.

Andosch, A., Affenzeller, M.J., Lütz, C., Lutz-Meindl, U., 2012. A freshwater green alga under cadmium stress: ameliorating calcium effects on ultrastructure and photosynthesis in the unicellular model Micrasterias. J. Plant Physiol. 169 (15), 1489–1500.

Anjum, S.A., Tanveer, M., Hussain, S., Bao, M., Wang, L., Khan, I., Ullah, E., Tung, S.A., Samad, R.A., Shahzad, B., 2015. Cadmium toxicity in Maize (*Zea mays* L.): consequences on antioxidative systems, reactive oxygen species and cadmium accumulation. Environ. Sci. Pollut. Res. 22 (21), 17022–17030.

Anjum, S.A., Tanveer, M., Hussain, S., Ashraf, U., Khan, I., Wang, L., 2017. Alteration in growth, leaf gas exchange, and photosynthetic pigments of maize plants under combined cadmium and arsenic stress. Water Air Soil Pollut. 228 (1), 13.

Anwar, S., Nawaz, M.F., Gul, S., Rizwan, M., Ali, S., Kareem, A., 2016. Uptake and distribution of minerals and heavy metals in commonly grown leafy vegetable species irrigated with sewage water. Environ. Monit. Assess. 188 (9), 541–543.

Apel, K., Hirt, H., 2004. Reactive oxygen species: metabolism, oxidative stress, and signal transduction. Annu. Rev. Plant Biol. 55, 373–399.

Assche, F.V., Clijsters, H., 1990. Effects of metals on enzyme activity in plants. Plant Cell Environ. 13 (3), 195–206.

Ausili, P., Borisov, A., Lindblad, P., Martensson, A., 2002. Cadmium affects the interaction between peas and root nodule bacteria. Acta Agric. Scand. Sect. B Soil Plant Sci. 52 (1), 8–17.

Bailly, C., El-Maarouf-Bouteau, H., Corbineau, F., 2008. From intracellular signaling networks to cell death: the dual role of reactive oxygen species in seed physiology. C. R. Biol. 331 (10), 806–814.

Baldantoni, D., Morra, L., Zaccardelli, M., Alfani, A., 2016. Cadmium accumulation in leaves of leafy vegetables. Ecotoxicol. Environ. Saf. 123, 89–94.

Balestrasse, K.B., Gardey, L., Gallego, S.M., Tomaro, M.L., 2001. Response of antioxidant defence system in soybean nodules and roots subjected to cadmium stress. Funct. Plant Biol. 28 (6), 497–504.

Balestrazzi, A., Botti, S., Zelasco, S., Biondi, S., Franchin, C., Calligari, P., Racchi, M., Turchi, A., Lingua, G., Berta, G., Carbonera, D., 2009. Expression of the PsMT A1 gene in white poplar engineered with the MAT system is associated with heavy metal tolerance and protection against 8-hydroxy-2′-deoxyguanosine mediated-DNA damage. Plant Cell Rep. 28 (8), 1179–1192.

Bansal, P., Sharma, P., 2000. Effect of $Pb^{2+}$ and $Cd^{2+}$ on respiration and mitochondrial electron transport chain in germinating pea seeds (*Pisum sativum* L.). Ind. J. Environ. Ecol. Plant (3), 249–254.

Bansal, P., Sharma, P., Goyal, V., 2002. Impact of lead and cadmium on enzyme of citric acid cycle in germinating pea seeds. Biol. Plant. 45 (1), 125–127.

Bartha, B., Kolbert, Z., Erdei, L., 2005. Nitric oxide production induced by heavy metals in *Brassica juncea* L. and *Pisum sativum* L. Acta Biol. Szeged 49 (2), 9–12.

Benavides, M.P., Gallego, S.M., Tomaro, M.L., 2005. Cadmium toxicity in plants. Braz. J. Plant Physiol. 17 (1), 21–34.

Bertoli, A.C., Cannata, M.G., Carvalho, R., Bastos, A.R.R., Freitas, M.P., Dos Santos Augusto, A., 2012. *Lycopersicon esculentum* submitted to Cd-stressful conditions in nutrition solution: nutrient contents and translocation. Ecotoxicol. Environ. Saf. 86, 176–181.

Boot, R.G., Mensink, M., 1990. Size and morphology of root systems of perennial grasses from contrasting habitats as affected by nitrogen supply. Plant Soil 129 (2), 291–299.

Boussen, S., Soubrand, M., Bril, H., Ouerfelli, K., Abdeljaouad, S., 2013. Transfer of lead, zinc and cadmium from mine tailings to wheat (*Triticum aestivum*) in carbonated Mediterranean (Northern Tunisia) soils. Geoderma 192, 227–236.

Bui, A.T., Nguyen, H.T., Nguyen, M.N., Tran, T.H.T., Vu, T.V., Nguyen, C.H., Reynolds, H.L., 2016. Accumulation and potential health risks of cadmium, lead and arsenic in vegetables grown near mining sites in Northern Vietnam. Environ. Monit. Assess. 188 (9), 525.

Cabala, R., El Zohri, M., Frank, H., 2011. Accumulation and translocation of Cd metal and the Cd-induced production of glutathione and phytochelatins in *Vicia faba* L. Acta Physiol. Plant. 33 (4), 1239–1248.

Cambier, S., Gonzalez, P., Durrieu, G., Bourdineaud, J.P., 2010. Cadmium-induced genotoxicity in zebrafish at environmentally relevant doses. Ecotoxicol. Environ. Saf. 73 (3), 312–319.

Carpena, R.O., Vázquez, S., Esteban, E., Fernández-Pascual, M., de Felipe, M.R., Zornoza, P., 2003. Cadmium-stress in white lupin: effects on nodule structure and functioning. Plant Physiol. Biochem. 41 (10), 911–919.

Chaney, R.L., 2015. How does contamination of rice soils with Cd and Zn cause high incidence of human Cd disease in subsistence rice farmers. Curr. Pollut. Rep. 1 (1), 13–22.

Chang, Y.S., Chang, Y.J., Lin, C.T., Lee, M.C., Wu, C.W., Lai, Y.H., 2013. Nitrogen fertilization promotes the phytoremediation of cadmium in *Pentas lanceolata*. Int. Biodeterior. Biodegrad. 85, 709–714.

Chen, F., Wang, F., Zhang, G., Wu, F., 2008. Identification of barley varieties tolerant to cadmium toxicity. Biol. Trace Elem. Res. 121 (2), 171–179.

Chugh, L.K., Sawhney, S.K., 1996. Effect of cadmium on germination, amylases and rate of respiration of germinating pea seeds. Environ. Pollut. 92 (1), 1–5.

Ci, D., Jiang, D., Wollenweber, B., Dai, T., Jing, Q., Cao, W., 2010. Cadmium stress in wheat seedlings: growth, cadmium accumulation and photosynthesis. Acta Physiol. Plant. 32 (2), 365–373.

Claire-Lise, M., Nathalie, V., 2012. The use of the model species *Arabidopsis halleri* towards phytoextraction of cadmium polluted soils. New Biotechnol. 30 (1), 9–14.

Cooper, J., Bolbot, J.A., Saini, S., Setford, S.J., 2007. Electrochemical method for the rapid on site screening of cadmium and lead in soil and water samples. Water Air Soil Pollut. 179 (1–4), 183–195.

Dakora, F.D., 1995. A functional relationship between leghaemoglobin and nitrogenase based on novel measurements of the two proteins in legume root nodules. Ann. Bot. 75 (1), 49–54.

Degryse, F., Buekers, J., Smolders, E., 2004. Radio-labile cadmium and zinc in soils as affected by pH and source of contamination. Eur. J. Soil Sci. 55 (1), 113–122.

Dhingra, H.R., Priefer, U.B., 2006. Impact of cadmium on structural and functional aspects of pea (*Pisum sativum* L.) root nodules. J. Plant Biol. 33 (3), 201.

Di Baccio, D., Castagna, A., Tognetti, R., Ranieri, A., Sebastiani, L., 2014. Early responses to cadmium of two poplar clones that differ in stress tolerance. J. Plant Physiol. 171 (18), 1693–1705.

Dominguez, M.T., Maranon, T., Murillo, J.M., Redondo-Gomez, S., 2011. Response of Holm oak (*Quercus ilex* subsp. ballota) and mastic shrub (*Pistacia lentiscus* L.) seedlings to high concentrations of Cd and Tl in the rhizosphere. Chemosphere 83 (8), 1166–1174.

Dong, J., Wu, F., Zhang, G., 2006. Influence of cadmium on antioxidant capacity and four microelement concentrations in tomato seedlings (*Lycopersicon esculentum*). Chemosphere 64 (10), 1659–1666.

Durrett, T.P., Gassmann, W., Rogers, E.E., 2007. The FRD3-mediated efflux of citrate into the root vasculature is necessary for efficient iron translocation. Plant Physiol. 144 (1), 197–205.

Ertani, A., Mietto, A., Borin, M., Nardi, S., 2017. Chromium in agricultural soils and crops: a review. Water Air Soil Pollut. 228 (5), 190.

Feng, J., Shi, Q., Wang, X., Wei, M., Yang, F., Xu, H., 2010. Silicon supplementation ameliorated the inhibition of photosynthesis and nitrate metabolism by cadmium (Cd) toxicity in *Cucumis sativus* L. Sci. Hortic. 123 (4), 521–530.

Fernandez, R., Bertrand, A., Reis, R., Mourato, M.P., Martins, L.L., Gonzalez, A., 2013. Growth and physiological responses to cadmium stress of two populations of *Dittrichia viscosa* (L.) Greuter. J. Hazard. Mater. 244, 555–562.

Fernandez-Pascual, M., De Lorenzo, C., De Felipe, M.R., Rajalakshmi, S., Gordon, A.J., Thomas, B.J., Minchin, F.R., 1996. Possible reasons for relative salt stress tolerance in nodules of white lupin cv. Multolupa. J. Exp. Bot. 47 (11), 1709–1716.

Finger-Teixeira, A., Ferrarese, M.D.L.L., Soares, A.R., da Silva, D., Ferrarese-Filho, O., 2010. Cadmium-induced lignification restricts soybean root growth. Ecotoxicol. Environ. Saf. 73 (8), 1959–1964.

Fischerova, Z., Tlustos, P., Szakova, J., Sichorova, K., 2006. A comparison of phytoremediation capability of selected plant species for given trace elements. Environ. Pollut. 144 (1), 93–100.

Florijn, P.J., Nelemans, J.A., Van Beusichem, M.L., 1993. Evaluation of structural and physiological plant characteristics in relation to the distribution of cadmium in maize inbred lines. Plant Soil 154 (1), 103–109.

Fotiadis, E., Lolas, P.C., 2011. Phytoremediation of Cd contaminated soil through certain weed and crop species. J. Agric. Sci. Technol. 1, 811–817.

Gallego, S.M., Pena, L.B., Barcia, R.A., Azpilicueta, C.E., Iannone, M.F., Rosales, E.P., Zawoznik, M.S., Groppa, M.D., Benavides, M.P., 2012. Unravelling cadmium toxicity and tolerance in plants: insight into regulatory mechanisms. Environ. Exp. Bot. 83, 33–46.

Gapper, C., Dolan, L., 2006. Control of plant development by reactive oxygen species. Plant Physiol. 141 (2), 341–345.

Gharaibeh, M.A., Marschner, B., Heinze, S., 2015. Metal uptake of tomato and alfalfa plants as affected by water source, salinity, and Cd and Zn levels under greenhouse conditions. Environ. Sci. Pollut. Res. 22 (23), 18894–18905.

Gichner, T., Patkova, Z., Szakova, J., Znidar, I., Mukherjee, A., 2008. DNA damage in potato plants induced by cadmium, ethyl methanesulphonate and $\gamma$-rays. Environ. Exp. Bot. 62 (2), 113–119.

Gill, R.A., Zang, L., Ali, B., Farooq, M.A., Cui, P., Yang, S., Ali, S., Zhou, W., 2015. Chromium-induced physio-chemical and ultrastructural changes in four cultivars of *Brassica napus* L. Chemosphere 120, 154–164.

Gouia, H., Suzuki, A., Brulfert, J., Ghorbal, M.H., 2003. Effects of cadmium on the co-ordination of nitrogen and carbon metabolism in bean seedlings. J. Plant Physiol. 160 (4), 367–376.

Gratao, P.L., Polle, A., Lea, P.J., Azevedo, R.A., 2005. Making the life of heavy metal-stressed plants a little easier. Funct. Plant Biol. 32 (6), 481–494.

Gratão, P.L., Monteiro, C.C., Rossi, M.L., Martinelli, A.P., Peres, L.E., Medici, L.O., Lea, P.J., Azevedo, R.A., 2009. Differential ultrastructural changes in tomato hormonal mutants exposed to cadmium. Environ. Exp. Bot. 67 (2), 387–394.

Gratao, P.L., Monteiro, C.C., Tezotto, T., Carvalho, R.F., Alves, L.R., Peters, L.P., Azevedo, R.A., 2015. Cadmium stress antioxidant responses and root-to-shoot communication in grafted tomato plants. BioMetals 28 (5), 803–816.

Groppa, M.D., Rosales, E.P., Iannone, M.F., Benavides, M.P., 2008. Nitric oxide, polyamines and Cd-induced phytotoxicity in wheat roots. Phytochemistry 69 (14), 2609–2615.

Guerrero-Campo, J., Palacio, S., Perez-Rontome, C., Montserrat-Marti, G., 2006. Effect of root system morphology on root-sprouting and shoot-rooting abilities in 123 plant species from eroded lands in north-east Spain. Ann. Bot. 98 (2), 439–447.

Hao, X., Chang, Q., Li, X., 2009. Synthesis, characterization, and properties of polymeric flocculant with the function of trapping heavy metal ions. J. Appl. Polym. Sci. 112 (1), 135–141.

Hasan, S.A., Hayat, S., Ali, B., Ahmad, A., 2008. 28-Homobrassinolide protects chickpea (*Cicer arietinum*) from cadmium toxicity by stimulating antioxidants. Environ. Pollut. 151 (1), 60–66.

Hassan, M., Mansoor, S., 2014. Oxidative stress and antioxidant defense mechanism in mung bean seedlings after lead and cadmium treatments. Turk. J. Agric. For. 38 (1), 55–61.

Hassan, W., Bano, R., Bashir, S., Aslam, Z., 2016. Cadmium toxicity and soil biological index under potato (*Solanum tuberosum* L.) cultivation. Soil Res. 54 (4), 460–468.

Hassinen, V., Vallinkoski, V.M., Issakainen, S., Tervahauta, A., Karenlampi, S., Servomaa, K., 2009. Correlation of foliar MT2b expression with Cd and Zn concentrations in hybrid aspen (*Populus tremula* × *tremuloides*) grown in contaminated soil. Environ. Pollut. 157 (3), 922–930.

Hediji, H., Djebali, W., Cabasson, C., Maucourt, M., Baldet, P., Bertrand, A., Zoghlami, L.B., Deborde, C., Moing, A., Brouquisse, R., Chaibi, W., 2010. Effects of long-term cadmium exposure on growth and metabolomic profile of tomato plants. Ecotoxicol. Environ. Saf. 73 (8), 1965–1974.

Hediji, H., Djebali, W., Belkadhi, A., Cabasson, C., Moing, A., Rolin, D., Brouquisse, R., Gallusci, P., Chaibi, W., 2015. Impact of long-term cadmium exposure on mineral content of *Solanum lycopersicum* plants: consequences on fruit production. S. Afr. J. Bot. 97, 176–181.

Hernandez-Allica, J., Becerril, J.M., Garbisu, C., 2008. Assessment of the phytoextraction potential of high biomass crop plants. Environ. Pollut. 152 (1), 32–40.

Horvath, E., Pal, M., Szalai, G., Paldi, E., Janda, T., 2007. Exogenous 4-hydroxybenzoic acid and salicylic acid modulate the effect of short-term drought and freezing stress on wheat plants. Biol. Plant. 51 (3), 480–487.

Hossain, M.A., Hasanuzzaman, M., Fujita, M., 2010. Up-regulation of antioxidant and glyoxalase systems by exogenous glycinebetaine and proline in mung bean confer tolerance to cadmium stress. Physiol. Mol. Biol. Plants 16 (3), 259–272.

Huang, B., Xin, J., Dai, H., Liu, A., Zhou, W., Yi, Y., Liao, K., 2015. Root morphological responses of three hot pepper cultivars to Cd exposure and their correlations with Cd accumulation. Environ. Sci. Pollut. Res. 22 (2), 1151–1159.

Irfan, M., Hayat, S., Ahmad, A., Alyemeni, M.N., 2013. Soil cadmium enrichment: allocation and plant physiological manifestations. Saudi J. Biol. Sci. 20 (1), 1–10.

Jin, X., Yang, X., Islam, E., Liu, D., Mahmood, Q., 2008. Effects of cadmium on ultrastructure and antioxidative defense system in hyperaccumulator and non-hyperaccumulator ecotypes of *Sedum alfredii* Hance. J. Hazard. Mater. 156 (3), 387–397.

Jinadasa, N., Collins, D., Holford, P., Milham, P.J., Conroy, J.P., 2016. Reactions to cadmium stress in a cadmium-tolerant variety of cabbage (*Brassica oleracea* L.): is cadmium tolerance necessarily desirable in food crops. Environ. Sci. Pollut. Res. 23 (6), 5296–5306.

Khan, A., Khan, S., Alam, M., Khan, M.A., Aamir, M., Qamar, Z., Rehman, Z.U., Perveen, S., 2016. Toxic metal interactions affect the bioaccumulation and dietary intake of macro-and micro-nutrients. Chemosphere 146, 121–128.

Kong, X., He, W., Qin, N., Liu, W., Yang, B., Yang, C., Xu, F., Mooij, W.M., Koelmans, A.A., 2017. Integrated ecological and chemical food web accumulation modeling explains PAH temporal trends during regime shifts in a shallow lake. Water Res. 119, 73–82.

Kranner, I., Colville, L., 2011. Metals and seeds: biochemical and molecular implications and their significance for seed germination. Environ. Exp. Bot. 72 (1), 93–105.

Kupper, H., Lombi, E., Zhao, F.J., McGrath, S.P., 2000. Cellular compartmentation of cadmium and zinc in relation to other elements in the hyper-accumulator *Arabidopsis halleri*. Planta 212 (1), 75–84.

Kuriakose, S.V., Prasad, M.N.V., 2008. Cadmium stress affects seed germination and seedling growth in *Sorghum bicolor* (L.) Moench by changing the activities of hydrolyzing enzymes. Plant Growth Regul. 54 (2), 143–156.

Lee, K., Bae, D.W., Kim, S.H., Han, H.J., Liu, X., Park, H.C., Lim, C.O., Lee, S.Y., Chung, W.S., 2010. Comparative proteomic analysis of the short-term responses of rice roots and leaves to cadmium. J. Plant Physiol. 167 (3), 161–168.

Lefevre, I., Marchal, G., Corréal, E., Zanuzzi, A., Lutts, S., 2009. Variation in response to heavy metals during vegetative growth in *Dorycnium pentaphyllum* Scop. Plant Growth Regul. 59 (1), 1–11.

León, A.M., Palma, J.M., Corpas, F.J., Gómez, M., Romero-Puertas, M.C., Chatterjee, D., Mateos, R.M., Luis, A., Sandalio, L.M., 2002. Antioxidative enzymes in cultivars of pepper plants with different sensitivity to cadmium. Plant. Physiol. Biochem. 40 (10), 813–820.

Li, Y., Chen, Z., Xu, S., Zhang, L., Hou, W., Yu, N., 2015. Effect of combined pollution of Cd and B [a] P on photosynthesis and chlorophyll fluorescence characteristics of wheat. Pol. J. Environ. Stud. 24 (1), 121–131.

Li, X., Zhou, Q., Sun, X., Ren, W., 2016. Effects of cadmium on uptake and translocation of nutrient elements in different welsh onion (*Allium fistulosum* L.) cultivars. Food Chem. 194, 101–110.

Lima, A.I.G., Pereira, S.I.A., Figueira, E.M.D.A.P., Caldeira, G.C.N., de Matos Caldeira, H.D.Q., 2006. Cadmium detoxification in roots of *Pisum sativum* seedlings: relationship between toxicity levels, thiol pool alterations and growth. Environ. Exp. Bot. 55 (2), 149–162.

Lin, R., Wang, X., Luo, Y., Du, W., Guo, H., Yin, D., 2007. Effects of soil cadmium on growth, oxidative stress and antioxidant system in wheat seedlings (*Triticum aestivum* L.). Chemosphere 69 (1), 89–98.

Liu, D., Kottke, I., 2004. Subcellular localization of cadmium in the root cells of *Allium cepa* by electron energy loss spectroscopy and cytochemistry. J. Biosci. 29 (3), 329–335.

Liu, W., Sun, L., Zhong, M., Zhou, Q., Gong, Z., Li, P., Tai, P., Li, X., 2012. Cadmium-induced DNA damage and mutations in *Arabidopsis* plantlet shoots identified by DNA fingerprinting. Chemosphere 89 (9), 1048–1055.

Lopez-Millan, A.F., Sagardoy, R., Solanas, M., Abadia, A., Abadia, J., 2009. Cadmium toxicity in tomato (*Lycopersicon esculentum*) plants grown in hydroponics. Environ. Exp. Bot. 65 (3), 376–385.

Lu, Y.L., Liang, L., Yang, H., 2013a. Joint ecotoxicology of cadmium and metsulfuron-methyl in wheat (*Triticum aestivum*). Environ. Monit. Assess. 185 (4), 2939–2950.

Lu, Z., Zhang, Z., Su, Y., Liu, C., Shi, G., 2013b. Cultivar variation in morphological response of peanut roots to cadmium stress and its relation to cadmium accumulation. Ecotoxicol. Environ. Saf. 91, 147–155.

Lunackova, L., Sottnikova, A., Masarovicova, E., Lux, A., Stresko, V., 2003. Comparison of cadmium effect on willow and poplar in response to different cultivation conditions. Biol. Plant. 47 (3), 403–411.

Lux, A., Martinka, M., Vaculik, M., White, P.J., 2010. Root responses to cadmium in the rhizosphere: a review. J. Exp. Bot. 62 (1), 21–37.

Mahmood, T., Gupta, K.J., Kaiser, W.M., 2009. Cadmium stress stimulates nitric oxide production by wheat roots. Pak. J. Bot. 41 (3), 6.

Maksimovic, I., Kastori, R., Krstic, L., Lukovic, J., 2007. Steady presence of cadmium and nickel affects root anatomy, accumulation and distribution of essential ions in maize seedlings. Biol. Plant. 51 (3), 589–592.

Marschner, H., 2012. Marschner's Mineral Nutrition of Higher Plants. vol. 89. Academic Press, London.

Metwally, A., Safronova, V.I., Belimov, A.A., Dietz, K.J., 2004. Genotypic variation of the response to cadmium toxicity in *Pisum sativum* L. J. Exp. Bot. 56 (409), 167–178.

Mitani-Ueno, N., Yamaji, N., Ma, J.F., 2018. Transport system of mineral elements in rice. In: Rice Genomics, Genetics and Breeding. Springer, Singapore, pp. 223–240.

Moller, I.M., Jensen, P.E., Hansson, A., 2007. Oxidative modifications to cellular components in plants. Annu. Rev. Plant Biol. 58, 459–481.

Mombo, S., Foucault, Y., Deola, F., Gaillard, I., Goix, S., Shahid, M., Schreck, E., Pierart, A., Dumat, C., 2016. Management of human health risk in the context of kitchen gardens polluted by lead and cadmium near a lead recycling company. J. Soils Sediments 16 (4), 1214–1224.

Monteiro, M.S., Santos, C., Soares, A.M.V.M., Mann, R.M., 2009. Assessment of biomarkers of cadmium stress in lettuce. Ecotoxicol. Environ. Saf. 72 (3), 811–818.

Nedjimi, B., Daoud, Y., 2009. Cadmium accumulation in *Atriplex halimus* subsp. schweinfurthii and its influence on growth, proline, root hydraulic conductivity and nutrient uptake. Flora 204 (4), 316–324.

Ozdener, Y., Kutbay, H.G., 2009. Toxicity of copper, cadmium, nickel, lead and zinc on seed germination and seedling growth in *Eruca sativa*. Fresenius Environ. Bull. 18 (1), 26.

Perfus-Barbeoch, L., Leonhardt, N., Vavasseur, A., Forestier, C., 2002. Heavy metal toxicity: cadmium permeates through calcium channels and disturbs the plant water status. Plant J. 32 (4), 539–548.

Pietrini, F., Zacchini, M., Iori, V., Pietrosanti, L., Ferretti, M., Massacci, A., 2010. Spatial distribution of cadmium in leaves and its impact on photosynthesis: examples of different strategies in willow and poplar clones. Plant Biol. 12 (2), 355–363.

Poghosyan, G.H., Mukhaelyan, Z.H., Vardevanyan, P.H., 2014. Influence of cadmium ions on growth and antioxidant system activity of wheat (*Triticum aestivum* L.) Seedlings. Int. J. Sci. Res. Environ. Sci. 2 (10), 371.

Polle, A., Klein, T., Kettner, C., 2013. Impact of cadmium on young plants of *Populus euphratica* and *P.× canescens*, two poplar species that differ in stress tolerance. New For. 44 (1), 13–22.

Puschenreiter, M., Turktas, M., Sommer, P., Wieshammer, G., Laaha, G., Wenzel, W.W., Hauser, M.T., 2010. Differentiation of metallicolous and non-metallicolous *Salix caprea* populations based on phenotypic characteristics and nuclear microsatellite (SSR) markers. Plant Cell Environ. 33 (10), 1641–1655.

Qian, H., Li, J., Pan, X., Jiang, H., Sun, L., Fu, Z., 2010. Photoperiod and temperature influence cadmium's effects on photosynthesis-related gene transcription in *Chlorella vulgaris*. Ecotoxicol. Environ. Saf. 73 (6), 1202–1206.

Qureshi, M.I., D'Amici, G.M., Fagioni, M., Rinalducci, S., Zolla, L., 2010. Iron stabilizes thylakoid protein-pigment complexes in Indian mustard during Cd-phytoremediation as revealed by BN-SDS-PAGE and ESI-MS/MS. J. Plant Physiol. 167 (10), 761–770.

Rehman, Z.U., Khan, S., Brusseau, M.L., Shah, M.T., 2017. Lead and cadmium contamination and exposure risk assessment via consumption of vegetables grown in agricultural soils of five-selected regions of Pakistan. Chemosphere 168, 1589–1596.

Rizwan, M., Ali, S., Abbas, T., Zia-ur-Rehman, M., Hannan, F., Keller, C., Al-Wabel, M.I., Ok, Y.S., 2016a. Cadmium minimization in wheat: a critical review. Ecotoxicol. Environ. Saf. 130, 43–53.

Rizwan, M., Meunier, J.D., Davidian, J.C., Pokrovsky, O.S., Bovet, N., Keller, C., 2016b. Silicon alleviates Cd stress of wheat seedlings (*Triticum turgidum* L. cv. Claudio) grown in hydroponics. Environ. Sci. Pollut. Res. 23 (2), 1414–1427.

Robinson, B.H., Mills, T.M., Petit, D., Fung, L.E., Green, S.R., Clothier, B.E., 2000. Natural and induced cadmium-accumulation in poplar and willow: implications for phytoremediation. Plant Soil 227 (1-2), 301–306.

Rodriguez-Serrano, M., Romero-Puertas, M.C., Pazmino, D.M., Testillano, P.S., Risueño, M.C., Luis, A., Sandalio, L.M., 2009. Cellular response of pea plants to cadmium toxicity: cross talk between reactive oxygen species, nitric oxide, and calcium. Plant Physiol. 150 (1), 229–243.

Sabeen, M., Mahmood, Q., Irshad, M., Fareed, I., Khan, A., Ullah, F., Hussain, J., Hayat, Y., Tabassum, S., 2013. Cadmium phytoremediation by *Arundo donax* L. from contaminated soil and water. Biomed. Res. Int. 3, 121–127.

Sanchez-Pardo, B., Carpena, R.O., Zornoza, P., 2013. Cadmium in white lupin nodules: impact on nitrogen and carbon metabolism. J. Plant Physiol. 170 (3), 265–271.

Sandalio, L.M., Dalurzo, H.C., Gomez, M., Romero-Puertas, M.C., Del Rio, L.A., 2001. Cadmium-induced changes in the growth and oxidative metabolism of pea plants. J. Exp. Bot. 52 (364), 2115–2126.

Sarwar, N., Imran, M., Shaheen, M.R., Ishaque, W., Kamran, M.A., Matloob, A., Rehim, A., Hussain, S., 2017. Phytoremediation strategies for soils contaminated with heavy metals: modifications and future perspectives. Chemosphere 171, 710–721.

Scebba, F., Arduini, I., Ercoli, L., Sebastiani, L., 2006. Cadmium effects on growth and antioxidant enzymes activities in *Miscanthus sinensis*. Biol. Plant. 50 (4), 688–692.

Schutzendubel, A., Polle, A., 2002. Plant responses to abiotic stresses: heavy metal-induced oxidative stress and protection by mycorrhization. J. Exp. Bot. 53 (372), 1351–1365.

Semane, B., Dupae, J., Cuypers, A., Noben, J.P., Tuomainen, M., Tervahauta, A., Kärenlampi, S., Van Belleghem, F., Smeets, K., Vangronsveld, J., 2010. Leaf proteome responses of *Arabidopsis thaliana* exposed to mild cadmium stress. J. Plant Physiol. 167 (4), 247–254.

Seth, C.S., Misra, V., Chauhan, L.K.S., Singh, R.R., 2008. Genotoxicity of cadmium on root meristem cells of *Allium cepa*: cytogenetic and Comet assay approach. Ecotoxicol. Environ. Saf. 71 (3), 711–716.

Shafi, M., Bakht, J., Razuddin, Hayat, Y., Zhang, G.P., 2011. Genotypic difference in the inhibition of photosynthesis and chlorophyll fluorescence by salinity and cadmium stresses in wheat. J. Plant Nutr. 34 (3), 315–323.

Shah, K., Dubey, R.S., 1995. Effect of cadmium on RNA level as well as activity and molecular forms of ribonuclease in growing rice seedlings. Plant Physiol. Biochem. 33, 577–584.

Shamsi, I.H., Wei, K., Zhang, G.P., Jilani, G.H., Hassan, M.J., 2008. Interactive effects of cadmium and aluminum on growth and antioxidative enzymes in soybean. Biol. Plant. 52 (1), 165–169.

Shamsi, I.H., Zhang, G., Hu, H., Xue, Q., Hussain, N., Ali, E., Shen, Q., Zheng, W., Zhang, Q., Liu, X., Jabeen, Z., 2014. Assessment of the hazardous effects of Cd on physiological and biochemical characteristics of soybean genotypes. Int. J. Agric. Biol. 16, 41–48.

Sharma, R.K., Agrawal, M., Agrawal, S.B., 2010. Physiological, biochemical and growth responses of lady's finger (*Abelmoschus esculentus* L.) plants as affected by Cd contaminated soil. Bull. Environ. Contam. Toxicol. 84 (6), 765–770.

Sipos, G., Solti, A., Czech, V., Vashegyi, I., Tóth, B., Cseh, E., Fodor, F., 2013. Heavy metal accumulation and tolerance of energy grass (*Elymus elongatus* subsp. ponticus cv. Szarvasi-1) grown in hydroponic culture. Plant Physiol. Biochem. (68), 96–103.

Smiri, M., Chaoui, A., El Ferjani, E., 2009. Respiratory metabolism in the embryonic axis of germinating pea seed exposed to cadmium. J. Plant Physiol. 166 (3), 259–269.

Solis-Dominguez, F.A., González-Chávez, M.C., Carrillo-González, R., Rodriguez-Vazquez, R., 2007. Accumulation and localization of cadmium in *Echinochloa polystachya* grown within a hydroponic system. J. Hazard. Mater. 141 (3), 630–636.

Stanova, A., Durisova, E., Banasova, V., Gurinova, E., Nadubinska, M., Kenderesova, L., Ovecka, M., Ciamporova, M., 2012. Root system morphology and primary root anatomy in natural non-metallicolous and metallicolous populations of three Arabidopsis species differing in heavy metal tolerance. Biology 67 (3), 505–516.

Sun, J., Wang, R., Zhang, X., Yu, Y., Zhao, R., Li, Z., Chen, S., 2013. Hydrogen sulfide alleviates cadmium toxicity through regulations of cadmium transport across the plasma and vacuolar membranes in *Populus euphratica* cells. Plant Physiol. Biochem. 65, 67–74.

Sun, H., Wang, X., Shang, L., Zhou, Z., Wang, R., 2017. Cadmium accumulation and its effects on nutrient uptake and photosynthetic performance in cucumber (*Cucumis sativus* L.). Philipp. Agric. Sci. 100 (3), 263–270.

Vaculík, M., Konlechner, C., Langer, I., Adlassnig, W., Puschenreiter, M., Lux, A., Hauser, M.T., 2012. Root anatomy and element distribution vary between two *Salix caprea* isolates with different Cd accumulation capacities. Environ. Pollut. 163, 117–126.

Valentovicova, K., Haluskova, L.U., Huttova, J., Mistrak, I., Tamas, L., 2010. Effect of cadmium on diaphorase activity and nitric oxide production in barley root tips. J. Plant Physiol. 167 (1), 10–14.

Varalakshmi, L.R., Ganeshamurthy, A.N., 2013. Phytotoxicity of cadmium in radish and its effects on growth, yield, and cadmium uptake. Commun. Soil Sci. Plant Anal. 44 (9), 1444–1456.

Vazquez, M.D., Poschenrieder, C.H., Barceld, Y., 1992. Cadmium in bean roots. New Phytol. 120 (2), 215–226.

Volland, S., Andosch, A., Milla, M., Stoger, B., Lutz, C., Lutz-Meindl, U., 2011. Intracellular metal compartmentalization in the green algal model system *Micrasterias denticulata* (Streptophyta) measured by transmission electron microscopy-coupled electron energy loss spectroscopy. J. Phycol. 47 (3), 565–579.

Volland, S., Lutz, C., Michalke, B., Lutz-Meindl, U., 2012. Intracellular chromium localization and cell physiological response in the unicellular alga Micrasterias. Aquat. Toxicol. 109, 59–69.

Volland, S., Bayer, E., Baumgartner, V., Andosch, A., Lütz, C., Sima, E., Lutz-Meindl, U., 2014. Rescue of heavy metal effects on cell physiology of the algal model system Micrasterias by divalent ions. J. Plant Physiol. 171 (2), 154–163.

Wahid, A., Ghani, A., Ali, I., Ashraf, M.Y., 2007. Effects of cadmium on carbon and nitrogen assimilation in shoots of mungbean [*Vigna radiata* (L.) Wilczek] seedlings. J. Agron. Crop Sci. 193 (5), 357–365.

Wan, Y., Luo, S., Chen, J., Xiao, X., Chen, L., Zeng, G., Liu, C., He, Y., 2012. Effect of endophyte-infection on growth parameters and Cd-induced phytotoxicity of Cd-hyperaccumulator *Solanum nigrum* L. Chemosphere 89 (6), 743–750.

Wang, M.Y., Siddiqi, M.Y., Ruth, T.J., Glass, A.D., 1993. Ammonium uptake by rice roots (II. Kinetics of 13NH4+ influx across the plasma lemma). Plant Physiol. 103 (4), 1259–1267.

Wang, M., Zou, J., Duan, X., Jiang, W., Liu, D., 2007. Cadmium accumulation and its effects on metal uptake in maize (*Zea mays* L.). Bioresour. Technol. 98 (1), 82–88.

Wang, Y., Qian, Y., Hu, H., Xu, Y., Zhang, H., 2011. Comparative proteomic analysis of Cd-responsive proteins in wheat roots. Acta Physiol. Plant. 33 (2), 349–357.

Wang, L., Xu, Y., Sun, Y., Liang, X., Lin, D., 2014. Identification of pakchoi cultivars with low cadmium accumulation and soil factors that affect their cadmium uptake and translocation. Front. Environ. Sci. Eng. 8 (6), 877–887.

Wang, Y., Ye, G., Chen, H., Hu, X., Niu, Z., Ma, S., 2015. Functionalized metal–organic framework as a new platform for efficient and selective removal of cadmium (II) from aqueous solution. J. Mater. Chem. A 3 (29), 15292–15298.

Wang, P., Deng, X., Huang, Y., Fang, X., Zhang, J., Wan, H., Yang, C., 2016. Root morphological responses of five soybean [*Glycine max* (L.) Merr] cultivars to cadmium stress at young seedlings. Environ. Sci. Pollut. Res. 23 (2), 1860–1872.

Wani, P.A., Khan, M.S., Zaidi, A., 2007a. Cadmium, chromium and copper in greengram plants. Agron. Sustain. Dev. 27 (2), 145–153.

Wani, P.A., Khan, M.S., Zaidi, A., 2007b. Impact of heavy metal toxicity on plant growth, symbiosis, seed yield and nitrogen and metal uptake in chickpea. Aust. J. Exp. Agric. 47 (6), 712–720.

Wei, S., Li, Y., Zhan, J., Wang, S., Zhu, J., 2012. Tolerant mechanisms of *Rorippa globosa* (Turcz.) Thell. hyperaccumulating Cd explored from root morphology. Bioresour. Technol. 118, 455–459.

White, P.J., 2000. Calcium channels in higher plants. Biochim. Biophys. Acta Biomembr. 1465 (1), 171–189.

White, P.J., Brown, P.H., 2010. Plant nutrition for sustainable development and global health. Ann. Bot. 105 (7), 1073–1080.

Xin, J., Huang, B., 2014. Subcellular distribution and chemical forms of cadmium in two hot pepper cultivars differing in cadmium accumulation. J. Agric. Food Chem. 62 (2), 508–515.

Xin, J., Huang, B., Dai, H., Liu, A., Zhou, W., Liao, K., 2014. Characterization of cadmium uptake, translocation, and distribution in young seedlings of two hot pepper cultivars that differ in fruit cadmium concentration. Environ. Sci. Pollut. Res. 21 (12), 7449–7456.

Xu, D., Chen, Z., Sun, K., Yan, D., Kang, M., Zhao, Y., 2013. Effect of cadmium on the physiological parameters and the subcellular cadmium localization in the potato (*Solanum tuberosum* L.). Ecotoxicol. Environ. Saf. 97, 147–153.

Xue, Z., Gao, H., Zhao, S., 2014. Effects of cadmium on the photosynthetic activity in mature and young leaves of soybean plants. Environ. Sci. Pollut. Res. 21 (6), 4656–4664.

Yamaji, N., Ma, J.F., 2017. Node-controlled allocation of mineral elements in Poaceae. Curr. Opin. Plant Biol. 39, 18–24.

Yannarelli, G.G., Fernandez-Alvarez, A.J., Santa-Cruz, D.M., Tomaro, M.L., 2007. Glutathione reductase activity and isoforms in leaves and roots of wheat plants subjected to cadmium stress. Phytochemistry 68 (4), 505–512.

Yourtchi, M.S., Bayat, H.R., 2013. Effect of cadmium toxicity on growth, cadmium accumulation and macronutrient content of durum wheat (Dena CV.). Int. J. Agric. Crop Sci. 6 (15), 1199.

Yousaf, B., Liu, G., Wang, R., Imtiaz, M., Zia-ur-Rehman, M., Munir, M.A.M., Niu, Z., 2016. Bioavailability evaluation, uptake of heavy metals and potential health risks via dietary exposure in urban-industrial areas. Environ. Sci. Pollut. Res. 23 (22), 22443–22453.

Zhang, X., Xia, H., Li, Z., Zhuang, P., Gao, B., 2010. Potential of four forage grasses in remediation of Cd and Zn contaminated soils. Bioresour. Technol. 101 (6), 2063–2066.

Zhao, Y., Gong, L., Zhou, B., Huang, Y., Liu, C., 2016. Detecting tomatoes in greenhouse scenes by combining AdaBoost classifier and colour analysis. Biosyst. Eng. 148, 127–137.

Zhi, Y., He, K., Sun, T., Zhu, Y., Zhou, Q., 2015. Assessment of potential soybean cadmium excluder cultivars at different concentrations of Cd in soils. J. Environ. Sci. 35, 108–114.

## FURTHER READING

Boominathan, R., Doran, P.M., 2003. Cadmium tolerance and antioxidative defenses in hairy roots of the cadmium hyperaccumulator, *Thlaspi caerulescens*. Biotechnol. Bioeng. 83 (2), 158–167.

Chiang, H.C., Lo, J.C., Yeh, K.C., 2006. Genes associated with heavy metal tolerance and accumulation in Zn/Cd hyperaccumulator *Arabidopsis halleri*: a genomic survey with cDNA microarray. Environ. Sci. Technol. 40 (21), 6792–6798.

De Michele, R., Vurro, E., Rigo, C., Costa, A., Elviri, L., Di Valentin, M., Careri, M., Zottini, M., di Toppi, L.S., Schiavo, F.L., 2009. Nitric oxide is involved in cadmium-induced programmed cell death in *Arabidopsis* suspension cultures. Plant Physiol. 150 (1), 217–228.

Di Toppi, L.S., Vurro, E., Rossi, L., Marabottini, R., Musetti, R., Careri, M., Maffini, M., Mucchino, C., Corradini, C., Badiani, M., 2007. Different compensatory mechanisms in two metal-accumulating aquatic macrophytes exposed to acute cadmium stress in outdoor artificial lakes. Chemosphere 68 (4), 769–780.

Garnier, L., Simon-Plas, F.R., Thuleau, P., Agnel, J.P., Blein, J.P., Ranjeva, R., Montillet, J.L., 2006. Cadmium affects tobacco cells by a series of three waves of reactive oxygen species that contribute to cytotoxicity. Plant Cell Environ. 29 (10), 1956–1969.

Hassan, M.J., Shao, G., Zhang, G., 2005. Influence of cadmium toxicity on growth and antioxidant enzyme activity in rice cultivars with different grain cadmium accumulation. J. Plant Nutr. 28 (7), 1259–1270.

He, S., He, Z., Yang, X., Stoffella, P.J., Baligar, V.C., 2015. Soil biogeochemistry, plant physiology, and phytoremediation of cadmium-contaminated soils. Adv. Agron. 134, 135–225.

Lane, T.W., Saito, M.A., George, G.N., Pickering, I.J., Prince, R.C., Morel, F.M., 2005. Biochemistry: a cadmium enzyme from a marine diatom. Nature 435 (7038), 42.

Liang, C.H.E.N., Xiao-Hua, L.O.N.G., Zhang, Z.H., Zheng, X.T., Rengel, Z., Zhao-Pu, L.I.U., 2011. Cadmium accumulation and translocation in two Jerusalem artichoke (*Helianthus tuberosus* L.) cultivars. Pedosphere 21 (5), 573–580.

Liu, M.Q., Yanai, J., Jiang, R.F., Zhang, F., McGrath, S.P., Zhao, F.J., 2008. Does cadmium play a physiological role in the hyperaccumulator *Thlaspi caerulescens*? Chemosphere 71 (7), 1276–1283.

Lombi, E., Zhao, F.J., McGrath, S.P., Young, S.D., Sacchi, G.A., 2001. Physiological evidence for a high-affinity cadmium transporter highly expressed in a *Thlaspi caerulescens* ecotype. New Phytol. 149 (1), 53–60.

Maksymiec, W., Wojcik, M., Krupa, Z., 2007. Variation in oxidative stress and photochemical activity in *Arabidopsis thaliana* leaves subjected to cadmium and excess copper in the presence or absence of jasmonate and ascorbate. Chemosphere 66 (3), 421–427.

McGrath, S.P., Zhao, F.J., 2003. Phytoextraction of metals and metalloids from contaminated soils. Curr. Opin. Biotechnol. 14 (3), 277–282.

Mittler, R., 2002. Oxidative stress, antioxidants and stress tolerance. Trends Plant Sci. 7 (9), 405–410.

Monteiro, C.C., Rolao, M.B., Franco, M.R., Peters, L.P., Cia, M.C., Capaldi, F.R., Carvalho, R.F., Gratão, P.L., Rossi, M.L., Martinelli, A.P., Peres, L.E., 2012.

Biochemical and histological characterization of tomato mutants. An. Acad. Bras. Cienc. 84 (2), 573–585.

Nelson, M.T., 1986. Interactions of divalent cations with single calcium channels from rat brain synaptosomes. J. Gen. Physiol. 87 (2), 201–222.

Qiu, R.L., Zhao, X., Tang, Y.T., Yu, F.M., Hu, P.J., 2008. Antioxidative response to Cd in a newly discovered cadmium hyperaccumulator, *Arabis paniculata* F. Chemosphere 74 (1), 6–12.

Riaz, S., Iqbal, M., Hussain, I., Rasheed, R., Ashraf, M.A., Mahmood, S., Younas, M., Iqbal, M.Z., 2014. Chronic cadmium induced oxidative stress not the DNA fragmentation modulates growth in spring wheat (*Triticum aestivum*). Int. J. Agric. Biol. 16, 789–794.

Semane, B., Cuypers, A., Smeets, K., Van Belleghem, F., Horemans, N., Schat, H., Vangronsveld, J., 2007. Cadmium responses in *Arabidopsis thaliana*: glutathione metabolism and antioxidative defence system. Physiol. Plant. 129 (3), 519–528.

Sharma, S.S., Dietz, K.J., 2009. The relationship between metal toxicity and cellular redox imbalance. Trends Plant Sci. 14 (1), 43–50.

Siddiqui, S., Meghvansi, M.K., Wani, M.A., Jabee, F., 2009. Evaluating cadmium toxicity in the root meristem of *Pisum sativum* L. Acta Physiol. Plant. 31 (3), 531.

Sun, R.L., Zhou, Q.X., Sun, F.H., Jin, C.X., 2007. Antioxidative defense and proline/phytochelatin accumulation in a newly discovered Cd-hyperaccumulator, *Solanum nigrum* L. Environ. Exp. Bot. 60 (3), 468–476.

Uraguchi, S., Watanabe, I., Yoshitomi, A., Kiyono, M., Kuno, K., 2006. Characteristics of cadmium accumulation and tolerance in novel Cd-accumulating crops, *Avena strigosa* and *Crotalaria juncea*. J. Exp. Bot. 57 (12), 2955–2965.

Uraguchi, S., Mori, S., Kuramata, M., Kawasaki, A., Arao, T., Ishikawa, S., 2009. Root-to-shoot Cd translocation via the xylem is the major process determining shoot and grain cadmium accumulation in rice. J. Exp. Bot. 60 (9), 2677–2688.

Ursache, R., Andersen, T.G., Marhavy, P., Geldner, N., 2018. A protocol for combining fluorescent proteins with histological stains for diverse cell wall components. Plant J. 93 (2), 399–412.

Wang, Z., Zhang, Y., Huang, Z., Huang, L., 2008. Antioxidative response of metal-accumulator and non-accumulator plants under cadmium stress. Plant Soil 310 (2), 137.

Zhou, W., Qiu, B., 2005. Effects of cadmium hyperaccumulation on physiological characteristics of *Sedum alfredii* Hance (Crassulaceae). Plant Sci. 169 (4), 737–745.

Zhou, Q., Guo, J.J., He, C.T., Shen, C., Huang, Y.Y., Chen, J.X., Guo, J.H., Yuan, J.G., Yang, Z.Y., 2016. Comparative transcriptome analysis between low-and high-cadmium-accumulating genotypes of pakchoi (*Brassica chinensis* L.) in response to cadmium stress. Environ. Sci. Technol. 50 (12), 6485–6494.

Zornoza, P., Vazquez, S., Esteban, E., Fernandez-Pascual, M., Carpena, R., 2002. Cadmium-stress in nodulated white lupin: strategies to avoid toxicity. Plant Physiol. Biochem. 40 (12), 1003–1009.

CHAPTER 4

# Adaptive and Tolerance Mechanisms in Herbaceous Plants Exposed to Cadmium

**Bodin Mongkhonsin*, Woranan Nakbanpote[†], Orapan Meesungnoen[†], Majeti Narasimha Vara Prasad[‡]**
*Department of General Science, Faculty of Education, Kalasin University, Kalasin, Thailand
[†]Department of Biology, Faculty of Science, Mahasarakharm University, Khamriang, Thailand
[‡]Department of Plant Sciences, University of Hyderabad, Hyderabad, India

## Abstract

Herbaceous plants are plants that have no persistent woody stem above ground. They are classified following life-cycle classification as annuals, biennials or perennials. Native herbaceous species usually colonize on remediated soils with potential use in phytoremediation of heavy metals. In case of cadmium, many herbaceous species are suitable for phytoextraction and phytostabilization. Their mechanisms involving are immobilization, exclusion and compartimentalization. They can adapt and synthesize phytochelatins, metallothioneins, stress proteins and phenolic compounds to tolerate cadmium and other metals. In addition, endophytic microorganism also involves in the plants tolerant mechanism. Growing herbal plants to cover and remediate cadmium contaminated areas must be concerned. Especially shoot parts of hyperaccumulative plants, spreading of cadmium in top soil can be increased from falling leaves of their life cycles. The cadmium also enters the food chain via insects, birds, herbivores, etc. Moreover, application of herbal plants collected from contaminated area for folk medicines and raw materials of drug and food/feed should be prohibited or controlled base on reliable researches and knowledges. Finally, phytomanagement and harvesting processes for cadmium phytoremediation are depending on the tolerant mechanisms and life cycles of each herbal plant species.

**Keywords:** Cadmium, Herbaceous plant, Herb, Tolerance mechanism, Endophyte

## 1 INTRODUCTION

Cadmium (Cd) is a nonessential element that negatively affects plant growth and development (Kabata-Pendias, 2011). Overt symptoms induced by the Cd content of plants are growth retardation, chlorosis, and root damage (Das et al., 1997; López-Millán et al., 2009; Panitlertumpai et al., 2013;

*Cadmium Toxicity and Tolerance in Plants*
https://doi.org/10.1016/B978-0-12-814864-8.00004-8

Mongkhonsin et al., 2016). An excessive Cd concentration affecting growth overrules some nutrients, such as Ca, Mg, P, N, Fe, and K; leading to growth inhibition, a decrease in biomass, and death of the plant. In addition, the metal inhibits physiological processes, such as respiration, photosynthesis, cell elongation, the plant-to-water relationship, and metabolism (Prasad, 1995; López-Millán et al., 2009; He et al., 2017). In the case of human toxicity, Cd is absorbed more efficiently by the lungs (30%–60%) than by the gastrointestinal tract; the latter being a saturable process (Nordberg et al., 1985). It is transported in the blood and widely distributed in the body, but accumulates primarily in the liver and kidneys. Metabolic transformations of Cd are limited to its binding to protein and nonprotein sulfhydryl groups, and various macromolecules, such as metallothionein, which is especially important in the kidneys and liver (Fowler, 2009).

Phytoremediation is a biological treatment process that utilizes natural processes harbored in (or stimulated by) plants to enhance degradation and removal of contaminants in environments (Alvarez and Illman, 2006). Phytoremediation utilizes physical, chemical, and biological processes of plants to remove, degrade, transform, or stabilize contaminants within soil. Some hyperaccumulators, which are usually found in metal contaminated sites, have the ability to accumulate high concentrations of metals without any impact on their growth and development (Lasat, 2002). The mechanisms for metal remediation are phytoextraction, phytostabilization, and phytovolatilization (Raskin and Ensley, 2000; Prasad, 2011). *Phytoextraction* aims to extract and translocate metals, in soluble form, from shallow contaminated soil to plant tissues (Prasad, 2011). *Phytostabilization/immobilization* aims to prevent the dispersion of contaminated sediments and soil by using plants (mainly grasses) to minimize erosion by wind or rain (Prasad, 2011). *Phytovolatilization* is the natural ability of a plant to volatilize a contaminant, which is taken up through its roots, via leaf stomata (Prasad, 2011; ITRC, 2009). The interface between microbes and plant roots (rhizosphere) is considered to greatly influence the growth and survival of plants. Therefore, the discovery of rhizosphere bacteria that are metal resistant and able to promote plant growth have raised high hopes for ecologically friendly and cost-effective strategies for the reclamation of metal-polluted soils (Ma et al., 2011).

Herbaceous plants are plants that have no persistent woody stem above ground, and they are classified according to their life-cycle as annuals, biennials, or perennials. Herbaceous plants include both forbs and graminoids, which are plants with a grass-like morphology. Annual herbaceous plants die completely when they have flowered and fruited, and then they grow again from seed. Herbaceous perennial and biennial plants have stems that die at the end of the growing season, but parts of the plant survive under or close to the ground from season to season, and they grow from living tissues remaining on or under the ground, including roots, a caudex, or various types of underground stems, such as bulbs, corms, stolons, rhizomes, and tubers (Stern, 2006). Native herbaceous species usually colonize on remediated soils, and can potentially be used in the phytoremediation of metals (Leguizamo et al., 2017). Many herbaceous species are suitable for phytoextraction and phytostabilization. Their mechanisms include immobilization, exclusion, and compartmentalization. Herbaceous plants can adapt and synthesize phytochelatins, metallothioneins, stress proteins, and phenolic compounds to tolerate Cd and other metals. Growing the plants to cover and remediate Cd-contaminated areas must be considered. Especially in the shoot parts of hyperaccumulative plants, the spread of Cd in top soil can increase from falling leaves during their life cycles. Therefore, phytomanagement and the harvesting processes for Cd phytoremediation depend on the tolerant mechanisms and life cycles of each herbal plant species. An application of herbaceous herbal plants collected from contaminated areas for folk medicines and raw materials of drugs and food/feed should be prohibited or controlled, based on reliable research and knowledge.

## 2 TOLERANCE AND ACCUMULATION MECHANISM OF HERBACEOUS PLANTS EXPOSED TO CADMIUM

The first stage of metal uptake in plants is the interaction with the soil in order to solubilize metals for absorption; in this stage, plants need to interact with the soil in the rhizosphere. The presence of microbes, a decrease in pH, a change in the redox potential, and/or the exudation of ligands are factors that can increase the metal uptake. In addition, two strategies have been identified for explanation of metal uptake and transport in plants. In the first, the plants (dicotyledons and nongrass monocotyledons) solubilize Fe (III) by reduction of the more soluble Fe(II) and a decrease in pH by $H^+$ excretion. In the second, the plants (graminaceous monocotyledons) produce specific ligands (phytosiderophores, PS) from the mugineic acid family (Callahan et al., 2006). Plant ligands play a role in the sequestration of metals from soils, as they are transported via the apoplastic and symplastic systems to the aboveground tissue.

## 2.1 Uptake and Transport Cd

Roots of most vascular perennial plants can enhance metal detoxification and tolerance of plants by selective metal exclusion and metal chelator secretion (Shanker et al., 2005; Luo et al., 2016). Root epidermis, which is the first plant structure in contact with metals, contains suberin and low-methylesterified pectin in the walls of root endodermal cells. The two compounds prevent passage of water and other molecules. Suberin in the root endodermis and exodermis cell walls serves as a barrier to control the movement of water and mineral ions, and it controls nutritional mineral accumulation and transportation (Baxter et al., 2009). Whereas, polysaccharides in the low-methylesterified pectin can bind with divalent and trivalent metal ions. Most essential and nonessential metals commonly enter through plants as divalent cations ($Zn^{2+}$, $Cd^{2+}$, $Fe^{2+}$, $Mn^{2+}$ or $Cu^{2+}$). Thus, the root cell walls enhance the binding of metal cations into root cells and contribute to reducing the inadvertent uptake of toxic metals by plants (Jentschke and Godbold, 2000). The metals enter plant roots through apoplastic and symplastic pathways (Fig. 1). The metal cations taken up via the apoplast system are further transported through the plasma-membrane into the cytoplasm. The symplast is formed by the interior of cells interconnected by the plasmodesmata. The root cell walls are the first barrier against metals' entry into root cells because functional groups of pectin and cellulose can bind some metal ions (Shanker et al., 2005; Luo et al., 2016).

Cd uptake occurs through transmembrane carriers engaged in the uptake of nutritional ions, such as $Fe^{2+}$, $Mg^{2+}$, $Mn^{2+}$, $Ca^{2+}$, $Cu^{2+}$, and $Zn^{2+}$ (Clemens, 2006; Gallego et al., 2012). $Zn^{2+}$ and $Cd^{2+}$ share transporters of zinc-regulated transporter protein (ZIP), which are responsible for the uptake of metals into cytosol (Krämer, 2010; Lin and Aarts, 2012). The *ZIP* gene appears to be an important factor to control $Zn^{2+}$ and $Cd^{2+}$ import, as it encodes a plasma membrane-localized Zn and Cd-specific transporter. In addition to *ZIP1*, *ZIP3*, *ZIP4*, *ZIP5*, *ZIP9*, *ZIP10*, *ZIP12*, and *IRT3* genes are involved in Zn/Cd-uptake(Krämer, 2010; Lin and Aarts, 2012).

Cd in the cytosol of root cells can be transported to other cellular compartments, such as vacuoles, mitochondria, and chloroplasts, and translocated to neighboring cells via plasmodesmata, or transporters on the plasma membrane, to reach the xylem vessels (Broadley et al., 2007; Choppala et al., 2014). Some perennial plants are nonhyperaccumulators or hyperaccumulators. Cd hyperaccumulative plants often exhibit high rates of Cd xylem loading to transport more metals to aerial parts of the plants. In plant hyperaccumulators, the efflux system is directed toward loading

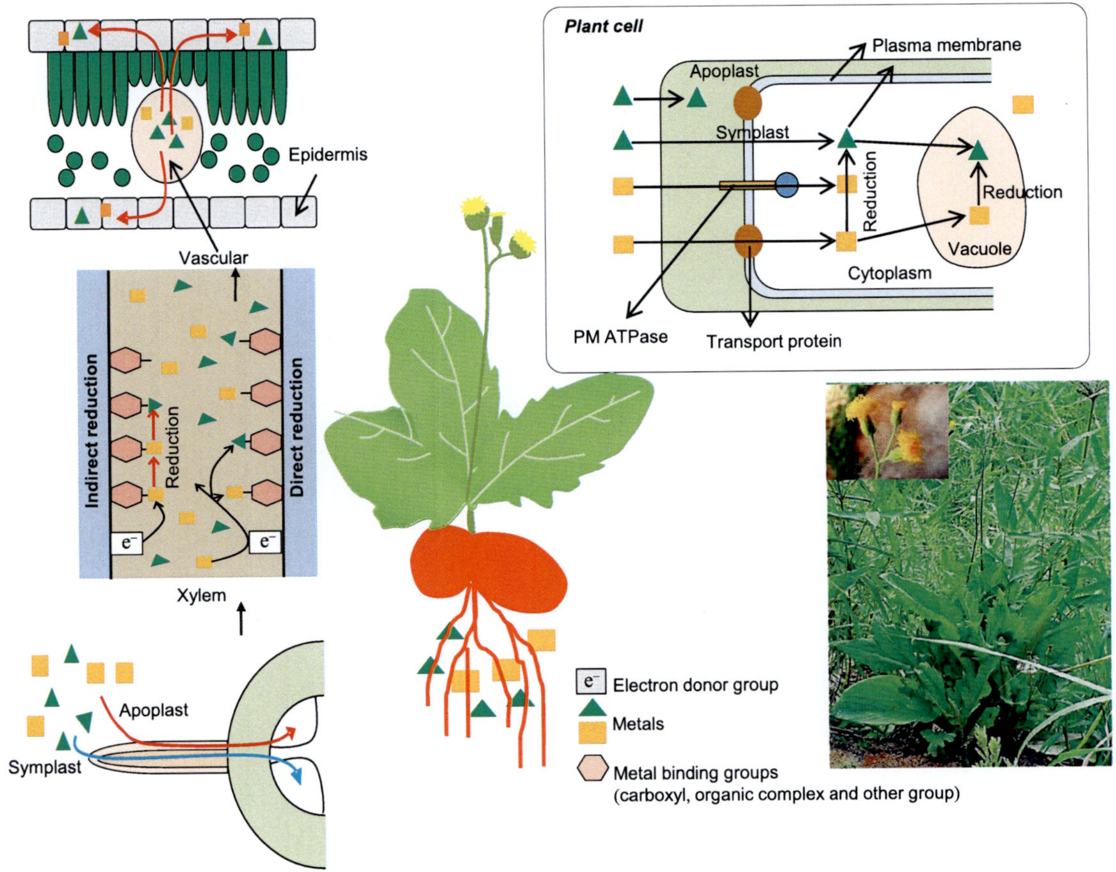


**Fig. 1** A hypothetical model of heavy metals (HMs) transportation, speciation and accumulation in herbaceous plant, occurring in root, xylem, leaves and plant cell. HMs are absorbed by root epidermal cells into the vascular system and transported from the roots to leaves. HMs are transported from root via endodermis to xylem via symplastic system and translocated to leaves via xylem and vein with apoplastic system. In the xylem and vain, HMs are accumulated in the xylem bond with oxygen ligands in the lignocellulosic structure of xylem and translocated to the leaves. In the leaf tissues, HMs could be changed to HMs-organic complexes, the complexes are not adsorbed in xylem and translocated to leaves. HMs in cytosol can be transported to organelles, especially vacuoles within the cells via transporters.

of metals into the xylem on their way to the shoot. The important cellular metal efflux transporter family is the P1B-type ATPase family. One prominent Zn efflux transporter is HMA4 (Heavy Metal ATPase 4). This plasma membrane-localized transporter is normally responsible for loading Zn into the xylem (Krämer, 2010). Moreover, other efflux transporters are involved in Zn or Cd redistribution, translocation, and detoxification, such as the ones encoded by the plant cadmium resistance 1 and 2 (*PCR1* and *PCR2*) genes. Activity of the plasma membrane-localized Cd efflux transporter, PCR1, enhances Cd tolerance by exporting Cd out of the cell and thus reducing Cd contents. Also, PCR2 functions as a Zn efflux transporter, which contributes to Zn distribution and detoxification in plants (Krämer, 2010; Lin and Aarts, 2012; Luo et al., 2016). The distributions of Cd in the shoot (stem and leaves) of *Gynura pseudochina* were examined by synchrotron-basemicro-X-ray fluorescence (μ-XRF) imaging. The μ-XRF imaging showed that the transport process regulating Cd movement during Cd treatment could be apoplast and symplastic from the root to the vascular tissue (Nakbanpote et al., 2010; Mongkhonsin et al., 2016). High Cd restricted movement of ions from the cortex into the stele is expected due to the presence of the Casparian strip, which blocks the apoplastic movement of ions, including $Cd^{2+}$(Tester and Leigh, 2001). Amounts of Cd were sequestered in the symplast of cortical cells and bound with xylem cell walls. Cd arriving in the xylem was then transferred to the shoots' tissue via the vascular system (Fig. 1) (Krämer, 2010; Akhter et al., 2014). Vollenweider et al. (2006) found that the main Cd sinks in the leaves of *Salix viminalis* were in the veins, and Cd storage in the veins was an active detoxification process. However, once the metals exceeded the tolerance threshold of the plant, they would be transported to mesophyll and epidermal cells (Hu et al., 2009; Mongkhonsin et al., 2016). Cd accumulated in substantial amounts in the mesophyll and epidermal cells, most likely in their vacuoles, particularly when the external supply of Cd was high (Küpper et al., 2000; Qiu et al., 2011).

## 2.2 Detoxification and Speciation

Several perennial plants are hyperaccumulators. Amounts of Cd are accumulated in plant cells after uptake and transport. Phytochelatins play an important role in detoxification of metals, and they are effective in binding metal ions by formation of stable metal-phytochelatins complexes (Sytar et al., 2013). *G. pseudochina* has properties that enable it to tolerate a high Cd and Zn concentration, related to the sulfur proteins and phytochelatins (Panitlertumpai et al., 2013), which are responsible for Cd or Zn and metal chelator complex (Cd-phytochelatin) accumulation into the vacuole (Figs. 1 and 2) (Lin and Aarts, 2012). Sytar et al. (2013) explained that phytochelatins

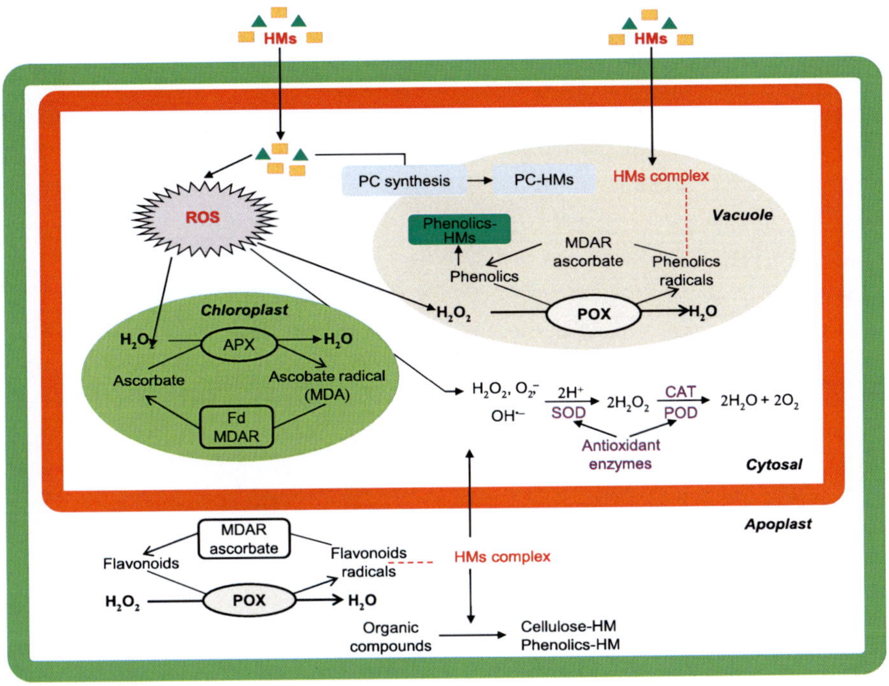


**Fig. 2** A schematic model for herbaceous responding to heavy metal stress. The consequence of *HMs* toxicity is the excessive accumulation of *reactive oxygen species* (*ROS*), ROS accumulation is counteracted by enzymatic antioxidant systems that include a variety of scavengers, such as superoxide dismutase (SOD), ascorbate peroxidases (APX) and catalase (CAT). These enzymes can also scavenged the major ROS, especially SOD, which is the major enzyme activity that modulates the relative amounts of $O_2^{\bullet -}$ and $H_2O_2$. After that CAT and several classes of peroxidase, such as APX, eliminate $H_2O_2$ by breaking it down directly to form $H_2O$ and $O_2$. The synthesis of flavonoids is correlated with the enzymatic and nonenzymatic system in a strategy against ROS. In the chloroplast and cytosol, ascorbate is oxidized to the *malondialdehyde* (*MDA*) radical by APX to detoxify $H_2O_2$. The MDA radical can be reduced to ascorbate by the nonenzymatic reaction of ferredoxin (Fd) or enzymatic reaction of monodehydroascorbate reductase (MDAR). Similarly in vacuoles, the ascorbate/phenolics/peroxidase (POX) system can efficiently reduce $H_2O_2$ without any accumulation of oxidized phenolic products. It has been proposed that the vacuoles and apoplast can function as sinks of $H_2O_2$ in plant cells, which the flavonoids/phytochelatins have effective processes for scavenging free radicals and play a role in resistance to HMs toxicity in plants by chelating complexes.

(PCs) are oligomers of GSH, synthesized by the enzyme phytochelatin synthase (PCS). The synthesis of PCs was initially found to be essential for tolerance of Cd, as concluded from an analysis of PCs (cad) mutants defective in $PC_S$, but later it was also shown to be relevant for tolerance to excess Zn. When Cd enters plant cells, low molecular weight (LMW) PCs will first form a PC–Cd complex. This LMW PC–Cd complex is subsequently sequestered into vacuoles by ATP-binding cassette (ABC) transporters. For example, two ABCC-type transporters from *Arabidopsis thaliana*, ABCC1 and ABCC2, can enhance Cd tolerance and accumulation through vacuolar sequestration of PC-Cd. LMW PC-Cd complexes bind sulfides to form a stable, high molecular weight (HMW) PC-Cd complex, which is stored in the vacuole (Sanitá di Toppi and Gabbrielli, 1999).

In leaves' tissue, once toxic metals are imported to the mesophyll, photosynthesis is damaged (Hu et al., 2009). Moreover, the corresponding distributions of Cd in the stem and leaves may be due to the metal binding phytochelatin and accumulation in the vacuole (Sanitá di Toppi and Gabbrielli, 1999; Panitlertumpai et al., 2013; Sytar et al., 2013). The speciation of Cd in stems and leaves of *G. pseudochina* indicated that Cd in samples were Cd ion ($Cd^{2+}$), and the first shell coordinated with $Cd^{2+}$ trended to be sulfur (S) and oxygen (O) (Mongkhonsin et al., 2016). Küpper et al. (2000) reported that *Arabidopsis halleri*, Cd may be complexed with O-donor ligands as being indicative of binding to the cell wall.

The detoxification process includes antioxidative defense system repair of Cd-induced toxic effects of reactive oxygen species (ROS), which includes phenols and flavonoids (Michalak, 2006; Posmyk et al., 2009). The Cd-induced antioxidative mechanism in plant cells mainly acts through enhancing the activity or presence of antioxidant enzymes, or increasing the accumulation of phenolics (Lin and Aarts, 2012). The enhanced accumulation of phenolics under metal stress can be synthetized from cinnamic acid, which is formed from phenylalanine by the action of phenyloalanine ammonia-lyase(PAL) (Sakihama et al., 2002; Grace, 2005; Michalak, 2006; Kováčik et al., 2009). In addition, the synthesis of flavonoids is correlated with the enzymatic and nonenzymatic system in a strategy against ROS. In the chloroplast and cytosol, ascorbate is oxidized to the malondialdehyde (MDA) radical by ascorbate peroxidase (APX) to detoxify $H_2O_2$. The MDA radical can be reduced to ascorbate by the nonenzymatic reaction of ferredoxin (Fd), or enzymatic reaction of MDA reductase (MDAR). Similarly in vacuoles, the ascorbate/phenolics/peroxidase (POX) system can efficiently reduce $H_2O_2$ without any accumulation of oxidized phenolic products. It has been proposed that the vacuoles and apoplast can function as sinks of $H_2O_2$ in plant cells, which allow the delocalized detoxification mechanism against $H_2O_2$ produced in other compartments during stress and development (Sakihama et al., 2002; Michalak, 2006) (Fig. 2). Moreover, the antioxidant action of phenolic compounds is due to their high tendency to chelate metals (Agati et al., 2012). The roots of many plants exposed to metals exude high levels of phenolics. The general chelating ability of phenolic compounds is probably related to the high nucleophilic character of the aromatic rings, rather than to specific chelating groups within the molecule. The antioxidant action of phenolics depends on the structure of molecules and number of hydroxyl groups in the molecules (Wei and Guo, 2014). The structures of phenolics and flavonoids are important determinants for the antioxidant potential of flavonoids (Fig. 3): (i) the *orto*3′, 4′-dihydroxy structure in the B ring, (ii) the 2,3-double bond in conjunction with the 4-oxo group in the C ring (which allows conjunction between the A and B ring, or electron delocalization), and (iii) the presence of a 3-OH group in the C ring and a 5-OH group in the A ring. Among them, the 3-OH group is the most significant determinant of electron-donating activity (Pietta, 2000; Wei and Guo, 2014).

Phenolic compounds, especially quercetin or rutin, have a greater ability to delay the oxidation process by reacting with metal ions (Souza and Giovani, 2004; Kováčik and Klejdus, 2008). Quercetin and rutin have bulked sugar substitution linked onto the 3-hydroxyl group via an ether bond, which binds metals preferentially at the 5-hydroxyl-4-keto site (Wei and Guo, 2014). In addition, metals were found to bind to the flavonoids, such as flavonol, chrysin, kaempferol, myricetin, and morin. The flavonoids have effective processes for scavenging free radicals, and play a role in resistance to Cd and Zn toxicity in plants by chelating complexes (Brown et al., 1998; Bai et al., 2004; Kostyuk et al., 2007; Medvidović-Kosanović et al., 2011). Confocal laser scanning microscopy (CLSM) images of *G. pseudochina* leaves showed flavonoid compounds localized in the cell wall (Mongkhonsin et al., 2016). Cell walls of plants consist mainly of three organic compounds, including cellulose, hemicellulose, and lignin. These compounds are also major components of natural lignocellulosic materials (Pérez et al., 2002; Agbor et al., 2011; Chen et al., 2014). Lignin, a complex of phenylpropanoid polymer in the plant cell wall, is predominantly deposited in the cell walls of secondary-thickened cells (Vanholme et al., 2010; Chen et al., 2014). Moreover, the cell wall is a major component of phenolic molecules such as flavonoids and anthocyanin (Sytar et al., 2013). For

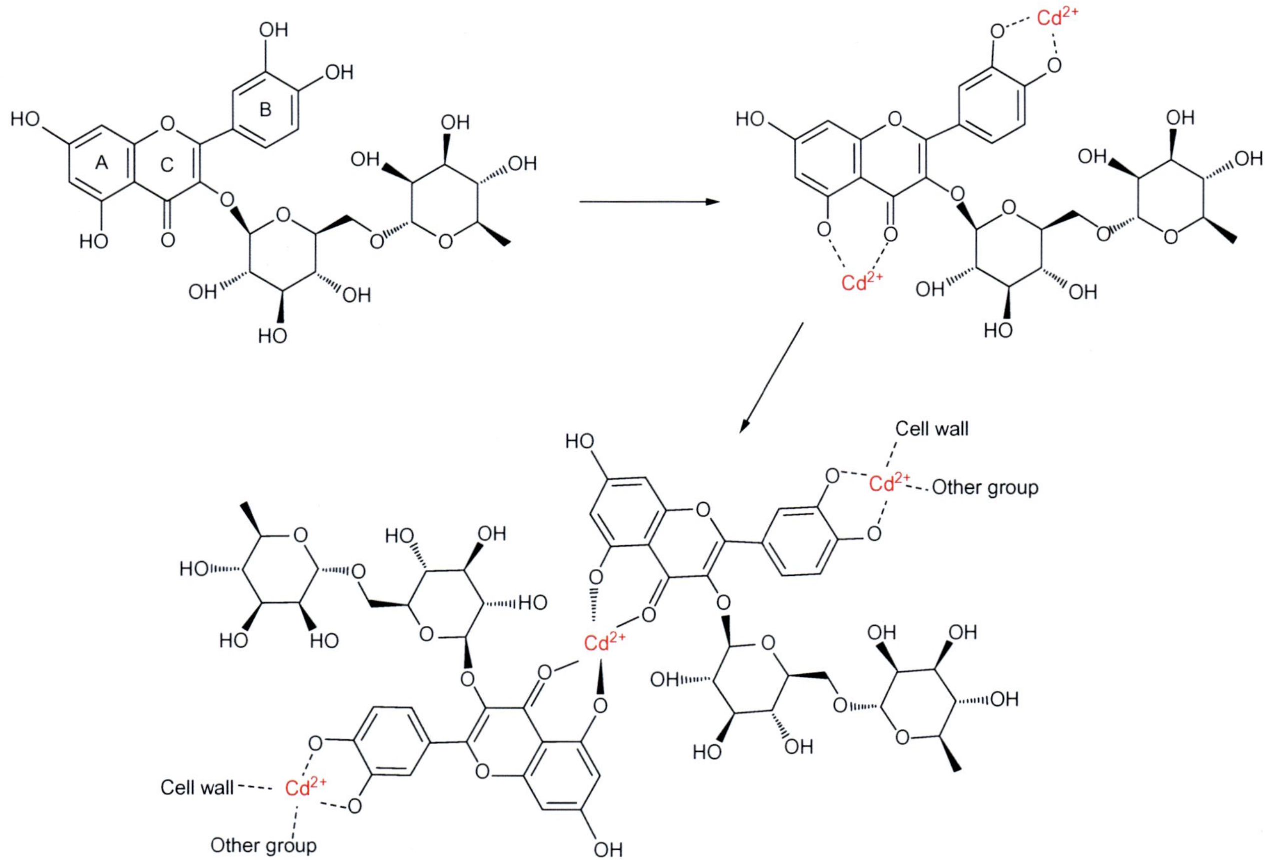


**Fig. 3** A schematic model of rutin, a polyphenolic compound in plant, binding with cadmium ion. HMs binding induces distinct changes in the proton resonances on the flavonoid rings, providing useful information to locate the HMs-binding sites. No HMs-binding are observed with flavone which lacks a chelation site. HMs are found to bind to the 3-hydroxyl-4-keto, catechol and 5-hydroxyl-4-keto chelation sites of flavonol, 3′,4′-dihydroxylflavone and chrysin, respectively. Phenolic compounds chelate HMs at the 3-hydroxyl-4-keto site, especially rutin binds HMs preferentially at the 5-hydroxyl-4-keto site.

example, the flavone tricin was fully compatible with lignification reactions, and is an authentic lignin monomer because it can only start a lignin chain function as a nucleation site for lignification in plants (del Río et al., 2012; Lan et al., 2015). Phenolic esters play a prominent role in many steps of lignin synthesis (Humphreys and Chapple, 2002). Plant phenolic compound precursors have been recognized largely as beneficial antioxidants that can scavenge harmful active oxygen species (Sytar et al., 2013). Flavonoid structures can chelate with metals, then the metal chelates mainly accumulate in the cell wall (Kováčik and Klejdus, 2008; Gallego et al., 2012). The distribution of Cd and Zn has correlations with the localization of flavonoid fluorescence in the cell walls, especially the cell wall of vascular tissue (Mongkhonsin et al., 2016).

Moreover, the excess Cd concentration affects the morphology of the root. The dark brown color of the root system could be caused by different processes. The dark brown color of roots tips, seen after exposure to high concentrations of Cd, indicates chronic exposure and stress-induced morphogenic responses (Potters et al., 2007). Even through *G. pseudochina* is a Zn/Cd hyperaccumulative plant, the *G. pseudochina* treated with a high Cd concentration (15 $mg L^{-1}$) showed the brown solution, which might be an exudate of phenolic compounds released into Murashige and Skoog (MS) agar (Mongkhonsin et al., 2016). Effects of Cd, Cu, and Ni at high concentrations on dry weight and phenolic compounds were expressed in the metal treated *Matricaria chamomilla*(Kováčik and Klejdus, 2008; Kováčik et al., 2008a,b, 2009), and the consequences of leaf chlorosis, necrosis, and root brown were occurring to *Lycopersicon esculentum*(López-Millán et al., 2009). The dark brown color of the root exposure to high Cd doses or alteration of the phenolic metabolism and brown-colored root systems indicated enhancement of phenylalanine ammonia-lyase(PAL) activity and lignin synthesis (Kováčik and Bačkor, 2007; Kováčik and Klejdus, 2008). In general, lignification in a metal-enriched environment is considered part of the defense reaction, which reduces metal entry into plant tissue (Ederli et al., 2004). In addition, Cd-induced changes in auxin levels near the root tip are generally decreased following exposure to Cd, mirroring the cessation of cell division (Potters et al., 2007).

## 2.3 Role of Amino Acid in Metal Tolerance

Amino acids play an important role in metal tolerance in plants (Sharma and Dietz, 2006; Verslues and Sharma, 2010). These organic compounds act as antioxidants and metal chelators during metal stress. Amino acid-dependent metal tolerance is very active at higher pH values compared with acidic pH, at which low molecular organic acids take part in metal chelation. The rise in cellular pool of amino acids often occurs in response to metal stress because of switching of signal transduction, halt in protein synthesis, and protein degradation, respectively (Mehta and Gaur, 1999). Most of the studies on changes in amino acids content during metal stress pointed accumulation of proline and histidine in plants.

Proline is a proteinogenic amino acid that plays an important role in plants as an antioxidant, an osmolyte, electron sink, the stabilizer of cellular macromolecules, and structural component of cell walls (Siripornadulsil et al., 2002). Synthesis of proline can be from either glutamate via exergonic reactions consuming nicotinamide adenine dinucleotide phosphate (NADPH), or ornithine, with the help of ornithine-d-aminotransferase (Verslues and Sharma, 2010). The consumption of NADPH during proline synthesis indicates that this amino acid acts as electrons sinks. Accumulation of proline is very common during water stress, and the accumulation of proline is found to be more abundant in plants parts with higher levels of metal accumulation. For example, the accumulation of proline was found to be higher in the leaves of the *Silene vulgaris*, where Cd accumulation was more pronounced compared with the root (Schat et al., 1997). The time taken to accumulate proline under Cd stress was 6–24 h, which indicates accumulation of proline is related to a water deficit in plants (Yamada et al., 2005). Therefore, the major reason for accumulation of proline under metal stress is impairment of water relationships in the plant cell. In contrast, metal-induced retardation in proline oxidation is also as the reason for the increase in the accumulation of proline during metal stress (Costa and Morel, 1994). Also, increased activities of D1-pyrroline-5-carboxylate reductase and ornithine-d-aminotransferase in detached rice leaves induced proline accumulation under metal stress (Chen et al., 2001). Activation of proline synthesis in plants also occurred as a result of Cd-induced membrane damage that resulted in leakage of osmolytes in the cell (Sharma and Dietz, 2006). Accumulation of proline in response to Cd stress helps in the regulation of osmotic pressure, redox reactions, metal chelation, and scavenging of free radicals. Proline-induced closure of stomata suppresses the uptake and translocation of Cd in several plant species because of inhibition of transpiration (Sharma and Dietz, 2006). This result indicates the indirect action of proline in conferring metal tolerance. The formation of a proline—Cd complex has been reported to protect glucose-6-phosphate dehydrogenase and nitrate reductase activity in plant cells (Sharma et al., 1998). The proline-mediated cytoplasmic sequestration of Cd was also revealed, with the help of Extended X-Ray Absorption Fine Structure (EXAFS) studies in *Chlamydomonas reinhardtii*(Siripornadulsil et al., 2002). Binding of proline with reactive oxygen species, such as hydroxyl radicals ($OH^-$) and singlet oxygen $^1O_2$, help to detoxify reactive oxygen species generated during Cd stress (Alia and Matysik, 2001). On the other hand, proline is believed to mediate the toxic effect of Cd on rice seedlings because of the parallelism observed between proline accumulation and the growth inhibition of rice seedlings (Chen and Kao, 1995).

Histidine helps metal chelation in plants (Anjum et al., 2015). The role of histidine in metal hyperaccumulation was revealed during studies with Ni-hyperaccumulating plants such as *Alyssum lesbiacum* and *Brassica juncea*, where a higher amount of Ni accumulation is associated with the rise of histidine levels in xylem (Kerkeb and Krämer, 2003). Amino acids such as asparagine, alanine, and glutamine have also been implicated in metal tolerance (Bottari and Festa, 1996). The possible role of these amino acids is metal chelation, which confers metal tolerance in plants. Cysteine also plays a very important role in Cd detoxification. This amino acid not only binds with Cd, but also takes part in the synthesis of glutathione, phytochelatin, and peptides that are known to bind with metals such as Cd (Zagorchev et al., 2013). Exogenous cysteine application increased Cd tolerance in rice seedlings via regulation of cellular redox status with the help of glutathione (Wang et al., 2017). It is noteworthy that redox regulation during Cd stress is also achieved with the activity of antioxidant enzymes.

## 3 ANTIOXIDANT ENZYMES INDUCED BY METAL STRESSES

After plants uptake and transport, Cd is accumulated in plant cells. Accumulated nonessential and excessive essential Cd can cause detrimental effects on plant growth and development (Gallego et al., 2012). The toxic symptoms of Cd include stunted growth, chlorosis, decreased photosynthesis, induced lipid peroxidation, enhanced proteolysis, disturbed balance between ROS and antioxidants, and eventually cell death (Gratão et al., 2005; Gallego et al., 2012). During evolution, plants developed several strategies to reduce the toxicity of Cd detoxification. Plants have important detoxification mechanisms to cope with normal ROS levels, which are activated to repair

Cd-induced toxic effects of ROS. The detoxification defense system includes nonantioxidative enzymes (phenolics, glutathione (GSH), ascorbic acid, and carotenoids) and antioxidative enzymes (superoxide dismutase (SOD), catalase (CAT), peroxidase (POD), ascorbate peroxidase (APX), and glutathione reductase (GR)) (Gill and Tuteja, 2010; Gallego et al., 2012; Lin and Aarts, 2012). The detoxification processes include cytosolic metal chelation and induction of the antioxidative systems to scavenge enhanced ROS production. Excess Cd can affect a variety of processes in plants, of which the major consequence is the enhanced production of ROS; specifically, Cd exposure induces overproduction of the active oxygen forms $O_2^{\bullet -}$, $H_2O_2$, and $^{\bullet}OH$, and increases lipid peroxidation of plant leaves and chlorophyll breakdown (Sivaci et al., 2007, 2008; Kováčik et al., 2008a,b; López-Millán et al., 2009; Cherif et al., 2011). The ROS accumulation is counteracted by enzymatic antioxidant systems that include a variety of scavengers, such as SOD, CAT, POD, APX, and GR (Gill and Tuteja, 2010; Sytar et al., 2013). These enzymes can also scavenge the major ROS, especially SOD, which is the major enzyme activity that modulates the relative amounts of $O_2^{\bullet -}$ and $H_2O_2$(Cakmak, 2000). After that, CAT and several classes of peroxidase, such as APX, and GR, eliminate $H_2O_2$ by breaking it down directly to form $H_2O$ and $O_2$(Michalak, 2006; Sytar et al., 2013; Lin and Aarts, 2012). In metal-sensitive plant species, ROS levels will rise substantially if there are not sufficient antioxidant enzymes available. Consequently, ROS-induced cellular damage will induce locally programmed cell death, and will generally affect plant growth and development. Metal-tolerant species normally produce high levels of ROS scavenging antioxidant enzymes, providing a sufficiently efficient antioxidant defense mechanism against metal-induced oxidative stress (Sharma and Dietz, 2009).

The increase of enzyme activity under Cd stress, as in the plants *Kandelia candel*, *Bruguiera gymnorrhiza*, *Solanum lycopersicum*, *Ceratophyllum demersum*, and *Morus alba*, had a consequence of more tolerance and radical scavenging in plant cells (Aravind and Prasad, 2003, 2005; Cherif et al., 2011; Tewari et al., 2008; Zhang et al., 2007). In *G. pseudochina*, Cd concentration over the threshold levels decreased phenolic, flavonoid, and antioxidant activities of the treated plant extracts (Mongkhonsin et al., 2016). Table 1 shows the responses of antioxidative enzymes (SOD, CAT, APX, GR, GSH) and phenolics levels in different herbaceous plants exposed to various toxic concentrations of Cd.

**Table 1** Summary of the responses of antioxidative enzymes (POD, POX, SOD, CAT, APX, GPX, GR) and non-antioxidative enzymes (GSH, phenolics) levels in different herbaceous plants exposed to Cd

| Plant species | Combination metals | Antioxidant systems[a] | | | | | | | | References |
|---|---|---|---|---|---|---|---|---|---|---|
| | | POX | SOD | CAT | APX | GPX | GR | GSH | Phenolics | |
| ***Dicotyledon*** | | | | | | | | | | |
| *Arabidopsis* | Cd | | + | + | | | | + | | Jozefczak et al., 2014 |
| *Arabidopsis thaliana* | Cd | | + | + | + | | + | + | | Cuypers et al., 2011 |
| *Arachis hypogaea* | Cd | | + | + | + | | + | | | Dinakar et al., 2008 |
| *Brassica juncea* | Cd | | + | +/− | +/− | | +/− | | | Mobin and Khan, 2007 |
| *Brassica juncea* | Cd | − | | − | + | − | | | | Markovska et al., 2009 |
| *Brassica juncea* | Cd | | − | + | − | − | +/− | + | | Nouairi et al., 2009 |
| *Brassica napus* | Cd | | − | − | − | + | + | − | | Nouairi et al., 2009 |
| *Brassica juncea* | Cd | | + | + | + | | + | + | | Iqbal et al., 2010 |
| *Brassica juncea* | Cd | | | + | + | | | | | Armas et al., 2015 |
| *Camellia sinensis* | Cd | | | | | | | | + | Zagoskina et al., 2007 |
| *Carthamus tinctorius* | Cd | | + | − | + | − | + | + | | Namdjoyan et al., 2011 |

*Continued*

**Table 1** Summary of the responses of antioxidative enzymes (POD, POX, SOD, CAT, APX, GPX, GR) and non-antioxidative enzymes (GSH, phenolics) levels in different herbaceous plants exposed to Cd—cont'd

| Plant species | Combination metals | Antioxidant systems | | | | | | | | References |
|---|---|---|---|---|---|---|---|---|---|---|
| | | POX | SOD | CAT | APX | GPX | GR | GSH | Phenolics | |
| *Crotalaria juncea* | Cd | | | + | | | + | | | Pereira et al., 2002 |
| Elodea | Cd | | +/− | +/− | | | | | | Maleva et al., 2012 |
| *Erica andevalensis* | Cd | | | | | | | | +/− | Márquez-García et al., 2012 |
| *Gynura pseudochina* | Cd, Zn, Zn + Cd | | | | | | | | +/− | Mongkhonsin et al., 2016 |
| *Hibiscus cannabinus* | Cd | Cd | Cd+ | + | | | + | | | Li et al., 2013b |
| *Lepidium sativum* | Cd | | + | + | + | | + | + | | Gill et al., 2012 |
| *Matricaria chamomilla* | Cd, Cu | | | | | | | | + | Kováčik et al., 2008a,b |
| *Matricaria chamomilla* | Cd, Cu | | | | + | | | | + | Kováčik et al., 2009 |
| Mung Bean | Cd | | + | | + | | | − | | Anjum et al., 2011 |
| *Nicotiana tabacum* | Cd | + | +/− | | | | | | | Martins et al., 2011 |
| *Orthosiphon stamineus* | Cd | | | | | | | | + | Ibrahim et al., 2017 |

| | | | | | | | | | | |
|---|---|---|---|---|---|---|---|---|---|---|
| *Pistia stratiotes* | Cd | | + | | | | | | | Li et al., 2013a |
| *Sinapis arvensis* | Cd, Pb | | | + | + | + | | | | Heidari and Sarani, 2011 |
| *Solanum nigrum* | Cd | | + | + | + | | | | | Fidalgo et al., 2011 |
| *Solanum lycopersicum* | Cd, Zn, Cd + Zn | | + | +/− | +/− | | +/− | | | Cherif et al., 2011 |
| Strawberry | Cd | | + | + | +/− | | | | | Muradoglu et al., 2015 |
| *Suaeda fruticosa* | Cd, Cu | | | − | − | − | | − | | Bankaji et al., 2015 |
| *Thlaspi caerulescens* | Cd | | | | | | | − | | Wójcik et al., 2005 |
| Tomato | Cd | +/− | | +/− | +/− | | +/− | | | Gratão et al., 2015 |
| *Vaccinium corymbosum* | Cd | | | | | | | | +/− | Manquián-Cerda et al., 2016 |
| *Vaccinium myrtillus* | Cd, Zn, Pb, Fe | | | | | | | | + | Białońska et al., 2007 |
| *Vaccinium myrtillus* | Cd + Fe + Mn + Pb + Zn | | | | | − | | +/− | | Kandziora-Ciupa et al., 2013 |
| *Vitis vinifera* | Cd | | | | | | | | +/− | Çetin et al., 2014 |

*Continued*

**Table 1** Summary of the responses of antioxidative enzymes (POD, POX, SOD, CAT, APX, GPX, GR) and non-antioxidative enzymes (GSH, phenolics) levels in different herbaceous plants exposed to Cd—cont'd

| Plant species | Combination metals | Antioxidant systems | | | | | | | | References |
|---|---|---|---|---|---|---|---|---|---|---|
| | | POX | SOD | CAT | APX | GPX | GR | GSH | Phenolics | |
| ***Monocotyledon*** | | | | | | | | | | |
| Maize | Cd | | | | | | | – | | Artiushenko et al., 2014 |
| *Myriophyllum spicatum* L. and *Myriophyllum triphyllum* | Cd | | | | | | | | +/– | Sivaci et al., 2007 |
| *Nigella sativa* | Cd | + | | | | | | | | Espanany et al., 2016 |
| *Phyllostachys pubescens* | Cd | | – | | | | | | | Li et al., 2016 |
| Rice | Cd | | | +/– | + | + | | | + | Roychoudhury et al., 2012 |
| *Triticum aestivum* | Cd | + | + | + | | | | | | Shafi et al., 2009 |

[a]Key of symbol: + increase, – decrease.

## 4 MICROORGANISMS INVOLVED IN PLANT TOLERANCE TO Cd EXPOSURE

The interface between microbes and plant roots (rhizosphere) is considered to greatly influence the growth and survival of plants. Phytoextraction could benefit from rhizosphere processes, primarily through plant-microbe induced solubilization prior to uptake by plants, such as the exudation of organic compounds that can decrease the pH, and thus solubilize metal cations (Wenzel et al., 2004). Phytostabilization can be supported with hyphae of mycorrhizal fungi and some root exudates (sugar and mucilage). The exudates can mechanically bind together soil particles and make the soil more resistant to erosion (Miller and Jastrow, 1990; Wenzel et al., 2004; Prasad, 2011). In addition, metal cations, oxyanions of metals, and metalloids may be immobilized by the release of exude compounds into the rhizosphere (Wenzel et al., 2004). Phytovolatilization probably relates to specific enzymes produced by rhizospheric microorganisms, which can transform mercury (Hg), selenium (Se), and arsenic (As) to volatile compounds (Terry and Zayed, 1994; Azaizeh et al., 1997; Wenzel et al., 2004).

Symbiotic, endophytic, and rhizospheric microorganisms enhance plant growth in heavily polluted soil by (i) supplying growth factors such as indole-3-acetic acid (IAA), cytokinins, and gibberellins, (ii) providing nutrients and essential elements such as nitrogen by nitrogen fixation, while phosphorus, magnesium, and insoluble minerals are transformed into soluble elements by acidification, chelation, and redox reaction before uptake to the plant, (iii) act as biocontrol of plant pathogen by producing antibiotic agents to suppress plant pathogens, and activate the induced systemic resistance of plant defense mechanisms (Lebeua et al., 2008; Rajkumar et al., 2010, 2012; Ma et al., 2011). In addition, preventing metal accumulation in plants is one a plant tolerance enhanced by microorganisms, including secretion of extracellular polymeric substances (EPS), biosurfactances, and glycoprotein to capture metal ions with ligands, or functional groups in their structures, such as thiol, carboxyl, hydroxyl, or amide groups (Nies, 1999; Meesungnoen et al., 2012). Siderophores increase metal mobility and availability; therefore, the compounds enhance metal uptake and accumulate in plants or phytoextraction. The siderophore-metal complex uptakes into plant roots via three mechanisms, including ligand exchange, siderophore-metal complex direct uptake, and siderophore degradation (Rajkumar et al., 2010). Chelation between siderophores and metal (Al, Cd, Cu and Ni) makes the metals less able to inhibit *Streptomyces* spp. to synthesis the auxins, which in turn enhances the plant growth (Dimkpa et al., 2008). Siderophores could indirectly influence the availability and amount of Cd taken up from the rhizosphere of plants (Dimkpa et al., 2009a), and supply plants with Fe in the presence of levels of metals (Al, Cu, Mn, Ni, and U) that inhibit Fe acquisition, leading to enhanced chlorophyll content, and circumventing lipid peroxidation effects on leaves (Dimkpa et al., 2009b).

Consequently, alternative phytoremediation methods that exploit rhizosphere bacteria to reduce metal toxicity to plants have been investigated. Furthermore, the discovery of rhizosphere bacteria that are metal resistant and able to promote plant growth have raised high hopes for ecologically friendly and cost-effective strategies for reclamation of metal-polluted soils (Jing et al., 2007; Ma et al., 2011). Meesungnoen et al. (2012) isolated plant growth-promoting bacteria *Pseudomonas aeruginosa* PDMZnCd2003 from rhizosphere of *G. pseudochina* growing in Zn- and Cd-contaminated soil. This bacterium tolerated Zn and Cd, and it had plant growth-promoting abilities of IAA production, nitrogen fixation, and phosphate solubilization even under the metals' stress. The Zn/Cd-tolerant mechanisms of the isolate were a production of siderophores and EPS under the metals' stress. Interestingly, Zn and/or Cd in nutrient broth induced *P. aeruginosa* PDMZnCd200 to secrete parrot and yellow-green chemicals, which probably indicated siderophores, such as pyocyanin and pyoverdine, respectively. Munjit et al. (2015) isolated Zn- and Cd-tolerant bacteria from the rhizosphere of *Zea mays* growing in Zn/Cd-contaminated agricultural soil. The biochemical test, and Gram's staining, indicated that the selected eight isolates belonged to the genus *Pseudomonas* and *Serratia*, and they were able to produce IAA under the Zn/Cd stress. Endophytic microorganisms are also involved in the plants' tolerance mechanisms. *Cupurvidus taiwanensis* KKU 2500-3 and *P. aeruginosa*KKU2500-8, -9, and -20, which were Cd-resistant bacteria isolated from Zn/Cd-contaminated soil in a rice field, increased Cd tolerance and decreased the accumulation of Cd in Thai jasmine rice, and *C. taiwanensis* KKU 2500-3 could possibly transform toxic soluble $CdCl_2$ into insoluble CdS (Siripornadulsil and Siripornadulsil, 2013). *Bacillus pumilus* E2S2 and *Bacillus* sp. E1S2 isolated from Zn/Cd hyperaccumulator *Sedum plumbizincicola* had plant growth-promoting properties of ACC deaminase, phosphate solubilizing, IAA, and siderophores activity (Ma et al., 2015). In addition, re-inoculation of *B. pumilus* E2S2 in the host plant increased plants' biomass and Cd uptake. Fungal tissues have a role in metal sorption and alleviation of metal toxicity exposure to the plant host

(Jentschke and Godbold, 2000; Hu et al., 2013). *Glomus mosseae* was able to adsorb Cd up to 0.5 mg Cd/g and support phytostabilization (Joner et al., 2000). Whereas, *Glomus geosporum* improved phytoextraction by enhancing Cu accumulation in the roots of *Aster tripolium* L. (Carvalho et al., 2006). *Glomus etunicatum* supported nutrient uptake (P, S, and Fe), then promoted growth of *Calopogonium mucunoides* under Pb stress (Souza et al., 2012). Table 2 shows the effects of microbial inoculation and their metabolites on Cd-exposed plants.

## 5 GROWING HERBACEOUS PLANTS FOR REHABILITATION, PHYTOREMEDIATION, AND SUSTAINABLE DEVELOPMENT

In a case study of the Padaeng Industry Public Company Limited (PDI), it was indicated that PDI's Mae Sot mine covers an area of 2078 rai, of which 79%, or 1643 rai, was used for mining purposes, and required environmental restoration after the termination of the mining business in 2016. The company began initial rehabilitation of the land in 2003. Rehabilitation started in places where mining and related activities were completed, as follows: (i) Fertile topsoil from outside the mine covered bare soil up to at least 30 cm thick before planting, (ii) Bare soil was covered mostly with vetiver and some ruzi grass. It is aimed at increasing more organic matter into the soil, preventing soil erosion, reducing the velocity of runoff water, and conserving moisture in the soil. (iii) Local trees species, such as teak (*Tectona grandis*), iron wood (*Ostrya virginiana*), Siamese sal (*Shorea obtusa* Wall. ex Blume), local cork tree (*Duabanga grandiflora*), orchid tree (*Bauhinia purpurea*), and so forth were planted. In addition, fast growing trees, such as Eucalyptus (*Eucalyptus globulus*), Royal Poinciana (*Delonix regia*) and Leucaena (*Leucaena leucocephala*) were also planted (Prasad et al., 2014, 2016). Moreover, native herbaceous plants should be reintroduced by growing between the local trees to rehabilitate the area. Figs. 4 and 5 show some native herbaceous plants in a conserved area, which has no mining activity, and a mining area of PDI's Mae Sot mine. The autochthonous species and allochthonous plants incorporate naturalized plants that establish themselves in the reproduction ecology and plant-plant interactions (Pugnaire, 2010; Ennajeh et al., 2010; Leguizamo et al., 2017). The rehabilitation will continue until 2023, then the company will return the area to the Royal Forest Department.

Agricultural fields in the Mae Sot District, which have received irrigation from the Mae Tao Creek passing through sources of Zn mineralization from

**Table 2** The study of microorganism/metabolites effects on Cd exposure plants

| Microorganism/ metabolites | Microorganism species | Effects of Cd tolerance in plant | Reference |
|---|---|---|---|
| Siderophores | *Streptomyces tendae* F4 | Siderophores reduced Cd uptake in bacteria while increased Cd uptake in *Helianthus annuus* | Dimkpa et al., 2009b |
| Bacteria | *Methylobacterium oryzae CBMB20 Burkholderia* sp. *CBMB40* | In gnotobiotic assay of *Lycopersicon esculentum* L. seedling, inoculation of these bacteria reduced the ethylene emission and increased tolerance index at different concentrations of $NiCl_2/CdCl_2$. In pot experiments, they reduced Ni(II) and Cd(II) accumulations in roots and shoots and increase in the plant growth. | Madhaiyan et al., 2007 |
| | *Burkholderia cepacia* | Bacteria significantly increased plant growth, Zn and Cd uptake in shoot and metal translocation of *Sedum alfredii*. Cd stimulated organic acid production in roots of *S. alfredii*. The secretion involved the metal resistance mechanism by chelating the metal ions and reducing their uptake | Li et al., 2007 |
| | *Bacillus* sp. J119 | *Bacteria* improved plant growth and Cd uptake in rape, maize, Sudan grass and tomato | Sheng et al., 2008 |
| | *Strain LKR01, LSE02, LSE03 and LSE04* | Inoculation of bacterial strains increased growth and Cd accumulation in *Solanum nigrum* L. | Luo et al., 2011 |
| | *Ralstonia eutropha*(B1) *Chrysiobacterium humi*(B2) | Inoculation of bacterial strains, especially *C. humi*(B2), *Helianthus annuus* enhanced the short term stabilization, decreased biomass loss and reduced aboveground tissue contamination. | Marques et al., 2013 |
| | *Serratia nematodiphila LRE07, Enterobacter aerogenes LRE17, Enterobacter sp. LSE04, Acinetobacter sp. LSE06* | Inoculation of these bacteria increased growth and Cd accumulation in the root, stem and leaf tissue of *Solanum nigrum* L. | Chen et al., 2010 |
| | *Burkholderia* sp. *Actinobacterium* | *Burkholderia* sp. reduced Zn and Cd uptake in roots while *Actinobacterium* increased Zn and Cd accumulations in leaf of *Salix caprea*. Both bacteria did not promote Plant growth. | Kuffner et al., 2010 |

| | | | |
|---|---|---|---|
| Mycorrhiza | *Glomus intraradices*, *Glomus constrictum* *Glomus mosseae* | Mycorrhiza improved growth and the reactive oxygen species (ROS) scavenging and reduced Cd concentration and accumulation in *T. erecta* L. | Liu et al., 2011 |
| | *Glomus constrictum* *Glomus claroideum* *Glomus mosseae* *Glomus microcarpum* *Sclerocystis sinuosa* *Glomus* sp. | Mycorrhiza might colonize ryegrass (*Lolium perenne* L.) and decreased Cd, Ni and Zn uptake. | Takács and Vörös, 2003 |
| | *Glomus versiforme* BGC GD01C | Inoculation with *G. versiforme* increased growth and Cd concentrations of *Solanum nigrum* at 25 and 50 mg Cd $kg^{-1}$ soils, but decreased at 100 mg Cd $kg^{-1}$ soil. | Liu et al., 2015 |
| | *Glomus intraradices* | Inoculation of *mycorrhiza* increased plant growth while Cd accumulation was lower than non-inoculated pots of *Medicago truncatula*. Mycorrhiza inoculation increased available Cd and reduced Cd in leachates. | Redon et al., 2008 |
| | *Glomus caledonium*90,036 (*Gc*) *Glomus mosseae* M47V (*Gm*) | Mycorrhiza increased soil acid phosphatase activities, and available P, and P absorptions in *Sedum alfredii* Hance and *Lolium perenne* L., shoot biomasses, and Cd absorptions. While mycorrhiza significantly decreased Cd phytoavailable by elevating soil pH. | Hu et al., 2013 |
| | *Glomus mosseae* Sel 85N and P792 | Mycorrhiza increased biomass and decreased Zn and Cd contents in plant tissues. Moreover, they reduced oxidative stress by increased production of total non-protein thiols and antioxidant enzymatic upregulation. | Garg and Kaur, 2012 |

*Continued*

**Table 2** The study of microorganism/metabolites effects on Cd exposure plants—cont'd

| Microorganism/ metabolites | Microorganism species | Effects of Cd tolerance in plant | Reference |
|---|---|---|---|
| Consortia | *Glomus caledonium*90,036 *Gigaspora margarita* ZJ37 *Gigaspora decipens* ZJ38 *Scutellospora gilmori* ZJ39 *Acaulospora* spp. *Glomus* spp. | Mycorrhiza consortia increased growth and P, Cu, Zn, Pb concentrations while Cd concentrations did not altered in *Elsholtzia splendens*. | Wang et al., 2005 |
| | Cd resistant microorganisms | Inoculation of microbial consortium increased Cd concentration in *Phragmites australis* stems while more than 48% Cd was remained in its belowground of *Juncus maritimus*. Both plants did not show significant signs of Cd toxicity. | da Silva et al., 2014 |

**Fig. 4** Some native herbaceous plants (A–I) in conserved area, which was no mining activity, of PDI's Mae Sot mine, taken during 2003–05.

**Fig. 5** Some native herbaceous plants (A–I) in mining area of PDI's Mae Sot mine, taken during 2003–05.

the mountains, have elevated levels of Cd, Zn, and Pb contamination (Simmons et al., 2005; Watcharamai and Saenton, 2013; Sebastian et al., 2016). The biomass produced during phytoremediation has been utilized to meet demands in green chemistry. For example, biofuel production is a sustainable way to convert the biomass produced in a phytoremediation program to a value-added product. Metal accumulating plants also showed an increase in accumulation of phytochemicals (Okema et al., 2014). Plants very often responded to Cd accumulation with an increase in phenolics. For example, plants used in phytoremediation, such as *G. pseudochina*, accumulated phenolics in response to Cd and Zn accumulation (Mongkhonsin et al., 2016). Phenolics are a useful commodity because of their color, taste, and health properties. These compounds also act as reducing agents in green chemistry. The ability of phenolics to screen UV light made these compounds attractive in the cosmetic industry (Działo et al., 2016). Therefore, a focus on extracting phenolics from plants used in phytoremediation is an important aspect. The well-known phenolics that respond to Cd stress are caffeic acid, coumarins, and flavonoids (Mongkhonsin et al., 2016). An increase in enzyme activity in the shikimic acid pathway and malonic acid pathway were responsible for the Cd-induced increase in phenolics in plants (Caretto et al., 2015). Besides, metal-induced upregulation of transferases enzymes that participate in chemical modifications such as methylation, acylation, and glycosylation results in diversity in the production of phenolics in plants under Cd stress C (Rohn et al., 2002). The difference in the genetic makeup of plants, as well as environmental conditions, also causes variations in the type of phenolics produced in response to Cd stress. It can be concluded that metal tolerance in plants not only helps to remove metals such as Cd from the soil, but also help to produce value-added products that ensure sustainable phytoremediation. However, the metals also enter the food chain via insects, birds, herbivores, and so forth, and application of herbal plants collected from contaminated areas for folk medicines and raw materials of drug and food/feed should be prohibited or controlled based on reliable research and knowledge.

In a case study of *G. pseudochina*, this Zn/Cd hyperaccumulative plant is a medicinal herbaceous plant. The interaction of Zn and Cd reduced the adverse effect of Cd on the plant growth. The combination treatments of Zn and Cd decreased the translocation factor when compared with the treatment of either Zn or Cd alone (Mongkhonsin et al., 2016). The leaf extraction showed that total phenolic and flavonoid contents corresponded to antioxidant activity. The treatment with Zn plus Cd still caused a higher antioxidant activity than only Cd in the extracts. Treatments with higher Zn and Cd concentrations were found in the leaf extracts at 21%–33% and 2%–3%, respectively. Therefore, the concentration of Zn and Cd in the extracts was lower when compared with Zn and Cd accumulated in the leaf biomass as 67%–79% and 97%–98%, respectively. Sukadeetad et al. (2018) showed that a *G. pseudochina* leaf extracted with ethanol contained chlorogenic acid (CGA), caffeic acid (CA), p-coumaric acid (PCA), and rutin (RUT), a xanthone derivative, a phenylpropanoid, a phenolic glycoside compound, a glycerol-phospholipid, and a low concentration of pyrrolizidine alkaloids (PAs). The crude extracts could inhibit the RelB canonical pathway and IL-8 production. Therefore, it is possible to apply *G. pseudochina* leaf extracts for psoriasis alleviation. Based on our research over the past 10 years, *G. pseudochina* is a useful case study for growing in a Zn/Cd contaminated area to support phytoremediation, especially the agricultural area near the Mae Taw Channel (Fig. 6). The precious phenolic compounds containing antioxidant properties could be extracted from the plant's leaves, and then the unneeded biomass should be a byproduct for the zinc phytomining process. However, rhizobacteria and endophytic bacteria probably affect the form of Cd accumulated in the plant cells, including effects of environment and soil properties. Therefore, the research and phytochemical investigation focusing on safety of the extracts should be further studied for an additional 2–5 years before supporting this idea.

**Fig. 6** A research field of growing *Gynura pseudochina* in Zn/Cd contaminated area near Mae Taw cannel (A), and growth of the plants in clay-loam soil (B) and clay soil with rice husk (C).

## ACKNOWLEDGMENTS

The authors gratefully acknowledge the Thailand Research Fund (TRF) (no. DBG5080016) and the Research and Researcher for Industries (RRi) under TRF (no. PHD58I0023) for financial support. B. Mongkhonsin and O. Meesungnoen would like to thank the Human Resource Development in Science Project (Science Achievement Scholarship of Thailand, SAST). Finally, we thank S. Khaokaew and Kasetsart University for supporting the field experiment in the Phatat Pha Daeng sub-district, Mae Sot, Tak Province, Thailand, and appreciate to PDI's Mae Sot mine for permission of taking photographs, useful information, and sincerity in environmental rehabilitation.

## REFERENCES

Agati, G., Azzarello, E., Pollastri, S., Tattini, M., 2012. Flavonoids as antioxidants in plants: location and functional significance. Plant Sci. 196, 67–76.

Agbor, V.B., Cicek, N., Sparling, R., Berlin, A., Levin, D.B., 2011. Biomass pretreatment: fundamentals toward application. Biotechnol. Adv. 29, 675–685.

Akhter, M.F., Omelon, C.R., Gordon, R.A., Moser, D., Macfie, S.M., 2014. Localization and chemical speciation of cadmium in the roots of barley and lettuce. Environ. Exp. Bot. 100, 10–19.

Alia, M.P., Matysik, J., 2001. Effect of proline on the production of singlet oxygen. Amino Acids 21, 195–200.

Alvarez, P.J.J., Illman, W.A., 2006. Bioremediation and Natural Attenuation: Process Fundamentals and Mathematical Models. John Wiley and Sons, New Jersey.

Anjum, N.A., Umar, S., Iqbal, M., Khan, N.A., 2011. Cadmium causes oxidative stress in mung bean by affecting the antioxidant enzyme system and ascorbate-glutathione cycle metabolism. Russ. J. Plant Physiol. 58, 92–99.

Anjum, N.A., Hasanuzzaman, M., Hossain, M.A., Thangavel, P., Roychoudhury, A., Gill, S.S., Rodrigo, M.A., Adam, V., Fujita, M., Kizek, R., Duarte, A.C., Pereira, E., Ahmad, I., 2015. Jacks of metal/metalloid chelation trade in plants—an overview. Front. Plant Sci. 6, 192. https://doi.org/10.3389/fpls.2015.00192.

Aravind, P., Prasad, M.N.V., 2003. Zinc alleviates cadmium-induced oxidative stress in *Ceratophyllum demersum* L.: a free-floating freshwater macrophyte. Plant Physiol. Biochem. 41, 391–397.

Aravind, P., Prasad, M.N.V., 2005. Cadmium-Zinc interaction in a hydroponic system using *Ceratophyllumdemersum* L.: adaptive ecophysiology, biochemistry and molecular toxicology. Braz. J. Plant Physiol. 17, 3–20.

Armas, T., Pinto, A.P., Varennes, A.D., Mourato, M.P., Martins, L.L., Gonçalves, M.L.S., Mota, A.M., 2015. Comparison of cadmium-induced oxidative stress in *Brassica juncea* in soil and hydroponic cultures. Plant Soil 388, 297–305.

Artiushenko, T., Syshchykov, D., Gryshko, V., Čiamporová, M., Fiala, R., Repka, V., Martinka, M., Pavlovkin, J., 2014. Metal uptake, antioxidant status and membrane potential in maize roots exposed to cadmium and nickel. Biologia 69, 1142–1147.

Azaizeh, H.A., Gowthaman, S., Terry, N., 1997. Microbial selenium volatilization in rhizosphere and bulk soils from a constructed wetland. J. Environ. Qual. 26, 666–672.

Bai, Y., Song, F., Chen, M., Xing, J., Liu, Z., Liu, S., 2004. Characterization of the rutin-metal complex by electrospray ionization tandem mass spectrometry. Anal. Sci. 20, 1147–1151.

Bankaji, I., Caçador, I., Sleimi, N., 2015. Physiological and biochemical responses of *Suaeda fruticosa* to cadmium and copper stresses: growth, nutrient uptake, antioxidant enzymes, phytochelatin, and glutathione levels. Environ. Sci. Pollut. Res. 22, 13058–13069.

Baxter, I., Hosmani, P.S., Rus, A., Lahner, B., Borevitz, J.O., Muthukumar, B., Mickelbart, M.V., Schreiber, L., Franke, R.B., Salt, D.E., 2009. Root suberin forms an extracellular barrier that affects water relations and mineral nutrition in Arabidopsis. PLoS Genet. 5(5). e1000492.

Białońska, D., Zobel, A.M., Kuraś, M., Tykarska, T., Sawicka-Kapusta, K., 2007. Phenolic compounds and cell structure in bilberry leaves affected by emissions from a Zn-Pb Smelter. Water Air Soil Pollut. 181, 123–133.

Bottari, E., Festa, M.R., 1996. Asparagine as a ligand for cadmium(II), lead(II) and zinc(II). Chem. Speciat. Bioavailab. 8, 75–83.

Broadley, M.R., White, P.J., Hammond, J.P., Zelko, I., Lux, A., 2007. Zinc in plants. New Phytol. 173, 677–702.

Brown, J.E., Khodr, H., Hider, R.C., Rice-Evans, C.A., 1998. Structural dependence of flavonoid interactions with $Cu^{2+}$ ions: implications for their antioxidant properties. Biochem. J. 330, 1173–1178.

Cakmak, I., 2000. Possible roles of zinc in protecting plant cells from damage by reactive oxygen species. New Phytol. 146 (2), 185–205. Tansley Review No. 111.

Callahan, D.L., Baker, A.J.M., Kolev, S.D., Wedd, A.G., 2006. Metal ion ligands in hyperaccumulating plants. J. Biol. Inorg. Chem. 11, 2–12.

Caretto, S., Linsalata, V., Colella, G., Mita, G., Lattanzio, V., 2015. Carbon fluxes between primary metabolism and phenolic pathway in plant tissues under stress. Int. J. Mol. Sci. 16, 26378–26394.

Carvalho, L.M., Caçador, I., Martins-Loução, M.A., 2006. Arbuscular mycorrhizal fungi enhance root cadmium and copper accumulation in the roots of the salt marsh plant *Aster tripolium* L. Plant Soil 285, 161–169.

Çetin, E.S., Babalik, Z., Hallaç-Türk, F., Göktürk-Baydar, N., 2014. The effects of cadmium chloride on secondary metabolite production in *Vitis vinifera* cv. cell suspension cultures. Biol. Res. 47, 47–53.

Chen, S.L., Kao, C.H., 1995. Cd induced changes in proline level and peroxidase activity in roots of rice seedlings. Plant Growth Regul. 17, 67–71.

Chen, C.T., Chen, L.M., Lin, C.C., Kao, C.H., 2001. Regulation of proline accumulation in detached rice leaves exposed to excess copper. Plant Sci. 160, 283–290.

Chen, L., Luo, S., Xiao, X., Guo, H., Chen, J., Wan, Y., Li, B., Xu, T., Xi, Q., Rao, C., Liu, C., Zeng, G., 2010. Application of plant growth-promoting endophytes (PGPE) isolated from *Solanum nigrum* L. for phytoextraction of Cd-polluted soils. Appl. Soil Ecol. 46, 383–389.

Chen, J., Mangelinckx, S., Ma, L., Wang, Z., Li, W., De, K.N., 2014. Caffeoylquinic acid derivatives isolated from the aerial parts of *Gynura divaricata* and their yeast α-glucosidase and PTP1B inhibitory activity. Fitoterapia 99, 1–6.

Cherif, J., Mediouni, C., Ammar, W.B., Jemal, F., 2011. Interactions of zinc and cadmium toxicity in their effects on growth and in antioxidative systems in tomato plants (*Solanum lycopersicum*). J. Environ. Sci. 23 (5), 837–844.

Choppala, G., Saifullah, Bolan, N., Bibi, S., Iqbal, M., Rengel, Z., Kunhikrishnan, A., Ashwath, N., Ok, Y.S., 2014. Cellular mechanisms in higher plants governing tolerance to cadmium toxicity. Crit. Rev. Plant Sci. 33, 374–391.

Clemens, S., 2006. Toxic metal accumulation, responses to exposure and mechanisms of tolerance in plants. Biochimie 88, 1707–1719.

Costa, G., Morel, J.L., 1994. Water relations, gas exchange and amino acid content in Cd-treated lettuce. Plant Physiol. Biochem. 32, 561–570.

Cuypers, A., Smeets, K., Ruytinx, J., Opdenakker, K., Keunen, E., Remans, T., Horemans, N., Vanhoudt, N., Van Sanden, S., Van Belleghem, F., Guisez, Y.,

Colpaert, J., Vangronsveld, J., 2011. The cellular redox state as a modulator in cadmium and copper responses in *Arabidopsis thaliana* seedlings. J. Plant Physiol. 168, 309–316.
da Silva, M.N., Mucha, A.P., Rocha, A.C., Teixeira, C., Gomes, C.R., Almeida, C.M.R., 2014. A strategy to potentiate Cd phytoremediation by saltmarsh plants-autochthonous bioaugmentation. J. Environ. Manag. 134, 136–144.
Das, P., Samantaray, S., Rout, G.R., 1997. Studies on cadmium toxicity in plants: a review. Environ. Pollut. 98, 29–36.
del Río, J.C., Rencoret, J., Prinsen, P., Martínez, A.T., Ralph, J., Gutiérrez, A., 2012. Structural characterization of wheat straw lignin as revealed by analytical pyrolysis, 2D-NMR, and reductive cleavage methods. J. Agric. Food Chem. 60, 5922–5935.
Dimkpa, C.O., Svatos, A., Dabrowska, P., Schmidt, A., Boland, W., Kothe, E., 2008. Involvement of siderophores in the reduction of metal-induced inhibition of auxin synthesis in *Streptomyces* spp. Chemosphere 74, 19–25.
Dimkpa, C.O., Merten, D., Svatoa, A., Bachel, G., Kothe, E., 2009a. Siderophores mediate reduced and increased uptake of cadmium by *Streptomyces tendae* F4 and sunflower (*Helianthus annuus*), respectively. J. Appl. Microbiol. 107, 1687–1696.
Dimkpa, C.O., Merten, D., Svatos, A., Buchel, G., Kothe, E., 2009b. Metal-induced oxidative stress impacting plant growth in contaminated soil is alleviated by microbial siderophores. Soil Biol. Biochem. 41, 154–162.
Dinakar, N., Nagajyothi, P.C., Suresh, S., Udaykiran, Y., Damodharam, T., 2008. Phytotoxicity of cadmium on protein, proline and antioxidant enzyme activities in growing *Arachis hypogaea* L. seedling. J. Environ. Sci. 20, 199–206.
Działo, M., Mierziak, J., Korzun, U., Preisner, M., Szopa, J., Kulma, A., 2016. The potential of plant phenolics in prevention and therapy of skin disorders. Int. J. Mol. Sci. 17, 160. https://doi.org/10.3390/ijms17020160.
Ederli, L., Reale, L., Ferranti, F., Pasqualini, S., 2004. Responses induced by high concentration of cadmium in *Phragmites australis* roots. Physiol. Plant. 121, 66–74.
Ennajeh, M., Cochard, H., Khemira, H., 2010. Re-introduction success of an autochthonous plant species, *Periploca angustifolia*, in the Natural Reserve of Oued Dekouk, Tunisia. Span. J. Agric. Res. 8 (4), 1005–1011.
Espanany, A., Fallah, S., Tadayyon, A., 2016. Seed priming improves seed germination and reduces oxidative stressin black cumin (*Nigella sativa*) in presence of cadmium. Ind. Crop. Prod. 79, 195–204.
Fidalgo, F., Freitas, R., Ferreira, R., Pessoa, A.M., Teixeira, J., 2011. *Solanum nigrum* L. antioxidant defense system isozymes are regulated transcriptionally and post-translationally in Cd induced stress. Environ. Exp. Bot. 72, 312–319.
Fowler, B.A., 2009. Monitoring of human populations for early markers of cadmium toxicity: a review. Toxicol. Appl. Pharmacol. 238, 294–300.
Gallego, S.M., Pena, L.B., Barcia, R.A., Azpilicueta, C.E., Iannone, M.F., Rosales, E.P., Zawoznik, M.S., Groppa, M.D., Benavides, M.P., 2012. Unravelling cadmium toxicity and tolerance in plants: insight into regulatory mechanisms. Environ. Exp. Bot. 83, 33–46.
Garg, N., Kaur, H., 2012. Response of antioxidant enzymes, phytochelatins and glutathione production towards Cd and Zn stresses in *Cajanus cajan*(L.) Millsp. genotypes colonized by arbuscular mycorrhizal fungi. J. Agron. Crop Sci. 199, 118–133.
Gill, S.S., Tuteja, N., 2010. Reactive oxygen species and antioxidant machinery in abiotic stress tolerance in crop plants. Plant Physiol. Biochem. 48, 909–930.
Gill, S.S., Khana, N.A., Tuteja, N., 2012. Cadmium at high dose perturbs growth, photosynthesis and nitrogen metabolism while at low dose it up regulates sulfur assimilation and antioxidant machinery in garden cress (*Lepidium sativum* L.). Plant Sci. 182, 112–120.
Grace, S.C., 2005. Phenolics as antioxidants. In: Smirnoff, N. (Ed.), Antioxidants and Reactive Oxygen Species in Plants. Blackwell Publishing Ltd., Oxford, pp. 141–168

Gratão, P.L., Polle, A., Lea, P.J., Azevedo, R.A., 2005. Making the life of heavy metal-stressed plants a little easier. Funct. Plant Biol. 32, 481–494.

Gratão, P.L., Monteiro, C.C., Tezotto, T., Carvalho, R.F., Alves, L.R., Peters, L.P., Azevedo, R.A., 2015. Cadmium stress antioxidant responses and root-to-shoot communication in grafted tomato plants. Biometals 28, 803–816.

He, S., Yang, X., He, Z., Baligar, V.C., 2017. Morphological and physiological responses of plants to cadmium toxicity: a review. Pedosphere 27 (3), 421–438.

Heidari, M., Sarani, S., 2011. Effects of lead and cadmium on seed germination, seedling growth and antioxidant enzymes activities of mustard (*Sinapis arvensis* L.). J. Agric. Biol. Sci. 6, 44–47.

Hu, P.J., Qiu, R.L., Senthilkumar, P., Jiang, D., Chen, Z.W., Tang, Y.T., Liu, F.J., 2009. Tolerance accumulation and distribution of zinc and cadmium in hyperaccumulator *Potentilla griffithii*. Environ. Exp. Bot. 66, 317–325.

Hu, J., Wu, S., Wu, F., Leung, H.M., Lin, X., Wong, M.H., 2013. Arbuscular mycorrhizal fungi enhance both absorption and stabilization of Cd by Alfred stonecrop (*Sedum alfredii* Hance) and perennial ryegrass (*Lolium perenne* L.) in a Cd-contaminated acidic soil. Chemosphere 93, 1359–1365.

Humphreys, J.M., Chapple, C., 2002. Rewriting the lignin roadmap. Curr. Opin. Plant Biol. 5, 224–229.

Ibrahim, M.H., Ismail, A., Omar, H., Zain, N.A.M., 2017. Application effects of cadmium and humic acid on the growth, chlorophyll fluorescence, leaf gas exchange and secondary metabolites in Misai Kucing (*Orthosiphon stamineus* Benth). Annu. Res. Rev. Biol. 18, 1–14.

Iqbal, N., Masood, A., Nazar, R., Syeed, S., Khan, N.A., 2010. Photosynthesis, growth and antioxidant metabolism in mustard (*Brassica juncea* L.) cultivars differing in Cd tolerance. Agric. Sci. China 9, 519–527.

ITRC (Interstate Technology & Regulatory Council), 2009. Phytotechnology Technical and Regulatory Guidance and Decision Trees. Interstate Technology & Regulatory Council. Phytotechnologies Team, Washington, DC.

Jentschke, G., Godbold, D.L., 2000. Metal toxicity and ectomycorrhizas. Physiol. Plant. 109, 107–116.

Jing, Y., He, Z., Yang, X., 2007. Role of soil rhizobacteria in phytoremediation of heavy metal contaminated soil. J Zhejiang Univ Sci B 8 (3), 192–207.

Joner, E.J., Briones, R., Leyval, C., 2000. Metal-binding capacity of arbuscular mycorrhizal mycelium. Plant Soil 226, 227–234.

Jozefczak, M., Keunen, E., Schat, H., Bliek, M., Hernández, L.E., Carleer, R., Remans, T., Bohler, S., Vangronsveld, J., Cuypers, A., 2014. Differential response of *Arabidopsis* leaves and roots to cadmium: glutathione-related chelating capacity vs antioxidant capacity. Plant Physiol. Biochem. 83, 1–9.

Kabata-Pendias, A., 2011. Trace Elements in Soils and Plants, fourth ed. Taylor and Francis Group, Boca Raton, Florida.

Kandziora-Ciupa, M., Ciepał, R., Nadgórska-Socha, A., Barczyk, G., 2013. A comparative study of heavy metal accumulationand antioxidant responses in *Vaccinium myrtillus* L. leaves in polluted and non-polluted areas. Environ. Sci. Pollut. Res. 20, 4920–4932.

Kerkeb, L., Krämer, U., 2003. The role of free histidine in xylem loading of nickel in *Alyssum lesbiacum* and *Brassica juncea*. Plant Physiol. 131 (2), 716–724.

Kostyuk, V.A., Potapovich, A.I., Kostyuk, T.V., Cherian, M.G., 2007. Metal complexes of dietary flavonoids: evaluation of radical scavenger properties and protective activity against oxidative stress *in vivo*. Cell. Mol. Biol. 53, 61–68.

Kováčik, J., Bačkor, M., 2007. Phenylalanine ammonia-lyase and phenolic compounds in chamomile tolerance to cadmium and copper excess. Water Air Soil Pollut. 185, 185–193.

Kováčik, J., Klejdus, B., 2008. Dynamics of phenolic acids and lignin accumulation in metal-treated*Matricaria chamomilla* roots. Plant Cell Rep. 27, 605–615.
Kováčik, J., Bačkor, M., Kaduková, J., 2008a. Physiological responses of *Matricaria chamomilla* to cadmium and copper excess. Environ. Toxicol. 23, 123–130.
Kováčik, J., Grúz, J., Bačkor, M., Tomko, J., Strnad, M., Repčák, M., 2008b. Phenolic compounds composition and physiological attributes of *Matricaria chamomilla* grown in copper excess. Environ. Exp. Bot. 62, 145–152.
Kováčik, J., Klejdus, B., Hedbavny, J., Štork, F., Bačkor, M., 2009. Comparison of cadmium and copper effect on phenolic metabolism, mineral nutrients and stress-related parameters in *Matricaria chamomilla* plants. Plant Soil 3 (20), 231–242.
Krämer, U., 2010. Metal hyperaccumulation in plants. Annu. Rev. Plant Biol. 6, 517–534.
Kuffner, M., De Maria, S., Puschenreiter, M., Fallmann, K., Wieshammer, G., Gorfer, M., Strauss, J., Rivelli, A.R., Sessitsch, A., 2010. Culturable bacteria from Zn- and Cd-accumulating*Salix caprea* with differential effects on plant growth and heavy metal availability. J. Appl. Microbiol. 108, 1471–1484.
Küpper, H., Lombi, E., Zhao, F.J., McGrath, S.P., 2000. Cellular comport mentation of cadmium and zinc in relation to other elements in the hyperaccumulator *Arabidopsis halleri*. Planta 212, 75–84.
Lan, W., Lu, F., Regner, M., Zhu, Y., Rencoret, J., Ralph, S.A., Zakai, U.I., Morreel, K., Boerjan, W., Ralph, J., 2015. Tricin, a flavonoid monomer in monocot lignification. Plant Physiol. 167, 1284–1295.
Lasat, M.M., 2002. Phytoextraction of toxic metals: a review of biological mechanisms. J. Environ. Qual. 31, 109–120.
Lebeua, T., Braud, A., Jézéquel, K., 2008. Performance of bioaugmentation-assisted phytoextraction applied to metal contaminated soils: a review. Environ. Pollut. 153, 497–522.
Leguizamo, M.A.O., Gómez, W.D.F., Sarmiento, M.C.G., 2017. Native herbaceous plant species with potential use in phytoremediation of heavy metals, spotlight on wetlands—a review. Chemosphere 168, 1230–1247.
Li, W.C., Ye, Z.H., Wong, M.H., 2007. Effects of bacteria on enhanced metal uptake of the Cd/Zn-hyperaccumulating plant, *Sedum alfredii*. J. Exp. Bot. 58, 4173–4182.
Li, Y., Zhang, S., Jiang, W., Liu, D., 2013a. Cadmium accumulation, activities of antioxidant enzymes, and malondialdehyde (MDA) content in *Pistia stratiotes* L. Environ. Sci. Pollut. Res. 20, 1117–1123.
Li, F., Qi, J., Zhang, G., Lin, L., Fang, P., Tao, A., Xu, J., 2013b. Effect of cadmium stress on the growth, antioxidative enzymes and lipid peroxidation in two Kenaf (*Hibiscus cannabinus* L.) plant seedlings. J. Integr. Agric. 12, 610–620.
Li, S., Chen, J., Islam, E., Wang, Y., Wu, J., Ye, Z., Yan, W., Peng, D., Liu, D., 2016. Cadmium-induced oxidative stress, response of antioxidants and detection of intracellular cadmium in organs of moso bamboo (*Phyllostachys pubescens*) seedlings. Chemosphere 153, 107–114.
Lin, Y.F., Aarts, M.G.M., 2012. The molecular mechanism of zinc and cadmium stress response in plants. Cell. Mol. Life Sci. 69, 3187–3206.
Liu, L.Z., Gong, Z.Q., Zhang, Y.L., Li, P.J., 2011. Growth, cadmium accumulation and physiology of marigold (*Tagetes erecta* L.) as affected by arbuscular mycorrhizal fungi. Pedosphere 21, 319–327.
Liu, H., Yuan, M., Tan, S., Yang, X., Lan, Z., Jiang, Q., Ye, Z., Jing, Y., 2015. Enhancement of arbuscular mycorrhizal fungus (*Glomus versiforme*) on the growth and Cd uptake by Cd-hyperaccumulator*Solanum nigrum*. Appl. Soil Ecol. 89, 44–49.
López-Millán, A.F., Sagardoy, R., Solanas, M., Abadía, A., Abadía, J., 2009. Cadmium toxicity in tomato (*Lycopersicon esculentum*) plants grown in Hydroponics. Environ. Exp. Bot. 65, 376–385.

Luo, S., Chen, L., Chen, J., Xiao, X., Xu, T., Wan, Y., Rao, C., Liu, C., Liu, Y., Lai, C., Zeng, G., 2011. Analysis and characterization of cultivable heavy metal-resistant bacterial endophytes isolated from Cd-hyperaccumulator*Solanum nigrum* L. and their potential use for phytoremediation. Chemosphere 85, 1130–1138.

Luo, Z.B., He, J., Polle, A., Rennenberg, H., 2016. Heavy metal accumulation and signal transduction in herbaceous and woody plants: paving the way for enhancing phytoremediation efficiency. Biotechnol. Adv. 34, 1131–1148.

Ma, Y., Prasad, M.N.V., Rajkumar, M., Freitas, H., 2011. Plant growth promoting rhizobacteria and endophytes accelerate phytoremediation of metalliferous soils. Biotechnol. Adv. 29, 248–258.

Ma, Y., Oliveira, R.S., Nai, F., Rajkumar, M., Luo, Y., Rocha, I., Freitas, H., 2015. The hyperaccumulator *Sedum plumbizincicola* harbors metal-resistant endophytic bacteria that improve its phytoextraction capacity in multi-metal contaminated soil. J. Environ. Manag. 156, 62–69.

Madhaiyan, M., Poonguzhali, S., Sa, T., 2007. Metal tolerating methylotrophic bacteria reduces nickel and cadmium toxicity and promotes plant growth of tomato (*Lycopersicon esculentum* L.). Chemosphere 69, 220–228.

Maleva, M.G., Nekrasova, G.F., Borisova, G.G., Chukina, N.V., Ushakova, O.S., 2012. Effect of heavy metals on photosynthetic apparatus and antioxidant status of Elodea. Russ. J. Plant Physiol. 59, 190–197.

Manquián-Cerda, K., Escudey, M., Zúñiga, G., Arancibia-Miranda, N., Molina, M., Cruces, E., 2016. Effect of cadmium on phenolic compounds, antioxidant enzyme activity and oxidative stress in blueberry (*Vaccinium corymbosum* L.) plantlets grown in vitro. Ecotoxicol. Environ. Saf. 133, 316–326.

Markovska, Y.K., Gorinova, N.I., Nedkovska, M.P., Miteva, K.M., 2009. Cadmium-induced oxidative damage and antioxidant responses in *Brassica juncea*plants. Biologia 53, 151–154.

Marques, A.P.G.C., Moreira, H., Franco, A.R., Rangel, A.O.S.S., Castro, P.M.L., 2013. Inoculating *Helianthus annuus* (sunflower) grown in zinc and cadmium contaminated soils with plant growth promoting bacteria-effects on phytoremediation strategies. Chemosphere 92, 74–83.

Márquez-García, B., Fernández-Recamales, M.Á., Córdoba, F., 2012. Effects of cadmium on phenolic composition and antioxidant activities of *Erica andevalensis*. J. Bot. Article ID 936950. https://doi.org/10.1155/2012/936950.

Martins, L.L., Mourato, M.P., Cardoso, A.I., Pinto, A.P., Mota, A.M., Gonçalves, M.L.S., Varennes, A., 2011. Oxidative stress induced by cadmium in *Nicotiana tabacum* L.: effects on growth parameters, oxidative damage and antioxidant responses in different plant parts. Acta Physiol. Plant. 33, 1375–1383.

Medvidović-Kosanović, M., Samardžić, M., Malatesti, N., Sak-Bosnar, M., 2011. Electroanalytical characterization of a copper(II)-rutin complex. Int. J. Electrochem. Sci. 6, 1075–1084.

Meesungnoen, O., Nakbanpote, W., Thiwthong, R., Thumanu, K., 2012. Zinc and cadmium resistance mechanism of *Pseudomonas aeruginosa* PDMZnCd2003. Curr. Res. J. Biol. Sci. 7, 4–13.

Mehta, S.K., Gaur, J.P., 1999. Heavy metal-induced proline accumulation and its role in ameliorating metal toxicity in *Chlorella vulgaris*. New Phytol. 143, 253–259.

Michalak, A., 2006. Phenolic compounds and their antioxidant activity in plants growing under heavy metal stress. Pol. J. Environ. Stud. 15 (4), 523–530.

Miller, R.M., Jastrow, J., 1990. Hierarchy of root and mycorrhizal interactions with soil aggregation. Soil Biol. Biochem. 22, 579–584.

Mobin, M., Khan, N.A., 2007. Photosynthetic activity, pigment composition and antioxidative response of two mustard (*Brassica juncea*) cultivars differing in photosynthetic capacity subjected to cadmium stress. J. Plant Physiol. 164, 601–610.

Mongkhonsin, B., Nakbanpote, W., Hokura, A., Nuengchamnong, N., Maneechai, S., 2016. Phenolic compounds responding to zinc and/or cadmium treatments in *Gynura pseudochina*(L.) DC. extracts and biomass. Plant Physiol. Biochem. 109, 549–560.

Munjit, R., Nakbanpote, W., Sangdee, A., 2015. Screening of zinc and cadmium tolerance and plant-growth promoting rhizobacteria from *Zea mays* L. rhizosphere. Proceeding of the 27th Annual Meeting of the Thai Society for Biotechnology and International Conference, November 17–20, 2015. Mandarin Hotel Bangkok by Centre Point, Bangkok, Thailand.

Muradoglu, F., Gundogdu, M., Ercisli, S., Encu, T., Balta, F., Jaafar, H.Z.E., Zia-Ul-Haq, M., 2015. Cadmium toxicity affects chlorophyll a and b content, antioxidant enzyme activities and mineral nutrient accumulation in strawberry. Biol. Res. 48, 1–7.

Nakbanpote, W., Panitlertumpai, N., Sukadeetad, K., Meesungneon, O., Noisa-nguan, W., 2010. Advances in phytoremediation research: a case study of *Gynura pseudochina*(L.) DC. In: Fürstner, I. (Ed.), Advanced Knowledge Application in Practice. SCIYO, Rijeka, Croatia, pp. 353–378.

Namdjoyan, S.H., Khavari-Nejad, R.A., Bernard, F., Nejadsattari, T., Shaker, H., 2011. Antioxidant defense mechanisms in response to cadmium treatments in two safflower cultivars. Russ. J. Plant Physiol. 58, 467–477.

Nies, D.H., 1999. Microbial heavy-metal resistance. Appl. Microbiol. Biotechnol. 51, 730–750.

Nordberg, G.F., Kjellstrom, T., Nordberg, M., 1985. Kinetics and metabolism. In: Cadmium and Health: A Toxicological and Epidemiological Appraisal. CRC Press, Boca Raton, pp. 103–178.

Nouairi, I., Ammar, W.B., Youssef, N.B., Miled, D.D.B., Ghorbal, M.H., Zarrouk, M., 2009. Antioxidant defense system in leaves of Indian mustard (*Brassica juncea*) and rape (*Brassica napus*) under cadmium stress. Acta Physiol. Plant. 31, 237–247.

Okema, A., Southway, C., Stirka, W.A., Street, R.A., Finniea, J.F., Van Stadena, J., 2014. Heavy metal contamination in South African medicinal plants: a cause for concern. S. Afr. J. Bot. 93, 125–130.

Panitlertumpai, N., Nakbanpote, W., Sangdee, A., Thumanu, K., Izumi Nakai, I., Hokura, A., 2013. Zinc and/or cadmium accumulation in *Gynura pseudochina*(L.) DC. studied *in vitro* and the effect on crude protein. J. Mol. Struct. 1036, 279–291.

Pereira, G.J.G., Molina, S.M.G., Lea, P.J., Azevedo, R.A., 2002. Activity of antioxidant enzymes in response to cadmium in *Crotalaria juncea*. Plant Soil 239, 123–132.

Pérez, J., Muñoz-Dorado, J., de la Rubia, T., Martínez, J., 2002. Biodegradation and biological treatments of cellulose, hemicellulose and lignin: an overview. Int. Microbiol. 5, 53–63.

Pietta, P.G., 2000. Flavonoids as antioxidants. J. Nat. Prod. 63, 1035–1042.

Posmyk, M.M., Kontek, R., Janas, K.M., 2009. Antioxidant enzymes activity and phenolic compounds content in red cabbage seedlings exposed to copper stress. Ecotoxicol. Environ. Saf. 72, 596–602.

Potters, G., Pasternak, T.P., Guisez, Y., Palme, K.J., Jansen, M.A.K., 2007. Stress-induced morphogenic responses: growing out of trouble? Trends Plant Sci. 12 (3), 98–105.

Prasad, M.N.V., 1995. Cadmium toxicity and tolerance in vascular plants. Environ. Exp. Bot. 35 (4), 525–545.

Prasad, M.N.V., 2011. Description of mechanisms involved in bioremediation. In: Prasad, M.N.V. (Ed.), A State-of-the-Art Report on Bioremediation, Its Applications to Contaminated Sites in India. Ministry of Environment & Forests Government of India, Hyderabad, India, pp. 7–22.

Prasad, M.N.V., Nakbanpote, W., Sebastian, A., Panitlertumpai, N., Phadermrod, C., 2014. Phytomanagement of padaeng zinc mine waste, Mae Sot District, Tak Province, Thailand. In: Hakeem, K.R., Sabir, M., Ozturk, M., Murmut, A. (Eds.), Soil Remediation and Plants: Prospects and Challenges. Elsevier, London, pp. 661–687 (Chapter 23).

Prasad, M.N.V., Nakbanpote, W., Phadermrod, C., Rose, D., Suthari, S., 2016. Mulberry and vetiver for phytostabilizaion of mine over burden: cogeneration of economic products. In: Prasad, M.N.V. (Ed.), Bioremediation and Bioeconomy. Elsevier, Amsterdam, pp. 295–328.

Pugnaire, F.I., 2010. Positive Plant Interactions and Community Dynamics. CRC Press, Florida, p. 156.

Qiu, R.L., Thangavel, P., Hu, P.J., Senthilkumar, P., Ying, R.R., Tang, Y.T., 2011. Interaction of cadmium and zinc on accumulation and sub-cellular distribution in leaves of hyperaccumulator *Potentilla griffithii*. J. Hazard. Mater. 186, 1425–1430.

Rajkumar, M., Ae, N., Prasad, M.N.V., Freitas, H., 2010. Potential of siderophore-producing bacteria for improving heavy metal phytoextraction. Trends Biotechnol. 28, 142–149.

Rajkumar, M., Sandhya, S., Prasad, M.N.V., Freitas, H., 2012. Perspectives of plant-associated microbes in heavy metal phytoremediation. Biotechnol. Adv. 30, 1562–1574.

Raskin, I., Ensley, B.D., 2000. Phytoremediation of Toxic Metals: Using Plants to Clean up the Environment. John Wiley and Sons, New York.

Redon, P.O., Béguiristain, T., Leyval, C., 2008. Influence of Glomus intraradices on Cd partitioning in a pot experiment with *Medicago truncatula* in four contaminated soils. Soil Biol. Biochem. 40, 2710–2712.

Rohn, S., Rawel, H.M., Kroll, J., 2002. Inhibitory effects of plant phenols on the activity of selected enzymes. J. Agric. Food Chem. 50 (12), 3566–3571.

Roychoudhury, A., Basu, S., Sengupta, D.N., 2012. Antioxidants and stress-related metabolites in the seedlings of two Indica rice varieties exposed to cadmium chloride toxicity. Acta Physiol. Plant. 34, 835–847.

Sakihama, Y., Cohen, M.F., Grace, S.C., Yamasaki, H., 2002. Plant phenolic antioxidant and prooxidant activities: phenolics-induced oxidative damage mediated by metals in plants. Toxicology 177, 67–80.

Sanitá di Toppi, L., Gabbrielli, R., 1999. Response to cadmium in higher plants. Environ. Exp. Bot. 41, 105–130.

Schat, H., Sharma, S.S., Vooijs, R., 1997. Heavy metal-induced accumulation of free proline in a metal-tolerant and a non-tolerant ecotype of *Silene vulgaris*. Physiol. Plant. 101, 477–482.

Sebastian, A., Panitlertumpai, N., Nakbanpote, W., Prasad, M.N.V., 2016. Fate of trace elements in rice paddies. In: Rinklebe, J., Knox, A.S., Paller, M. (Eds.), Trace Elements in Waterlogged Soils and Sediments. CRC Press, Florida, pp. 313–328.

Shafi, M., Bakht, J., Hassan, M.J., Raziuddin, M., Zhang, G., 2009. Effect of cadmium and salinity stresses on growth and antioxidant enzyme activities of wheat (*Triticum aestivum* L.). Bull. Environ. Contam. Toxicol. 82, 772–776.

Shanker, A.K., Cervantes, C., Loza-Tavera, H., Avudainayagam, S., 2005. Chromium toxicity in plants. Environ. Int. 31 (5), 739–753.

Sharma, S.S., Dietz, K.J., 2006. The significance of amino acids and amino acid-derived molecules in plant responses and adaptation to heavy metal stress. J. Exp. Bot. 57, 711–726.

Sharma, S.S., Dietz, K.J., 2009. The relationship between metal toxicity and cellular redox imbalance. Trends Plant Sci. 14 (1), 43–50.

Sharma, S.S., Schat, H., Vooijs, R., 1998. *In vitro* alleviation of heavy metal-induced enzyme inhibition by proline. Phytochemistry 49, 1531–1535.

Sheng, X., He, L., Wang, Q., Ye, H., Jiang, C., 2008. Effects of inoculation of biosurfactant-producing*Bacillus* sp. J119 on plant growth and cadmium uptake in a cadmium-amended soil. J. Hazard. Mater. 155, 17–22.

Simmons, R.W., Pongsakul, P., Saiyasitpanich, D., Klinphoklap, S., 2005. Elevated levels of cadmium and zinc in paddy soils and elevated levels of cadmium in rice grain downstream of a zinc mineralized area in Thailand: implications for public health. Environ. Geochem. Health 27, 501–511.

Siripornadulsil, S., Siripornadulsil, W., 2013. Cadmium-tolerant bacteria reduce the uptake of cadmium in rice: potential for microbial bioremediation. Ecotoxicol. Environ. Saf. 94, 94–103.

Siripornadulsil, S., Traina, S., Verma, D.P.S., Sayre, R.T., 2002. Molecular mechanisms of proline-mediated tolerance to toxic heavy metals in transgenic microalgae. Plant Cell 14, 2837–2847.

Sivaci, A., Sivaci, E.R., Sökmen, M., 2007. Changes in antioxidant activity, total phenolic and abscisic acid constituents in the aquatic plants *Myriophyllum spicatum* L. and *Myriophyllum triphyllum* Orchard exposed to cadmium. Ecotoxicology 16, 423–428.

Sivaci, A., Elmas, E., Gümüş, F., Sivaci, E.R., 2008. Removal of cadmium by *Myriophyllum heterophyllum* Michx. and *Potamogeton crispus* L. and its effect on pigments and total phenolic compounds. Arch. Environ. Contam. Toxicol. 54, 612–618.

Souza, R.F.V., Giovani, W.F., 2004. Antioxidant properties of complexes of flavonoids with metal ions. Redox Rep. 9, 97–104.

Souza, L.A., Andrade, S.A.L., Souza, S.C.R., Schiavinato, M.A., 2012. Arbuscular mycorrhiza confers Pb tolerance in *Calopogonium mucunoides*. Acta Physiol. Plant. 34, 523–531.

Stern, K.R., 2006. Introductory Plant Biology, 10th ed. McGraw-Hill Higher Education, Boston, p. 624.

Sukadeetad, K., Nakbanpote, W., Heinrich, M., Nuengchamnong, N., 2018. Effect of drying methods and solvent extraction on the phenolic compounds of *Gynura pseudochina*(L.) DC. leaf extracts and their anti-psoriatic property. Ind. Crop. Prod. 120, 34–46.

Sytar, O., Kumar, A., Latowski, D., Kuczynska, P., Strzałka, K., Prasad, M.N.V., 2013. Heavy metal-induced oxidative damage, defense reactions, and detoxification mechanisms in plants. Acta Physiol. Plant. 35, 985–999.

Takács, T., Vörös, I., 2003. Effect of metal non-adapted arbuscular mycorrhizal fungi on Cd, Ni and Zn uptake by ryegrass. Acta Agron. Hung. 51, 347–354.

Terry, N., Zayed, A.M., 1994. Selenium volatilization in plants. In: Frankenberger, W.T., Benson, S. (Eds.), Selenium in the Environment. Dekker, New York, pp. 343–367.

Tester, M., Leigh, R.A., 2001. Partitioning of nutrient transport processes in roots. J. Exp. Bot. 52, 445–457.

Tewari, R.K., Kumar, P., Sharma, P.N., 2008. Morphology and physiology of zinc-stressed mulberry plants. J. Plant Nutr. Soil Sci. 171, 286–294.

Vanholme, R., Demedts, B., Morreel, K., Ralph, J., Boerjan, W., 2010. Lignin biosynthesis and structure. Plant Physiol. 153, 895–905.

Verslues, P.E., Sharma, S., 2010. Proline metabolism and its implications for plant-environment interaction. Arabidopsis Book 8. e0140.

Vollenweider, P., Cosio, C., Madeleine, S., Gunthardt-Goerg, M.S., Keller, C., 2006. Localization and effects of cadmium in leaves of a cadmium-tolerant willow (*Salix viminalis* L.). Part II. Microlocalization and cellular effects of cadmium. Environ. Exp. Bot. 58, 25–40.

Wang, F., Lin, X., Yin, R., 2005. Heavy metal uptake by arbuscular mycorrhizas of *Elsholtzia splendens* and the potential for phytoremediation of contaminated soil. Plant Soil 269, 225–232.

Wang, W., Cang, L., Zhou, D., Yu, Y., 2017. Exogenous amino acids increase antioxidant enzyme activities and tolerance of rice seedlings to cadmium stress. Environ. Prog. Sustain. Energy 36, 155–161.

Watcharamai, T., Saenton, S., 2013. Leaching of heavy metals from Mae Tao Watershed's agricultural soils, Mae Sot District, Tak Province. Proceeding International Graduate Research Conference 2013, Chiang Mai University, Thailand, pp. ST50–ST54.

Wei, Y., Guo, M., 2014. Zinc-binding sites on selected flavonoids. Biol. Trace Elem. Res. 161, 223–230.

Wenzel, W.W., Lombi, E., Adriano, D.C., 2004. Root and rhizosphere process in metal hyperaccumulation and phytoremediation technology. In: Prasad, M.N.V. (Ed.), Heavy Metal Stress in Plants from Biomolecules to Ecosystems. second ed. Springer-Verlag, Berlin Heidelberg, pp. 313–344.

Wójcik, M., Vangronsveld, J., Tukiendorf, A., 2005. Cadmium tolerance in *Thlaspi coerulescens* I. Growth parameters, metal accumulation and phytochelatin synthesis in response to cadmium. Environ. Exp. Bot. 53, 151–161.

Yamada, M., Morishita, H., Urano, K., Shiozaki, N., Yamaguchi-Shinozaki, K., Shinozaki, K., Yoshiba, Y., 2005. Effects of free proline accumulation in petunia under drought stress. J. Exp. Bot. 56, 1975–1981.

Zagorchev, L., Seal, C.E., Kranner, I., Odjakova, M., 2013. A central role for thiols in plant tolerance to abiotic stress. Int. J. Mol. Sci. 14 (4), 7405–7432.

Zagoskina, N.V., Goncharuk, E.A., Alyavina, A.K., 2007. Effect of cadmium on the phenolic compounds formation in the callus cultures derived from various organs of the tea plant. Russ. J. Plant Physiol. 4, 237–243.

Zhang, F.Q., Wang, Y.S., Lou, Z.P., Dong, J.D., 2007. Effect of heavy metal stress on antioxidative enzymes and lipid peroxidation in leaves and roots of two mangrove plant seedlings (*Kandelia candel* and *Bruguiera gymnorrhiza*). Chemosphere 67 (1), 44–50.

## FURTHER READING

Márquez-García, B., Ferńndez, M.A., Córdoba, F., 2009. Phenolics composition in *Erica* sp. differentially exposed to metal pollution in the Iberian Southwestern Pyritic Belt. Bioresour. Technol. 100, 446–451.

Mourato, M.P., Moreira, I.N., Leitão, I., Pinto, F.R., Sales, J.R., Martins, L.L., 2015. Effect of heavy metals in plants of the genus *Brassica*. Int. J. Mol. Sci. 16, 17975–17998.

Page, V., Feller, U., 2015. Heavy metals in crop plants: transport and redistribution processes on the whole plant level. Agronomy 5, 447–463.

# CHAPTER 5

# Cadmium-Induced Anatomical Abnormalities in Plants

**Kavita Shah*, Sareeta Nahakpam[†,‡], Vyomendra Chaturvedi*, Prerna Singh[‡]**
*Institute of Environment and Sustainable Development (IESD), Banaras Hindu University, Varanasi, India
†Department of Biochemistry, Institute of Science, Banaras Hindu University, Varanasi, India
‡Department of Biochemistry and Crop Physiology, Bihar Agricultural University, Bhagalpur, India

## Abstract

The continuous addition of the heavy metal cadmium (Cd) in the biosphere results from industrial processes, from use of fertilizers or other crop protection agents, from municipal waste, and through atmospheric fallout. To ensure the protection of crops from this dangerous abiotic stress, it is important to understand and assess the availability, accumulation, uptake, and localization of cadmium in crop plants, in addition to its toxic responses. The uptake of Cd by plants results in various anatomical, histochemical, and biochemical alterations inside the plant cells. The Cd-induced toxicity resulting from its uptake and distribution in plants is dependent largely upon its uptake concentration in the plant and plant species, and even among varieties of the same species. Cadmium localization varies in different plants, causing anatomical abnormalities within the plant cells. This chapter is an update on the uptake, localization, and distribution of cadmium and the corresponding anatomical changes observed in crops grown in Cd-rich soil. This chapter also discusses the major abnormalities in different plant species based on the amount of Cd uptake and localization, in particular.

**Keywords:** Anatomy, Cadmium, Crop, Heavy metal, Histochemical, Localization, Plant

## 1 INTRODUCTION

Cadmium is widespread in soil, water, and the atmosphere, with primary sources being waste incinerators, the cement and metallurgic industries, phosphate fertilizers, automobile exhaust, and heating systems (Gallego et al., 2012). Higher Cd toxicity is due to its solubility in water and high mobility in organisms (Singh and Shah, 2014). Bioaccumulation in the food chain is dangerous, leading to diseases based on the solubility of cadmium increases via root exudates, which make it more available for plants (Tudoreanu and Phillips, 2001). It is the genetic characteristics of plants,


https://doi.org/10.1016/B978-0-12-814864-8.00005-X

and the availability of Cd in the soil that determine the absorption of cadmium in agriculture crop roots (Roberts, 2014). Due to difficulty in controlling environmental Cd-accumulation, rice crops grown in Cd contaminated soil tend to accumulate Cd, resulting in dietary Cd intake in rice-eating populations (Meharg et al., 2013). The cadmium phytotoxicity is caused by Cd-induced oxidative stress that affects nucleic acids, proteins, and lipids; thereby causing growth inhibition, or even cell death (Loix et al., 2017).

Plants exposed to varied concentrations of Cd have varied responses, and induce anatomical abnormalities. Exposure of plants to varied Cd doses ranging between 1 μM and 1 mM is shown in Table 1. Different plant species show varied effects of these cadmium doses on growth, depending on the duration of exposure and type of plant tissue under study. This is a consequence of the tissue-specific and time-specific amount of Cd uptake, and its location and distribution within the plant parts. For the past few decades, researchers have studied the effect of varied concentrations (1 μM–1 mM) of cadmium on the growth of different plant species, which is shown in Table 1. Different doses of Cd in the soil have different impacts on the growth of plant species, which largely depends on the type of plant under study.

Most plants exhibit visible Cd toxicity symptoms such as stunted growth, chlorosis, necrosis, root browning, and even death (Gratão et al., 2009; Chang et al., 2013). Being highly mobile in phloem, Cd can accumulate in any part of the plant, causing stunted growth, chlorosis, and leaf epinasty (Gallego et al., 2012). Excess Cd accumulation in plants can profoundly interfere with a series of physiological processes, such as photosynthesis and respiration, uptake, transport, and assimilation of mineral nutrients and uptake of water. In addition, Cd stress alters gene and protein expression, induces or inhibits enzymes, enhances reactive oxygen species (ROS) accumulation and lipid peroxidation, and disturbs metabolisms. Cadmium toxicity is often described as decreased root length and dry mass, as well as enlarged root diameter (Gratão et al., 2009). Cadmium inhibits seed imbibition and reduces water content in seedlings, thereby impacting germination and plant stands. For a seed to germinate, the water potential of the embryo needs to meet a critical threshold. A hydrothermal time model showed that seed water potential was closely correlated with the changes in germination speed and rate, and affected the speed of dormancy alleviation (Alvarado and Bradford, 2005). It also induces biochemical changes in roots and leaves, such as lignification of cell walls in root tissues and leaf main

**Table 1** An update effect of varied duration and cadmium doses on growth of different plant species grown under hydroponics, sand cultures, or soil from literature (2010–16)

| Growth condition | Plant species | Level of cadmium exposure | Changes due to Cd toxicity | References |
|---|---|---|---|---|
| **(A)** Grown under hydroponics | | | | |
| **1.** Under hydroponics is for less than a week | *Helianthus annuus* | 20 μM | Root and leaf fresh weight, chlorophyll and carotenoid content | Saidi et al. (2014) |
| | *Hordeum vulgare* | 50 μM | Meristem, elongation and mature zone and Cd uptake sites. Cytoplasm was Cd storage sites | Shi et al. (2016) |
| | *Populus nigra* | 50 μM | Decreased quantum yields of PSII and protein contents | Lomaglio et al. (2015) |
| **2.** Under hydroponics for a week (7 days) | *Brassica juncea* | 50 or 200 μM | Shoot length, root and shoot fresh weight, and chlorophyll *a* and *b* content. Chlorosis also observed | Mohamed et al. (2012) |
| | *Ceratopteris pteridoides* | 5, 10, 20, and 40 μM | Root mass, leaf mass, total mass, and leaf width | Deng et al. (2014) |
| | *Daucus carota* | 36 μM | Increased lateral root formation, cell-to-cell separation events leading to schizogenous oil duct formation | Sanità di Toppi et al. (2012) |
| | *Gossypium hirustum* L. | 25, 50, and 100 μM | Growth inhibition, plant height, biomass and leaf area decreased; chlorophyll reduction, PSII damage, gas exchange, chlorophyll flourescence | Liu et al. (2014) |

*Continued*

**Table 1** An update effect of varied duration and cadmium doses on growth of different plant species grown under hydroponics, sand cultures, or soil from literature (2010–16)—cont'd

| Growth condition | Plant species | Level of cadmium exposure | Changes due to Cd toxicity | References |
|---|---|---|---|---|
| | *Groenlandia densa* | 0–97 μM | Photosynethic pigmentation, chlorophyll *a* and *b*, total chlorophyll ratios | Yilmaz and Parlak (2011) |
| | *Hibiscus cannabinus* | 30 μmol/L | Root and shoot length, and root and shoot biomass | Li et al. (2013) |
| | *Phaseolus vulgaris* | 1 mM | Root and shoot length, leaf area per plant, and plant dry mass, number of pods per pot, pod yield per pot, and pod protein | Howladar (2014) |
| 3. Under hydroponics for 2 weeks (8–15 days) | *Solanum lycopersicum* | 10 μmol/L | Phytosynthetic pigmentation | Cherif et al. (2011) |
| | *Solanum lycopersicum* | 0.5 mM | Plant growth | Monteiro et al. (2011) |
| | *Arabidopsis thaliana* | 50–100 μM | Chlorophyll and carotenoid levels | Martínez-Peñalver et al. (2012) |
| | *Brassica juncea* | 483 and 966 μM | Root and shoot fresh weight | Ahmad et al. (2011) |
| | *Brassica napus* | 50 μM; 100 and 500 mM | Plant height, root length leaf area, and number of leaves; fresh and dry biomass, root diameter, root surface area, number of root tips, and root volume | Ali et al. (2013) and Ehsan et al. (2014) |
| | *Hordeum vulgare* | 5 μM | Plant height, root length, and biomass. Leaf necrosis and root browning were also observed | Chen et al. (2010) |
| | *Juncus effusus* | 200–500 μM | Fresh and dry biomass | Najeeb et al. (2011) |

| | | | | |
|---|---|---|---|---|
| | *Nicotiana tabacum* | 100 and 500 μM | Decreased plant growth, chlorophyll, nitrates and iron content | Iannone et al. (2015) |
| | *Pteris vittata* | 30, 60, and 100 μM | Size reduction in root apical meristem, decrease in root hairs at higher concentration, reddish precipitates localized in vacuoles inside root hairs | Balestri et al. (2014) |
| | *Sedum alfredii* | 400 μM | Root elongation and plasma membrane integrity | Tian et al. (2011) |
| | *Tagetes patula* | 10, 25, and 50 μM | Root and shoot fresh and dry weight, biomass, growth rate, and chlorophyll content | Liu et al. (2011) |
| **(B)** Grown under sand cultures for more than 2 weeks | *Gossypium hirsutum* L. | 0.5, 1, 5, 10, and 20 mg/kg | Decrease in plant height, white Cd deposits in roots, stem, root, leaf transpiration affected | Chen et al. (2015) |
| | *Hordeum vulgare* | 0, 3, 6, 9, and 12 mg/L | More accumulation in roots than leaves, decrease in fresh biomass, endoderm cell wall thickening and lignification, lack of passage cells in endodermis, pericycle cell wall thickening | Alle et al. (2016) |

*Continued*

**Table 1** An update effect of varied duration and cadmium doses on growth of different plant species grown under hydroponics, sand cultures, or soil from literature (2010–16)—cont'd

| Growth condition | Plant species | Level of cadmium exposure | Changes due to Cd toxicity | References |
|---|---|---|---|---|
| | *Lactuca sativa* | 1, 2, 4, and 6 mg/L | More accumulation in leaves, decrease in fresh biomass, necrosis of leaves, chlorotic spots | Alle et al. (2016) |
| | *Oryza sativa* | 10, 50, 100, and 500 µM | Accumulation in roots, deposition in parenchyma cells, increase protease activity, nucleoli prominence of root cell, root browning | Shah et al. (2013) |
| **(C)** Grown in soil for more than 4 weeks (1 month) | *Solanum tuberosum* | 1, 5, and 25 mg/kg | High accumulation in roots and leaves, decrease chlorophyll content (i.e., Chl *a* and Chl *b*) and carotenoid (Cx + *c*), white deposits in the tubers, increased malondialdehyde (MDA) content | Xu et al. (2013) |
| | *Triticum aestivum* | 200 mg/kg | Reduced Rubisco activity, chlorophyll contents, gas exchange parameters, leaf nitrogen content and plant growth | Khan et al. (2015) |

veins, alters the chloroplast ultrastructure, and reduces net photosynthetic rate, stomatal conductance, and leaf transpiration (Souza et al., 2011).

Cadmium can cause DNA damage/oxidation indirectly via ROS production. The hydroxyl radical ($HO^{\bullet}$) is highly electrophilic and can oxidize DNA by reacting with pyrimidine aromatic rings. The $HO^{\bullet}$ radical preferentially attacks position 5 of cytosine and thymine and can also attack the hydrogen atom of the thymine methyl group, resulting in the formation of an allylic radical (which can in turn form peroxyl radicals by reacting with oxygen) (Nzengue et al., 2011).

It has been reported that Cd induces premature senescence in the leaves of several plant species, including wheat, peas, and *Elodea canadensis*. The senescence of leaves is mainly characterized by cessation of photosynthesis, disintegration of organelle structures, an increase in lipid peroxidation, and membrane damage. Herein, an attempt is made to present and update on the Cd uptake and distribution in crop plants, together with the corresponding anatomical changes, to help explain the abnormalities that develop in plants grown in Cd rich soil.

## 2 CADMIUM IN THE ENVIRONMENT

Cd is extremely rare in the earth's crust. It is a silvery-white, and lustrous, but tarnishable soft and ductile metal in the earth's crust. It has a relatively high vapor pressure (USGS, 2001). It is present in Zn ores through isomorphic replacement, and is usually divalent. Geochemically, cadmium behaves in a similar way to zinc. It is the 65th most abundant element in the earth's crust, with a lithosphere concentration of nearly 0.1–0.2 mg/kg. The only known Cd-based ore is greenockite (CdS) (Shahid et al., 2017). Cd is used in electroplating, paints, pigments, plastics, silver-cadmium batteries; and for coating machinery, photography, and television phosphors (Naja and Volesky, 2009); and in atmospheric depositions, wastewater and industrial discharge, and pesticide and insecticide applications (Rao and Kashifuddin, 2016; Shahid et al., 2017). Naturally, Cd in the earth's crust is mobilized through rock-weathering or volcanic activities (Choppala et al., 2014; Zhao et al., 2015). There are reports that Cd mostly occurs as guest metals with lead coal and zinc (Baker et al., 1990). Geogenic Cd is largely present due to black shale exposed to air and water (Liu et al., 2017). Significant Cd released into surface environments causes Cd contamination in soil water and food crops.

Cd never occurs in isolation in natural environments, but mostly as a guest metal in Pb:Zn mineralization. A global natural concentration of cadmium in soil lies in the range 0.07–1.1 mg/kg, which may vary with soil type, or soils impacted by Cd phosphate fertilizers; apart from soil and coal, Cd concentration in the different parts of the earth's crust generally does not exceed 0.6 g/kg (Liu et al., 2012). Cadmium concentrations higher than 10 mg Cd/kg in soil can be toxic to plants. Varied Cd concentrations among soils reflect the differences in the concentrations in the parent material from which these soils are formed. It is recommended that soils containing >1 mg Cd/kg dry soil not be used for agricultural or horticultural purposes so that the risk of elevated Cd concentration and its phytotoxicity is minimized (Shahid et al., 2017).

The residence time of Cd in soils of temperate climate is estimated to be 75 to 380 years (Kabata-Pendias and Henryk, 2001). Workers reported that cadmium usually averages 0.2 mg/kg in the lithosphere; whereas surface soil and plant food stuffs have 0.53 and <0.66 mg/kg dry weight, respectively. In soils of tropical rainforests, the rate of leaching of the elements is much shorter, and is calculated at about 40 years. All similar estimations have clearly indicated that the complete removal of metallic contaminants from soils is nearly impossible. The distribution of Cd in forest ecosystems is of special environmental concern. Cd is likely to be concentrated in the outer parts of conifer trees (spruce and pine), phloem, and bark (up to 1.3 ppm), which leads to its accumulation in wood ash up to 30 ppm.

## 3 UPTAKE OF CADMIUM IN PLANTS

Soil properties such as soil particle size, temperature, pH, and cation exchange capacity affect Cd phyto uptake. In addition, plant physiology, such as the root surface area, rate of root exudation, and transpiration also play an important role in the mobilization of Cd inside the plant. The majority of plant species accumulate Cd in root tissues, with only very small portions being transferred to the shoots (Podazza et al., 2012).

Direct uptake of Cd from the atmosphere by higher plants is very low compared with that from soil and water. The uptake depends on Cd availability and its concentration, and several plant species are known to have a high tolerance to this toxic element. Such plants involve different mechanisms, such as active excretion, metal exclusion, restricted distribution in tissues, binding of metal to the cell wall, organic chelation, and localization in vacuoles to avoid, neutralize, or adapt to Cd metal (Shah and Nahakpam, 2012; Gallego et al., 2012).

Mclaughlin et al. (2011) reported an increasing order of Cd accumulation from soil into vegetable species such as French beans, beetroots, radishes, peas, carrots, broccoli < potatoes, tomatoes, zucchini, and sweet corn < onions, leeks, parsnips, < turnips < cabbage, kale < lettuce, and spinach; whereas Rosén et al. (2012) suggested considerable Cd accumulation in tobacco leaves. Cd is reported to accumulate in the seeds or grains of Indian mustard (Qureshi et al., 2010), and in tuberous plants such as potatoes (Baghour et al., 2001). How different plant species relate to metal accumulation, compartmentalization, and detoxification is still under intensive investigation. Different workers have reported changes in morphological traits of plants, altered internal structures of roots, and changes in xylem pipes under heavy metal stress, and have suggested that such changes are species- and tissue-specific (Hose et al., 2001; Enstone et al., 2002; Hermle et al., 2006; Probst et al., 2009; Lux et al., 2011a). Although the response of plants to Cd treatment has been intensively studied, the knowledge concerning root anatomical structure in fixing and transport of metal ions is unclear. The physiological anatomical and physical mechanism underlying prevention of Cd uptake by the roots and subsequent transport to xylem, therefore, needs extensive study to help understand the particular role played by root structures in monocotyledons and dicotyledons (Lux et al., 2011a,b).

Anatomical structures and changes therein are important to understand in the uptake of metal ions because studies in monocots and dicots have indicated the accumulation of different amounts of Cd from soil (Martin et al., 2006; Osvalde and Paegle, 2007; Abe et al., 2008; Chang et al., 2014), suggesting that morphologically different inner cell layers of the cortex—root endodermis play an important role in the transport of soil solutions carrying the heavy metals inside the plant system. Accelerated endodermis and exodermis formation of hypodermal periderm, lignification of the cell wall, and formation of Cd chelates are known to occur in the presence of heavy metals (Hose et al., 2001; Enstone et al., 2002; Lux et al., 2011a,b; Zelko et al., 2008). Thickening of the abaxial and adaxial sclerenchyma, as well as pericycle tissues, in the presence of heavy metals reported in *Brachiaria decumbens* may be related to metal absorption in cell walls as an alternative pathway to prevent their translocation to photosynthetic tissues (Gomes et al., 2011). Similarly, roots of *Salix viminalis* grown in Cd-rich soil had thickened cell walls with a greater concentration of metals therein, as compared with other tissues (Vollenweider et al., 2006). In monocots, it is the casparian strips that hold up heavy metals (Gomes et al., 2011). The roots where secondary growth is absent in the endodermis are further differentiated and characterized with thickened and lignified casparian strips, and suberization followed by irregular thickening of endodermal cell wall. Monocots, largely, and a few dicots exhibit the irregular thickenings of the endodermis cell wall (Evert, 2006). Most of the field crops are monocots, bearing differentiated endodermises, and therefore it is quite likely that they play a substantial role in preventing a free penetration of pollutants through the apoplast; therefore, insights on differences in root system functioning and capability of heavy metal accumulation among the plant species can be obtained by gaining scientific knowledge about root structure modification. The mechanism of cadmium uptake by plants is represented in Fig. 1, and largely follows the following steps:

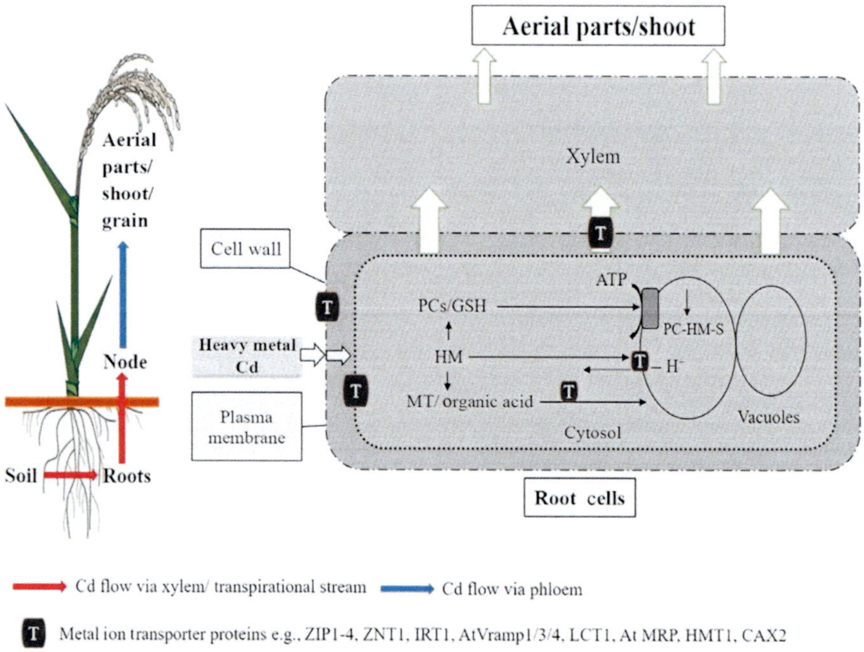


**Fig. 1** Schematic representation of the possible mechanisms of different subcellular organs for transportation, detoxification, and deposition of heavy metals in plant root cells. *(Based on Ahsan, N., Lee, B., Komatsu, S., 2011. Organellar proteomics: a high-throughput approach for better understanding of heavy metal accumulation and detoxification in plants. In: Sherameti, I., Varma, A. (Eds.), Detoxification of Heavy Metals. Springer, Berlin, Heidelberg, pp. 273–287; Shah, K., 2011. Cadmium metal detoxification and hyperaccumulators. In: Sherameti, I., Varma, A. (Eds.), Detoxification of Heavy Metals. Springer, Berlin, Heidelberg, pp. 181–203.)*

**(i)** Uptake of cadmium by the roots.
**(ii)** Transfer of cadmium from roots to shoots.
**(iii)** Transport of cadmium to grains.

## 3.1 Cadmium Uptake by Roots

Subcellular accumulation of Cd in plants varies greatly, not only among plant species, but also among cultivars and genotypes of the same species (Xin et al., 2013), and the plants' Cd detoxification and tolerance mechanisms. For most plants, Cd tends to accumulate in the roots, with only a small portion being translocated to aerial parts (Ahmad et al., 2011). Generally, Cd content in plants decreases in order of roots > stems > leaves > fruits > seeds (Benavides et al., 2005). Some species can sequester high concentrations of Cd in their roots with reduced translocation to shoots, or remobilize Cd from leaves to roots through the phloem (Mendoza-Cózatl et al., 2011). These are referred to as low Cd-accumulating plants. Recently, Zhou et al. (2015) reported 60%–90% accumulation of Cd in roots of 32 hybrid rice (*O. sativa*) cultivars. The concentration of Cd in plant roots can be tenfold higher than in the shoot (Wang et al., 2009), suggesting that roots serve as an effective barrier to Cd translocation. The efficacy of the root cell walls in trapping Cd is attributed to the number of available functional groups (e.g., hydroxyl, carboxyl) that are able to form complexes with trace metals (Nishizono et al., 1987). Cadmium accumulation in plant roots varies with plant type and rhizosphere Cd concentration. Similarly, Cd is mainly concentrated in the apical cells of plant roots at low soil Cd concentration, or in the proximal subapical region (Seregin et al., 2004). Cadmium mainly accumulates in nuclei and vacuoles, while a lesser amount of Cd may also be found in the plastids and cytoplasm of plant roots (Liu et al., 2007).

## 3.2 Cadmium Transfer From Roots to Shoots

Cadmium movement from roots to aerial tissues occurs via transpiration-driven xylem loading, with symplastic and/or apoplastic transport (Hasan et al., 2009; Kranner and Colville, 2011) of free $Cd^{2+}$, or Cd complexed with various chelates (Maestri et al., 2010; Lux et al., 2011a,b). Plants respond to exposure to toxic concentrations of Cd with the production of a class of Cd-binding polypeptides called phytochelatins. Phytochelatins are ubiquitous in the plant kingdom, and are believed to function as detoxifying agents for Cd by virtue of their ability to bind Cd, and thereby prevent it from reacting with the sulphydryl groups of vital enzymes and proteins. These compounds contain a high proportion of cysteine, and bind Cd by metal thiol coordination (Grant et al., 1998). Xylem loading plays an important role in Cd accumulation within aerial plant parts. Low Cd loading into the xylem sap in the roots has been associated with subsequent low Cd transport to aerial parts (Hasan et al., 2009; Verbruggen et al., 2009; Santos-Echeandía et al., 2010).

A number of important membrane ion transporter gene families identified and characterized in recent years are reported to be involved in Cd loading into xylem sap and transport. Some of the genes identified for Cd metal ion transport in plants are—ZIP1-4, ZNT1, IRT1, AtVramp1/3/4; LCT1, At MRP, HMT1 (in Cytosol), and CAX2 (in Vacuoles) (Shah, 2011). These gene families include the ATP binding cassette (ABC) superfamily, HMA (heavy metal ATPase), ZIP (ZRT, IRT-like protein), NRAMP (natural resistance-associated macrophage protein), YSL (yellow-stripe-like transporter), NAS (nicotinamine synthase), SAMS (*S*-adenosyl-methionine synthetase), FER (ferritin Fe (III) binding), CDF (cation diffusion facilitator), NRT (nitrate transporter), and IREG (iron regulated transporter) (Memon and Schröder, 2009; Maestri et al., 2010). Several studies have reported the involvement of P1B-ATPase, and also ZIP family transporters, in Cd transfer across the plasmalemma into the shoots (Hanikenne et al., 2008; Wong and Cobbett, 2009), using energy provided by ATP hydrolysis to transport Cd against its electrochemical gradient. The IRT1 gene, which is transcriptionally responsive to Fe deficiency, can also transport Cd to plant shoot tissues (Cohen et al., 1998).

Free Cd ions act as opportunistic hitchhikers and enter the root system through mechanisms that take advantage of transporters for essential nutrients such as Fe, Zn, and calcium (Ca). For example, the Fe depletion results in upregulation of Fe uptake systems, subsequently leading to a higher Cd influx as well (Wagner, 1993). The best-studied transporters include members of the ZIP (zinc regulated transporters, iron-regulated transporter-like protein) and the Nramp (natural resistance-associated macrophage protein) families. Cadmium enters root cells as $Cd^{2+}$ through ZIP (zinc-regulated transporters) transporters, such as orthologues of AtIRT1 and TcZNT1/TcZIP4, through orthologues of the wheat TaLCT1 transporter, or via cation channels (Verbruggen et al., 2009). Or it might also enter root cells as Cd-chelates by YSL (yellow-stripe1-like) proteins, which belong to the OPT superfamily of oligopeptide transporters (Curie et al., 2009).

In *N. caerulescens*, ZNT1 transporters mediate high-affinity Zn and low-affinity Cd uptake (Pence et al., 2000). Finally, it was shown that Cd is able to permeate *Arabidopsis thaliana* guard cells through Ca channels, thereby disrupting proper Ca signaling (Perfus-Barbeoch et al., 2002).

### 3.3 Cadmium Uptake in Rice Grains

The key Cd transport processes from soil to rice grains at first involve root uptake and xylem loading (root-to-shoot translocation), followed by redirection of transport through intervascular transfer at nodes and subsequent remobilization from leaf blades via phloem (Fig. 1). Most research is conducted on *Oryza sativa*, because it is both a model organism and a major source of Cd within the human diet (Clemens and Ma, 2016). There are seven members of the OsNramp family; however, the function of OsNramp2, OsNramp6, and OsNramp7 is still unknown (Wu et al., 2016). It was shown that OsNramp5 is a major transporter of both Mn and Cd into the interior of root cells (Sasaki et al., 2012). OsNramp1 could transport Fe and Cd in yeast, and is proposed to be involved in Cd accumulation in plants as well.

In rice, Cd absorbed from soils is finally transported into the stele after passing the endodermis and casparian strip with help from OsNramp5. In the stele, OsHMA2, an efflux-type metal transporter on the plasma membrane, is suggested to facilitate Cd and zinc (Zn) loading to the xylem. After Cd is transported from the roots to the shoots, phloem transport mediates Cd deposition into grains and governs grain Cd levels. Nodes (present in the culm of rice) control nutrient transport and distribution through complex vascular bundles that connect roots, leaves, and upper nodes. During grain maturation, most Cd transported through xylem is transferred to phloem at nodes, and this results in preferential Cd transport into upper nodes, and finally into grains rather than into leaf blades. In leaf blades, the stored Cd is remobilized through phloem, and transported into grains after redirection at the nodes (Uraguchi and Fujiwara, 2012).

## 4 LOCALIZATION OF CADMIUM IN PLANT CELLS

The heavy metal transportation, detoxification, and deposition in plant cells thus is dependent on the direct involvement of a number of organ-specific pathways, such as entering cells via their cell wall binding potential and entering the cytosol via the efflux pumping activity of the plasma membrane. Within the cytosol, heavy metals are transported, detoxified, and deposited through the involvement of several low molecular weight thiols, peptides, and organic acids; and thus transported or deposited, and become localized in the vacuoles. Fig. 2 shows the localization of cadmium in the cross-sections of different plants tissues exposed to Cd toxicity.

## 4.1 Cadmium (Cd) Localization in Tissues of Cotton (*Gossypium hirsutum* L.)

Chen et al. (2015) showed that cotton plants grew well and bore fibers when soil Cd concentrations were in the range of 0.26–20.26 mg/kg; however, beyond this, the growth of cotton plants were inhibited to a certain extent, and the plant height decreased during seedling and developing periods. In varied soil conditions, Cd concentrations of 0.26–1.26 mg/kg Cd BCFs (bioconcentration factor) in various cotton organs were leaf > stem > shell > root > seed > fiber, and TF values were > 1. Energy dispersive X-ray microanalysis indicated cotton leaf transpiration to play a significant role in extracting soil Cd, while roots and stems were the main compartments of Cd storage. Many white Cd deposits were found inside the intercellular space of secondary phloem in the roots, in the vessel and near the vascular cylinder (Fig. 2(i)). This indicated root periderm and secondary phloem to have an important part in the adsorption and reservation of Cd in cotton roots through the active or passive infiltration of metal ions. The work also revealed cotton leaves to have the greatest ability to accumulate Cd in soils, perhaps through transpiration, which would transport Cd from the root to the leaf via the stem. SEM-EDX (scanning electron microscopy and energy dispersive X-ray) microanalysis in the leaf further revealed Cd to be mainly distributed in the palisade cells adjacent to the upper epidermis, while less Cd was found in the spongy tissue near the lower epidermis. Fu et al. (2011) deduced that the vacuole were the predominant sink for Cd in cotton. Localization of Cd in root cells indicated Cd to likely combine with sulfur-rich peptides or other constituents (such as organic acids) of the cell sap in the vacuole of the cotton root (Monteiro et al., 2009).

## 4.2 Localization of Cadmium in Potatoes (*Solanum tuberosum* L.)

The subcellular distribution of Cd in different organs was analyzed in potato plants exposed to a Cd concentration of 25 mg/kg (Xu et al., 2013). Granular white deposits were observed inside the intercellular space, the interior surface of the plasma membrane, and on the surface of the elliptical starch granules in the tubers (Fig. 2(ii)). The localization of Cd in the tuber cells

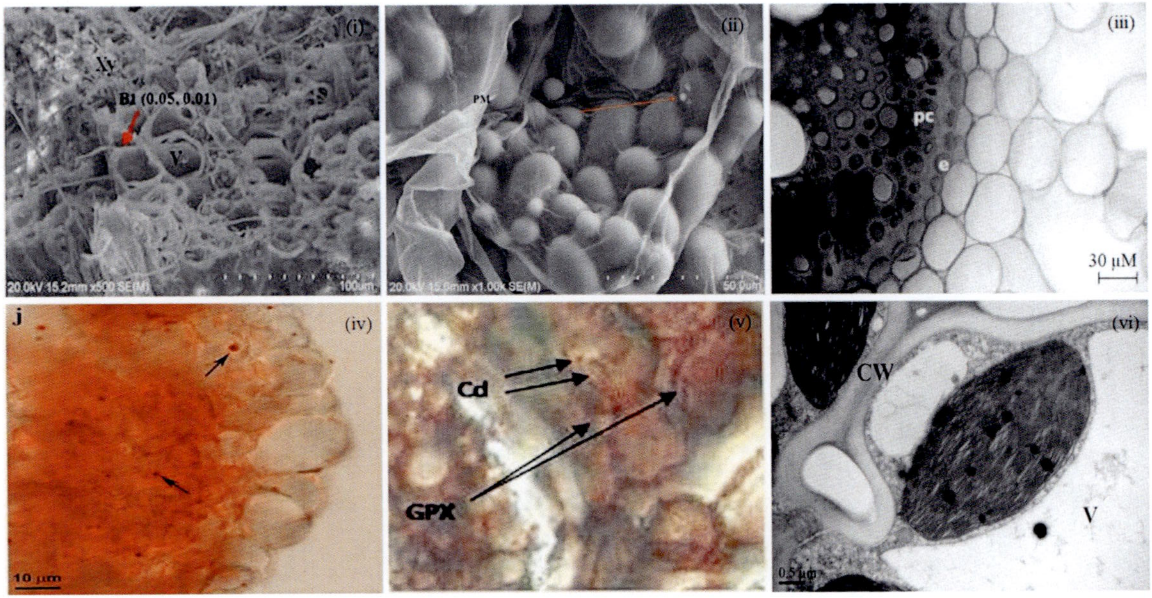


**Fig. 2** Cross sections of plant tissues from plants exposed to cadmium. (i) *Gossypium hirsurum* L. root exposed to 20.26 μM Cd (Chen et al., 2015). (ii) Stem of *Solanum tuberosum* L. exposed to 25 μM Cd (Xu et al., 2013). (iii) *Hordeum vulgare* L. root exposed to Cd scale bar 30 μM (Alle et al., 2016). (iv) Root hairs of *Pteris vittata* exposed to 100 μM Cd (Balestri et al., 2014). (v) *Oryza sativa* L. root exposed to 100 μM Cd (Shah et al., 2013). (vi) *Miscanthus floridulus* L. exposed to 10 μM Cd (Guo et al., 2016). *Cd*, Cadmium; *CW*, cell wall; *GPX*, peroxidase enzyme; *PC*, pericycle; *V*, vacuole; *Xy*, xylem.

indicated that the Cd was likely coordinated with the sulfur-rich peptides and organic acids of the cell sap in the vacuole of the tuber. A significant amount of Cd did not move from the root to the tuber because sulfhydryl binding or/and complexation to other constituents of the sieve tubes may easily reduce the mobility of Cd from the phloem to the tuber. However, Cd was detected in the cortex and the adjacent phloem, which suggests that a proportion of Cd might be transported from the xylem to the phloem. These researchers, when studying Cd in the phloem by EDX, determined that the spectrum revealed that the deposits of Cd contained sulfur (S) and phosphorus (P). Cd was found mainly distributed in the leaf tissues of the potato and outside of the leaf's stoma. The white deposits of Cd were detected in the spongy tissue, the palisade tissue, the stoma, and the adjacent cells outside of the stoma. These bindings should influence the physiological metabolism of the potato, such as the change of physiological parameters. A decrease in the chlorophyll content suggested the accumulation of Cd in the spongy tissue (Xu et al., 2013). They concluded that the complexation of Cd might be ascribed to two parts: Cd complexation with the products of photosynthesis, and Cd coordination with the compartments of the cell walls (including cellulose, hemicelluloses, and pectin). The Cd complexation could reduce the migration of Cd from the above-ground portion of the potato to the tuber, and then decrease the toxicity of Cd to the tuber of the potato.

## 4.3 Cadmium Localization in Barley *(Hordeum vulgare)*

The studies on changes of the root's anatomical structure in barely exposed to Cd (Alle et al., 2016) revealed the presence of lignified pericycle cell walls, and the lack of passage cells in the endodermis, which were otherwise present in control groups and proximal pericycle cell walls, which remained unlignified and maintained parenchymatic character (Fig. 2(iii)). This evidence suggests that passage cells of the endodermis allow the passing solutions heading toward the vascular cylinder by symplast to play an active role in ion uptake, and could be important for Cd uptake and transfer (Peterson and Enstone, 1996; Alle et al., 2016). Penetration of the soil solution is ensured by the transpiration in tracheary elements of the radial vascular bundle of the root, and the passage cells, in their turn, are located opposite to xylem elements, and ensure the penetration in the vascular cylinder only via the symplast in barley (Gomes et al., 2011).The EDX analysis of *Hordeum vulgare* for analyzing the localization of cadmium by Shi et al. (2016) revealed that the root hairs of barley play an important role in the acquisition of Cd from soil. Cd ion deposits were observed in elongation, meristem, and the mature zone in root tips, which indicated that these are the main sites of Cd uptake. From a transverse section point of view, the Cd level in the mature zone was high in comparison with the meristem zone, due to its perfect conducting system, whereas cytoplasm and the cell wall were the main sites of cadmium in the root tips of barley.

## 4.4 Localization of Cadmium in Ferns *Pteris vittata*

In situ localization of cadmium-treated *Pteris vittata* revealed the presence of Cd as reddish precipitates inside the root hairs, localized mainly in vacuoles (Balestri et al., 2014) (Fig. 2 (iv)). In the fronds, they were localized in several tissues: in the epidermal cells, including guard cells and scales, largely diffused on the dorsal surface of the pinna, and in the mesophyll cells close to vascular bundles. In *P. vittata,* Cd induced a change in the size of the root apical meristem, which became progressively shorter with increasing Cd concentration, suggesting that Cd influences mitotic activity. A change in the number of root hairs in *P. vittata* indicates Cd-induced changes in the differentiation pattern of some rhizodermal cells.

## 4.5 Localization of Cadmium in Rice (*Oryza sativa*)

A study conducted by Shah et al. (2013) on rice cultivars grown under cadmium revealed Cd uptake and localization in the root tissues as purple granules on the parenchyma cells below the exodermis (Fig. 2(v)). The precipitates of cadmium seen as purple stains were localized abundantly in vacuoles, but were difficult to distinguish in cell walls. Some small amounts of $Cd^{2+}$ deposits were also seen in the parenchyma cells of the aerenchyma. Peroxidase (GPX) were also present in the vascular cylinder, on the exodermal walls and the walls of the cortical cells (Fig. 2(v)). Cadmium was also found to be associated and located in the root elongation zone of the root hairs (Fig. 3) in the study.

## 4.6 Localization of Cd in *Miscanthus* spp.

Guo et al. (2016) studied the localization of cadmium in *Miscanthus* grown under 10 μM Cd levels. Microscopic studies revealed the presence of dense Cd granules in the cell wall, and chloroplast of *Miscanthus* leaves (Fig. 2 (vi)), in addition to its presence in the vacuoles.

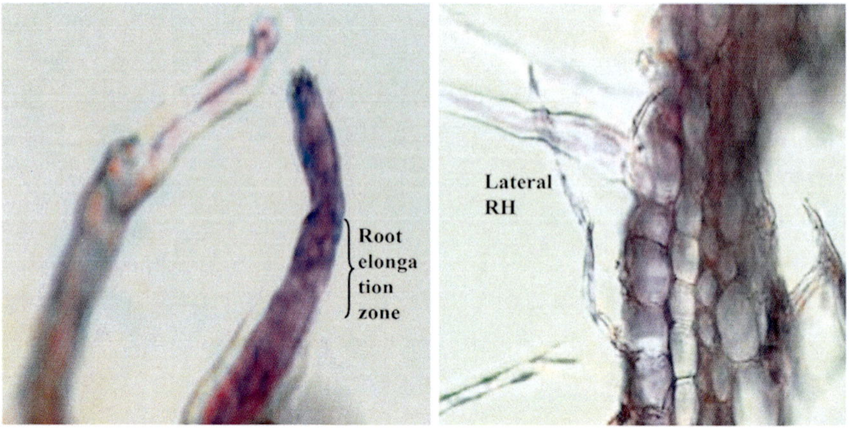


**Fig. 3** Photograph of rice roots showing uptake and translocation in 15 days old 500 μM Cd $(NO_3)_2$ in the growth medium. *RH*, root hair. *(Reproduced with permission from Nahakpam, S., 2009. Biochemical Changes in Rice (Oryza sativa L.) Seedlings Subjected to Heat and Metal Induced Oxidative Stress (Ph.D. thesis). Banaras Hindu University, Varanasi.)*

## 5 CADMIUM-INDUCED MAJOR ANATOMICAL CHANGES AND ABNORMALITIES IN PLANTS

In a study conducted by Daud et al. (2013) on two transgenic cotton cultivars (BR001, GK30) and their parent line (Coker 312), the deposition of electron dense precipitates (most probably Cd) were observed. Increasing Cd levels resulted in a starch accumulation in leaf mesophyll cells in the form of starch granules. Under nonstressed conditions, whole mesophyll cells showed a clean and thin cell with a well-shaped nucleus and a number of lipid bodies. The chloroplasts were elliptical-shaped, possessing properly arranged thylakoids with few starch granules. In Cd treated cells, there was an increase in the number of lipid bodies and nucleoli. Chloroplasts were misshaped, with little swollen and disrupted thylakoids, while there was an increase in the number of starch granules and plastoglobuli. The electron-dense granules were observed in the vacuolar, cytosolic regions, as well as attached to the cellular walls in Cd-treated leaf mesophyll cells, which showed that Cd was mostly accumulated in the dead parts, as well as they cytosol, of the cell.

A 10 μM concentration of exogenous Cd caused the granal and stroma lamella of chloroplasts to condense, and a loss of connection between the lamellae with small osmiophilic globules were observed in *Miscanthus*

(Guo et al., 2016). Similarly, cadmium-induced severe ultrastructural changes in root meristematic and leaf mesophyll cells of *S. alfredii*, which were the major site of damage induced by Cd stress, were the chloroplasts that led to imbalanced lamellae formation, leading to early senescence of leaves (Jin et al., 2008).

Cadmium stress resulted in greater root diameter, owing to increased parenchyma cell size and enlarged cortical tissues, which have a functional role in increasing plant resistance to radial flows of water and solutes (Maksimović et al., 2007). Cadmium-induced inhibition of root elongation might be attributed to the depolymerization of microtubules of the cell cytoskeleton, and the formation of chromosome aberrations, which result in lower mitotic activities of meristematic cells (Seth et al., 2008).

The histochemical staining of $Cd^{2+}$ using QAI revealed the purple granules of $Cd^{2+}$ in root tissues, which were restricted mainly to the parenchyma cells, the vacuoles, and the nuclei, supporting the hypothesis that localized accumulation could be a possible strategy for the defense response in rice plants (Singh and Shah, 2014). Cadmium exerts its toxicity through membrane damage and inactivation of enzymes (Shah, 2011). TEM analysis revealed an intact membrane in cv. Bh-1 under 100 μM $Cd^{2+}$ levels; however, slight tissue disorganization was observed in cv. DR-92. The cell wall ingrowths, as observed in the hypodermis of cv. DR-92, were otherwise absent in cv. Bh-1, suggesting that this might serve as a mode to avoid damage to the cell membrane. Root anatomical structures of four rice breeding materials (maintainer lines Yixiang B and E2B, restorer $2^+$ lines R892 and Mianhui725), grown under different Cd levels, were observed, and the root resistance to $2^+$ Cd ions was evaluated. For four rice breeding materials, the desquamation in the epidermis and new roots in the cortex were observed under low Cd stress. While under high Cd stress, root growth was inhibited in maintainer lines, and new roots could not develop or break off, and a few black spots appeared in the cortex and pericycle. In this study, the prominent changes in rice root were noted in the epidermis and cortex. It displayed desquamation in the epidermis, and damage in the exodermis. Also, the new roots appeared in the cortex. The relatively small changes appeared in the vascular cylinder and xylem pipe. These results were consistent with the conclusions as described by Küpper et al. (1999). Moreover, there were black spots in the cortex and pericycle, exhibiting deep staining in the pericycle, and thickening the cell wall after Cd treatment.

Cadmium treatment accelerated root maturation in *P. vittata*, inducing morphogenic changes that can be linked to adaptive strategies. These changes involve the increase in size of the root apical meristem, the development of root hairs, and the earlier differentiation of xylem elements and suberin lamellae. But *P. vittata* seems to be unable to survive for a longer period at a very high concentration of cadmium (e.g., 100 μM), because, in such conditions, the observed extreme morpho-anatomical and cytological changes at the root level unavoidably limit and alter the root function (Balestri et al., 2014). Following their uptake from the external solutions, cadmium accumulated in the apoplast, and did not readily enter uninjured cells. Thus, the absorbed Cd primarily concentrated in the apoplast of the periblem cells, and the distribution of the metal followed the pattern characteristic of the positively charged compounds of diverse origin, unable to readily cross the plasmalemma (Seregin et al., 2004). Structural alterations produced by Cd were detected in *Z. mays* and *H. vulgare* plants. A negative effect of Cd in the vacuoles has been observed. Cd caused a decline in vacuolar content, not only in cells from the leaves of *Z. mays,* but also in root cells of *H. vulgare*, probably due to the inhibition caused by Cd in water uptake. *H. vulgare* root cells presented deformation of the cell wall with a high Cd retention capacity.

## 6 INTERVENTIONS FOR INTEGRATED MANAGEMENT OF CADMIUM TOXICITY IN CROP PLANTS FOR FOOD SECURITY

Food security is a multidimensional effort that covers sufficiency, safety, and nutritional quality. There is a need to identify suitable amendments for improved crop production and effective amendments for soil remediation to meet the sustainable development goals that focus on attaining food security for all. Soil amendments can alter the physico-chemical composition of soil, or provide substrates for the adsorption of metals, or form stable complexes; thereby reducing the bioavailability or mobility of metal ions (Sharma and Nagpal, 2017) and their uptake by plants.

Cadmium (Cd) contamination has posed an increasing challenge to environmental quality and food security. Sensitivity or resistance of plants to Cd is highly dependent on the effectiveness of these Cd toxicity-coping mechanisms. Cd-sensitive plants generally exhibit reduced growth and productivity, which adversely affects their yield. It is possible that plants can be protected from Cd stress through development of an efficient antioxidative system comprised of expression and activation of antioxidative enzymes in plant cells, metal-chelation in the extracellular spaces and cytoplasm, or compartmentalization of toxic metals away from essential metabolic activities of the cell, either through biotechnological advancements, or management of soil-Cd levels. There are possibilities for employing metal-tolerant plants that can generally take up and accumulate high levels of Cd. Some plants may exhibit a reduction in metal translocation to other organs from the roots, mainly inferred from the high capacities of Cd chelation and sequestration in root cells, or in above-ground vegetative parts. In this way, these organs could be protected from Cd toxicity, even though there might be a high soil Cd level (Ashrafzadeh and Leung, 2016). In a quantitative trait loci (QTL) study of 39 chromosomes of brown rice (*Oryza sativa*), the regions on chromosomes 3, 6, and 8 were found to be associated with Cd-accumulation control. Furthermore, development of genotypic cultivars comprised of Cd resistance in plants can also limit Cd uptake.

To eliminate the long-term risk of Cd entering the food chain, soils contaminated with Cd should be cleaned up using cost-effective remediation measures. Soil organic matter plays an important role in retaining soil Cd through its high CEC, which in turn renders Cd less available to plants. Another possibility is to increase the contents of organic matter, or other organic adsorptive phases, that can promote Cd fixation, and thereby reduce its uptake by plants (Hu et al., 2016).

## 7 FUTURE PROSPECTS

The amount of Cd accumulated and translocated in the plant depends on the plant genotype, which differs among species, and among cultivars within a species. Concentration of Cd in the edible portions of plants can be influenced by genetic manipulation, and by soil and crop management practices. Application of Ca or Zn may increase or decrease Cd uptake by plants, depending on the levels of Cd, Ca, and Zn in the soil, and the relative importance of competition in soil sorption or plant uptake sites. Other factors such as liming, tillage, and residue management, crop rotation, and environmental conditions may also impact Cd accumulation in plants.

Hyperaccumulators generally have well-established detoxification mechanisms, and can therefore accumulate high levels of Cd without showing any physio-morphological toxicity symptoms. The hyperaccumulator species and tolerant species can be explored to study the possibility of using them as a tool to remove Cd. The knowledge on genes involved in hyperaccumulation mechanisms will open the opportunity in biotechnology to transfer specific genes to other cultivars to develop a safe variety.

Development of low Cd accumulating rice varieties has increased transgenic and plant breeding approaches. These tolerant/metal hyperaccumulator plants can be employed together with nontolerant varieties in mixed cropping patterns to avoid cadmium toxicity, and as part of sustainable agricultural practices.

From the discussions, it has been concluded that no single method or management practice could be helpful in increasing crop yield and production in Cd-contaminated soil. An integrated approach employing different strategies is required to reduce the uptake of cadmium by crop plants and its migration into the food chain. Biotechnology, soil amendments, crop rotation, and metal immobilization are some of the tools and techniques that can be helpful in sustainable agriculture and food security in metal-contaminated regions through the globe.

## REFERENCES

Abe, T., Fukami, M., Ogasawara, M., 2008. Cadmium accumulation in the shoots and roots of 93 weed species. Soil Sci. Plant Nutr. 54, 566–573.

Ahmad, P., Nabi, G., Ashraf, M., 2011. Cadmium-induced oxidative damage in mustard [*Brassica juncea* (L.) Czern. & Coss.] plants can be alleviated by salicylic acid. S. Afr. J. Bot. 77, 36–44.

Ali, H., Khan, E., Sajad, M.A., 2013. Phytoremediation of heavy metals-concepts and applications. Chemosphere 91, 869–881.

Alle, V., Kondratovics, U., Osvalde, A., Vikmane, M., 2016. Differences in cadmium accumulation and induced changes in root anatomical structures in plants used for food. Agron. Res. 14, 1249–1260.

Alvarado, V., Bradford, K.J., 2005. Hydrothermal time analysis of seed dormancy in true (botanical) potato seeds. Seed Sci. Res. 15, 77–88.

Ashrafzadeh, S., Leung, D.W.M., 2016. Development of cadmium-safe crop cultivars: a mini review. J. Crop Improv. 30, 107–117.

Baghour, M., Moreno, D.A., Víllora, G., Hernández, J., Castilla, N., Romero, L., 2001. Phytoextraction of Cd and Pb and physiological effects in potato plants (*Solanum Tuberosum* var. spunta): importance of root temperature. J. Agric. Food Chem. 49, 5356–5363.

Baker, A.J.M., Ewart, K., Hendry, G.A.F., Thorpe, P.C., Walker, P.L., 1990. The evolutionary basis of cadmium tolerance in higher plants. 4th International Conference on Environmental Contamination, Barcelona, Spain, pp. 23–29.

Balestri, M., Ceccarini, A., Forino, L.M.C., Zelko, I., Martinka, M., Lux, A., Ruffini Castiglione, M., 2014. Cadmium uptake, localization and stress-induced morphogenic response in the fern *Pteris vittata*. Planta 239, 1055–1064.

Benavides, M.P., Gallego, S.M., Tomaro, M.L., 2005. Cadmium toxicity in plants. Braz. J. Plant Physiol. 17, 21–34.

Chang, Y.S., Chang, Y.J., Lin, C.T., Lee, M.C., Wu, C.W., Lai, Y.H., 2013. Nitrogen fertilization promotes the phytoremediation of cadmium in *Pentas lanceolata*. Int. Biodeterior. Biodegrad. 85, 709–714.

Chang, C.Y., Yu, H.Y., Chen, J.J., Li, F.B., Zhang, H.H., Liu, C.P., 2014. Accumulation of heavy metals in leaf vegetables from agricultural soils and associated potential health risks in the Pearl River Delta, South China. Environ. Monit. Assess. 186, 1547–1560.

Chen, F., Wang, F., Wu, F., Mao, W., Zhang, G., Zhou, M., 2010. Modulation of exogenous glutathione in antioxidant defense system against Cd stress in the two barley genotypes differing in Cd tolerance. Plant Physiol. Biochem. 48, 663–672.

Chen, Z., Zhao, Y., Fan, L., Xing, L., Yang, Y., 2015. Cadmium (Cd) localization in tissues of cotton (*Gossypium hirsutum* L.), and its phytoremediation potential for Cd-contaminated soils. Bull. Environ. Contam. Toxicol. 95, 784–789.

Cherif, J., Mediouni, C., Ben Ammar, W., Jemal, F., 2011. Interactions of zinc and cadmium toxicity in their effects on growth and in antioxidative systems in tomato plants (*Solanum lycopersicum*). J. Environ. Sci. 23, 837–844.

Choppala, G., Saifullah, Bolan, N., Bibi, S., Iqbal, M., Rengel, Z., Kunhikrishnan, A., Ashwath, N., Ok, Y.S., 2014. Cellular mechanisms in higher plants governing tolerance to cadmium toxicity. Crit. Rev. Plant Sci. 33, 374–391.

Clemens, S., Ma, J.F., 2016. Toxic heavy metal and metalloid accumulation in crop plants and foods. Annu. Rev. Plant Biol. 67, 489–512.

Cohen, C.K., Fox, T.C., Garvin, D.F., Kochian, L.V., 1998. The role of iron-deficiency stress responses in stimulating heavy-metal transport in plants. Plant Physiol. 116, 1063–1072.

Curie, C., Cassin, G., Couch, D., Divol, F., Higuchi, K., Le Jean, M., Misson, J., Schikora, A., Czernic, P., Mari, S., 2009. Metal movement within the plant: contribution of nicotianamine and yellow stripe 1-like transporters. Ann. Bot. 103, 1–11.

Daud, M.K., Ali, S., Variath, M.T., Zhu, S.J., 2013. Differential physiological, ultramorphological and metabolic responses of cotton cultivars under cadmium stress. Chemosphere 93, 2593–2602.

Deng, G., Li, M., Li, H., Yin, L., Li, W., 2014. Exposure to cadmium causes declines in growth and photosynthesis in the endangered aquatic fern (*Ceratopteris pteridoides*). Aquat. Bot. 112, 23–32.

Ehsan, S., Ali, S., Noureen, S., Mahmood, K., Farid, M., Ishaque, W., Shakoor, M.B., Rizwan, M., 2014. Citric acid assisted phytoremediation of cadmium by *Brassica napus* L. Ecotoxicol. Environ. Saf. 106, 164–172.

Enstone, D.E., Peterson, C.A., Ma, F., 2002. Root endodermis and exodermis: structure, function, and responses to the environment. J. Plant Growth Regul. 21, 335–351.

Evert, R.F., 2006. Structure and development of the plant body—an overview. In: Esau's Plant Anatomy. John Wiley & Sons, Hoboken, NJ, pp. 1–13.

Fu, X., Dou, C., Chen, Y., Chen, X., Shi, J., Yu, M., Xu, J., 2011. Subcellular distribution and chemical forms of cadmium in *Phytolacca americana* L. J. Hazard. Mater. 186, 103–107.

Gallego, S.M., Pena, L.B., Barcia, R.A., Azpilicueta, C.E., Iannone, M.F., Rosales, E.P., Zawoznik, M.S., Groppa, M.D., Benavides, M.P., 2012. Unravelling cadmium toxicity and tolerance in plants: insight into regulatory mechanisms. Environ. Exp. Bot. 83, 33–46.

Gomes, M.P., de Sá e Melo Marques, T.C.L.L., de Oliveira Gonçalves Nogueira, M., de Castro, E.M., Soares, Â.M., 2011. Ecophysiological and anatomical changes due to uptake and accumulation of heavy metal in *Brachiaria decumbens*. Sci. Agric. 68, 566–573.

Grant, C.A., Buckley, W.T., Bailey, L.D., Selles, F., 1998. Cadmium accumulation in crops. Can. J. Plant Sci. 78, 1–17.

Gratão, P.L., Monteiro, C.C., Rossi, M.L., Martinelli, A.P., Peres, L.E.P., Medici, L.O., Lea, P.J., Azevedo, R.A., 2009. Differential ultrastructural changes in tomato hormonal mutants exposed to cadmium. Environ. Exp. Bot. 67, 387–394.

Guo, H., Hong, C., Chen, X., Xu, Y., Liu, Y., Jiang, D., Zheng, B., 2016. Different growth and physiological responses to cadmium of the three *Miscanthus* species. PLoS One 11, 1–16.

Hanikenne, M., Talke, I.N., Haydon, M.J., Lanz, C., Nolte, A., Motte, P., Kroymann, J., Weigel, D., Krämer, U., 2008. Evolution of metal hyperaccumulation required cis-regulatory changes and triplication of HMA4. Nature 453, 391–395.

Hasan, S.A., Fariduddin, Q., Ali, B., Hayat, S., Ahmad, A., 2009. Cadmium: toxicity and tolerance in plants. J. Environ. Biol. 30, 165–174.

Hermle, S., Günthardt-Goerg, M.S., Schulin, R., 2006. Effects of metal-contaminated soil on the performance of young trees growing in model ecosystems under field conditions. Environ. Pollut. 144, 703–714.

Hose, E., Clarkson, D.T., Steudle, E., Schreiber, L., Hartung, W., 2001. The exodermis: a variable apoplastic barrier. J. Exp. Bot. 52, 2245–2264.

Howladar, S.M., 2014. A novel *Moringa oleifera* leaf extract can mitigate the stress effects of salinity and cadmium in bean (*Phaseolus vulgaris* L.) plants. Ecotoxicol. Environ. Saf. 100, 69–75.

Hu, Y., Cheng, H., Tao, S., 2016. The challenges and solutions for cadmium-contaminated rice in China: a critical review. Environ. Int. 92–93, 515–532.

Iannone, M.F., Groppa, M.D., Benavides, M.P., 2015. Cadmium induces different biochemical responses in wild type and catalase-deficient tobacco plants. Environ. Exp. Bot. 109, 201–211.

Jin, X., Yang, X., Islam, E., Liu, D., Mahmood, Q., 2008. Effects of cadmium on ultrastructure and antioxidative defense system in hyperaccumulator and non-hyperaccumulator ecotypes of *Sedum alfredii* Hance. J. Hazard. Mater. 156, 387–397.

Kabata-Pendias, A., Henryk, P., 2001. Trace Elements in Soils and Plants, third ed. CRC Press, Boca Raton, FL.

Khan, M.I.R., Nazir, F., Asgher, M., Per, T.S., Khan, N.A., 2015. Selenium and sulfur influence ethylene formation and alleviate cadmium-induced oxidative stress by improving proline and glutathione production in wheat. J. Plant Physiol. 173, 9–18.

Kranner, I., Colville, L., 2011. Metals and seeds: biochemical and molecular implications and their significance for seed germination. Environ. Exp. Bot. 72, 93–105.

Küpper, H., Jie Zhao, F., McGrath, S., 1999. Cellular compartmentation of zinc in leaves of the hyperaccumulator *Thlaspi caerulescens*. Plant Physiol. 119, 305–312.

Li, F.T., Qi, J.M., Zhang, G.Y., Lin, L.H., Fang, P.P., Tao, A.F., Xu, J.T., 2013. Effect of cadmium stress on the growth, antioxidative enzymes and lipid peroxidation in two kenaf (*Hibiscus cannabinus* L.) plant seedlings. J. Integr. Agric. 12, 610–620.

Liu, D., Kottke, I., Adam, D., 2007. Localization of cadmium in the root cells of *Allium cepa* by energy dispersive X-ray analysis. Biol. Plant. 51, 363–366.

Liu, C., Guo, J., Cui, Y., Lü, T., Zhang, X., Shi, G., 2011. Effects of cadmium and salicylic acid on growth, spectral reflectance and photosynthesis of castor bean seedlings. Plant Soil 344, 131–141.

Liu, X., Lou, C., Xu, L., Sun, L., 2012. Distribution and bioavailability of cadmium in ornithogenic coral-sand sediments of the Xisha archipelago, South China Sea. Environ. Pollut. 168, 151–160.

Liu, L., Sun, H., Chen, J., Zhang, Y., Li, D., Li, C., 2014. Effects of cadmium (Cd) on seedling growth traits and photosynthesis parameters in cotton (*Gossypium hirsutum* L.). Plant Omi. J. 7, 284–290.

Liu, Y., Xiao, T., Perkins, R.B., Zhu, J., Zhu, Z., Xiong, Y., Ning, Z., 2017. Geogenic cadmium pollution and potential health risks, with emphasis on black shale. J. Geochem. Explor. 176, 42–49.

Loix, C., Huybrechts, M., Vangronsveld, J., Gielen, M., Keunen, E., Cuypers, A., 2017. Reciprocal interactions between cadmium-induced cell wall responses and oxidative stress in plants. Front. Plant Sci. 8, 1–19.

Lomaglio, T., Rocco, M., Trupiano, D., de Zio, E., Grosso, A., Marra, M., Delfine, S., Chiatante, D., Morabito, D., Stefania, G., 2015. Effect of short-term cadmium stress on *Populus nigra* L. detached leaves. J. Plant Physiol. 182, 40–48.

Lux, A., Martinka, M., Vaculík, M., White, P.J., 2011a. Root responses to cadmium in the rhizosphere: a review. J. Exp. Bot. 62, 21–37.

Lux, A., Vaculík, M., Martinka, M., Lišková, D., Kulkarni, M.G., Stirk, W.A., Van Staden, J., 2011b. Cadmium induces hypodermal periderm formation in the roots of the monocotyledonous medicinal plant *Merwilla plumbea*. Ann. Bot. 107, 285–292.

Maestri, E., Marmiroli, M., Visioli, G., Marmiroli, N., 2010. Metal tolerance and hyperaccumulation: costs and trade-offs between traits and environment. Environ. Exp. Bot. 68, 1–13.

Maksimović, I., Kastori, R., Krstić, L., Luković, J., 2007. Steady presence of cadmium and nickel affects root anatomy, accumulation and distribution of essential ions in maize seedlings. Biol. Plant. 51, 589–592.

Martin, D., Vollenweider, P., Buttler, A., Günthardt-Goerg, M.S., 2006. Bioindication of heavy metal contamination in vegetable gardens. For. Snow Landsc. Res. 80, 169–180.

Martínez-Peñalver, A., Graña, E., Reigosa, M.J., Sánchez-Moreiras, A.M., 2012. The early response of *Arabidopsis thaliana* to cadmium- and copper-induced stress. Environ. Exp. Bot. 78, 1–9.

Mclaughlin, M.J., Smolders, E., Degryse, F., Rietra, R., 2011. Uptake of metals from soil into vegetables. In: Swartjes, F.A. (Ed.), Dealing With Contaminated Sites: From Theory Towards Practical Application. Springer, Dordrecht, pp. 325–367.

Meharg, A.A., Norton, G., Deacon, C., Williams, P., Adomako, E.E., Price, A., Zhu, Y., Li, G., Zhao, F.J., McGrath, S., Villada, A., Sommella, A., De Silva, P.M.C.S., Brammer, H., Dasgupta, T., Islam, M.R., 2013. Variation in rice cadmium related to human exposure. Environ. Sci. Technol. 47, 5613–5618.

Memon, A.R., Schröder, P., 2009. Implications of metal accumulation mechanisms to phytoremediation. Environ. Sci. Pollut. Res. 16, 162–175.

Mendoza-Cózatl, D.G., Jobe, T.O., Hauser, F., Schroeder, J.I., 2011. Long-distance transport, vacuolar sequestration, tolerance, and transcriptional responses induced by cadmium and arsenic. Curr. Opin. Plant Biol. 14, 554–562.

Mohamed, A.A., Castagna, A., Ranieri, A., Sanità di Toppi, L., 2012. Cadmium tolerance in *Brassica juncea* roots and shoots is affected by antioxidant status and phytochelatin biosynthesis. Plant Physiol. Biochem. 57, 15–22.

Monteiro, M.S., Santos, C., Soares, A.M.V.M., Mann, R.M., 2009. Assessment of biomarkers of cadmium stress in lettuce. Ecotoxicol. Environ. Saf. 72, 811–818.

Monteiro, C.C., Carvalho, R.F., Gratão, P.L., Carvalho, G., Tezotto, T., Medici, L.O., Peres, L.E.P., Azevedo, R.A., 2011. Biochemical responses of the ethylene-insensitive never ripe tomato mutant subjected to cadmium and sodium stresses. Environ. Exp. Bot. 71, 306–320.

Naja, G.M., Volesky, B., 2009. Metals in the environment: toxicity and sources. In: Wang, L.-K., Chen, J.P., Hung, Y.T., Shammas, N.K. (Eds.), Handbook on Heavy Metals in the Environment. Taylor & Francis and CRC Press, Boca Raton, FL, pp. 13–62.

Najeeb, U., Jilani, G., Ali, S., Sarwar, M., Xu, L., Zhou, W., 2011. Insights into cadmium induced physiological and ultra-structural disorders in *Juncus effusus* L. and its remediation through exogenous citric acid. J. Hazard. Mater. 186, 565–574.

Nishizono, H., Ichikawa, H., Suziki, S., Ishii, F., 1987. The role of the root cell wall in the heavy metal tolerance of *Athyrium yokoscense*. Plant Soil 101, 15–20.

Nzengue, Y., Candéias, S.M., Sauvaigo, S., Douki, T., Favier, A., Rachidi, W., Guiraud, P., 2011. The toxicity redox mechanisms of cadmium alone or together with copper and zinc homeostasis alteration: its redox biomarkers. J. Trace Elem. Med. Biol. 25, 171–180.

Osvalde, A., Paegle, G., 2007. Cadmium accumulation, responses to exposure and possible protection mechanisms in plants. Proc. Latv. Acad. Sci. Sect. B Nat. Exact Appl. Sci. 61, 212–218.

Pence, N.S., Larsen, P.B., Ebbs, S.D., Letham, D.L.D., Lasat, M.M., Garvin, D.F., Eide, D., Kochian, L.V., 2000. The molecular physiology of heavy metal transport in the Zn/Cd hyperaccumulator *Thlaspi caerulescens*. Proc. Natl. Acad. Sci. 97, 4956–4960.

Perfus-Barbeoch, L., Leonhardt, N., Vavasseur, A., Forestier, C., 2002. Heavy metal toxicity: cadmium permeates through calcium channels and disturbs the plant water status. Plant J. 32, 539–548.

Peterson, C.A., Enstone, D.E., 1996. Functions of passage cells in the endodermis and exodermis of roots. Physiol. Plant. 97, 592–598.

Podazza, G., Arias, M., Prado, F.E., 2012. Cadmium accumulation and strategies to avoid its toxicity in roots of the citrus rootstock Citrumelo. J. Hazard. Mater. 215–216, 83–89.

Probst, A., Liu, H., Fanjul, M., Liao, B., Hollande, E., 2009. Response of *Vicia faba* L. to metal toxicity on mine tailing substrate: geochemical and morphological changes in leaf and root. Environ. Exp. Bot. 66, 297–308.

Qureshi, M.I., D'Amici, G.M., Fagioni, M., Rinalducci, S., Zolla, L., 2010. Iron stabilizes thylakoid protein-pigment complexes in Indian mustard during Cd-phytoremediation as revealed by BN-SDS-PAGE and ESI-MS/MS. J. Plant Physiol. 167, 761–770.

Rao, R.A.K., Kashifuddin, M., 2016. Adsorption studies of Cd(II) on ball clay: comparison with other natural clays. Arab. J. Chem. 9, S1233–S1241.

Roberts, T.L., 2014. Cadmium and phosphorous fertilizers: the issues and the science. Procedia Eng. 83, 52–59.

Rosén, K., Eriksson, J., Vinichuk, M., 2012. Uptake and translocation of $^{109}$Cd and stable Cd within tobacco plants (*Nicotiana sylvestris*). J. Environ. Radioact. 113, 16–20.

Saidi, I., Chtourou, Y., Djebali, W., 2014. Selenium alleviates cadmium toxicity by preventing oxidative stress in sunflower (*Helianthus annuus*) seedlings. J. Plant Physiol. 171, 85–91.

Sanità di Toppi, L., Vurro, E., De Benedictis, M., Falasca, G., Zanella, L., Musetti, R., Lenucci, M.S., Dalessandro, G., Altamura, M.M., 2012. A bifasic response to cadmium stress in carrot: early acclimatory mechanisms give way to root collapse further to prolonged metal exposure. Plant Physiol. Biochem. 58, 269–279.

Santos-Echeandía, J., Vale, C., Caetano, M., Pereira, P., Prego, R., 2010. Effect of tidal flooding on metal distribution in pore waters of marsh sediments and its transport to water column (Tagus estuary, Portugal). Mar. Environ. Res. 70, 358–367.

Sasaki, A., Yamaji, N., Yokosho, K., Ma, J.F., 2012. Nramp5 is a major transporter responsible for manganese and cadmium uptake in rice. Plant Cell 24, 2155–2167.

Seregin, I.V., Shpigun, L.K., Ivanov, V.B., 2004. Distribution and toxic effects of cadmium and lead on maize roots. Russ. J. Plant Physiol. 51, 525–533.

Seth, C.S., Misra, V., Chauhan, L.K.S., Singh, R.R., 2008. Genotoxicity of cadmium on root meristem cells of *Allium cepa*: cytogenetic and Comet assay approach. Ecotoxicol. Environ. Saf. 71, 711–716.

Shah, K., 2011. Cadmium metal detoxification and hyperaccumulators. In: Sherameti, I., Varma, A. (Eds.), Detoxification of Heavy Metals. Springer, Berlin, Heidelberg, pp. 181–203.

Shah, K., Nahakpam, S., 2012. Heat exposure alters the expression of SOD, POD, APX and CAT isozymes and mitigates low cadmium toxicity in seedlings of sensitive and tolerant rice cultivars. Plant Physiol. Biochem. 57, 106–113.

Shah, K., Singh, P., Nahakpam, S., 2013. Effect of cadmium uptake and heat stress on root ultrastructure, membrane damage and antioxidative response in rice seedlings. J. Plant Biochem. Biotechnol. 22, 103–112.

Shahid, M., Dumat, C., Khalid, S., Niazi, N.K., Antunes, P.M.C., 2017. Cadmium bioavailability, uptake, toxicity and detoxification in soil-plant system. In: de Voogt, P. (Ed.), Reviews of Environmental Contamination and Toxicology. In: vol. 241. Springer International Publishing, Cham, pp. 73–137.

Sharma, A., Nagpal, A.K., 2017. Soil amendments: a tool to reduce heavy metal uptake in crops for production of safe food. Rev. Environ. Sci. Biotechnol. 17, 1–17.

Shi, Q., Wang, J., Zou, J., Jiang, Z., Wang, J., Wu, H., Jiang, W., Liu, D., 2016. Cd subcellular localization in root tips of *Hordeum vulgare*. Pol. J. Environ. Stud. 25, 903–908.

Singh, P., Shah, K., 2014. Evidences for reduced metal-uptake and membrane injury upon application of nitric oxide donor in cadmium stressed rice seedlings. Plant Physiol. Biochem. 83, 180–184.

Souza, V.L., De Almeida, A.A.F., Lima, S.G.C., Júlio, J.C., Da, C., Silva, D., Mangabeira, P.A.O., Gomes, F.P., 2011. Morphophysiological responses and programmed cell death induced by cadmium in *Genipa americana* L. (Rubiaceae). Biometals 24, 59–71.

Tian, S., Lu, L., Zhang, J., Wang, K., Brown, P., He, Z., Liang, J., Yang, X., 2011. Calcium protects roots of *Sedum alfredii* H. against cadmium-induced oxidative stress. Chemosphere 84, 63–69.

Tudoreanu, L., Phillips, C.J.C., 2001. Modeling cadmium uptake and accumulation in plants. Adv. Agron. 84, 121–157.

Uraguchi, S., Fujiwara, T., 2012. Cadmium transport and tolerance in rice: perspectives for reducing grain cadmium accumulation. Rice 5, 1–8.

USGS, 2001. Cadmium statistics and information. U.S. Geological Survey Mineral Commodity Summaries, pp. 38–39.

Verbruggen, N., Hermans, C., Schat, H., 2009. Mechanisms to cope with arsenic or cadmium excess in plants. Curr. Opin. Plant Biol. 12, 364–372.

Vollenweider, P., Cosio, C., Günthardt-Goerg, M.S., Keller, C., 2006. Localization and effects of cadmium in leaves of a cadmium-tolerant willow (*Salix viminalis* L.). Part II microlocalization and cellular effects of cadmium. Environ. Exp. Bot. 58, 25–40.

Wagner, G.J., 1993. Accumulation of cadmium in crop plants and its consequences to human health. Adv. Agron. 51, 1–2.

Wang, H.C., Wu, J.S., Chia, J.C., Yang, C.C., Wu, Y.J., Juang, R.H., 2009. Phytochelatin synthase is regulated by protein phosphorylation at a threonine residue near its catalytic site. J. Agric. Food Chem. 57, 7348–7355.

Wong, C.K.E., Cobbett, C.S., 2009. HMA P-type ATPases are the major mechanism for root-to-shoot cd translocation in *Arabidopsis thaliana*. New Phytol. 181, 71–78.

Wu, D., Yamaji, N., Yamane, M., Kashino-Fujii, M., Sato, K., Feng Ma, J., 2016. The HvNramp5 transporter mediates uptake of cadmium and manganese, but not iron. Plant Physiol. 172, 1899–1910.

Xin, J., Huang, B., Yang, Z., Yuan, J., Zhang, Y., 2013. Comparison of cadmium subcellular distribution in different organs of two water spinach (*Ipomoea aquatic* Forsk.) cultivars. Plant Soil 372, 431–444.

Xu, D., Chen, Z., Sun, K., Yan, D., Kang, M., Zhao, Y., 2013. Effect of cadmium on the physiological parameters and the subcellular cadmium localization in the potato (*Solanum tuberosum* L.). Ecotoxicol. Environ. Saf. 97, 147–153.

Yilmaz, D.D., Parlak, K.U., 2011. Changes in proline accumulation and antioxidative enzyme activities in *Groenlandia densa* under cadmium stress. Ecol. Indic. 11, 417–423.

Zelko, I., Lux, a., Czibula, K., 2008. Difference in the root structure of hyperaccumulator *Thlaspi caerulescens* and non-hyperaccumulator *Thlaspi arvense*. Int. J. Environ. Pollut. 33, 123.

Zhao, F.J., Ma, Y., Zhu, Y.G., Tang, Z., McGrath, S.P., 2015. Soil contamination in China: current status and mitigation strategies. Environ. Sci. Technol. 49, 750–759.

Zhou, H., Zeng, M., Zhou, X., Liao, B.H., Peng, P.Q., Hu, M., Zhu, W., Wu, Y.J., Zou, Z.J., 2015. Heavy metal translocation and accumulation in iron plaques and plant tissues for 32 hybrid rice (*Oryza sativa* L.) cultivars. Plant Soil 386, 317–329.

## FURTHER READING

Agami, R.A., Mohamed, G.F., 2013. Exogenous treatment with indole-3-acetic acid and salicylic acid alleviates cadmium toxicity in wheat seedlings. Ecotoxicol. Environ. Saf. 94, 164–171.

Ahsan, N., Lee, B., Komatsu, S., 2011. Organellar proteomics: a high-throughput approach for better understanding of heavy metal accumulation and detoxification in plants. In: - Sherameti, I., Varma, A. (Eds.), Detoxification of Heavy Metals. Springer, Berlin, Heidelberg, pp. 273–287.

Arao, T., Kawasaki, A., Baba, K., Mori, S., Matsumoto, S., 2009. Effects of water management on cadmium and arsenic accumulation and Dimethylarsinic acid concentrations in Japanese rice. Environ. Sci. Technol. 43, 9361–9367.

Arao, T., Ishikawa, S., Murakami, M., Abe, K., Maejima, Y., Makino, T., 2010. Heavy metal contamination of agricultural soil and countermeasures in Japan. Paddy Water Environ. 8, 247–257.

Asgher, M., Khan, N.A., Khan, M.I.R., Fatma, M., Masood, A., 2014. Ethylene production is associated with alleviation of cadmium-induced oxidative stress by sulfur in mustard types differing in ethylene sensitivity. Ecotoxicol. Environ. Saf. 106, 54–61.

Bian, R., Chen, D., Liu, X., Cui, L., Li, L., Pan, G., Xie, D., Zheng, J., Zhang, X., Zheng, J., Chang, A., 2013. Biochar soil amendment as a solution to prevent Cd-tainted rice from China: results from a cross-site field experiment. Ecol. Eng. 58, 378–383.

Chaney, R.L., Reeves, P.G., Ryan, J.A., Simmons, R.W., Welch, R.M., Angle, J.S., Scott Angle, J., 2004. An improved understanding of soil Cd risk to humans and low cost methods to phytoextract Cd from contaminated soils to prevent soil Cd risks. Biometals 17, 549–553.

Cui, L., Li, L., Zhang, A., Pan, G., Bao, D., Chang, A., 2011. Biochar amendment greatly reduces rice CD uptake in a contaminated paddy soil: a two-year field experiment. Bioresources 6, 2605–2618.

Delmail, D., Labrousse, P., Hourdin, P., Larcher, L., Moesch, C., Botineau, M., 2011. Physiological, anatomical and phenotypical effects of a cadmium stress in different-aged chlorophyllian organs of *Myriophyllum alterniflorum* DC (Haloragaceae). Environ. Exp. Bot. 72, 174–181.

Farooq, M.A., Ali, S., Hameed, A., Ishaque, W., Mahmood, K., Iqbal, Z., 2013. Alleviation of cadmium toxicity by silicon is related to elevated photosynthesis, antioxidant enzymes; suppressed cadmium uptake and oxidative stress in cotton. Ecotoxicol. Environ. Saf. 96, 242–249.

Fidalgo, F., Freitas, R., Ferreira, R., Pessoa, A.M., Teixeira, J., 2011. *Solanum nigrum* L. antioxidant defence system isozymes are regulated transcriptionally and post-translationally in Cd-induced stress. Environ. Exp. Bot. 72, 312–319.

Gill, S.S., Khan, N.A., Tuteja, N., 2012. Cadmium at high dose perturbs growth, photosynthesis and nitrogen metabolism while at low dose it up regulates sulfur assimilation and antioxidant machinery in garden cress (*Lepidium sativum* L.). Plant Sci. 182, 112–120.

Hasan, S.A., Hayat, S., Ahmad, A., 2011. Brassinosteroids protect photosynthetic machinery against the cadmium induced oxidative stress in two tomato cultivars. Chemosphere 84, 1446–1451.

Hayat, S., 2012. Foliar spray of brassinosteroid enhances yield and quality of *Solanum lycopersicum* under cadmium stress. Saudi J. Biol. Sci. 19, 325–335.

Hseu, Z.Y., Su, S.W., Lai, H.Y., Guo, H.Y., Chen, T.C., Chen, Z.S., 2010. Remediation techniques and heavy metal uptake by different rice varieties in metal-contaminated soils of Taiwan: new aspects for food safety regulation and sustainable agriculture. Soil Sci. Plant Nutr. 56, 31–52.

Nahakpam, S., 2009. Biochemical Changes in Rice (*Oryza sativa* L.) Seedlings Subjected to Heat and Metal Induced Oxidative Stress (Ph.D. thesis). Banaras Hindu University, Varanasi.

Rady, M.M., Hemida, K.A., 2015. Ecotoxicology and environmental safety modulation of cadmium toxicity and enhancing cadmium-tolerance in wheat seedlings by exogenous application of polyamines. Ecotoxicol. Environ. Saf. 119, 178–185.

Sergeant, K., Kieffer, P., Dommes, J., Hausman, J., Renaut, J., 2014. Proteomic changes in leaves of poplar exposed to both cadmium and low-temperature. Environ. Exp. Bot. 106, 112–123.

Sun, L.M., Zheng, M.M., Liu, H.Y., Peng, S.B., Huang, J.L., Cui, K.H., Nie, L.X., 2014. Water management practices affect arsenic and cadmium accumulation in rice grains. Sci. World J. 1–6.

Wang, F., Chen, F., Cai, Y., Zhang, G., Wu, F., 2011a. Modulation of exogenous glutathione in ultrastructure and photosynthetic performance against Cd stress in the two barley genotypes differing in Cd tolerance. Biol. Trace Elem. Res. 144, 1275–1288.

Wang, M.Y., Chen, A.K., Wong, M.H., Qiu, R.L., Cheng, H., Ye, Z.H., 2011b. Cadmium accumulation in and tolerance of rice (*Oryza sativa* L.) varieties with different rates of radial oxygen loss. Environ. Pollut. 159, 1730–1736.

Zhang, Z.C., Chen, B.X., Qiu, B.S., 2010. Phytochelatin synthesis plays a similar role in shoots of the cadmium hyperaccumulator *Sedum alfredii* as in non-resistant plants. Plant Cell Environ. 33, 1248–1255.

# CHAPTER 6

# Cadmium Contamination in Water and Soil

**Qaisar Mahmood*, Mahnoor Asif*, Shahida Shaheen*, Malik Tahir Hayat*, Shafaqat Ali†**

*Department of Environmental Sciences, COMSATS, Abbottabad, Pakistan
†Department of Environmental Sciences and Engineering, Government College University, Faisalabad, Pakistan

## Abstract

With an increasing population rate, economic development, and climatic changes, water scarcity and access to safe water are becoming urgent challenges for the whole world. Cadmium (Cd) and its compounds are distributed in settled, dissolved, adsorbed, or bio-accumulated forms in aquatic ecosystems. Poor quality surface water, marine water, and ground water due to Cd toxicity have disturbed the ecological balance of the environment. Cd is a toxic heavy metal, and its chronic exposure leads to antagonistic changes in living organisms when it enters the food chain. It has gained the attention of scientists after a breakout of Itai-Itai disease in Japan that resulted from wastewater irrigation of agricultural lands. Cd enters the environment through natural and anthropogenic sources. Its bioavailability in soil depends on physicochemical properties of soil. Due to high Cd mobility in the soil, it causes toxicity to plants, and humans, as well as soil microbes. In plants, high Cd concentrations exert toxic effects on plants' physiological and metabolic processes. In humans, Cd intake occurs through three routes, which include food intake, occupational exposure, and smoking. However, Cd through food intake is a major route of exposure.

**Keywords:** Cadmium contamination, Drinking water, Cadmium in food, Soil deterioration, Stream contamination

## 1 CADMIUM

The discovery of cadmium (Cd) dates back to 1817, and its widespread use in various industrial products has been reported since the end of the 19th century (Tolcin, 2009). Cd is a group IIB member of the periodic table, with mass number 112.48 and atomic number 48 (Banaei et al., 2016). The most common oxidation state of cadmium is +2, and some compounds of the +1 oxidation state are also present in natural ecosystems, but the metallic state

*Cadmium Toxicity and Tolerance in Plants*
https://doi.org/10.1016/B978-0-12-814864-8.00006-1

(CdO) is insoluble, and rarely present in water. Due to its chemical resemblance with zinc, it ensures its occurrence naturally in zinc and lead sulfide ore-containing reservoirs (Beheshti et al., 2015).

## 2 CADMIUM AS A CONTAMINANT

Cadmium is a nonessential toxic metal with no biological functions (Vig et al., 2003), and it is found in soil ($0.53\,mg\,kg^{-1}$), the lithosphere ($0.2\,mg\,kg^{-1}$), and sedimentary rocks ($0.3\,mg\,kg^{-1}$). Cd is highly carcinogenic, even in low concentrations (Khan et al., 2017). Its half-life in soil varies from 15 to 1100 years (Vig et al., 2003). Cadmium contamination gained the attention of scientists in the 1960s after the reporting of an osteoporosis-like bone disease called "Itai-Itai" in Japan. This disease was caused by consumption of Cd-polluted rice and water by people living near a historical zinc-lead mine (Pan et al., 2010).

## 3 SOURCES OF CADMIUM IN SOIL

Cd is a significant soil pollutant, which is a matter of health concern (Fig. 1; Jinadasa et al., 2016). Cd in soil can be present in different forms with varying mobility, bioavailability, chemical interaction, and toxicity (Tang et al., 2016). Cd enters into the soil through natural, as well as anthropogenic, sources. Volcanic eruptions, forest fires, sea sprays, and weathering of rocks are natural sources of Cd in the environment. A high concentration of Cd is present in mafic and ultramafic rocks. Mudstone, silt stone, and black shale

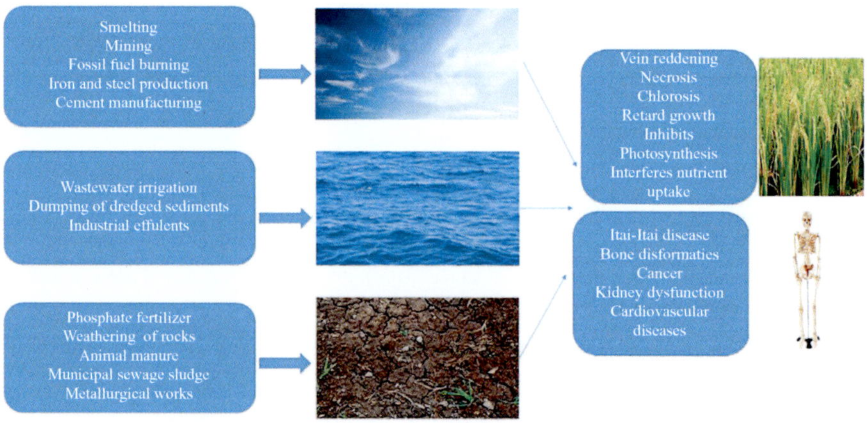


**Fig. 1** Sources of cadmium in air, water, and soil along with toxic effects of cadmium on plants and humans.

also contain significant amounts of Cd, which, upon weathering, release it into the soil (Khan et al., 2017). Smelting, ash from fossil fuel combustion, manure, application of municipal sewage sludge, waste from cement manufacture and use of phosphate fertilizer, and metallurgical works are major anthropogenic sources of Cd contamination of soil (Järup, 2003; Pan et al., 2010). Other sources of Cd include plastic stabilizers, solar panels, motor oil, pigments, batteries, fungicides, polyvinyl chloride plastic, and rubber and textile manufacturing (Khan et al., 2017). Cd pollution in river water also contaminates the soil through irrigation, flooding, or dumping of dredged sediments (Fig. 1; Pan et al., 2010). Wastewater irrigation and use of phosphate fertilizers are major contributors of Cd contamination in agricultural soil (Khan et al., 2017). Cd accumulation in soil depends on soil properties and the level of pollution (Franz et al., 2008). Plants can uptake more Cd in acidic soil (Järup, 2003). Cd in soil form complexes with hydroxyl, chlorides, thiol, and sulfhydryls groups, and biological activity of Cd is governed by these complexes (Vig et al., 2003).

Among different countries of the world, France has the highest Cd concentration in soil ($16.7\,mg\,kg^{-1}$), with Belgium as second highest ($7.61\,mg\,kg^{-1}$), followed by China ($7.43\,mg\,kg^{-1}$) (Khan et al., 2017). According to the US Geological Survey, 2017, 23,000 metric tons of Cd were produced worldwide in the 2016. In China, 743 metric tons of Cd were emitted in 2009 (Khan et al., 2017). China is facing the problem of Cd contaminated soil in >11 provinces and 25 districts, and as a result, $\sim 1.2 \times 10^{10}$ kg of grain have been found to be polluted by Cd. Rice uptakes a high Cd concentration, and China is the leader in rice production. However, in South Korea, Cd concentration in rice grains produced in soil contaminated by heavy metals near metal mines was recorded to be greater than permissible limits, that is, $0.2\,mg$ Cd $kg^{-1}$, set by the Korean Food and Drug administration. In Japan, rice is the leading source of Cd ingestion by humans. The Environment Protection Administration of Taiwan (TEPA) conducted different projects and found background concentrations of Cd of $<0.4\,mg\,kg^{-1}$; however, after establishment of the electroplating industry, it reached $30\,mg\,kg^{-1}$ (Rafiq et al., 2014).

## 4 FACTORS AFFECTING Cd BIOAVAILABILITY IN SOIL

According to Rafiq et al. (2014), Cd uptake by rice grains is greatly influenced by the type of soil. Cd bioavailability in soil depends upon physicochemical properties of soil, such as soil pH, redox potential, organic matter, salinity, Fe oxides, calcium carbonate, and cation exchange capacity. Therefore, it is necessary to select the proper soil type for agricultural purposes. However, pH is an important factor that influences Cd bioavailability. When soil pH decreases, there is increase in the conversion of Cd from an immobile form, to a mobile form, thus increasing Cd phytoavailability. Redox potential also influences Cd mobility in soil, when conditions changes from suboxic ($+350\,mV < Eh < +100\,mV$) to anoxic ($Eh < +100\,mV$), Cd becomes available to plants for uptake. Similarly, salinity also increases phytoavailability of Cd in soil, as high chloride ion concentrations in soil results in the formation of stable $CdCl^+$ and $CdCl_2$ compounds, which enhances mobility of Cd in soil (Tang et al., 2016).

Cd availability in soil is affected by organic matter, causing modifications in physicochemical properties of soil. Organic matter does not reduce Cd concentration in soil, but changes bioavailability of Cd in soil, making it immobile through the adsorption process. Instead of using a synthetic chelating agent, natural organic molecules, such as compost, sewage sludge, and manure can be used as organic amendments. Organic matter decreases soil pH, and as a result, Cd accumulation in the soil decreases, thus decreasing Cd uptake by plants (Tang et al., 2016; Khan et al., 2017). Fe (II)-oxidizing bacteria produces bacteriogenic Fe oxides, and Fe (III)-reducing bacteria produces secondary Fe minerals that result in immobilization of Cd in soil. Soils containing high $CaCO_3$ are able to adsorb $Cd^{2+}$ from soil, thus reducing mobility and bioavailability of Cd in soil (Tang et al., 2016).

## 5 Cd CONTAMINATION IN SOIL

In agriculture, the soil concentration of Cd should be $<3\,\mu g\,g^{-1}$ in dry soil to avoid its transfer through the food chain. In nonpolluted soil, Cd concentration increases with the concentration of clay particles, but is usually $<1\,\mu g\,g^{-1}$ in dry soil. Topsoil contains high Cd content, where it binds with organic matter. Cd is found in soil solutions primarily as $Cd^{2+}$, but it also exists in the form of Cd chelates (Lux et al., 2011). Because of high mobility of Cd in the soil, it is toxic to plants and humans. In humans, Cd accumulation generally occurs through the food chain cycle (Rizwan et al., 2017). Most commonly used soil remediation techniques are soil washing, phytoremediation, solidification, stabilization, excavation, and organic amendments (Khan et al., 2017).

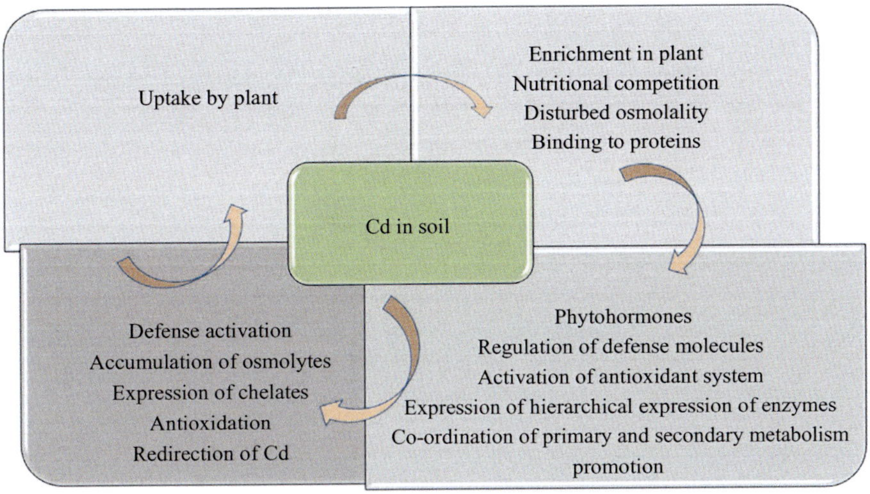


**Fig. 2** Soil environment contaminated with cadmium initiates physiological limitations in plants grown on that contaminated soil.

## 5.1 Effects of Cd Contaminated Soil on Plants

High Cd concentrations have harmful effects on plants physiological and metabolic processes, as shown in Fig. 2. Cd leads to a decrease in leaf length and fresh weight, seed germination, chlorophyll content, stomatal conductance, and adenosine triphosphate ATP concentration (Khan et al., 2017; Rizwan et al., 2017). It inhibits photosynthesis, and causes mitochondrial degradation, thus leading to vein reddening, necrosis, chlorosis, retarding the growth of roots and shoots, and decreases nutrient uptake (Khan et al., 2017). It negatively affects plant growth by interfering with plant metabolic processes, inhibiting proton pumping, reducing root growth, and damaging photosynthetic process (Rafiq et al., 2014). Cd enhances the production of ROS, resulting in oxidative stress (Fig. 2). Cd contamination interferes with the uptake and translocation of essential nutrients such as Ca, Mg, Zn, Fe, and K (Khan et al., 2017, 2010b). Cd toxicity results in the inhibition of photochemical reactions through decreasing the activity of electron transport (Jinadasa et al., 2016).

Seedlings of *Arabidopsis thaliana* treated with Cd show reduced root axis length and enhanced lateral root diameter as compared with control groups (Vitti et al., 2013). Cd stress of 60 mg $kg^{-1}$ reduced the dry weights of potato seedlings; and root, shoot, and seedling length as compared with a control group (Hassan et al., 2016). Jinadasa et al. (2016) reported that high Cd content accumulated in the leaves of cabbage (*Brassica oleracea* L.) without showing any toxicity symptoms, having devastating effects on human health. Hédiji et al. (2010) found that Cd treatment decreases chlorophyll and carotenoid contents in tomato leaves as compared with the control group. Cd treatment decreases the concentration of Zn, Ca, K, and Mn in different parts of tomato seedlings (Bertoli et al., 2012). A high concentration of Cd considerably reduced the mineral nutrients in lettuce, potatoes, and tomatoes, depending on the type of species (Khan et al., 2016). Hossain et al. (2010) found that under Cd stress, hydrogen peroxide ($H_2O_2$) production is enhanced in mung beans. A high Cd concentration results in overproduction of ROS in soybeans, compared with the control group (Shamsi et al., 2014).

## 5.2 Effects of Cd on Soil Microbes

Soil is a habitat for a large number of diversified microbes (Vig et al., 2003). Soil microbial activity is an indicator of soil quality (Su, 2014). Cd contamination in soil has toxic effects on soil microbes, reducing their number and activities (Belimov et al., 2005; Khan et al., 2017, 2010a,b). Cd's effect on soil microbes depends on the soil type and Cd concentration in soil (Vig et al., 2003). According to Su (2014), microbial biomass in soil near a mine was reported to be lower than soil away from the mine. Sensitivity of microbes to heavy metals depends on their species, strains of the same species, and activities of the same species (Giller et al., 1998). Microbial biomass carbon, basal respiration, and dehydrogenase activity are indicators of Cd toxicity in soil (Liao et al., 2005). Khan et al. (2010a,b) found that soil amended with Cd, especially at higher concentrations, changed the structure of the bacterial community. Adding Cd in the paddy soils greatly enhances the inhibition of functions of the microbial community, and Cd bioavailability in paddy soil changes with time (Liao et al., 2005). Many soil microbes can tolerate heavy metals, and play a significant role in their mobilization and immobilization. When *Brassica napus* was inoculated with metal-resistant PGPR, it enhanced the growth of plants grown in soil contaminated with Cd. Similarly, barley plants, when inoculated with auxin-producing and nitrogen-fixing PGPR, immobilized Cd and enhanced the growth of the barley plants cultivated in soil with toxic Cd concentrations (Belimov et al., 2005).

## 5.3 Food Intake of Cd

Cd accumulation and distribution varies in different species, and different cultivars within species, called CSCs (Cd safe cultivars) (Liu et al., 2010). A high bioaccumulation index of Cd was observed in plants grown in Cd contaminated soil, with no harmful effects on crops. The soil crop system is involved in Cd entry through the food chain, causing serious health problems in humans (Liu et al., 2016). There is a growing interest in using crops that can accumulate low concentrations of heavy metals in their edible parts for treatment of heavy metal-contaminated agricultural soils (Murakami et al., 2009). Different mechanisms at different levels might be involved in reducing Cd uptake in low Cd cultivars. It depends on the translocation of Cd from roots to shoots, Cd immobilization in cell walls of different parts of plant, and phloem loading (Rizwan et al., 2017). The maximum tolerable intake of Cd recommended by WHO/FAO is 400–500 $\mu g\,week^{-1}$, or equal to 70 $\mu g\,day^{-1}$ (Irfan et al., 2013).

Humans are exposed to Cd through different routes, but food intake is the major route of exposure. The EFSA (European Food Safety Authority) (2012) analyzed Cd occurrence in different food products, and found grains and grain products (26.9%), starchy roots and tubers (13.2%), and vegetables and vegetable products (16.0%) are the major contributors to Cd intake in Europe. Xiong et al. (2014) studied the concentration of Cd in different parts of lettuce, radishes, and parsley introduced via roots and foliar pathways, and their human accessibility, and found higher accessibility through root transfer.

Different cropping systems, such as crop rotation and intercropping, have been used in agricultural soil for Cd remediation (Liu et al., 2016). Liu et al. (2016) studied the use of different cropping systems to remediate Cd-contaminated soils, and found that edible parts accumulated less Cd than inedible parts. Plants can uptake Cd more readily than other metals, like Cu, Zn, Pb, and Ni (Prokop et al., 2003). The soil-to-plant transfer is generally measured by

$$\mathrm{PTF} = C_{\mathrm{plant}}/C_{\mathrm{soil}}$$

Defined as a ratio between Cd concentration in the plant to Cd concentration in the soil,

where:

PTF is the plant transfer factor.

$C_{plant}$ represents the Cd concentration in plants.

$C_{soil}$ represents the Cd concentrations in soil (Khan et al., 2010a,b).

Molluscs and crustaceans such as oysters and crabs contains high levels of Cd. A high concentration of Cd is present in the livers and kidneys of animals, and in cocoa beans, mushrooms, and oil seeds. Cereals such as rice, and vegetables such as lettuce, potatoes, and carrots contains higher Cd concentrations than dairy products, fish, milk, eggs, and meat. Omnivores have less Cd intake than vegetarians and shellfish consumers (Järup and Åkesson, 2009).

## 6 Cd CONTAMINATION IN WATER

Due to the unavailability of safe water, the spread of waterborne diseases and the death of children $<5$ years old is common in developing countries (Konar et al., 2016). Heavy metals such as Cd, Cr, Pb, Al, and Hg are harmful water pollutants, and pollutants of ecological systems, if their intake exceeds the permissible limit decided by the NEQS policies (Gill et al., 2013).

Major natural sources of Cd are forest fires and volcanic eruptions on the land and in the deep sea, which released about 60% of the total natural discharge into the environment; hence, it's intermixing with water bodies increased with surface runoff (Selinus and Alloway, 2013). Other natural sources of Cd are geological weathering of the mineral rocks in soil (Chrastný et al., 2015). Anthropogenic water contamination by Cd is possible by the following activities: mining, smelting, smoking, industrial waste water discharge, and airborne particulate matter, as shown in Fig. 2 (Fuge, 2013). According to the Safe Drinking Water Act by the EPA, an allowable maximum concentration of Cd in drinking water is 0.005 $mg L^{-1}$ (https://www.epa.gov/sites/production/files/2016-03/.../Cd-final-factsheet.pdf).

According to the Clean Water Act (1997) set by USEPA, Cd and its compounds such as Cd acetate, Cd bromide, and Cd chloride are designated as hazardous substances, and guidelines have been established for water bodies (Kenneth et al., 2008).

Similarly, the Resource Conservation and Recovery Act (1976) also formulated the toxicity characteristics and leaching procedure limit for Cd of about 1.0 $mg L^{-1}$ (Royston, 2013). Cd and its compounds are highly water soluble compared with other metals, which promotes their high bioavailability and bioaccumulation. On an industrial scale, the most commonly used water-soluble Cd salts by purity percentage are Cd chloride (98% to 99.999%), Cd sulfide (98% to 99.999%), Cd sulfate (98% to 99.999%),

Cd oxide (99% to 99.9999%), and Cd nitrate (>99%) (Lewis, 2016). Surface water contamination (40%–56%) by Cd is due to the use of phosphate fertilizers (Pan et al., 2010).

Chronologically, Cd toxicity in aquatic organisms causes antagonistic effects by decreasing growth, reproduction, and immune and endocrine system development (McGeer et al., 2012). The public has been warned that communities in which people regularly eat shellfish and fish meats may have higher Cd exposure (ATSDR, 2010) http://www.atsdr.cdc.gov/csem/cadmium/cdbiologic_fate.html.

Seafood, including finfish, shellfish, crab, and seaweed, and seafood products (fish oil and sauces) that contain higher Cd have become a significant threat for seafood consumers all over the world (da Araújo et al., 2016).

## 6.1 Cd in Fresh Water

Heavy metal contamination of water bodies disrupts the biodiversity of aquatic systems by decreasing the reproduction rate of aquatic organisms, which may result in the low population rate or extinction of many freshwater species (Closs et al., 2015; Hamilton et al., 2016). The biodiversity in aquatic ecosystems makes them an interesting environment for exploring conservation and restoration strategies in an ecosystem exposed to heavy metals toxicity (Ellis et al., 2017). Algae, insects, and fish are good biomarkers for pollutants bioaccumulation in an aquatic ecosystem (Stankovic et al., 2014).

The rate of heavy metals bioaccumulation biomagnification in an aquatic ecosystem depends on the chemistry of water, seasonal variations, and consumption and excretion of fish (Abdel-Khalek, 2015). Sediment analysis of water bodies is very important for the screening of environmental pollutants, as these can intermix with water and pose a threat to aquatic life (Uluturhan et al., 2011). In artificial lakes and ponds of southern Poland (Silesian plateau), the contamination index of Cd in sediments was found to be higher than the earth's crust due to the geochemical background of Poland, and the geo-environmental variations of water bodies (Rzętała, 2016).

The solubility of Cd in water depends on its pH; the more acidic conditions support the solubility of Cd more than basic conditions (Honma et al., 2016). In fresh water bodies, Cd and its compounds can exist in abiotic conditions such as water and sediment, and biotic components of aquatic ecosystem environments, such as organisms; but their rate of contamination causes antagonistic effects on flora and fauna if their uptake surpasses the toxicity tolerance limits (Shi et al., 2016). Cd is easily absorbed, and accumulates in tissues, which is very harmful if it becomes part of the food chain (Olmedo et al., 2013). A comparative study of Cd bioaccumulation in bull trouts (*S. confluentus*) and brook trouts (*S. fontinalis*) showed that the effect of Cd toxicity was different for these two species, and that bull trouts suffered more during exposure (Meador, 2015).

Fresh water fish (specifically the gills, tissues, and digestive tracks) were investigated for Cd toxicity absorbed in sublethal concentrations, and it was found that the Cd concentration was highest in tissues of *C. carpio,* which was alarming in respect to aquatic systems (Afshan et al., 2014). A transcriptomic analysis of Cd-exposed wild yellow perch (*P. flavescens*) showed evolutionary toxicological changes by coding SNPs genes when compared with noncontaminated freshwater fish (Bélanger-Deschênes et al., 2013). A molecular study of freshwater ecology is an important tool for the investigation of the ill effects of Cd toxicity and tolerance by aquatic organisms, and their level of transmission through the food chain to higher organisms. A micro-electrophoretic study of fresh water fish*, C. catla and C. mrigala,* depicted that Cd toxicity caused genotoxicity by DNA damage expression (Hussain et al., 2017).

The contaminated aquatic ecosystems adsorbed a number of heavy metals in sediments and dissolved forms. Therefore, the uptake and accumulations of heavy metals by aquatic flora (rooted, submerged, or floated) is an important factor for the bioavailability of toxins through food chains (Mebane, 2006). Algae are the primary producers of water and heavy metals accumulators and transmitters in an aquatic ecosystem when they enter into the food chain (Liu et al., 2008). *Chlorella vulgaris* is a type of freshwater green algae that can be used as a biomarker for the ecological screening of Cd-contaminated waters (Çelekli et al., 2016).

Trancriptomic and metabolomics studies of freshwater alga *C. reinhardtii.* Were analyzed for various Cd stresses (0, 8.1, and 114.8 μM) after 48 and 72 h, and it was revealed that the transcriptomic responses of algae were more pronounced than the metabolomics responses (Jamers et al., 2013). Cd stress, along with other heavy metals stress, was also analyzed in diatom *P. lanceolatum,* and it was observed that diatom was hypersensitive against various concentrations (0.25, 0.35, and 0.62 $mg L^{-1}$) of Cd (Sbihi et al., 2012). Rivers, when flowing through natural water bodies, also add the toxic heavy metals that settle down into the sediment. An investigation of Lake Symsar in Northern Poland depicted that a number of heavy metals

(Zn, Mn, Cd, and Ni) were accumulated in the top sedimentation layer of the lake, and were posing stress to the natural ecosystem (Kuriata-Potasznik et al., 2016).

## 6.2 Cd in Drinking Water

Groundwater contamination became an alarming issue because of its associated health risks when utilized as drinking water (Razak et al., 2015), and the possible risk to human health has been studied widely. Cd, like other heavy metals, enters into surface water and groundwater by the weathering of rocks, or through mining or smelting sites (Fetter et al., 2017). The level of Cd contamination of drinking water increased by the corrosion of galvanized water supply pipes, water storage tanks that are mostly platted with Cd and Zn, or by using paint-coated utensils (Salehi et al., 2017). Another source of drinking water contamination is landfill leachate, because landfills are made up of municipal waste, Cd-processing industrial waste, or other metallic processing waste materials (El-Salam and Abu-Zuid, 2015).

Other sources of ground water contamination with Cd are the application of organic fertilizers and animal manure (Wu and Sun, 2016). The Waikato Regional Council of New Zealand has monitored the ground water, surface water, and sediment of Peat Lake, where the Cd levels were higher in sediments due to the geological contamination of parent materials; whereas it was normal in water samples (Taylor et al., 2017). An analysis of drinking water from shallow wells of Cd-contaminated soils has also shown great health risks (Muhammad et al., 2017). In the Red Sea, in Jizan, Saudi Arabia, the qualitative analysis of water, sediment, and white shrimp (*L. vannamei*) was carried out, and the results showed that the level of Cd and other heavy metals was higher than the WHO standards, which can pose health risks for the community (Mortuza and Al-Misned, 2017).

Another case study of Cd-contaminated surface and ground water was conducted in two towns of South Africa, with the application of multivariate statistical tools, which highlighted the casual agents of pollution (Elumalai et al., 2017). The observations from this case study are similar to those of the spread of Itai-Itai disease in Japan (1912), in which the major symptoms of the toxicity-linked disease were demineralization of bones, inflammation of joints, osteoporosis, and bones fractures. It has been reported that Japan and China are the top most victimized countries of the world, due to environmental Cd exposure, which is comparatively higher in these two countries than in any other country (Han et al., 2009; Egeonu and Herat, 2016).

## 6.3 Cd in Rain Water

Due to an increasing urbanization, rainwater runoff contains a variety of pollutants, organic matter, heavy metals, oil derivatives, chlorides, and other xenobiotics, and is a significant threat for the aquatic environment, if there are improper storm drainage systems (St-Hilaire et al., 2016). Due to the high demand of water and water scarcity, the rainwater harvesting approach is a valuable technique. The quality of rainwater from free-fall to storage needs to be safe for domestic use (Owusu-Boateng et al., 2015). Rainwater quality analysis for heavy metals contamination has been done by many researchers, which revealed the atmospheric quality of a specific region (Santos et al., 2011; Cerqueira et al., 2014).

Cd contamination in rain water was observed to be higher due to the corrosion of galvanized steel roofs than the rainwater from cemented or Al-Zn coated steel roofs. Similarly, the material of catchment and storage tanks is also a factor in rainwater pollution by Cd (Asiima et al., 2015). The Australian government initialized household rain water tanks as an alternative water resource in 2005 to address the water scarcity issue. An investigation in Melbourne urban roof tanks indicated that the level of contamination of Pb, Cd, Al, Zinc, and Fe exceeded the Australian Drinking Water Guidelines (ADWG), and was deemed unsafe for daily use (Magyar et al., 2008).

## 6.4 Cd Pollution in Lakes and Streams

Contamination of lakes and streams is common only where these water bodies are located near the source of emissions; whereas the lakes and streams of remote areas are being disturbed due to the geochemistry of those specific regions (Schindler and Kamber, 2013). Sediment analysis is an important tool for the screening of ecological conditions of water bodies because sediment provides the basin for atmospheric contaminants (Bindler et al., 2011). For the heavy metals screening of shallow lakes, a multicore analysis can play a key role for the improvement of environmental conditions of an ecological system (Engstrom and Rose, 2013).

Trace heavy metals (Cu, Cd, Pb, Fe, Mn, As, Zn, Ni, and Cr) values were high in Léré Lake, in Western Mayo-Kebbi (Chad), and sediment and fish tissues exhibited contamination that was higher than the guidelines of the WHO, which created significant health risks to the surrounding communities (Mahamat et al., 2017). Dam Lake Aswan (Egypt) was checked for heavy metals contamination in dam water, sediment, fish (*T. nilotica*) (specifically in the vertebral column and fins of fish of various weights and ages), and aquatic macrophytes (*N. armeta*), and the value of Cd, Pb, and Ni were found to be detrimental in fish tissues (Rashed, 2001). In Korea, when the level of toxic elements of a stream's sediment near Shihwa Lake were analyzed, there was a high concentration of Zn > Cu > Pb > Cr > Ni > Co > As > Cd > Hg, which indicated that the geo-accumulation index was high near industrial areas, and the level of contaminants and the geo-accumulation index were lower in rural areas (Jeong et al., 2016).

## 6.5 Marine Water

The Cd cycle in marine water is very interesting because of its distribution in various isotopic forms, their role in nutrition cycles, and as tracer elements in oceans. Fractionation of Cd isotopes (106Cd to 116Cd) in ocean water is an important aspect of its distribution in oceans (Abouchami et al., 2013). Rivers are thought to be the most important source of marine Cd, where it acts as micronutrient for biota (Lambelet et al., 2013). Cd toxicity is relatively lower in marine ecosystems than in freshwater due to the difference in the environmental status of both water bodies (McGeer et al., 2012).

The marine water organisms show more tolerance and adoptability toward Cd toxicity due to the physiological functions (defense systems) of inhabitants, or other biotic and abiotic factors (Rocha et al. (2015). Seaweed is a promising bio-indicator of heavy metals in marine ecosystems. *Ulva lactuca* marine green algae responded to various Cd concentrations (0, 2.5, 5, and 10 $mg L^{-1}$), and DNA damage was observed after high exposure (Babu et al., 2014). Cd contaminated marine biota increased the rate of health disorders through dietary consumption. Shellfish and other fish, when consuming Cd-containing algae or water, accumulate a considerable amount of Cd in their bodies, which can then transfer to humans through dietary consumption, and may cause many health issues (Fig. 3). Oysters can accumulate alarmingly high amounts of Cd in their tissues. Oysters' (*C. gigas*) accumulation of high amounts of Cd can result in DNA damage and expression of stress-responsive proteins and antioxidants (Meng et al., 2017).

In Malwathu Oya (Sri Lanka), an investigation about food toxicology revealed that different body parts of fish species (gills, kidney, liver, and muscle) surpassed the allowable safe limit of Cd during cropping season, which may be due to the surface run-off of phosphates containing fertilizers (Perera et al., 2016). A toxicological investigation of marine mussels (*Mytilus galloprovincialis*) indicated the expression of DNA-damaged chromosome aberration when exposed to (Cd, Zn, Cu, Pb) (Zhang et al., 2017).

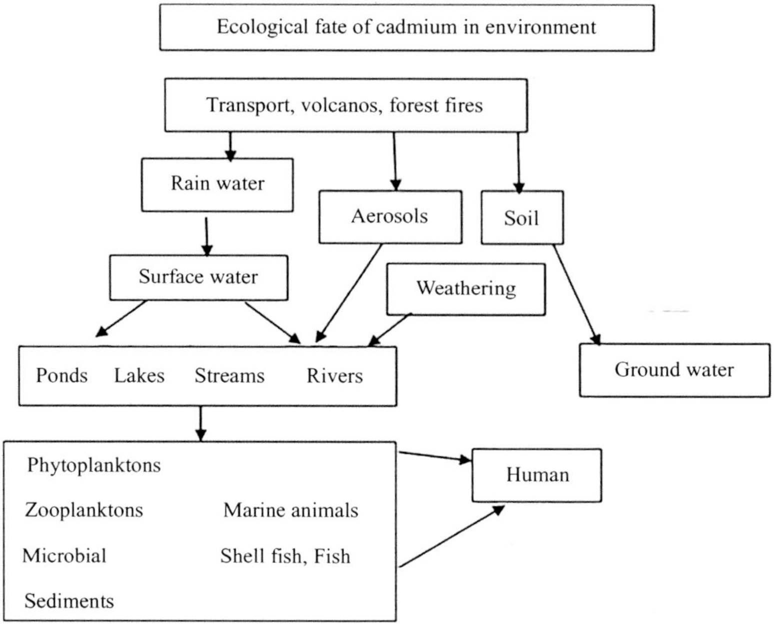


**Fig. 3** Ecological fate of Cd in environment.

# 7 CASE STUDIES

## 7.1 Case Study-1

A case study was performed in a town in southeastern China that is popular for its ceramic industry. Cd contamination in the soil was based on impacts from the ceramic industry from 2004 to 2014. In this study, samples of ceramics and pigment were collected from a selected factory, and sediment core samples were collected from a nearby river. A geo-accumulation index indicated that the sediment was highly polluted by Cd. It was concluded, by principal component analysis, that Cd pollution in a nearby river was a result of proximity to the ceramic factory. Sediment dating showed that the Cd pollution began shortly after the ceramic factory's establishment (2002–04). It was inferred from the preceding results that the ceramic industry in the town was responsible for a high Cd concentration in soil (Liao et al., 2015).

## 7.2 Case Study-2

This study was performed in Japan. Samples from cases of Itai-Itai disease were collected to assess chronic Cd hepatotoxicity. To assess Cd accumulation in different organs, 89 cases of Itai-Itai disease, with 27 non-Cd polluted cases, were taken as control groups. For a histopathological assessment, 80 cases of Itai-Itai disease and 70 control cases were studied. Immunohistochemistry was performed for metallothionein, which is involved in Cd detoxification. In the Itai-Itai disease group, the hepatic Cd content was found to be higher than the Cd content in all other organs. In the Itai-Itai disease group, fibrosis in the liver was found to be considerably higher than in the control group. Similarly, high metallothionein expression was observed in the Itai-Itai disease group than the control group. A high hepatic accumulation of Cd due to prolonged exposure led to fibrosis. However, it was determined that high metallothionein expression lowers Cd hepatotoxicity (Baba et al., 2013).

## 8 CONCLUSION

Cadmium is a toxic heavy metal with no biological functions. It enters into the environment through natural, as well as anthropogenic sources. Poor quality of surface water, marine water, and ground water due to Cd toxicity has disturbed the ecological balance of the environment. High Cd concentrations have harmful effects on plants' physiological and metabolic processes. Cd contamination in soil has toxic effects on soil microbes, reducing their number and activities; however, many soil microbes can tolerate heavy metals, and play a significant role in their mobilization and immobilization. In humans, Cd accumulation generally occurs through the food chain cycle. Cd uptake and accumulation by aquatic flora (rooted, submerged, or floating) is an important factor for the bioavailability of toxins through food chains.

## REFERENCES

Abdel-Khalek, A.A., 2015. Risk assessment, bioaccumulation of metals and histopathological alterations in Nile tilapia (*Oreochromis niloticus*) facing degraded aquatic conditions. Bull. Environ. Contam. Toxicol. 94 (1), 77–83.

Abouchami, W., Galer, S.J., Horner, T.J., Rehkämper, M., Wombacher, F., Xue, Z., Lambelet, M., Gault-Ringold, M., Stirling, C.H., Schönbächler, M., Shiel, A.E., 2013. A common reference material for cadmium isotope studies—NIST SRM 3108. Geostand. Geoanal. Res. 37 (1), 5–17.

Agency for Toxic Substances and Disease Registry (ATSDR), 2010. ATSDR Fact Sheet; Case Studies in Environmental Medicine (CSEM) Cadmium Toxicity What Is the Biological Fate of Cadmium in the Body? https://www.atsdr.cdc.gov/csem/cadmium/cdbiologic_fate.html (accessed 14 January 2018).

Afshan, S., Ali, S., Ameen, U.S., Farid, M., Bharwana, S.A., Hannan, F., Ahmad, R., 2014. Effect of different heavy metal pollution on fish. Res. J. Chem. Environ. Sci 2 (1), 74–79.

Asiima, A.R., Mbabazi, J., Tebandeke, E., Ntale, M., 2015. Trace metal and nutrient constitution of rain water and sediment/sludge harvested in various storage tanks from galvanized iron roof tops in Kampala City. Carib. J. Sci. Tech. 3, 718–730.

Baba, H., Tsuneyama, K., Yazaki, M., Nagata, K., Minamisaka, T., Tsuda, T., Nomoto, K., Hayashi, S., Miwa, S., Nakajima, T., Nakanishi, Y., Nakanishi, Y., 2013. The liver in itai-itai disease (chronic cadmium poisoning): pathological features and metallothionein expression. Mod. Pathol. 26 (9), 1228–1234.

Babu, M.Y., Palanikumar, L., Nagarani, N., Devi, V.J., Kumar, S.R., Ramakritinan, C.M., Kumaraguru, A.K., 2014. Cadmium and copper toxicity in three marine macroalgae: evaluation of the biochemical responses and DNA damage. Environ. Sci. Pollut. Res. Int. 21 (16), 9604–9616.

Banaei, A., Saadat, A., Mohammad Goli, M., McArdle, P., Pourbasheer, E., Pargolghasemi, P., 2016. Synthesis, characterization, and molecular structures of Ni (II) and Cd (II) complexes derived from dithiophosphonate. Heteroat. Chem. 27 (6), 353–360.

Beheshti, A., Lalegani, A., Behvandi, F., Safaeiyan, F., Sarkarzadeh, A., Bruno, G., Rudbari, H.A., 2015. Synthesis and structural characterization of dinuclear Cd 2+, Hg 2+ and Fe 2+ complexes with neutral bi and tetradentate flexible pyrazole-based ligands. J. Mol. Struct. 1082, 143–150.

Bélanger-Deschênes, S., Couture, P., Campbell, P.G., Bernatchez, L., 2013. Evolutionary change driven by metal exposure as revealed by coding SNP genome scan in wild yellow perch (*Perca flavescens*). Ecotoxicology 22 (5), 938–957.

Belimov, A.A., Hontzeas, N., Safronova, V.I., Demchinskaya, S.V., Piluzza, G., Bullitta, S., Glick, B.R., 2005. Cadmium-tolerant plant growth-promoting bacteria associated with the roots of Indian mustard (*Brassica juncea* L. Czern.). Soil Biol. Biochem. 37 (2), 241–250.

Bertoli, A.C., Cannata, M.G., Carvalho, R., Bastos, A.R.R., Freitas, M.P., dos Santos Augusto, A., 2012. *Lycopersicon esculentum* submitted to Cd-stressful conditions in nutrition solution: nutrient contents and translocation. Ecotoxicol. Environ. Saf. 86, 176–181.

Bindler, R., Rydberg, J., Renberg, I., 2011. Establishing natural sediment reference conditions for metals and the legacy of long-range and local pollution on lakes in Europe. J. Paleolimnol. 45, 519–531.

Çelekli, A., Gültekin, E., Bozkurt, H., 2016. Morphological and biochemical responses of *Spirogyra setiformis* exposed to cadmium. Clean: Soil, Air, Water 44 (3), 256–262.

Cerqueira, M.R.F., Pinto, M.F., Derossi, I.N., Esteves, W.T., Santos, M.D.R., Matos, M.A.C., Lowinsohn, D., Matos, R.C., 2014. Chemical characteristics of rainwater at a southeastern site of Brazil. Atmos. Pollut. Res. 5 (2), 253–261.

Chrastný, V., Čadková, E., Vaněk, A., Teper, L., Cabala, J., Komárek, M., 2015. Cadmium isotope fractionation within the soil profile complicates source identification in relation to Pb–Zn mining and smelting processes. Chem. Geol. 405, 1–9.

Clean Water Act, 1997. Clean Water Act. EPA's Office.

Closs, G.P., Angermeier, P.L., Darwall, W.R., Balcombe, S.R., 2015. Why are freshwater fish so threatened. In: Closs, G.P., Krkosek, M., Olden, J. (Eds.), Conservation of Freshwater Fishes. Cambridge University Press, pp. 37–75.

da Araújo, C.F.S., Lopes, M.V., Vasquez, M.R., Porcino, T.S., Ribeiro, A.S.V., Rodrigues, J.L.G., do Prado Oliveira, S.S., Menezes-Filho, J.A., 2016. Cadmium and lead in seafood from the Aratu Bay, Brazil and the human health risk assessment. Environ. Monit. Assess. 188 (4), 259.

Egeonu, N., Herat, S., 2016. E-waste: a problem or an opportunity? Review of issues, challenges and solutions in African countries. Int. J. Environ. Waste Manag. 17 (3–4), 318–339.

Ellis, J.I., Clark, D., Atalah, J., Jiang, W., Taiapa, C., Patterson, M., Hewitt, J., 2017. Multiple stressor effects on marine infauna: responses of estuarine taxa and functional traits to sedimentation, nutrient and metal loading. Sci. Rep. 7, 12013.

El-Salam, M.M.A., Abu-Zuid, G.I., 2015. Impact of landfill leachate on the groundwater quality: a case study in Egypt. J. Adv. Res. 6 (4), 579–586.

Elumalai, V., Brindha, K., Lakshmanan, E., 2017. Human exposure risk assessment due to heavy metals in groundwater by pollution index and multivariate statistical methods: a case study from south africa. Water 9 (4), 234.

Engstrom, D.R., Rose, N.L., 2013. A whole-basin, mass-balance approach to Paleolimnology. J. Paleolimnol. 49, 333–347.

European Food Safety Authority, 2012. Cadmium dietary exposure in the European population. EFSA J. 10, 2551.

Fetter, C.W., Boving, T., Kreamer, D., 2017. Contaminant Hydrogeology. Waveland Press, Long Grove, USA.

Franz, E., Römkens, P., van Raamsdonk, L.E.O., Van Der Fels-Klerx, I., 2008. A chain modeling approach to estimate the impact of soil cadmium pollution on human dietary exposure. J. Food Prot. 71 (12), 2504–2513.

Fuge, R., 2013. Anthropogenic sources. In: Essentials of Medical Geology. Springer, Dordrecht, pp. 59–74.

Gill, S.S., Hasanuzzaman, M., Nahar, K., Macovei, A., Tuteja, N., 2013. Importance of nitric oxide in cadmium stress tolerance in crop plants. Plant Physiol. Biochem. 63, 254–261.

Giller, K.E., Witter, E., Mcgrath, S.P., 1998. Toxicity of heavy metals to microorganisms and microbial processes in agricultural soils: a review. Soil Biol. Biochem. 30 (10), 1389–1414.

Hamilton, P.B., Cowx, I.G., Oleksiak, M.F., Griffiths, A.M., Grahn, M., Stevens, J.R., Carvalho, G.R., Nicol, E., Tyler, C.R., 2016. Population-level consequences for wild fish exposed to sublethal concentrations of chemicals—a critical review. Fish Fish. 17 (3), 545–566.

Han, J.X., Shang, Q., Du, Y., 2009. Effect of environmental cadmium pollution on human health. Health 1 (3), 159–166.

Hassan, W., Bano, R., Bashir, S., Aslam, Z., 2016. Cadmium toxicity and soil biological index under potato (*Solanum tuberosum L.*) cultivation. Soil. Res. 54 (4), 460–468.

Hédiji, H., Djebali, W., Cabasson, C., Maucourt, M., Baldet, P., Bertrand, A., Zoghlami, L.B., Deborde, C., Moing, A., Brouquisse, R., Chaïbi, W., 2010. Effects of long-term cadmium exposure on growth and metabolomic profile of tomato plants. Ecotoxicol. Environ. Saf. 73 (8), 1965–1974.

Honma, T., Ohba, H., Kaneko-Kadokura, A., Makino, T., Nakamura, K., Katou, H., 2016. Optimal soil Eh, pH, and water management for simultaneously minimizing arsenic and cadmium concentrations in rice grains. Environ. Sci. Technol. 50 (8), 4178–4185.

Hossain, M.A., Hasanuzzaman, M., Fujita, M., 2010. Up-regulation of antioxidant and glyoxalase systems by exogenous glycinebetaine and proline in mung bean confer tolerance to cadmium stress. Physiol. Mol. Biol. Plants 16 (3), 259–272.

Hussain, B., Sultana, T., Sultana, S., Al-Ghanim, K.A., Masood, S., Ali, M., Mahboob, S., 2017. Microelectrophoretic study of environmentally induced DNA damage in fish and its use for early toxicity screening of freshwater bodies. Environ. Monit. Assess. 189 (3), 115.

Irfan, M., Hayat, S., Ahmad, A., Alyemeni, M.N., 2013. Soil cadmium enrichment: allocation and plant physiological manifestations. Saudi J. Biol. Sci. 20 (1), 1–10.

Jamers, A., Blust, R., De Coen, W., Griffin, J.L., Jones, O.A., 2013. An omics based assessment of cadmium toxicity in the green alga *Chlamydomonas reinhardtii*. Aquat. Toxicol. 126, 355–364.

Järup, L., 2003. Hazards of heavy metal contamination. Br. Med. Bull. 68 (1), 167–182.

Järup, L., Åkesson, A., 2009. Current status of cadmium as an environmental health problem. Toxicol. Appl. Pharmacol. 238 (3), 201–208.

Jeong, H., Kim, K.T., Kim, E.S., Ra, K., Lee, S.Y., 2016. Sediment quality assessment for heavy metals in streams around the Shihwa lake. J. Korean Soc. Mar. Environ. Energy 19 (1), 25–36.

Jinadasa, N., Collins, D., Holford, P., Milham, P.J., Conroy, J.P., 2016. Reactions to cadmium stress in a cadmium-tolerant variety of cabbage (*Brassica oleracea* L.): is cadmium tolerance necessarily desirable in food crops? Environ. Sci. Pollut. Res. 23 (6), 5296–5306.

Kenneth, J.A., William, J.B., Jagdish, B., Michael, J.B., Robert, C., et al., 2008. Clean Water Act Brief. Stanford University, USA.

Khan, S., Hesham, A.E.L., Qiao, M., Rehman, S., He, J.Z., 2010a. Effects of Cd and Pb on soil microbial community structure and activities. Environ. Sci. Pollut. Res. 17 (2), 288–296.

Khan, S., Rehman, S., Khan, A.Z., Khan, M.A., Shah, M.T., 2010b. Soil and vegetables enrichment with heavy metals from geological sources in Gilgit, northern Pakistan. Ecotoxicol. Environ. Saf. 73 (7), 1820–1827.

Khan, S., Munir, S., Sajjad, M., Li, G., 2016. Urban park soil contamination by potentially harmful elements and human health risk in Peshawar City, Khyber Pakhtunkhwa, Pakistan. J. Geochem. Explor. 165, 102–110.

Khan, M.A., Khan, S., Khan, A., Alam, M., 2017. Soil contamination with cadmium, consequences and remediation using organic amendments. Sci. Total Environ. 601, 1591–1605.

Konar, M., Evans, T.P., Levy, M., Scott, C.A., Troy, T.J., Vörösmarty, C.J., Sivapalan, M., 2016. Water resources sustainability in a globalizing world: who uses the water? Hydrol. Process. 30 (18), 3330–3336.

Kuriata-Potasznik, A., Szymczyk, S., Skwierawski, A., Glińska-Lewczuk, K., Cymes, I., 2016. Heavy metal contamination in the surface layer of bottom sediments in a flow-through lake: a case study of Lake Symsar in Northern Poland. WaterSA 8 (8), 358.

Lambelet, M., Rehkämper, M., van de Flierdt, T., Xue, Z., Kreissig, K., Coles, B., Porcelli, D., Andersson, P., 2013. Isotopic analysis of Cd in the mixing zone of Siberian rivers with the Arctic Ocean—new constraints on marine Cd cycling and the isotope composition of riverine Cd. Earth Planet. Sci. Lett. 361, 64–73.

Lewis, R.A., 2016. Hawley's Condensed Chemical Dictionary. John Wiley & Sons, Hoboken, NJ.

Liao, M., Luo, Y.K., Zhao, X.M., Huang, C.Y., 2005. Toxicity of cadmium to soil microbial biomass and its activity: effect of incubation time on Cd ecological dose in a paddy soil. J. Zhejiang Univ. Sci. B 6 (5), 324.

Liao, Q.L., Liu, C., Wu, H.Y., Jin, Y., Hua, M., Zhu, B.W., Chen, K., Huang, L., 2015. Association of soil cadmium contamination with ceramic industry: a case study in a Chinese town. Sci. Total Environ. 514, 26–32.

Liu, H., Li, L., Yin, C., Shan, B., 2008. Fraction distribution and risk assessment of heavy metals in sediments of Mushui lake. J. Environ. Sci. 20, 390–397.

Liu, W., Zhou, Q., An, J., Sun, Y., Liu, R., 2010. Variations in cadmium accumulation among Chinese cabbage cultivars and screening for Cd-safe cultivars. J. Hazard. Mater. 173 (1), 737–743.

Liu, Y., Liu, K., Li, Y., Yang, W., Wu, F., Zhu, P., Zhang, J., Chen, L., Gao, S., Zhang, L., 2016. Cadmium contamination of soil and crops is affected by intercropping and rotation systems in the lower reaches of the Minjiang River in south-western China. Environ. Geochem. Health 38 (3), 811–820.

Lux, A., Martinka, M., Vaculík, M., White, P.J., 2011. Root responses to cadmium in the rhizosphere: a review. J. Exp. Bot. 62 (1), 21–37.

Magyar, M.I., Mitchell, V.G., Ladson, A.R., Diaper, C., 2008. In: Lead and other heavy metals: common contaminants of rainwater tanks in Melbourne. Proceedings of Water Down Under, 2008, p. 409.

Mahamat, T.N., Kidjakebo, S., Abel, S.K., Bertrand, N.G., 2017. Heavy metals pollution in water, sediments and fish of Léré lake, Western Mayo Kebbi (Chad). J. Environ. Prot. 5 (1), 1.

McGeer, J.C., Niyogi, S., Smith, D.S., 2012. Cadmium. In: Wood, C.M., Farrell, A.P., Brauner, C.J. (Eds.), Homeostasis and Toxicology of Non-Essential Metals. Fish Physiology. vol. 31, pp. 125–184 (Part B).

Meador, J.P., 2015. Tissue concentrations as the dose metric to assess potential toxic effects of metals in field-collected fish: copper and cadmium. Environ. Toxicol. Chem. 34 (6), 1309–1319.

Mebane, C.A., 2006. Cadmium risks to freshwater life: derivation and validation of low-effect criteria values using laboratory and field studies (Report no. 2006–5245). US Geological Survey.

Meng, J., Wang, W., Li, L., Yin, Q., Zhang, G., 2017. Cadmium effects on DNA and protein metabolism in oyster (*Crassostrea gigas*) revealed by proteomic analyses. Sci. Rep. 7, 11716.

Mortuza, M.G., Al-Misned, F.A., 2017. Environmental contamination and assessment of heavy metals in water, sediments and shrimp of Red Sea Coast of Jizan, Saudi Arabia. J. Aquat. Pollut. Toxicol. 1, 1.

Muhammad, N., Banoori, N., Akbar, A., Azizullah, A., Khan, M., Qasim, M., Rahman, H., 2017. Microbial and toxic metal contamination in well drinking water: potential health risk in selected areas of Kohat, Pakistan. Urban Water J. 14 (4), 394–400.

Murakami, M., Nakagawa, F., Ae, N., Ito, M., Arao, T., 2009. Phytoextraction by rice capable of accumulating Cd at high levels: reduction of Cd content of rice grain. Environ. Sci. Technol. 43, 5878–5883.

Olmedo, P., Pla, A., Hernández, A.F., Barbier, F., Ayouni, L., Gil, F., 2013. Determination of toxic elements (mercury, cadmium, lead, tin and arsenic) in fish and shellfish samples. Risk assessment for the consumers. Environ. Int. 59, 63–72.

Owusu-Boateng, G., Gadogbe, M.K., KNUST, K., 2015. Domestic rainwater harvesting in a water-stressed community and variation in rainwater quality from source to storage. Int. J. Sustain. Dev. 14 (2), 225–243.

Pan, J., Plant, J.A., Voulvoulis, N., Oates, C.J., Ihlenfeld, C., 2010. Cadmium levels in Europe: implications for human health. Environ. Geochem. Health 32 (1), 1–12.

Perera, P.C.T., Sundarabarathy, T.V., Sivananthawerl, T., Kodithuwakku, S.P., Edirisinghe, U., 2016. Arsenic and cadmium contamination in water, sediments and fish is a consequence of paddy cultivation: evidence of river pollution in Sri Lanka. Achiev. Life Sci. 10 (2), 144–160.

Prokop, Z., Cupr, P., Zlevorova-Zlamalikova, V., Komarek, J., Dusek, L., Holoubek, I., 2003. Mobility, bioavailability, and toxic effects of cadmium in soil samples. Environ. Res. 91 (2), 119–126.

Rafiq, M.T., Aziz, R., Yang, X., Xiao, W., Rafiq, M.K., Ali, B., Li, T., 2014. Cadmium phytoavailability to rice (*Oryza sativa L.*) grown in representative Chinese soils. A model to improve soil environmental quality guidelines for food safety. Ecotoxicol. Environ. Saf. 103, 101–107.

Rashed, M.N., 2001. Cadmium and lead levels in fish (*Tilapia nilotica*) tissues as biological indicator for lake water pollution. Environ. Monit. Assess. 68 (1), 75–89.

Razak, N.H.A., Praveena, S.M., Aris, A.Z., Hashim, Z., 2015. Drinking water studies: a review on heavy metal, application of biomarker and health risk assessment (a special focus in Malaysia). J. Epidemiol. Glob. Health 2015 (5), 297–310.
Resource Conservation and Recovery Act, 1976. EPA's Office.
Rizwan, M., Ali, S., Adrees, M., Ibrahim, M., Tsang, D.C., Zia-ur-Rehman, M., Rizwan, M., Ali, S., Adrees, M., Ibrahim, M., Tsang, D.C., Zia-ur-Rehman, M., Zahir, Z.A., Rinklebe, J., Tack, F.M., Ok, Y.S., 2017. A critical review on effects, tolerance mechanisms and management of cadmium in vegetables. Chemosphere 182, 90–105.
Rocha, T.L., Gomes, T., Mestre, N.C., Cardoso, C., Bebianno, M.J., 2015. Tissue specific responses to cadmium-based quantum dots in the marine mussel *Mytilus galloprovincialis*. Aquat. Toxicol. 169, 10–18.
Royston, M.G., 2013. Pollution Prevention Pays. Elsevier, Beccles, London.
Rzętała, M.A., 2016. Cadmium contamination of sediments in the water reservoirs in Silesian Upland (southern Poland). J. Soils Sediments 16 (10), 2458–2470.
Salehi, M., Li, X., Whelton, A.J., 2017. Metal accumulation in representative plastic drinking water plumbing systems. J. Am. Water Works Assoc. 109 (11), E479–E493.
Santos, P.S., Otero, M., Santos, E.B., Duarte, A.C., 2011. Chemical composition of rainwater at a coastal town on the southwest of Europe: what changes in 20years? Sci. Total Environ. 409 (18), 3548–3553.
Sbihi, K., Cherifi, O., El Gharmali, A., Oudra, B., Aziz, F., 2012. Accumulation and toxicological effects of cadmium, copper and zinc on the growth and photosynthesis of the freshwater diatom *Planothidium lanceolatum* (Brébisson) Lange-Bertalot: a laboratory study. J. Mater. Environ. Sci. 3 (3), 497–506.
Schindler, M., Kamber, B.S., 2013. High-resolution lake sediment reconstruction of industrial impact in a world-class mining and smelting center, Sudbury, Ontario, Canada. Appl. Geochem. 37, 102–116.
Selinus, O., Alloway, B.J., 2013. Environmental pathology. In: Centeno, J.A., Finkelman, R.B., Fuge, R., Lindh, U., Smedley, P. (Eds.), Essentials of Medical Geology. Springer, New York, NY, p. 820.
Shamsi, I.H., Zhang, G., Hu, H., Xue, Q., Hussain, N., Ali, E., Shen, Q., Zheng, W., Zhang, Q., Liu, X., Jabeen, Z., 2014. Assessment of the hazardous effects of Cd on physiological and biochemical characteristics of soybean genotypes. Int. J. Agric. Biol. 16, 41–48.
Shi, Y., Wang, R., Lu, Y., Song, S., Johnson, A.C., Sweetman, A., Jones, K., 2016. Regional multi-compartment ecological risk assessment: establishing cadmium pollution risk in the northern Bohai Rim, China. Environ. Int. 94, 283–291.
Stankovic, S., Kalaba, P., Stankovic, A.R., 2014. Biota as toxic metal indicators. Environ. Chem. Lett. 12 (1), 63–84.
St-Hilaire, A., Duchesne, S., Rousseau, A.N., 2016. Floods and water quality in Canada: a review of the interactions with urbanization, agriculture and forestry. Can. Water Resour. J. 41 (1–2), 273–287.
Su, C., 2014. A review on heavy metal contamination in the soil worldwide: situation, impact and remediation techniques. Environ. Skept. Crit. 3 (2), 24.
Tang, X., Li, Q., Wu, M., Lin, L., Scholz, M., 2016. Review of remediation practices regarding cadmium-enriched farmland soil with particular reference to China. J. Environ. Manag. 181, 646–662.
Taylor, M., Caldwell, J., Sneath, G., 2017. In: Current state and trend of cadmium levels in soil, freshwater and sediments across the Waikato region. Science and Policy: Nutrient Management Challenges for the Next Generation. Occasional Report No. 30. Fertilizer and Lime Research Centre, Massey University, Palmerston North, p. 11.

Tolcin, A.C., 2009. Cadmium. In Mineral Commodity Summaries. U.S. Geological Survey, http://minerals.usgs.gov/minerals/pubs/commodity/cadmium/mcs-2009-cadmi.pdf.

Uluturhan, E., Kontas, A., Can, E., 2011. Sediment concentrations of heavy metals in the Homa Lagoon (Eastern Aegean Sea): assessment of contamination and ecological risks. Mar. Pollut. Bull. 62 (9), 1989–1997.

Vig, K., Megharaj, M., Sethunathan, N., Naidu, R., 2003. Bioavailability and toxicity of cadmium to microorganisms and their activities in soil: a review. Adv. Environ. Res. 8 (1), 121–135.

Vitti, A., Nuzzaci, M., Scopa, A., Tataranni, G., Remans, T., Vangronsveld, J., Sofo, A., 2013. Auxin and cytokinin metabolism and root morphological modifications in *Arabidopsis thaliana* seedlings infected with Cucumber mosaic virus (CMV) or exposed to cadmium. Int. J. Mol. Sci. 14 (4), 6889–6902.

Wu, J., Sun, Z., 2016. Evaluation of shallow groundwater contamination and associated human health risk in an alluvial plain impacted by agricultural and industrial activities, Mid-west China. Expo. Health 8, 311–329.

Xiong, T., Leveque, T., Shahid, M., Foucault, Y., Mombo, S., Dumat, C., 2014. Lead and cadmium phytoavailability and human bioaccessibility for vegetables exposed to soil or atmospheric pollution by process ultrafine particles. J. Environ. Qual. 43 (5), 1593–1600.

Zhang, Y.F., Chen, S.Y., Qu, M.J., Adeleye, A.O., Di, Y.N., 2017. Utilization of isolated marine mussel cells as an in vitro model to assess xenobiotics induced genotoxicity. Toxicol. In Vitro 44, 219–229.

## FURTHER READING

Henson, M.C., Chedrese, P.J., 2004. Endocrine disruption by cadmium, a common environmental toxicant with paradoxical effects on reproduction. Exp. Biol. Med. (Maywood) 229 (5), 383–392.

Nookabkaew, S., Rangkadilok, N., Prachoom, N., Satayavivad, J., 2016. Concentrations of trace elements in organic fertilizers and animal manures and feeds and cadmium contamination in herbal tea (Gynostemma pentaphyllum Makino). J. Agric. Food Chem. 64 (16), 3119–3126.

Rahman, M.M., Rahman, M.M., Chongling, Y., Islam, K.S., 2011. Accumulation, distribution and toxicological effects induced by cadmium on the development of mangrove plant *Kandelia candel (L.)*. Contemp. Probl. Ecol. 4 (2), 133–139.

UNEP, 2010. Final Review of Scientific Information on Cadmium. United Nations Environment Programme, Nairobi. Available from, www.unep.org/hazardoussubstances/Portals/9/Lead_Cadmium/docs/Interim_reviews/UNEP_GC26_INF_11_Add_2_Final_UNEP_Cadmium_review_and_apppendix_Dec_2010.pdf.

WHO, 2013. Ten Chemicals of Major Health Concern. Retrieved from, www.who.int/ipcs/assessment/public_health/chemicals_phc/en/index.html.

# CHAPTER 7

# Environmental Hazards of Cadmium: Past, Present, and Future

**Malik Tahir Hayat*, Muhammad Nauman*, Nida Nazir*, Shafaqat Ali†, Nazneen Bangash‡**

*Department of Environmental Sciences, COMSATS, Abbottabad, Pakistan
†Department of Environmental Sciences and Engineering, Government College University, Faisalabad, Pakistan
‡Department of Bio Sciences, COMSATS, Islamabad, Pakistan

## Abstract

Cadmium (Cd) is a toxic element that is not required by all living organisms. It poses many environmental threats that can ultimately cause disorders in plants and diseases in human beings. It can accumulate in different organs of the human body due to its persistent nature. Cd is released into the environment through various natural and anthropogenic activities such as mining, smelting, and refining. It is also being used in many other industrial applications, such as plastics, pigments, enamels, ceramics, and steel plating. Cd exposure can be through water, air, and soil, which results in Cd toxicity. Some industrial processes produce Cd as a byproduct. Using emissions control equipment, the emission from the incineration of municipal solid waste can be controlled up to 95%. Cd toxicity can affect bones, the respiratory system, kidneys, reproduction, and can potentially lead to cancer in humans. Low concentrations of cadmium are highly toxic, and carcinogenic to plants. Cd has no part in plants' growth and other metabolic processes. It is mainly accumulated in the roots, shoots, and edible parts of the plants. Some plants can tolerate a small concentration of Cd, while some show changes in their phenotype, such as a decrease in weight, roots, and shoot lengths. Cytotoxicity, inflammation, and deterioration have also been observed in plants due to Cd toxicity. Different types of food may have different levels of cadmium, such as green leafy vegetables, fish, and meat. So, if the present trend of Cd utilization continues, worse consequences will result. Restoration of Cd-contaminated sites and proper disposal of Cd-containing materials must be followed to prevent Cd toxicity.

**Keywords:** Cadmium toxicity, Food chain, Accumulation, Industrial waste, Agricultural waste, Urban health

*Cadmium Toxicity and Tolerance in Plants*
https://doi.org/10.1016/B978-0-12-814864-8.00007-3

## 1 INTRODUCTION

The element cadmium (Cd) was discovered in 1817 by F. Strohmeier, a German chemist. Cadmium (atomic number 48) is a simple element, and it lies deep beneath the earth's surface. It is a highly toxic element, and its commercial use was banned until the end of the 19th century (Fthenakis, 2004). Its texture is like soft metal, and is bluish-white. Its chemical properties are similar to two stable metals, mercury and zinc, which are present in group 12 on the periodic table, with 321°C melting and 767°C boiling points, respectively. It is usually not found as a pure element in the environment, and is usually used in the formation of lead and copper ore, some complex oxides, sulfides, and carbonates in zinc.

Cadmium is not required by the body, and is toxic in nature, and has unhealthy effects on the human body, especially on bones. Cadmium can accumulate in the kidneys and liver, and may be a source of contamination in the future as an inhaled carcinogen (Joseph, 2009).

Cadmium in the environment is also toxic for micro-organisms, plants, and animals, as cadmium is a chemically simple element, and is persistent in nature, so it is not easily broken into less toxic compounds in the environment.

## 2 SOURCES OF CADMIUM

Cadmium is commonly seen as an impurity present in zinc ores; mostly it is isolated as a byproduct during the production of zinc (Zn). Roasting the ores of zinc sulfide in the presence of oxygen converts zinc sulfide to oxide. Cadmium is produced by mining, refining, or smelting of zinc, and is a lesser byproduct of the production of copper and lead (Bi et al., 2006). Therefore, cadmium is produced in relation to zinc, instead of through its direct production. Cadmium production almost doubled from 1950 to 1990, about 20,000 tons annually, while its consumption in this time period remained constant. Because of the improvement of refining technologies, production of cadmium in Europe has decreased, resulting in a significant reduction in cadmium emissions. Recycling accounts for 18% of the world's cadmium production (Frishberg, 2009).

Mostly, cadmium is used in nickel-cadmium production. About 80% of total cadmium in 2004 was used for cadmium production, but refined cadmium is also used in plastics pigments, plastics stabilizers, enamels, ceramics, and iron and steel plating; and is also used for alloying tin, copper, and lead alloys (Lambert, 2009). However, its consumption has significantly decreased in the production of stabilizers, pigments, and alloys since 1990 (Cavalcante et al., 2009).

## 3 EXPOSURE OF CADMIUM

Natural and anthropogenic (man-made) emissions are two major sources from which cadmium can enter the environment. Mainly, the primary deposition of cadmium is into three environmental sections: air, soil, and water; but after initial deposition into these elements, considerable transfer is observed. Emissions of cadmium are more mobile in air than in water, and least so in soils (Kaličanin, 2009).

### 3.1 Natural Exposure

Generally, the concentration of cadmium in the earth is around 0.1–0.5 ppm, but sedimentary rocks may accumulate higher levels. Phosphorites (a nondetrital sedimentary rock) and marine phosphates have been observed with cadmium concentration as much as 500 ppm (Jackson and Macgillivray, 1995). Erosion and weathering of rock materials are causing cadmium to be transported in large quantities by rivers (an estimated 15,000 mt (metric tons) of cadmium) into the world's oceans per annum. Volcanic eruption and forest fires are also major sources of cadmium emissions naturally, and around 820 metric tons of cadmium is annually released into the atmosphere due to volcanic activities, and up to 70 metric tons of cadmium's annual emissions are due to forest fires (Nriagu, 1990).

## 4 ANTHROPOGENIC CADMIUM EMISSIONS

The anthropogenic emissions of cadmium in the air mainly come from coal burning, fuel combustion, smelting of nonferrous metal, smelting of steel and iron, incineration of municipal solid waste, and from industries of phosphate fertilizer and cement. Cadmium can enter the environment either from its producer, user, or through disposal of cadmium-containing products. Many studies have attempted to provide an overview of man-made emissions of cadmium into water, air, and soil, and their sources (Bro-Rasmussen, 1996; Jackson and Macgillivray, 1995).

## 5 POINT SOURCES VERSUS DIFFUSED SOURCE

Two types of sources are potentially available for the emission of cadmium: point sources and diffused sources. Large production agencies are considered point sources, while disposal materials emissions are diffused emissions. Considerable work has been done in the past 20 years to control emissions from point sources, and a constant decrease has been observed (Jackson and Macgillivray, 1995) (Ramos et al., 1994). Scientists have focused on the control of diffused emissions by developing different strategies (Shafer, 1997).

## 6 EMISSIONS OF CADMIUM FROM MUNICIPAL SOLID WASTE INCINERATION

The incineration of municipal waste is a huge source of cadmium emissions in the environment, and ultimately poses a threat to human health. The ratio for incinerating municipal waste varies between 10% and 90% in different countries. State-of-the-art emission control equipment can capture 50%–99% of cadmium emissions. Furthermore, the different programs are being developed to create new methods of reducing the emission of cadmium from waste incineration.

Different products found in the municipal solid waste are responsible for cadmium emissions, including steel, and iron, cement, gypsum, and other construction materials and nonferrous alloys and metals products, residues of fossil fuels, and other natural substances, such as food, plants, and grasses in which some concentration of cadmium may found (Hasselriis and Licata, 1996).

### 6.1 Air

Several other processes resulting in the yield of cadmium include the burning of petroleum products, for example, coal or oil, and the burning of metropolitan waste, including plastics and nickel-cadmium batteries, which can also be deposited as solid waste (Sahmoun et al., 2005).

It is correspondingly released into the atmosphere as a result of industrial waste, which is expelled out of the chimneys of steel and iron factories.

Cadmium, when expelled into the environment from purifying, mining, and other processes, may diffuse in respirable-sized particles, thus reaching far-flung areas, which results in environmental pollution. It is then brought to earth in the form of rain. On the earth, it moves naturally into the layers of soil, and is then absorbed by plants, for example, green vegetables, root crops, grains, and cereals (ATSDR, 1999). Cadmium oxide additionally exists as tiny particles in air, caused by fumes and smog sometimes. These are the aftereffects of refining, patching, and other industrial and mechanical processes.

## 6.2 Water

Cadmium is still discharged in large amount into the environment due to human activities, and in water, cadmium is released from ship fuel and paints. Due to the high toxicity of cadmium, it has an adverse impact on water quality. It is reported that on average, <5 ng/L of cadmium is present in oceans (Bro-Rasmussen, 1996). Cadmium is one of the most prolific constituents of seawater. Concentration of cadmium is also found in drinking water supplies. Higher concentrations of cadmium are also reported in coastal areas. Scientists have examined the level and toxicity of cadmium in the Haihe basin in north China (Tang et al., 2017). Cadmium from drainage lines causes the increased cadmium in domestic drinking channels.

## 6.3 Food Chain

Different types of food may have different levels of cadmium. Green leafy vegetables such as spinach and lettuce, and some basic staple foods such as cereals and potatoes show high values, from 30 to 150 ppb. Sunflowers, peanuts, and soybean seeds naturally show high levels of cadmium, with no apparent negative health effects. Normally, fish and meat have a cadmium content of <5–40 ppb. Very high concentrations of cadmium were reported in animal organs such as the liver and kidneys (Satarug et al., 2003). Because of different agricultural practices, the level of cadmium in food may widely vary in some areas; some farmers use phosphate fertilizers, some use animal manure, and some use waste sludge as fertilizers. With the change in the varieties of crops, we see different levels of atmospheric cadmium deposition. Many studies show that the level of cadmium (about 95%) in nonsmokers mainly comes from food sources (Jarup, 2002). There is significant evidence that the levels of cadmium in food have declined significantly in the past few decades due to the gradual control of cadmium emissions in the environment.

## 7 CASE STUDY IN ENVIRONMENTAL MEDICINE 1990

A woman between the ages of 55–60 visited a doctor's office with protestations of low back agony, causing her a great deal of pain when walking. The pain had continuously increased in the 5 years prior to menopause. The pain is particularly noticeable after a continued period of sitting.

The patient's history uncovers that she has been married for 38 years. Her significant other, who is healthy, owns and works in a little sales shop in their home. The patient used to make adornments available for purchase in their shop for around 35 years. Their family consists of two grown children in good health.

The patient has no individual or family history of kidney ailments, diabetes, hypertension, or cardiovascular ailments. Additionally, she has no history of back injury or significant weight reduction. She habitually smoked about one to two packs of cigarettes every day for 40 years. She has not taken any estrogens, vitamins, calcium supplements, or other drugs.

Upon a proper checkup, the doctors observed the patient to have stooped posture and a waddling stride. Her blood pressure was 120 over 70. The patient's teeth had yellow staining, and her fingernails were likewise stained with nicotine. She was anemic The stomach and cardiovascular analysis were typical. Tapping her lower lumbar are created discomfort. Yet raising her straightened leg was not painful. Her profound ligament reflexes were in place. The rest of the physical and neurologic examination were ordinary. Sensation was typical in her legs and feet. The level of movement was typical in her hips and knees.

A urine analysis indicated 3+ proteinuria and glycosuria. BUN, creatinine, and egg white levels were ordinary. Roentgenograms of the lumbosacral spine and pelvis presented pseudo fractures and other proof of serious osteomalacia and mellow osteoporosis. No osteolytic or osteoblastic injuries were present.

The patient creates adornments in the storm cellar of her home, where she utilizes rough processors, etching hardware, patching devices, and different crude materials. She wears face mask while working with the processor. The work place is dusty, having two little windows close to the highest point of one divider, fit for giving aeration. There are no nearby or general machine-driven fumes. She confessed to smoking, and also to eating in the work area. The patient and her husband used to grow vegetables in the garden in the backyard. An adjacent treatment plant of wastewater provides free manure, which her husband used to apply to the garden at regular intervals.

The garden had always been irrigated and watered with the cadmium-infused water from the city.

The patient's problems also include backache, glycosuria, proteinuria, osteomalacia, yellow teeth, and much more. Smoking also caused increased in cadmium toxicity in the patient. Laboratory findings also led to the diagnosis of renal dysfunction, unhealthy lungs, and an infected stomach.

## 8 PRESENT SCENARIO AND USES OF Cd

Having a soft texture and a white-silver color, cadmium traces were first found in paint pigments used during World War 1. Cadmium is basically produced as a byproduct of processes such as mining and smelting of raw metals such as copper and zinc. A huge amount of cadmium is used in the manufacture of Ni-Cad batteries. Nowadays, around a quarter of all cadmium production is used in electrodes, which are the most important part of alkaline batteries. The rest of the proportion is used in paints and pigments, and in manufacturing of coatings and fertilizer products. It is also used as a neutron absorbent in nuclear fission (Žaltauskaitė and Sodienė, 2014).

### 8.1 Nickle-Cadmium Batteries

Batteries includes a positive electrode plate (nickel) and a negative electrode (cadmium) immersed in a potassium-hydroxide electrolytic solution. Each panel has a steel frame made of a nickel-plated steel sheet containing the active material (positive plate: mainly nickel, graphite and cobalt; negative plate: mainly iron oxide and cadmium). Ni-Cd batteries have great utility in rail and aircraft enterprises, and emergency electricity in customer appliances, which includes cordless strength tools, cell phones, camcorders, portable PCs, and toys (Morrow and Keating, 1997). Once collected, nickel-cadmium batteries can be recycled almost 100%. Currently, there are nine major nickel and cadmium recycling plants in the United States, Japan, and Europe that can recover about 20,000 tons of consumer and industrial nickel-Cd batteries and their waste. There is enough capacity to recover all nickel and cadmium from the batteries currently being collected.

### 8.2 Cadmium Pigments

Cadmium as a paint pigment was used to add glaze in many colors, such as red and deep pink, in the manufacturing of enamels, plastic pigments, glasses, and ceramics. As they can persist in the environment without fading for a very long time, and with-stand high pressure and temperatures, they are applied to areas where high pressure and temperature process are required.

## 8.3 Cadmium Stabilizers

Polyvinylchloride (PVC) stabilizers containing cadmium compounds can endure very harsh environments, so products containing these stabilizers are created for harsh weather, and temperature- and ultra-violet light-resistance. Around 1.7 tons of cadmium per year have been used as cadmium stabilizers since 1991–92. But no cadmium stabilizers were ever used in toys or food in contact with polyvinylchlorides. In 2000, some PVC manufacturing industries voluntarily signed an agreement to not use any cadmium-containing stabilizers by 2001. Later, some amendments were made to this agreement, and 0.01% of cadmium was allowed in all polyvinylchloride products. Somewhere, the level of cadmium could be 0.1% of weight, and there would be no penalty.

## 8.4 Cadmium Coatings

Cadmium coatings are important for metals because of their excellent quality of corrosion resistance, especially in moist and alkaline conditions. Like zinc, cadmium also provides excellent protection to metals and other alloys from being corroded, and acts as a substitute for steel, especially when its damaged coating exposes areas of metals. Cadmium coatings, in addition, provide good characteristics of torque because of a low friction coefficient and good lubricity, which restricts galvanic corrosion caused by electrical contact between metals and good conductivity of electricity. It also reduces the risk of jam in mechanical operations due to corrosion remains (Morrow, 1996).

## 8.5 Alloys and Minor Uses

Cadmium alloys encompass

1. Electrical conductivity alloys
2. Heat conductivity alloys
3. Electrical contact alloys

Different small-scale uses of cadmium are found in solar cells in the form of cadmium sulfate and telluride. Cadmium is also used in electronic appliances such as semiconductors.

# 9 POTENTIAL EFFECTS OF CADMIUM ON HUMANS

Cadmium was discovered in 1817 by F. Strohmeier, a German chemist. The initial observations made by clinical medicine were based on the toxic effect of cadmium. Delayed respiratory symptoms and acute gastrointestinal symptoms were observed in workers that used calcium carbonate as a polishing agent, due to inhalation and oral intake in 1858. In 1940, some acute intestinal effects were reported, including diarrhea and vomiting in persons consuming cadmium-contaminated drinks and foods (Njcaud et al., 1942). Osteomalacia was observed in some cadmium workers (Nicaud et al., 1942). There is generally a higher concentration in women's bodies in the blood, kidneys, cortex, and urine (Vahter et al., 2007). Increased intestinal absorption is the main reason of higher concentrations of cadmium in women (Fig. 1).

## 9.1 Effect on Bones

Itai-Itai is a bone disease that occurred after World War II in the Fuchu Toyama territory located in Japan. Itai-Itai ("ouch-ouch") manifests as distortions and fractures in the long bones in the skeleton, triggering unbearable pain in affected people. In postmenopausal women, Cd was initially wrongly associated with the density of bone mineral. An increase in baseline

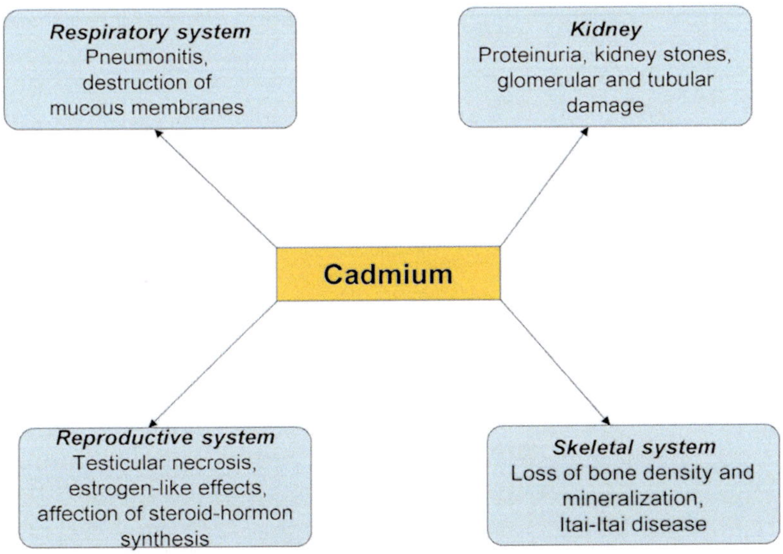


**Fig. 1** Effects of cadmium on humans.

extraction of urinary cadmium was linked with a 74% increased risk of bone fractures in women (Godt et al., 2006). A study shows that those who have eaten rice grown in cadmium-polluted irrigated fields have shown symptoms of this disease. Also, virtual fractures characteristic of osteomalacia (softening of bones) and severe softening of skeletal bones might be observed.

## 9.2 Renal Effects

Kidney problems are the major disorder resulting from the acute exposure of cadmium. Cadmium-metallothionein (CdMT) is the form in which cadmium enters the kidney. Cd-MT is reabsorbed in the proximal tubules after it is filtered through the glomerulus. After that, cadmium is accumulated in the tubules cells. During one's lifespan, the cadmium concentration increases in the tubulus cells of the kidneys (Godt et al., 2006). Renal tubular damage of 4% is related to 2.4 μg of cadmium per gram creatinine excretion in urine. This shows the correlation of cadmium with the level of kidney damage induced (Järup et al., 1998). Urinary excreted β2-microglobulin, retinol-binding proteins (RBP), and *N*-acetyl-α-D-glucosaminidase (NAG) are the principal indicators of kidney damage (Bernard, 2004). In China, a cadmium study presented considerably high levels of urinary β2-microglobulin and RBP, mainly in individuals having a high concentration of blood cadmium as compared with people with normal values (Jin et al., 2002). Both tubular and glomerular damages where observed in first group. It has been debated whether tubular damage is reversible or not, and the general opinion is that tubular damage is irreversible (Godt et al., 2006).

## 9.3 Respiratory System

Cigarette smoke is the main source of inhalative cadmium intoxication. About 40%–60% of cadmium in cigarette smoke is absorbed by human lungs (Elinder et al., 1976).

Cadmium accumulation of 15 mg was observed in the average 50-year-old person who did not smoke. While the similar lifespan of a smoker shows 30 mg of cadmium. Smokers generally have 4–5 times more cadmium in their blood than a nonsmoker. Usually, inhalatively resorbed cadmium in the form of cadmium-cysteine complexes reaches the blood circulation system, which results in shortness of breath, destruction of mucous membranes, and lung edema (Zalups and Ahmad, 2003).

## 9.4 Reproductive System

The toxic effects of cadmium on the male reproductive system have been widely observed in animals, but inadequate information on humans is available (Nordberg, 2009). The production of progesterone and testosterone was observed. High dosages inhibit ovarian progesterone biosynthesis, whereas low dosages of cadmium stimulate it. An increase in spontaneous abortion and low birth weight are associated with maternal exposure to cadmium (Frery et al., 1993).

## 9.5 Carcinogenity

Cadmium has been categorized as a group I carcinogen for human beings by the International Agency for research on Cancer (IARC, 1993), while the European commission categorized some cadmium compounds as possibly carcinogenic. Cd carcinogenity in humans was first reported in 1967. Both Kipling and Waterhouse reported prostate cancer in Cd workers (1967), but no increase in lung cancer was reported. An increase in lung cancer because of Cd exposure was stated in studies by (Lemen et al., 1976). Molecular mechanisms of cadmium carcinogenity are not fully understood yet, but some factors may contribute to it, such as upregulation of mitogenic signaling, acquisition of apoptotic resistance by cadmium exposure, and perturbance of DNA-repairing mechanisms.

Cadmium contamination causes human renal and prostate cancers and lung cancer. However, cadmium carcinogenesis in other body parts ofhumans, such as the liver, pancreas, and stomach are still unclear. But in an animal, cadmium causes cancers at different sites, and by various routes and mechanisms, as shown in Fig. 2, where we can see the mechanism of cadmium as a carcinogen.

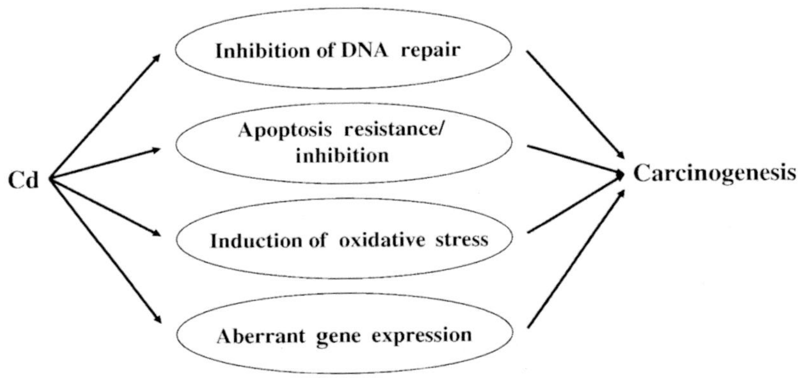


**Fig. 2** Mechanisms of cadmium carcinogenesis.

Cd is persistent in the environment, as well as in biological tissues, causing serious concerns, especially in the involvement of Cd in cancer in humans. Cd generally exists in the environment, some food products, and in particular work places.

In spite of the progress made in understanding different mechanisms of cadmium carcinogenesis, more research is required to entirely understand the contribution of Cd in the occurrence of human carcinogenesis, especially in populations where people are exposed to a high concentration of Cd.

## 10 POTENTIAL EFFECTS OF CADMIUM ON PLANTS

Even a low concentration of cadmium is highly toxic and carcinogenic to plants (Benavides et al., 2005: Khan et al., 2015). Cadmium is a nonessential element that is not involved in the growth and development of plants, animals, and humans. Soil organisms can also bear toxic effects due to high cadmium concentrations, which easily transfer into the vegetative cover; thus it eventually enters the food chain (Wahid et al., 2009; Li et al., 2016). Cd plays no metabolic function in plants, but is found to be stored in the roots, shoots, and edible parts (Rascio and Navari-Izzo, 2011; Zarcinas et al., 2004). Uptake of cadmium by plants from soil mainly depends on its bioavailability and concentration. However, a low concentration is directly taken from the deposition of contaminated dust from the atmosphere (Clemens, 2006).

Some plants are able to tolerate some amounts of Cd, but its higher concentration leads to phytotoxicity. The concentration of Cd causing phytotoxicity depends on the plant species. Its concentrations of $>5–10\,\mu g\,Cd\,g^{-1}$ DM result in toxicity in plants (White and Brown, 2010).

### 10.1 Effect on the Phenotype

Toxicity of Cd has a powerful effect on the phenotype of plants, including a decrease in fresh weight and leaf length of wild garlic (Street et al., 2010).

### 10.2 Effect on Chlorophyll Content

The concentration of ATP and chlorophyll content were decreased by cadmium (Hédiji et al., 2010). The synthesis of chlorophyll in plants inhibits the photosynthesis in leaves, and affects the functioning of photochemical centers (Jing et al., 2005). The dry weight of shoots and roots (wheat) was decreased by cadmium toxicity (Zhang et al., 2002).

### 10.3 Effect on Physiological and Metabolic Processes

Cadmium toxicity disturbs the functions of mitochondria, resulting in cytotoxicity, inflammation, and deterioration. Cd also causes chlorosis and cell death of plant tissues. It is also involved in the growth hindrance of shoots and roots of the plants. The decrease in nutrient uptake and vein reddening is also caused by cadmium toxicity (Júnior et al., 2014; Khan et al., 2016). Tomato plants are affected by cadmium, which changes physiological and metabolic processes and inhibits growth (López-Millán et al., 2009).

### 10.4 Reactive Oxygen Species

One of the main effects of cadmium on cells is the production of reactive oxygen species that damage the membrane (Belkhadi et al., 2010). Nutrient deficiency is seen in the plants grown in contaminated soil of cadmium, which causes a negative impact on the nutritional value and growth of the plants (Khan et al., 2015).

## 11 CADMIUM INTAKE IN HUMANS

### 11.1 Ingestion

Mostly, cadmium that enters the human body comes from land-based foods. We are either getting food from plants or animals grown up in or on cadmium-contaminated soil. Cadmium present in the soil becomes part of our food chain, which is the primary cause of cadmium intake in the human body. Some studies show that 90% of cadmium in our body comes from land-based foods, while aquatic food and drinking water are responsible for only 2% of the intake of cadmium in the human body. As mentioned herein, the level of cadmium present in crops significantly depends on the soil, and agriculture trends such as usage of fertilizers and sludge (Van Assche et al., 1988).

### 11.2 Cadmium Consumption From Meals

Many scientists are trying to determine the average daily intake of cadmium in foods, and generally, these studies show that the daily average dietary intake of nonsmokers in noncontaminated areas is currently between 10 and 25 μg of cadmium (Elinder et al., 1985; OECD, 1994; ATSDR, 1997). Generally, in many countries, a decreasing trend of cadmium in blood levels has been confirmed (Ducoffre et al., 1992; Hoffmann et al., 2001). WHO and other health organizations have developed a weekly intake value of cadmium, according to body weight, that a human body can tolerate. The weekly cadmium intake level is 7 μg/kg in relation to the daily intake that the human body can tolerate, which is on average 70 μg/kg for 70 kg for men and 60 μg/kg for 60 kg for women, per day. This clearly shows the maximum tolerable cadmium intake from food sources (Van Assche and Ciarletta, 1992).

## 11.3 Inhalation

Inhaled cadmium contributes a significantly smaller amount of cadmium, as observed in workers exposed to high concentrations of cadmium, and smokers. Nowadays, in a professional working environment, the chance of cadmium inhalation is much more controlled. Cadmium exposure is also well-controlled in working environments that are directly related to cadmium and nonferrous production industries. On the other hand, emissions form thermal power plants, and other main industries where cadmium may have low levels of cadmium present in form of impurities, such as the steel and iron industries, may be important because of their emissions.

Cadmium amounts in cigarette smokers is 1–3 μg, which is significantly more as compared with nonsmokers. Some reports illustrate that one cigarette, which contains around 1–2 μg of Cd, is about 10% of inhaled cadmium content. The amount of cadmium exposure can be increased or decreased by the type of tobacco and filter used in cigarette production. A daily cigarette smoker can inhale almost double the average daily intake of a nonsmoker. There are different anthropogenic sources of cadmium inhalation listed, with percentage of exposure, in Table 1.

**Table 1** Anthropogenic sources of cadmium

| Sources | Exposure % |
|---|---|
| Phosphate fertilizers | 41.3 |
| Fossil fuel combustion | 22 |
| Iron and steel production | 16.7 |
| Natural sources | 8 |
| Nonferrous metals | 6.3 |
| Cement production | 2.5 |
| Cadmium products | 2.5 |
| Incineration | 1 |

The anthropogenic sources of cadmium, including fossil gasoline combustion, some industrial activities, and phosphate fertilizers, contribute more to a human's exposure to cadmium than their use, production, and disposal of cadmium yields in centration of all cadmium-incorporated materials. Though, proven previously, this remark must be positioned inside the potential of the average of each day. Cadmium intakes below those vital limits ensure human fitness. Thus, no action appears required to lessen cadmium from its present levels.

## 12 CADMIUM MANAGEMENT IN THE BODY

The greater part of cadmium is transported to proteins, for example, albumin and metallothionein, immediately after it enters the blood.

After uptake in GI-blood, the liver is the most important organ to be affected. In the liver, cadmium activates the generation of protein known as metallothionein. Then, through successive hepatocyte putrefaction and apoptosis, cadmium metallothionein edifices sweep away into the sinusoidal blood. Here, components of the consumed Cd enter the entero-hepatical cycle by route of secretion into the biliary tract in the form of cadmium-glutathione conjugates. Enzymatically debased to cadmium-cysteine homes inside the biliary tree, cadmium re-enters the small digestive organs (Järup and Åkesson, 2009).

The kidney is the most important organ for long-haul cadmium accumulation. Here, in the kidney, the half-life time of cadmium is nearly 10 years. A deep-rooted admission can accordingly prompt a cadmium deposition in the kidneys, therefore bringing about tubules cell rot.

The blood grouping of cadmium fills as a solid marker, while urinary presence is displayed in secretion and renal collection (Klaassen et al., 2009). Discharge of cadmium occurs with the aid of defecation and urination. The process of treatment of cadmium in the human body is shown in Fig. 3.

## 13 FUTURE PERSPECTIVES OF CADMIUM

Cadmium will remain in our societies, in all types of products, both cheap and expensive. Today, it is well controlled, so there are no compelling reasons to reduce or boycott cadmium products, and the transmission of cadmium from objects to humans is very small, so this concern is not very important.

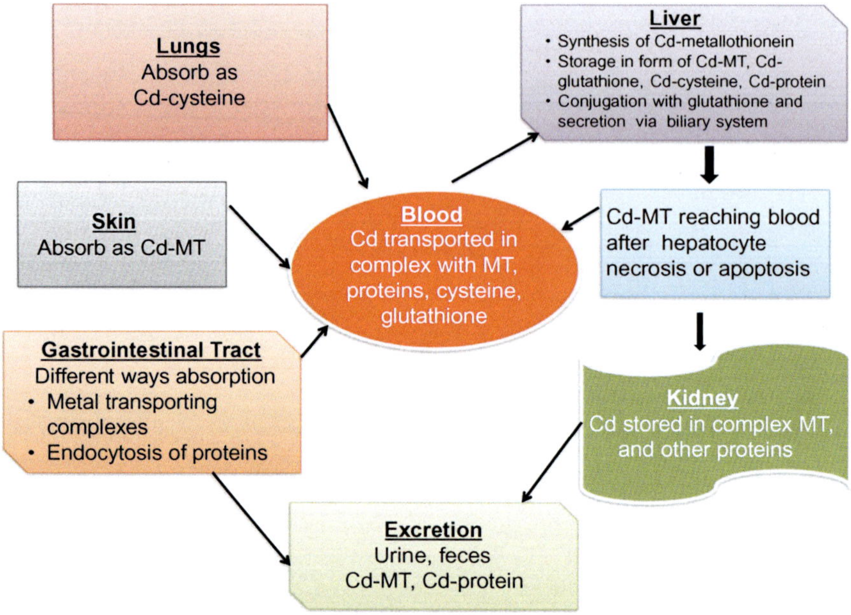


**Fig. 3** Process of treatment of cadmium in the human body.

## 13.1 Recycling of Cd Containing Products

Nickel-cadmium batteries are imperative in numerous objects, especially those requiring high electricity, prolonged lifecycles, and super low- or high-temperatures operation. Essential nonrechargeable batteries can be replaced by rechargeable Ni-Cd batteries. In this way, we can decrease the waste considerably. The recyclable materials of Ni-Cd batteries can be 99% reused to produce new batteries. All battery producers from Europe, Japan, Australia, North America, and OECD increased their battery production from recycled materials, which enhances the rate of reusing Ni-Cd. Ni-Cd batteries are additionally influencing critical commitments to the advancement of the electric auto in Europe, in this manner adding to the change of urban air quality (Satarug et al., 2003).

## 13.2 Worse Consequences if the Present Trend Continues

In some plastics, glass, chinaware, and decorations, cadmium hue and stabilizers are inevitable additives that produce delicate shapes and long-lasting battery life for demanding applications. From a biological prospective, it is essential to generate and sustain functional materials through extended management lives, and eventually reduce the contribution to waste streams in the world (Jin et al., 2002). The production of low-cost alternatives that reduce the life of the products will eventually result in increasing waste around the world. It must also be emphasized that the cadmium used in these applications is a very rigid form of insoluble material that meshes with objects.

Cadmium covers the industry in such a way as to provide ultra-low corrosion factors, low resistance electric equipment, large electroplating values, a large coating range, a special variety of protection that can cover a wide range of surfaces, excellent durability, and soldering ability. As a result, cadmium-covered elements are favored for a variety of basic applications by providing benefits in areas such as aviation, atomic energy, barriers, electricity, mining, and ocean work. More importantly, the cadmium-plated silverware and articles are reused without interruption (Godt et al., 2006).

### 13.3 Restoration and Proper Disposal

The cadmium recovery project ensures that cadmium flows from waste are retained to help prevent the contamination of earth. This allows low levels of human cadmium consumption under WHO standards, and is a relatively low commitment to reducing cadmium. Such measures will undermine efforts to collect and recycle cadmium on a global scale (Nordberg, 2004). It will also reduce the EU's impact on the world market through the specialization of plants and labor camps. In Japan and North America, these assumptions have just been reached. The OECD program to reduce the risk of cadmium also received comparative results.

The view was expressed that the European Union should cooperate with well-informed enterprise management activities for the collection and reuse of products containing cadmium, which may increase rational and safe usage of cadmium in the modern world (Misra and Pandey, 2005).

## 14 CONCLUSION

As discussed herein, cadmium (Cd) is a highly toxic element that is not required by the body. It is released into the environment through various natural and anthropogenic sources that can be further categorized into point and diffuse sources. The incineration of municipal waste is a huge source of Cd emission in the environment. Humans are exposed to Cd through air, water, and food contamination, and it causes many diseases that can impact the gastrointestinal, renal, reproductive, and respiratory systems, including

Itai-Itai, osteomalacia diseases, and in severe cases, cancer. It is being used in various industrial applications as pigments, stabilizers, and coatings and alloys. Plants are also affected by Cd exposure, which can result in various disorders such as reduced chlorophyll content, less fresh and leaf weight, and physiological and metabolic dysfunctions under stressed conditions. Therefore, it is crucial to prevent the use of Cd-containing products to reduce its harmful impacts on humans, as well as other living organisms.

## REFERENCES


ATSDR, 1997. 1997 CERCLA priority list of hazardous substances that will be the subject of toxicological profiles and support document. Division of Toxicology, Agency for Toxic Substances and Disease Registry. U.S. Department of Health and Human Services, Atlanta, GA.

Agency for Toxic Substances and Disease Registry (ATSDR), 1999. Toxicological Profile for Cadmium (update), Agency for Toxic Substances and Disease Registry. Atlanta, Georgia, pp. 1–397.

Belkhadi, A., Hediji, H., Abbes, Z., Nouairi, I., Barhoumi, Z., Zarrouk, M., Djebali, W., 2010. Effects of exogenous salicylic acid pre-treatment on cadmium toxicity and leaf lipid content in *Linum usitatissimum* L. Ecotoxicol. Environ. Saf. 73 (5), 1004–1011.

Benavides, M.P., Gallego, S.M., Tomaro, M.L., 2005. Cadmium toxicity in plants. Braz. J. Plant Physiol. 17 (1), 21–34.

Bernard, A., 2004. Renal dysfunction induced by cadmium: biomarkers of critical effects. Biometals 17 (5), 519–523.

Bi, X., Feng, X., Yang, Y., Qiu, G., Li, G., 2006. Quantitative assessment of cadmium emission from zinc smelting and its influences on the surface soils and mosses in Hezhang County, Southwestern China. Atmos. Environ. 40 (22), 4228–4233. https://doi.org/10.1016/j.atmosenv.2006.02.019.

Bro-Rasmussen, F., 1996. Contamination by persistent chemicals in food chain and human health. Sci. Total Environ. 188 (Suppl), S45–S60. https://doi.org/10.1016/0048-9697(96)05276-X.

Cavalcante, P.M.T., Dondi, M., Guarini, G., Raimondo, M., Baldi, G., 2009. Colour performance of ceramic nano-pigments. Dyes Pigments 80 (2), 226–232. https://doi.org/10.1016/j.dyepig.2008.07.004.

Clemens, S., 2006. Toxic metal accumulation, responses to exposure and mechanisms of tolerance in plants. Biochimie 88 (11), 1707–1719.

Ducoffre, G., Claeys, F., Sartor, F., 1992. Decrease in blood cadmium levels over time in Belgium. Arch. Environ. Health 47 (5), 354–356. https://doi.org/10.1080/00039896.1992.9938374.

Elinder, C.G., Kjellström, T., Friberg, L., Linnman, B.L.L., 1976. Cadmium in kidney cortex, liver, and pancreas from Swedish autopsies. Arch. Environ. Health 31 (6), 292–302.

Elinder, C.G., Kjellström, T., Hogstedt, C., et al., 1985. Cancer mortality of cadmium workers. Occup. Environ. Med. 42, 651–655.

Frery, N., Nessmann, C., Girard, F., Lafond, J., Moreau, T., Blot, P., Huel, G., 1993. Environmental exposure to cadmium and human birthweight. Toxicology 79 (2), 109–118.

Frishberg, I.V., 2009. Production of zinc, cadmium and their alloy powders. In: Neikov, O.D., Naboychenko, S.S., Murashova, I.V., Gopienko, V.G., Frishberg, I.V., Lotsko, D.V. (Eds.), Handbook of Non-Ferrous Metal Powders. Elsevier, Oxford, pp. 409–422 (Chapter 19).

Fthenakis, V.M., 2004. Life cycle impact analysis of cadmium in CdTe PV production. Renew. Sust. Energ. Rev. 8 (4), 303–334.

Godt, J., Scheidig, F., Grosse-Siestrup, C., Esche, V., Brandenburg, P., Reich, A., Groneberg, D.A., 2006. The toxicity of cadmium and resulting hazards for human health. J. Occup. Med. Toxicol. 1 (1), 22.

Hasselriis, F., Licata, A., 1996. Analysis of heavy metal emission data from municipal waste combustion. J. Hazard. Mater. 47 (1), 77–102. https://doi.org/10.1016/0304-3894(95)00107-7.

Hédiji, H., Djebali, W., Cabasson, C., Maucourt, M., Baldet, P., Bertrand, A., Chaïbi, W., 2010. Effects of long-term cadmium exposure on growth and metabolomic profile of tomato plants. Ecotoxicol. Environ. Saf. 73 (8), 1965–1974.

Hoffmann, K., Krause, C., Seifert, B., 2001. The German Environmental Survey 1990/92 (GerES II): primary predictors of blood cadmium levels in adults. Arch. Environ. Health 56 (4), 374–379. https://doi.org/10.1080/00039890109604471.

IARC, 1993. Beryllium, Cadmium, Mercury, and Exposures in the Glass Manufacturing Industry, vol. 58. IARC Working Group on the Evaluation of the Carcinogenic Risk of Chemicals to Humans. World Health Organization.

Jackson, T., Macgillivray, A., 1995. Accounting for cadmium: tracking emissions of cadmium from the global economy. Chem. Ecol. 11 (3), 137–181. https://doi.org/10.1080/02757549508039067.

Järup, L., 2002. Cadmium overload and toxicity. Nephrol. Dial. Transplant. 17 (2), 35–39.

Järup, L., Åkesson, A., 2009. Current status of cadmium as an environmental health problem. Toxicol. Appl. Pharmacol. 238 (3), 201–208.

Järup, L., Berglund, M., Elinder, C.G., Nordberg, G., Vanter, M., 1998. Health effects of cadmium exposure—a review of the literature and a risk estimate. Scand. J. Work Environ. H, 1–51.

Jin, T., Nordberg, M., Frech, W., Dumont, X., Bernard, A., Ye, T.T., Li, Y., 2002. Cadmium biomonitoring and renal dysfunction among a population environmentally exposed to cadmium from smelting in China (ChinaCad). Biometals 15 (4), 397–410.

Jing, D., Fei-bo, W.U., Guo-ping, Z., 2005. Effect of cadmium on growth and photosynthesis of tomato seedlings. J. Zhejiang Univ. Sci. B 6 (10), 974–980.

Joseph, P., 2009. Mechanisms of cadmium carcinogenesis. Toxicol. Appl. Pharmacol. 238 (3), 272–279. https://doi.org/10.1016/j.taap.2009.01.011.

Júnior, C.A.L., Mazzafera, P., Arruda, M.A.Z., 2014. A comparative ionomic approach focusing on cadmium effects in sunflowers (*Helianthus annuus* L.). Environ. Exp. Bot. 107, 180–186.

Kaličanin, B.M., 2009. Determination of very toxic metal—cadmium in natural water samples. Desalination 249 (1), 58–62. https://doi.org/10.1016/j.desal.2009.03.006.

Khan, A., Khan, S., Khan, M.A., Qamar, Z., Waqas, M., 2015. The uptake and bioaccumulation of heavy metals by food plants, their effects on plants nutrients, and associated health risk: a review. Environ. Sci. Pollut. Res. Int. 22 (18), 13772–13799.

Khan, A., Khan, S., Alam, M., Khan, M.A., Aamir, M., Qamar, Z., Perveen, S., 2016. Toxic metal interactions affect the bioaccumulation and dietary intake of macro-and micronutrients. Chemosphere 146, 121–128.

Klaassen, C.D., Liu, J., Diwan, B.A., 2009. Metallothionein protection of cadmium toxicity. Toxicol. Appl. Pharmacol. 238 (3), 215–220.

Lambert, P., 2009. 6—Sustainability of metals and alloys in construction. In: Khatib, J.M. (Ed.), Sustainability of Construction Materials. Woodhead Publishing, pp. 148–170.

Lemen, R.A., Lee, J.S., Wagoner, J.K., Blejer, H.P., 1976. Cancer mortality among cadmium production workers. Ann. N. Y. Acad. Sci. 271 (1), 273–279.

Li, Y., Tang, H., Hu, Y., Wang, X., Ai, X., Tang, L., Qiu, J., 2016. Enrofloxacin at environmentally relevant concentrations enhances uptake and toxicity of cadmium in the earthworm *Eisenia fetida* in farm soils. J. Hazard. Mater. 308, 312–320.

López-Millán, A.F., Sagardoy, R., Solanas, M., Abadía, A., Abadía, J., 2009. Cadmium toxicity in tomato (*Lycopersicon esculentum*) plants grown in hydroponics. Environ. Exp. Bot. 65 (2), 376–385.

Misra, V., Pandey, S.D., 2005. Hazardous waste, impact on health and environment for development of better waste management strategies in future in India. Environ. Int. 31 (3), 417–431.

Morrow, H., 1996. Questioning the need to develop alternatives for cadmium coatings. In: Proceedings of 2nd Annual Cadmium Alternatives Conference, National Defence Centre for Environmental Excellence, Johnstown, Pennsylvania, pp. 13–15.

Morrow, H., Keating, J., 1997. Overview paper on effective recycling of Ni-Cd batteries. In: Proceedings of the OECD Workshop on the Effective Collection and Recycling of Nickel-Cadmium Batteries, Lyon, France, pp. 23–25.

Njcaud, P., Lafitte, A., Gros, A., 1942. Les troubles de l'intoxication chronique par le cadmium [Symptoms of chronic cadmium intoxication]. Arch. Mal. Prof. Med. Trav. Secur. Soc. 4, 192–202.

Nordberg, G.F., 2004. Cadmium and health in the 21st century—historical remarks and trends for the future. Biometals 17 (5), 485–489.

Nordberg, G.F., 2009. Historical perspectives on cadmium toxicology. Toxicol. Appl. Pharmacol. 238 (3), 192–200.

Nriagu, J.O., 1990. Global metal pollution: poisoning the biosphere? Environ. Sci. Policy Sustain. Dev. 32 (7), 7–33. https://doi.org/10.1080/00139157.1990.9929037.

OECD, 1994. Risk Reduction Monograph No. 5. Cadmium. Environment Directorate OECD, Paris.

Ramos, L., Hernandez, L.M., Gonzalez, M.J., 1994. Sequential fractionation of copper, lead, cadmium and zinc in soils from or near Doñana National Park. J. Environ. Qual. 23 (1), 50–57. https://doi.org/10.2134/jeq1994.00472425002300010009x.

Rascio, N., Navari-Izzo, F., 2011. Heavy metal hyperaccumulating plants: how and why do they do it? And what makes them so interesting? Plant Sci. 180 (2), 169–181.

Sahmoun, A.E., Case, L.D., Jackson, S.A., Schwartz, G.G., 2005. Cadmium and prostate cancer: a critical epidemiologic analysis. Cancer Investig. 23 (3), 256–263.

Satarug, S., Baker, J.R., Urbenjapol, S., Haswell-Elkins, M., Reilly, P.E., Williams, D.J., Moore, M.R., 2003. A global perspective on cadmium pollution and toxicity in non-occupationally exposed population. Toxicol. Lett. 137 (1–2), 65–83.

Shafer, W.H., 1997. Sanitary engineering, water pollution and resources. In: Shafer, W.H. (Ed.), Masters Theses in the Pure and Applied Sciences: Accepted by Colleges and Universities of the United States and Canada. In: vol. 40. Springer US, Boston, MA, pp. 309–316.

Street, R.A., Kulkarni, M.G., Stirk, W.A., Southway, C., Van Staden, J., 2010. Effect of cadmium on growth and micronutrient distribution in wild garlic (*Tulbaghia violacea*). S. Afr. J. Bot. 76 (2), 332–336.

Tang, W., Zhang, W., Zhao, Y., Zhang, H., Shan, B., 2017. Basin-scale comprehensive assessment of cadmium pollution, risk, and toxicity in riverine sediments of the Haihe Basin in North China. Ecol. Indic. 81 (Suppl. C), 295–301. https://doi.org/10.1016/j.ecolind.2017.06.011.

Vahter, M., Åkesson, A., Lidén, C., Ceccatelli, S., Berglund, M., 2007. Gender differences in the disposition and toxicity of metals. Environ. Res. 104 (1), 85–95.

Van Assche, F., Ciarletta, P., 1992. Cadmium in the environment: levels, trends, and critical pathways. In: Seventh International Cadmium Conference, pp. 51–56.

Van Assche, F., Cardinaels, C., Clijsters, H., 1988. Induction of enzyme capacity in plants as a result of heavy metal toxicity; dose-response relations in *Phaseolus vulgaris* L., treated with zinc and cadmium. Environ. Pollut. 52, 103–115.

Wahid, A., Arshad, M., Farooq, M., 2009. Cadmium phytotoxicity: responses, mechanisms and mitigation strategies: a review. In: Organic Farming, Pest Control and Remediation of Soil Pollutants. Springer, Netherlands, pp. 371–403.
White, P.J., Brown, P.H., 2010. Plant nutrition for sustainable development and global health. Ann. Bot. 105 (7), 1073–1080.
Žaltauskaitė, J., Sodienė, I., 2014. Effects of cadmium and lead on the life-cycle parameters of juvenile earthworm *Eisenia fetida*. Ecotoxicol. Environ. Saf. 103, 9–16.
Zalups, R.K., Ahmad, S., 2003. Molecular handling of cadmium in transporting epithelia. Toxicol. Appl. Pharmacol. 186 (3), 163–188.
Zarcinas, B.A., Ishak, C.F., McLaughlin, M.J., Cozens, G., 2004. Heavy metals in soils and crops in Southeast Asia. Environ. Geochem. Health 26 (3), 343–357.
Zhang, G., Fukami, M., Sekimoto, H., 2002. Influence of cadmium on mineral concentrations and yield components in wheat genotypes differing in Cd tolerance at seedling stage. Field Crop Res. 77 (2), 93–98.

## FURTHER READING

Grant, C.A., Buckley, W.T., Bailey, L.D., Selles, F., 1998. Cadmium accumulation in crops. Can. J. Plant Sci. 78 (1), 1–17. https://doi.org/10.4141/p96-100.

# CHAPTER 8

# Role of Phytochelatins in Cadmium Stress Tolerance in Plants

**Javed Ahmad*, Arlene Asthana Ali*, Mohd Affan Baig*, Muhammad Iqbal†, Inamul Haq*, M. Irfan Qureshi***
*Department of Biotechnology, Jamia Millia Islamia, New Delhi, India
†Department of Botany, Jamia Hamdard, Hamdard Nagar, New Delhi, India

## Abstract

Cadmium (Cd), a nonessential element, is readily taken up by plants, and negatively affects plant metabolism, growth, and development. Due to their sessile nature, plants have evolved different mechanisms to mitigate metal toxicity. Such mitigation takes place by immobilization of heavy metal ions, their exclusion, and the chelation of metal ions, as well as their compartmentalization. Primarily, in an adaptive response to Cd, plants strongly produce S-rich molecules, phytochelatins (PCs), which are well known for mediating a diverse array of Cd stresses in plants. PCs are enzymatically synthesized peptides and they have been measured in plants grown in environments with high and low metal concentrations. This chapter reviews the recent advances in PC research dealing with their biosynthetic regulation, roles in cadmium detoxification and/or accumulation, and PC synthase gene expression, to provide a better understanding of the mechanism involved in improving the phytoremediation efficiency of plants.

**Keywords:** Cadmium, Phytochelatin, Metal-binding ligands, Cd detoxification, Phytochelatin synthase, Antioxidants, Cd tolerance

## 1 INTRODUCTION

Heavy metal toxicity has affected the agricultural soils of many parts of the world. Soil contamination by the addition of heavy metals, such as arsenic (As), lead (Pb), cadmium (Cd), mercury (Hg), zinc (Zn), cobalt (Co), copper (Cu), chromium (Cr), and nickel (Ni) is a crucial environmental issue because of the potentially harmful effects the metals pose to the environment (Ahmad et al., 2018; Jalmi et al., 2018). Such poisonous substances are widely spread in the soil and impart acute, chronic toxicity to plants.

*Cadmium Toxicity and Tolerance in Plants*
https://doi.org/10.1016/B978-0-12-814864-8.00008-5

Photosynthetic organisms are the primary source of metals introduced into the food chain for animals and human beings (Srivastava et al., 2017). Non-essential metals such as Cd are highly toxic, and cause toxic effects, even at very low concentrations. The minimum regulatory cadmium (Cd) concentration in the soil utilized for agricultural activities is 100 mg $kg^{-1}$ of soil (Salt et al., 1995). But this threshold is continuously being exceeded because of various human activities. Cd causes toxicity due to its binding nonspecifically with sulfhydryl groups of proteins, and replacement of the essential metal ions in metalloenzymes. However, a variety of mechanisms have been developed in plants that regulate the uptake and accumulation of heavy metals present in soil (Viehweger, 2014). Plant responses to heavy metal toxicity involve the immobilization of heavy metal ions, their exclusion, and the chelation of metal ions, as well as their compartmentalization, along with generalized mechanisms, including synthesis of stress metabolites and stress proteins (Ahmad et al., 2018; Hasan et al., 2017). The most extensively used mechanism for detoxification of heavy metals in plants is the chelation of heavy metals by a ligand, often followed by the compartmentalization of the ligand-metal complex (Liu et al., 2015). Plants contain phytochelatins (PCs) and metallothioneins (MTs) as common ligands that bind heavy metals. PCs are found in numerous varieties of plants, including the dicots, monocots, gymnosperms, and algae, indicating that the pathway to produce PCs evolved quite early, so as to impart tolerance to the vascular plants against the adverse toxic environment (Shukla et al., 2016). PCs are synthesized enzymatically from a tripeptide glutathione (GSH), using the enzyme "phytochelatin synthase (PCS)" and trapping the heavy metals by binding to them. The PC-heavy metal complexes are produced within the cytosol, and eventually transported and sequestered into the cell vacuole (Merlos et al., 2014). Recent advances in the study of PC-related aspects are made mainly on the molecular and biochemical fronts, using the hyperaccumulator plant species. Heavy metal interaction with functional proteins at the molecular level, with regard to the activation of PC synthase by a heavy metal such as Cd, followed by synthesis of PCs, has attracted the attention of researchers. Herein is an attempt to gather and collate the latest advancement made in the study of phytochelatins, with references to cadmium detoxification in plants. A better understanding of detoxification mechanisms will facilitate the selection of hyperaccumulator plant species, and the genetic manipulation of such plants may further improve their phytoremediation ability (Iqbal et al., 2014).

## 2 STRUCTURE, BIOSYNTHESIS, AND REGULATION OF PHYTOCHELATIN

Phytochelatins are represented by the oligomeric form of glutathione, having the characteristic recurrence of γ-glutamate (Glu)-cysteine (Cys) dipeptide followed by a glycine (Gly). The general structure of PCs (Fig. 1A) is symbolized by (γ-Glu-Cys)*n*-Gly [(PC)*n*], where $n=2–11$. However, n usually corresponds to 5 (Cobbett and Goldsbrough, 2002; Pochodylo and Aristilde, 2017). PCs have been identified in different groups of organisms, including vascular plants, algae, fungi, microorganisms, and animals. Some plant species showed structural variations of PCs, such as (γ-Glu-Cys)*n*-γ-Ala, (γ-Glu-Cys)*n*-Ser and (γ-Glu-Cys)*n*-Glu (Grill et al., 2007). GSH acts as a substrate for PC biosynthesis, as shown by different molecular and biochemical studies. GSH biosynthesis involves two enzymes, viz. a rate-limiting enzyme γEC synthetase, and a GSH synthetase (Fig. 1B) (Merlos et al., 2014; Rea, 2012). Cadmium stress increases the activity of γEC synthetase, whereas the treatment with buthionine sulfoximine (BSO) has a suppressive effect (Cobbett, 2000). A mutant of

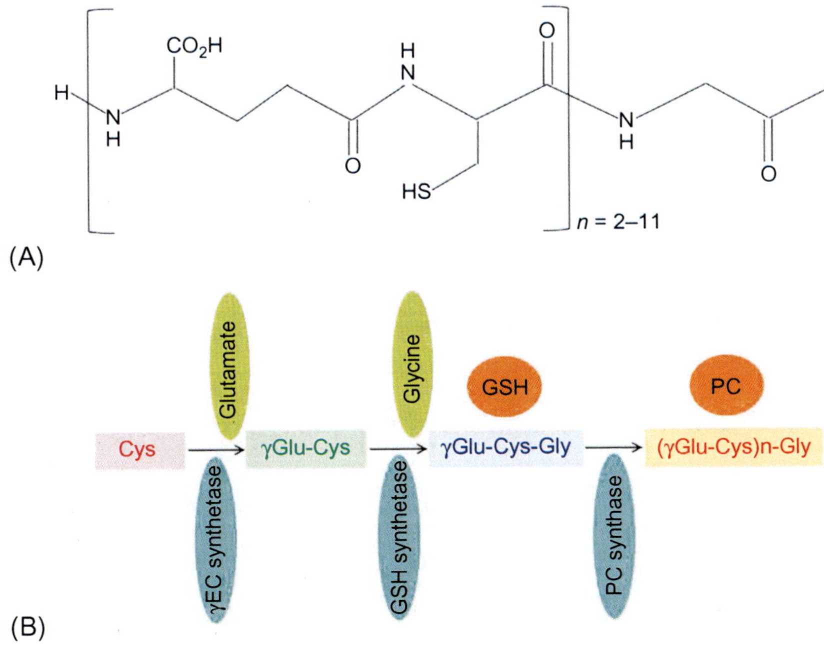


**Fig. 1** (A) The general structure of PCs. (B) Scheme of the biosynthesis of phytochelatins.

*Schizosacharomyces pombe* (PC deficient) (Mutoh and Hayashi, 1988) and a mutant of *Arabidopsis thaliana* (Cd-hypersensitive) (Cobbett et al., 1998; Howden et al., 1995) did not show γEC synthetase or GSH synthetase activity. On the other hand, a *sulfate transporter* 1;2 knockout mutant of *A. thaliana* (Yamaguchi et al., 2017) demonstrated an enhanced synthesis of phytochelatins upon exposure to Cd. Furthermore, a *cad1-3* mutant of *A. thaliana* (De Benedictis et al., 2018) (knockout mutant for *phytochelatin synthase1)* showed a defective deposition of callose upon the exposure of Cd, indicating the involvement of AtPCS1 not only in the synthesis of PCs, but also in the callose deposition. In addition, a mutant of *A. thaliana*, which was deficient of phytochelatin synthase, showed high sensitivity to Cd. However, this mutant was grown successfully as a wild type plant at the normal Zn and Cu level, signifying the positive role of phytochelatin in metal toxicity. The syntheses of PCs are mediated by phytochelatin synthase (PCS) via a posttranslational process. It is a γ-glutamylcysteine dipeptidyl (trans) peptidase that catalyzes the synthesis of PCs from GSH, or GSH and previously synthesized PCs, or from the previously synthesized PCs alone to produce polymers consisting of 2–11 γ-Glu-Cys repeats (López-Climent et al., 2014). The occurrence of a γ bond between cysteine and glutamic acid shows that the synthesis of PCs cannot take place by the ribosomes (Kang et al., 2007). Phytochelatin synthase (PCS) genes were isolated from *A. thaliana* (*AtPCS1*), *S. pombe* (*SpPCS*), *Triticum aestivum* (*TaPCS1*), *Arabidopsis halleri* (*AhPCS1*), *Thlaspi caerulescens* (*TcPCS1*), *Ceratophyllum demersum* (*CdPCS1*), *Malus hupehensis* (*MhPCS*), *Suaeda salsa* (*SsPCS*) by a PCR-based method (Clemens et al., 1999; Cong et al., 2016; Ha et al., 1999; Jiang et al., 2013; Meyer et al., 2011; Shukla et al., 2012; Vatamaniuk et al., 1999). These isolated genes encode 50–55 kDa polypeptides showing ~50% homology to each other. By the alignment of PCS, the proteins sequence indicates a high level of similarity in the N-terminal region, while the C-terminal domain shows enormous variability. The partial proteolysis and mutant studies have established that the N-terminal core domain is enough for PCS activity; as a result, it can be considered as a catalytic domain (Ruotolo et al., 2004; Tsuji et al., 2005). Nevertheless, the C-terminal domain is also important, and responsible for ensuring high PCS activity, and maintains protein stability and responds to a broad spectrum of heavy metal ions (Ruotolo et al., 2004; Vivares et al., 2005). The PCS gene is constitutively expressed; however, the expression level is too high in the presence of cadmium ions, as compared with other metal ions.

Exposure to heavy metal does not influence the gene of PC synthase, which has a constitutive expression in higher plants (Hirata et al., 2005).

The study of AtPCS1 with the northern and RT-PCR technique has confirmed this, demonstrating that the exposure of metals such as Cd, Pb, Hg, or of the oxidative stress does not control the transcription of AtPCS1 (Cazalé and Clemens, 2001). However, the level of AtPCS1 mRNA at the initial stage of plant development in Cd-treated *Arabidopsis* was two times greater than in the nontreated ones (Lee and Korban, 2002). Therefore, it appears that the expression of the PC synthase gene may be regulated in specific organs, and at the developmental stages that require a strong defense against the exposure of toxic heavy metals. The activity of PC synthase is not only modulated by Cd, As, or other heavy metals; but also affected by intracellular amounts of GSH. It is very clear that the GSH production is controlled by oxidative stress (Ahmad et al., 2017). The externally applied, as well as internally released $H_2O_2$ enhances the level of GSH in different plants (Rahantaniaina et al., 2017). Using the molecular and physiological approaches, Rizwan et al. (2018) demonstrated that the exogenous supply of GSH can reverse the unfavorable effects of Cd, as reflected by plant growth and development, together with the induction of PCs from GSH and genes related to Cd tolerance, including the phytochelatin synthetase-like protein, MYB, and WRKY transcription factors. Previously, such regulatory mechanisms were used with *Dunaliella tertiolecta* for the induction of PC synthesis in response to Cd stress (Tsuji et al., 2003). Moreover, the activity of PCS is regulated by phosphorylation and dephosphorylation. In this context, Wang et al. (2009) demonstrated by their in-vitro experiments that enzymes such as casein kinase 2 (CK2) and alkaline phosphatase regulate the PCS activity by phosphorylation and de-phosphorylation, respectively. That the amino acid Thr 49 is the site of phosphorylation was demonstrated by the site-directed mutagenesis experiments with AtPCS1. The phosphorylation does not occur in the mutant AtPCS1 (T49A), and the activity of wild-type protein (PCS1) is significantly higher than that of the mutant AtPCS1 (T49A).

## 3 PHYTOCHELATINS ACT AS A MARKER OF Cd STRESS

Various studies have reported the induction of PCs in response to Cd in many plant species, including freshwater green algae (*Scenedesmus vacuolatus*, Le Faucheur et al., 2005), marine diatom (*Thalassiosira nordenskioeldii*, Wang and Wang, 2011), tobacco cell cultures (*Nicotiana tabacum*, Zitka et al., 2011a, 2011b), rice (*Oryza sativa*, Nocito et al., 2011), flax (*Linum usitatissimum*, Najmanova et al., 2012), mouse-ear cress (*A. thaliana*, Gielen et al., 2017), and garden vetch (*Vicia sativa*, Zhang et al., 2018). Such studies evidently demonstrate that phytochelatins act as a protector of plants against metal poisoning, and maintain the metal homeostasis in the cell. Therefore, PCs can be used as a biomarker for the early detection of heavy metal in plants, especially under Cd stress. Several metals can cause toxicity in plants at the same time, and may act synergistically, antagonistically, and independently; depending on the interaction and chemistry of different metals (Son et al., 2014; Sun et al., 2005a,b). The synthesis of phytochelatins and glutathione was monitored as an indicator of heavy metal stress in wheat plants under Cd and Pb (Sun et al., 2005a,b). In this study focused on concentrations of PCs and GSH in roots, stems, and leaves of plants, Cd proved to be more effective than Pb for the production of phytochelatins and glutathione, but their combined application suppressed the formation of GSH in the growing plants. These findings suggest that the mutual interactions of metals are also crucial with reference to the performance of PCs and GSH as the important stress markers. Interaction of Cd with PCs is stronger, as compared with other metals; for example, T-DNA line and *Tos 17* mutants of rice showed less Cd accumulation and high As accumulation (Uraguchi et al., 2017), indicating that the interaction of both these metals with PCs have different mechanisms. PCs are synthesized and accumulated in the underground, as well as the aerial plant parts. Many studies suggest that they are produced and accumulated first in the roots. Yan et al. (2017) demonstrated that Cd exposure induced the expression of the PC synthase gene, which was found primarily in the roots of Lvbao-701 and Chixin-4 cultivars of *Brassica parachinensis*, with a relatively greater induction in the Lvbao-701 than in the Chixin-4 roots. Hence, the higher accumulation of Cd and over-expressed PCs synthesis in the roots could be the reason why Lvbao-701 cultivars show a reduced Cd translocation to shoots, and better tolerance to Cd stress, than Chixin-4 cultivars. On the other hand, Szalai et al. (2013) reported that longer exposure of maize plants to Cd resulted in a low accumulation of PCs in roots and a high abundance of phytochelatin synthase in leaves. Similarly, Mohamed et al. (2012) found ample accumulation of PCs in both roots and shoots of *Brassica juncea* under Cd stress, but shoots exhibited a relatively higher accumulation, indicating that substrate reduction and feedback regulation could be responsible for this difference. Different plants have different levels of tolerance against metal toxicity that directs the PCs synthesis, accumulation, and metal translocation in plants. Zhang et al. (2010) suggested that Cd detoxification in the hyperaccumulator plant *Sedum alfredii* is equivalent to that of the nonresistant plants mediated by PCs action, mainly in shoots. Tian et al. (2017) demonstrated that the hyper-tolerance feature of *S. alfredii* is due to its efficient performance concerning the root-to-shoot translocation of soil contaminants, and suggested that it may be mediated by high Cd signal in the vascular tissue of young stems. Moreover, vacuoles of the shoot parenchyma cells showed efficient storage of Cd mediated by PCs, which resisted its movement into the protoplasts. This may represent a crucial development in metal homeostasis and tolerance in shoots of the Cd hyperaccumulator *S. alfredii*. In the Cd-hyperaccumulators *T. aestivum* (Hentz et al., 2012) and *B. juncea* (Rahman et al., 2016), PC-mediated metal detoxification was seen largely in roots. Dong et al. (2017) observed no variation in the accumulation of Cd between the roots of tall fescue and Kentucky bluegrass, possibly showing the occurrence of the same mechanism of tolerance, that is, the action of PCs in roots of both the species.

## 4 SYNERGISTIC ROLE OF PHYTOCHELATINS WITH ANTIOXIDANTS

Cd detoxification in plants mediated by phytochelatins and antioxidant systems has been widely studied (Bashir et al., 2015; Mohamed et al., 2012; Seth et al., 2008). Previously, Chen et al. (2008) verified the coupling of phytochelatins with antioxidative systems in his study. They observed an increased biosynthesis of PCs, along with increased activity of antioxidants in *Brassica chinensis*, which jointly enhanced the plant tolerance to Cd stress. The improved level of GSH under Cd stress could be possible due to an increase of enzymes related to GSH biosynthesis, including γ-glutamylcysteine synthetase, glutathione synthetase, and glutathione reductase, and also the enzymes involved in PC biosynthesis. The peroxidases, including ascorbate peroxidase (APX), glutathione peroxidase (GPX), and guaiacol peroxidase (GOPX) are auspicious and talented antioxidant enzymes that have a specific role in metal stress tolerance, and help in evaluating the sublethal metal toxicity in plants (Ahmad et al., 2018; Radotić et al., 2000). This group of enzymes efficiently nullifies the hydrogen peroxide or lipid peroxides that are produced by the Cd-led oxidative stress. Much like phytochelatins (cysteine-rich oligomers), peroxidases can have the redox active cysteine or selenocysteine molecules that exhibit connectivity at the cysteine level. Srivastava et al. (2004) reported that the induction of PC synthesis and modulation of antioxidant systems takes place simultaneously in *Cuscuta reflexa* grown under Cd stress, indicating that the Cd-mediated ROS activate both the systems concurrently. The glyoxalase system is also active in plants under Cd stresses and tries to reduce the adverse effects of metals. The high levels of Gly I and Gly II maintain the GSH level and the GSH/GSSG ratio using the glyoxalase system, which probably maintains the synthesis of PCs under Cd stress (Hasanuzzaman et al., 2017; Nahar et al., 2016). The tolerance of safflower seedlings to Cd also increased by the combined action of antioxidant enzymes and PCs, indicating that both are crucial, and perform coordinately for Cd detoxification (Namdjoyan et al., 2012). On the other hand, poplar plants showed a high resistance to Cd stress with the supplementation of nitrogen, thus suggesting that nitrogen possibly activates the defense system by utilizing the glutathione and PC synthetic pathways (Lin et al., 2016). A similar association of antioxidant enzymes and PC biosynthesis for combating Cd stress was observed in *Bacopa monnieri* (Mishra et al., 2006). The photosynthesis and other physiological processes were not reduced much, and protein synthesis was induced. Such a response may emerge due to enhanced activity of GR and a high concentration of GSH, which may provide sufficient GSH, not only for PC synthesis, but also for antioxidant functions. Accordingly, plants might well neutralize the ROS by the enhanced activities of SOD, APX, GPX, GSH, and GR, while the amount of GSH could suffice for PC production (Mishra et al., 2006). Indian mustard (*B. juncea*), known for its high tolerance capacity against Cd stress, was evaluated by Mohamed et al. (2012) for its tolerance to Cd stress in terms of the antioxidant and Cd sequestering mechanisms. They found that roots were more efficient in free-radical scavenging than shoots, as evident by the high level of peroxidase, catalase, and ascorbate activities. Simultaneously, Garg and Kaur (2013) substantiated the findings of Mohamed et al. (2012) by demonstrating the increased concentration of nonprotein thiols, as well as activity of antioxidant enzymes (SOD, CAT, POD) in *Cajanus cajan* genotypes that act synergistically in fighting against the Cd-induced oxidative stress. *B. juncea* also showed a citric acid-mediated cadmium tolerance and phytoremediation process by the synchronized alliances of antioxidant defense, metal chelation, and glyoxalase systems. Citric acid could enhance the accessibility of Cd by reducing the pH. It acts as a metal chelator like PCs. Therefore, increased availability of Cd promotes its binding with citric acid and PCs during the detoxification process (Mahmud et al., 2018).

## 5 PC-Cd COMPLEXES ARE SEQUESTERED IN VACUOLES

Studies have shown that plants pump the hyperaccumulated metals into vacuoles, rather than synthesizing the metal-specific ligands for detoxification purposes. Recent studies have identified the metal compartmentation as a vital resistance mechanism (Iqbal et al., 2014; Jabeen et al., 2009). It causes minimum damage to photosynthesis when metals are preferentially sequestered in compartments (vacuoles) and cell types (epidermal storage cells) (Küpper et al., 2004; Psaras and Manetas, 2001). After a brief exposure to Cd, the vacuoles of tobacco leaves were found to contain Cd and PCs (Vögeli-Lange and Wagner, 1990). Vacuoles were the centers of Cd detoxification in *Thlaspi* leaves (Wójcik et al., 2005). In the bean roots, the vacuole and nuclei showed the presence of Cd, whereas the cell wall, cytoplasm, and plastids were devoid of it (Vázquez et al., 1992). Both the age and tissue-dependent variation in metal complexation (Cd and Zn) were observed in the hyperaccumulator *T. caerulescens* (Ganges ecotype). With the increasing leaf age, an inverse variation was observed between the fraction of ligands and the total metal found in the epidermal vacuoles (Küpper et al., 2004). The sequestration of total metal in the epidermal vacuoles might be the probable explanation for a decrease in the amount of ligands. The young and mature tissues (leaves, petioles, and stems) contained a higher percentage of sulfur (S) ligands (e.g., phytochelatins-bound Cd) than the senescent leaves. The results indicate that plants find it more profitable to synthesize strong ligands in young leaves until the sequestration in the vacuole becomes the main detoxification mechanism. On the other hand, it is likely that the mesophyll tissue containing the metal-susceptible photosynthetic machinery (Prasad and Strzałka, 1999) is protected from the harsh effects of metal toxicity with strong ligands, whereas the metals in the epidermis are constantly present in vacuoles in the form of free ions or weak organic acid complexes (Küpper et al., 2004). The relatively low percentage of weak metal complexes in the stem and petioles could be attributed to the lower ratio of epidermal vacuole volume, and hence, the low quantity of the sequestered complexes in these organs, in comparison with the total tissue volume. Vacuolar localization of heavy metal ions, an important strategy employed by the naturally evolved heavy-metal-tolerant plants, and the specifically increased sequestration of Cd into the mesophyll vacuoles of tolerant plants, could be accomplished by a greater uptake capacity due to more vacuoles per unit leaf area (thicker leaves) and a speedy transfer of detoxified complexes into the vacuole in tolerant plants (Chardonnens et al., 1998).

In phytochelatins forming the metal-phytochelatin (M-PC), the metal is attached to the SH group of the cysteine unit (Potesil et al., 2005). It is well documented that heavy metals are shipped from the cytosol to the vacuoles, where the metal either gets detached from the cysteine and may bind to other molecules, or form stable PC-MS complexes (Reese et al., 1992;

Strasdeit et al., 1991). An occurrence of low and high molecular weight (LMW and HMW) PC-Cd complexes has been described in *B. juncea* and *A. thaliana* (Howden et al., 1995; Speiser et al., 1992a, 1992b). One conventional model of heavy metal detoxification proposes that heavy metal introduction stimulates PC production, which chelate Cd to form LMW complexes in the cytoplasm. The vacuoles are the sites of LMW complex sequestration, thereby leading to the formation of HMW complexes, and Cd is compartmentalized (Cobbett, 2000; Zhang et al., 2010) (Fig. 2). Vögeli-Lange and Wagner (1990) suggested that Cd in the form of a Cd-(γ-EC)nG [Cadmium poly(γ-glutamylcysteinyl) glycine] complex moves across the tonoplast. Once inside the vacuole, the complex tends to dissociate, depending on the concentration of other chelators and their affinity to Cd, as well as on the chain length of the Cd-(γ-EC)nGs. When there is a dearth of high affinity chelators to Cd, only the low molecular complexes will readily dissociate at a normal vacuole pH of about 5 (Matsumoto et al., 1990). Accumulation of intact Cd-(γ-EC)nGs in the vacuole may lead to detrimental effects, as this would withdraw a big chunk

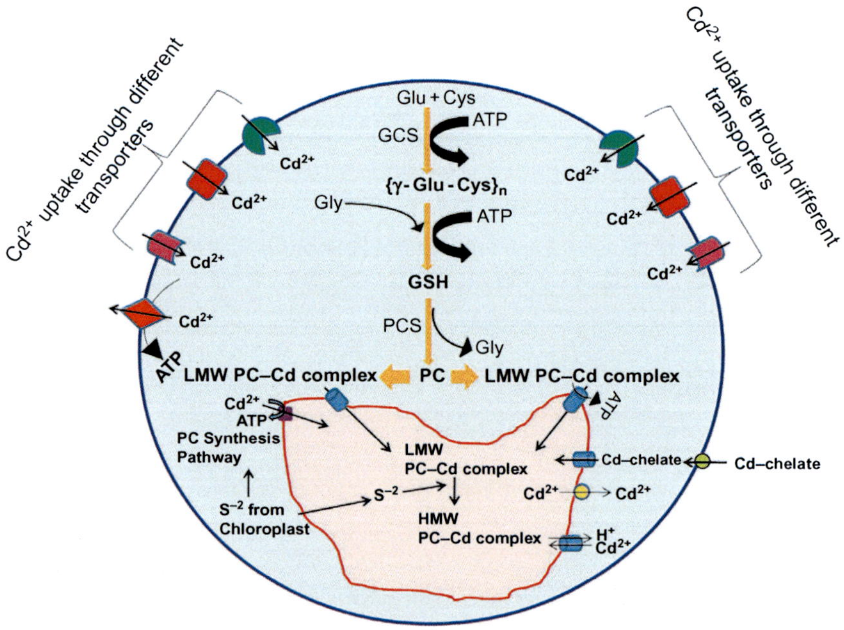


**Fig. 2** Pictorial presentation of the key steps involved in the uptake of cadmium ($Cd^{2+}$) and their detoxification by phytochelatins (PCs) in plant cells.

of the cytoplasmic glutathione pool, or other low molecular sulfur compounds. Therefore, re-shuttling of (γ-EC)nG seems to be an essential step for an effective compartmentalization of Cd. This leads to the speculation that the rate of PC breakdown and the PC concentration could depend on the rate of transport across the tonoplast (de Knecht et al., 1994). X-ray cryo-microanalytical techniques have illuminated the process of ultimate speciation of heavy metals in the vacuoles. At high Cd concentrations, globular deposits containing Cd, potassium (K), and phosphorus (P), first detected in root cells of *Agrostis gigantea*, were found in mature cells, whereas sheet-like deposits with Cd and sulfur (S) were found in immature cells (Rauser and Ackerley, 1987; Van Steveninck et al., 1990). It can be hypothesized that tolerance is achieved due to an increased ability to transport metals into the vacuole, which depends on (a) a higher concentration of substances in the cytosol-forming metal complexes, such as (γ-EC)nGs and possibly malic acid, which can be transported over to the tonoplast, (b) an efficient (higher affinity/capacity) metal-transporting coordination in the tonoplast, and (c) a higher metal-storage capacity in the vacuole (Ernst et al., 1992).

Transcriptomic studies in hyperaccumulator and nonhyperaccumulator species show that genes encoding proteins involved in heavy metals shuttle across the vacuolar membrane or plasma membrane and express constitutively, which may be a probable means of metal sequestration. Overexpression studies of CAX genes (in *A. halleri* and *T. caerulescens*) and the "cation diffusion facilitator" (CDF) family members, also known as "metal transporter proteins" (MTPs) enhance cation exchange/efflux from the cytosol, enabling the Cd sequestration and hyperaccumulation of heavy metals (Cracium et al., 2006; van de Mortel et al., 2008; Rascio and Navari-Izzo, 2011). In fission yeast, the sequestration of PC-Cd complexes to the vacuole has been clearly demonstrated through studies of the Cd-sensitive mutant, *hmt1*. ATP-binding cassettes (ABC) are membrane transport proteins present in the tonoplast, which are encoded by *Hmt1* gene (Ortiz et al., 1992). There has also been a report on an energy-consuming, proton gradient-independent mechanism comparable to *Hmt1* that performs the task of conveying PC-Cd complexes and PCs to oat roots vacuoles (Cobbett, 2000; Salt and Rauser, 1995). According to Wagner (1993), PC-mediated Cd detoxification occurs only when plants are exposed to higher concentrations of Cd (which is rare in the environment), whereas at lower Cd concentrations generally found in the soil ($\leq$0.3 μM), the metal would preferentially bind to vacuolar citrate. Contrary to this notion,

Howden and colleagues observed that *Arabidopsis* mutants deficient in PC are prone to Cd levels as low as 0.6 μM (Howden et al., 1995).

However, how the metal-tolerant plants exactly bring about a successful vacuolar compartmentalization is vague, even today. Analyzing isolated vacuoles and tonoplasts from metal-sensitive and metal-tolerant plants may add to our existing knowledge of heavy-metal sequestration.

## 6 SULFIDE IONS AND PC FUNCTION WITH REFERENCE TO Cd

There have been reports of the presence of sulfide ions in metal complexes of *S. pombe* and *Candida glabrata* and in the phytochelatin of *Chlorella* and homophytochelatins of *Glycine* max (Grill et al., 1986; Hayashi et al., 1986; Gekeler et al., 1988; Hunter and Mehra, 1998). Evidence such as the release of $H_2S$ when purified PC complexes undergo acidification and higher molecular mass displayed by the sulfide ion-containing complexes has confirmed the presence of sulfide ions (Steffens, 1990). The significance of sulfide ions in heavy metal detoxification came to light when a number of Cd-sensitive mutants believed to be affected in aspects of sulfide metabolism were identified in yeast species. The *S. pombe* mutants affected in the enzymes of the adenine biosynthetic pathway were hypersensitive to Cd and provided proof for the importance of sulfide ions in PC-Cd complexes (Speiser et al., 1992a, 1992b; Juang et al., 1993). Cd-sensitive, *hem2* mutants of *C. glabrata*, deficient in porphobilinogen synthase (an enzyme that synthesizes the cofactor siroheme of sulfite reductase) also highlighted the significance of sulfide ions in the PCs function (Hunter and Mehra, 1998; Pal and Rai, 2010). This deficiency may have led to Cd-sensitivity in these two mutants. Studies reveal that sulfide ions are a part of the high molecular weight (HMW) complexes of PCs in vacuoles, which provide thermodynamic and acid stability to the complexes, and resistance against proteolytic degradation, as well as determining the number of cadmium ions sequestered per PC peptide (Gill and Tuteja, 2011; Hu et al., 2001; Nocito et al., 2006; Reddy and Prasad, 1990; Steffens, 1990) (Fig. 2). The peptide composition of HMW complexes and their molecular weight in *C. glabrata* seemed to vary depending on the sulfide content and composition of the growth media (Mehra et al., 1994; Mendoza-Cózatl et al., 2005). A build-up of phytochelatin complexes under low pH could be observed, depending on the degree of sulfide incorporation. It has also been found that the amount of labile sulfur present at low pH values

(5–3.5) is a determining factor for displacement of Cd from PCs (Dameron et al., 1989). Sulfide stabilizes the pH, at which point displacement of metal occurs. It has been demonstrated that sulfide-deficient complexes half dissociate at a slightly higher pH than the Cd complexes rich in sulfide (Steffens, 1990). Integration of high sulfide in the vacuolar PC-Cd complexes of yeasts led to the formation of dense aggregates of 20A° diameter CdS nanocrystallite, and small CdS crystallites in tomatoes. These particles might lead to a way of storing Cd in the vacuoles (de Knecht et al., 1994; Reese et al., 1992).

Sulfide in metal complexes can be detected by the characteristic UV transition in the 305–318 nm regions, whose maximum is dependent on the amount of sulfide in the complex. Reese and colleagues highlighted that Cd concentrations to which tomato plants are exposed determined the S:Cd ratio in PC-Cd complexes in the roots (Reese et al., 1992). The ratio was $<0.1$ below 50 μM Cd in the Cd-sensitive and Cd-tolerant *Silene vulgaris* (de Knecht et al., 1994). In the absence of sulfide, the Cys-Cd ratio has been measured to be 2:1 to 4:1, which drops to ~1:1 when sulfide ions are added (Strasdeit et al., 1991). There was an increase in Cd/SCys ratio from 0.6 to 1.6 after incorporation of sulfide ions in Cd-PC complex in the Cd-stressed *Phaeodactylum tricornutum* cells, which indicated an increase in the number of Cd ions per CdS crystallite (Morelli et al., 2002). During Cd stress, intracellular sulfide levels were augmented by five- to sevenfold in *S. pombe* (Perego et al., 1997); hence, there has been speculation about the source of these inorganic sulfide ions. Results obtained from a *S. pombe* mutant sensitive to Cd have suggested that the protein sulfide oxidoreductase maintained an adequate equilibrium of sulfide produced during heavy metal stress (Weghe and Ow, 1999). Sulfur-containing compounds formed from cysteine sulfinate (generated from cysteine by the action of cysteine dioxygenase) are the direct source of intracellular sulfide ions to be incorporated into the HMW complexes (Juang et al., 1993). The varied roles of sulfide ions in the phytochelatin-metal complex may amplify the usefulness of these metal complexes for detoxification of toxic heavy metals. Pulse chase studies may provide some insight into the possible role of sulfides present in phytochelatin complexes in cysteine biosynthesis. Although the plant is benefitted by sulfide, excess sulfide concentrations are detrimental for the PC-mediated metal tolerance (Pal and Rai, 2010). Also, little is known about the utility of sulfur in sequestration of other metals by PCs. Further studies are needed to unravel the mechanism of sulfide metabolism, particularly in plants not known to have Cd-sensitive sulfide metabolism mutants.

## 7 DETECTION OF PHYTOCHELATINS IN PLANTS

As the phytochelatins in plants are the probable candidates involved in stress tolerance, it calls for an accurate detection of phytochelatins, and other thiol-containing proteins, for example, glutathione, metallothioneins, and metalloenzymes. Variations in the molecular size and lack of chromophores of these peptides make their analysis difficult. Classical electrochemical methods, especially the differential pulse and cyclic voltammetry, have been widely used (Kizek et al., 2004; Vacek et al., 2004; Vitecek et al., 2006). Another method, chronopotentiometric analysis with a constant current, has been able to detect thiols, even at low amounts (Paleček et al., 2004; Strouhal et al., 2003; Supalkova et al., 2007). Numerous analytical approaches have been used to study the allocation and quantification of PCs induced by metal toxicity in different plants. The analytical quantification and separation of PCs is significant. However, sample preparation is also important for the integrality of PCs, for example, standardization of extraction and clean-up protocols are crucial for an efficient transfer of PCs from complex plant tissues to a simpler aqueous solution that can be used for analysis.

### 7.1 Sample Preparation

Plant samples are first harvested and rinsed with water that contains additives such as EDTA/$CaCl_2$/$CaNO_3$ for efficient removal of surface metal ions, and blotted dry with paper. The samples are then divided into the underground (roots) and above-ground (shoots, i.e., stem and leaves) parts, which are either stored at −80°C or immediately used for analysis. The plant material is ground in mortar and pestle with liquid nitrogen. The powdered plant material is extracted in water or phosphate/HEPES buffer (pH 7.5), and the homogenate is centrifuged at 10,000 × *g*. The supernatant is filtered through a 0.2 μm membrane filter to remove the cell debris. Consequently, the filtrate can be analyzed by reversed phase liquid chromatography (RPLC) or size-exclusion chromatography (SEC). The filtrate can be acidified by a solution of hydrochloric acid (HCl) and diethylene tri-amine penta-acetic acid (DTPA), which is responsible for the dissociation of metal-phytochelatin (M-PC) complexes. Further, the total content of derivatized (by monobromobimane) nonprotein thiols is measured by spectrophotometer. Different preconcentration and cleanup methods are important for different plant extracts, which are analyzed by liquid chromatography (Bardarov et al., 2015; Minocha et al., 2008; Serrano et al., 2015) (see Fig. 3.)

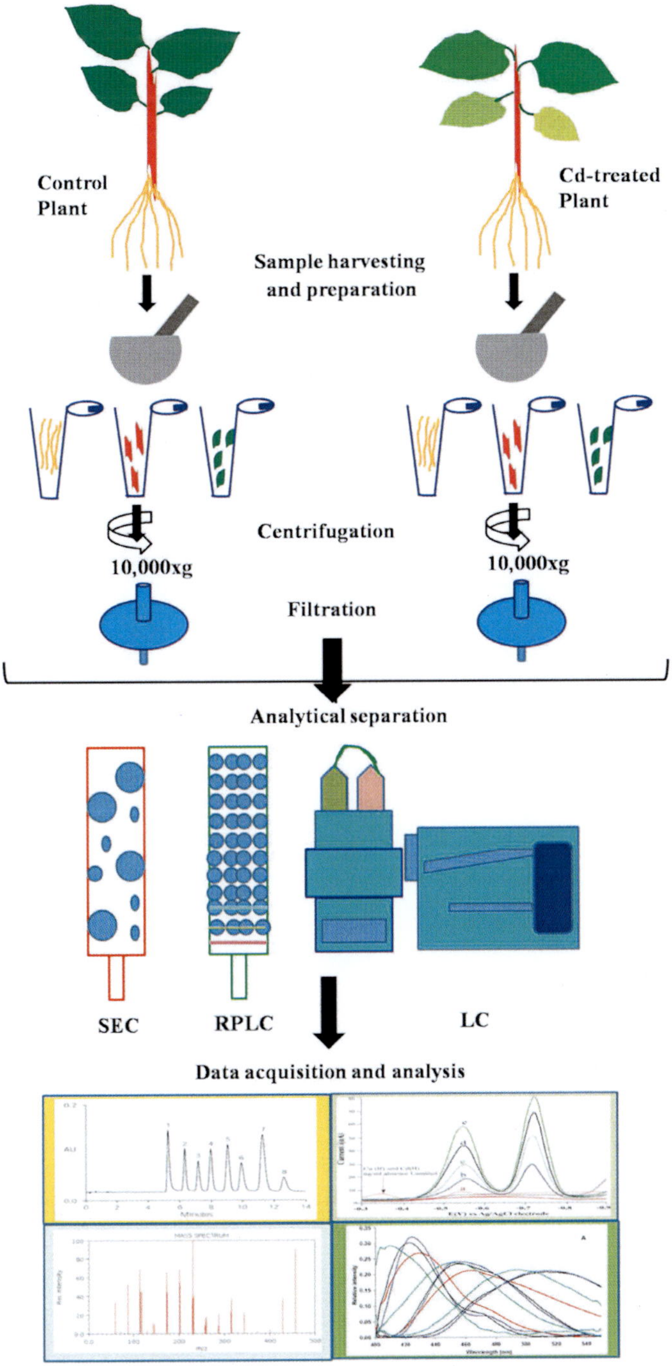


**Fig. 3** Schematic workflow for detection of phytochelatins in plants.

## 7.2 Analysis

Liquid chromatography (LC) is a powerful technique for the separation of induced PCs and related compounds from the complex plant extracts (Liu et al., 2015). Size exclusion chromatography (SEC) can be used for the separation of PCs by using the Tris/phosphate/HEPES buffer (pH 7) (Miszczak et al., 2013). Similarly, reversed-phase high-performance liquid chromatography (RP-HPLC) efficiently separates the induced PCs at acidic pH by use of a mobile phase of water with a small fraction of acetonitrile (ACN) or methanol (MeOH) with trifluoroacetic acid (TFA) (El-Zohri et al., 2005). Various atomic spectrometers, including atomic absorption, atomic fluorescence, and inductively coupled plasma mass spectrometers, can be used to detect the metal content in chromatographic fractions. On the other hand, detection of peptides mediated through thiols derivatized with the common chemical monobromobimane takes place efficiently by the fluorescence spectrophotometer (Serrano et al., 2015). In addition, Ellman's reagent is also important for the determination of nonprotein thiols by UV-Vis absorption spectrophotometer (Winther and Thorpe, 2014). The coupling of MS to LC has been especially useful in the study of the peptidic fractions. Electrochemical detection can also be possible by using electrodes that involve slow oxidation of thiol groups. Porous carbon electrodes and glassy carbon electrodes are commonly used for this purpose (Rodrigo et al., 2013). Synthetic mixtures of metal ions and PCs are specially employed in various experiments for better understanding of the chelating mechanisms of PCs. Chronopotentiometric and voltammetric methods produce complex information constituting many overlapping signals that are generally resolved by the sophisticated chemometric methods (Fojta et al., 2006). Among them, multivariate curve resolution, along with alternating least squares (MCR-ALS), is well-suited in relatively simple systems to give pure signals, which provide a clear picture of the comparative stability of the complexes and their concentrations, thus deciphering their stoichiometry. Such information enables the formulation of the complexation model in numerous cases (Cavanillas et al., 2012). In more complex cases, an increase in the voltammetric matrix with data obtained from corresponding optical techniques, such as circular dichroism (CD), can provide a better differentiation between similar species (Gusmao et al., 2010). Gaussian peak adjustment (GPA) provides satisfactory results for nonlinear datasets containing signals moving through the potential axis, which is not detected by MCRALS. Electrodes based on mercury were used mostly for chemometric and electrochemical study of metal complexation by the thiol-rich peptides (Serrano et al., 2015). Mercury hazards and policies against its use led to the development of modified electrodes containing bismuth, for example, bismuth film electrodes (BiFE) or ex situ bismuth screen-printed electrodes (ex situ BiSPCE), which are potential alternatives to mercury (Alberich et al., 2009). MS is a powerful technique for chromatographic analysis of PC extracts, and is a valuable tool for the study of metal-PC complexes, using electrospray ionization (ESI-MS). Mass spectra obtained from a direct injection of the synthetic metal ion-PCs mixtures or SEC fractions from natural extracts provide valuable information that allows identification of complex formation. Such identification considers isotopic distribution of atoms (e.g., Cd) that can be quite helpful. However, ionization can introduce artifacts in the interpretation of complexation process; for example, PC complexes having several PC units are very rarely detected, which could be due to the breaking of bonds of the metal ions joining the different units. Huge differences in the sensitivity of MS could enhance the importance of species that are negligible in the solution. These problems can be prevented by comparing the MS results with those arising from different techniques such as voltammetry, CD, or calorimetry. The isothermal titration calorimetry (ITC) technique, which is rarely applied to the study of these systems, has great potential for accurate determination of metal-PC stoichiometries, stability constants, and other thermodynamic parameters, for example, enthalpies or entropies (Chekmeneva et al., 2007; Serrano et al., 2015).

Column chromatography (gel filtration and HPLC), followed by metal detection by UV or florescence spectrophotometry, polarography, inductively coupled plasma-atomic emission spectroscopy (ICP-AES), inductively coupled plasma mass spectrometry (ICP-MS), or electrospray ionization mass spectrometry (ESI-MS) are the popular analytical methods for analyzing the phytochelatins and their metal complexes (Grill et al., 1985; Hogstrand and Haux, 1992; Maitani et al., 1996; Vacchina et al., 2000; Iglesia-Turiño et al., 2006). Two chromatographic separation methods commonly used for measuring the PCs are (1) separation with reversed phase-HPLC (RP-HPLC) and UV detection, using postcolumn derivatization with DTNB [5,5′-dithiobis (2-nitrobenzoic acid)], also known as Ellman's reagent (de Knecht et al., 1994; Grill et al., 1987), and (2) precolumn derivatization with monobromobimane (mBBr), followed by the separation with RP-HPLC and fluorescence detection (Stolt et al., 2003; Landberg and Greger, 2004; Zhang et al., 2008). The latter is more sensitive, ensures a higher stability of the monobromobimane derivatives, and a negligible fluorescence loss over a long duration of time (20 months approx.) (Sneller et al., 2000; Sun et al., 2005a, 2005b). While experimenting with Cd-affected plants for detection of PCs, Doring and colleagues demonstrated that PCs can be concentrated by freeze-drying the samples and by preventing further dilution of the derivatized products (Döring et al., 2000). In a study of PCs of Cd-sensitive and tolerant *S. vulgaris*, phenylisothiocyanate was used to derivatize the amino acids, separation was performed by a reversed-phase Pico-Tag column (Waters) based HPLC, and PC-Cd complexes were examined using the FPLC-Mono Q anion-exchange column (de Knecht et al., 1994). For detecting and quantifying the metal(loid) complexes, besides the free PCs, a popular method of ICP-MS is used in which HPLC is coupled to a metal-specific detector. However, running the standards is necessary because ICP-MS fails to provide the structural information (Bluemlein et al., 2009). Another element-specific method is the extended X-ray absorption fine structure (EXAFS), which facilitates analysis of metal complexes in vivo. This can also be performed on whole frozen tissues (Salt et al., 1995), which is an advantage over the other spectroscopic methods. In vivo ligand background of Cd and Zn in the Cd/Zn hyperaccumulator *T. caerulescens* (Ganges ecotype) was evaluated by the EXAFS method, a more sensitive method than NMR (Küpper et al., 2004). The Cd EXAFS data disclosed a primarily tetrahedral coordination of cadmium by sulfur in the phytochelatins of maize exposed to low Cd levels (Pickering et al., 1999). Denaturing polyacrylamide gel electrophoresis (Tricine-SDS-urea PAGE) coupled with the fluorescent tag, mBrB, was also used for extensive profiling of thiol-rich peptides, including the phytochelatins, and the results were confirmed by the NMR and HPLC-tandem MS analyses (Fan et al., 2004). An optimized high-performance liquid chromatography coupled with an electrochemical detector (HPLC-ED) method was found suitable for studying the effect of different Cd concentrations on the content of various thiol compounds up to the femtomole levels, including PC2 in maize kernels (Potesil et al., 2005). Zitka and colleagues determined PC2 by HPLC-ED, following the PCS activation in tobacco BY-2 cells, by treating them with different doses of cadmium ions (Zitka et al., 2011a,b). ESI-tandem MS (ESI-MS/MS) has been established as a powerful and sensitive technique for protein and peptide identification, with a wide range of detection levels ($10^{-15}$ to $10^{-18}$ mol), and it comes with many advantages, such as the minimum amount of sample requirements (few microliters), absence of a liquid-liquid or a solid-phase step, and omission of the derivatization step (Allen and Hutchens, 1992; Yates et al., 1996; Iglesia-Turiño et al., 2006). PC synthesis was assayed in response to Cd stress in *S. alfredii* with the help of RP-HPLC, HPLC-mass spectrometry (HPLC-MS), and HPLC-tandem mass spectrometry (HPLC-MS/MS) instruments with the electrospray interface (ESI) source (Zhang et al., 2008).

Gel filtration, although preferable for PC analysis, involves an acidification step, which dissembles any PC-metal complex, thereby preventing the latter's detection. Thus, the low resolution of gel filtration is not able to provide us with details about the nature, type, structure, number, and stoichiometry of PC-metal complexes. To overcome this shortcoming, nano-electrospray tandem mass spectrometry (nano-ESI-MS/MS) and capillary liquid chromatography/electrospray ionization tandem mass spectrometry (capillary LC/ESI-MS/MS) was used for the detection of Cd-induced PCs and PC-Cd complexes in *Datura innoxia* (Yen et al., 1999). Nano-ESI/MS/MS is able to maintain the required resolution and sensitivity for proper detection and identification of PCs and PC-metal complexes, with less tedious sample preparation. Capillary LC/ESI-MS/MS is complementary to nano-ESI/MS/MS in providing information or detecting PC-metal complexes that are missed by nano-ESI/MS/MS when the analytes studied are scarce in the mixture. Thus, the combination of these two techniques creates an advanced, sensitive tool for evaluating the size, stoichiometry, and number of PC-metal complexes in plant tissues in a better way.

## 8 CONCLUDING REMARKS

Extraction, identification, and biosynthesis of phytochelatins in the resistant and nonresistant plant species have given a clear view of metal chelation and allocation in different plant parts. Plants can deal with Cd by means of GSH, sulfide biosynthesis, and PCs compartmentalization. For efficient phytoremediation, we need to know the exact nature of PCs synthesis, and the degree of interaction between PCs and metals, such as the strong interaction between PCs and Cd. Further, the expression level of phytochelatin synthase genes is implicated in the differential metal stress tolerance. The study of metal binding with PCs analyzed by the combination of various analytical methods provides a holistic view of the metal-PCs complexity. Future studies with a proper combination of different analytical techniques will likely give a deeper insight into metal homeostasis by PCs.

## REFERENCES

Ahmad, J., Bashir, H., Bagheri, R., Baig, A., Al-Huqail, A., Ibrahim, M.M., Qureshi, M.I., 2017. Drought and salinity induced changes in ecophysiology and proteomic profile of *Parthenium hysterophorus*. PLOS One 12 (9) p.e0185118.

Ahmad, J., Baig, M.A., Ali, A.A., Al-Huqail, A.A., Ibrahim, M.M., Qureshi, M.I., 2018. Differential antioxidative and biochemical responses to aluminium stress in *Brassica juncea* cultivars. Hortic. Environ. Biotechnol. 59 (5), 615–627.

Alberich, A., Serrano, N., Ariño, C., Díaz-Cruz, J.M., Esteban, M., 2009. Bismuth film electrodes for the study of metal thiolate complexation: an alternative to mercury electrodes. Talanta 78 (3), 1017–1022.

Allen, M.H., Hutchens, T.W., 1992. Electrospray-ionization mass spectrometry for the detection of discrete peptide/metal-ion complexes involving multiple cysteine (sulfur) ligands. Rapid Commun. Mass Spectrom. 6 (4), 308–312.

Bardarov, K., Naydenov, M., Djingova, R., 2015. HPLC-HRMS method for fast phytochelatins determination in plants. Application to analysis of *Clinopodium vulgare* L. Talanta 142, 20–27.

Bashir, H., Ibrahim, M.M., Bagheri, R., Ahmad, J., Arif, I.A., Baig, M.A., Qureshi, M.I., 2015. Influence of sulfur and cadmium on antioxidants, phytochelatins and growth in Indian mustard. AoB Plants 7, 1–13. https://doi.org/10.1093/aobpla/plv 001.

Bluemlein, K., Raab, A., Feldmann, J., 2009. Stability of arsenic peptides in plant extracts: off-line versus on-line parallel elemental and molecular mass spectrometric detection for liquid chromatographic separation. Anal. Bioanal. Chem. 393 (1), 357–366.

Cavanillas, S., Gusmão, R., Ariño, C., Díaz-Cruz, J.M., Esteban, M., 2012. Voltammetric analysis of phytochelatin complexation in ternary metal mixtures supported by multivariate analysis and ESI-MS. Electroanalysis 24 (2), 309–315.

Cazalé, A.C., Clemens, S., 2001. *Arabidopsis thaliana* expresses a second functional phytochelatin synthase. FEBS Lett. 507 (2), 215–219.

Chardonnens, A.N., Ten Bookum, W.M., Kuijper, L.D., Verkleij, J.A., Ernst, W.H., 1998. Distribution of cadmium in leaves of cadmium tolerant and sensitive ecotypes of *Silene vulgaris*. Physiol. Plant. 104 (1), 75–80.

Chekmeneva, E., Díaz-Cruz, J.M., Arino, C., Esteban, M., 2007. Binding of $Cd^{2+}$ and $Zn^{2+}$ with the Phytochelatin (γ-Glu-Cys) 4-Gly: a voltammetric study assisted by multivariate curve resolution and electrospray ionization mass spectrometry. Electroanalysis 19 (2–3), 310–317.

Chen, L., Guo, Y., Yang, L., Wang, Q., 2008. Synergistic defensive mechanism of phytochelatins and antioxidative enzymes in *Brassica chinensis* L. against Cd stress. Chin. Sci. Bull. 53 (10), 1503.

Clemens, S., Kim, E.J., Neumann, D., Schroeder, J.I., 1999. Tolerance to toxic metals by a gene family of phytochelatin synthases from plants and yeast. EMBO J. 18 (12), 3325–3333.

Cobbett, C.S., 2000. Phytochelatins and their roles in heavy metal detoxification. Plant Physiol. 123 (3), 825–832.

Cobbett, C., Goldsbrough, P., 2002. Phytochelatins and metallothioneins: roles in heavy metal detoxification and homeostasis. Annu. Rev. Plant Biol. 53 (1), 159–182.

Cobbett, C.S., May, M.J., Howden, R., Rolls, B., 1998. The glutathione-deficient, cadmium-sensitive mutant, cad2–1, of *Arabidopsis thaliana* is deficient in γ-glutamylcysteine synthetase. Plant J. 16 (1), 73–78.

Cong, M., Zhao, J., Lü, J., Ren, Z., Wu, H., 2016. Homologous cloning, characterization and expression of a new halophyte phytochelatin synthase gene in *Suaeda salsa*. Chin. J. Oceanol. Limnol. 34 (5), 1034–1043.

Cracium, A.R., Courbot, M., Bourgis, F., Salis, P., Saumitou-Laprade, P., Verbruggen, N., 2006. Comparative cDNA-AFLP analysis of Cd-tolerant and-sensitive genotypes derived from crosses between the Cd hyperaccumulator *Arabidopsis halleri* and *Arabidopsis lyrata* ssp. petraea. J. Exp. Bot. 57 (12), 2967–2983.

Dameron, C.T., Reese, R.N., Mehra, R.K., Kortan, A.R., Carroll, P.J., Steigerwald, M.L., Brus, L.E., Winge, D.R., 1989. Biosynthesis of cadmium sulphide quantum semiconductor crystallites. Nature 338 (6216), 596–597.

De Benedictis, M., Brunetti, C., Brauer, E.K., Andreucci, A., Popescu, S.C., Commisso, M., Guzzo, F., Sofo, A., Ruffini Castiglione, M., Vatamaniuk, O.K., Sanita'di Toppi, L., 2018. The *Arabidopsis thaliana* knockout mutant for phytochelatin synthase 1 (cad1-3) is defective in callose deposition, bacterial pathogen defense and auxin content. Front. Plant Sci. 9, 19. https://doi.org/10.3389/fpls.2018.00019.

de Knecht, J.A., van Dillen, M., Koevoets, P.L., Schat, H., Verkleij, J.A., Ernst, W.H., 1994. Phytochelatins in cadmium-sensitive and cadmium-tolerant *Silene vulgaris* (chain length distribution and sulfide incorporation). Plant Physiol. 104 (1), 255–261.

Dong, Q., Xu, P., Wang, Z., 2017. Differential cadmium distribution and translocation in roots and shoots related to hyper-tolerance between tall fescue and Kentucky bluegrass. Front. Plant Sci. 8, 113. https://doi.org/10.3389/fpls.2017.00113.

Döring, S., Korhammer, S., Oetken, M., Markert, B., 2000. Analysis of phytochelatins in plant matrices by pre-column derivatization, high-performance liquid chromatography and fluorescence-detection. Fresenius J. Anal. Chem. 366 (3), 316–318.

El-Zohri, M.H.A., Cabala, R., Frank, H., 2005. Quantification of phytochelatins in plants by reversed-phase HPLC-ESI-MS-MS. Anal. Bioanal. Chem. 382 (8), 1871–1876.

Ernst, W.H.O., Verkleij, J.A.C., Schat, H., 1992. Metal tolerance in plants. Acta Bot. Neer. 41 (3), 229–248.

Fan, T.W.M., Lane, A.N., Higashi, R.M., 2004. An electrophoretic profiling method for thiol-rich phytochelatins and metallothioneins. Phytochem. Anal. 15 (3), 175–183.

Fojta, M., Fojtová, M., Havran, L., Pivoňková, H., Dorčák, V., Šestáková, I., 2006. Electrochemical monitoring of phytochelatin accumulation in *Nicotiana tabacum* cells exposed to sub-cytotoxic and cytotoxic levels of cadmium. Anal. Chim. Acta 558 (1–2), 171–178.

Garg, N., Kaur, H., 2013. Response of antioxidant enzymes, phytochelatins and glutathione production towards Cd and Zn stresses in *Cajanus cajan* (L.) Millsp. genotypes colonized by arbuscular mycorrhizal fungi. J. Agron. Crop Sci. 199 (2), 118–133.

Gekeler, W., Grill, E., Winnacker, E.L., Zenk, M.H., 1988. Algae sequester heavy metals via synthesis of phytochelatin complexes. Arch. Microbiol. 150 (2), 197–202.

Gielen, H., Vangronsveld, J., Cuypers, A., 2017. Cd-induced Cu deficiency responses in *Arabidopsis thaliana*: are phytochelatins involved? Plant Cell Environ. 40 (3), 390–400.

Gill, S.S., Tuteja, N., 2011. Cadmium stress tolerance in crop plants: probing the role of sulfur. Plant Signal. Behav. 6 (2), 215–222.

Grill, E., Winnacker, E.L., Zenk, M.H., 1985. Phytochelatins: the principal heavy-metal complexing peptides of higher plants. Science 230 (4726), 674–676.

Grill, E., Gekeler, W., Winnacker, E.L., Zenk, H.H., 1986. Homo-phytochelatins are heavy metal-binding peptides of homo-glutathione containing Fabales. FEBS Lett. 205 (1), 47–50.

Grill, E., Winnacker, E.L., Zenk, M.H., 1987. Phytochelatins, a class of heavy-metal-binding peptides from plants, are functionally analogous to metallothioneins. Proc. Natl. Acad. Sci. 84 (2), 439–443.

Grill, E., Mishra, S., Srivastava, S., Tripathi, R.D., 2007. Role of phytochelatins in phytoremediation of heavy metals. In: Environ. Bioremed. Technol. Springer, Berlin, Heidelberg, pp. 101–146.

Gusmao, R., Arino, C., Díaz-Cruz, J.M., Esteban, M., 2010. Cadmium binding in mixtures of phytochelatins and their fragments: a voltammetric study assisted by multivariate curve resolution and mass spectrometry. Analyst 135 (1), 86–95.

Ha, S.B., Smith, A.P., Howden, R., Dietrich, W.M., Bugg, S., O'Connell, M.J., Goldsbrough, P.B., Cobbett, C.S., 1999. Phytochelatin synthase genes from Arabidopsis and the yeast *Schizosaccharomyces pombe*. Plant Cell 11 (6), 1153–1163.

Hasan, M., Cheng, Y., Kanwar, M.K., Chu, X.Y., Ahammed, G.J., Qi, Z.Y., 2017. Responses of plant proteins to heavy metal stress—a review. Front. Plant Sci. 8, 1492.

Hasanuzzaman, M., Nahar, K., Hossain, M.S., Mahmud, J.A., Rahman, A., Inafuku, M., Oku, H., Fujita, M., 2017. Coordinated actions of glyoxalase and antioxidant defense systems in conferring abiotic stress tolerance in plants. Int. J. Mol. Sci. 18 (1), 200. https://doi.org/10.3390/ijms18010200.

Hayashi, Y., Nakagawa, C.W., Murasugi, A., 1986. Unique properties of Cd-binding peptides induced in fission yeast, *Schizosaccharomyces pombe*. Environ. Health Perspect. 65, 13–19.

Hentz, S., McComb, J., Miller, G., Begonia, M., Begonia, G., 2012. Cadmium uptake, growth and phytochelatin contents of *Triticum aestivum* in response to various concentrations of cadmium. World Environ. 2 (3), 44–50.

Hirata, K., Tsuji, N., Miyamoto, K., 2005. Biosynthetic regulation of phytochelatins, heavy metal-binding peptides. J. Biosci. Bioeng. 100 (6), 593–599.

Hogstrand, C., Haux, C., 1992. Evaluation of differential pulse polarography for the quantification of metallothionein—a comparison with RIA. Anal. Biochem. 200 (2), 388–392.

Howden, R., Andersen, C.R., Goldsbrough, P.B., Cobbett, C.S., 1995. A cadmium-sensitive, glutathione-deficient mutant of *Arabidopsis thaliana*. Plant Physiol. 107 (4), 1067–1073.

Hu, S., Lau, K.W., Wu, M., 2001. Cadmium sequestration in *Chlamydomonas reinhardtii*. Plant Sci. 161 (5), 987–996.

Hunter, T.C., Mehra, R.K., 1998. A role for HEM2 in cadmium tolerance 1. J. Inorg. Biochem. 69 (4), 293–303.

Iglesia-Turiño, S., Febrero, A., Jauregui, O., Caldelas, C., Araus, J.L., Bort, J., 2006. Detection and quantification of unbound phytochelatin 2 in plant extracts of *Brassica napus* grown with different levels of mercury. Plant Physiol. 142 (2), 742–749.

Iqbal, M., Ahmad, A., Ansari, M.K., Qureshi, M.I., Aref, I.M., Khan, P.R., Hegazy, S.S., El-Atta, H., Husen, A., Hakeem, K.R., 2014. Improving the phytoextraction capacity of plants to scavenge metal (loid)-contaminated sites. Environ. Rev. 23 (1), 44–65.

Jabeen, R., Ahmad, A., Iqbal, M., 2009. Phytoremediation of heavy metals: physiological and molecular mechanisms. Bot. Rev. 75 (4), 339–364.

Jalmi, S.K., Bhagat, P.K., Verma, D., Noryang, S., Tayyeba, S., Singh, K., Sharma, D., Sinha, A.K., 2018. Traversing the links between heavy metal stress and plant signaling. Front. Plant Sci. 9, 12.

Jiang, Q.Q., Sun, X.L., Cao, H., Zou, Y.M., Shu, H.R., 2013. Isolation and expression analysis of a phytochelatin synthase gene (MhPCS) from *Malus hupehensis*. J. Fruit Sci. 3, 341–347.

Juang, R.H., McCue, K.F., Ow, D.W., 1993. Two purine biosynthetic enzymes that are required for cadmium tolerance in *Schizosaccharomyces pombe* utilize cysteine sulfinate in vitro. Arch. Biochem. Biophys. 304 (2), 392–401.

Kang, S.H., Singh, S., Kim, J.Y., Lee, W., Mulchandani, A., Chen, W., 2007. Bacteria metabolically engineered for enhanced phytochelatin production and cadmium accumulation. Appl. Environ. Microbiol. 73 (19), 6317–6320.

Kizek, R., Vacek, J., Trnková, L., Jelen, F., 2004. Cyclic voltammetric study of the redox system of glutathione using the disulfide bond reductant tris (2-carboxyethyl) phosphine. Bioelectrochemistry 63 (1–2), 19–24.

Küpper, H., Mijovilovich, A., Meyer-Klaucke, W., Kroneck, P.M., 2004. Tissue-and age-dependent differences in the complexation of cadmium and zinc in the cadmium/zinc hyperaccumulator *Thlaspi caerulescens* (Ganges ecotype) revealed by X-ray absorption spectroscopy. Plant Physiol. 134 (2), 748–757.

Landberg, T., Greger, M., 2004. No phytochelatin (PC2 and PC3) detected in *Salix viminalis*. Physiol. Plant. 121 (3), 481–487.

Le Faucheur, S., Behra, R., Sigg, L., 2005. Phytochelatin induction, cadmium accumulation, and algal sensitivity to free cadmium ion in *Scenedesmus vacuolatus*. Environ. Toxicol. Chem. 24 (7), 1731–1737.

Lee, S., Korban, S., 2002. Transcriptional regulation of *Arabidopsis thaliana* phytochelatin synthase (AtPCS1) by cadmium during early stages of plant development. Planta 215 (4), 689–693.

Lin, T., Wan, X., Zhang, F., 2016. The short-term responses of glutathione and phytochelation synthetic pathways genes to additional nitrogen under cadmium stress in poplar leaves. Russ. J. Plant Physiol. 63 (6), 754–762.

Liu, P., Cai, W.J., Yu, L., Yuan, B.F., Feng, Y.Q., 2015. Determination of phytochelatins in rice by stable isotope labeling coupled with liquid chromatography-mass spectrometry. J. Agric. Food Chem. 63 (25), 35–42.

López-Climent, M.F., Arbona, V., Pérez-Clemente, R.M., Zandalinas, S.I., Gómez-Cadenas, A., 2014. Effect of cadmium and calcium treatments on phytochelatin and glutathione levels in citrus plants. Plant Biol. 16 (1), 79–87.

Mahmud, J.A., Hasanuzzaman, M., Nahar, K., Bhuyan, M.B., Fujita, M., 2018. Insights into citric acid-induced cadmium tolerance and phytoremediation in *Brassica juncea* L.: coordinated functions of metal chelation, antioxidant defense and glyoxalase systems. Ecotoxicol. Environ. Saf. 147, 990–1001.

Maitani, T., Kubota, H., Sato, K., Yamada, T., 1996. The composition of metals bound to class III metallothionein (phytochelatin and its desglycyl peptide) induced by various metals in root cultures of *Rubia tinctorum*. Plant Physiol. 110 (4), 1145–1150.

Matsumoto, Y., Okada, Y., Min, K.S., Onosaka, S., Tanaka, K., 1990. Amino acids and peptides. XXVII: Synthesis of phytochelatin-related peptides and examination of their heavy metal-binding properties. Chem. Pharm. Bull. 38 (9), 2364–2368.

Mehra, R.K., Mulchandani, P., Hunter, T.C., 1994. Role of CdS quantum crystallites in cadmium resistance in *Candida glabrata*. Biochem. Biophys. Res. Commun. 200 (3), 1193–1200.

Mendoza-Cózatl, D., Loza-Tavera, H., Hernández-Navarro, A., Moreno-Sánchez, R., 2005. Sulfur assimilation and glutathione metabolism under cadmium stress in yeast, protists and plants. FEMS Microbiol. Rev. 29 (4), 653–671.

Merlos, M.A., Micháleka, P., Kryštofováa, O., Zítkaa, O., Adama, V., Kizeka, R., 2014. The role of phytochelatins in plant and animals: a review. J. Met. Nanotechnol. 4, 22–27.

Meyer, C.L., Peisker, D., Courbot, M., Craciun, A.R., Cazalé, A.C., Desgain, D., Schat, H., Clemens, S., Verbruggen, N., 2011. Isolation and characterization of *Arabidopsis halleri* and *Thlaspi caerulescens* phytochelatin synthases. Planta 234 (1), 83–95.

Minocha, R., Thangavel, P., Dhankher, O.P., Long, S., 2008. Separation and quantification of monothiols and phytochelatins from a wide variety of cell cultures and tissues of trees and other plants using high performance liquid chromatography. J. Chromatogr. A 1207 (1–2), 72–83.

Mishra, S., Srivastava, S., Tripathi, R.D., Govindarajan, R., Kuriakose, S.V., Prasad, M.N.V., 2006. Phytochelatin synthesis and response of antioxidants during cadmium stress in *Bacopa monnieri* L. Plant Physiol. Biochem. 44 (1), 25–37.

Miszczak, A., Rosłon, M., Zbroja, G., Brama, K., Szalacha, E., Gawrońska, H., Pawlak, K., 2013. SEC ICP MS and CZE ICP MS investigation of medium and high molecular weight complexes formed by cadmium ions with phytochelatins. Anal. Bioanal. Chem. 405 (14), 4667–4678.

Mohamed, A.A., Castagna, A., Ranieri, A., di Toppi, L.S., 2012. Cadmium tolerance in *Brassica juncea* roots and shoots is affected by antioxidant status and phytochelatin biosynthesis. Plant Physiol. Biochem. 57, 15–22.

Morelli, E., Cruz, B.H., Somovigo, S., Scarano, G., 2002. Speciation of cadmium-γ-glutamyl peptides complexes in cells of the marine microalga *Phaeodactylum tricornutum*. Plant Sci. 163 (4), 807–813.

Mutoh, N., Hayashi, Y., 1988. Isolation of mutants of *Schizosaccharomyces pombe* unable to synthesize cadystin, small cadmium-binding peptides. Biochem. Biophys. Res. Commun. 151 (1), 32–39.

Nahar, K., Hasanuzzaman, M., Alam, M.M., Rahman, A., Suzuki, T., Fujita, M., 2016. Polyamine and nitric oxide crosstalk: antagonistic effects on cadmium toxicity in mung bean plants through upregulating the metal detoxification, antioxidant defense and methylglyoxal detoxification systems. Ecotoxicol. Environ. Saf. 126, 245–255.

Najmanova, J., Neumannova, E., Leonhardt, T., Zitka, O., Kizek, R., Macek, T., Mackova, M., Kotrba, P., 2012. Cadmium-induced production of phytochelatins and speciation of intracellular cadmium in organs of *Linum usitatissimum* seedlings. Ind. Crop. Prod. 36 (1), 536–542.

Namdjoyan, S., Namdjoyan, S., Kermanian, H., 2012. Induction of phytochelatin and responses of antioxidants under cadmium stress in safflower (*Carthamus tinctorius*) seedlings. Turk. J. Bot. 36 (5), 495–502.

Nocito, F.F., Lancilli, C., Crema, B., Fourcroy, P., Davidian, J.C., Sacchi, G.A., 2006. Heavy metal stress and sulfate uptake in maize roots. Plant Physiol. 141 (3), 1138–1148.

Nocito, F.F., Lancilli, C., Dendena, B., Lucchini, G., Sacchi, G.A., 2011. Cadmium retention in rice roots is influenced by cadmium availability, chelation and translocation. Plant Cell Environ. 34 (6), 994–1008.

Ortiz, D.F., Kreppel, L., Speiser, D.M., Scheel, G., McDonald, G., Ow, D., 1992. Heavy metal tolerance in the fission yeast requires an ATP-binding cassette-type vacuolar membrane transporter. EMBO J. 11 (10), 3491–3499.

Pal, R., Rai, J.P.N., 2010. Phytochelatins: peptides involved in heavy metal detoxification. Appl. Biochem. Biotechnol. 160 (3), 945–963.

Paleček, E., Masařík, M., Kizek, R., Kuhlmeier, D., Hassmann, J., Schülein, J., 2004. Sensitive electrochemical determination of unlabeled MutS protein and detection of point mutations in DNA. Anal. Chem. 76 (19), 5930–5936.

Perego, P., Weghe, J.V., Ow, D.W., Howell, S.B., 1997. Role of determinants of cadmium sensitivity in the tolerance of *Schizosaccharomyces pombe* to cisplatin. Mol. Pharmacol. 51 (1), 12–18.

Pickering, I.J., Prince, R.C., George, G.N., Rauser, W.E., Wickramasinghe, W.A., Watson, A.A., Dameron, C.T., Dance, I.G., Fairlie, D.P., Salt, D.E., 1999. X-ray absorption spectroscopy of cadmium phytochelatin and model systems. Biochim. Biophys. Acta Protein Struct. Mol. Enzymol. 1429 (2), 351–364.

Pochodylo, A.L., Aristilde, L., 2017. Molecular dynamics of stability and structures in phytochelatin complexes with Zn, Cu, Fe, Mg, and Ca: implications for metal detoxification. Environ. Chem. Lett. 15 (3), 495–500.

Potesil, D., Petrlova, J., Adam, V., Vacek, J., Klejdus, B., Zehnalek, J., Trnkova, L., Havel, L., Kizek, R., 2005. Simultaneous femtomole determination of cysteine reduced and oxidized glutathione, and phytochelatin in maize (*Zea mays* L.) kernels using high-performance liquid chromatography with electrochemical detection. J. Chromatogr. A 1084 (1–2), 134–144.

Prasad, M.N.V., Strzałka, K., 1999. Impact of heavy metals on photosynthesis. In: Heavy Metal Stress in Plants. Springer, Berlin Heidelberg, pp. 117–138.

Psaras, G.K., Manetas, Y., 2001. Nickel localization in seeds of the metal Hyperaccumulator *Thlaspi pindicum* Hausskn. Ann. Bot. 88 (3), 513–516.

Radotić, K., Dučić, T., Mutavdžić, D., 2000. Changes in peroxidase activity and isoenzymes in spruce needles after exposure to different concentrations of cadmium. Environ. Exp. Bot. 44 (2), 105–113.

Rahantaniaina, M.S., Li, S., Chatel-Innocenti, G., Tuzet, A., Issakidis-Bourguet, E., Mhamdi, A., Noctor, G., 2017. Cytosolic and chloroplastic DHARs cooperate in oxidative stress-driven activation of the salicylic acid pathway. Plant Physiol. 174 (2), 956–971.

Rahman, M., Haq, N., Williams, I., 2016. Phytoaccumulation of arsenic, cadmium and lead by *Brassica juncea* parents and their F1 hybrids. J. Environ. Prot. 7, 613–622.

Rascio, N., Navari-Izzo, F., 2011. Heavy metal hyperaccumulating plants: how and why do they do it? And what makes them so interesting? Plant Sci. 180 (2), 169–181.

Rauser, W.E., Ackerley, C.A., 1987. Localization of cadmium in granules with in differentiating and mature root cells. Can. J. Bot. 65 (4), 643–646.

Rea, P.A., 2012. Phytochelatin synthase: of a protease a peptide polymerase made. Physiol. Plant. 145 (1), 154–164.

Reddy, G.N., Prasad, M.N.V., 1990. Heavy metal-binding proteins/peptides: occurrence, structure, synthesis and functions. A review. Environ. Exp. Bot. 30 (3), 251–264.

Reese, R.N., White, C.A., Winge, D.R., 1992. Cadmium-sulfide crystallites in Cd-(γEC) nG peptide complexes from tomato. Plant Physiol. 98 (1), 225–229.

Rizwan, M., Ali, S., ur Rehman, M.Z., Rinklebe, J., Tsang, D.C., Bashir, A., Maqbool, A., Tack, F.M.G., Ok, Y.S., 2018. Cadmium phytoremediation potential of *Brassica* crop species: a review. Sci. Total Environ. 631, 1175–1191.

Rodrigo, M.A.M., Zitka, O., Kominkova, M., Adam, V., Beklova, M., Kizek, R., 2013. Analysis of cadmium-phytochelatins 2 complexes using flow injection analysis coupled with electrochemical detection mass spectrometry. Int. J. Electrochem. Sci. 8, 4409–4421.

Ruotolo, R., Peracchi, A., Bolchi, A., Infusini, G., Amoresano, A., Ottonello, S., 2004. Domain organization of phytochelatin synthase functional properties of truncated enzyme species identified by limited proteolysis. J. Biol. Chem. 279 (15), 14686–14693.

Salt, D.E., Rauser, W.E., 1995. MgATP-dependent transport of phytochelatins across the tonoplast of oat roots. Plant Physiol. 107 (4), 1293–1301.

Salt, D.E., Prince, R.C., Pickering, I.J., Raskin, I., 1995. Mechanisms of cadmium mobility and accumulation in Indian mustard. Plant Physiol. 109 (4), 1427–1433.

Serrano, N., Díaz-Cruz, J.M., Ariño, C., Esteban, M., 2015. Recent contributions to the study of phytochelatins with an analytical approach. Trends Anal. Chem. 73, 129–145.

Seth, C.S., Chaturvedi, P.K., Misra, V., 2008. The role of phytochelatins and antioxidants in tolerance to Cd accumulation in *Brassica juncea* L. Ecotoxicol. Environ. Saf. 71 (1), 76–85.

Shukla, D., Kesari, R., Mishra, S., Dwivedi, S., Tripathi, R.D., Nath, P., Trivedi, P.K., 2012. Expression of phytochelatin synthase from aquatic macrophyte *Ceratophyllum demersum* L. enhances cadmium and arsenic accumulation in tobacco. Plant Cell Rep. 31 (9), 1687–1699.

Shukla, D., Trivedi, P.K., Nath, P., Tuteja, N., 2016. Metallothioneins and phytochelatins: role and perspectives in heavy metal(loid)s stress tolerance in crop plants. Abiotic Stress Response in Plants. In 237–264.

Sneller, F.E.C., van Heerwaarden, L.M., Koevoets, P.L., Vooijs, R., Schat, H., Verkleij, J.A., 2000. Derivatization of Phytochelatins from *Silene vulgaris*, induced upon exposure to arsenate and cadmium: comparison of derivatization with Ellman's reagent and monobromobimane. J. Agric. Food Chem. 48 (9), 4014–4019.

Son, J.A., Narayanankutty, D.P., Roh, K.S., 2014. Influence of exogenous application of glutathione on rubisco and rubisco activase in heavy metal-stressed tobacco plant grown in vitro. Saudi J. Biol. Sci. 21 (1), 89–97.

Speiser, D.M., Abrahamson, S.L., Banuelos, G., Ow, D.W., 1992a. *Brassica juncea* produces a phytochelatin-cadmium-sulfide complex. Plant Physiol. 99 (3), 817–821.

Speiser, D.M., Ortiz, D.F., Kreppel, L., Scheel, G., McDonald, G., Ow, D.W., 1992b. Purine biosynthetic genes are required for cadmium tolerance in *Schizosaccharomyces pombe*. Mol. Cell. Biol. 12 (12), 5301–5310.

Srivastava, S., Tripathi, R.D., Dwivedi, U.N., 2004. Synthesis of phytochelatins and modulation of antioxidants in response to cadmium stress in *Cuscuta reflexa*—an angiospermic parasite. J. Plant Physiol. 161 (6), 665–674.

Srivastava, V., Sarkar, A., Singh, S., Singh, P., de Araujo, A.S., Singh, R.P., 2017. Agroecological responses of heavy metal pollution with special emphasis on soil health and plant performances. Front. Environ. Sci. 5, 64.

Steffens, J.C., 1990. The heavy metal-binding peptides of plants. Annu. Rev. Plant Biol. 41 (1), 553–575.

Stolt, J.P., Sneller, F.E.C., Bryngelsson, T., Lundborg, T., Schat, H., 2003. Phytochelatin and cadmium accumulation in wheat. Environ. Exp. Bot. 49 (1), 21–28.

Strasdeit, H., Duhme, A.K., Kneer, R., Zenk, M.H., Hermes, C., Nolting, H.F., 1991. Evidence for discrete Cd (SCys) 4 units in cadmium phytochelatin complexes from EXAFS spectroscopy. J. Chem. Soc. Chem. Commun. (16), 1129–1130.

Strouhal, M., Kizek, R., Vacek, J., Trnková, L., Němec, M., 2003. Electrochemical study of heavy metals and metallothionein in yeast *Yarrowia lipolytica*. Bioelectrochemistry 60 (1–2), 29–36.

Sun, Q., Wang, X.R., Ding, S.M., Yuan, X.F., 2005a. Effects of interaction between cadmium and plumbum on phytochelatins and glutathione production in wheat (*Triticum aestivum* L.). J. Integr. Plant Biol. 47 (4), 435–442.

Sun, Q., Ye, Z.H., Wang, X.R., Wong, M.H., 2005b. Increase of glutathione in mine population of *Sedum alfredii*: a Zn hyperaccumulator and Pb accumulator. Phytochemistry 66 (21), 2549–2556.

Supalkova, V., Huska, D., Diopan, V., Hanustiak, P., Zitka, O., Stejskal, K., Baloun, J., Pikula, J., Havel, L., Zehnalek, J., Adam, V., 2007. Electroanalysis of plant thiols. Sensors 7 (6), 932–959.

Szalai, G., Krantev, A., Yordanova, R., Popova, L.P., Janda, T., 2013. Influence of salicylic acid on phytochelatin synthesis in *Zea mays* during Cd stress. Turk. J. Bot. 37 (4), 708–714.

Tian, S., Xie, R., Wang, H., Hu, Y., Hou, D., Liao, X., Brown, P.H., Yang, H., Lin, X., Labavitch, J.M., Lu, L., 2017. Uptake, sequestration and tolerance of cadmium at cellular levels in the hyperaccumulator plant species *Sedum alfredii*. J. Exp. Bot. 68 (9), 2387–2398.

Tsuji, N., Hirayanagi, N., Iwabe, O., Namba, T., Tagawa, M., Miyamoto, S., Miyasaka, H., Takagi, M., Hirata, K., Miyamoto, K., 2003. Regulation of phytochelatin synthesis by zinc and cadmium in marine green alga, *Dunaliella tertiolecta*. Phytochemistry 62 (3), 453–459.

Tsuji, N., Nishikori, S., Iwabe, O., Matsumoto, S., Shiraki, K., Miyasaka, H., Takagi, M., Miyamoto, K., Hirata, K., 2005. Comparative analysis of the two-step reaction catalyzed by prokaryotic and eukaryotic phytochelatin synthase by an ion-pair liquid chromatography assay. Planta 222 (1), 181–191.

Uraguchi, S., Tanaka, N., Hofmann, C., Abiko, K., Ohkama-Ohtsu, N., Weber, M., Kamiya, T., Sone, Y., Nakamura, R., Takanezawa, Y., Kiyono, M., 2017. Phytochelatin synthase has contrasting effects on cadmium and arsenic accumulation in rice grains. Plant Cell Physiol. 58 (10), 1730–1742.

Vacchina, V., Lobinski, R., Oven, M., Zenk, M.H., 2000. Signal identification in size-exclusion HPLC-ICP-MS chromatograms of plant extracts by electrospray tandem mass spectrometry (ES MS/MS). J. Anal. At. Spectrom. 15 (5), 529–534.

Vacek, J., Petřek, J., Kizek, R., Havel, L., Klejdus, B., Trnková, L., Jelen, F., 2004. Electrochemical determination of lead and glutathione in a plant cell culture. Bioelectrochemistry 63 (1–2), 347–351.

van de Mortel, J.E., Schat, H., Moerland, P.D., van Themaat, E.V.L., Van Der Ent, S.J.O.E.R.D., Blankestijn, H., Ghandilyan, A., Tsiatsiani, S., AARTS, M.G., 2008. Expression differences for genes involved in lignin, glutathione and sulphate metabolism in response to cadmium in *Arabidopsis thaliana* and the related Zn/Cd-hyperaccumulator *Thlaspi caerulescens*. Plant Cell Environ. 31 (3), 301–324.

Van Steveninck, R.F.M., Van Steveninck, M.E., Fernando, D.R., Edwards, L.B., Wells, A.J., 1990. Electron probe X-ray microanalytical evidence for two distinct mechanisms of Zn and Cd binding in a Zn tolerant clone of *Lemna minor* L. C. R. Acad. Sci. III Sci. Vie 310 (13), 671–678.

Vatamaniuk, O.K., Mari, S., Lu, Y.P., Rea, P.A., 1999. AtPCS1, a phytochelatin synthase from Arabidopsis: isolation and in vitro reconstitution. Proc. Natl. Acad. Sci. U. S. A. 96 (12), 7110–7115.

Vázquez, M.D., Poschenrieder, C., Barceló, J., 1992. Ultrastructural effects and localization of low cadmium concentrations in bean roots. New Phytol. 120, 215–226.

Viehweger, K., 2014. How plants cope with heavy metals. Bot. Stud. 55 (1), 35.

Vitecek, J., Petrlova, J., Petrek, J., Adam, V., Potesil, D., Havel, L., Mikelova, R., Trnkova, L., Kizek, R., 2006. Electrochemical study of S-nitrosoglutathione and nitric oxide by carbon fibre NO sensor and cyclic voltammetry possible way of monitoring of nitric oxide. Electrochim. Acta 51 (24), 5087–5094.

Vivares, D., Arnoux, P., Pignol, D., 2005. A papain-like enzyme at work: native and acyl-enzyme intermediate structures in phytochelatin synthesis. Proc. Natl. Acad. Sci. U. S. A. 102 (52), 18848–18853.

Vögeli-Lange, R., Wagner, G.J., 1990. Subcellular localization of cadmium and cadmium-binding peptides in tobacco leaves: implication of a transport function for cadmium-binding peptides. Plant Physiol. 92 (4), 1086–1093.

Wagner, G.J., 1993. Accumulation of cadmium in crop plants and its consequences to human health. Adv. Agron. 51, 173–212.

Wang, M.J., Wang, W.X., 2011. Cadmium sensitivity, uptake, subcellular distribution and thiol induction in a marine diatom: exposure to cadmium. Aquat. Toxicol. 101 (2), 377–386.

Wang, H.C., Wu, J.S., Chia, J.C., Yang, C.C., Wu, Y.J., Juang, R.H., 2009. Phytochelatin synthase is regulated by protein phosphorylation at a threonine residue near its catalytic site. J. Agric. Food Chem. 57 (16), 7348–7355.

Weghe, J.G.V., Ow, D.W., 1999. A fission yeast gene for mitochondrial sulfide oxidation. J. Biol. Chem. 274 (19), 13250–13257.

Winther, J.R., Thorpe, C., 2014. Quantification of thiols and disulfides. Biochim. Biophys. Acta. (BBA)-General Subjects. 1840 (2), 838–846.

Wójcik, M., Vangronsveld, J., D'Haen, J., Tukiendorf, A., 2005. Cadmium tolerance in *Thlaspi caerulescens*: II. Localization of cadmium in *Thlaspi caerulescens*. Environ. Exp. Bot. 53 (2), 163–171.

Yamaguchi, C., Ohkama-Ohtsu, N., Shinano, T., Maruyama-Nakashita, A., 2017. Plants prioritize phytochelatin synthesis during cadmium exposure even under reduced sulfate uptake caused by the disruption of SULTR1; 2. Plant Signal. Behav. 12(5) e1325053. https://doi.org/10.1080/15592324.2017.1325053.

Yan, S.U.N., Hui, Y.E., Zebin, W.E.I., Xuhui, K.O.N.G., Qi-Tang, W.U., 2017. Root cell wall and phytochelatins in low-cadmium cultivar of *Brassica parachinensis*. Pedosphere 1–12 (In Press).

Yates, J.R., McCormack, A.L., Link, A.J., Schieltz, D., Hays, L., 1996. Tutorial review. Future prospects for the analysis of complex biological systems using micro-column liquid chromatography-electrospray tandem mass spectrometry. Analyst 121 (7), 65R–76R.

Yen, T.Y., Villa, J.A., DeWitt, J.G., 1999. Analysis of phytochelatin-cadmium complexes from plant tissue culture using nano-electrospray ionization tandem mass spectrometry and capillary liquid chromatography/electrospray ionization tandem mass spectrometry. J. Mass Spectrom. 34 (9), 930–941.

Zhang, Z., Gao, X., Qiu, B., 2008. Detection of phytochelatins in the hyperaccumulator *Sedum alfredii* exposed to cadmium and lead. Phytochemistry 69 (4), 911–918.

Zhang, Z.C., Chen, B.X., Qiu, B.S., 2010. Phytochelatin synthesis plays a similar role in shoots of the cadmium hyperaccumulator *Sedum alfredii* as in non-resistant plants. Plant Cell Environ. 33 (8), 1248–1255.

Zhang, X., Rui, H., Zhang, F., Hu, Z., Xia, Y., Shen, Z., 2018. Overexpression of a functional *Vicia sativa* PCS1 homologue increases cadmium tolerance and phytochelatins synthesis in *Arabidopsis*. Front. Plant Sci. 9, 107. https://doi.org/10.3389/fpls.2018.00107.

Zitka, O., Krystofova, O., Sobrova, P., Adam, V., Zehnalek, J., Beklova, M., Kizek, R., 2011a. Phytochelatin synthase activity as a marker of metal pollution. J. Hazard. Mater. 192 (2), 794–800.

Zitka, O., Skutkova, H., Krystofova, O., Sobrova, P., Adam, V., Zehnalek, J., Havel, L., Beklova, M., Hubalek, J., Provaznik, I., Kizek, R., 2011b. Rapid and ultrasensitive method for determination of phytochelatin using high performance liquid chromatography with electrochemical detection. Int. J. Electrochem. Sci. 6 (5), 1367–1381.

## FURTHER READING

Chekmeneva, E., Díaz-Cruz, J.M., Ariño, C., Esteban, M., 2009. Binding of $Hg^{2+}$ with phytochelatins: study by differential pulse voltammetry on rotating Au-disk electrode, electrospray ionization mass-spectrometry, and isothermal titration calorimetry. Environ. Sci. Technol. 43 (18), 7010–7015.

Gusmão, R., Cavanillas, S., Ariño, C., Díaz-Cruz, J.M., Esteban, M., 2010. Circular dichroism and voltammetry, assisted by multivariate curve resolution, and mass spectrometry of the competitive metal binding by phytochelatin PC5. Anal. Chem. 82 (21), 9006–9013.

Vacchina, V., Chassaigne, H., Łobiński, R., Oven, M., Zenk, M.H., 1999. Characterisation and determination of phytochelatins in plant extracts by electrospray tandem mass spectrometry. Analyst 124 (10), 1425–1430.

# CHAPTER 9

# Cadmium-Induced Oxidative Stress in Plants, Cadmium Toxicity, and Tolerance in Plants: From Physiology to Remediation

**Anna Grobelak, Jakub Świątek, Aneta Murtaś, Marta Jaskulak**
Department of Environmental Engineering, Faculty of Infrastructure and Environment, Czestochowa University of Technology, Czestochowa, Poland

## Abstract

Cadmium enters the environment mainly through the ground, and it is found in manure and pesticides. This trace element is a byproduct of zinc, lead, and copper production. Cadmium toxicity has been extensively studied in plants; however, induced oxidative stress and its biochemical mechanism has not yet been well established, and is an ongoing interest of scientists. The harmful effects of free ROS radicals are termed "oxidative stress." Indicators of cadmium-induced oxidative stress include δ-aminolevulinic acid dehydratase (ALA-D) activity, S-adenosyl-l-methinone (SAM), thiobarbituric acid reactive substances (TBARS), GSH levels, δ-aminolevulinic acid (ALA) and porphobilinogen (PBG), superoxide dismutase (SOD), guaiacol peroxidase (GPOX), proteins, and gene expression. Cadmium treatments cause inhibition of specific enzymes' activity and gene expression. Studies have also confirmed that different soil amendments and treatments significantly decrease the cadmium-induced oxidative stress in plants.

**Keywords:** Cadmium, Oxidative stress, Indicators, Plants, Soil, Amendments

## 1 INTRODUCTION: CADMIUM IN THE ENVIRONMENT

Cadmium (Cd) can mostly be found in the earth's layers, and always occurs in combination with zinc. Cadmium is also an unavoidable residual product of zinc, lead, and copper mining. Cadmium enters the environment mainly through the ground, because it is found in artificial fertilizers, manure, and pesticides (Hasanuzzaman et al., 2012). A frequent result of soil acidification processes is increased mobility of all Cd cations due to their increased desorption into the soil solution (Grobelak et al., 2018). The creation of

*Cadmium Toxicity and Tolerance in Plants*
https://doi.org/10.1016/B978-0-12-814864-8.00009-7

mobile metal complexes with plant root exudates, along with the process of desorption of metals from the soil, may be the dominant mechanisms associated with the migration of metals into the soil profile (Dessureault-Rompré et al., 2015). Due to natural processes, a very large amount of Cd is released into the environment. About half of this Cd is released into rivers through weathering of rocks, and some Cd is released into the air through forest fires and volcanoes. The rest of the Cd is released through anthropogenic activities, such as industry. Cadmium enters the environment mainly through the ground, and it is found in manure and pesticides. This trace element is an inevitable byproduct of zinc, lead, and copper production (Aravind and Prasad, 2005). Cadmium in jewelry, plastics, and brazing sticks was banned in the EU starting in December 2011 (http://europa.eu/rapid/press-release_IP-11-620_en.htm).

## 2 OXIDATIVE STRESS

Cadmium toxicity has been extensively studied in plants; however, induced oxidative stress and its biochemical mechanism has not yet been well established, and has garnered the ongoing attention of scientists. The main mechanism of Cd phytotoxicity is the induction of oxidative stress. The harmful effect of free ROS radicals is termed "oxidative stress." Indicators of Cd-induced oxidative stress include δ-aminolevulinic acid dehydratase (ALA-D) activity, S-adenosyl-l-methinone (SAM), thiobarbituric acid reactive substances (TBARS), GSH levels, δ-aminolevulinic acid (ALA) and porphobilinogen (PBG), superoxide dismutase (SOD), guaiacol peroxidase (GPOX), proteins, and gene expression (Foyer and Noctor, 2005).

Plant organisms are exposed to the continuous production and action of harmful particles such as reactive oxygen species (ROS), which are the product of oxygen metabolism. During oxidative stress, increased production of ROS is observed, leading to tissue damage, and even death (Borisova et al., 2016). Low concentrations of reactive oxygen species fulfill signal functions related to control and regulation of the growth and development processes, response to biotic and biotic stress factors, and the process of programmed cell death. For this purpose, it is necessary to maintain a balance between the level of ROS produced and the possibility of their inactivation by antioxidative mechanisms in plant cells (Jaskulak et al., 2018). In the basic energetic state, called triplet oxygen, the $O_2$ molecule is not very active, and the total oxygen reduction leads to the formation of a water molecule. Reactive oxygen species are, in turn, products of one-electron oxygen reduction, or its photochemical excitation. They are formed in the mitochondria, cell membranes, chloroplasts, peroxisomes, and apoplasts (Asada, 2006). Reactive oxygen species (ROS) produced in all eukaryotic cells include singlet oxygen, superoxide anion radicals, hydrogen peroxide, and hydroxyl radicals (Medrano and Parry, 2008). Some ROS have an unpaired electron, which makes them free radicals, in which dismutation leads to the formation of hydrogen peroxide, according to the following reaction (Sharkey, 2005):

$$O_2^- + O_2^- + 2H_2O \rightarrow O_2 + 2H_2O_2$$

In chloroplasts and mitochondria of plants, reactive oxygen species are formed during electron flow through the electron chain in the thylakoid membrane of chloroplasts, and through the mitochondrial respiratory chain located in the inner membrane of the mitochondrion. A significant source of the formation of reactive oxygen species is also peroxisomes. During natural physiological processes, ROS are accompanied by processes such as closing stomata, papilla formation in legumes, and aging of cells (Mittler and Blumwald, 2015). Excessive production of reactive oxygen species associated with the occurrence of stress factors results in oxidative damage of nucleic acids, lipids, and proteins in cells. Oxidative changes in protein structures include their carbonylation, oxidation of thiol groups with the formation of disulfide bridges, fragmentation of the polypeptide chain, or protein aggregation as a result of the formation of additional S-S connections. Other sources of ROS may be reactions involving peroxisomal oxidases, cytochrome P-450 enzymes, NAD (P)H oxidases, or xanthine oxidase. Direct measurement of ROS levels in cells is a very difficult and inaccurate task due to the short half-life, high reactivity, and low concentration. Oxidative stress is indirectly determined, based on the effects caused by ROS (Inzé and Van Montagu, 1995).

The harmful effect of free ROS is evident in biological systems when there is either an excessive production of ROS and/or a deficiency of enzymatic and nonenzymatic antioxidants. The redox stress/oxidative stress is a complex process. Its impact on the organism depends on the type of oxidant, on the site and intensity of its production, on the composition and activities of various antioxidants, and on the ability of repair systems.

Plant cells have numerous defense mechanisms that allow for the optimization of damage caused by ROS overproduction. These systems include oxidative enzymes and nonenzymatic antioxidants. Oxidative enzymes are directly involved in the removal of ROS, and in oxidation reactions and the reduction of ascorbic acid and glutathione (Hasanuzzaman et al., 2018).

# 3 INDICATORS OF CADMIUM-INDUCED OXIDATIVE STRESS

## 3.1 Enzymatic Indicators

Cadmium toxicity has been widely studied in plants; nevertheless, its biological mechanism of action has not yet been closely examined. Hence, proteins containing a redox-active Fe atom, such as catalase (CAT), class I and class III peroxidases (PODs), and NADPH oxidase have a crucial role in both the production and scavenging of reactive oxygen species (ROS) (Tchounwou et al., 2012). The protection mechanisms consist of enzymes such as superoxide dismutase (SOD), CAT, and ascorbate peroxidase (APX), and different metabolites; for example, glutathione (GSH) and ascorbate (AsA). Moreover, GSH functions as a precursor for phytochelatins (PCs). These particles have the capability for chelating Cd, thus decreasing its harmfulness (Jozefczak et al., 2015).

### *3.1.1 Superoxide Dismutase (SOD)*

Superoxide peroxide (SOD) is an enzyme catalyzing the reactions of dismutation of two superoxide anion radicals to the hydrogen peroxide and oxygen molecules.

$$O_2 + O_2 + 2H^+ \rightarrow H_2O_2 + O_2$$

This reaction takes place in two stages, and includes the reduction and oxidation of the metal enzyme that is contained in the catalytic center. Depending on the type of cofactor, the ferric dismutases are: FeSOD, occurring mainly in plastids, manganese dismutases: MnSOD, occurring in the mitochondria and copper-zinc dysmutases: Cu/ZnSOD, present in the cytosol, peroxisomes, apoplast, chloroplasts, and mitochondria. SOD is a significant antioxidant defense in almost all living cells exposed to oxygen (Zhang et al., 2009).

*Peroxidase (POD)* is directly involved in the removal of hydrogen peroxide. Peroxidase activity is important in the functioning of plant cells by taking part in germination of seedlings, cell aging, leaf drooping, and fruit ripening, or reactions to emerging adverse environmental factors and pathogens. These include ascorbate peroxidase (APX), which is a key enzyme in the course of an ascorbate-glutathione cycle, where when attaching two particles of ascorbate reduces $H_2O_2$ to water molecules to form monodehydroascorbate (MDHA) (Gill and Tuteja, 2010). Glutathione peroxidase plays an important role in the removal of ROS, in which the main function is to protect cellular structures against oxidation. In addition, this enzyme is characterized by tolerance to high temperatures, while retaining partial activity, even at 100°C.

Among the peroxidases, we also separate guaiacol peroxidase present in the cytosol, intercellular spaces, vacuoles, and cell walls that plays an important role in the process of lignification, ethylene biosynthesis, and protection against pathogens (Batish et al., 2006; Chen et al., 2008).

*Catalase (CAT)* carries out reactions of hydrogen peroxide dismutation to the oxygen and water molecule in accordance with the reaction:

$$2H_2 + O_2 \rightarrow 2H_2O + O$$

Catalases occur in peroxisomes and glyoxysomes, as well as cytosol and mitochondria. In all eukaryotes, catalases are tetramers composed of four molecules, with NADPH molecules and heme groups in active centers. In relation to peroxidases, they react with hydrogen peroxide up to 10,000 times faster. Among the catalases, there are three main isomeric forms: CAT1, CAT2, and CAT3. In places of significant production of hydrogen peroxide, such as cytosol, peroxisomes, and glyoxysomes, CAT2 dominates. The third isomeric form, CAT3, dominates in young tissues of plants and seeds, catalyzing the decomposition of $H_2O_2$ resulting from the degradation of fatty acids. The concentration of $H_2O_2$ in the cells determines the type of catalase activity. At a high concentration of hydrogen peroxide, catalases exhibit catalase activity allowing the removal of $H_2O_2$, while at low concentrations, the dominant activity of catalase is peroxidase activity (Azevedo et al., 2005).

*δ-Aminolevulinic acid dehydratase (ALA-D) and δ-aminolevulinic acid (ALA) activity* is the next defense mechanism when exposed to stress factors such as the presence of cadmium. ALA can be formed through two paths: the enzymatic condensation of glycine and succinyl-CoA by ALA synthase (ALA-S), and the C5 pathway transforming glutamate to ALA. The C5 pathway is predominant in plants, algae, and most bacteria; whereas ALA-S is primarily limited to animal mitochondria, some fungi, and the α-group of purple eubacteria (*Rhizobium* and *Rhodobacter*). ALA undergoes enolization and further metal-catalyzed aerobic oxidation at physiological pH to yield superoxide radical ($O_2^-$), hydrogen peroxide ($H_2O_2$), and hydroxyl radical ($HO^\bullet$). Thus, accumulation of ALA in the cell might endogenously increase reactive oxygen species (ROS), leading to oxidative stress (Noriega et al., 2007). From in vitro tests using leaf discs exposed to ALA or $Cd^{2+}$, it was noted that TBARS levels were boosted, while GSH content and SOD and GPOX activities and expressions were reduced. The defensive role of SAM against oxidative stress generated by $Cd^{2+}$ and ALA was also confirmed. Accumulation of ALA in soybean plants due to treatment with $Cd^{2+}$ is mainly responsible for oxidative stress generation (Noriega et al., 2007).

*Glutathione (GSH)* is the most abundant antioxidant in aerobic cells, present in micromolar (μM). Glutathione, a tripeptide with the sulfhydryl (—SH) group, is an essential compound predominantly involved in redox equilibrium preservation of the cellular environment. Glutathione is known as a key signaling molecule underlying the higher PAL (phenylalanine ammonia lyase) activity, indicating the significance of PAL in Cd plants' tolerance. The higher activity of PAL was noted in plants under Cd exposition (DalCorso et al., 2010). In the study of Mohanpuria et al. (2007), Cd exposure of *Camellia sinensis* L. decreased chlorophyll and protein, but the transcript levels of glutathione biosynthetic genes viz. gamma-glutamylcysteine synthetase (gamma-ECS) and glutathione synthetase (GSHS) enlarged. Additionally, transcript levels of glutathione reductase (GR), an enzyme involved in reduction of oxidized glutathione (GSSG) to reduced glutathione (GSH), also indicated upregulation on Cd exposure. The research also confirmed that the upregulation in the transcript of glutathione metabolic genes is connected with the defense of plants from Cd exposition (Mohanpuria et al., 2007). The decline in GSH contents was also noted for *Lycopersicon esculentum* after Cd exposure (Ammar et al., 2008).

The *malondialdehye (MDA)* or *thiobarbituric acidreactive-substances (TBARS)* assay are used widely to assess peroxidation of lipids in plants. Malondialdehyde is produced through auto-oxidation and enzymic degradation of polyunsaturated fatty acids in plants cells. This secondary end product of the oxidation of polyunsaturated fatty acids reacts with two molecules of thiobarbituric acid (TBA) via an acid-catalyzed nucleophilic-addition reaction. This marker is popular due to its simplicity and low cost of analysis (Hodges et al., 1999).

*Dithiothreitol (DTT, $HSCH_2(CH(OH))_2CH_2SH$)* is generally used as a cell-free degree of the oxidative potential of particles (Verma et al., 2011). The redox-active chemicals in particulate matter (PM) oxidize supplemented DTT to its disulfide formula. The DTT linear loss is used to measure the oxidative properties of the PM. Then, the redox-active species in PM provide an electron for oxygen dissolution and superoxide generation, which can distribute ROS to other plants. In addition, the DTT assay delivers a quantitative measure of oxidation, but does not provide the information about specific ROS (Charrier and Anastasio, 2012).

## 3.2 Nonenzymatic Proteins and Small-Molecule Antioxidants

An important role in the removal of excess ROS is also played by nonenzymatic proteins and small-molecule antioxidants. *Ascorbate (AsA)* is a strong antioxidant that reacts with singlet oxygen, organic peroxide radicals, hydrogen peroxide, superoxide, and hydroxyl radicals. It occurs in the cytosol, vacuoles, chloroplasts, mitochondria, and peroxisomes. AsA is a main antioxidant, but also shares biosynthesis constituents with cell wall precursors. Moreover, AsA and Class III lignifying POD enzymes contribute within a complex redox system, in which hydrogen peroxide ($H_2O_2$) is put forward as a main sensor of the redox balance, being both a signaling particle to induce lignin biosynthesis, and a substrate for lignifying PODs (Loix et al., 2017).

*Tocopherol* is a hydrophobic antioxidant present in biological membranes, mainly reducing lipid peroxide radicals and singlet oxygen. During the reaction of tocopherol with radicals, less reactive tocopherol radicals are formed, and removed by other antioxidants—glutathione and ascorbate.

*Carotenoids* effectively expire molecules that are in the triplet excited state (e.g., chlorophyll), as well singlet oxygen, by quenching triplet chlorophyll, which prevents excessive production of ROS. They also react with lipid radicals. Carotenoids, and among them particularly β-carotene, can act (in vivo testing) as chemical quenchers of $O_2$, with β-carotene endoperoxide being a key oxidation product produced from this activity in plants. The latter product delivers a new internal probe of $O_2$ production, which could be used for the primary recognition of oxygen stress in plants. Other carotenoid metabolites produced by $O_2$ oxidation of carotenoids, such as aldehydes, can be likewise formed enzymatically by CCDs, and are thus not specific to $O_2$ (Ramel et al., 2012).

*Flavonoids* inhibit the activity of lipoxygenases; they are a substrate for other enzymes, including peroxidases for the decomposition of hydrogen peroxide. The antioxidant functions in plant cells also have substances such as proline, mannitol, and sorbitol. In addition, the mitochondrial inner membrane contains a decoupling protein (UCP) that has the ability to reduce ROS production by scattering the proton electrochemical gradient (Tchounwou et al., 2012).

## 3.3 Interactions Between Cadmium-Induced Cell Wall Responses and Oxidative Stress

The intracellular defense system only has the ability to deal with negative effects of Cd to a certain extent. The plant cell wall is an important plant component, as it represents the first contact point of plants with Cd, and has an important function in keeping excess Cd out of the cell. The main mechanisms are associated with the stimulation of primary cell wall remodeling and its relative uptake of crucial plant nutrients and the initiation of lignin biosynthesis (Loix et al., 2017). The significant role is recognized to AsA as it is functioning as a key antioxidant; but moreover, bonds biosynthesis particles with cell wall precursors. It was found that AsA and Class III lignifying POD enzymes contribute within a complex redox system, in which hydrogen peroxide ($H_2O_2$) is delivered forward as a main sensor of the redox equilibrium. Thus, it is working both as a signaling particle to generate lignin biosynthesis, and as a substrate for lignifying PODs (Loix et al., 2017).

# 4 CADMIUM AFFECTS PLANTS GROWTH—CASE STUDIES

## 4.1 Soil pH Changes

The collection of metals in plants is closely related to the properties of the substrate, including its pH, the content of organic substances, and the concentration of all ions. Acidic soil pH causes the breakdown of chemical compounds into ions—forms in which metals easily penetrate into plant cells. One of the ways to limit the availability of trace metals for plants is soil liming leads to an increase in pH. At the alkaline pH, a significant amount of trace metals becomes inaccessible to plants. In general, most metals show relatively higher mobility in acidification and oxidizing conditions, while basic and reducing conditions favor the reduction of metal mobility (Placek et al., 2017). According to Kabata-Pendias and Pendias (2001), the bioavailability of Cd is changing from very high at pH 2 to very little at pH 7. It was found that for Cd contaminated soil treated with soil amendments influencing pH, the significant change in Cd bioavailability was noted (Fig. 1). The liming process can reduce the bioavailability, and thus toxicity, of Cd-contaminated soil. Moreover, these soil amendments seems to be the easiest and cheapest to apply.

In some research studies, the combined treatments of nutrients and lime were tested (Fig. 2). The largest biomass yield of grass was obtained for a treatment of the combined use of potassium phosphate and sewage sludge, and for a treatment of combined use of superphosphates and sewage sludge. However, the use of only sewage sludge in a dose of 3% d.w. gave similar results, as well as combinations of K6 and K7 with additional inorganic fertilizers application. The liming of the Cd-contaminated soil (K) yielded very little grass biomass in comparison with nutrient treatments.

In studies on the content of Cd in the biomass, a very large reduction in the concentration of Cd in biomass comparison to control (untreated soil)

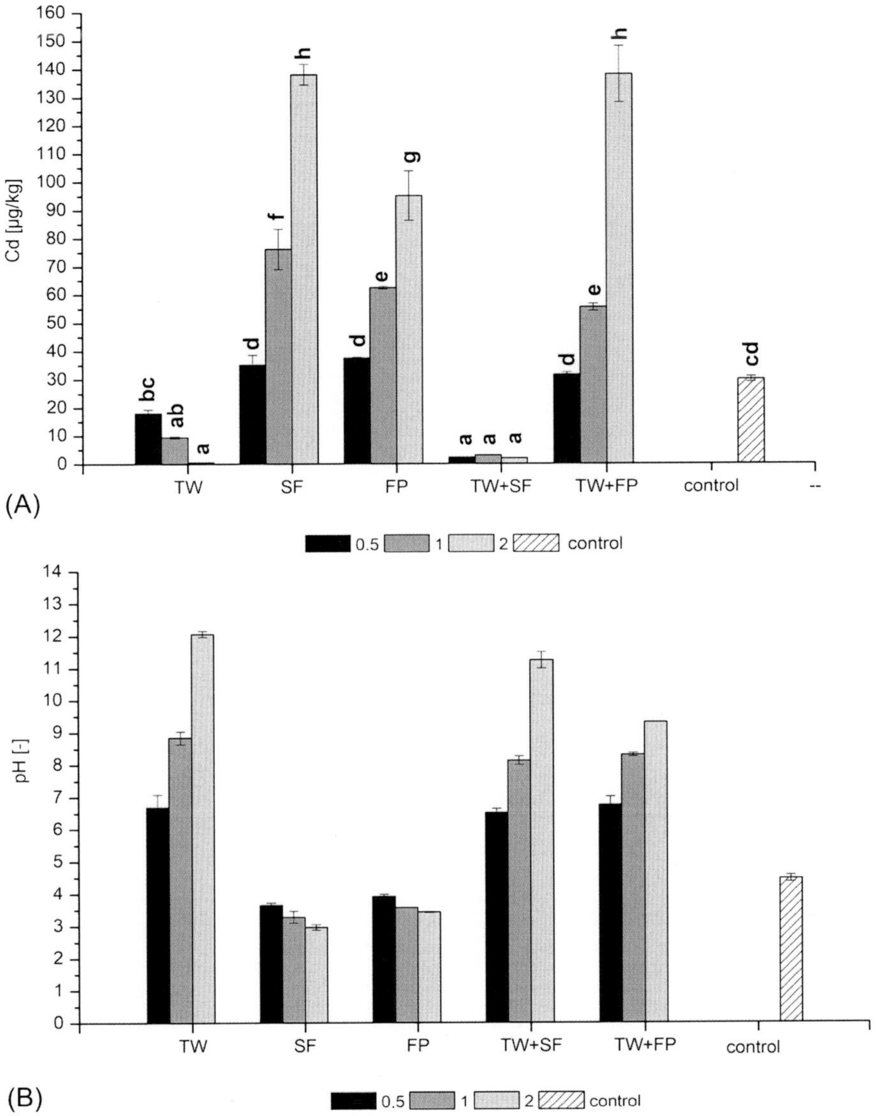


**Fig. 1** (A) Cd content extracted with water and pH. (B) Value of soil material; 0.5%–2% of inorganic additive; TW—lime fertilizer, SF—superphosphate, FP—potassium phosphate.

was obtained after the application of inorganic fertilizers (Fig. 3). The best results, however, were obtained for the treatments of the combined use of sewage sludge and inorganic additives. In the treatments used, the most Cd accumulated mainly in the plants' roots.

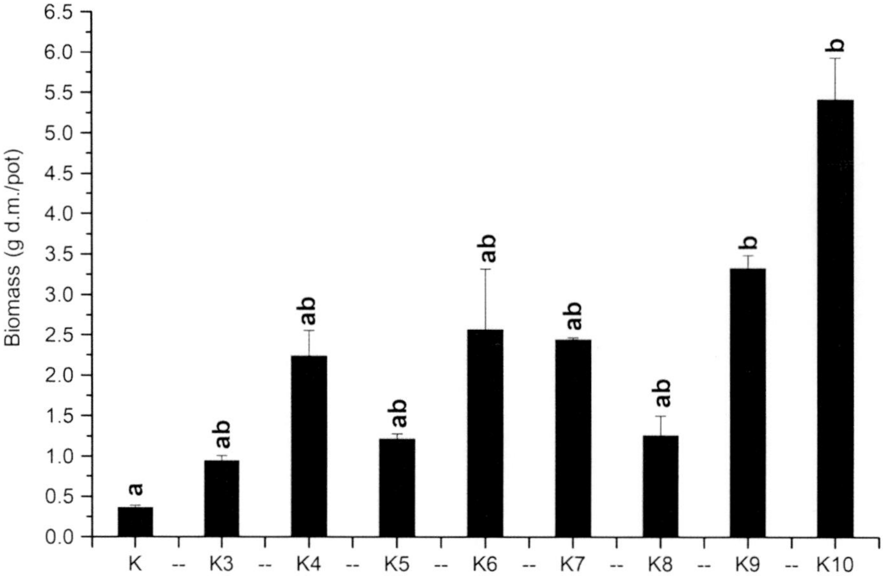


**Fig. 2** Biomass growth after 10 weeks of the experiment; K—control, K3—sewage sludge 1%, K4—sewage sludge 3%, K5—TW + SF 0.8%, K6—TW + SF 0.8% + 1% sewage sludge, K7—TW + SF 0.8% + 3% sewage sludge, K8—TW + FP 0.8%, K9—TW + FP 0.8% + 1% sewage sludge, K10—TW + FP 0.8% + 3% sewage sludge.

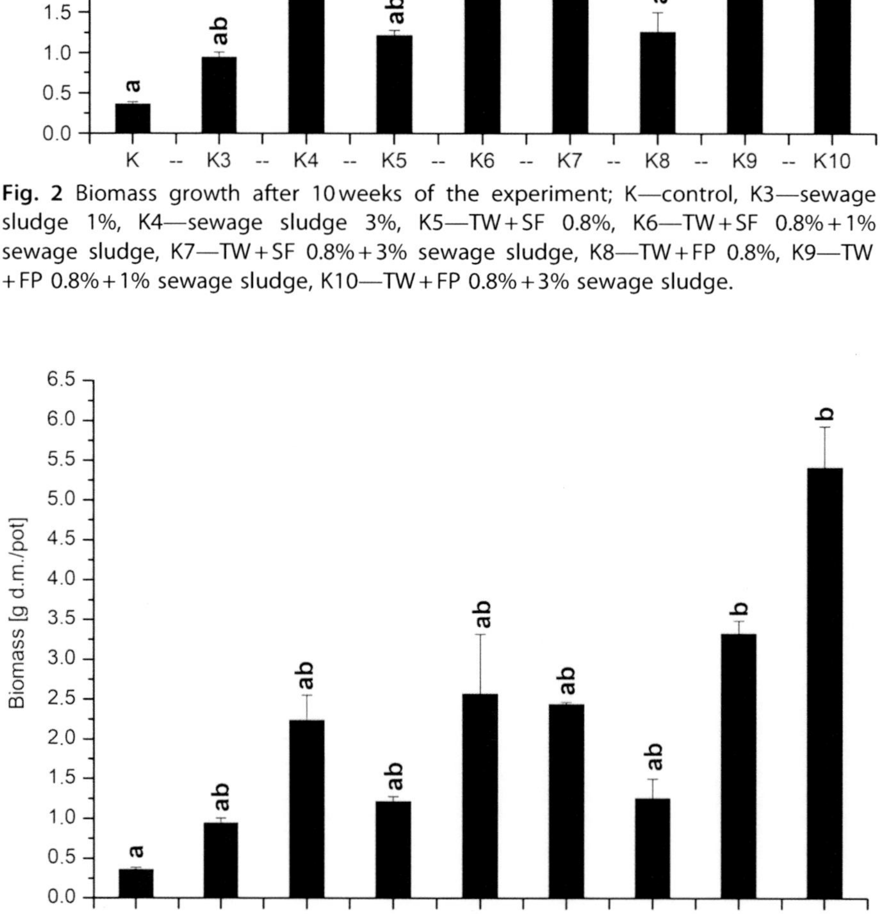


**Fig. 3** Cd in biomass of fescue grass after 10 weeks of the experiment; K—control, K3—sewage sludge 1%, K4—sewage sludge 3%, K5—TW + SF 0.8%, K6—TW + SF 0.8% + 1% sewage sludge, K7—TW + SF 0.8% + 3% sewage sludge, K8—TW + FP 0.8%, K9—TW + FP 0.8% + 1% sewage sludge, K10—TW + FP 0.8% + 3% sewage sludge.

## 4.2 Nutrients Uptake

Plants exposed to Cd have been shown to exhibit excessive root destruction, inhibition of the photosynthesis system, modifications of enzymes, and a decreased biomass. According to Hédiji et al. (2015), these biological abnormalities might be an indirect effect initiated by a deficiency of mineral nutrition within the plants due to Cd uptake. One possible mechanism by which Cd could affect the uptake and translocation of nutrients is cell wall remodeling. The chlorosis effect was noted after Cd exposition, mainly as a result of Cd-induced Fe deficiency in the shoots of *A. thaliana* (Xu et al., 2015). Similarities in physicochemical features among Cd and other divalent cations, for example, Ca, Zn, and magnesium (Mg), can antagonize Cd uptake. The Cd-induced responses on the uptake of plant nutrients are very complex (Cannata et al., 2014). In the conducted study where plants were growing on Cd-contaminated soil (7 mg/kg) amended with sewage sludge, the different morphological features (Fig. 4) were observed in comparison with plants growing in unamended soil.

The vivid green color of plants, lush growth, and intensive propagation are all observed in grass species growing on soils with organic soil additives (Fig. 4). In terms of morphology, the best grass growth was obtained for *Festuca arundinacea* Schreb. Moreover, the root system of plants cultivated on the substrate with the addition of sludge was very developed, and the whole lump of soil was overgrown with root mass. On substrates without added sludge, plant growth was inhibited. Only a seedling of a few centimeters with 2–3 chlorotic leaves were obtained, devoid of, or with a very poor, root system. Only for *F. arundinacea* Schreb the plants developed roots a few centimeters long. Similar results were found for *Lupinus luteus,* grown in Cd-contaminated soil (Fig. 5). Plants grown in soil without horse manure application were very small, with reduced roots in comparison with plants grown in amended soil.

## 4.3 Oxidative Stress Markers

Analysis of *guaiacol peroxidase enzyme (GPX) activity* was carried out in plants of *Sinapis alba* (Fig. 6) in order to determine the remediation effect in Cd (7 mg/kg) contaminated soil (anthropogenic contamination) and treated with organic amendments. Peroxidases are one of the most important defense mechanisms against free oxygen radicals. The lowest level of GPX activity therefore means a lower level of abiotic stress. It was found that in plants grown in organic amended soil, the GPX activity was much lower than for untreated soil.

**Fig. 4** Morphology of grass plants grown on sewage sludge-amended soil contaminated with Cd (7 mg/kg) and unamended soil Dg—*Dactylis glomerata* L., Lw—*Lolium westrwoldicum* L., Fr—*Festuca rubra* L., Lp—*Lolium perenne* L., Fr—*Festuca arundinacea* Schreb., Ac—*Agrostis capillaris.*

**Fig. 5** Shoots (4 weeks culture) of *Lupinus luteus*; from left: reference soil (clean), contaminated soil (7 mg/kg Cd), 10% v/v horse manure amendment; 20% v/v horse manure amendment.

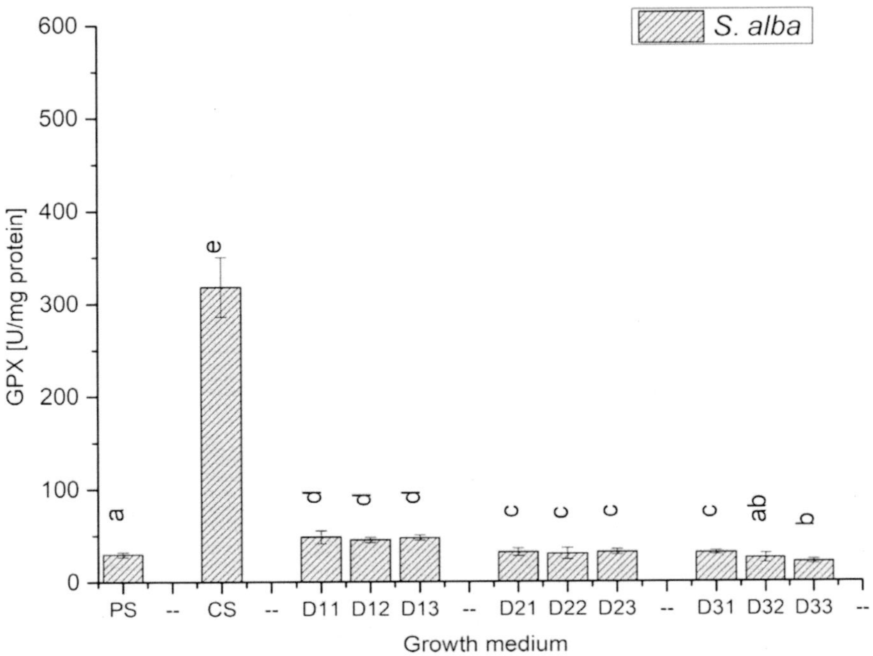


**Fig. 6** Guaiacol peroxidase activity in green biomass of *S. alba* after 3 weeks of culture in Cd (7 mg/kg) contaminated soil, aboveground parts; Peat—reference soil, M.Sl—soil, D1—soil + cow manure (10%, 15%, and 20%/kg d.w.), D2—soil + horse manure (10%, 15% and 20%), D3—soil + vermicompost (10%, 15% and 20%).

The *chlorophyll content of aboveground parts* of plants can be determined to monitor the physiological condition of plants in the process of phytoremediation of soil contaminated with Cd (Fig. 7).

Measurement of the chlorophyll content in the aboveground parts of the studied plants is an effective marker of the abiotic stress level in the studied plant species. The lowest chlorophyll content in both vase and in vitro tests was recorded in plants grown on soil contaminated without any additives. With the use of selected organic fertilizers, the chlorophyll content in the aboveground parts of the studied plants increased, which is an improvement on the physiological status of the cultivated plants. This result applies to both chlorophyll A, B, and their sum. The highest content of chlorophyll was recorded on the reference soil free of impurities (peat).

The production of *phenolic compounds* is one of the most important defense mechanisms of plants against the impact of adverse environmental conditions. Along with improving the quality of degraded soils, the content

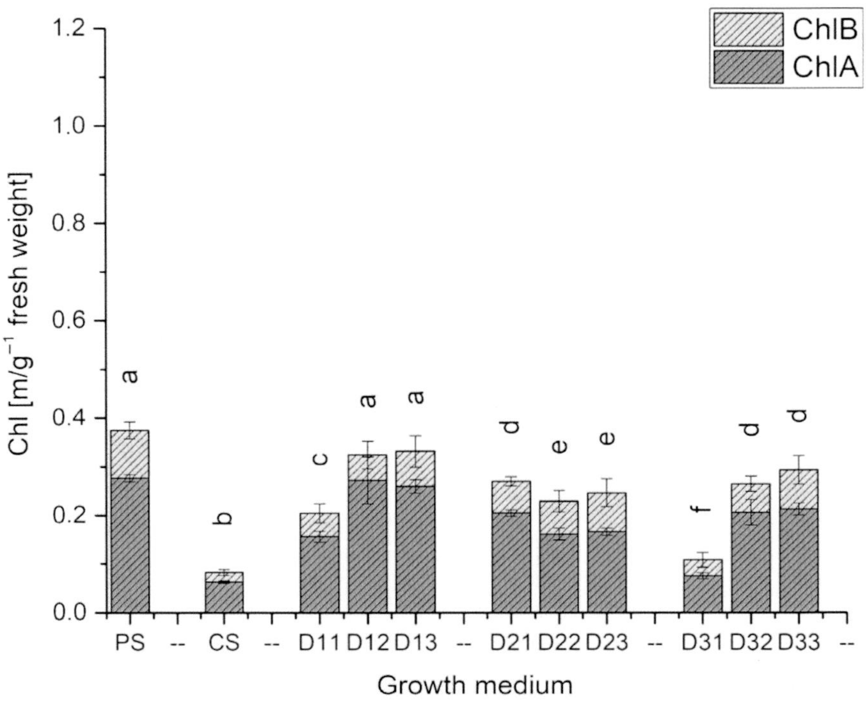


**Fig. 7** Chlorophyll content in aboveground parts of *S. alba*; 3 weeks of culture on soil contaminated with Cd (7 mg/kg); Peat—reference soil, M.Sl—soil, D1—soil + cow manure (10%, 15% and 20%/kg. d.d), D2—soil + horse manure (10%, 15% and 20%), D3—soil + vermicompost (10%, 15% and 20%).

of phenolic compounds in the aboveground parts of the tested plants decreases, which indicates the correct course of the remediation process of a given soil. Determination of the content of phenolic compounds in aboveground parts of plants grown in Cd-contaminated soil showed high sensitivity to changes caused by an organic soil amendment to contaminated soil. The highest (Fig. 8) content of phenolic compounds was recorded on control (contaminated) soil without amendments (M.Sl).

The addition of organic fertilizers caused a significant decrease in phenolic substances in the aboveground parts of the studied plants. In the studies by Jaskulak et al. (2018), in the case of phenolic compounds, all three studied plant species: *S. alba, Robinia pseudoacacia*, and *L. luteus* showed a similar content of phenolic compounds in above-ground parts—approx. 20–35 μL/g after Cd exposure (7 mg/kg). In the study investigating the regulatory function of exogenous addition of calcium (Ca) in enhancing the antioxidant defense after Cd treatment in rice, it was found that exposure to Cd caused chlorosis, leaf rolling symptoms, and growth inhibition. Calcium application

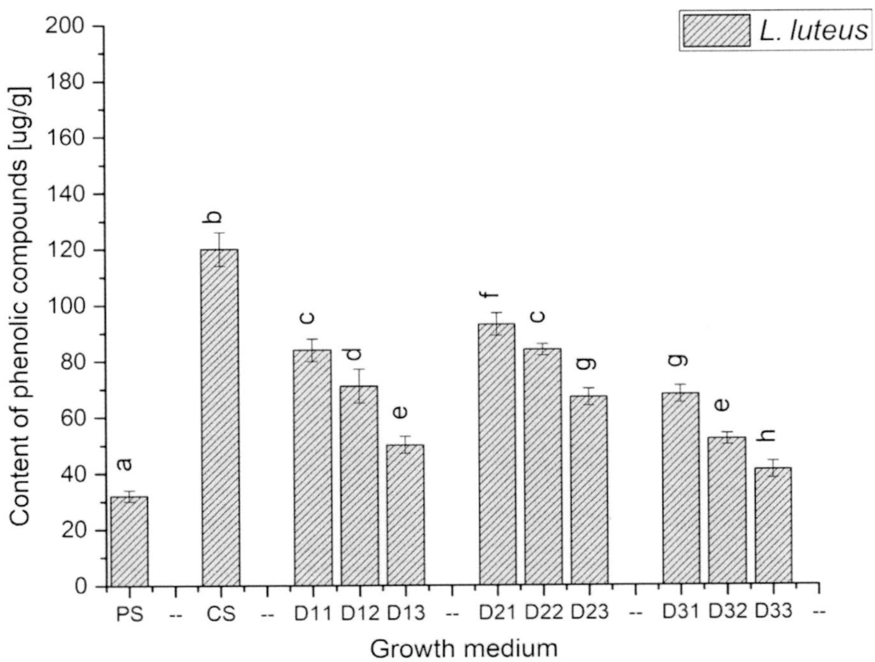


**Fig. 8** The content of phenolic compounds in aboveground parts of *L. luteus*; 3 weeks of culture on soil contaminated with Cd (7 mg/kg); peat—reference soil, M.Sl—soil, D1—soil + cow manure (10%, 15% and 20%/kg. d.w.), D2—soil + horse manure (10%, 15% and 20%), D3—soil + vermicompost (10%, 15% and 20%).

reduced Cd uptake, increased the AsA level, and increased the activities of superoxide dismutase (SOD) and catalase (CAT) (Rahman et al., 2016).

The study (Fig. 9) of the level of gene expression using the qPCR (quantitative real time PCR) technique of rbLc encoding a large RuBisCO subunit in relation to the reference genes used: 18S and actin showed the high sensitivity of this marker to changing conditions of the medium under Cd contamination.

The highest expression of RuBisCO was demonstrated by samples grown on soil without Cd contamination (peat) and with the 20% addition of bovine manure (D11) and vermicompost (D31). The lowest level of expression, and thus the highest toxic effect of Cd from the soil, was recorded for contaminated soil without any additives. The level of expression grew along with the improvement of the quality of the substrate by the addition of organic fertilizers, and all differences between the concentrations used are statistically significant.

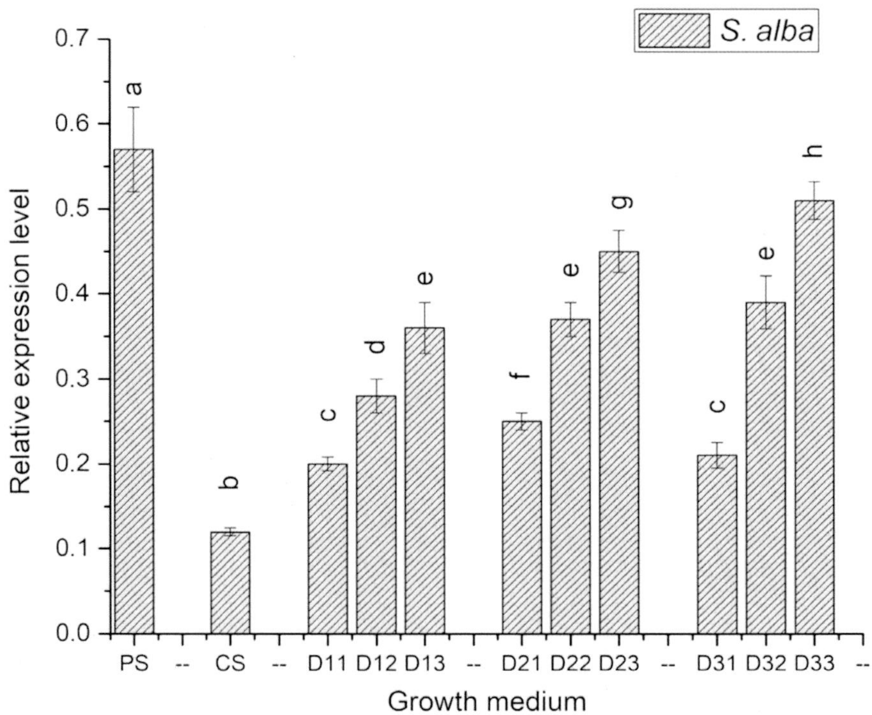


**Fig. 9** Relative level of expression of the *rbLc* gene, *S. alba*, in vitro tests, Peat—reference soil, M.Sl—soil, D1—soil + cow manure (10%, 15% and 20%/kg d.w.), D2—soil + horse manure (10%, 15% and 20%), D3—soil + vermicompost (10%, 15% and 20%).

## 5 SUMMARY

The optimized oxidative stress markers may be used to determine the effect of selected soil amendments on plants' development, in planning phytoremediation processes, carbon sequestration, or toxicity tests in all plant species, regardless of their taxonomy. The current studies indicate that oxidative stress, induced by cadmium, plays a role in decreasing the biomass of plants, and that soil supplementation by organic fertilizers might be useful in reversing the oxidative stress effects. Exogenous nutrients, organic and inorganic (Ca- and Fe-rich substances) soil amendments, and application controls the antioxidant protection and reduces the Cd toxicity for plants.

## ACKNOWLEDGMENT

The authors declare that they have no conflict of interest. Research was funded from internal grant BS PB 401/304/11,BS/MN-401-301/17 and MNISW/2017/89 DiR/NN2.

## REFERENCES

Ammar, B., Mediouni, W.C., Tray, B., et al., 2008. Glutathione and phytochelatin contents in tomato plants exposed to cadmium. Biol. Plant. 52, 314. https://doi.org/10.1007/s10535-008-0065-9.

Aravind, P., Prasad, M.N.V., 2005. Modulation of cadmium-induced oxidative stress in *Ceratophyllum demersum* by zinc involves ascorbate-glutathione cycle and glutathione metabolism. Plant Physiol. Biochem. 43 (2), 107–116.

Asada, K., 2006. Production and scavenging of reactive oxygen species in chloroplasts and their functions. Plant Physiol. 141, 391–396.

Azevedo, H., Gomes, C., Pinto, G., Santos, C., 2005. Cadmium effects in sunflower: membrane permeability and changes in catalase and peroxidase activity in leaves and calluses. J. Plant Nutr. 28, 2233–2241.

Batish, D.R., Singh, H.P., Setia, N., Kaur, S., Kohli, R.K., 2006. 2-Benzoxazolinone (BOA) induced oxidative stress, lipid peroxidation and changes in some antioxidant enzyme activities in mung bean (*Phaseolus aureus*). Plant Physiol. Biochem. 44, 819–827.

Borisova, G., Chukina, N., Maleva, M., Kumar, A., Prasad, M.N.V., 2016. Thiols as biomarkers of heavy metal tolerance in the aquatic macrophytes of Middle Urals, Russia. Int. J. Phytoremediation 18 (10), 1037–1045.

Cannata, M.G., Bertoli, A.C., Carvalho, R., Bastos, A.R.R., Freitas, M.P., Augusto, A.S., 2014. Effects of cadmium on the content, accumulation, and translocation of nutrients in bean plant cultivated in nutritive solution. Commun. Soil Sci. Plant Anal. 45, 223–235.

Charrier, J.G., Anastasio, C., 2012. On dithiothreitol (DTT) as a measure of oxidative potential for ambient particles: evidence for the importance of soluble transition metals. Atmos. Chem. Phys. 12 (5), 11317–11350.

Chen, L., Guo, Y., Yang, L., Wang, Q., 2008. Synergistic defensive mechanism of phytochelatins and antioxidative enzymes in *Brassica chinensis* L. against Cd stress. Chin. Sci. Bull. 53, 1503–1511.

DalCorso, G., Farinati, S., Furini, A., 2010. Regulatory networks of cadmium stress in plants. Plant Signal. Behav. 5, 663–667. https://doi.org/10.4161/psb.5.6.11425.

Dessureault-Rompré, j., Zebarth, J.B., Burton, D.L., Georgallas, A., 2015. Predicting soil nitrogen supply from soil properties. Can. J. Soil Sci. 95, 63–75.

Foyer, C.H., Noctor, G., 2005. Oxidant and antioxidant signalling in plants: a re-evaluation of the concept of oxidative stress in a physiological context. Plant Cell Environ. 28, 1056–1071.

Gill, S.S., Tuteja, N., 2010. Reactive oxygen species and antioxidant machinery in abiotic stress tolerance in crop plants. Plant Physiol. Biochem. 48, 909–930.

Grobelak, A.R., Kokot, P., Singh, B.R., Kacprzak, M., 2018. Mine waste rehabilitation case studies from Poland. In: Prasad, M.N.V., de Campos Favas, P.J., Maiti, S.K. (Eds.), Bio-Geotechnologies for Mine Site Rehabilitation. first ed. Elsevier, Amsterdam, pp. 515–529 730 pp.

Hasanuzzaman, M., Hossain, M.A., Fujita, M., 2012. Exogenous selenium pretreatment protects rapeseed seedlings from cadmium-induced oxidative stress by upregulating antioxidant defense and methylglyoxal detoxification systems. Biol. Trace Elem. Res. 149 (2), 248–261.

Hasanuzzaman, M., Nahar, K., Rahman, A., Mahmud, J.A., Fujita, M., 2018. Exogenous glutathione attenuates lead-induced oxidative stress in wheat by improving antioxidant defense and physiological mechanisms. J. Plant Interact. 13 (1), 203–212.

Hédiji, H., Djebali, W., Belkadhi, A., Cabasson, C., Moing, A., Rolin, D., et al., 2015. Impact of long-term cadmium exposure on mineral content of *Solanum lycopersicum* plants: consequences on fruit production. S. Afr. J. Bot. 97, 176–181.

Hodges, D., DeLong, J., Forney, C., et al., 1999. Improving the thiobarbituric acid-reactive-substances assay for estimating lipid peroxidation in plant tissues containing anthocyanin and other interfering compounds. Planta 207 (4), 604–611.

Inzé, D., Van Montagu, M., 1995. Oxidative stress in plants. Curr. Opin. Biotechnol. 6 (2), 153–158.

Jaskulak, M., Rorat, A., Grobelak, A., Kacprzak, M., 2018. Antioxidative enzymes and expression of rbcL gene as tools to monitor heavy metal-related stress in plants. J. Environ. Manag. 218, 71–78.

Jozefczak, M., Bohler, S., Schat, H., Horemans, N., Guisez, Y., Remans, T., et al., 2015. Both the concentration and redox state of glutathione and ascorbate influence the sensitivity of Arabidopsis to cadmium. Ann. Bot. 116, 601–612.

Kabata-Pendias, A., Pendias, H., 2001. Trace Elements in Soils and Plants. CRC Press, Boca Raton, FL. https://doi.org/10.1201/b10158-25.

Loix, C., Huybrechts, M., Vangronsveld, J., Gielen, M., Keunen, E., Cuypers, A., 2017. Frontiers reciprocal interactions between cadmium-induced cell wall responses and oxidative stress in plants. Front. Plant Sci. 8, 1867.

Medrano, M., Parry, J., 2008. Long term water stress inactivates Rubisco in subterranean clover. Ann. Appl. Biol. 131 (3), 491–501.

Mittler, R., Blumwald, E., 2015. The roles of ROS and ABA in systemic acquired acclimation. Plant Cell 27 (1), 64–70.

Mohanpuria, P., Rana, N.K., Yadav, S.K., 2007. Cadmium induced oxidative stress influence on glutathione metabolic genes of *Camellia sinensis* (L.) O. Kuntze. Environ. Toxicol. 22 (4), 368–374.

Noriega, G.O., Balestrasse, K.B., Batlle, A., et al., 2007. Cadmium induced oxidative stress in soybean plants also by the accumulation of delta-aminolevulinic acid. Biometals 20, 841.

Placek, A., Grobelak, A., Hiller, J., Stępień, W., Jelonek, P., Jaskulak, M., Kacprzak, M., 2017. The role of organic and inorganic amendments in carbon sequestration and immobilization of heavy metals in degraded soils. J. Sustain. Dev. Energy Water Environ. Syst. 5 (4), 509–517.

Rahman, A., Mostofa, M.G., Nahar, K., Hasanuzzaman, M., Fujita, M., 2016. 2016. Exogenous calcium alleviates cadmium-induced oxidative stress in rice (*Oryza sativa* L.) seedlings by regulating the antioxidant defense and glyoxalase systems. Braz. J. Bot. 39, 393.

Ramel, F., Birtic, S., Cuiné, S., Triantaphylidès, C., Ravanat, J.L., Havaux, M., 2012. Chemical quenching of singlet oxygen by carotenoids in plants. Plant Physiol. 158 (3), 1267–1278.

Sharkey, T., 2005. Effects of moderate heat stress on photosynthesis: importance of thylakoid reactions, rubisco deactivation, reactive oxygen species, and thermotolerance provided by isoprene. Plant Cell Environ. 28, 269–277.

Tchounwou, P.B., Yedjou, C.G., Patlolla, A.K., Sutton, D.J., 2012. Heavy metals toxicity and the environment. In: Luch, A. (Ed.), Molecular, Clinical and Environmental Toxicology. Springer, Basel, pp. 133–164.

Verma, V., Pakbin, P., Cheung, K.L., Cho, A.K., Schauer, J.J., Shafer, M.M., Kleinman, M.T., Sioutas, C., 2011. Physicochemical and oxidative characteristics of semi-volatile components of quasi-ultrafine particles in an urban atmosphere. Atmos. Environ. 45, 1025–1033.

Xu, S.S., Lin, S.Z., Lai, Z.X., 2015. Cadmium impairs iron homeostasis in Arabidopsis thaliana by increasing the polysaccharide contents and the iron-binding capacity of root cell walls. Plant Soil 392, 71–85.

Zhang, H., Lian, C., Shen, Z., 2009. Proteomic identification of small, copper-responsive proteins in germinating embryos of *Oryza sativa*. Ann. Bot. 103, 923–930.

CHAPTER 10

# Cadmium-Induced Oxidative and Nitrosative Stress in Plants

**Susana Mabel Gallego, María Patricia Benavides**
Universidad de Buenos Aires—Consejo Nacional de Investigaciones Cientificas y Tecnicas, Instituto de Química y Fisicoquímica Biológicas "Profesor Alejandro C. Paladini" (IQUIFIB), Facultad de Farmacia y Bioquímica, Buenos Aires, Argentina

## Abstract

The numerous deleterious health effects upon exposure to toxic heavy metals in the environment are a matter of serious concern and a global issue. Cadmium is classified as a no-redox metal, but in biological systems, it is characterized by developing redox cell imbalance. It is well documented that in many plants, acute and chronic Cd exposure increases reactive oxygen and nitrogen species (ROS and RNS, respectively) in the cell. ROS and RNS could be acting as deleterious molecules inducing an oxidative/nitrosative injury that could result in growth inhibition, accelerated senescence, or cell death; or as essential components in the signaling cascade through plants' response to cope with metal stress. An understanding of the mechanisms by which plants perceive, transduce, and counterbalance Cd-induced oxidative or nitrosative stress signals to initiate acclimation or adaptive responses can be helpful in developing plants with increased tolerance to metal stress.

**Keywords:** Antioxidant system, Nitric oxide, ROS, RNS, Nitrosative, Cell death, Tolerance

## 1 INTRODUCTION

The steady increase in heavy metal pollution problems in the world has raised new issues that need an urgent solution. Cadmium (Cd) is a highly toxic trace pollutant for animals and plants, and its presence in soil comes from anthropogenic sources in more than 90% of cases (Caldas and Machado, 2004). Its increasing concentration in cultivable soils has been widely recognized as a serious concern over the past 40 years (Sanitá di Toppi and Gabbrielli, 1999; Benavides et al., 2005; Dal Corso et al., 2008; Azevedo et al., 2012; Gallego et al., 2012; Gong et al., 2017).

Considering that plants can absorb Cd with no apparent effort, they are a significant source of this metal exposure for humans through the food chain (Lux et al., 2011; Gallego et al., 2012; Uraguchi and Fujiwara, 2013). Also,

*Cadmium Toxicity and Tolerance in Plants*
https://doi.org/10.1016/B978-0-12-814864-8.00010-3

Cd shows a long biological half-life of 10–30 years, is efficiently retained, and results in an additive increase in the human body over time, which is becoming a global problem and serious health concern (Gong et al., 2017). For that reason, research is growing in view of understanding not only the levels of Cd accumulation in the different plant organs and the effects of this heavy metal on plant physiology (Azevedo et al., 2012; Clemens et al., 2013), but also the mechanisms involved in Cd-induced tolerance (Cuypers et al., 2011; Farinati et al. 2011; Gratão et al., 2012).

Cadmium is nonessential for plant growth and development, but once absorbed from the soil by accompanying essential nutrients, it can produce complex changes at genetic, biochemical, and physiological levels that result in substantial growth reductions (Dal Corso et al., 2008; Azevedo et al., 2012; Gallego et al., 2012). Cd can induce severe disturbances in the plant, such as in photosynthesis, water relations, and mineral uptake (Azevedo et al., 2012 and references therein). For that reason, it is crucial that we focus on the reduction of Cd accumulation in biological systems, and consequently, the attenuation of the toxic effects of the metal on cultivated plants.

Usually, $Cd^{2+}$ enters plant cells through transmembrane carriers engaged in the uptake of $Ca^{2+}$, $Mg^{2+}$, $Fe^{2+}$, $Zn^{2+}$, and $Cu^{2+}$ (Zargar et al., 2015). The first barriers to most of the stress factors are plant cell walls and cell membranes. Once accumulated in the plant, Cd usually induces a series of toxicity symptoms. As consequence, a complex biochemical pathway within the cell can be triggered concomitantly with transcription regulation of Cd-responsive genes, such as induction of these genes involved in the antioxidant systems or those related to metal tolerance and accumulation, Cd binding, transport, or detoxification and cellular Cd homeostasis (Cuypers et al., 2011; Gratão et al., 2012; Li et al., 2016; Shi et al., 2016).

Several studies have shown that cadmium causes stimulation of membrane-localized NADPH oxidase and, consequently, an increase in ROS production (Heyno et al., 2008; De Michele et al., 2009; Arasimowicz-Jelonek et al., 2012). Cd can disrupt cellular redox homeostasis to burst the reactive oxygen species (ROS) in plants, resulting in the generation of oxidative stress. Oxidative damage can produce inhibition of growth and development, disturbances in physiological processes, alteration in nitrogen metabolism, and increased rates of DNA damage and cell death (Groppa et al., 2008; Lehotai et al., 2011; Rosales et al., 2011; Gallego et al., 2012; Wu et al., 2015; Méndez et al., 2016). Fortunately, higher plants have evolved a series of protective mechanisms, including antioxidative enzymes and nonenzymatic antioxidants to counteract this oxidative stress, as a sophisticated way for controlling ROS levels through the Foyer-Halliwell-Asada cycle (Foyer and Noctor, 2011).

In addition to the accumulation of ROS, Cd induces the formation of reactive nitrogen species (RNS), among which nitric oxide (NO) is the main one (Groppa et al., 2008; Besson-Bard and Wendehenne, 2009; Arasimowicz-Jelonek et al., 2012; Gill et al., 2012; Chmielowska-Bak et al., 2014; Alemayehu et al., 2015; Liu et al., 2015). NO has been recognized as a key multifunctional signaling molecule in various plant physiological processes, including growth regulation, root formation, flowering, senescence, cell death, defense against abiotic and biotic stresses, as well as the alleviation of heavy metal toxicity (Laspina et al., 2005; Rodríguez-Serrano et al., 2009; Yu et al., 2014; Domingos et al., 2015; Corpas et al., 2017).

The main objective of this chapter is to present the current knowledge of the formation, signaling, and metabolism of crucial components of the ROS/RNS network and the posttranslational modifications derived from their interactions with proteins, with the aim to enhance our understanding of the interplay between ROS and RNS under Cd stress, and the mechanisms involved in plant responses to Cd stress.

## 2 REACTIVE OXYGEN SPECIES IN PLANTS

The term "ROS" describes a family of oxygen-containing chemical species produced in biological systems that are characterized by their chemical reactivity, half-life, and lipid solubility (Halliwell and Gutteridge, 1999). These small oxidant molecules include chemical forms produced by the partial reduction of molecular oxygen ($O_2$), such as the free radical superoxide anion ($O_2^{\bullet-}$), and hydroxyl radical ($HO^{\bullet}$), and the active nonradical hydrogen peroxide ($H_2O_2$). Other biologically relevant ROS are formed when photosensitizers, such as chlorophyll, absorb light energy and transfer some of that energy to $O_2$, producing the electronically excited form singlet oxygen ($^1O_2$) (Halliwell, 2006; Pospíšil, 2012).

The generation of ROS in living organisms is an unavoidable consequence of aerobic metabolism (Fischer et al., 2016). They are byproducts in the normal transport of electrons in mitochondria and chloroplast, where electrons released to the cellular environment can react with $O_2$ (Turrens, 2003; Singh et al., 2016). Moreover, in the chloroplast membrane, insufficient energy dissipation during photosynthesis can lead to formation of a chlorophyll triplet state that can transfer its excitation energy onto ground-state $O_2$ to make $^1O_2$ (Triantaphylidès and Havaux, 2009).

The endogenous ROS production can also derive from regulated enzymatic processes in different cell compartments for signaling, biosynthetic reactions, chemical defense, and detoxification functions (Halliwell, 2006). In the plant cell, a variety of metabolic processes generate ROS enzymatically. In plant peroxisomes, fatty acid degradation, photorespiration, purine metabolism, sulfite detoxification, sarcosine metabolism, and polyamine catabolism pathways often include an oxidative step generating ROS as byproducts (Sandalio and Romero-Puertas, 2015; Del Río and López-Huertas, 2016; Reumann and Bartel, 2016; Corpas et al., 2017). In the endoplasmic reticulum, specific oxidases produce ROS for proper oxidative protein folding (Sevier and Kaiser, 2008; Zito, 2015), and NADPH-oxidases (NOXs) located in the plasma membrane catalyze the production of $O_2^{\bullet -}$ (Sagi and Fluhr, 2006; Qu et al., 2017). In plants, the respiratory burst oxidase homolog (*rboh*) gene family encodes the key enzymatic subunit of the NADPH oxidase (Torres et al., 1998) with 10 *rboh* homologs identified in Arabidopsis (ATRBOH A-J) and differentially expressed in the plant tissues (Sagi and Fluhr, 2006). Other enzymatic sources of ROS include cell wall peroxidases, oxalate-oxidases, and amino-oxidases in the apoplast (Apel and Hirt, 2004; Mittler et al., 2004; Wrzaczek et al., 2013). The presence of redox-active metals, such as iron ($Fe^{2+}/Fe^{3+}$) or copper ($Cu^{+}/Cu^{2+}$), also contribute to ROS formation through the Haber-Weiss and Fenton reactions (Kehrer, 2000).

ROS have become one of the main players that modulate a wide variety of physiological processes (Mittler et al., 2011; Sies, 2015), not only in the whole cell, but also inside each cell compartment (Go and Jones, 2008). Oxidation-reduction (redox) reactions in cells are utilized in fundamental regulatory processes, collectively termed "redox signaling" and "redox control." For the regulation of ROS metabolism, a highly efficient and sophisticated ROS scavenging system has evolved in aerobic organisms (Ahmad et al., 2008). This antioxidant defense machinery includes enzymes such as superoxide dismutase (SOD), catalase (CAT), ascorbate peroxidase (AP), glutathione reductase (GR), monodehydroascorbate reductase (MDHAR), dehydroascorbate reductase (DHAR), glutathione peroxidase (GPX), guaicol peroxidase (GPOX), and glutathione-S transferase (GST); and, low molecular weight antioxidants, such as glutathione (GSH), ascorbic acid, tocopherols, carotenoids, and flavonoids (Flora, 2009; Gill and Tujeta, 2010; Hasanuzzaman et al., 2017; Rahantaniaina et al., 2017). This antioxidant system is present in plant cells, cytosol, mitochondria, and peroxisomes; as well as in the chloroplast, and constitutes a series of coupled redox reactions called the ascorbate-glutathione cycle, or "Foyer-Halliwell-Asada" pathway (Foyer and Noctor, 2011). Glutathione (γ-L-glutamyl-L-cysteinylglycine) comprises the major nonprotein thiol in plant cells, and besides its antioxidant function, GSH is involved in the protection against heavy metals as the precursor to the synthesis of phytochelatins (PCs), molecules involved in the complexation and detoxification of Cd ions (Cobbett and Goldsbrough, 2002).

The ROS gene network of plants comprises the total genes that encode ROS-producing and ROS-removing proteins involved in the mantainance of cell redoxtasis (Inupakutika et al., 2016). In Arabidopsis, the basic ROS gene network is constituted by more than 150 genes (Mittler et al., 2004). Under physiological conditions, moderate levels of ROS allow reversible interactions with biomolecules to mediate the regulation of cellular processes (Mittler, 2017; Sies, 2017). In plants, ROS are related to growth and development (Kocsy et al., 2013; Singh et al., 2016), cell differentiation (Schmidt and Schippers, 2015), flowering (Schippers et al., 2016), plant immunity (Qi et al., 2017; Segal and Wilson, 2018), and responses to the environment (Raja et al., 2017). Moreover, ROS acts as plant redox systemic signaling, thereby propagating the response to environmental cues (Schmitt et al., 2014; Gaupels et al., 2017; Raja et al., 2017). However, this cellular redox equilibrium can be disturbed by adverse environmental conditions, resulting in ROS overaccumulation. This "imbalance between oxidants and antioxidants in favor of the oxidants, leading to a disruption of redox signaling and control and/or molecular damage" is defined as oxidative stress (Jones, 2006; Sies and Jones, 2007).

Due to their abundance and ubiquity in the plant cell, proteins are one of the main macromolecules susceptible to oxidative modifications (Davies, 2005, 2016; Sies, 2015). Protein oxidation involves covalent reversible or irreversible modifications of amino acid residues, induced either directly by ROS or indirectly by reactions with secondary byproducts of oxidative stress that would result in alterations of protein structure, enzymatic activity, or gene transcription (Møller et al., 2007; Samardzic and Rodgers, 2017). Unregulated protein modifications caused by ROS can derive from oxidative damage to proteins. The sulfur-containing amino acids cysteine (Cys) and methionine (Met) are the main targets of redox modification in protein (Schöneich, 2017). The direct and specific oxidation of the thiol residue of Cys to sulfenic acids and disulfides constitutes "redox sulfur switches" that lead to modulation of the protein functional state (Kemp et al., 2008; Poole, 2015; Boronat et al., 2017). The redox reversible modifications of Cys residues are controlled by thioredoxins, glutaredoxins, and glutathione (Poole, 2015). Cysteine oxidation beyond a reversible state, such as sulfinic (R-$SO_2H$) and sulfonic acid (R-$SO_3H$) species, are considered irreversible modifications that lead to a permanently damaged and possibly nonfunctional protein (Klomsiri et al., 2011). Nowadays, advances in redox proteomics allow a better understanding of cellular regulation by thiol switching in living organisms (Akter et al., 2015; Go et al., 2015; Nietzel et al., 2017). Owing to the presence of sulfur, Met belongs to the most easily oxidized amino acids. Methionine oxidation yields a racemic mixture of the two methionine sulfoxide (Met-SO) stereoisomers, Met-S-(o) and Met-R-(o) forms. Met sulfoxide can be reduced back to Met by the thioredoxin-dependent enzyme family methionine sulfoxide reductases (Msrs) (Drazic and Winter, 2014; Kim et al., 2014). There are two classes of Msr enzymes that act stereo-specifically; MsrA reduces the S-epimer specifically, while MsrB reduces the R-epimer (Boschi-Muller and Branlant, 2014). Msrs are present in all aerobic organisms and, though both types of Msrs share common catalytic mechanisms, they are similar neither at the primary amino acid sequence, nor at the structural level (Zhang and Weissbach, 2008). The highest level of Met oxidation, R-$SO_2CH_3$ (sulfone) appears to be irreversible.

The formation of carbonyl derivatives (reactive aldehydes and ketones) on amino acid residues is other ROS-induced posttranslational protein modification, an irreversible process used as a global marker of protein oxidation (Møller et al., 2007; Madian and Regnier, 2010; Augustyniak et al., 2015). Side chain oxidation occurs on a large number of amino acids, and it is carried out by different mechanisms (Berlett and Stadtman, 1997; Rao and Møller, 2011; Chondrogianni et al., 2014). This nonrepairable protein carbonylation results in protein misfolding, breakdown, and/or crosslinking, which are toxic to cells (Nystrom, 2005). Carbonylated protein can be selectively removed from the cytoplasm and nucleus by the multicatalytic protease (proteasome) (Poppek and Grune, 2006; Jung et al., 2014). In particular, the unbound 20S proteasome core is the main protease responsible for degrading oxidized proteins, in an independent ubiquitin and ATP manner (Raynes et al., 2016). The proteasome itself is also subjected to oxidative stress derived posttranslational modifications that affect its activity (Lefaki et al., 2017). Current knowledge of mammalian cells has resulted in the addition of specific proteolytic systems to degrade oxidatively damaged proteins in lysosomes, mitochondria, and peroxisomes (Chondrogianni et al., 2014; Samardzic and Rodgers, 2017). Although protein carbonylation is often associated with negative consequences of environmental stress (Braconi et al., 2011), modifications in the abundance and carbonylation patterns of proteins have been connected to plant physiology, like life cycle progression (Johansson et al., 2004), fruit senescence (Qin et al., 2009), seed viability (Yin et al. 2017), and leaf senescence (Have et al., 2015). In humans, evidence connects protein carbonylation with signal transduction under stress conditions (Cai and Yan, 2013).

## 3 CADMIUM-INDUCED OXIDATIVE STRESS

Cadmium toxicity is closely associated with oxidative stress generation. The capacity of this metal to induce oxidative stress in a huge variety of plant species, organs, or cell suspensions, at different metal concentrations and exposure times, has been extensively reported (Gallego et al., 2012). But, unlike redox active metals, cadmium is unable to perform single electron transfer reactions, nor can it catalyze ROS generation directly. Nowadays, it is recognized that Cd-induced oxidative stress comes from the disturbance of the ROS network. Cd produces an imbalance in ROS-producer and ROS-scavenger systems. Moreover, protein oxidative modifications and adjustments of the mechanism to reverse, repair, or degrade oxidative damaged proteins are also well documented in response to cadmium.

It is widely known that cadmium induces ROS accumulation in plants. Histochemical staining using nitroblue tetrazolium (NBT) and 3,3′ diaminobenzidine (DAB), and fluorogenic dyes, such as dihydroethidium (DHE) or 2′,7′,dichlorodihydrofluorescein diacetate ($H_2DCFDA$), for $O_2^{\bullet -}$ and $H_2O_2$/peroxides, respectively, as well as analytical chemical determinations has been used to demonstrate ROS presence in plant tissues. Several papers have reported an increase in ROS levels during cadmium exposure, as in tomato seedlings (Singh et al., 2018), leaves and roots of maize seedlings (Kováčik and Babula, 2017; Li et al., 2017), rice seedlings (Srivastava et al., 2014), fronds of duckweed (*Lemna minor*) (Lu et al., 2018), leaves of cowpea plants (Pereira et al., 2018); root tips of wheat (Pena et al., 2012), *Arabidopsis thaliana* seedlings (Cho and Seo, 2005), rice roots (Yeh et al., 2007) soybean roots (Chmielowska-Bąk et al., 2017), and cotton leaves (Daud et al., 2015).

Chloroplasts are one of the main sources of ROS during metal stress (Azpilicueta et al., 2007, Méndez et al., 2016). Ultrastructural disorganization and dysfunction of the photosynthetic apparatus induced by cadmium are closely associated to lipids and protein oxidative damage (Wang et al.,

2014; Bashir et al., 2015; Lysenko et al., 2015; Arena et al., 2017; Pereira de Araújo et al., 2017). Thus, an efficient antioxidant system to scavenge ROS is essential to maintain chloroplast integrity during Cd stress (Wu et al., 2015; Guo et al., 2016). Also, ROS accumulation was observed in mitochondria and peroxisomes (Romero-Puertas et al., 2004; Corpas and Barroso, 2014). At the root level, $O_2^-$ of mitochondrial origin contributes to explain the overall redox imbalance produced by cadmium (Garnier et al., 2006; Heyno et al., 2008).

At the biochemical level, cadmium indirectly modulates the activity of the constitutive form of the plasma membrane NADPH oxidase. The activity of NOX as a ROS producer in response to short-term Cd stress has been reported in different plant tissues and species, for example, soybean roots (Peréz-Chaca et al., 2014; Chmielowska-Bąk et al., 2017), pea seeds (Smiri et al., 2010), and lupine roots (Arasimowicz-Jelonek et al., 2012). Cadmium-triggered ROS production by NOX was also demonstrated by using the enzyme inhibitor diphenyleneiodonium chloride (DPI), for example, in tobacco cell cultures (Olmos et al., 2003), pea leaves (Romero-Puertas et al., 2004), rice roots (Yeh et al., 2007), and soybean roots (Chmielowska-Bąk et al. 2017). Garnier et al. (2006) reported that an increase in cytosolic free calcium concentration originated from internal pools precedes and is required for NOX activation by cadmium. The analysis of NADPH oxidase transcripts levels (Smeets et al., 2009), and the response of defective Arabidopsis mutants (Gupta et al., 2017) indicate that RBOHs can differentially regulate ROS metabolism under cadmium stress. Thus, specific signatures from different RBOHs initiate ROS signaling and control in response to mild stress (Gupta et al., 2017), allowing the plant to successfully cope with a severe stress (Mittler, 2017). Nevertheless, cadmium ions can also decrease NOX activity by competing with $Ca^{2+}$ ions, as it was reported in isolated plasma membranes (Heyno et al., 2008). NOX inhibition by $Cd^{2+}$ was also reported in sunflower leaves by Groppa et al. (2012).

Cadmium can interfere with the antioxidant defense system and indirectly lead to the production of ROS. A vast amount of research has studied the response of the ROS-scavenger system during Cd exposure in different tissues and plant species. Recently, proteomic profiling has been adopted to study the influence of cadmium stress in gene expression. Analysis of the protein functional category indicates a central role for the oxidative stress response in plant adaptation to the metal. Cadmium up-regulated the antioxidant enzymes CAT and MDHAR in *Arabidosis thaliana* cell suspensions. Moreover, it also resulted in an increase in proteins involved in amino acid biosynthesis required for GSH and PCs formation, glutamine and glutamate synthetases (GS and GOGAT), γ-glutamylcysteine synthase, glutamate dehydrogenase (GDH); and GSH metabolism, glutathione S-transferase (GST) (Sarry et al., 2006). GST protein abundance also increased in roots of *Arabidopsis thaliana* upon $Cd^{2+}$ exposure (Roth et al., 2006). The activation of PC synthesis generates a strong sink for reduced sulfur, which correlates with the up-regulation of ATP sulphurylase (APS), an enzyme that catalyzes the first step in S assimilation, in roots (Roth et al., 2006). Proteome changes in Arabidopsis leaves coincided with the incremental increase in GS and GST proteins (Semane et al., 2010). Moreover, SOD was up-regulated in leaves of *Arabidopsis thaliana* (Semane et al., 2010; Carneiro et al., 2017) and in the metallophyte *Arabidopsis halleri* (Farinati et al., 2009). An increase of AP, SOD, and GST proteins was reported in leaves of *Triticum aestivum* seedlings treated with cadmium (Wang et al., 2011), while in *Triticum polonicum* L., sulfate and GSH metabolism-related proteins (sulfurtransferase, APS, GS and GST) were enhanced in roots (Wang et al., 2016). In rice plants, AP, GR, and GST were up-regulated in roots, while peroxidase increased in leaves (Lee et al., 2010). More recent reports comparing two genotypes of rice with different Cd accumulation abilities in grains showed differences in the ROS detoxifying enzymes AP, DHAR, and SOD (Xue et al., 2014). An abundance of peroxidases and enzymes involved in sulfur reduction, APS and 5′-adenylylsulfate reductase, together with cysteine synthase (CS), were reported in alfalfa roots (Dai et al., 2017). Peroxidase was also found to be up-regulated in cotton leaves (Daud et al., 2015). Besides, SOD, peroxidase, CAT, GR, and GST were induced in the roots of radishes (*Raphanus sativus* L.) (Xu et al., 2017). In *Brachypodium distachyon* L., an experimental model for studying small-grain cereals, SOD, AP and GST increased under $Cd^{2+}$ stress in seedling leaves (Cheng et al., 2018); whereas AP is among the proteins whose expression was significantly induced in tobacco (*Nicotiana tabacum*) (Xie et al., 2014). GS was found to be up-regulated specifically in cell suspensions of the Cd-tolerant fiber flax (*Linum usitatissimum* L.) cv. Jitka (Hradilová et al., 2010). Proteins involved in glutathione homeostasis, CS and GS, were up-regulated in the leaf of the Cd-hyperaccumulator *Brassica juncea* (D'Alessandro et al., 2013), while the induction of proteins related to thiol metabolism (cysteine and methionine synthesis), antioxidative defense (AP, SOD, and MDHAR), and the detoxification enzyme GST (Alvarez et al., 2009) were detected in roots. Proteomic studies of the high Cd accumulation cultivar (Pen) and the low Cd accumulation cultivar (Nan) of the vegetable *Amaranthus gangeticus* showed up-regulation of the proteins from GSH-related metabolism in Pen compared with Nan (He et al., 2018). Proteins involved in sulfur and GSH metabolism, together with GST, were differentially expressed under Cd stress in roots of *A. hybridus* (Jin et al., 2016) and leaves of *Phytolacca americana* (Zhao et al., 2011). Furthermore, the hyperaccumulator *Arabis paniculata* enhances shoot GST, AP, and SOD proteins in response to Cd (Zeng et al., 2011). Peroxidases, GS and GST, were up-regulated in response to cadmium in different species of the genus *Populus* (Kieffer et al., 2008, 2009; Durand et al., 2010; Marmiroli et al., 2013; Sergeant et al., 2014; Yang et al., 2015a,b).

From the overall literature, it is clear that the central role of GSH takes place during Cd stress. The metal induces thiol metabolism in plants, not only by stimulating sulfur assimilation pathways in roots, but also by triggering GSH and PCs synthesis (Gill and Tuteja, 2011; Jozefczak et al., 2012; Hasanuzzaman et al., 2017). Depletion of the GSH pool during Cd stress has consequences for the cell redox balance, leading to an increase in ROS levels that accounts for oxidative damage to cell components (Ortega-Villasante et al., 2005, Jozefczak et al., 2014; Flores-Cáceres et al., 2015). Thus, adaptation to growing in the presence of Cd is linked to the capacity to maintain high intracellular GSH concentrations (Gallego et al., 2005).

Adaptive redox homeostasis allows multiple antioxidative pathways to activate to maintain cellular metabolism, and to recycle damaged macromolecules. In general, all the proteins involved in the glutathione-ascorbate cycle might help plant cells to re-establish redox homeostasis under Cd stress. Moreover, Msr systems could be considered alternative enzymes involved in maintaining cell redox balance, acting outside of the ascorbate-glutathione cycle, and thus being involved in plant Cd tolerance. In this context, up-regulation of Msr was detected in the proteomic studies of *Brassica juncea* (Alvarez et al., 2009) and radishes (Xu et al., 2017). Likewise, Méndez et al. (2016) reported that cadmium treatment significantly increased mRNA accumulation of Msrs in leaves of *Arabidosis thaliana*. One possible mechanism that mediates Msrs induction is through the enhancement of ROS production that modulates the differential expression of the system, as was suggested in *Chlamydomonas reinhardtii* (Chang et al., 2014).

Proteomic analysis has shown that several enzymes involved in the ubiquitin (Ub)/26S proteasoma pathway and other peptidases were induced by Cd in plants (Roth et al. 2006; Kieffer et al., 2008; Alvarez et al., 2009; Durand et al., 2010; Semane et al., 2010; Sergeant et al., 2014; Xie et al. 2014; Jin et al., 2016; Xu et al., 2017). Kurepa et al. (2008) linked the increased accumulation of 20SP to the increased oxidized protein turnover, and placed proteasome function in an essential role for oxidative stress tolerance in plants. The 20S proteasome proteolytic pathway was found to be up-regulated by cadmium at both RNA and activity levels in Arabidopsis leaves (Polge et al., 2009). Moreover, Pena et al. (2007) suggested a threshold response of the 20S proteasome to Cd stress was mediated through oxidative protein modification of the proteasome itself, thus avoiding accumulation of carbonylated proteins in the cell.

## 4 NITRIC OXIDE IN PLANTS

Fifteen years after discovery, NO is now seen as an essential molecule in many biological systems. After being identified in mammalian systems in 1987 (Palmer et al., 1987), the interest in NO and its roles in plants have steadily increased, and it is a matter of debate among plant scientists. After the first reports were published in 1998 (Delledonne et al., 1998; Durner et al., 1998), a number of papers have been written each year in this field.

Nitric oxide is a free radical with an unpaired electron shared between the nitrogen and the oxygen atom. It is neither a strong oxidant (Eo' $^{\bullet}NO/NO^{-} = 0.39\,V$) nor a strong reductant (Eo' $^{\bullet}NO/NO^{+} = 1.21\,V$) (Radi, 1996), and therefore is not reactive with most of biological molecules, despite its radical nature. It does not react with itself, nor does it react with most biological molecules that have no unpaired electrons; but it tends to combine with other radicals and transition metals (Heinrich et al., 2013). NO is also a small and lipophilic molecule that can easily cross lipid membranes, which is associated with its hydrofobicity and small size (Thomas, 2015). In its radical form, $^{\bullet}NO$, it has a single electron in its $2p$-$\pi$ antibonding orbital, and it can react far away from its site of formation (100 μm to 1 mm), and its half life is assumed to be on the order of seconds (1–10 s), which is a long half life compared with other free radicals (Ignarro, 1990; Radi, 2004; Bartesaghi and Radi, 2018).

Nitrosonium cation ($NO^{+}$) is formed by the loss of this electron, while the addition of an extra electron to this orbital leads to nitroxyl radical ($NO^{-}$), with both compounds displaying distinct properties and reactivities (Astier et al., 2011). Nitric oxide may undergo autoxidation reactions in the presence of oxygen, leading to nitrogen dioxide $^{\bullet}NO_2$ formation, a strong oxidizing and nitrating agent (E°' $^{\bullet}NO_2/NO_2^{-} = 0.99\,V$) (Nottingham and Sutter, 1986). The formation of RNS from $^{\bullet}NO$ requires the presence of oxidants such as superoxide radical, hydrogen peroxide, and transition metal centers, the concentration of which can be increased either by $^{\bullet}NO$ itself or by the same mediators that up-regulate $^{\bullet}NO$ production (Bartesaghi and Radi, 2018). Nitrogen dioxide may further react with $^{\bullet}NO$, forming dinitrogen trioxide ($N_2O_3$), an unstable species that mainly mediates nitrosation and deamination reactions (Bartesaghi et al., 2011). Nitric oxide does react fast with other oxygen-, carbon-, sulfur-, and nitrogen-centered radicals, such as $^{\bullet}OH$, tyrosyl ($Tyr^{\bullet}$), thiyl ($RS^{\bullet}$), and peroxyl ($ROO^{\bullet}$) radicals (Bartesaghi and Radi, 2018). These molecules have much higher reactivity than NO, and can play a dual role in biological systems, both as signaling molecules and toxic compounds (Hughes, 2008; Thomas, 2015).

In mammals, NO is synthetized by a family of enzymes called nitric oxide synthases (NOS) from L-arginine (L-Arg), NADPH, $O_2$, FAD and FMN, using tetrahydrobiopterin (BH4) and calmodulin as cofactors (Campbell et al., 2014). NOS proteins catalyze the NADP-dependent oxidation of L-Arg to NO and citrulline. Three NOS isoforms have been isolated and characterized: neuronal NOS (nNOS), inducible NOS (iNOS), and endothelial NOS (eNOS) (Li and Poulos, 2005). A novel NOS was identified in *Arabidopsis*, which had the same co-factor requirements as mammalian NOS, but exhibited no significant sequence homology to the mammalian form (Guo et al., 2003). Although the mutant *atnos1* has been used to show the roles of NO in floral development (He et al., 2004), or the interaction of NO with ROS (Zhao et al., 2007), subsequently AtNOS1 was found not to possess NOS activity, was established to be a GTPase (Moreau et al., 2008), and was renamed AtNOA1 (Nitric Oxide Associated1; Zemojtel et al., 2006). Recently, a related NOS protein, which possessed 45% similarity to human NOS, has been discovered in the single-cell green algae *Ostreococcus tauri* (Foresi et al., 2010), but this gene does not seem to be present in higher plants. Up to date, no protein or gene was identified in plants that had any sequence similarity to the complete animal NOS proteins (Crawford, 2006; Mur et al., 2013a; Corpas and Barroso, 2017; Santolini et al., 2017), despite the completion of numerous genome projects (Yu et al., 2014), representing a significant impediment to the further development of the field. In this context, Corpas and Barroso (2017) postulated that the L-Arg-dependent NO synthase-like activity detected in higher plants could be the result of cooperation between discrete proteins, which together can generate NO; this would explain the absence of findings at the molecular level on the presence of a protein complex similar to that for animal NOS.

The main controversy around NO generation in plants is the finding of the precise sources for NO formation (Morot-Gaudry-Talarmain et al., 2002; Domingos et al., 2015; Santolini et al., 2017). In this scenario, there are several proposed routes, which depend upon either reductive or oxidative chemistry (Gupta et al., 2011; Yu et al., 2014). Oxidative mechanisms include the production of NO from L-Arg (Delledonne et al., 2001; Corpas et al., 2006), hydroxylamines (Rumer et al., 2009), or polyamines, the latter being a potentially interesting source of NO in plants by a mechanism that is still unresolved (Tun et al., 2006; Groppa et al., 2008; Recalde et al., 2018). Regarding the enzymatic origin of NO by a reductive chemical reaction, the best described enzymatic source of NO in plants is the reductive pathway involving NAD(P)H-dependent nitrate reductase (NR), a cytosolic enzyme associated with nitrogen assimilation, whose primary function is the reduction of nitrate to nitrite. However, both in vitro and in vivo analyses provided strong evidence that NR also generates NO from nitrite (Yamasaki, 2000), but with lower efficiency (Rockel et al., 2002; Planchet et al., 2006). Other reductive sources are a plasma membrane-bound nitrite-NO reductase (Ni-NOR) (Stohr et al., 2001; Stohr and Ullrich, 2002) and mitochondrial nitrite reduction (Tischner et al., 2004; Planchet et al., 2005). In *Arabidopsis thaliana*, NR has two genes, NIA1 and NIA2, which catalyze mainly the reduction of nitrate to nitrite, with NIA2 being the gene encoding the enzyme responsible for the majority of NR activity (Wilkinson and Crawford, 1991).

A great amount of data demonstrate that NO is a component of many regulatory pathways in plants, and the mechanisms underlying its functions have received much attention. It is now well established that NO is a key player in the regulation of different plant developmental processes, including photomorphogenesis, plant defense, stomatal aperture, hormone interactions, and leaf senescence (Lamattina et al., 2003; He et al., 2004; Mur et al., 2013b; Baudouin and Hancock, 2014; Sanz et al., 2014; Yu et al., 2014). Among others, NO modulates root development (Correa-Aragunde et al., 2004, 2008; Lombardo et al., 2006; Groppa et al., 2008; Lombardo and Lamattina, 2012; Recalde et al., 2018), it is involved in the control of protein kinases activity (Courtois et al., 2008), it is linked to lipid and cGMP signaling (Gaupels et al., 2011), and interacts with ROS, mainly superoxide and hydrogen peroxide (Yoshioka et al., 2011). Nitric oxide has been implicated in the defense against several plant pathogens (Delledonne et al. 1998; Wendehenne et al., 2004; Mur et al. 2005, Yun et al., 2012), and it can influence the activity of transcription factors and intervene in upstream signaling cascades, mRNA stability, and translation (Yun et al., 2012; Corpas and Barroso, 2015). In spite of the extensive studies on NO function in different processes in plants, the whole picture of NO impact on living cells, including production, activity and metabolism of NO, still has to be completed. Although the mechanisms of NO generation in plants come to be established, the mechanisms of NO signaling are gradually being identified. In this scenario, the conversion of NO accumulation into a defined biological function is crucial, delimiting the wide range of interactions that can be established between NO and its biological targets, in a concentration and redox-dependent fashion.

In the present information era, omics analysis has become a high-throughput approach to gaining insight into plant responses and tolerance to stress. Proteomics became popular in the 1990s, and has greatly evolved to a mature stage today, becoming a very useful technique in the study of the posttranslational modification of proteins (PTM), those that can occur as a consequence of NO activity in plants (Lindermayr et al., 2005, 2006; Astier and Lindermayr, 2012; Zaffagnini et al., 2016; Gong et al., 2017; Liu et al., 2018). Protein modification by NO is a key mechanism when considering down-stream effects of this signaling molecule in plants. Most of the work being developed is focused on three types of protein modifications, which are specific to NO signaling: tyrosine nitration, thiol protein S-nitrosylation, and the interaction of NO moiety with metalloproteins in so-called metal nitrosylation events (Astier and Lindermayr, 2012; Mur et al., 2012; Kovacs and Lindermayr, 2013). To date, more than 200 different proteins have been found to be regulated through these NO-dependent processes, which are involved in a large variety of cellular functions and signaling (Astier et al., 2011). However, the fundamental knowledge of NO production, sensing, and transduction in plants and the PTMs of proteins that could be potential targets for NO remains inadequately characterized. Almost no reports at the proteomics level can be found to reveal the global and systematic mechanisms related to NO-induced Cd stress tolerance in plants. In this chapter, we will refer particularly to the first two PMTs mentioned herein, from a general point of view and in relation to the knowledge related to Cd stress.

## 4.1 Protein Tyr Nitration

Different modes of action mechanisms of NO signaling have been reported in plants. One of the molecular footprints left by the reactions of reactive nitrogen species with biomolecules is tyrosine (Tyr) nitration, the reaction of a nitrating agent with a tyrosine residue of a target protein that leads to the covalent addition of a nitro group ($—NO_2$) in the ortho position of the phenolic hydroxyl group, resulting in the formation of 3-nitrotyrosine (Schopfer et al., 2003; Radi, 2004). The nitro group, $NO_2^+$, originates from nitrogen dioxide ($NO_2$), a potent oxidizing agent resulting from the reaction between NO and $O_2$, or from $ONOO^-$. Therefore, peroxynitrite derives from NO, and consequently it belongs to the group of NO originated molecules, the RNS (Radi, 2004; Kolbert et al., 2017). Under physiological conditions, $ONOO^-$ reacts with $CO_2$ to give the carbonate radical ($CO_3^{\bullet-}$) and $NO_2$, acting as a main nitrating agent. Peroxynitrite itself is also a powerful oxidant produced following the interaction between •NO and $O_2^{\bullet-}$ (Astier et al., 2011; Radi, 2013, Yeo et al., 2015). Contrary to tyrosine phosphorylation, Tyr nitration does not involve enzymatic activity (Kolbert et al., 2017).

This well-established PTM attracts a great deal of interest in plant research because it can alter protein function, and consequently, it can produce remarkable changes in several plants' physiological issues. The in vivo occurrence of protein Tyr nitration (PTN) was first described by Ischiropoulos et al. (1992). Tyr nitration is restricted to specific target Tyr residues (Astier and Lindermayr, 2012), and though this modification, was first thought to be irreversible. A growing body of evidence has now demonstrated that denitration processes may occur, both in enzymatic or nonenzymatic ways, suggesting that it is a reversible mechanism (Romero-Puertas et al., 2007; Vandelle and Delledonne, 2011). However, though denitrase activity has recently been characterized in animals (Deeb et al., 2013), the denitrase enzyme has not yet been identified in plants; thus the reversibility of Tyr nitration still remains uncertain in plants (Kolbert et al., 2017).

The biological significance of PTN has not been well established in plants yet. In this sense, only 1%–2% of the total tyrosine pool may be the target of in vivo nitration (Radi 2013; Bartesaghi and Radi, 2018), suggesting that the process is highly specific (Kolbert et al., 2017). This fact is reinforced by the low number of $YNO_2$ sites in plant enzymes containing several Tyr amino acids, as methionine synthase or monodehydroascorbate reductase (Lozano-Juste et al., 2011; Begara-Morales et al., 2015). In this sense, the biological relevance of PTN is questioned (Souza et al., 2008). Tyrosine nitration has been observed, to a certain degree, in healthy or unstressed plants (Begara-Morales et al., 2013), indicating that these plants have a physiological nitroproteome that supposes regulatory function for nitration (Kolbert et al., 2017). This issue leads us to wonder if Tyr nitration is an unavoidable consequence of stress, or if it regulates protein pool size in an active form.

Although the existence of physiological nitroproteomes have been reported for many species (Tanou et al., 2012; Signorelli et al., 2013; Chaki et al., 2015; Lehotai et al., 2016), nitrated poteins have been identified only in some of them. The corresponding proteins are related to essentially all major cellular processes, ranging from primary/secondary metabolism to signaling. Lozano-Juste et al. (2011) identified 127 nitroproteins in wild-type *Arabidopsis thaliana* grown under normal conditions, and 21 proteins were found to be nitrated in sunflower hypocotyls (Chaki et al., 2009), and 26 nitrated proteins were found in the roots of nonstressed *Citrus* plants (Tanou et al., 2012). Cecconi et al. (2009) reported the identification of eleven proteins that are undergoing Tyr nitration during the development of the hypersensitive response, whereas using the symbiotic model involving root nodule formation between *Medicago truncatula* and *Sinorhizobium meliloti*, Melo et al. (2011) have shown that one of the guanylate cyclases (MtGS1a) is modified by Tyr nitration

Regarding the consequences of Tyr nitration, it is likely that it interferes with signaling based on phosphorylation/dephosphorylation, especially if both modifications involve the same Tyr residues. In plant cells, the available data show that PTN usually causes functional loss of the particular enzyme protein, including metionine synthase, cytosolic L-ascorbate peroxidase, peroxisomal hydroxypyruvate reductase, leghemoglobin-1, and monodehydroascorbate reductase I (Kolbert et al., 2017 and references therein). On the other hand, stress-induced Tyr nitration has been widely proven in different plant species (reviewed by Corpas et al., 2013), which attributes nitration as a biomarker of secondary nitrosative stress (Kolbert et al., 2017). Recent results indicate a general occurrence of tyrosine nitration as the effect of stress situations, as those produced by salt stress in sunflower seedlings (David et al., 2016), by cadmium in soybean roots (Gzyl et al., 2016), or by zinc in Brassica species (Feigl et al., 2016).

## 4.2 Protein *S*-Nitrosylation

The identification of S-nitrosylated proteins has gained interest in recent years in relation to the identification of plant proteins regulated by NO at the posttranslational level, providing new insights of how NO mediates its activities at the molecular levels (Lee et al., 2011; Zaffagnini et al., 2016; Kovacs and Lindermayr, 2013). S-nitrosylation is a redox-based post-translation modification, which refers to the covalent binding of an NO group to Cys residues of targets proteins, thus leading to S-nitrosothiols

(SNOs) (Hess et al., 2005; Astier and Lindermayr, 2012; Hess and Stamler, 2012). S-nitrosoglutathione (GSNO), formed by S-nitrosylation of the antioxidant molecule GSH, is the major low-molecular-weight S-nitrosothiol, considered to be an NO reservoir in cells (Barroso et al., 2006; Leitner et al., 2009; Lindermayr and Durner, 2009).

S-nitrosylation is a reversible mechanism, because S-NO is extremely labile, light- and redox-sensitive, and can be specifically and enzymatically broken down by thioredoxins (Kneeshaw et al., 2014; Yu et al., 2014); and can also be descomposed nonenzymatically by ascorbate or glutathione. Furthermore, S-nitrosoglutathione reductase (GSNOR) decomposes GSNO and indirectly controls S-NO levels (Feechan et al., 2005). Moreover, the study of S-nitrosylated proteins is slowed by the lack of conserved consensus motif surrounding the nitrosylated Cys residue. Although animal studies suggested that an acid-base motif is located near the S-nitrosylated Cys, thereby contributing to trans-nitrosylation (Marino and Gladyshev, 2010; Seth and Stamler, 2011), this structural requirement is not stringent enough to unequivocally identify putative target Cys in silico. In addition, whether such features are conserved in plant proteins is currently unknown, mainly due to the low number of S-nitrosylated sites reported up until now (Puyaubert et al., 2014; Zaffagnini et al., 2016). Owing to its specificity and reversibility, protein nitrosylation is a posttranslational modification with a fundamental role in signaling networks. Whether NO and RNS damage the cell by an indiscriminate modification of biological molecules, or trigger signaling cascades through S-nitrosylation of specific Cys, will mainly depend on the local concentration of NO that dynamically vary inside the cells as a result of different NO biosynthetic pathways, complex NO chemistry, and a wide variety of NO scavenging systems (Zaffagnini et al., 2016).

The majority of all NO-affected proteins seem to be regulated by S-nitrosylation of a single critical Cys residue, which occurs by oxygen-dependent chemical reactions, or by the transfer of NO from a nitrosothiol to a protein sulfhydryl group (transnitrosylation) (Lindermayr et al., 2005; Lee et al., 2011). This redox modification is a central route for NO bioactivity, as it can affect protein activity (activation or inhibition), translocation, and protein function, serving as a molecular switch, enabling the target protein to be directly responsive to changes in the cellular redox status. S-nitrosylation has been shown to modulate enzyme activity (Lindermayr et al., 2005; Belenghi et al., 2007; Wang et al., 2009; Yun et al., 2012; Hu et al., 2015; Yang et al., 2015a,b), protein localization

(Tada et al., 2008), and protein-protein interactions (Hara et al., 2005). Due to their reactivity with intracellular reducing agents, as ascorbic acid or GSH, and with reduced metal ions, especially $Cu^{1+}$, nitrosothiols are exceptionally labile. This lability results in tissue half-lives of seconds to a few minutes, and therefore provides a very sensitive mechanism for regulating cellular processes. Therefore, this redox-based posttranslational modification can be considered to be analogous to other, more well-established protein modifications, as phosphorylation (Lindermayr et al., 2005).

Despite most proteins possessing Cys residues, only some of them are specifically regulated by S-nitrosylation in vivo and, therefore, contribute to the NO-dependent modulation of protein function (Belenghi et al., 2007; Palmieri et al., 2010; Begara-Morales et al., 2014). This specificity is governed by several factors. Among others, S-nitrosothiol formation can be favored in more ionizable Cys residues, such as those surrounded by acidic and basic amino acid residues (Astier et al., 2011), or might be promoted by the presence of hydrophobic compartments surrounding the resident Cys residue into proteins (Seth and Stamler, 2011). Moreover, the colocalization of NO sources with key proteins targeted for S-nitrosylation is assumed to contribute to the specificity of S-nitrosylation (Hess et al., 2005).

In plants, the first report about the biological function of S-nitrosylation came out through a genetic approach, which revealed a central role for SNOs in plant disease resistance (Feechan et al., 2005). Furthermore, the exogenous addition of NO donors to plant protein extracts also demonstrated the in vitro formation of plant SNOs (Lindermayr et al., 2005). The list of S-nitrosylated plant proteins has grown exponentially in the past few years, reportedly due to the application of the biotin-switch technique (Jaffrey and Snyder, 2001), a very efficient tool for identifying S-nitrosylated proteins in vivo. The list includes proteins specifically S-nitrosylated during plant immune function (Wang et al., 2009; Lindermayr et al., 2010), heavy metal exposure (De Michele et al., 2009), temperature (Airaki et al., 2012), and salinity (Tanou et al., 2009; Leterrier et al., 2012; Camejo et al., 2013).

Identification of plant S-nitrosylated proteins was first been achieved on a proteome-wide scale from plant tissues, cell suspensions, cellular extracts, or organelles treated with nitrosylating agents, thus leading to elevated levels of S-nitrosylated proteins (Astier and Lindermayr, 2012). A pioneering work by Lindermayr and co-workers in 2005 identified plant proteins sensitive to this posttranslational process. They treated extracts of *Arabidopsis* cell cultures with the NO donor GSNO, resulting in an effective formation of nitrosothiols and S-nitrosylated proteins. The proteomic analyses made by nanoLC/MS/MS resulted in the identification of thousands of proteins belonging to stress-related proteins, signaling/regulating proteins, redox-related proteins, cytoskeleton proteins, metabolic proteins, and others, many of them being already characterized as targets for *S*-nitrosylation in animals (Lindermayr et al., 2005). Within the stress-related proteins, several proteins were identified that already undergo S-nitrosylation in animal systems. They also identified S-nitrosylated proteins belonging to a cluster of signaling and regulating factors such as elongation (eEF-1 and eEF-2) and initiation (4A-1 and 5A-4-related) factors, which suggest that NO interferes with protein synthesis (Lindermayr et al., 2005). Other candidates for S-nitrosylation in *Arabidopsis* were cytoskeleton proteins such as tubulin $\alpha$, tubulin $\beta$, an actin-depolymerizing factor, actin, and proteins related to enzymes of the antioxidant system (Lindermayr and Durner, 2009). Thioredioxin and glutaredoxin function as redox-regulators of target enzymes and transcription factors (Rouhier et al., 2001). Both proteins harbor two highly conserved Cys residues within their active site and represent possible targets for S-nitrosylation. Another important group of metabolic enzymes undergoing S-nitrosylation includes enzymes involved in sulfur metabolism, such as Cys synthase, *S*-adenosylhomocysteinase, vitamin B12-independent Met synthase, and S-adenosylmethione (SAM) synthetase (Lindermayr et al., 2005). Similarly, in *Brassica juncea* (Abat and Deswal, 2009), as well as in citrus (*Citrus aurantium* L.) (Tanou et al., 2009), about a thousand S-nitrosylated proteins were identified from leaf protein extracts exposed to GSNO. Tanou et al. (2009) suggested that S-nitrosylation of critical Cys residues might prevent irreversible protein carbonylation triggered by ROS, by promoting conformational changes. In accordance with this hypothesis, pretreatment of citrus plants with the NO donor sodium nitroprusside (SNP) dramatically reduced the extent of protein carbonylation, as well as the deleterious physiological effects induced by salt stress.

Fares et al. (2011) used the combined BSM and ICAT labeling to investigate quantitatively the role of nitrosylation events in the early response to salt stress in *Arabidopsis* plants, in the absence of an NO donor. The author reported a set of sites detected to be salt-sensitive under the control of the cell nitrosylation/denitrosylation machinery (Fares et al., 2011). The main classes of nitrosylated proteins detected in Arabidopsis were involved in protein (or amino acid) metabolism or in energy metabolism, followed by proteins participating in cell structure or photosynthesis, and a lower percentage (20%) related to miscellaneous classes, as lipid metabolism or transport proteins, such as the phosphate transporter PHT3;1 (At5g14040) and an adenylate translocator (At3g08580), that are proteins not previously characterized as targets of NO in plants. It is worth mentioning that the reported proteins belonged to the same major classes of nitrosylated proteins identified in human or animal cells (Liu et al., 2010). However, among the 135 Arabidopsis accessions previously quoted as nitrosylables, only a minor part was identified by Fares et al. (2011).

## 5 CADMIUM-INDUCED NITROSATIVE STRESS

Plant adaptation to changes in environmental conditions is crucial to preventing potential damage. Sometimes, these changes are intense enough to cause stress in plants that is not detectable at the phenotype level, but has a molecular response to adapt to metabolic pathways. As we mentioned herein, it is very important to reduce the accumulation and toxic effects of Cd on crops. For that reason, efforts have been made to understand the importance of NO signaling in controlling Cd stress tolerance, but on a larger scale, global and systematic studies are necessary to unravel the involvement of NO in plant tolerance to Cd-induced toxicity.

There is growing evidence indicating that cadmium stress modulates NO generation in plants (Chmielowska-Bak et al., 2014). However, the time and intensity of NO generation seems to be strictly dependent on the form and concentration of metal used, the duration of stress treatment, the plant species and developmental phase of the model plant, as well as the plant tissue or organ analyzed (Xiong et al., 2010; Azevedo et al., 2012).

Some of the defense mechanisms that concern NO formation related to Cd stress have been widely documented. A dose-dependent and rapid production of NO was exhibited in soybean cells treated with 4 or 7 μM Cd for 72 h (Kopyra et al., 2006), whereas a long-term treatment with 1 μM Cd for 4 weeks produces a 2.4-fold increase in NO content in wheat roots (Mahmood et al., 2009). NO content increased in the root apices of the wheat seedlings when treated with 0.1 mM $CdCl_2$ for 5 days (Corpas et al., 2008), and was involved in Cd-induced disturbances in root growth of 7-d old wheat (Groppa et al., 2008) and *Arabidopsis* (Besson-Bard et al., 2008) plants. An endogenous NO accumulation in response to Cd was found to be involved in plant cell death (PCD) induction in both a cell suspension culture (De Michele et al., 2009), and a whole plant model system (Arasimowicz-Jelonek et al., 2012; Ye et al., 2013). Exogenous application of NO can alleviate Cd toxicity by inhibiting Cd transport from the root to the shoot in rice (Singh and Shah, 2014) and peanuts (Dong et al., 2016), or by modulating the antioxidant system in sunflower plants (Laspina et al., 2005), whereas it can partially reverse Cd-induced structural abnormalities in barley roots (Alemayehu et al., 2015), suggesting that NO can improve Cd stress tolerance at multiple physiological levels.

At the molecular level, many signaling molecules and plant growth regulators are involved in cadmium sensing and downstream plant response, including modifications at both transcriptional and posttranslation levels. As mentioned earlier in this chapter, the signaling mode of NO action at the molecular level includes protein modification by binding to critical Cys residues, heme or iron-sulfur centers, and Tyr residue nitration through peroxynitrite formation (Arasimowicz-Jelonek and Floryszak-Wieczorek, 2011). Corpas and Barroso (2014) showed that $ONOO^-$ is overproduced under Cd stress, suggesting that the metabolism of ROS in peroxisomes could participate in the mechanism of the response to this metal. However, very little is known about the interaction between RNS and ROS during plant-metal interactions. Liu et al. (2018) reported that Cd provokes endogenous $^{\bullet}NO$ generation, but a reduction in GSNO (increase in SNOs) in *Arabidopsis thaliana*, suggesting that these $^{\bullet}NO$-derived molecules in an oxidative environment can mediate the generation of $ONOO^-$ between $^{\bullet}NO$ and $O_2^{\bullet-}$, regulating PTMs through the nitration of Tyr residues, indicating that Cd provoked oxidative and nitrosative stress. One possible explanation is that Tyr-$NO_2$ is mediated by $ONOO^-$, which inhibits electron transport on the acceptor side of photosystem II or S-adenosyl-L-homocysteine hydrolase activity (Szabó et al., 2007; Chaki et al., 2009).

The phenomenon of Tyr nitration as a regulatory modification induced by RNS during a plant response has been scarcely reported on for Cd as the stress-induced agent. Few reports are available regarding Cd-induced Tyr nitration in plants, though $ONOO^-$ formation could provide an important regulatory loop for NO bioactivity under Cd stress conditions (Gzyl et al., 2016). In soybeans, five of the 10 identified proteins undergoing Tyr nitration in roots were down-regulated during both moderate and high Cd exposure (Gzyl et al., 2016). Recent data obtained by immunoelectron microscopy show the presence of Tyr-nitrated proteins in pea leaf peroxisomes under Cd stress (Barroso et al., 2006), and proteomic analysis of isolated pea leaf peroxisomes has shown that peroxisomal NADH-dependent hydroxypyruvate reductase is a target of nitration by peroxynitrite, resulting in a loss of function of the protein (Corpas et al., 2013).

Proteins might be a good starting point in the search of nitrosative stress markers in plants (Corpas et al., 2007). At least three processes influence the function of the target protein: (i) inhibition of function, (ii) gain of function, and (iii) nonimpact on function (Radi, 2004). In addition, several publications have reported that Tyr-$NO_2$ may prevent/stimulate further phosphorylation (Szabó et al., 2007). In this case, an increase in protein nitration was observed under several types of abiotic stresses, such as salinity, low or high temperature, and wounding (Liu et al., 2018 and references therein); however, no specific proteins have been identified.

In *Arabidopsis* plants, Liu et al. (2018) reported an increase in the nitration of at least three polypeptides of 75 kDa, 48 kDa, and 32 kDa in Cd-stressed roots, although no bands were observed under control conditions. However, Tyr-$NO_2$ of leaves was evident in Cd-induced and control plants, while immunoreactive bands of 48 kDa were observed in both treated and control roots. In soybean roots, a high Cd dose (170 μM Cd) induced elevated levels of $ONOO^-$, whereas moderate (85 μM Cd) and high Cd levels provoked a lowered expression of 5 proteins undergoing Tyr nitration (Gzyl et al., 2016). The nitrated proteins included proteins involved in oxidation-reduction processes, transferase activity, glycolytic processes, and phosphorylation.

Most enzymes of the ascorbate-GSH cycle have been identified as potential nitrated proteins, showing that $^{\bullet}$NO-PTM could compromise this cycle functioning (Lin et al., 2012; Tanou et al., 2012; Corpas et al., 2016). It has been reported that cytosolic APX in peas is inactivated by $ONOO^-$ as consequence of Tyr-$NO_2$, and as a result of the detoxification of $H_2O_2$ by this cycle (Begara-Morales et al., 2014, 2015). Tyrosine nitration provokes inhibition of carbonic anhydrase (β-CA) activity (Chaki et al., 2013), whereas nitrosative stress provokes the inhibition of ferredoxin–NADP reductase activity in sunflowers under high temperature stress (Chaki et al., 2011). Glutathione reductase has also been identified as a Tyr-$NO_2$ target (Chaki et al., 2009), but chloroplastic and cytosolic pea GR activities are not affected by $ONOO^-$-mediated $NO_2$-Tyr (Begara-Morales et al., 2016), while an increase in GR in Cd-exposed Arabidopsis has been described (Liu et al., 2018), implying that the regulation of Tyr-$NO_2$ on target proteins may be directed to maintain GSH level and thus, the cellular redox state (Liu et al., 2018). In this context, Cd treatment is accompanied by a rapid increase in NO and phytochelatin (PC) synthesis, and NO is involved in Cd-induced programmed cell death in *Arabidopsis* cell suspensions by modulating the concentration and function of PC through S-nitrosylation (De Michele et al., 2009). Another interesting finding is that

NADPH-generating dehydrogenases were unaffected under Cd stress because, as with other stresses, such as high salinity (Leterrier et al., 2012) and low temperature (Airaki et al., 2012), these enzymes were usually increased. Therefore, this evidence supports our hypothesis that the ascorbate-GSH cycle is regulated by ˙NO-PTMs such as S-nitrosylation Tyr-$NO_2$ (Liu et al., 2018), but the type of abiotic stress involved should be considered before drawing any general conclusions.

S-nitrosylation of the elements involved in signal transduction pathways may lead to alterations in their functioning, as has been observed in the case of the AtMYB2 TF (Serpa et al., 2007). S-nitrosylation is involved in the ameliorating effect of SNP against Cd toxicity in leaves of *Boehmeria nivea* (L.) Gaud (Wang et al., 2015); whereas in *Pisum sativum* L. plants under Cd stress, Ortega-Galisteo et al. (2012) identified six peroxisomal proteins as putative S-nitrosylation targets involved in photorespiration, β-oxidation, and ROS detoxification in pea plants, among which CAT, glycolate oxidase (GO), and MDH were inhibited by NO donors.

The information related to the specific impact of this modification on the structure of target proteins as those mentioned herein, and the role of the tyrosine or Cys target of nitration/nitrosylation, respectively, is necessary to better understand the cross-talk between NO and ROS in the antioxidant defense against nitrosative stress.

## 6 THE INTERPLAY BETWEEN ROS AND RNS IN PLANT RESPONSES TO Cd STRESS

Reactive nitrogen species and ROS signaling pathways in Cd - plant interactions are closely connected (Fig. 1). Therefore, the prediction of informative bio-indicators from the pool of oxidated, nitrated, or nitrosylated proteins, the expression of which are modulated by stress factors, should represent a new challenge in plant oxidative or nitrosative biology, becoming important elements that may allow us to assess the early stages of disorders in plant metabolism. Moreover, the identification of PTMs proteins altered by oxidative or nitrosative processses whose expression is altered during Cd stress may contribute to our understanding of the mechanism involved in the tolerance response to Cd toxicity.

As developed earlier in this chapter, it is well known that Cd stress induces a serious ROS burst in the plant cell. To show the intrincate relationship between ROS and RNS when analyzing plant responses to Cd stress, Gong et al. (2017) observed that the cucumber plant has evolved defense mechanisms to control the accumulation of Cd and to minimize the damage from

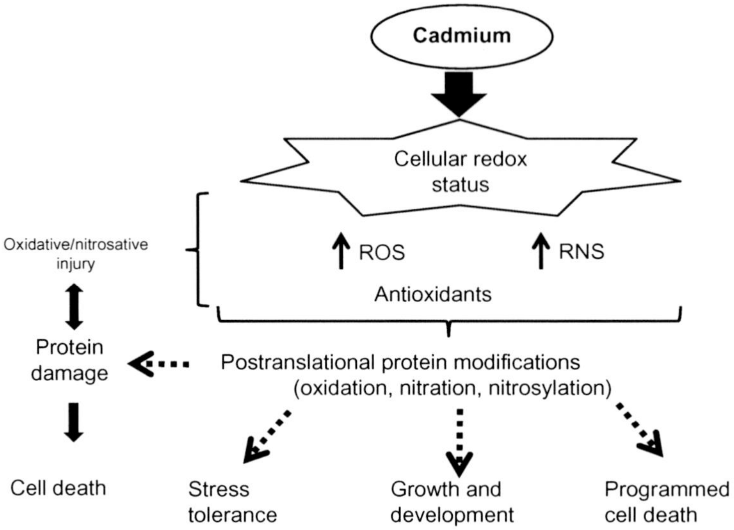


**Fig. 1** Schematic diagram showing the role of redox protein postranslational modifications as intermediaries in the cell recognition and plant response to cadmium stress. *ROS*, reactive oxygen species; *RNS*, reactive nitrogen species.

ROS-derived Cd toxicity, which are not powerful enough to fight against Cd stress. However, using a metabolic pathway analysis, they identified 1691 proteins that can be used to establish the role of NO in regulating molecular changes of the proteome in cucumber leaves exposed to Cd. The authors demonstrated that several Cd-disruptive pathways were markedly reversed by NO treatments, which can confer a regulatory network of $Ca^{2+}$ signaling activation, Cd transport inhibition, Cd chelation, localization, detoxification, and ROS scavenging, and were responsible for the mitigative effects of plant growth inhibition and chlorosis, as well as cell death.

On the other hand, Liu et al. (2018) assessed that the high reactivity of ROS and $^{\bullet}$NO suggest a "collaborative change" after the application of modulators in *Arabidopsis* plants, considering that Cd enhanced both ROS and $^{\bullet}$NO signals, while the presence of cPTIO (a NO trapping agent)/L-NAME (a NOS inhibitor) depleted the in vivo $^{\bullet}$NO signal and increase in ROS; the opposite was found after the application of SNP alone, or in co-administration with Cd. Similar responses in terms of $^{\bullet}$NO changes were also observed in soybeans (Kopyra et al., 2006), and confirmed the unequivocal action of the given compounds, despite differences in species, concentration, and duration of the treatment. The impacts of SNP and cPTIO/L-NAME on the Cd-induced effects were observed in many species, and despite different concentrations used, a similar action was observed: SNP ameliorates negative Cd toxicity, but cPTIO or L-NAME reverses this action in terms of oxidative stress-related parameters (Xiong, et al., 2010; Signorelli et al., 2013). Furthermore, Liu et al. (2018) reported that the observed effects of NO-trapping agents used together with Cd in *Arabidopsis* were opposite of those reported in *Pisum sativum* (Rodríguez-Serrano et al., 2009), where Cd-induced oxidative stress was related to depletion of $^{\bullet}$NO, and this process was evoked by changes in signal molecules (e.g., $Ca^{2+}$ signal). Taking these reports into account, it can be suggested that ROS and $^{\bullet}$NO/RNS are involved in signal transduction pathways that activate the response to stress in plants, and RNS plays a vital role in the regulation of metal cytotoxicity.

Aside from the complexity of nitrosylation signaling networks and associated oxidative regulatory pathways, their participation in signal transduction and environmental adaptation clearly requires signal integration, considering other multiple signaling pathways and molecules. Cysteine residues undergo many other types of posttranslational modifications in addition to nitrosylation, among which oxidoreduction of disulfide bonds have been shown to play a major role in redox signaling (Paulsen and Carroll, 2013). These redox modifications are interconnected, and influence each other (Lounifi et al., 2013) using simple molecules such as ROS and RNS to form a highly complex redox network with numerous proteins regulated by several modifications (Zaffagnini et al., 2016). The redox signaling network, which remains mostly unexplored, is probably a major component of signal integration, and constitutes the molecular signature of the ROS/RNS crosstalk, whose importance in cell signaling has been widely recognized (Zaninotto et al., 2006; Sevilla et al., 2015). Moreover, this network probably involves spatial and temporal regulation of several redox posttranslational modifications on hundreds of proteins in a highly dynamic manner. Although most studies on redox PTMs have been focused on protein carbonylation, tyrosine nitration, nitrosylation, and oxidoreduction of disulfide bonds, other redox modifications, as those mentioned previously in this chapter, exist and should be considered. These modifications include *S*-sulfhydration, sulfenic/sulfonic/sulfinic acids, methionine oxidation, and diverse types of S-thiolation, such as cysteinylation (Zaffagnini et al., 2016, and references therein). The importance of these additional redox modifications is emerging, and future studies should provide new insights into their functional role and their integration within the redox signaling network.

## REFERENCES

Abat, J.K., Deswal, R., 2009. Differential modulation of S-nitrosoproteome of *Brassica juncea* by low temperature: change in S-nitrosylation of Rubisco is responsible for the inactivation of its carboxylase activity. Proteomics 9, 4368–4380.

Ahmad, P., Sarwat, M., Sharma, S., 2008. Reactive oxygen species, antioxidants and signaling in plants. J. Plant Biol. 51 (3), 167–173.

Airaki, M., Leterrier, M., Proteomics Mateos, R.M., Valderrama, R., Chaki, M., Barroso, J.B., del Río, L.A., Palma, J.M., Corpas, F.J., 2012. Metabolism of reactive oxygen species and reactive nitrogen species in pepper (*Capsicum annuum* L.) plants under low temperature stress. Plant Cell Environ. 35, 281–295.

Akter, S., Huang, J., Waszczak, C., Jacques, S., Gevaert, K., Van Breusegem, F., Messens, J., 2015. Cysteines under ROS attack in plants: a proteomics view. J. Exp. Bot. 66 (10), 2935–2944.

Alemayehu, A., Zelinová, V., Bočcová, B., Huttová, J., Mistrík, I., Tamás, L., 2015. Enhanced nitric oxide generation in root transition zone during the early stage of cadmium stress is required for maintaining root growth in barley. Plant Soil 390, 213–222.

Alvarez, S., Berla, B.M., Sheffield, J., Cahoon, R.E., Jez, J.M., Hicks, M., 2009. Comprehensive analysis of the *Brassica juncea* root proteome in response to cadmium exposure by complementary proteomic approaches. Proteomics 9, 2419–2431.

Apel, K., Hirt, H., 2004. Reactive oxygen species: metabolism, oxidative stress, and signal transduction. Annu. Rev. Plant Biol. 55, 373–399.

Arasimowicz-Jelonek, M., Floryszak-Wieczorek, J., 2011. Understanding the fate of peroxynitrite in plant cell. From physiology to pathophysiology. Phytochemistry 72 (8), 681–688.

Arasimowicz-Jelonek, M., Floryszak-Wieczorek, J., Deckert, J., Rucińska-Sobkowiak, R., Gzyl, J., Pawlak-Sprada, S., Abramowski, D., Jelonek, T., Gwóźdź, E.A., 2012. Nitric oxide implication in cadmium-induced programmed cell death in roots and signaling response of yellow lupine plants. Plant Physiol. Biochem. 58, 124–134.

Arena, C., Figlioli, F., Sorrentino, M.C., Izzo, L.G., Capozzi, F., Giordano, S., Spagnuolo, V., 2017. Ultrastructural, protein and photosynthetic alterations induced by Pb and Cd in *Cynara cardunculus* L., and its potential for phytoremediation. Ecotoxicol. Environ. Safety 145, 83–89.

Astier, J., Lindermayr, C., 2012. Nitric oxide-dependent posttranslational modification in plants: an update. Int. J. Mol. Sci. 13 (11), 15193–15208.

Astier, J., Rasul, S., Koen, E., Manzoor, H., Besson-Bard, A., Lamotte, O., Jeandroz, S., Durner, J., Lindermayr, C., Wendehenne, D., 2011. S-nitrosylation: an emerging post-translational protein modification in plants. Plant Sci. 181, 527–533.

Augustyniak, E., Adam, A., Wojdyla, K., Rogowska-Wrzesinska, A., Willetts, R., Korkmaz, A., Atalay, M., Weber, D., Grune, T., Borsa, C., Gradinaru, D., Chand Bollineni, R., Fedorova, M., Griffiths, H.R., 2015. Validation of protein carbonyl measurement: a multi-centre study. Redox Biol. 4, 149–157.

Azevedo, R.A., Gratão, P.L., Monteiro, C.C., Carvalho, R.F., 2012. What is new in the research on cadmium-induced stress in plants? Food Energy Sec. 1 (2), 133–140.

Azpilicueta, C.E., Benavides, M.P., Tomaro, M.L., Gallego, S.M., 2007. Mechanism of CATA3 induction by cadmium in sunflower leaves. Plant Physiol. Biochem. 45 (8), 589–595.

Barroso, J.B., Corpas, F.J., Carreras, A., Rodríguez-Serrano, M., Esteban, F.J., Fernández-Ocaña, A., Chaki, M., Romero-Puertas, M.C., Valderrama, R., Sandalio, L.M., 2006. Localization of S-nitrosoglutathione and expression of S-nitrosoglutathione reductase in pea plants under cadmium stress. J. Exp. Bot. 57, 1785–1793.

Bartesaghi, S., Radi, R., 2018. Fundamentals on the biochemistry of peroxynitrite and protein tyrosine nitration. Redox Biol. 14, 618–625.

Bartesaghi, S., Romero, N., Radi, R., 2011. Nitric oxide and derived oxidants. In: - Pantapoulos, K., Schipper, H.M. (Eds.), Principiles of Free Radical Biomedicine. Nova Science Publishers Inc, New York, USA.

Bashir, H., Qureshi, M.I., Ibrahim, M.M., Iqbal, M., 2015. Chloroplast and photosystems: impact of cadmium and iron deficiency. Photosynthetica 53, 321–335.

Baudouin, E., Hancock, J.T., 2014. Nitric oxide signaling in plants. Front. Plant Sci. 4, 553.

Begara-Morales, J.C., Chaki, M., Sanchez-Calvo, B., Mata-Perez, C., Leterrier, M., Palma, J.M., Barroso, J.B., Corpas, F.J., 2013. Protein tyrosine nitration in pea roots during development and senescence. J. Exp. Bot. 64, 1121–1134.

Begara-Morales, J.C., Sánchez-Calvo, B., Chaki, M., Mata-Pérez, C., Valderrama, R., Padilla, M.N., López-Jaramillo, J., Luque, F., Corpas, F.J., Barroso, J.B., 2015. Differential molecular response of monodehydroascorbate reductase and glutathione reductase by nitration and S-nitrosylation. J. Exp. Bot. 66, 5983–5996.

Begara-Morales, J.C., Sánchez-Calvo, B., Chaki, M., Valderrama, R., Mata-Pérez, C., López-Jaramillo, J., Padilla, M.N., Carreras, A., Corpas, F.J., Barroso, J.B., 2014. Dual regulation of cytosolic ascorbate peroxidase (APX) by tyrosine nitration and S-nitrosylation. J. Exp. Bot. 65, 527–538.

Begara-Morales, J.C., Sánchez-Calvo, B., Chaki, M., Valderrama, R., Mata-Pérez, C., Padilla, M.N., Corpas, F.J., Barroso, J.B., 2016. Antioxidant systems are regulated by nitric oxide-mediated post-translational modifications (NO-PTMs). Front. Plant Sci. 16 (7), 152.

Belenghi, B., Romero-Puertas, M.C., Vercammen, D., Brackenier, A., Inze, D., Delledonne, M., Van Breusegem, F., 2007. Metacaspase activity of *Arabidopsis thaliana* is regulated by S-nitrosylation of a critical cysteine residue. J. Biol. Chem. 282, 1352–1358.

Benavides, M.P., Gallego, S.M., Tomaro, M.L., 2005. Cadmium toxicity in plants. Braz. J. Plant Physiol. 17 (1), 21–34.

Berlett, B.S., Stadtman, E.R., 1997. Protein oxidation in aging, disease and oxidative stress. J. Biol. Chem. 272, 20313–20316.

Besson-Bard, A., Courtois, C., Gauthier, A., Dahana, J., Dobrowolska, G., Jeandroz, S., Pugin, A., Wendehenne, D., 2008. Nitric oxide in plants: production and cross-talk with $Ca^{2+}$ signaling. Mol. Plant. 1 (2), 218–228.

Besson-Bard, A., Wendehenne, D., 2009. NO contributes to cadmium toxicity in *Arabidopsis thaliana* by mediating an iron deprivation response. Plant Signal. Behav. 4, 252–254.

Boronat, S., Domènech, A., Hidalgo, E., 2017. Proteomic characterization of reversible thiol oxidations in proteomes and proteins. Antioxid. Redox Signal. 26 (7), 329–344.

Boschi-Muller, S., Branlant, G., 2014. Methionine sulfoxide reductase: chemistry, substrate binding recycling process and oxidase activity. Bioorg. Chem. 57, 222–230.

Braconi, D., Bernardini, G., Santucci, A., 2011. Linking protein oxidation to environmental pollutants: redox proteomic approaches. J. Proteomics. 74 (11), 2324–2337.

Cai, Z., Yan, L.J., 2013. Protein oxidative modifications: beneficial roles in disease and health. J. Biochem. Pharmacol. Res. 1 (1), 15–26.

Caldas, E.D., Machado, L.L., 2004. Cadmium, mercury and lead in medicinal herbs in Brazil. Food Chem. Toxicol. 42 (4), 599–603.

Camejo, D., Romero-Puertas, M.C., Rodriguez-Serrano, M., Sandalio, L.M., Lazaro, J.J., Jimenez, A., Sevilla, F., 2013. Salinity-induced changes in S-nitrosylation of pea mitochondrial proteins. J. Proteomics 79, 87–99.

Campbell, M.G., Smith, B.C., Potter, C.S., Carragher, B., Marletta, M.A., 2014. Molecular architecture of mammalian nitric oxide synthases. Proc. Natl. Acad. Sci. USA 111, E3614–E3623.

Carneiro, J.M.T., Chacón-Madrid, K., Galazzi, R.M., Campos, B.K., Arruda, S.C.C., Azevedo, R.A., Arruda, M.A.Z.J., 2017. Evaluation of silicon influence on the mitigation of cadmium-stress in the development of *Arabidopsis thaliana* through total metal content, proteomic and enzymatic approaches. Trace Elem. Med. Biol. 44, 50–58.

Cecconi, D., Orzetti, S., Vandelle, E., Rinalducci, S., Zolla, L., Delledonne, M., 2009. Protein nitration during defense response in *Arabidopsis thaliana*. Electrophoresis 30, 2460–2468.

Chaki, M., Valderrama, R., Fernández-Ocaña, A.M., Carreras, A., Gómez-Rodríguez, M.V., López-Jaramillo, J., Begara-Morales, J.C., Sánchez-Calvo, B., Luque, F., Leterrier, M., 2011. High temperature triggers the metabolism of S-nitrosothiols in sunflower mediating a process of nitrosative stress which provokes the inhibition of ferredoxin-NADP reductase by tyrosine nitration. Plant Cell Environ. 34, 1803–1818.

Chaki, M., Alvarez de Morales, P., Ruiz, C., Begara-Morales, J.C., Barroso, J.B., Corpas, F.J., Palma, J.M., 2015. Ripening of pepper (*Capsicum annuum*) fruit is characterized by an enhancement of protein tyrosine nitration. Ann. Bot. 116 (4), 637–647.

Chaki, M., Carreras, A., López-Jaramillo, J., Begara-Morales, J.C., Sánchez-Calvo, B., Valderrama, R., Corpas, F.J., Barroso, J.B., 2013. Tyrosine nitration provokes inhibition of sunflower carbonic anhydrase (β-CA) activity under high temperature stress. Nitric Oxide 29, 30–33.

Chaki, M., Valderrama, R., Fernández-Ocaña, A.M., Carreras, A., López-Jaramillo, J., Luque, F., Palma, J.M., Pedrajas, J.R., Begara-Morales, J.C., Sánchez-Calvo, B., Gómez-Rodríguez, M.V., Corpas, F.J., Barroso, J.B., 2009. Protein targets of tyrosine nitration in sunflower (*Helianthus annuus* L.) hypocotyls. J. Exp. Bot. 60, 4221–4234.

Chang, H.L., Tseng, Y.L., Ho, K.L., Shie, S.C., Wu, P.S., Hsu, Y.T., Lee, T.M., 2014. Reactive oxygen species modulate the differential expression of methionine sulfoxide reductase genes in *Chlamydomonas reinhardtii* under high light illumination. Physiol. Plant. 150, 550–564.

Cheng, Z.W., Chen, Z.Y., Yan, X., Bian, Y.W., Deng, X., Yan, Y.M., 2018. Integrated physiological and proteomic analysis reveals underlying response and defense mechanisms of *Brachypodium distachyon* seedling leaves under osmotic stress, cadmium and their combined stresses. J. Proteomics. 170, 1–13.

Chmielowska-Bąk, J., Arasimowicz-Jelonek, M., Izbiańska, K., Frontasyeva, M., Zinicovscaia, I., Guiance-Varela, C., Deckert, J., 2017. NADPH oxidase is involved in regulation of gene expression and ROS overproduction in soybean (*Glycine max* L.) seedlings exposed to cadmium. Acta Soc. Bot. Pol. 86 (2), 3551.

Chmielowska-Bak, J., Gzyl, J., Rucińska-Sobkowiak, R., Arasimowicz-Jelonek, M., Deckert, J., 2014. The new insights into cadmium sensing. Front. Plant Sci. 5, 245. https://doi.org/10.3389/fpls.2014.00245.

Cho, U.-H., Seo, N.-H., 2005. Oxidative stress in *Arabidopsis thaliana* exposed to cadmium is due to hydrogen peroxide accumulation. Plant Sci. 168, 113–120.

Chondrogianni, N., Petropoulos, I., Grimm, S., Georgila, K., Catalgol, B., Friguet, B., Grune, T., Gonos, E.S., 2014. Protein damage, repair and proteolysis. Mol. Aspects Med. 35, 1–71.

Clemens, S., Aarts, M.G.M., Thomine, S., Verbruggen, N., 2013. Plant science: the key to preventing slow cadmium poisoning. Trends Plant Sci. 18, 92–99.

Cobbett, C., Goldsbrough, P., 2002. Phytochelatins and metallothioneins: roles in heavy metal detoxification and homeostasis. Annu. Rev. Plant Biol. 53, 159–182.

Corpas, F.J., Carreras, A., Esteban, F.J., Chaki, M., Valderrama, R., del Rio, L.A., Barroso, J.B., 2008. Localization of S-nitrosothiols and assay of nitric oxide synthesis and S-nitrosoglutathione reductase activity in plants. Meth. Enzymol. 437, 561–574.

Corpas, F.J., Barroso, J.B., 2015. Nitric oxide from a "green" perspective. Nitric Oxide 45, 15–19.
Corpas, F.J., Barroso, J.B., 2014. Peroxynitrite ($ONOO^-$) is endogenously produced in *Arabidopsis* peroxisomes and is overproduced under cadmium stress. Ann. Bot. 113 (1), 87–96.
Corpas, F.J., Barroso, J.B., 2017. Nitric oxide synthase-like activity in higher plants. Nitric Oxide 68, 5–6.
Corpas, F.J., Barroso, J.B., Carreras, A., Valderrama, R., Palma, J.M., Leon, A.M., Sandalio, L.M., del Rio, L.A., 2006. Constitutive arginine-dependent nitric oxide synthase activity in different organs of pea seedlings during plant development. Planta 224, 246–254.
Corpas, F.J., Barroso, J.B., Palma, J.M., Rodriguez-Ruiz, M., 2017. Plant peroxisomes: a nitro-oxidative cocktail. Redox Biology 11, 535–542.
Corpas, F.J., Chaki, M., Begara-Morales, J.C., Valderrama, R., Sánchez-Calvo, B., Barroso, J.B., 2016. Functional implications of S-nitrosothiols under nitrooxidative stress induced by abiotic conditions. Adv. Bot. Res. 77, 79–96.
Corpas, F.J., del Río, L.A., Barroso, J.B., 2007. Need of biomarkers of nitrosative stress inplants. Trends Plant Sci. 12, 436–438.
Corpas, F.J., Palma, J.M., del Río, L.A., Barroso, J.B., 2013. Protein tyrosine nitration in higher plants grown under natural and stress conditions. Front. Plant Sci. 4, 29. https://doi.org/10.3389/fpls.2013.00029.
Correa-Aragunde, N., Graziano, M., Lamattina, L., 2004. Nitric oxide plays a central role in determining lateral root development in tomato. Planta 218, 900–905.
Correa-Aragunde, N., Lombardo, C., Lamattina, L., 2008. Nitric oxide: an active nitrogen molecule that modulates cellulose synthesis in tomato roots. New Phytol. 179, 386–396.
Courtois, C., Besson, A., Dahan, J., Bourque, S., Dobrowolska, G., Pugin, A., Wendehenne, D., 2008. Nitric oxide signalling in plants: interplays with $Ca^{2+}$ and protein kinases. J. Exp. Bot. 59, 155–163.
Crawford, N.M., 2006. Mechanisms for nitric oxide synthesis in plants. J. Exp. Bot. 57, 471–478.
Cuypers, A., Smeets, K., Ruytinx, J., Opdenakker, K., Keunen, E., Remans, T., Horemans, N., Vanhoudt, N., Van Sanden, S., Van Belleghem, F., Guisez, Y., Colpaert, J., Vangronsveld, J., 2011. The cellular redox state as a modulator in cadmium and copper responses in *Arabidopsis thaliana* seedlings. J. Plant Physiol. 168, 309–316.
Dai, C., Cui, W., Pana, J., Xie, Y., Wang, J., Shen, W., 2017. Proteomic analysis provides insights into the molecular bases of hydrogen gas-induced cadmium resistance in *Medicago sativa*. J. Proteomics 152, 109–120.
Dal Corso, G., Farinati, S., Maistri, S., Furini, A., 2008. How plants cope with cadmium: staking all on metabolism and gene expression. J. Int. Plant Biol. 50, 1268–1280.
D'Alessandro, A., Taamalli, M., Gevi, F., Timperio, A.M., Zolla, L., Ghnaya, T., 2013. Cadmium stress responses in *Brassica juncea*: hints from proteomics and metabolomics. J. Proteome Res. 12 (11), 4979–4997.
Daud, M.K., Quiling, H., Lei, M., Ali, B., Zhu, S.J., 2015. Ultrastructural, metabolic and proteomic changes in leaves of upland cotton in response to cadmium stress. Chemosphere 120, 309–320.
David, A., Yadav, S., Baluska, F., Bhatla, S.C., 2016. Nitric oxide accumulation and protein tyrosine nitration as a rapid and long distance signalling response to salt stress in sunflower seedlings. Nitric Oxide 50, 28–37.
Davies, M.J., 2005. The oxidative environment and protein damage. Biochim. Biophys. Acta 1703, 93–109.
Davies, M.J., 2016. Protein oxidation and peroxidation. Biochem. J. 473, 805–825.

De Michele, R., Vurro, E., Rigo, C., Costa, A., Elviri, L., Valentin, M., Zottini, M., Sanita di Toppi, L., Schiavo, F.L., 2009. Nitric oxide is involved in cadmium-induced programmed cell death in *Arabidopsis suspension* cultures. Plant Physiol. 150, 217–228.

Deeb, R.S., Nuriel, T., Cheung, C., Summers, B., et al., 2013. Characterization of a cellular denitrase activity that reverses nitration of cyclooxygenase. Am. J. Physiol. Heart Circ. Physiol. 305, H687–H698.

Del Río, L.A., López-Huertas, E., 2016. ROS generation in peroxisomes and its role in cell signaling. Plant Cell Physiol. 57 (7), 1364–1376.

Delledonne, M., Xia, Y., Dixon, R.A., Lamb, C., 1998. Nitric oxide functions as a signal in plant disease resistance. Nature 394, 585–588.

Delledonne, M., Zeier, J., Marocco, A., Lamb, C., 2001. Signal interactions between nitric oxide and reactive oxygen intermediates in the plant hypersensitive disease resistance response. Proc. Nat. Acad. Sci. USA 98, 13454–13459.

Domingos, P., Prado, A.M., Wong, A., Gehring, C., Feijo, J.A., 2015. Nitric oxide: a multitasked signaling gas in plants. Mol. Plant. 8, 506–520.

Dong, Y., Chen, W., Xu, L., Kong, J., Liu, S., He, Z., 2016. Nitric oxide can induce tolerance to oxidative stress of peanut seedlings under cadmium toxicity. Plant Growth Regul. 79, 19–28.

Drazic, A., Winter, J., 2014. The physiological role of reversible methionine oxidation. Biochim. Biophys. Acta. 1844 (8), 1367–1382.

Durand, T.C., Sergeant, K., Planchon, S., Carpin, S., Label, P., Morabito, D., Hausman, J.F., Renaut, J., 2010. Acute metal stress in *Populus tremula* x *P. alba* (717-1B4 genotype): leaf and cambial proteome changes induced by cadmium$^{2+}$. Proteomics 10, 349–368.

Durner, J., Wendehenne, D., Klessig, D.F., 1998. Defense gene induction in tobacco by nitric oxide, cyclic GMP, and cyclic ADP-ribose. Proc. Natl. Acad. Sci. U.S.A. 95, 10328–10333.

Fares, A., Rossignol, M., Peltier, J.B., 2011. Proteomics investigation of endogenous S-nitrosylation in *Arabidopsis*. Biochem. Biophys. Res. Comm. 416, 331–336.

Farinati, S., DalCorso, G., Bona, E., Corbella, M., Lampis, S., Cecconi, D., Polati, R., Berta, G., Vallini, G., Furini, A., 2009. Proteomic analysis of *Arabidopsis halleri* shoots in response to the heavy metals cadmium and zinc and rhizosphere microorganisms. Proteomics 9, 4837–4850.

Farinati, S., Dal Corso, G., Panigati, M., Furini, A., 2011. Interaction between selected bacterial strains and *Arabidopsis halleri* modulates shoot proteome and cadmium and zinc accumulation. J. Exp. Bot. 62, 3433–3447.

Feechan, A., Kwon, E., Yun, B.-W., Wang, Y., Pallas, J.A., Loake, G.J., 2005. A central role for S-nitrosothiols in plant disease resistance. Proc. Natl. Acad. Sci. USA. 102, 8054–8059.

Feigl, G., Kolbert, Z., Lehotai, N., Molnar, A., Ördög, A., Borde, A., Laskay, G., Erdei, L., 2016. Different zinc sensitivity of Brassica organs is accompanied by distinct responses in protein nitration level and pattern. Ecotoxicol. Environ. Saf. 125, 141–152.

Fischer, W.W., Hemp, J., Valentine, J.S., 2016. How did life survive Earth's great oxygenation? Curr. Opin. Chem. Biol. 31, 166–178.

Flora, S.J.S., 2009. Structural, chemical and biological aspects of antioxidants for strategies against metal and metalloid exposure. Oxid. Med. Cell. Longev. 2, 191–206.

Flores-Cáceres, M.L., Hattab, S., Hattab, S., Boussetta, H., Banni, M., Hernández, L.E., 2015. Specific mechanisms of tolerance to copper and cadmium are compromised by a limited concentration of glutathione in alfalfa plants. Plant Sci. 233, 165–173.

Foresi, N., Correa-Aragunde, N., Parisi, G., Calo, G., Salerno, G., Lamattina, L., 2010. Characterization of a nitric oxide synthase from the plant kingdom: NO generation from the green alga *Ostreococcus tauri* is light irradiance and growth phase dependent. Plant Cell 22, 3816–3830.

Foyer, C.H., Noctor, G., 2011. Ascorbate and glutathione: the heart of the redox hub. Plant Physiol. 155, 2–18.

Gallego, S.M., Kogan, M.J., Azpilicueta, C.E., Peña, C., Tomaro, M.L., 2005. Glutathione mediated antioxidative mechanisms in sunflower (*Helianthus annuus* L.) cells in response to cadmium stress. Plant Growth Regul. 46, 267–276.

Gallego, S.M., Pena, L.B., Barcia, R.A., Azpilicueta, C.E., Iannone, M.F., Rosales, E.P., Zawoznik, M.S., Groppa, M.D., Benavides, M.P., 2012. Unravelling cadmium toxicity and tolerance in plants: insight into regulatory mechanisms. Environ. Exp. Bot. 83, 33–46.

Garnier, L., Simon-Plas, F., Thuleau, P., Agnel, J.P., Blein, J.P., Ranjeva, R., Montillet, J.L., 2006. Cadmium affects tobacco cells by a series of three waves of reactive oxygen species that contribute to cytotoxicity. Plant Cell Environ. 29 (10), 1956–1969.

Gaupels, F., Durner, J., Kogel, K.H., 2017. Production, amplification and systemic propagation of redox messengers in plants? The phloem can do it all!. New Phytol. 214 (2), 554–560.

Gaupels, F., Kuruthukulangarakoola, G.T., Durner, J., 2011. Upstream and downstream signals of nitric oxide in pathogen defence. Curr. Opin. Plant Biol. 14, 707–714.

Gill, S.S., Hasanuzzaman, M., Nahar, K., Macovei, A., Tuteja, N., 2012. Importance of nitric oxide in cadmium stress tolerance in crop plants. Plant Physiol. Biochem. 63, 254–261.

Gill, S.S., Tujeta, N., 2010. Reactive oxygen species and antioxidant machinery in abiotic stress tolerance in crop plants. Plan Physiol. Biochem. 48, 909–930.

Gill, S.S., Tuteja, N., 2011. Cadmium stress tolerance in crop plants. Probing the role of sulfur. Plant Signal. Behav. 6 (2), 215–222.

Go, Y.M., Jones, D.P., 2008. Redox compartmentalization in eukaryotic cells. Biochim. Biophys. Acta 1780, 1273–1290.

Go, Y.M., Chandler, J.D., Jones, D.P., 2015. The cysteine proteome. Free Rad. Biol. Med. 84, 227–245.

Gong, B., Nie, W., Yan, Y., Gao, Z., Shi, Q., 2017. Unravelling cadmium toxicity and nitric oxide induced tolerance in *Cucumis sativus*: insight into regulatory mechanisms using proteomics. J. Hazard. Mat. 336, 202–213.

Gratão, P.L., Monteiro, C.C., Carvalho, R.F., Tezotto, T., Piotto, F.A., Peres, L.E.P., Azevedo, R.A., 2012. Biochemical dissection of diageotropica and Never ripe tomato mutants to Cd stressful conditions. Plant Physiol. Biochem. 56, 79–96.

Groppa, M.D., Ianuzzo, M.P., Rosales, E.P., Vázquez, S.C., Benavides, M.P., 2012. Cadmium modulates NADPH oxidase activity and expression in sunflower leaves. Biol. Plant. 56 (1), 167–171.

Groppa, M.D., Rosales, E.P., Iannone, M.F., Benavides, M.P., 2008. Nitric oxide, polyamines and Cd-induced phytotoxicity in wheat roots. Phytochemistry 69, 2609–2615.

Guo, F.Q., Okamoto, M., Crawford, N.M., 2003. Identification of a plant nitric oxide synthase gene involved in hormonal signaling. Science 302, 100–103.

Guo, H., Hong, C., Chen, X., Xu, Y., Liu, Y., Jiang, D., Zheng, B., 2016. Different growth and physiological responses to cadmium of the three Miscanthus species. PLoS ONE 11 (4), e0153475.

Gupta, D.K., Pena, L.B., Romero-Puertas, M.C., Hernández, A., Inouhe, M., Sandalio, L.M., 2017. NADPH oxidases differentially regulate ROS metabolism and nutrient uptake under cadmium toxicity. Plant Cell Environ. 40 (4), 509–526.

Gupta, K.J., Fernie, A.R., Kaiser, W.M van Dongen, J.T., 2011. On the origins of nitric oxide. Trends Plant Sci. 16 (3), 160–168.

Gzyl, J., Izbianska, K., Floryszak-Wieczorek, J., Jelonek, T., Arasimowicz-Jelonek, M., 2016. Cadmium affects peroxynitrite generation and tyrosine nitration in seedling roots of soybean (*Glycine max* L.). Environ. Exp. Bot. 131, 155–163.

Halliwell, B., 2006. Reactive species and antioxidants. Redox biology is a fundamental theme of aerobic life. Plant Physiol. 141, 312–322.

Halliwell, B., Gutteridge, J.M.C., 1999. Free Radicals in Biology and Medicine. Oxford University Press, Oxford.

Hara, M.R., Agrawal, N., Kim, S.F., Cascio, M.B., Fujimuro, M., Ozeki, Y., Takahashi, M., Cheah, J.H., Tankou, S.K., Hester, L.D., Ferris, C.D., Hayward, S.D., Snyder, S.H., Sawa, A., 2005. S-nitrosylated GAPDH initiates apoptotic cell death by nuclear translocation following Siah1 binding. Nat. Cell Biol. 7, 665–674.

Hasanuzzaman, M., Nahar, K., Anee, T.I., Fujita, M., 2017. Glutathione in plants: biosynthesis and physiological role in environmental stress tolerance. Physiol. Mol. Biol. Plants. 23 (2), 249–268.

Have, M., Leitao, L., Bagard, M., Castell, J.-F., Repellin, A., 2015. Protein carbonylation during natural leaf senescence in winter wheat, as probed by fluorescein-5-thiosemicarbazide. Plant Biol. 17, 973–979.

He, C.-T., Zhou, Y.-H., Huang, Y.-Y., Fu, H.-L., Wang, X.-S., Gong, F.-Y., Tan, X., Yang, Z.-Y., 2018. Different proteomic processes related to the cultivar-dependent cadmium accumulation of *Amaranthus gangeticus*. J. Agric. Food Chem. 66 (5), 1085–1095.

He, Y., Tang, R.H., Hao, Y., Stevens, R.D., Cook, C.W., Ahn, S.M., Jing, L., Yang, Z., Chen, L., Guo, F., Fiorani, F., Jackson, R.B., Crawford, N.M., Pei, Z.M., 2004. Nitric oxide represses the *Arabidopsis* floral transition. Science 305, 1968–1971.

Heinrich, T.A., da Silva, R.S., Miranda, K.M., Switzer, C.H., Wink, D.A., Fukuto, J.M., 2013. Biological nitric oxide signalling: chemistry and terminology. Br. J. Pharmacol. 169 (7), 1417–1429.

Hess, D.T., Matsumoto, A., Kim, S.O., Marshall, H.E., Stamler, J.S., 2005. Protein S-nitrosylation: purview and parameters. Nat. Rev. Mol. Cell Biol. 6, 150–166.

Hess, D.T., Stamler, J.S., 2012. Regulation by S-nitrosylation of protein post-translational modification. J. Biol. Chem. 287, 4411–4418.

Heyno, E., Klose, C., Krieger-Liszkay, A., 2008. Origin of cadmium-induced reactive oxygen species production: mitochondrial electron transfer versus plasma membrane NADPH oxidase. New Phytol. 179, 687–699.

Hradilová, J., Rehulka, P., Rehulková, H., Vrbová, M., Griga, M., Brzobohatý, B., 2010. Comparative analysis of proteomic changes in contrasting flax cultivars upon cadmium exposure. Electrophoresis 31 (2), 421–431.

Hu, J., Huang, X., Chen, L., Sun, X., Lu, C., Zhang, L., Wang, Y., Zuo, J., 2015. Site-specific nitrosoproteomic identification of endogenously S-nitrosylated proteins in *Arabidopsis*. Plant Physiol. 167, 1731–1746.

Hughes, M.N., 2008. Chemistry of nitric oxide and related species. Methods Enzymol. 436, 3–19.

Ignarro, L.J., 1990. Biosynthesis and metabolism of endothelium-derived nitric oxide. Ann. Rev. Pharmacol. Toxicol. 30, 535–560.

Inupakutika, M.A., Sengupta, S., Devireddy, A.R., Azad, R.K., Mittler, R., 2016. The evolution of reactive oxygen species metabolism. J. Exp. Bot. 67 (21), 5933–5943.

Ischiropoulos, H., Zhu, L., Chen, J., Tsai, M., Martin, J.C., Smith, C.D., Beckman, J.S., 1992. Peroxynitrite-mediated tyrosine nitration catalyzed by superoxide dismutase. Arch. Biochem. Biophys. 298, 431–437.

Jaffrey, S.R., Snyder, S.H., 2001. The biotin switch method for the detection of S-nitrosylated proteins. Sci. STKE 2001, PL1.

Jin, H., Xu, M., Chen, H., Zhang, S., Han, X., Tang, Z., Sun, R., 2016. Comparative proteomic analysis of differentially expressed proteins in *Amaranthus hybridus* L. roots under cadmium stress. Water Air Soil Pollut. 227 (7), 1–12.

Johansson, E., Olsson, O., Nyström, T., 2004. Progression and specificity of protein oxidation in the life cycle of *Arabidopsis thaliana*. J. Biol. Chem. 279 (21), 22204–22208.
Jones, D.P., 2006. Redefining oxidative stress. Antiox. Redox Signal. 8 (9–10), 1865–1879.
Jozefczak, M., Keunen, E., Schat, H., Bliek, M., Hernández, L.E., Carleer, R., Remans, T., Bohler, S., Vangronsveld, J., Cuypers, A., 2014. Differential response of Arabidopsis leaves and roots to cadmium: glutathione-related chelating capacity vs antioxidant capacity. Plant Physiol. Biochem. 83, 1–9.
Jozefczak, M., Remans, T., Vangronsveld, J., Cuypers, A., 2012. Glutathione is a key player in metal-induced oxidative stress defenses. Int. J. Mol. Sci. 13 (3), 3145–3175.
Jung, T., Höhn, A., Grune, T., 2014. The proteasome and the degradation of oxidized proteins: Part II – protein oxidation and proteasomal degradation. Redox Biol. 2, 99–104.
Kehrer, J.P., 2000. The Haber-Weiss reaction and mechanisms of toxicity. Toxicology 149, 43–50.
Kemp, M., Go, Y.M., Jones, D.P., 2008. Non-equilibrium thermodynamics of thiol/disulfide redox systems: a perspective on redox systems biology. Free Radic. Biol. Med. 44 (6), 921–937.
Kieffer, P., Dommes, J., Hoffmann, L., Hausman, J.F., Renaut, J., 2008. Quantitative changes in protein expression of cadmium-exposed poplar plants. Proteomics 8, 2514–2530.
Kieffer, P., Schröder, P., Dommes, J., Hoffmann, L., Renaut, J., Hausman, J.F., 2009. Proteomic and enzymatic response of poplar to cadmium stress. J. Proteomics 72, 379–396.
Kim, G., Weiss, S.J., Levine, R.L., 2014. Methionine oxidation and reduction in proteins. Biochim. Biophys. Acta. 1840 (2), 901–905.
Klomsiri, C., Karplus, P.A., Poole, L.B., 2011. Cysteine-based redox switches in enzymes. Antioxid. Redox Signal. 14 (6), 1065–1077.
Kneeshaw, S., Gelineau, S., Tada, Y., Loake, G.J., Spoel, S.H., 2014. Selective protein denitrosylation activity of thioredoxin-h5 modulates plant immunity. Mol. Cell 56, 153–162.
Kocsy, G., Tari, I., Vanková, R., Zechmann, B., Gulyás, Z., Poór, P., Galiba, G., 2013. Redox control of plant growth and development. Plant Sci. 211, 77–91.
Kolbert, Z., Feigl, G., Bordé, A., Molnár, A., Erdei, L., 2017. Protein tyrosine nitration in plants:present knowledge, computational prediction and future perspectives. Plant Physiol. Biochem. 113, 56–63.
Kopyra, M., Stachon-Wilk, M., Gwozdz, E.A., 2006. Effects of exogenous nitric oxide on the antioxidant capacity of cadmium-treated soybean cell suspension. Acta Physiol. Plant. 28, 525–536.
Kováčik, J., Babula, P., 2017. Fluorescence microscopy as a tool for visualization of metal-induced oxidative stress in plants. Acta Physiol. Plant. 39, 157–164.
Kovacs, I., Lindermayr, C., 2013. Nitric oxide-based protein modification: formation and site-specificity of protein S-nitrosylation. Front. Plant Sci 4, 137. https://doi.org/10.3389/fpls.2013.00137.
Kurepa, J., Toh-E, A., Smalle, J.A., 2008. 26S proteasome regulatory particle mutants have increased oxidative stress tolerance. Plant J. 53, 102–114.
Lamattina, L., Garcia-Mata, C., Graziano, M., Pagnussat, G., 2003. Nitric oxide: the versatility of an extensive signal molecule. Annu. Rev. Plant Biol. 54, 109–136.
Laspina, N.V., Groppa, M.D., Tomaro, M.L., Benavides, M.P., 2005. Nitric oxide protects sunflower leaves against cadmium induced oxidative stress. Plant Sci. 167, 323–330.
Lee, K., Bae, D.W., Kim, S.H., Han, H.J., Liu, X., Park, H.C., Lim, C.O., Lee, S.Y., Chung, W.S., 2010. Comparative proteomic analysis of the short-term responses of rice roots and leaves to cadmium. J. Plant Physiol. 167, 161–168.
Lee, T.Y., Chen, Y.J., Lu, T.C., Huang, H.D., Chen, Y.J., 2011. SNOSite: exploiting maximal dependence decomposition to identify cysteine S-nitrosylation with substrate site specificity. PloS One 5, e21849.

Lefaki, M., Papaevgeniou, N., Chondrogianni, N., 2017. Redox regulation of proteasome function. Redox Biol. 13, 452–458.
Lehotai, N., Lyubenova, L., Schröder, P., Feigl, G., Ördög, A., Szilagyi, K., Erdei, L., Kolbert, Z., 2016. Nitro-oxidative stress contributes to selenite toxicity in pea (*Pisum sativum* L.). Plant Soil 400, 107–122.
Lehotai, N., Petö, A., Bajkán, S., Erdei, L., Tari, I., Kolbert, Z., 2011. *In vivo* and *in situ* visualization of early physiological events induces by heavy metals in pea root meristem. Acta Physiol. Plant. 33, 2199–2207.
Leitner, M., Vandelle, E., Gaupels, F., Bellin, D., Delledonne, M., 2009. NO signals in the haze: nitric oxide signalling in plant defence. Curr. Opin. Plant Biol. 12, 451–458.
Leterrier, M., Barroso, J.B., Valderrama, R., Palma, J.M., Corpas, F.J., 2012. NADP-dependent isocitrate dehydrogenase from *Arabidopsis* roots contributes in the mechanism of defence against the nitro-oxidative stress induced by salinity. Sci. World J. 2012, 694740.
Li, H., Poulos, T.L., 2005. Structure–function studies on nitric oxide synthases. J. Inorg. Biochem. 99, 293–305.
Li, J., Wei, X., Yu, P., Deng, X., Xu, W., Ma, M., Zhang, H., 2016. Expression of cadR enhances its specific activity for Cd detoxification and accumulation in *Arabidopsis*. Plant Cell Physiol. 57 (8), 1720–1731.
Li, M., Hao, P., Cao, F., 2017. Glutathione-induced alleviation of cadmium toxicity in *Zea mays*. Plant Physiol. Biochem. 119, 240–249.
Lin, A., Wang, Y., Tang, J., Xue, P., Li, C., Liu, L., Hu, B., Yang, F., Loake, G.J., Chu, C., 2012. Nitric oxide and protein S-nitrosylation are integral to hydrogen peroxide-induced leaf cell death in rice. Plant Physiol. 158, 451–464.
Lindermayr, C., Durner, J., 2009. S-nitrosylation in plants: pattern and function. J. Proteomics 73, 1–9.
Lindermayr, C., Saalbach, G., Bahnweg, G., Durner, J., 2006. Differential inhibition of *Arabidopsis* methionine adenosyltransferases by protein S-nitrosylation. J. Biol. Chem. 281, 4285–4291.
Lindermayr, C., Saalbach, G., Durner, J., 2005. Proteomic identification of S-nitrosylated proteins in *Arabidopsis*. Plant Physiol. 137, 921–930.
Lindermayr, C., Sell, S., Muller, B., Leister, D., Durner, J., 2010. Redox regulation of the NPR1-TGA1 system *of Arabidopsis thaliana* by nitric oxide. Plant Cell 22, 2894–2907.
Liu, M., Hou, J., Huang, L., Huang, X., Heibeck, T.H., Zhao, R., Pasa-Tolic, L., Smith, R.D., Li, Y., Fu, K., Zhang, Z., Hinrichs, S., Ding, S., 2010. Site-specific proteomics approach for study protein S-nitrosylation. Anal. Chem. 82, 7160–7168.
Liu, S., Yang, R., Tripathi, D.K., Li, X., He, W., Wua, M., Alie, S., Maa, M., Cheng, Q., Pan, Y., 2018. The interplay between reactive oxygen and nitrogen species contributes in the regulatory mechanism of the nitro-oxidative stress induced by cadmium in *Arabidopsis*. J. Hazard. Mat. 344, 1007–1024.
Liu, S., Yang, R., Pan, Y., Maa, M., Pan, J., Zhao, Y., Cheng, Q., Wu, M., Wang, M., Zhang, L., 2015. Nitric oxide contributes to minerals absorption, proton pumps and hormone equilibrium under cadmium excess in *Trifolium repens* L. plants. Ecotox. Environ. Safety 119, 35–46.
Lombardo, M.C., Graziano, M., Polacco, J.C., Lamattina, L., 2006. Nitric oxide functions as a positive regulator of root hair development. Plant Signal. Behav. 1, 28–33.
Lombardo, M.C., Lamattina, L., 2012. Nitric oxide is essential for vesicle formation and trafficking in Arabidopsis root hair growth. J. Exp. Bot. 63, 4875–4885.
Lounifi, I., Arc, E., Molassiotis, A., Job, D., Rajjou, L., Tanou, G., 2013. Interplay between protein carbonylation and nitrosylation in plants. Proteomics 13, 568–578.
Lozano-Juste, J., Colom-Moreno, R., Leon, J., 2011. *In vivo* protein tyrosine nitration in *Arabidopsis thaliana*. J. Exp. Bot. 62, 3501–3517.

Lu, Q., Zhang, T., Zhang, W., Su, C., Yang, Y., Hu, D., Xu, Q., 2018. Alleviation of cadmium toxicity in *Lemna minor* by exogenous salicylic acid. Ecotoxicol. Environ. Safety 147, 500–508.

Lux, A., Martinka, M., Vaculík, M., White, P.J., 2011. Root responses to cadmium in the rhizosphere: a review. J. Exp Bot. 62 (1), 21–37.

Lysenko, E.A., Klaus, A.A., Pshybytko, N.L., Kusnetsov, V.V., 2015. Cadmium accumulation in chloroplasts and its impact on chloroplastic processes in barley and maize. Photosynthesis Res. 125, 291–303.

Madian, A.G., Regnier, F.E., 2010. Proteomic identification of carbonylated proteins and their oxidation sites. J. Proteome Res. 9, 3766–3780.

Mahmood, T., Gupta, K.J., Kaiser, W.M., 2009. Cadmium stress stimulates nitric oxide production by wheat roots. Pak. J. Bot. 41, 1285–1290.

Marino, S.M., Gladyshev, V.N., 2010. Cysteine function governs its conservation and degeneration and restricts its utilization on protein surfaces. J. Mol. Biol. 404 (5), 902–916.

Marmiroli, M., Imperiale, D., Maestri, E., Marmiroli, N., 2013. The response of Populus spp to cadmium stress: chemical, morphological and proteomics study. Chemosphere 93, 1333–1344.

Melo, P.M., Silva, L.S., Ribeiro, I., Seabra, A.R., Carvalho, H.G., 2011. Glutamine synthetase is a molecular target of nitric oxide in root nodules of *Medicago truncatula* and is regulated by tyrosine nitration. Plant Physiol. 157, 1505–1517.

Méndez, A.A.E., Pena, L.B., Benavides, M.P., Gallego, S.M., 2016. Priming with NO controls redox state and prevents cadmium-induced general up-regulation of methionine sulfoxide reductase gene family in *Arabidopsis*. Biochimie 131, 128–136.

Mittler, R., 2017. ROS are good. Trends Plant Sci. 22 (1), 11–19.

Mittler, R., Vanderauwera, S., Gollery, M., Van Breusegem, F., 2004. Reactive oxygen gene network of plants. Trends Plant Sci. 9, 490–498.

Mittler, R., Vanderauwera, S., Suzuki, N., Miller, G., Tognetti, V.B., Vandepoele, K., Gollery, M., Shulaev, V., Van Breusegem, F., 2011. ROS signaling: the new wave? Trends Plant Sci. 16 (6), 300–309.

Møller, I.M., Jensen, P.E., Hansson, A., 2007. Oxidative modifications to cellular components in plants. Annu. Rev. Plant Biol. 58, 459–481.

Moreau, M., Lee, G.I., Wang, Y., Crane, B.R., Klessig, D.F., 2008. AtNOS/A1 is a functional *Arabidopsis thaliana* cGTPase and not a nitric oxide synthase. J. Biol. Chem. 283, 32957–32967.

Morot-Gaudry-Talarmain, Y., Rockel, P., Moureaux, T., Quillere, I., Leydecker, M.T., Kaiser, W.M., Morot-Gaudry, J.F., 2002. Nitrite accumulation and nitric oxide emission in relation to cellular signaling in nitrite reductase antisense tobacco. Planta 215, 708–715.

Mur, L.A., Sivakumaran, A., Mandon, J., Cristescu, S.M., Harren, F.J., Hebelstrup, K.H., 2012. Haemoglobin modulates salicylate and jasmonate/ethylene-mediated resistance mechanisms against pathogens. J. Exp. Bot. 63, 4375–4387.

Mur, L.A., Mandon, J., Persijn, S., Cristescu, S.M., Moshkov, I.E., Novikova, G.V., Hall, M.A., Harren, F.J., Hebelstrup, K.H., Gupta, K.J., 2013a. Nitric oxide in plants: an assessment of the current state of knowledge. AoB Plants 5, pls052. https://doi.org/10.1093/aobpla/pls052.

Mur, L.A., Prats, E., Pierre, S., Hall, M.A., Hebelstrup, K.H., 2013b. Integrating nitric oxide into salicylic acid and jasmonic acid/ethylene plant defense pathways. Front. Plant Sci. 4, 215. https://doi.org/10.3389/fpls.2013.00215.

Mur, L.A., Santosa, I.E., Laarhoven, L.J.J., Holton, N.J., Harren, F.J.M., Smith, A.R., 2005. Laser photoacoustic detection allows in planta detection of nitric oxide in tobacco following challenge with avirulent and virulent *Pseudomonas syringae* pathovars. Plant Physiol. 138, 1247–1258.

Nietzel, T., Mostertz, J., Hochgräfe, F., Schwarzländer, M., 2017. Redox regulation of mitochondrial proteins and proteomes by cysteine thiol switches. Mitochondrion 33, 72–83.

Nottingham, W.C., Sutter, J.R., 1986. Kinetics of the oxidation of nitric oxide by clorine and oxgyen in nonaqueous media. Int. J. Chem. Kinet. 18, 1289–1302.

Nystrom, T., 2005. Role of oxidative carbonylation in protein quality control and senescence. EMBO J. 24, 1311–1317.

Olmos, E., Martínez-Solano, J.R., Piqueras, A., Hellín, E., 2003. Early steps in the oxidative burst induced by cadmium in cultured tobacco cells (BY-2 line). J. Exp. Bot. 54, 291–301.

Ortega-Galisteo, A.P., Rodríguez-Serrano, M., Pazmiño, D.M., Gupta, D.K., Sandalio, L.M., Romero-Puertas, M.C., 2012. S-Nitrosylated proteins in pea (*Pisum sativum* L.) leaf peroxisomes: changes under abiotic stress. J. Exp. Bot. 63, 2089–2103.

Ortega-Villasante, C., Rellán-Álvarez, Z.Z., Del Campo, F.F., Carpena-Ruiz, R.O., Hernández, L.E., 2005. Cellular damage induced by cadmium and mercury in *Medicago sativa*. J. Exp. Bot. 56 (418), 2239–2251.

Palmer, R.M., Ferrige, A.G., Moncada, S., 1987. Nitric oxide release accounts for the biological activity of endothelium-derived relaxing factor. Nature 327, 524–526.

Palmieri, M.C., Lindermayr, C., Bauwe, H., Steinhauser, C., Durner, J., 2010. Regulation of plant glycine decarboxylase by S-nitrosylation and glutathionylation. Plant Physiol. 152, 1514–1528.

Paulsen, C.E., Carroll, K.S., 2013. Cysteine-mediated redox signaling: chemistry, biology, and tools for discovery. Chem. Rev. 113, 4633–4679.

Pena, L.B., Barcia, R.A., Azpilicueta, C.E., Méndez, A.A.E., Gallego, S.M., 2012. Oxidative post translational modifications of proteins related to cell cycle are involved in cadmium toxicity in wheat seedlings. Plant Sci. 196, 1–7.

Pena, L.B., Pasquini, L.A., Tomaro, M.L., Gallego, S.M., 2007. 20S proteasome and accumulation of oxidized and ubiquitinated proteins in maize leaves subjected to cadmium stress. Phytochemistry 68, 1139–1146.

Pereira de Araújo, R., Furtado de Almeida, A.A., Silva Pereira, L., Mangabeira, P.A.O., Olimpio Souza, J., Pirovani, C.P., Ahnert, D., Baligar, V.C., 2017. Photosynthetic, antioxidative, molecular and ultrastructural responses of young cacao plants to Cd toxicity in the soil. Ecotoxicol. Environ. Saf. 144, 148–157.

Pereira, T.S., Pereira, T.S., Souza, C.L.F.C., Lima, E.J.A., Batista, B.L., Lobato, A.K.S., 2018. Silicon deposition in roots minimizes the cadmium accumulation and oxidative stress in leaves of cowpea plants. Physiol. Mol. Biol. Plants. 24 (1), 99–114.

Peréz-Chaca, M.V., Rodrígues-Serrano, M., Molina, A.S., Pedranzani, H.E., Zirulnik, F., Sandalio, L.M., Romero-Puertas, M.C., 2014. Cadmium induces two waves of reactive oxygen species in *Glycine max* (L.) roots. Plant Cell Environ. 37, 1672–1687.

Planchet, E., Gupta, K., Sonoda, M., Kaiser, W.M., 2005. Nitric oxide emission from tobacco leaves and cell suspensions: rate limiting factors and evidence for the involvement of mitochondrial electron transport. Plant J. 41, 732–743.

Planchet, E., Sonoda, M., Zeier, J., Kaiser, W.M., 2006. Nitric oxide (NO) as an intermediate in the cryptogein-induced hypersensitive response: a critical re-evaluation. Plant Cell Environ. 29, 59–69.

Polge, C., Jaquinod, M., Holzer, F., Bourguignon, J., Walling, L., Brouquisse, R., 2009. Evidence for the existence in *Arabidopsis thaliana* of the proteasome proteolytic pathway: activation in response to cadmium. J. Biol. Chem. 284 (51), 35412–35424.

Poole, L.B., 2015. The basics of thiols and cysteines in redox biology and chemistry. Free Radic. Biol. Med. 80, 148–157.

Poppek, D., Grune, T., 2006. Proteasomal defense of oxidative protein modifications. Antiox. Redox Signal. 8 (1-2), 173–184.

Pospíšil, P., 2012. Molecular mechanisms of production and scavenging of reactive oxygen species by photosystem II. Biochim. Biophys. Acta 1817, 218–231.

Puyaubert, J., Fares, A., Reze, N., Peltier, J.B., Baudouin, E., 2014. Identification of endogenously S-nitrosylated proteins in Arabidopsis plantlets: effect of cold stress on cysteine nitrosylation level. Plant Sci. 215–216, 150–156.

Qi, J., Wang, J., Gong, Z., Zhou, J.-M., 2017. Apoplastic ROS signaling in plant immunity. Curr. Opin. Plant Biol. 3, 92–100.

Qin, G., Meng, X., Wang, Q., Tian, S., 2009. Oxidative damage of mitochondrial proteins contributes to fruit senescence: a redox proteomics analysis. J. Proteome 8 (5), 2449–2462.

Qu, Y., Yan, M., Zhang, Q., 2017. Functional regulation of plant NADPH oxidase and its role in signaling. Plant Signal. Behav. 12 (8), e1356970.

Radi, R., 1996. Reactions of nitric oxide with metalloproteins. Chem. Res. Toxicol. 9 (5), 828–835.

Radi, R., 2004. Nitric oxide, oxidants, and protein tyrosine nitration. Proc. Natl. Acad. Sci. U.S.A. 12, 4003–4008.

Radi, R., 2013. Protein tyrosine nitration: biochemical mechanisms and structural basis of functional effects. Acc. Chem. Res. 46, 550–559.

Rahantaniaina, M.S., Li, S., Chatel-Innocenti, G., Tuzet, A., Mhamdi, A., Vanacker, H., Noctor, G., 2017. Glutathione oxidation in response to intracellular $H_2O_2$: key but overlapping roles for dehydroascorbate reductases. Plant Signal. Behav. 12 (8), e1356531.

Raja, V., Majeed, U., Kang, H., Andrabi, K.I., John, R., 2017. Abiotic stress: interplay between ROS, hormones and MAPKs. Environ. Exp. Bot. 137, 142–157.

Rao, R.S., Møller, I.M., 2011. Pattern of occurrence and occupancy of carbonylation sites in proteins. Proteomics 11, 4166–4173.

Raynes, R., Pomatto, L.C.D., Davies, K.J.A., 2016. Degradation of oxidized proteins by the proteasome: distinguishing between the 20S, 26S, and immunoproteasome proteolytic pathways. Mol. Aspects Med. 50, 41–55.

Recalde, L., Vázquez, A., Groppa, M.D., Benavides, M.P., 2018. Reactive oxygen species and nitric oxide are involved in polyamine-induced growth inhibition in wheat plants. Protoplasma. https://doi.org/10.1007/s00709-018-1227-z (in press).

Reumann, S., Bartel, B., 2016. Plant peroxisomes: recent discoveries in functional complexity, organelle homeostasis, and morphological dynamics. Curr. Opin. Plant Biol. 34, 17–26.

Rockel, P., Strube, F., Rockel, A., Wildt, J., Kaiser, W.M., 2002. Regulation of nitric oxide (NO) production by plant nitrate reductase *in vivo* and *in vitro*. J. Exp. Bot. 53, 103–110.

Rodríguez-Serrano, M., Romero-Puertas, M.C., Pazmino, D.M., Testillano, P.S., Risueño, M.C., Del Río, L.A., Sandalio, L.M., 2009. Cellular response of pea plants to cadmium toxicity: cross talk between reactive oxygen species, nitric oxide, and calcium. Plant Physiol. 150, 229–243.

Romero-Puertas, M.C., Laxa, M., Mattè, A., Zaninotto, F., Finkemeier, I., Jones, A.M., Perazzolli, M., Vandelle, E., Dietz, K.J., Delledonne, M., 2007. S-nitrosylation of peroxiredoxin II E promotes peroxynitrite-mediated tyrosine nitration. Plant Cell 19, 4120–4130.

Romero-Puertas, M.C., Rodríguez-Serrano, M., Corpas, F.J., Gómez, M., del Río, L.A., Sandalio, L.M., 2004. Cadmium-induced subcellular accumulation of $O_2^-$ and $H_2O_2$ in pea leaves. Plant Cell Environ. 27, 1122–1134.

Rosales, E.P., Iannone, M.F., Groppa, M.D., Benavides, M.P., 2011. Nitric oxide inhibits nitrate reductase activity in wheat leaves. Plant Physiol. Biochem. 49 (2), 124–130.

Roth, U., von Roepenack-Lahaye, E., Clemens, S., 2006. Proteome changes in *Arabidopsis thaliana* roots upon exposure to $Cd^{2+}$. J. Exp. Bot. 57 (15), 4003–4013.

Rouhier, N., Gelhaye, E., Sautiere, P.E., Brun, A., Laurent, P., Tagu, D., Gerard, J., de Fay, E., Meyer, Y., Jacquot, J.P., 2001. Isolation and characterization of a new peroxiredoxin from poplar sieve tubes that uses either glutaredoxin or thioredoxin as a proton donor. Plant Physiol. 127, 1299–1309.

Rumer, S., Gupta, K.J., Kaiser, W.M., 2009. Plant cells oxidize hydroxylamines to NO. J. Exp. Bot. 60, 2065–2072.

Sagi, M., Fluhr, R., 2006. Production of reactive oxygen species by plant NADPH oxidases. Plant Physiol. 141, 336–340.

Samardzic, K., Rodgers, K.J., 2017. Oxidised protein metabolism: recent insights. Biol. Chem. 398 (11), 1165–1175.

Sandalio, L.M., Romero-Puertas, M.C., 2015. Peroxisomes sense and respond to environmental cues by regulating ROS and RNS signalling networks. Ann. Bot. 116, 475–485.

Sanitá di Toppi, L., Gabbrielli, R., 1999. Response to cadmium in higher plants. Env. Exp. Bot. 41, 105–130.

Santolini, J., André, F., Jeandroz, S., Wendehenne, D., 2017. Nitric oxide synthase in plants: where do we stand? Nitric Oxide 63, 30–38.

Sanz, L., Fernández-Marcos, M., Modrego, A., Lewis, D.R., Muday, G.K., Pollmann, S., Dueñas, M., Santos-Buelga, C., Lorenzo, O., 2014. Nitric oxide plays a role in stem cell niche homeostasis through its interaction with auxin. Plant Physiol. 166, 1972–1984.

Sarry, J.E., Kuhn, L., Ducruix, C., Lafaye, A., Junot, C., Hugouvieux, V., Jourdain, A., Bastien, O., Fievet, J.B., Vailhen, D., Amekraz, B., Moulin, C., Ezan, E., Garin, J., Bourguignon, J., 2006. The early responses of *Arabidopsis thaliana* cells to cadmium exposure explored by protein and metabolite profiling analyses. Proteomics 6 (7), 2180–2198.

Schippers, J.H.M., Foyer, C.H., van Dongen, J.T., 2016. Redox regulation in shoot growth. SAM maintenance and flowering. Curr. Opin. Plant Biol. 29, 121–128.

Schmidt, R., Schippers, J.H.M., 2015. ROS-mediated redox signaling during cell differentiation in plants. Biochim. Biophys. Acta. 1850, 1497–1508.

Schmitt, F.J., Renger, G., Friedrich, T., Kreslavski, V.D., Zharmukhamedov, S.K., Los, D.A., Kuznetsov, V.V., Allakhverdiev, S.I., 2014. Reactive oxygen species: re-evaluation of generation, monitoring and role in stress-signaling in phototrophic organisms. Biochim. Biophys. Acta. 1837 (6), 835–848.

Schöneich, C., 2017. Sulfur radical-induced redox modifications in proteins: analysis and mechanistic aspects. Antioxid. Redox Signal. 26 (8), 388–405.

Schopfer, F.J., Baker, P.R., Freeman, B.A., 2003. NO-dependent protein nitration: A cell signaling event or an oxidative inflammatory response? Trends Biochem. Sci. 28, 646–654.

Segal, L.M., Wilson, R.A., 2018. Reactive oxygen species metabolism and plant-fungal interactions. Fungal Genet. Biol. 110, 1–9.

Semane, B., Dupae, J., Cuypers, A., Noben, J.P., Tuomainen, M., Tervahauta, A., Kärenlampi, S., Van Belleghem, F., Smeets, K., Vangronsveld, J., 2010. Leaf proteome responses of *Arabidopsis thaliana* exposed to mild cadmium stress. J. Plant Physiol. 167 (4), 247–254.

Sergeant, K., Kieffer, P., Dommes, J., Hausman, J.-F., Renaut, J., 2014. Proteomic changes in leaves of poplar exposed to both cadmium and low-temperature. Environ. Exp. Bot. 106, 112–123.

Serpa, V., Vernal, J., Lamattina, L., Grotewold, E., Cassia, R., Terenzi, H., 2007. Inhibition of AtMYB2 DNA-binding by nitric oxide involves cysteine S-nitrosylation. Biochem. Biophys. Res. Commun. 361, 1048–1053.

Seth, D., Stamler, J.S., 2011. The SNO-proteome: causation and classifications. Curr. Opin. Chem. Biol. 15, 129–136.

Sevier, C.S., Kaiser, C.A., 2008. Ero1 and redox homeostasis in the endoplasmic reticulum. Biochim. Biophys. Acta. 1783, 549–556.

Sevilla, F., Camejo, D., Ortiz-Espín, A., Calderón, A., Lázaro, J.J., Jiménez, A., 2015. The thioredoxin/peroxiredoxin/sulfiredoxin system: current overview on its redox function in plants and regulation by reactive oxygen and nitrogen species. J. Exp. Bot. 66, 2945–2955.

Shi, X., Sun, H., Chen, Y., Pan, H., Wang, S., 2016. Transcriptome sequencing and expression analysis of cadmium (Cd) transport and detoxification related genes in Cd-accumulating *Salix integra*. Front. Plant Sci. 7, 1577.

Sies, H., 2015. Oxidative stress: a concept in redox biology and medicine. Redox Biol. 4, 180–183.

Sies, H., 2017. Hydrogen peroxide as a central redox signaling molecule in physiological oxidative stress: oxidative eustress. Redox Biol. 11, 613–619.

Sies, H., Jones, D.P., 2007. Oxidative stress. In: Fink, G. (Ed.), Encyclopedia of Stress. 2nd ed. Elsevier, New York, USA, pp. 45–48.

Signorelli, S., Corpas, F.J., Borsani, O., Barroso, J.B., Monza, J., 2013. Water stress induces a differential and spatially distributed nitro-oxidative stress response in roots and leaves of *Lotus japonicus*. Plant Sci. 201-202, 137–146.

Singh, P., Shah, K., 2014. Evidences for reduced metal-uptake and membrane injury upon application of nitric oxide donor in cadmium stressed rice seedlings. Plant Physiol. Biochem. 83, 180–184.

Singh, R., Singh, S., Parihar, P., Mishra, R.K., Tripathi, D.K., Singh, V.P., Chauhan, D.K., Prasad, S.M., 2016. Reactive oxygen species (ROS): beneficial companions of plants′ developmental processes. Front. Plant Sci. 7, 1299.

Singh, S., Singh, A., Srivastava, P.K., Prasad, S.M., 2018. Cadmium toxicity and its amelioration by kinetin in tomato seedlings vis-à-vis ascorbate-glutathione cycle. J. Photochem. Photobiol. B: Biol. 178, 76–84.

Smeets, K., Opdenakker, K., Remans, T., Van Sanden, S., Van Belleghem, F., Semane, B., Horemans, N., Guisez, Y., Vangronsveld, J., Cuypers, A., 2009. Oxidative stress-related responses at transcriptional and enzymatic levels after exposure to Cd or Cu in a multipollution context. J. Plant Physiol. 166 (18), 1982–1992.

Smiri, M., Chaoui, A., Rouhier, N., Gelhaye, E., Jacquot, J.P., El Ferjani, E., 2010. Oxidative damage and redox change in pea seeds treated with cadmium. C. R. Biol. 333 (11-12), 801–807.

Souza, J.M., Peluffo, G., Radi, R., 2008. Protein tyrosine nitration-Functional alteration or just a biomarker? Free Rad. Biol. Med. 45, 357–366.

Srivastava, R.K., Pandey, P., Rajpoot, R., Rani, A., Dubey, R.S., 2014. Cadmium and lead interactive effects on oxidative stress and antioxidative responses in rice seedlings. Protoplasma 251 (5), 1047–1065.

Stohr, C., Strube, F., Marx, G., Ullrich, W.R., Rockel, P., 2001. A plasma membrane-bound enzyme of tobacco roots catalyses the formation of nitric oxide from nitrite. Planta 212, 835–841.

Stohr, C., Ullrich, W.R., 2002. Generation and possible roles of NO in plant roots and their apoplastic space. J. Exp. Bot. 53, 2293–2303.

Szabó, C., Ischiropoulos, H., Radi, R., 2007. Peroxynitrite: biochemistry, pathophysiology and development of therapeutics. Nat. Rev. Drug Discov. 6, 662–680.

Tada, Y., Spoel, S.H., Pajerowska-Mukhtar, K., Mou, Z., Song, J., Wang, C., Zuo, J., Dong, X., 2008. Plant immunity requires conformational changes [corrected] of NPR1 via S-nitrosylation and thioredoxins. Science 321, 952–956.

Tanou, G., Filippou, P., Belghazi, M., Job, D., Diamantidis, G., Fotopoulos, V., Molassiotis, A., 2012. Oxidative and nitrosative-based signaling and associated post-translational modifications orchestrate the acclimation of citrus plants to salinity stress. Plant J. 72, 585–599.

Tanou, G., Job, C., Rajjou, L., Arc, E., Belghazi, M., Diamantidis, G., Molassiotis, A., Job, D., 2009. Proteomics reveals the overlapping roles of hydrogen peroxide and nitric oxide in the acclimation of citrus plants to salinity. Plant J. 60, 795–804.

Thomas, D.D., 2015. Breathing new life into nitric oxide signaling: a brief overview of the interplay between oxygen and nitric oxide. Redox Biol. 5, 225–233.

Tischner, R., Planchet, E., Kaiser, W.M., 2004. Mitochondrial electron transport as a source for nitric oxide in the unicellular green alga *Chlorella sorokiniana*. FEBS Lett. 576, 151–155.

Torres, M.A., Onouchi, H., Hamada, S., Machida, C., Hammond-Kosack, K.E., Jones, J.D., 1998. Six *Arabidopsis thaliana* homologues of the human respiratory burst oxidase (gp91phox). Plant J. 14 (3), 365–370.

Triantaphylidès, C., Havaux, M., 2009. Singlet oxygen in plants: production, detoxification and signaling. Trends Plant Sci. 14, 219–228.

Tun, N.N., Santa-Catarina, C., Begum, T., Silveira, V., Handro, W., Floh, E.I.S., Scherer, G.F.E., 2006. Polyamines induce rapid biosynthesis of nitric oxide (NO) in *Arabidopsis thaliana* seedlings. Plant Cell Physiol. 47, 346–354.

Turrens, J.F., 2003. Mitochondrial formation of reactive oxygen species. J. Physiol. 552, 335–344.

Uraguchi, S., Fujiwara, T., 2013. Rice breaks ground for cadmium-free cereals. Curr. Opin. Plant Biol. 16 (3), 28–34.

Vandelle, E., Delledonne, M., 2011. Peroxynitrite formation and function in plants. Plant Sci. 181 (5), 534–539.

Wang, D., Liu, Y., Tan, X., Liu, H., Zeng, G., Hu, X., Jian, H., Gu, Y., 2015. Effect of exogenous nitric oxide on antioxidative system and S-nitrosylation in leaves of *Boehmeria nivea* (L.) Gaud under cadmium stress. Environ. Sci. Pollut. Res. 22, 3489–3497.

Wang, Y., Hu, H., Xu, Y., Li, X.X., Zhang, H.J., 2011. Differential proteomic analysis of cadmium-responsive proteins in wheat leaves. Biol. Plant. 55, 586–590.

Wang, Y., Jiang, X., Li, K., Wu, M., Zhang, R., Zhang, L., Chen, G., 2014. Photosynthetic responses of *Oryza sativa* L. seedlings to cadmium stress: physiological, biochemical and ultrastructural analyses. Biometals 27, 389–401.

Wang, Y., Wang, X., Wang, C., Wang, R., Peng, F., Xiao, X., Zeng, J., Fan, X., Kang, H., Sha, L., Zhang, H., Zhou, Y., 2016. Proteomic profiling of the interactions of Cd/Zn in the roots of dwarf polish wheat (*Triticum polonicum* L.). Front. Plant Sci. 7, 1378.

Wang, Y.Q., Feechan, A., Yun, B.W., Shafiei, R., Hofmann, A., Taylor, P., Xue, P., Yang, F.Q., Xie, Z.S., Pallas, J.A., Chu, C.C., Loake, G.J., 2009. S-nitrosylation of AtSABP3 antagonizes the expression of plant immunity. J. Biol. Chem. 284, 2131–2137.

Wendehenne, D., Dürner, J., Klessig, D.F., 2004. Nitric oxide: a new player in plant signalling and defence responses. Curr. Opin. Plant Biol. 7, 449–455.

Wilkinson, J.Q., Crawford, N.M., 1991. Identification of the Arabidopsis CHL3 gene as the nitrate reductase structural gene NIA2. Plant Cell. 3, 461–471.

Wrzaczek, M., Brosché, M., Kangasjaärvi, J., 2013. ROS signaling loops-production, perception, regulation. Curr. Opin. Plant Biol. 16, 575–582.

Wu, Z., Zhao, X., Sun, X., Tan, Q., Tang, Y., Nie, Z., Qu, C., Chen, X., Hu, C., 2015. Antioxidant enzyme systems and the ascorbate–glutathione cycle as contributing factors to cadmium accumulation and tolerance in two oilseed rape cultivars (*Brassica napus* L.) under moderate cadmium stress. Chemosphere 138, 526–536.

Xie, L., He, X., Shang, S., Zheng, W., Liu, W., Zhang, G., Wu, F., 2014. Comparative proteomic analysis of two tobacco (*Nicotiana tabacum)* genotypes differing in Cd tolerance. BioMetals 27, 1277–1289.

Xiong, J., Fu, G., Tao, L., Zhu, C., 2010. Roles of nitric oxide in alleviating heavy metal toxicity in plants. Arch. Biochem. Biophys. 497, 13–20.

Xu, L., Wang, Y., Zhang, F., Tang, M., Chen, Y., Wang, J., Karanja, B.K., Luo, X., Zhang, W., Liu, L., 2017. Dissecting root proteome changes reveals new insight into cadmium stress response in radish (*Raphanus sativus* L.). Plant Cell Physiol. 58 (11), 1901–1913.

Xue, D., Jiang, H., Deng, X., Zhang, X., Wang, H., Xu, X., Hu, J., Zeng, D., Guo, L., Qian, Q., 2014. Comparative proteomic analysis provides new insights into cadmium accumulation in rice grain under cadmium stress. J. Hazard. Mat. 280, 269–278.

Yamasaki, H., 2000. Nitrite-dependent nitric oxide production pathway: implications for involvement of active nitrogen species in photoinhibition *in vivo*. Philos. Trans. Royal Soc. London, Series B: Biol. Sci. 355, 1477–1488.

Yang, H., Mu, J., Chen, L., Feng, J., Hu, J., Li, L., Zhou, J.M., Zuo, J., 2015a. S-Nitrosylation positively regulates ascorbate peroxidase activity during plant stress responses. Plant Physiol. 167, 1604–1615.

Yang, Y., Li, X., Yang, S., Zhou, Y., Dong, C., Ren, J., Sun, X., Yang, Y., 2015b. Comparative physiological and proteomic analysis reveals the leaf response to cadmium-induced stress in Poplar (*Populus yunnanensis*). PLoS ONE 10 (9), e0137396.

Ye, Y., Li, Z., Xing, D., 2013. Nitric oxide promotes MPK6-mediated caspase- 3-like activation in cadmium-induced *Arabidopsis thaliana* programmed cell death. Plant Cell Environ. 36, 1–15.

Yeh, C.M., Chien, P.S., Huang, J., 2007. Distinct signaling pathways for induction of MAP kinase activities by cadmium and copper in rice roots. J. Exp. Bot. 58, 659–671.

Yeo, W.S., Kim, Y.J., Kabir, M.H., Kang, J.W., Kim, K.P., 2015. Mass spectrometric analysis of protein tyrosine nitration in aging and neurodegenerative diseases. Mass Spectrom. Rev. 34, 166–183.

Yin, G., Xin, X., Fu, S., An, M., Wu, S., Chen, X., Zhang, J., He, J., Whelan, J., Lu, X., 2017. Proteomic and carbonylation profile analysis at the critical node of seed ageing in *Oryza sativa*. Sci. Rep. 7, 40611.

Yoshioka, H., Mase, K., Yoshioka, M., Kobayashi, M., Asai, S., 2011. Regulatory mechanisms of nitric oxide and reactive oxygen species generation and their role in plant immunity. Nitric Oxide 25, 216–221.

Yu, M., Lamattina, L., Spoel, S.H., Loake, G.J., 2014. Nitric oxide function in plant biology: a redox cue in deconvolution. New Phytol. 202, 1142–1156.

Yun, B.W., Spoel, S.H., Loake, G.J., 2012. Synthesis of and signalling by small, redox active molecules in the plant immune response. Biochim. Biophys. Acta 1820, 770–776.

Zaffagnini, M., De Mia, M., Morisse, S., Di Giacinto, N., Marchand, C.H., Maes, A., Lemaire, S.D., Trost, P., 2016. Protein S-nitrosylation in photosynthetic organisms: a comprehensive overview with future perspectives. Biochim. Biophys. Acta 1864, 952–966.

Zaninotto, F., La Camera, S., Polverari, A., Delledonne, M., 2006. Cross talk between reactive nitrogen and oxygen species during the hypersensitive disease resistance response. Plant Physiol. 141, 379–383.

Zargar, S.M., Fujiwara, M., Inaba, S., Kobayashi, M., Kurata, R., Ogata, Y., Fukao, Y., 2015. Correlation analysis of proteins responsive to Zn Mn, or Fe deficiency in *Arabidopsis* roots based on iTRAQ analysis. Plant Cell Rep. 34, 157–166.

Zemojtel, T., Frohlich, A., Palmieri, M.C., Kolanczyk, M., Mikula, I., Wyrwicz, L.S., Wanker, E.E., Mundlos, S., Vingron, M., Martasek, P., Durner, J., 2006. Plant nitric oxide synthase: a never-ending story? Trends Plant Sci. 11, 524–525.

Zeng, X.W., Qiu, R.L., Ying, R.R., Tang, Y.T., Tang, L., Fang, X.H., 2011. The differentially-expressed proteome in Zn/Cd hyperaccumulator *Arabis paniculata* Franch. in response to Zn and Cd. Chemosphere 82 (3), 321–328.

Zhang, X.H., Weissbach, H., 2008. Origin and evolution of the protein-repairing enzymes methionine sulphoxide reductases. Biol. Rev. Camb. Philos. Soc. 83 (3), 249–257.

Zhao, L., Sun, Y.L., Cui, S.X., Chen, M., Yang, H.M., Liu, H.M., Chai, T.Y., Huang, F., 2011. Cd-induced changes in leaf proteome of the hyperaccumulator plant *Phytolacca americana*. Chemosphere 85 (1), 56–66.

Zhao, M.G., Tian, Q.Y., Zhang, W.H., 2007. Nitric oxide synthase dependent nitric oxide production is associated with salt tolerance in *Arabidopsis*. Plant Physiol. 144, 206–217.

Zito, E., 2015. ERO1: a protein disulfide oxidase and $H_2O_2$ producer. Free Rad. Biol. Med. 83, 299–304.

## FURTHER READING

Crawford, N.M., Galli, M., Tischner, R., Heimer, Y.M., Okamoto, M., Mack, A., 2006. Response to Zemojtel et al.: plant nitric oxide synthase: back to square one. Trends Plant Sci. 1, 526–527.

Freschi, L., 2013. Nitric oxide and phytohormone interactions: current status and perspectives. Front. Plant Sci. 9 (4), 398.

Rinalducci, S., Murgiano, L., Zolla, L., 2008. Redox proteomics: basic principles and future perspectives for the detection of protein oxidation in plants. J. Exp. Bot. 59, 3781–3801.

# CHAPTER 11

# Damage and Protection of the Photosynthetic Apparatus Under Cadmium Stress

**Anelia G. Dobrikova, Emilia L. Apostolova**
Institute of Biophysics and Biomedical Engineering, Bulgarian Academy of Sciences, Sofia, Bulgaria

## Abstract

Cadmium (Cd) is one of the most toxic pollutants that may cause potentially harmful effects on crops. Accumulation of Cd in the plant tissues is influenced by the levels of Cd available in the soil in which the plants are grown. It is known that Cd ions cause alterations in the functionality of photosynthetic membranes by affecting the lipid composition, the chlorophyll metabolism, the chloroplast ultrastructure, and the activity of both photosystems. The study of signaling molecules that mediate Cd stress tolerance is an important step in better understanding how plants can acclimate to the adverse environment. Here we review the protection role and mechanisms of action of some plant growth regulators, such as brassinosteroids, salicylic acid, gibberellic acid, and gibberellin-insensitive DELLA proteins on the photosynthetic apparatus under Cd stress.

**Keywords:** Cadmium stress, Photosynthetic apparatus, Brassinosteroids, Salicylic acid, Gibberellic acid, DELLA proteins

## ABBREVIATIONS

| | |
|---|---|
| **24-EBL** | 24-epibrassinolide |
| **28-HBL** | 28-homobrassinolide |
| **BR** | brassinosteroid |
| **Cd** | cadmium |
| **GA** | gibberellic acid |
| **LHC** | light-harvesting complex |
| **OEC** | oxygen-evolving complex |
| **PSI** | photosystem I |
| **PSII** | photosystem II |
| **ROS** | reactive oxygen species |
| **SA** | salicylic acid |


https://doi.org/10.1016/B978-0-12-814864-8.00011-5

## 1 INTRODUCTION

Cadmium (Cd) represents one of the most toxic substances released into the environment, resulting in a serious human health hazard (Nordberg, 2009). About 70% of the Cd intake by humans originates from plant sources, as Cd is efficiently retained in the human body, and shows a long biological half-life of 10–30 years. Thus, a serious health threat can potentially arise, even from low-level chronic exposure. In the 1990s, Cd was classified as a group 1 carcinogen based on data obtained for workers occupationally exposed to Cd (Nordberg, 2009).

Accumulation of Cd in plant tissues is influenced by the levels of Cd available in the soil, as well as by the uptake of Cd from plant roots. Once transported to shoots and leaves, Cd ions cause alterations in the functionality of photosynthetic membranes by affecting the lipid composition, the chlorophyll metabolism, the chloroplast ultrastructure, and the activity of both photosystems. It should be noted that Cd-induced generation of reactive oxygen species (ROS) in thylakoid membranes can also lead to damage of the photosynthetic apparatus (Hasan et al., 2011). Cadmium inhibits both the "light" and "dark" reactions of the photosynthesis in higher plants, as the Calvin cycle is considered to be more sensitive to its activity (Krupa et al., 1993). Furthermore, it has been suggested that Cd initially destroys the photosynthetic carbon reduction cycle, and subsequently influences the photosynthetic electron transport and photosynthetic rate, $P_N$ (Burzyński and Źurec, 2007; Mobin and Khan, 2007; Ahammed et al., 2013).

Cd accumulation in leaves also leads to deleterious effects on the biosynthesis of chlorophyll (Wang and Zhou, 2006), the transpiration and respiration processes, and the stomatal opening (Hasan et al., 2009), which ultimately cause an inhibition of the photosynthetic rate (Krupa et al., 1993; Ci et al., 2010; Hasan et al., 2011). Similarly, Ahammed et al. (2013) have concluded that the Cd-induced inhibition of photosynthesis can be attributed to: (1) stomatal and nonstomatal limitation; (2) reduction in photosynthetic pigments, and (3) destruction of the photosynthetic apparatus.

The signaling molecules that mediate plant stress responses have attracted considerable attention in the past few years. The increasing research in this area aims to provide a better understanding of the protection mechanisms in plants, and to improve plant resistance to Cd stress. This chapter attempts to present current knowledge on the application of some exogenous plant growth regulators, such as brassinosteroids (BRs), salicylic acid (SA), gibberellic acid (GA), and modified GA-insensitive DELLA proteins, to understand their defense effects on the photosynthetic apparatus of plants under Cd stress.

## 2 ACCUMULATION OF CADMIUM IN HIGHER PLANTS

The bioavailability of soil Cd is determined by the concentration and speciation of Cd in soil solution. The most available to plants is free Cd ions ($Cd^{2+}$) (see in Grant et al., 1998). Soil solutions that contain Cd concentrations ranging from 0.04 to 0.32 μM are considered nonpolluted (Wanger, 1993). The uptake of Cd by higher plants depends on its concentration and its bioavailability, modulated by the temperature, the redox potential, the presence of organic matter, and the concentrations of other components (Toppi and Gabbrielli, 1999). It has been shown that the pH of soil influences the Cd phytoavailability, by influencing the distribution of Cd between the soil and the solute (Naidu et al., 1994). The sorption capacity of soil is another factor influencing the amount of Cd in the soil solution. Displacement of Cd from sorption sites in the soil by competing ions in the soil solution leads to an increase of its solution concentration (Grant et al., 1998). The absorption of Cd by plants also depends on size and uptake characteristics of the plant root system (Mullins et al., 1986).

Increased amounts of Cd lead to inhibition of the nitrate reductase activity in shoots, and so decrease the nitrate absorption (Hernandez et al., 1996). In addition, Cd inhibits the induction of root Fe(III) reductase and leads to Fe(II) deficiency, which in turn influences the photosynthesis (Alcántara et al., 1994). Plants' exposure to toxic concentrations of Cd leads to the synthesis of Cd-binding polypeptides, phytochelatins. They are believed to function as detoxifying agents by their ability to bind Cd (see in Grant et al., 1998). Synthesis of phytochelatins at low Cd concentrations has also been shown, but their binding to Cd has not been demonstrated (Tukendorf and Rauser, 1990; Ahner et al., 1995). The role of phytochelatins in the Cd accumulation in plants exposed to low Cd levels remains to be clarified.

Accumulation and distribution of Cd in plants vary among plant species, and among cultivars within a species (Kuboi et al., 1986; Alexander et al., 2006; Wang et al., 2013b; Dobrikova et al., 2017; Xin et al., 2017). Previously, Kuboi et al. (1986) divided the plants into three groups, depending on the amount of Cd that can accumulate: low accumulators, which include Leguminosae; moderate accumulators, which include

Gramineae, Liliaceae, Cucurbitaceae, and Umbelliferae; and high accumulators, which include Chenopodiaceae, Cruciferae, Solanaceae, and Compositae. The emerging molecular understanding of Cd uptake, and retention in root and root-to-shoot translocation will also enable the development of low Cd-accumulating crops (Clemens et al., 2013). The relationship between the roots and heavy metals is built basically through two strategies: exclusion and accumulation. Exclusion, or avoidance, leads to limitation of Cd uptake. Recent studies on Cd accumulating plants have led to important insights into the processes controlling the passage of Cd from the soil to edible plant organs, as plant tolerance mechanisms involve accumulation and storing of Cd by its binding to amino acids, proteins, and peptides (Pál et al., 2006; Xin et al., 2017). Previously, it has been found that plant hyperaccumulators can accumulate and tolerate, without visible symptoms, exceptionally high concentrations of heavy metals in their tissues (Reeves and Baker, 2000). These plant species, defined as "hyperaccumulators," usually have: (1) a bioaccumulation factor index (BF, the ratio of metal concentration in plants to that in soil) >1.0, sometimes reaching 50–100, and (2) a translocation factor index (TF, the ratio of metal concentration in shoots to that in roots) >1.0, which is used to measure the effectiveness of a plant in translocating a metal from its roots to its shoots. BF and TF values lower than 1 indicate that plants have an excluding effect on heavy metal uptake.

Cadmium is a mobile element, easily absorbed by the roots, and transported to the shoots, negatively affecting plant growth, as the expression of metal ion transporter genes correlates with the Cd concentration (Oono et al., 2016). It has been shown that in the case of poorly mobile elements, their levels in plants' organs decrease in the following order: roots > shoots > fruits > seeds (Sękara et al., 2005). It has been shown that red beets, pumpkin, chicory, and alfalfa accumulate more Cd in leaves compared with the roots, while common beans, wheat, barley, and maize accumulate more Cd in the roots than in the leaves (Sękara et al., 2005; Lysenko et al., 2015; Dobrikova et al., 2017). Furthermore, the plants have no similarity in Cd accumulation in leaves and in chloroplasts. A recent study by Lysenko et al. (2015) showed that maize plants accumulate more Cd in their leaves as compared with barley plants, whereas in chloroplasts, the Cd accumulation was higher in barley than in maize. Finally, the translocation of Cd from the roots to the shoots and leaves, and its accumulation in chloroplasts, will disturb photosynthesis.

# 3 IMPACT OF CADMIUM ON THE PHOTOSYNTHETIC APPARATUS

In higher plants, Cd stress decreases biomass production and photosynthetic pigments, and induces oxidative stress and alters the antioxidant enzyme activities, depending on the plant species, and thus reduces the effectiveness of photosynthetic performance (Parmar et al., 2013; Tran and Popova, 2013; Arivazhagan and Sharavanan, 2015). The Cd toxicity directly influences the photosynthetic electron transport and net $CO_2$ assimilation, as well as indirectly decreasing the chlorophyll content (by indirect inhibition of chlorophyll biosynthesis), or damping the photosynthetic apparatus on different levels of its organization; the chloroplast fine structure, and lipid and protein composition of thylakoid membranes (Burzyński and Kłobus, 2004; Sárvári et al., 2008). As already discussed, the Cd-induced effects on plants depend on its applied concentration, duration of action, and plant species. For example, Ali et al. (2014) showed that the 6-day treatment of *Brassica napus* with Cd at lower concentrations (100 μM) stimulates the seedling growth, but at higher concentrations (500 μM), inhibits the growth, decreases the pigment content, and increases the content of ROS. Unlike *Brassica species,* which are heavy-metal tolerant plants, the treatment of ryegrass seedlings with a lower concentration (100 μM) of $CdCl_2$ for 2 weeks resulted in toxicity symptoms such as chlorosis and necrotic spots on leaves (Wang et al., 2013a).

The Cd-induced alterations in the functions of photosynthetic membranes by affecting the lipid composition, the pigment content, the chloroplast ultrastructure, and the activity of both photosystems are reviewed here. The impact of Cd ions on different components of the photosynthetic apparatus, located in thylakoid membranes, is summarized and presented in Fig. 1.

## 3.1 Pigments, Lipids, and Structural Changes

Leaf chlorosis is one of the common signs of Cd toxicity (Baryla et al., 2001; Liu et al., 2011). The leaf pigment loss has been proposed to result from a degradation and/or a reduction in the formation of photosynthetic pigments as the primary target of Cd action (Fagioni et al., 2009; Hasan et al., 2011; Ahammed et al., 2013). It has been suggested that the decrease of chlorophyll content is a result of the replacement of Mg ions with Cd ions in chlorophyll molecules (Parmar et al., 2013), as the percent conversion of chlorophyll to pheophytin increases with an increase in Cd concentration

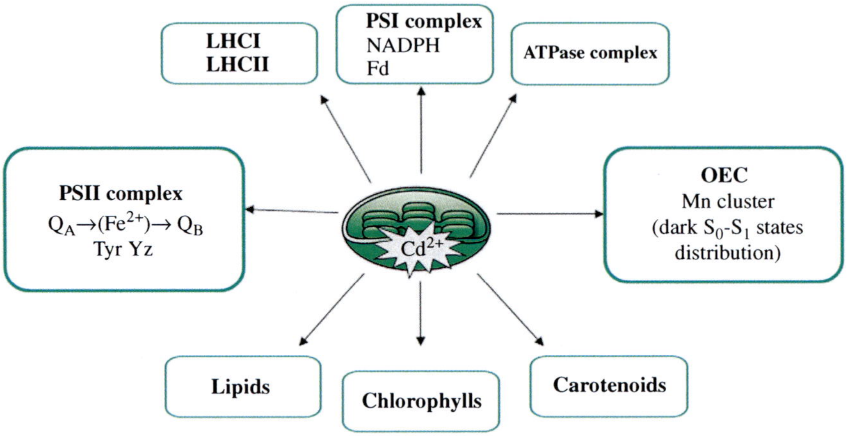


**Fig. 1** Schematic presentation of the target sites of Cd action in the thylakoid membranes of higher plants.

(Mobin and Khan, 2007). The inhibition of chlorophyll biosynthesis by interfering with protochlorophyllide production and activation of enzymatic degradation of chlorophylls by chlorophyllase plays a crucial role in the loss of chlorophyll content (Ci et al., 2010; Hasan et al., 2011; Ahammed et al., 2013). According to Baryla et al. (2001), the leaf chlorosis is due to a marked decrease in the chloroplast density, caused by a reduction in the number of chloroplasts per cell, and a change in cell size of oilseed rape plants. Subsequently, the decrease in chlorophyll content undoubtedly affects the photosynthetic efficiency. A previous study has shown that the reduction in both the chlorophyll content and the chlorophyll *a*/*b* ratio in long-term Cd treated plants is connected with the alteration in the ratio of chlorophyll-protein complexes, together with the reorganization of thylakoid membranes (Sárvári et al., 2008). The effect of Cd stress on the carotenoid content also strongly depends on Cd concentrations and plant species. The increase in carotenoid content has been observed in *Cucumis sativa* and *Zea mays*, while in *Pisum sativum* these pigments decreased under Cd stress (see Parmar et al., 2013). Generally, it has been found that carotenoids are less affected by Cd than chlorophylls, thus the carotenoid to chlorophyll ratio is higher in the Cd treated plants (López-Millán et al., 2009; Lysenko et al., 2015; Dobrikova et al., 2017 and refs. therein).

Furthermore, it has also been suggested that the treatment of *Brassica napus* with 50 μM Cd for a long period of time (15 days) drastically decreases the amounts of chloroplast membrane lipids: monogalactosyldiacylglycerol (MGDG), digalactosyldiacylglycerol (DGDG), sulfolipids (SL), and phosphatidylglycerol (PG) (Nouairi et al., 2006). The authors have also shown a decrease in the levels of polyunsaturated fatty acids, particularly linolenic acid (C18:3) and hexadecatrienoic acid (C16:3), as well as a decline of the *trans* palmitoleic acid (C16:1-*trans*) of PG, which regulates the oligomeric organization of the light-harvesting complex of PSII (LHCII). In barley plants, the observed increase in ethylene production, as well as the specific reduction of C18:3 fatty acid content, most probably resulted from Cd-induced lipid peroxidation in the thylakoid membranes, leading to their disruption (Vassilev et al., 2004). Similarly, studies on tomato plants also showed that the high Cd concentrations after a 10-day exposure induce a decrease in the content of polar acyl lipids in chloroplast membranes, especially MGDG, DGDG, and PG, accompanied by changes in their composition and fatty acid content, resulting in a lower degree of fatty acid unsaturation (Djebali et al., 2005). Moreover, a greater decrease in MGDG content than that of DGDG in Cd-stressed tomato chloroplasts has also been observed, resulting in a decline of the MGDG/DGDG ratio. It is important to note that MGDG and C16:1-*trans* are considered to be indispensable for PSII activity because they regulate the supramolecular structure of PSII complexes (Djebali et al., 2005). Considering all this, it could be concluded that the Cd toxicity is associated with structural changes in chloroplast membranes. The data from ultrastructural investigations have shown that Cd induces disorganization in leaf structure, as well as severe alterations in the chloroplast fine structure, which lead to a disturbed shape and dilation of thylakoid membranes (Djebali et al., 2005; Zhu et al., 2005; Gratão et al., 2009). Moreover, it has been shown that higher Cd concentrations decrease the number and the size of chloroplasts, as well as the stacking degree of thylakoid membranes, causing thylakoid swelling (Hakmaoui et al. 2007; Moussa and El-Gamal, 2010). Similarly, the accumulation of Cd in maize plants also led to a disorganization of grana membranes, and an increase in the number and the size of plastoglobuli in chloroplasts, as induced vesiculation (Rascio et al., 1993). Furthermore, it has been found that Cd exposure causes disorganization of the trimeric structure of LHCII, resulting in diminished numbers of aggregated LHCII complexes, and also causes disorganization of the PSII supramolecular structure in the thylakoid membranes (Janik et al., 2010). Further, Basa et al. (2014) have also shown marked changes in the organization of photosynthetic apparatus, including a decrease in the amount of the PSII supercomplexes, and an increase in the monomeric form of the LHCII antennae under Cd stress. All these reported Cd-induced structural changes, destacking of the grana thylakoids, and rearrangement of the complexes in the thylakoid membranes have been accompanied by an increase in the energy transfer from PSII to PSI, as well as an increase of the absorption cross-section of the PSI antenna (Dobrikova et al., 2017).

## 3.2 Functional Activity of Photosystems I and II

The Cd-induced inhibition of photochemical processes, such as down-regulation of PSII photochemistry and linear electron transport, has been suggested to be due to the limitation in the use of ATP and NADPH by the Calvin cycle and the accompanying increase of the pH gradient across the thylakoid membranes (Krupa et al., 1993). Furthermore, it has been found that Cd stress induces changes in both photosystems (Atal et al., 1991; Wodala et al., 2012), depending on the plant species and applied Cd concentrations (Prasad, 1995; Parmar et al., 2013). In addition, it has also been demonstrated that PSII-mediated electron transport is more affected by Cd than PSI-mediated electron transport (Atal et al., 1991; Sigfridsson et al., 2004; Vassilev et al., 2004; Wang et al., 2013b; Dobrikova et al., 2017). Previous study of the photosynthetic processes in cucumbers has demonstrated that the effects of low Cd concentrations (20 and 50 μM) after a short period of exposure (48 h) on the maximal efficiency of PSII photochemistry (Fv/Fm), the photochemical (qP) and nonphotochemical quenching (qN) were insignificant (Burzyński and Źurec, 2007). On the other hand, investigations with higher Cd concentration and/or longer periods of treatment have shown a decrease of the chlorophyll fluorescence parameters: Fv/Fm, qP and the quantum efficiency of PSII ($\Phi_{PSII}$) in different plant species subjected to Cd stress (López-Millán et al., 2009; Ci et al., 2010; Liu et al., 2011; Ahammed et al., 2013). The decrease in the ratio Fv/Fm indicates the negative effects of heavy metals on the photochemical reactions, which may affect the efficiency of PSII photochemistry by blocking electron transport (Ahammed et al., 2013). Liu et al. (2011) showed that the treatment with 50 μM $CdCl_2$ for a long period of time (12 days) significantly decreases the photosynthetic rate ($P_N$), the maximal efficiency of PSII photochemistry, and the effective quantum yield of PSII in castor beans.

Moreover, the action of Cd ions on the photosynthetic apparatus influences both the donor and the acceptor sides of the PSII complex, which leads to an increase in the amount of the inactive PSII centers (Atal et al., 1991;

Dobrikova et al., 2017). It has been proposed that the Cd competitive binding to the essential $Ca^{2+}$ cofactor in the $Mn_4Ca$ cluster of OEC leads to inhibition of the oxygen evolution (Faller et al., 2005). Sigfridsson et al. (2004) have also suggested that Cd rapidly inhibits the functional activity of PSII on the donor side between OEC and the first electron donor of P680 (Tyr $Y_Z$) via two mechanisms: it blocks the oxidation of $Y_Z$ in a $Ca^{2+}$-independent manner, and blocks the OEC in a $Ca^{2+}$-dependent manner. Detailed analysis of the flash-induced oxygen yields, obtained with a Joliot-type rate electrode using an extended version of Kok's model, has indicated that Cd causes modifications of the $Mn_4Ca$ cluster of the OEC, resulting in a higher number of PSII centers in the initial $S_0$ state, as well as in the blocked $S_B$ state, accompanied with an increase of misses and double hits (Dobrikova et al., 2017). An earlier study by Zhu et al. (2005) revealed that decreased oxygen evolution in response to Cd treatment is not attributed to the simple closing of the stomata, but to a lower photosynthetic yield. In addition, it has been found that Cd causes some modifications in the secondary quinone acceptor, $Q_B$, on the PSII acceptor side, thus delaying the electron flow from $Q_A$ to $Q_B$ (Faller et al., 2005), and modifying the conformation of the $Q_B$ pocket (see Sigfridsson et al., 2004).

However, in some plants exposed to Cd stress, PSI is more sensitive than PSII (Sárvári et al., 2008; Fagioni et al., 2009), as the Cd-induced inhibition of PSI was supposed to be a result of Cd interference with electron transport on the PSI reducing side, as well as a significant reduction of antenna pigment-proteins of PSI, LHCI. Recent investigations into wheat and pea plants showed that Cd stress affects PSI photochemistry (Wodala et al., 2012; Dobrikova et al., 2017), as indicated by a decrease in the quantum efficiency of PSI ($\Phi_{PSI}$) and the number of electrons in the intersystem chain (Wodala et al., 2012). The dark relaxation kinetics of $P700^+$ observed in wheat plants exposed to 100 μM $CdCl_2$ for 10 days revealed that Cd influences the two subpopulations of PSI, rapidly and slow operated PSI centers, located in the stroma lamellae and the grana margins, respectively (Dobrikova et al., 2017). On the other hand, it has been suggested that Cd-induced iron deficiency is another possible reason for damage of PSI (Sárvári et al., 2008), because the deficiency of the iron leads to generation of ROS in thylakoid membranes, which destroys iron-sulfur centers of PSI (Michel and Pistorius, 2004).

It has also been found that the different concentrations of Cd lead to activation of different oxidative responses, depending on plant species (Hoseini and Zargari, 2013). In addition, the same authors also showed the bigger

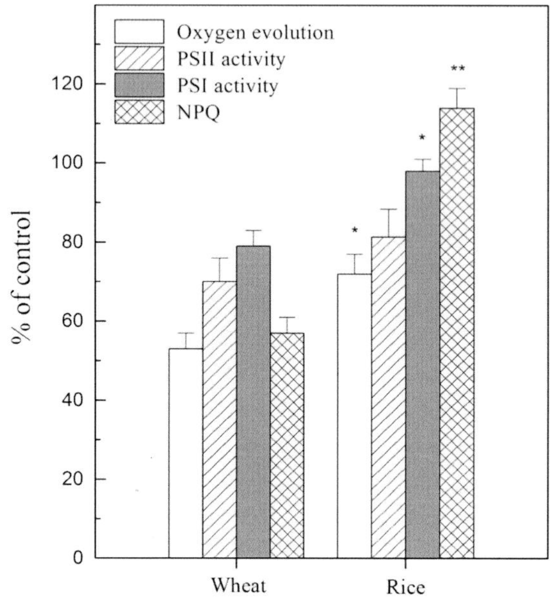


**Fig. 2** Comparison of the Cd-induced effects on: (1) the amplitude of oxygen evolution (measured with a Joliot-type electrode without artificial acceptor); (2) the photochemical activities of PSII ($H_2O \rightarrow BQ$) and PSI ($DCPIPH_2 \rightarrow MV$) and (3) nonphotochemical quenching (NPQ) in thylakoid membranes isolated from wheat and rice plants treated with 100 µM $CdCl_2$ for 10 days. The values are represented as a percentage of the untreated controls and the statistical differences are given between the values of the same parameters (*$P < .05$, **$P < .01$).

sensitivity of wheat plants in comparison with rice plants in regard to Cd-induced stress (Hoseini and Zargari, 2013). These findings are confirmed by our observations on the different sensitivity of the photosynthetic apparatus of wheat and rice plants to Cd toxicity (Fig. 2). The obtained results revealed that the treatment with 100 µM $CdCl_2$ for 10 days causes less prominent inhibition of the oxygen evolution and PSII-mediated electron transport in rice thylakoid membranes than in wheat. Moreover, the PSI-mediated electron transport is almost unchanged in the thylakoid membranes from rice leaves, while in wheat leaves, PSI activity decreases substantially after Cd exposure (Fig. 2). The most pronounced differences between rice and wheat plants are observed in the nonphotochemical quenching parameters (NPQ and $\Phi_{NO}$), as the NPQ values in rice leaves are slightly increased after the Cd treatment, whereas in wheat they are significantly reduced, more than that of the controls, which correlates with a stronger increase (by about 150%) of nonregulated energy dissipation in PSII

($\Phi_{NO}$, data not shown). Previously, Lysenko et al. (2015) demonstrated a similar set of differences in NPQ values between barley and maize plants subjected to Cd stress. They found that the parameter NPQ decreased exponentially with the increase of Cd concentration in barley plants, due to a decline in the fast relaxing component of NPQ (qE, pH-dependent energy dissipation). Conversely, in maize plants, neither NPQ nor its component qE were changed significantly (Lysenko et al., 2015), as maize is considered to be a heavy metal-tolerant plant (see Moravcová et al., 2018).

A moderate inhibition of PSII, and especially PSI photochemistry, ultimately results in a reduced number of electrons for $CO_2$ fixation and other assimilation processes, which probably contributes to inhibition of growth and other heavy metal stress-induced symptoms.

## 4 PROTECTION ROLE OF SIGNALING MOLECULES ON THE PHOTOSYNTHETIC APPARATUS UNDER CADMIUM STRESS

Signaling molecules (hormones, growth regulators, proteins, amino acids, nucleotides, etc.) are essential for the growth, development, and adaptation of plants, as well as for the activation of their antioxidant responses to a number of environmental stress factors, including heavy metals (see Gururani et al., 2015). Some tolerance mechanisms that plants have developed to cope with Cd toxicity are related to plant growth regulators, such as salicylic acid (SA), brassinosteroids (BRs), gibberellic acid (GA), jasmonic acid (JA), abscisic acid (ABA), auxins, and so forth. Here we review the recent studies on the defense role and the mechanisms of action of some exogenously added plant growth regulators, such as BRs, SA, GA, as well as GA-insensitive DELLA proteins, on the photosynthetic apparatus of higher plants under Cd stress.

### 4.1 Brassinosteroids

Brassinosteroids (BRs) are a class of plant steroid hormones involved in several physiological and morphological processes, as well as in plant responses to environmental stresses via activation of different protective mechanisms (reviewed in Vardhini, 2017). The exogenous application of BRs has been studied extensively in various crop plants under normal and abiotic stress conditions. It has been reported that BRs also minimize the toxic effects and symptoms generated by an excess of heavy metals, thus inducing plant protection against Cd stress (Ahammed et al., 2013). Moreover, many studies have demonstrated that the exogenously applied BRs enhance plant tolerance to heavy metal stress, and can be used to reduce their accumulation in plants (Ahammed et al., 2013; Fariduddin et al., 2014; Vardhini, 2015). Most likely, the growth-promoting activity of BRs may be associated with improved photosynthesis. Several studies have shown that the exogenous application of very low concentrations of 24-epibrassinolide (24-EBL, a biologically active BR) influences not only the plant growth and development processes, but also photosynthetic performance, as well 0.1 $mg L^{-1}$ 24-EBL has been found to be most effective concentration (Yu et al., 2004; Dobrikova et al., 2014 and refs. therein). Yu et al. (2004) examined the effect of 24-EBL spray application on the photosynthesis, and found an increase in the quantum yield of PSII electron transport, higher Fv/Fm, and increased Rubisco activity in *Cucumis sativus*. Our recent study (Dobrikova et al., 2014) reported that the foliar application of low concentrations of 24-EBL (0.1 $mg L^{-1}$) on pea plants improved the PSII activity, showing by stimulation of the photosynthetic parameters such as $\Phi_{PSII}$, qP, the PSII electron-transport rate and the oxygen evolution. These data also suggested that the defense action of 24-EBL on the photosynthetic apparatus could be due to changes in the structural organization of photosynthetic membranes, as well as altered kinetics of oxygen evolution, decreased ratio of functionally active PSII$\alpha$ to PSII$\beta$ centers, and an increased amount of PSII centers in the most reduced $S_0$ state, because these changes are prerequisites for plants' adaptation to stress environmental factors (Dobrikova et al., 2014).

Further, it has been reported that the interactive effects of Cd exposure and exogenous application of 24-EBL in tomato plants resulted in slightly improved photosynthetic activity and pigment content, compared with Cd treatment alone for 40 days (Ahammed et al., 2013). Moreover, these results suggested that the exogenous 24-EBL application protects plants against Cd-induced ROS accumulation and membrane lipid peroxidation. It has also been shown that the foliar application of 0.1 μM 24-EBL to Cd-exposed plants alleviated the photosynthetic inhibition and the oxidative stress by causing an enhancement in the activity of enzymes and related transcript levels of the antioxidant system, the secondary metabolism, and the xenobiotic detoxification system (Ahammed et al., 2013). Furthermore, after the 24-EBL application, the Cd accumulation in both leaves and roots was significantly reduced. These findings thus suggest an unambiguous role of 24-EBL in the alleviation of Cd stress.

Importantly, the 24-EBL application on heavy metal-exposed plants can significantly increase the chlorophyll content by stimulating the transcription of genes implicated in the biosynthesis of pigments, thus improving leaf elongation and the photosynthetic rate (Anuradha and Rao, 2009; Bajguz, 2011). Janeczko et al. (2005) showed that 24-EBL (0.1 μM) added to a nutrient solution alleviates Cd-induced inhibition of the primary photochemistry of PSII in rape seedlings by diminishing the damage of PSII reaction centers and oxygen-evolving centers, as well as by maintaining efficient photosynthetic electron transport. The ameliorating effect of 1 and 2 μM 24-EBL (seed soaking) against Cd toxicity on the photosynthetic pigment, the net photosynthetic rate, and enzyme activities of radish plants have also been observed by Anuradha and Rao (2009). Furthermore, it has been reported that the foliar application of 24-EBL leads to increasing the pigment content, the maximum photochemical efficiency of PSII (Fv/Fm), the quantum efficiency of PSII ($\Phi_{PSII}$), and the photochemical quenching coefficient (qP) of Cd-treated plants (Ahammed et al., 2013). The same authors also showed positive interactions of 24-EBL and phenanthrene when applied together for Cd stress alleviation. Additionally, the foliar 24-EBL application (0.1 μM) on cucumber plants induces transcripts for photosynthetic genes, and activates enzymes involved in the Calvin cycle, and thus improves $CO_2$ assimilation and photosynthetic electron transport (Jiang et al., 2012). Gao et al. (2013) have reported that the exogenous foliar spraying of 24-EBL enhanced plant growth and photosynthesis, and reduced Cd biological enrichment in *Helianthus tuberosus* seedlings subjected to Cd stress.

Foliage application of other BR, 28-HBL (0.01 μM) to Cd-treated *Brassica juncea* seedlings also showed decreased accumulation of Cd, increased plant growth, chlorophyll content, net photosynthetic rate, as well as antioxidant enzymes and proline content (Hayat et al., 2007). Other investigations have shown that the foliar spray of two tomato cultivars with two BRs (28-HBL or 24-EBL) were very effective in neutralizing the adverse effects generated by Cd that reflect better photosynthetic performance (i.e., $P_N$, $g_s$ and internal $CO_2$ concentration) (Hayat et al., 2010a). Similarly, Hasan et al. (2011) also reported that foliar spraying of 28-HBL or 24-EBL (0.01 μM) to two tomato cultivars subjected to Cd stress not only improved the activity of photosynthetic machinery and antioxidant defense system in both the cultivars, but also nullified the damaging effect of the metal on these parameters.

Recently, Jakubowska and Janicka (2017) have explored in more detail the mechanisms of action of BRs to define their role in plant adaptation to Cd stress. The authors showed a stimulating effect of exogenous BRs treatment (24-EBL in nutrient medium) on the activity of two plasma membrane enzymes, $H^+$-ATPase and NADPH oxidase, and proposed their key role in BR-induced Cd stress tolerance. Also, the relative expression of genes encoding both enzymes was affected by BRs. These findings contribute to knowledge about plant responses triggered in plants under Cd stress, as well as suggest that functional BR signaling and synthesis are needed for normal accumulation of a proton pump under Cd stress conditions as one of the plants' adaptation mechanisms.

## 4.2 Protection Role and Mechanisms of Action of Salicylic Acid

Salicylic acid (SA) is considered a hormone-like plant growth regulator with a phenolic nature, which is involved in the regulation of various physiological processes, as well as some defense mechanisms in plants (Pancheva et al., 1996; Rivas-San Vicente and Plasencia, 2011; Khan et al., 2015; Moravcová et al., 2018). Exogenous application of SA influences seed germination and fruit yield (Cutt and Klessing, 1992; Moravcová et al., 2018), stomatal closure (Larqué-Saavedra, 1979), membrane permeability (Barkosky and Einhellig, 1993), and plant growth and photosynthesis (Khan et al., 2003; Maslenkova et al., 2009). Accumulating evidence indicates that SA plays a key role in the protection of the photosynthetic apparatus of plants that are exposed to environmental stress (for review see Gururani et al., 2015).

The effects of exogenous SA treatment depend on numerous factors, such as the species and the developmental stage of plants, the mode of application, and its concentration (Horváth et al., 2007; Maslenkova et al., 2009). The concentration of exogenous SA is very important in the regulation of plant growth and photosynthesis under nonstress conditions, as the higher concentrations of SA can impose negative effects (Pancheva et al., 1996; Maslenkova et al. 2009; Singh et al., 2010; Yotsova et al., 2018). Moravcová et al. (2018) have indicated that the effect of exogenously applied SA is connected with the changes in the endogenous SA content of germinated maize seeds and developing roots. Moreover, it has also been shown that seeds pretreated with SA in lower concentrations (50 μM) increased the proportion of open PSII reaction centers in unstressed maize plants, as demonstrated by the increase in photochemical quenching of the variable chlorophyll fluorescence (qP). A lower concentration of SA (10 μM) improves the photosynthetic net $CO_2$ assimilation in mustard seedlings, which is accompanied with up-regulation of carboxylation efficiency, Rubisco activity, and chlorophyll content (Rivas-San Vicente and Plasencia, 2011). Our recent study revealed that the exogenous application of SA in low concentrations (10 μM) through the rooting medium of rice seedlings led to a stimulation of the plant growth, an increase of the chlorophyll content in leaves, as well as a stimulation of the photochemical activities of both photosystems and the cyclic electron flow around the PSI (Yotsova et al., 2018). In addition, it has been found that these changes are not accompanied with changes in the energy transfer between pigment-protein complexes in the photosynthetic membranes. Janda et al. (2012) have suggested an indirect effect of SA on the photosynthetic electron transport processes in tobacco plants, as they observed different effects of exogenous SA application (0.5–2 mM) on PSII charge separation and stabilization in isolated thylakoid membranes and intact leaves. Higher concentrations of SA (1–5 mM) lead to an inhibition of the photosynthetic rate (Pancheva et al., 1996) and to reduced chlorophyll content in cowpeas, wheat, and *Arabidopsis* (Rivas-San Vicente and Plasencia, 2011). On the other hand, Singh et al. (2010) have shown that low concentrations of SA (up to 50 μM) induce an enhancement in the growth characteristics and the chlorophyll content in cucumbers, while higher SA concentrations decrease these parameters. It has also been demonstrated that concentrations of exogenous SA higher than 500 μM applied through the rooting medium decrease the number of the active PSII centers and disturb the redox cycling of the rest of the operative oxygen-evolving centers, which is the result of structural changes in the catalytic Mn clusters of OEC at the protein level (Maslenkova et al. 2009 and refs. therein). In addition, the strong depletion of 33 kDa polypeptide of OEC has been shown, which is assumed to function as an Mn-binding protein. The changes in the membrane composition are accompanied with changes in the structure of granal thylakoid membranes (Usunova and Popova, 2000), which are accompanied with variations in the chlorophyll *a*/*b* ratio (Arfan et al., 2007). On the contrary, when barley plants were sprayed with 500 μM SA, the changes in the net $CO_2$ assimilation rate and the photochemistry of PSII did not register (Habibi, 2012).

Many investigations revealed the protective role of SA against environment stress conditions, including heavy metal toxicity (Metwally et al., 2003; Gunes et al., 2005; Arfan et al., 2007; Guo et al., 2007; Horváth et al., 2007; Hayat et al., 2010b; Mishra and Choudhuri, 1999; Liu et al., 2011; Khan et al., 2015; Moravcová et al., 2018). Salicylic acid has long been found as one of the growth regulators that reduces Cd toxicity in plants, as it has been reported that the SA application decreases the growth inhibition induced by Cd toxicity in barley (Metwally et al., 2003), soybeans (Drazic and Mihailovic, 2005), rice (Guo et al., 2007), maize (Krantev et al., 2008), wheat (Moussa and El-Gamal, 2010), and castor bean seedlings (Liu et al., 2011). The SA may indirectly alleviate Cd toxicity through different response mechanisms in plants, some of which include the regulation of the antioxidant machinery and lipid metabolism, leading to maintenance of membrane integrity and protection of membranes from degradation (Mishra and Choudhuri, 1999; Horváth et al., 2007; Krantev et al., 2008; Moussa and El-Gamal, 2010). It has been suggested that SA takes part in the signaling pathway of various antioxidant enzymes involved in ROS-scavenging, thus protecting plants from oxidative damage (Yang et al., 2004; Guo et al., 2007; Krantev et al., 2008; Habibi, 2012; Moravcová et al., 2018 and refs. therein). It has been shown that the presoaking of maize seeds with 500 μM SA before Cd stress increases antioxidant enzyme activities and diminishes the oxidative damage caused by Cd (Krantev et al., 2008). The authors also revealed an alleviation of the negative effect of Cd on the rate of $CO_2$ fixation and Rubisco activity. Therefore, the short-term treatment with SA (presoaking of seeds for 6 h) also decreased Cd accumulation in the shoots and the roots, alleviated the inhibitory effects of Cd on plant growth, chlorophyll levels and the photosynthesis in maize seedlings (Krantev et al., 2008). Similarly, Moussa and El-Gamal (2010) have shown that pretreatment of wheat seeds with SA (500 μM) for 20 h resulted in alleviating the Cd effect on chloroplast ultrastructure, chlorophyll content, and $CO_2$ fixation. Obtained results by Guo et al. (2007) have revealed that the pretreatment of rice plants with 10 μM SA (for 24 h) enhances the activities of antioxidant enzymes, as well as nonenzymatic antioxidants in roots, thus lowering the lipid peroxidation in Cd stressed plants. Studying different plant species, it has been suggested that the enhanced antioxidant capacity is the reason for SA-enhanced tolerance toward most kinds of abiotic stresses (Rivas-San Vicente and Plasencia, 2011).

## 4.3 Gibberellic Acid

Bioactive gibberellins, or gibberellic acid (GA), are plant hormones that control the growth and development throughout the life cycle of plants, including shoot and root growth, leaf morphogenesis, germination dormancy, seed production, and flowering (see Marzec, 2017). A central role of GA signaling in plant responses to abiotic stress is becoming increasingly evident (Colebrook et al., 2014). Reduction of GA levels and signaling has been shown to contribute to plant growth restriction under exposure to several stresses. Moreover, GA is also involved in plant adaptation to Cd stress by decreasing the negative effects of Cd on the growth parameters and photosynthesis (Ghorbanli et al., 2000; Zhu et al., 2012). Ghorbanli et al. (2000)

have shown that the exogenous addition of GA (10 mg $m^{-3}$) in the nutrient solution reduced the negative effects of Cd ions on the growth rate (of roots, shoots, leaves), the chlorophyll content, and the photosynthetic performance of soybean plants exposed to Cd stress. Another recent study (Zhu et al., 2012) has found that under Cd exposure, the application of 5 μM GA improves the root growth, as it reduces the Cd content and the lipid peroxidation in roots, indicating that GA can partially alleviate Cd toxicity. In addition, it has also been demonstrated that GA remarkably suppresses the up-regulation of the expression of iron-regulated transporter 1 (*IRT1*) in *Arabidopsis thaliana* (a transporter essential for iron uptake from the soil and for plant growth), which is involved in Cd absorption. These results showed that GA-alleviated Cd toxicity is mediated through the reduction of the Cd-dependent NO accumulation and expression of $Cd^{2+}$ uptake related gene *IRT1* in *Arabidopsis* (Zhu et al., 2012).

There is still less understanding of the relationship between GA signaling and Cd-induced inhibition of the leaf photosynthesis, beyond the well-established role of GA as a growth regulator, and the finding that it regulates the levels of ROS (Achard et al., 2008). Investigations on the influence of GA on photosynthesis (Yuan and Xu, 2001) have shown that the short-term application of GA in soybeans and broad bean plants results in increased net photosynthetic rate, $P_N$, increased stomatal conductance, improved photosynthetic oxygen evolution, and increased carboxylation efficiency, which is related to the increase in Rubisco content. Some researchers have reported the improved stress tolerance in plants exposed to heavy metal toxicity through increased photosynthetic activity with the application of GA (Ghorbanli et al., 2000; Ouzounidou and Ilias, 2005). The protective role of GA in sunflower plants exposed to copper (Cu) stress has been established by a detailed analysis of major photosynthetic parameters revealing that the GA treatment increases energy trapping in PSII reaction centers and the stability of the LHCII complex, as well as increasing the parameters $P_N$ and Fv/Fm (Ouzounidou and Ilias, 2005). Another study by Ghorbanli et al. (2000) has shown that the addition of GA to a nutrient solution of *Glycine max* plants exposed to Cd stress alleviates the growth-inhibitory effects of Cd toxicity, which has been evident by improved plant growth rate, increased chlorophyll content, and net $CO_2$ assimilation rate, indicating the role of GA in improved photosynthesis. In addition, Cd toxicity is considered to be associated with altered ultrastructural changes in thylakoids (Gratão et al., 2009), but there is no evidence to date whether GA alleviates changes in the architecture of chloroplasts under Cd stress.

## 4.4 DELLA Proteins

DELLA proteins are the negative growth regulators in the GA signaling pathway that are destabilized by bioactive GA (Wen et al., 2013). Reduced GA synthesis, or GA sensitivity, inhibits the DELLA degradation pathway, and this promotes the accumulation of DELLA proteins, contributed to a spectacular increase in the yield of rice (*Oryza sativa*) and wheat (*Triticum aestivum*) during the green revolution in the 1960s (for review see Van de Velde et al., 2017). DELLA proteins may also mediate interaction between GA and ABA pathways because one of their targets regulates ABA metabolism. Moreover, the DELLA accumulated proteins suppress plant growth at the cellular, organ, and whole-plant level, resulting in dwarf or semi-dwarf plants. In rice, semi-dwarf 1 (sd1) mutant plants contain a defective GA biosynthesis gene, whereas wheat mutants have reduced height genes, (*Rht1* and *Rht2*), encoding-modified DELLA proteins, which have a reduced affinity for the GA receptor, GID1, resulting in increased DELLAs accumulation compared with the wild type plants (Wen et al., 2013; Van de Velde et al., 2017). DELLA proteins have also been found to modulate the antioxidant defense system as a way of increasing plant tolerance to various abiotic stresses (Achard et al., 2008; Robert-Seilaniantz et al., 2010; Wen et al., 2013; Colebrook et al., 2014). In particular, the GA-insensitive DELLA proteins significantly increase plant stress tolerance by activating the antioxidant defense system, and thus reducing the levels of ROS (Achard et al., 2008; Robert-Seilaniantz et al., 2010).

Our recent study (Dobrikova et al., 2017) has demonstrated, for the first time, a role of the *Rht-B1c*-encoded DELLA proteins in protective mechanisms and tolerance of the photosynthetic apparatus in wheat plants exposed to Cd stress. Compared to the tall wild type, the dwarf Rht-B1c mutant has been found to be characterized by less pigment reduction, less inhibited photosynthesis, and better function of the photosynthetic apparatus (i.e., PSI and PSII activity and oxygen evolution) under Cd stress, despite the higher Cd accumulation in its shoots. It has also been suggested that the DELLA proteins affect plant tolerance by structural reorganizations in the photosynthetic membranes, leading to alterations on the PSII donor side (i.e., modification of the Mn cluster of the OEC), as well as to an increase of the rapidly operated PSI centers (i.e., those operated by cyclic electron transport) located in the stroma lamellae (Dobrikova et al., 2017). Furthermore, the PSI-dependent cyclic electron flow has been suggested to play a significant role in preventing oxidative damage to the photosynthetic apparatus during exposure of plants to Cd and Cu stress (Wodala et al., 2012).

Therefore, the enhanced capacity for cyclic electron transport observed in the DELLA mutants in comparison with the wild type plants is probably one of the protective mechanisms by which these plants alleviate oxidative damage under Cd stress.

To our knowledge, these results are so far the only evidence for the involvement of the DELLA proteins in the process of plant adaptation to Cd stress (Dobrikova et al., 2017), because other studies are conducted on other abiotic stresses, such as drought and salinity. Therefore, further research is necessary to provide a detailed understanding of the role of DELLA proteins in the regulation of activities of photosynthetic enzymes and hormone homeostasis in crop plants. Understanding the DELLA-regulatory mechanisms in crop plants under heavy metal stress holds great potential to improve yield.

## 5 CONCLUSION

Cadmium affects photosynthesis either directly or indirectly, thus influencing plant growth and development. Cd toxicity influences the pigments, lipids, and protein composition of the photosynthetic apparatus, and inhibits the functions of both photosystems, as this effect is stronger on PSII than on PSI. The essential role in the adaptation of higher plants to the heavy metal stress plays the signaling molecules. This chapter summarizes the knowledge on the defense effects and underlying mechanisms of exogenously applied signaling molecules involved in the adaptive response on the photosynthetic apparatus to Cd stress. It is shown that these molecules decrease the Cd accumulation, regulate the levels of ROS production, as well as cause modifications in the photosynthetic apparatus associated with improved photosynthesis. The study of signaling molecules that mediate the heavy metal stress resistance is important to a better understanding of how plants can tolerate the adverse environment, and in particular, Cd toxicity.

## REFERENCES

Achard, P., Renou, J.-P., Berthomé, R., Harberd, N.P., Genschik, P., 2008. Plant DELLAs restrain growth and promote survival of adversity by reducing the levels of reactive oxygen species. Curr. Biol. 18, 656–660.

Ahammed, G.J., Choudhary, S.P., Chen, S., Xia, X., Shi, K., Zhou, Y., Yu, J., 2013. Role of brassinosteroids in alleviation of phenanthrene-cadmium co-contamination-induced photosynthetic inhibition and oxidative stress in tomato. J. Exp. Bot. 64 (1), 199–213.

Ahner, B.A., Kong, S., Morel, F.M.M., 1995. Phytochelatin production in marine algae. 1. An interspecies comparison. Limnol. Oceanogr. 40, 649–657.

Alcántara, E., Romera, F.J., Canëte, M., De La Guardia, M.D., 1994. Effects of heavy metals on both induction and function of root Fe(III) reductase in Fe-deficient cucumber (*Cucumis sativus* L.) plants. J. Exp. Bot. 45, 1893–1898.

Alexander, P.D., Alloway, B.J., Dourado, A.M., 2006. Genotipic variations in the accumulation of Cd, Cu, Pb and Zn exhibited by six commonly grown vegetables. Environ. Pollut. 144, 736–745.

Ali, B., Qian, P., Jin, R., Ali, S., Khan, M., Aziz, R., Tian, T., Zhou, W., 2014. Physiological and ultra-structural changes in *Brassica napus* seedlings induced by cadmium stress. Biol. Plant. 58 (1), 131–138.

Anuradha, S., Rao, S.S.R., 2009. Effect of 24-epibrassinolide on the photosynthetic activity of radish plants under cadmium stress. Photosynthetica 47, 317–320.

Arfan, M., Athar, H.R., Ashraf, M., 2007. Does exogenous application of salicylic acid through the rooting medium modulate growth and photosynthetic capacity in two differently adapted spring wheat cultivars under salt stress. J. Plant Physiol. 164, 685–694.

Arivazhagan, V., Sharavanan, P.S., 2015. Effect of cadmium on photosynthetic responses and biochemical contents of maize plants. Am. J. Environ. Eng. Sci. 2, 32–36.

Atal, N., Saradhi, P.P., Mohanty, P., 1991. Inhibition of the chloroplast photochemical reactions by treatment of wheat seedlings with low concentrations of cadmium: analysis of electron transport activities and changes in fluorescence yield. Plant Cell Physiol. 32, 943–951.

Bajguz, A., 2011. Suppresion of Chlorella vulgaris growth by cadmium, lead and copper stress and its restoration by endogenous brassinolide. Arch. Environ. Contam. Toxicol. 60, 406–416.

Barkosky, R.R., Einhellig, F.A., 1993. Effects of salicylic acid on plant water relationship. J. Chem. Ecol. 19, 237–247.

Baryla, A., Crrier, P., Franck, F., Coulomb, C., Sahut, C., Havaux, M., 2001. Leaf chlorosis in oilseed rape plants (*Brassica napus*) grown on cadmium polluted soil: causes and consequences for photosynthesis and growth. Planta 212, 696–709.

Basa, B., Lattanzio, G., Solti, Á., Tóth, B., Abadía, J., Fodor, F., Sárvári, É., 2014. *Changes induced* by *cadmium stress* and *iron deficiency* in *the composition* and *organization* of *thylakoid complexes* in *sugar beet* (*Beta vulgaris* L.). Environ. Exp. Botany 101, 1–11.

Burzyński, M., Kłobus, G., 2004. Changes of photosynthetic parameters in cucumber leaves under Cu, Cd, and Pb stress. Photosynthetica 42, 505–510.

Burzyński, M., Źurec, A., 2007. Effect of copper and cadmium on photosynthesis in cucumber cotyledons. Photosynthetica 45, 239–244.

Ci, D., Jiang, D., Wollenweber, B., Dai, T., Jing, Q., Cao, W., 2010. Cadmium stress in wheat seedlings: growth, cadmium accumulation and photosynthesis. Acta Physiol. Plant. 32, 365–373.

Clemens, S., Aarts, M.G.M., Thomine, S., Verbruggen, N., 2013. Plant science: the key to preventing slow cadmium poisoning. Trends Plant Sci. 18 (2), 92–99.

Colebrook, E.H., Thomas, S.G., Phillips, A.L., Hedden, P., 2014. The role of gibberellin signalling in plant responses to abiotic stress. J. Exp. Biol. 217, 67–75.

Cutt, J.R., Klessing, D.F., 1992. Salicylic acid in plants: a changing perspective. Pharmaceut. Technol. 16, 25–34.

Djebali, W., Zarrouk, M., Brouquisse, R., El Kahoui, S., Limam, F., Ghorbel, M.H., Chaïbi, W., 2005. Ultrastructure and lipid alterations induced by cadmium in tomato (*Lycopersicon esculentum*) chloroplast membranes. Plant Biol. 7 (4), 358–368.

Dobrikova, A.G., Vladkova, R.S., Rashkov, G.D., Todinova, S.J., Krumova, S.B., Apostolova, E.L., 2014. Effects of exogenous 24-epibrassinolide on the photosynthetic membranes under non-stress conditions. Plant Physiol. Biochem. 80, 75–82.

Dobrikova, A.G., Yotsova, E.K., Börner, A., Landjeva, S.P., Apostolova, E.L., 2017. The wheat mutant DELLA-encoding gene (*Rht-B1c*) affects plant photosynthetic responses to cadmium stress. Plant Physiol. Biochem. 114, 10–18.

Drazic, G., Mihailovic, N., 2005. Modification of cadmium toxicity in soybean seedlings by salicylic acid. Plant Sci. 168, 511–517.

Fagioni, M., D'Amici, G.M., Timperio, A.M., Zolla, L., 2009. Proteomic analysis of multiprotein complexes in the thylakoid membrane upon cadmium treatment. J. Proteome Res. 8 (1), 310–326.

Faller, P., Kienzler, K., Krieger-Liszkay, A., 2005. Mechanism of $Cd^{2+}$ toxicity: $Cd^{2+}$ inhibits photoactivation of photosystem II by competitive binding to the essential $Ca^{2+}$ site. Biochim. Biophys. Acta 1706, 158–164.

Fariduddin, Q., Yusuf, M., Ahmad, I., Ahmad, A., 2014. Brassinosteroids and their role in response of plants to abiotic stresses. Biologia Plant. 58 (1), 9–17.

Gao, H., Liu, J., Zheng, Q., Hong, L., Wang, C., et al., 2013. Regulation of exogenous brassinosteroid on growth and photosynthesis of *Helianthus tuberosus* seedlings and cadmium biological enrichment under cadmium stress. Shengtai Xuebao 33, 1935–1943.

Ghorbanli, M., Kaveh, S.H., Sepehr, M.F., 2000. Effects of cadmium and gibberellin on growth and photosynthesis of Glycine max. Photosynthetica 37 (4), 627–631.

Gratão, P.L., Monteiro, C.C., Rossi, M.L., Martinelli, A.P., Peres, L.E.P., Medici, L.O., et al., 2009. Differential ultrastructural changes in tomato hormonal mutants exposed to cadmium. Environ. Exp. Bot. 67, 387–394.

Grant, C.A., Buckley, W.T., Bailey, L.D., Selles, F., 1998. Cadmium accumulation in crop. Can. J. Plant Sci. 78, 1–17.

Gunes, A., Inal, A., Alpaslan, M., Cicek, N., Guneri, E., Erasland, F., Guzelordu, T., 2005. Effects of exogenously applied salicylic acid on the induction of multiple stress tolerance and mineral nutrition in maize (*Zea mays* L.). Arch. Agron. Soil Sci. 51, 687–695.

Guo, B., Liang, Y.C., Zhu, Y.G., Zhao, F.J., 2007. Role of salicylic acid in alleviating oxidative damage in rice roots (*Orysa sativa*) subjected to cadmium stress. Environ. Pollut. 147, 743–749.

Gururani, M.A., Mohanta, T.K., Bae, H., 2015. Current understanding of the interplay between phytohormones and photosynthesis under environmental stress. Int. J. Mol. Sci. 16 (8), 19055–19085.

Habibi, G., 2012. Exogenous salicylic acid alleviates oxidative damage of barley plants under drought stress. Acta Biol. Szegedien. 56, 57–63.

Hakmaoui, A., Ater, M., Bóka, K., Barón, M., 2007. Copper and cadmium tolerance, uptake and effect on chloroplast ultrastructure. Studies on *Salix purpurea* and *Phragmites australis*. Z. Naturforsch. C 62 (5–6), 417–426.

Hasan, S.A., Hayat, S., Ahmad, A., 2011. Brassinosteroids protect photosynthetic machinery against the cadmium induced oxidative stress in two tomato cultivars. Chemosphere 84, 1446–1451.

Hasan, S.A., Hayat, S., Ahmad, A., 2009. Screening of tomato (*Lycopersicon esculentum*) cultivars against cadmium through shotgun approach. J. Plant Interact. 4, 187–201.

Hayat, Q., Hayat, S., Irfan, M., Ahmad, A., 2010b. Effect of exogenous salicylic acid under changing environment: a review. Environ. Exp. Bot. 68, 14–25.

Hayat, S., Ali, B., Hasan, S.A., Ahmad, A., 2007. Brassinosteroids enhanced the level of antioxidants under cadmium stress in *Brassica juncea*. Environ. Exp. Bot. 60, 33–41.

Hayat, S., Hasan, S.A., Hayat, Q., Ahmad, A., 2010a. Brassinosteroids protect *Lycopersicon esculentum* from cadmium toxicity applied as shotgun approach. Protoplasma 239, 3–14.

Hernandez, L.E., Carpena-Ruiz, R., Garate, A., 1996. Alterations in the mineral nutrition of pea seedlings exposed to cadmium. J. Plant Nutr. 19, 1581–1598.

Horváth, E., Szalai, G., Janda, T., 2007. Induction of abiotic stress tolerance by salicylic acid signaling. J. Plant Growth Regul. 26, 290–300.

Hoseini, S.M., Zargari, F., 2013. Cadmium in plants: a review. Int. J. Farm. Alli. Sci. 2, 579–581.

Jakubowska, D., Janicka, M., 2017. The role of brassinosteroids in the regulation of the plasma membrane H+-ATPase and NADPH oxidase under cadmium stress. Plant Sci. 264, 37–47.

Janda, K., Hideg, É., Szalai, G., Kovács, L., Janda, T., 2012. Salicylic acid may indirectly influence the photosynthetic electron transport. J. Plant Physiol. 169, 971–978.

Janeczko, A., Koscielniak, J., Pilipowicz, M., Szarek-Lukaszewska, G., Skoczowski, A., 2005. Protection of winter rape photosystem 2 by 24-epibrassinolide under cadmium stress. Photosynthetica 43, 293–298.

Janik, E., Maksymiec, W., Mazur, R., Garstka, M., Gruszecki, W.I., 2010. Structural and functional modifications of the major light-harvesting complex II in cadmium or copper-treated *Secale cereale*. Plant Cell Physiol. 51, 1330–1340.

Jiang, Y.P., Cheng, F., Zhou, Y.H., Xia, X.J., Shi, K., Yu, J.Q., 2012. Interactive effects of $CO_2$ enrichment and brassinosteroid on $CO_2$ assimilation and photosynthetic electron transport in *Cucumis sativus*. Environ. Exp. Bot. 75, 98–106.

Khan, W., Prithiviraj, B., Smith, D.L., 2003. Photosynthetic responses of corn and soybean to foliar application of salicylates. J. Plant Physiol. 160, 485–492.

Khan, M.I.R., Fatma, M., Per, T.S., Anjum, N.A., Khan, N.A., 2015. Salicylic acid-induced abiotic stress tolerance and underlying mechanisms in plants. Review. Front. Plant Sci. 6, 462.

Krantev, A., Yordanova, R., Janda, T., Szalai, G., Popova, L., 2008. Treatment with salicilic acid decreases the effect of cadmium on photosynthesis in maize plants. J. Plant Physiol. 165, 920–931.

Krupa, Z., Öquist, G., Hunter, N.P.A., 1993. The effects of cadmium on photosynthesis of *Phaseolus vulgaris* L.—a fluorescence analysis. Physiol. Plant 88, 626–630.

Kuboi, T., Noguchi, A., Yazaki, J., 1986. Family-dependent cadmium accumulation in higher plants. Plant Soil 92, 405–415.

Larqué-Saavedra, A., 1979. Stomatal closure in response to acetylsalicylic acid treatment. Z. Pflannzenphysiol. 93, 371–375.

Liu, C., Guo, J., Cui, Y., Lű, T., Zhang, X., Shi, G., 2011. Effect of cadmium and salicilic acid on growth, spectral reflectance and photosynthesis of castor bean seedling. Plant Soil 344, 131–141.

López-Millán, A.F., Sagardoy, R., Solanas, M., AbadHa, A., AbadHa, J., 2009. Cadmium toxicity in tomato (*Lycopersicon esculentum*) plants grown in hydroponics. Environ. Exp. Bot. 65, 376–385.

Lysenko, E.A., Klaus, A.A., Pshybytko, N.L., Kusnetsov, V.V., 2015. Cadmium accumulation in chloroplasts and its impact on chloroplastic processes in barley and maize. Photosynth. Res. 125, 291–303.

Marzec, M., 2017. Strigolactones and gibberellins: a new couple in the phytohormone world? Trends Plant Sci. 22 (10), 813–815.

Maslenkova, L., Peeva, V., Stojnova, Z., Popova, L., 2009. Salicylic acid-induced changes in photosystem II reactions in barley plants. Biotechnol. Biotechnol. Eq. 23, 297–300.

Metwally, A., Finkmemeier, I., Georgi, M., Dietz, K.J., 2003. Salicylic acid alleviates the cadmium toxicity in barley seedlings. Plant Physiol. 132, 272–281.

Michel, K.P., Pistorius, E.K., 2004. Adaptation of the photosynthetic electron transport chain in cyanobacteria to iron deficiency: the function of IdiA and IsiA. Physiol. Plant. 120, 36–50.

Mishra, A., Choudhuri, M.A., 1999. Effects of salicylic acid on heavy metal-induced membrane degradation mediated by lipoxygenase in rice. Biol. Plant. 42, 409–415.

Mobin, M., Khan, N.A., 2007. Photosynthetic activity, pigment composition and antioxidative response of two mustard (Brassica juncea) cultivars differing in photosynthetic capacity subjected to cadmium stress. J. Plant Physiol. 164 (5), 601–610.

Moravcová, Š., Tůma, J., Dučaiová, Z.K., Waligórski, P., Kula, M., Saja, D., Słomka, A., Bąba, W., Libik-Konieczny, M., 2018. Influence of salicylic acid pretreatment on seeds

germination and some defence mechanisms of *Zea mays* plants under copper stress. Plant Physiol. Biochem. 122, 19–30.
Moussa, H., El-Gamal, S., 2010. Effect of salicylic acid pretreatment on cadmium toxicity in wheat. Biol. Plant. 54, 315–320.
Mullins, G.L., Sommers, L.E., Barber, S.A., 1986. Modelling the plant uptake of cadmium and zinc from soils treated with sewage sludge. Soil Sci. Soc. Am. J. 50, 1245–1250.
Naidu, R., Bolan, N.S., Kookana, R.S., Tiller, K.G., 1994. Ionic-strength and pH effects on the sorption of cadmium and the surface charge of soils. Eur. J. Soil Sci. 45, 419–429.
Nordberg, G.F., 2009. Historical perspectives on cadmium toxicology. Toxicol. Appl. Pharmacol. 238, 192–200.
Nouairi, I., Ammar, W.B., Youssef, N.B., Daoud, D.B.M., Ghorbal, M.H., Zarrouk, M., 2006. Comparative study of cadmium effects on membrane lipid composition of *Brassica juncea* and *Brassica napus* leaves. Plant Sci. 170, 511–519.
Oono, Y., Yazava, T., Kanamori, H., Saaki, H., Mori, S., Handa, H., Matsumoto, T., 2016. Genome-wide transcriptome analysis of cadmium stress in rice. BioMed Res. Internat. 2016, https://doi.org/10.1155/2016/9739505. Article ID 9739505.
Ouzounidou, G., Ilias, I., 2005. Hormone-induced protection of sunflower photosynthetic apparatus against copper toxicity. Biol. Plant. 49, 223–228.
Pál, M., Horváth, E., Janda, T., Páldi, E., Szalai, G., 2006. Physiological changes and defense mechanisms induced by cadmium stress in maize. J. Plant Nutr. Soil Sci. 169, 239–246.
Pancheva, T., Popova, L., Uzunova, A., 1996. Effects of salicylic acid on growth and photosynthesis in barley plants. J. Plant Physiol. 149, 57–63.
Parmar, P., Kumari, N., Sharma, V., 2013. Structural and functional alterations in photosynthetic apparatus of plants under cadmium stress. Bot. Stud. 54, 45–51.
Prasad, M.N., 1995. Cadmium toxicity and tolerance in vascular plants. Environ. Exp. Bot. 35, 525–545.
Rascio, N., Dalla, V.D., Ferretti, M., Merlo, L., Ghisi, R., 1993. Some effects of cadmium on maize plants. Arch. Environ. Contam. Toxicol. 25, 244–249.
Reeves, R.D., Baker, A.J.M., 2000. Metal-accumulating plants. In: Raskin, I., Ensley, B. (Eds.), Phytoremediation of Toxic Metals: Using Plants to Clean up the Environment. Wiley, New York, pp. 193–229.
Rivas-San Vicente, M., Plasencia, J., 2011. Salicilic acid beyond defence: its role in plant growth and development. J. Exp. Bot. 62, 3321–3338.
Robert-Seilaniantz, A., Bari, R., Jones, J.D.G., 2010. A biotic or abiotic stress? In: Pareek, A., Sopory, S.K., Bohnert, H., Govindjee, (Eds.), Abiotic Stress Adaptation in Plants: Physiological, Molecular and Genomic Foundation. Springer, Dordrecht, The Netherlands, pp. 103–122.
Sárvári, É., Cseh, E., Balczer, T., Szigeti, Z., Záray, G., Fodor, F., 2008. Effect of Cd on the iron re-supply-induced formation of chlorophyll-protein complexes in cucumber. Acta Biol. Szeged 52 (1), 183–186.
Sękara, A., Poniedzialek, M., Ciura, J., Jędrszczyk, E., 2005. Cadmium and lead accumulation and distribution in the organs of nine crops: implications for phytoremediation. Polish J. Environ. Stud. 14, 509–516.
Sigfridsson, K.G.V., Bernát, G., Mamedov, F., Styring, S., 2004. Molecular interference of $Cd^{2+}$ with photosystem II. Biochim. Biophys. Acta 1659, 19–31.
Singh, P.K., Chaturvedi, V.K., Bose, B., 2010. Effects of salicylic acid on seedling growth and nitrogen metabolism in cucumber (*Cucumis sativus* L.). J. Stress Physiol. Biochem. 6, 102–113.
Toppi, L.S., Gabbrielli, R., 1999. Response to cadmium in higher plants. Environ. Exp. Bot. 41, 105–130.
Tran, T.A., Popova, L., 2013. Functions and toxicity of cadmium in plants: recent advances and future prospects. Turk. J. Bot. 37, 1–13.

Tukendorf, A., Rauser, W.E., 1990. Changes in glutathione and phytochelatins in roots of maize seedlings exposed to cadmium. Plant Sci. 70, 155–166.

Usunova, A.N., Popova, L.P., 2000. Effect of salicylic acid on leaf anatomy and chloroplast ultrastructure of barley plants. Photosynthetica 38, 243–250.

Van de Velde, K., Ruelens, P., Geuten, K., Rohde, A., Van Der Straeten, D., 2017. Exploiting DELLA signaling in cereals. Trends Plant Sci. 22 (10), 880–893.

Vardhini, B.V., 2015. Brassinosteroids are potential ameliorators of heavy metal stresses in plants. In: Ahmad, P. (Ed.), Plant Metal Interaction: Emerging Remediation Techniques. Elsevier, The Netherlands, pp. 209–237.

Vardhini, B.V., 2017. Modifications of morphological and anatomical characteristics of plants by application of brassinosteroids under various abiotic stress conditions—a review. Plant Gene 11, 70–89.

Vassilev, A., Lidon, F., Scotti, P., Da Graca, M., Yordanov, I., 2004. Cadmium-induced changes in chloroplast lipids and photosystem activities in barley plants. Biol. Plant. 48, 153–156.

Wang, M.E., Zhou, Q.X., 2006. Joint stress of chlorimuronethyl and cadmium on wheat *Triticum aestivum* at biochemical levels. Environ. Pollut. 144, 572–580.

Wang, Q., Liang, X., Dong, Y., Xu, L., Zhang, X., Kong, J., Liu, S., 2013a. Effect of exogenous salicylic acid and nitric oxide on physiological characteristics of perennial ryegrass under cadmium stress. J. Plant Growth. Reg. 32, 721–731.

Wang, X., Zhang, Z.W., Tu, S.H., Feng, W.Q., Xu, F., Zhu, F., et al., 2013b. Comparative study of four rice cultivars with different levels of cadmium tolerance. Biologia 68, 74–81.

Wanger, G.J., 1993. Accumulation of cadmium in crop plants and its consequences to human health. Adv. Agron. J. 51, 173–212.

Wen, W., Deng, Q., Jia, H., Wei, L., Wei, J., Wang, H., et al., 2013. Sequence variations of the partially dominant DELLA gene *Rht-B1c* in wheat and their functional impacts. J. Exp. Bot. 64, 3299–3312.

Wodala, B., Eitel, G., Gyula, T.N., Ördög, A., Horváth, F., 2012. Monitoring moderate Cu and Cd toxicity by chlorophyll fluorescence and P700 absorbance in pea leaves. Photosynthetica 50, 380–386.

Xin, J., Zhao, X., Tan, Q., Sun, X., Hu, C., 2017. Comparison of cadmium absorption, translocation, subcellular distribution and chemical forms between two radish cultivars (*Raphanus sativus* L.). Ecotoxicol. Environ. Safety 145, 258–265.

Yang, Y., Qi, M., Mei, C., 2004. Endogenous salicylic acid protects rice plants from oxidative damage caused by aging as well as biotic and abiotic stress. Plant J. 40, 909–919.

Yotsova, E., Dobrikova, A., Stefanov, M., Apostolova, E., 2018. Impact of salicylic acid on the growth and the activity of photosynthetic apparatus in rice under non-stress conditions. C. R. Acad. Bulg. Sci. 71 (3), 368–375. https://doi.org/10.7546/CRABS.2018.03.09.

Yu, J.Q., Huang, L.F., Hu, W.H., Zhou, Y.H., Mao, W.H., Ye, S.F., Nogués, S., 2004. A role for brassinosteroids in the regulation of photosynthesis in *Cucumis sativus*. J. Exp. Bot. 5, 1135–1143.

Yuan, L., Xu, D., 2001. Stimulation effect of gibberellic acid short-term treatment on leaf photosynthesis related to the increase in Rubisco content in broad bean and soybean. Photosynth. Res. 68, 39–47.

Zhu, R., Macfie, S.M., Ding, Z., 2005. Cadmium-induced plant stress investigated by scanning electrochemical microscopy. J. Exp. Bot. 56, 2831–2838.

Zhu, X.F., Jiang, T., Wang, Z.W., Lei, G.J., Shi, Y.Z., Li, G.X., et al., 2012. Gibberellic acid alleviates cadmium toxicity by reducing nitric oxide accumulation and expression of IRT1 in *Arabidopsis thaliana*. J. Hazard. Mater 239 (240), 302–307.

# CHAPTER 12

# Cadmium-Induced Imbalance in Nutrient and Water Uptake by Plants

**Asif Naeem*[,†], Mehwish Zafar[†], Hinnan Khalid[‡,§], Muhammad Zia-ur-Rehman[‡], Zahoor Ahmad[¶], Muhammad Ashar Ayub[†], M. Farooq Qayyum[||]**

[*]Soil and Environmental Sciences Division, Nuclear Institute for Agriculture and Biology (NIAB), Faisalabad, Pakistan
[†]Institute of Soil and Environmental Sciences, University of Agriculture, Faisalabad, Pakistan
[‡]Faculty of Agriculture, Institute of Soil and Environmental Sciences, University of Agriculture, Faisalabad, Pakistan
[§]Punjab Agriculture Department, Shahkot, Pakistan
[¶]Cholistan Institute of Desert Studies (CIDS), The Islamia University of Bahawalpur (IUB), Bahawalpur, Pakistan
[||]Department of Soil Science, Faculty of Agricultural Sciences & Technology, Bahaudin Zakariya University Multan, Multan, Pakistan

## Abstract

The inadequacy of fresh water availability in the agricultural sector has become a severe problem. Coupled with prevailing drought conditions, unavailability of plant essential nutrients has accelerated the food security threat significantly. However, unchecked loading of cadmium (Cd) into the environment due to the extensive use of adulterated fertilizers, sewage, and waste water, and dumping of industrial waste into agri-lands has made the plight of agricultural commodities more critical. Among different heavy metals, Cd is the most toxic, due to its high solubility and mobility in soil-plant systems. It inhibits plant growth and development through several mechanisms. Cd competes with plant essential nutrients (PENs) for uptake and translocation, and adversely affects plants' water status. Cadmium toxicity disturbs plant metabolism because of the restricted uptake and translocation of mineral nutrients. This chapter will discuss, in detail, the effects of Cd toxicity on nutrient uptake by roots and their onward translocation to aerial plant parts. Moreover, this chapter will also discuss possible strategies to offset the harm of Cd toxicity with the use of organic and inorganic amendments, incorporation of plant growth regulators, and the addition of microbial inoculums.

**Keywords:** Heavy metals, Cadmium, Nutrient uptake, Translocation, Water stress, Grain mineral status

*Cadmium Toxicity and Tolerance in Plants*
https://doi.org/10.1016/B978-0-12-814864-8.00012-7

# 1 INTRODUCTION

Heavy metals (HMs) pollution in the biosphere obtrudes irreversible threats to the sustainability of human life because of their toxicity and protracted persistence in most of the environment (Satarug, 2012). Although HMs are considered the backbone of industrialization, many HMs are not only metabolically nonessential for biological processes, but are also toxic for the biosphere (Zhang et al. 2014; Nouha et al., 2016). The most distinct anthropogenic activities, such as mining, pesticide use, sewage sludge, phosphatic fertilizers, and irrigating crops with waste effluents; and industrial processes, such as nickel-cadmium batteries production, and production of pigments, plastics, and metal plating are the most important sources of HMs accumulating in soil and water (Mutlu et al., 2012; Murtaza et al., 2015).

Water is a requirement of living organisms. Irregular rainfall and seasonal evapotranspiration in roots can cause freshwater depletion. The insufficiency of ground water for irrigation is a constantly rising problem, due to the utilization of sewage water in agriculture becoming a regular practice in three-fourths of the cities in Asia, America, and Africa (Gupta et al., 2008). Because the treatment of waste water is often painstaking and expensive, it is sometimes considered an unreasonable luxury, affordable only for prosperous countries (Paranychianakis et al., 2006; Levy et al., 2010). Continuous use of such waste water results in HMs contamination of water and soil.

Among the list of the most toxic heavy metals, cadmium (Cd) ranks 7th. Cadmium causes long-term deterioration of the environment (Mahar et al., 2015). Cd contamination in soils is mainly increasing due to the agricultural application of municipal effluents for different crops (Perkins and Mason, 2015). It has higher mobility and solubility within its surroundings compared with different heavy metals. Once Cd is introduced into the soil, it persists for an extended period of time, because it does not endure chemical or microbial degradation (Park et al., 2011). The long-term usage of waste water increases toxic levels of cadmium in the plants, and triggers a variety of physiological, as well as metabolic alterations (Villiers et al., 2011; Hasan et al., 2009). However, as each HM affects a specific site of the plant, visual toxic symptoms differ between HMs. The most prevalent visual responses of cadmium toxicity include stunted plant growth (Sharma and Dubey, 2007), turgor loss, necrosis, leaf chlorosis, reduced seed germination, and a damaged photosynthetic apparatus, correlated with senescence or plant death

(Rizwan et al., 2016a; Rizwan et al., 2016b). All these consequences are related to molecular, ultrastructural, and biochemical changes within plant cells and/or tissues caused by HMs presence, such as cadmium (Gamalero et al., 2009). Cadmium accumulation also alters mineral nutrients' uptake. Cadmium has been shown to interact with the availability of nutrient elements, and also, some of these nutrients have a protective role against the toxic effects of Cd stress (Nazar et al., 2012).

In humans, the chronic effects of cadmium (Cd) consist of bone fractures, hypertension, lung cancer, kidney dysfunction, pulmonary adenocarcinomas, and prostatic proliferative lesions (Colacino et al., 2014; Demenesku et al., 2014). Keeping in view the harmful effects of HMs in the ecosystem, their bioavailability should be restricted/minimized. Bioavailability is the amount of Cd available for plants' uptake. Different approaches, including physical, chemical, and biological can be used to reduce the HMs bioavailability. Some effective technologies include soil replacement (Yao et al., 2012), electro-kinetic techniques (Suèr et al., 2003), the vitrify technique (heating at 1400–2000°C), chemical extraction (Dermont et al., 2008), chemical immobilization/stabilization, cultivating hyper-accumulator plant species (Chen et al., 2013), low-HMs-accumulating species (Jamali et al., 2009), and endophytic bacteria inoculum (Ma et al., 2016b).

## 2 SOURCES OF Cd CONTAMINATION IN AGRICULTURE SOILS

Heavy metal toxicity reflects unavoidable impacts on soil health and properties. Heavy metal toxicity not only impairs plant growth and yielding parameters, but also disturbs the soil's physico-chemical properties, and soil micro-biota (Hinojosa et al., 2004; Huang et al., 2009). Toxicity of HMs cause the deficiency of plant's essential nutrients, and soil fertility of such agricultural lands becomes a question mark (Huang et al., 2009).

Cd has been declared more hazardous as compared with Pb due to high mobility and less attraction for soil particles. Cd is released into the environment in variable amounts from natural and anthropogenic activities.

### 2.1 Natural Sources

Volcanic eruptions, windblown dust, forest fires, and sea spray are among the natural sources of Cd into the atmosphere. Parent rocks weathering also contributes to the release of Cd into the environment (Khan et al., 2010;

Liu et al., 2013a,b). A high amount of Cd is present in mafic and ultramafic rocks, and when these rocks are weathered, they release a significant amount of Cd into soil (Shah et al., 2010). Up to 100 $mg kg^{-1}$ concentration of Cd is present in black shale, and the soil derived from their deposits is enriched with Cd (He et al., 2005). 4.6 $mg kg^{-1}$ mean Cd concentrations was observed in siltstone and mudstone, while it was observed 1.7 $mg kg^{-1}$ in carbonate rocks, which are collected from the Three Gorges Region of Jianping (Liu et al., 2013a,b). However, 10% of total contribution of natural sources is counted toward Cd contamination in soil, as compared with the total release of Cd from all sources.

## 2.2 Anthropogenic Sources

Anthropogenic sources, including the fossil fuel combustion, use of phosphate fertilizers, wastes from the cement industry, metallurgical works, sewage sludge, industrial and municipal wastes, and smelting, mining, and metals ore processing contribute more than 90% ($5.6–38 \times 10^6$ $kg year^{-1}$) of the release of Cd into the environment (Bi et al., 2006; Cloquet et al., 2006). In addition, to resist corrosion, Cd is also widely used in steel plating, pigments, plastic stabilizers, solar panels, and batteries. Other sources of Cd may include polyvinyl chloride plastic manufacturing, alloys, solders, motor oil, rubber, fungicides, and textile manufacturing (Tamaddon and Hogland, 1993). Based on the findings of the British Geological Survey, the total worldwide production of Cd in 2015 was approximately 24,900 metric tons (British Geological Survey, 2017). In 2016, the worldwide production of Cd, excluding the United States, was 23,000 metric tons (U.S. Geological Survey, 2017). Cd occurs in all types of zinc (Zn) ores as a guest element due to geochemical similarity. Thus, substantial amounts are deposited locally (Bi et al., 2006; Roy and Mcdonald, 2015). Urban soil is mainly polluted with Cd from industrial processes and vehicular emissions (Khan et al., 2016a). A large amount (67%) is used in the production of metallic Cd electroplate used in nickel-cadmium batteries (Greenwood and Earnshaw, 2001). In China for the year 2009, the total estimated emission of Cd was 743 metric tons, of which 57% was contributed by industrial processes (Cheng et al., 2014). Nriagu and Pacyna (1988) reported that the worldwide estimated total input of Cd to soil through different anthropogenic sources was 5 to $38 \times 106$ $kg year^{-1}$ with 12% from agriculture and animal wastes, including fertilizers. The contamination of agriculture soil with Cd from wastewater irrigation also contributes to a larger extent. In China,

Cd-enriched farmland area has reached $27.86 \times 104\,km^2$ (Cao et al., 1999). Among different sources, the use of phosphate fertilizers is generally considered to be one of the major sources of Cd input to agricultural soil (Smolders and Mertens, 2013). Concentrations in phosphate fertilizers used in Europe were observed in the range of 0.1–120 mg Cd (kg $P_2O_5$)$^{-1}$ (Nziguheba and Smolders, 2008). In addition, the application of contaminated manure also adds Cd to agricultural soil. According to Nicholson et al., 2003, the input of Cd to agricultural soil from manure, when applied at a rate of N equivalent to 250 kgN $ha^{-1}$ $year^{-1}$, ranged from 1.4 to 6.1 g Cd $ha^{-1}$ $year^{-1}$.

## 3 CADMIUM SPECIATION IN CONTAMINATED SOILS

In terrestrial ecosystems, soil plays an important role for cadmium cycling, and thus, accumulates more as compared with water and air (Ashraf et al., 2012; Alahabadi et al., 2017; Maghakyan et al., 2017). Therefore, knowledge of Cd speciation obtained by following the continuous metal extraction protocols is of primary concern for agricultural soils (Tessier et al., 1979). Various studies reveal that from a total metals concentration, their bio-available fraction is more important in agricultural soils. Ashraf et al. (2012) states that the increasing order for mobility of Cd can be described as: exchangeable fraction > carbonate bound fraction > bound to oxides > bound to organic compounds > residual metal concentration.

Oxide-bound Cd is found in nodules and cement-like structures present between the soil particles that become mobile in acidic and reduced conditions. Similarly, carbonate bound fraction become mobile upon the change of redox conditions toward the reductive and prolonged acidic environment. However, Cadmium bound to organic compounds remains very stable, and becomes available only in strongly oxidizing conditions (Tessier et al., 1979; Huang et al., 2007). Therefore, bio-available metal concentration is strongly related with various factors including soil properties, organic matter, plant factors, climatic conditions, and micro-organisms (Chang et al., 2014; Liu et al., 2017). Soil particles provide the binding sites to the Cd and sustain the equilibrium with the soil solution. However, solubility of the bound Cd depends upon various factors, including complex formation with soluble humic substances, activation of chemi-sorption processes with soil minerals such as clay and oxides, formation of insoluble precipitates such as sulfides, oxides, carbonates, and phosphates, and stimulation of co-precipitation processes with other minerals (Khan et al., 2017;

Saha et al., 2017). Therefore, variation in the described soil properties causes variation in Cd availability in different soil types.

## 4 UPTAKE, TRANSPORT, AND TOXICITY OF Cd TO PLANTS

Plants generally take up Cd from soil solutions, with the help of the root system. Therefore, the concentration of Cd in the soil solution determines the actual amount of Cd uptake by the plants (Wei et al., 2016; Rehman et al., 2017). Various HMs, including cadmium (Cd), lead (Pb), copper (Cu), and so forth use ion transporters such as $Fe^{2+}$, $Mn^{2+}$, $Zn^{2+}$, and $Ca^{2+}$, for their entry from the soil solution into the root system (Clemens, 2006). However, Cd uptake and accumulation ability varies by plant type and species (Liu et al., 2003; Grant et al., 2008; Ghani et al., 2014; Ji et al., 2016). Moreover, Cd uptake and accumulation efficiency depend on the following primary processes: uptake ability of plant roots, transporting efficiency of xylem sap, and ultimate re-translocation within seeds of plants. Roots play a key role for the uptake and accumulation of water, essential nutrients, and toxic HMs. In addition, a small fraction of Cd concentration present in the soil system get adsorbed onto the roots' surface, therefore increasing the surface area of plant roots, prolonging the exposure duration, and accumulating extended levels of Cd into various plant parts (Hentz et al., 2012; Wei et al., 2014; Rizwan et al., 2016a) (Fig. 1).

The Cd uptake ability of plant roots varies with varying root structures, surface areas, and apices (Kubo et al., 2011). It is also well documented that natural resistance associated macrophage proteins (NRAMPs) play a prominent role in Cd transport across the cell membrane (Nevo and Nelson, 2006; Song et al., 2014). However, the presence of the root endodermis, coupled with the exodermis, act as the solute flow barrier during Cd toxicity (Lux et al., 2011; Redjala et al., 2011). A higher level of Cd availability causes the competition for nutrient ion uptake by nutrient ion transporters such as proteins of iron, zinc, and calcium; and thus, an ample quantity of plant essential nutrients causes a hindrance in Cd uptake and accumulation by plants (Hart et al., 2002; Benavides et al., 2005; Lux et al., 2011) (Fig. 1). Generally, after uptake of Cd by roots, a maximum part of Cd becomes fractionized into the root cells; however, only a small fraction moves toward the shoots and grains (Gill et al., 2011; Khan et al., 2014; Song et al., 2014). Contrary results of various studies upon fractionation and accumulation of Cd in different plant parts also opens the window for varied mechanisms

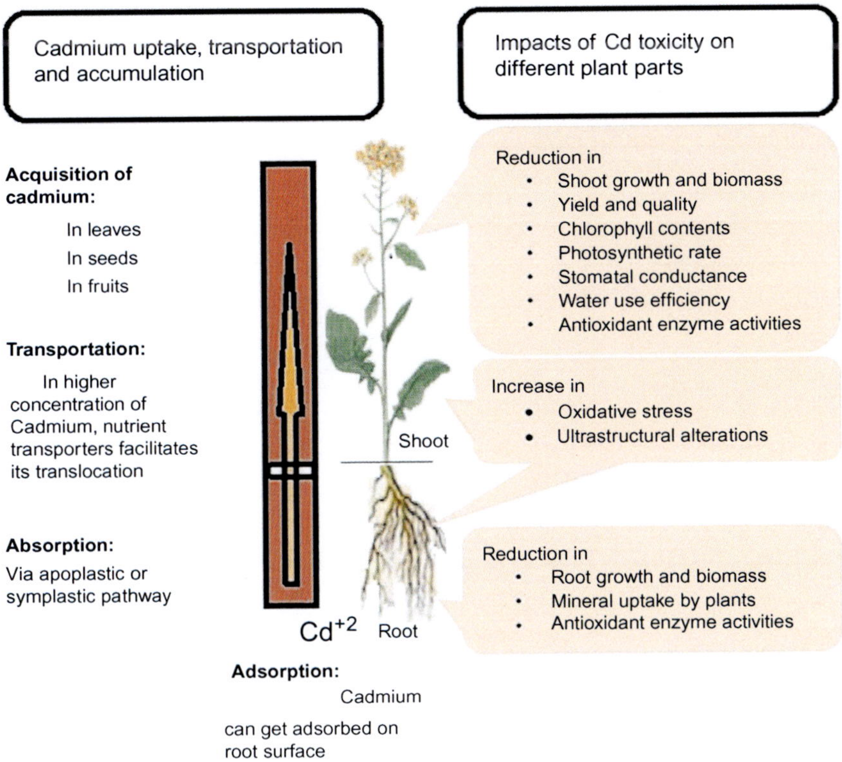


**Fig. 1** Suggested mechanism of Cd uptake, accumulation, and toxicity in crop plants. *(Modified from Rizwan, M., Ali, S., Zia ur Rehman, M., Rinklebe, J., Tsang, D.C.W., Bashir, A., Ok, Y.S., 2018. Cadmium phytoremediation potential of* Brassica *crop species: a review. Sci. Total Environ. 631–632, 1175–1191. 10.1016/j.scitotenv.2018.03.104.)*

with species variation (Abe et al., 2008; Dwivedi et al., 2014; Rehman et al., 2017). Once Cd enters into the roots, then transportation occurs, either with the help of an apoplastic or symplastic pathway. However, the maximum part of the absorbed Cd from different complexes with various ligands (such as organic acids and phyto-chelations) ultimately becomes part of the root vacuoles and/or nuclei (Hart et al., 2002; Lux et al., 2011).

Xylem is of primary concern for Cd movement from roots toward the shoots, leaves, and grains. Plasmalemma (basic component of endodermis) and casparian strips also support the central cylinder for upward metal flow within the plant parts (Clemens, 2006; Seregin et al., 2004; Tudoreanu and Phillips, 2001). Translocation of Cd from roots toward aerial parts becomes accomplished by both passive (transpiration process) and active (ion transport channels) mechanisms (Hart et al., 2002; Nakanishi et al., 2006).

However, the amount of Cd transfer and/or accumulation varies within plant species and genotype (Dunbar et al., 2003; Khan et al., 2014; Rehman et al., 2017). It has been reported that Cd accumulation into grains occurs through the cell sap movement by the phloem (Liu et al., 2007; Yoneyama et al., 2010). However, the amount of Cd translocated depends upon the concentration of Cd in the lower parts, as well as a fraction of the other nutrients into the cell sap. Plants accumulate Cd from soils, and thus affect plant growth and development due to its toxic effects on different physiological and morphological processes, including reduced growth, mutagenic effects, rolling of leaves, chlorosis, necrosis, and increased oxidative stress and mortality (Rizwan et al., 2016a). They also have a detrimental effect on seed germination, acceleration of nutrient imbalance, and yield of the plant (Naeem et al., 2016; Rizwan et al., 2016a).

## 5 EFFECT OF Cd TOXICITY ON NUTRIENT HOMEOSTASIS

Cd has significant negative impacts on plant nutritional values and growth rates, and most plants grown in Cd-contaminated soil are nutrient deficient (Khan et al., 2015). Yang et al. (1996) reported Cd effects on the uptake of various nutrients by plants. At the root region, Cd causes mineral deficiency by competing for absorption with minerals having similar chemical properties, such as Ca and Mg (Barceló and Poschenrieder, 1990). Thus, Cd toxicity affects phytoavailability of both macro- and micro-nutrients.

### 5.1 Effect of Cd Toxicity on Phytoavailability of Macronutrients

The primary nutrient elements (N, P, K) are an integral part of our agricultural system for optimizing crop productivity and for sustainable agriculture. Bioaccumulation of Cd causes the alteration of various physiological functions by affecting N metabolism (Chaffei et al., 2004). A significant decrease occurred in K and P concentrations in the edible parts of tomatoes and potatoes at different Cd application doses, while the content of vitamin C was increased (Khan et al., 2016b). Being an important part of genetic, structural, and metabolic components, N plays an important role in plant growth and metabolism (Hasan et al., 2008). In the tomatoes, a decrease in N contents (70, 9 and 34%) was observed at different doses (1.0, 2.5 and 5 $mg\,kg^{-1}$) of Cd (Khan et al., 2016b). In hydroponic conditions, a decrease in N content of soybeans was also reported under Cd stress, but the changes in N content were not consistent with increasing Cd stress. Similarly, a decrease in

P content of the soybeans was also observed because of Cd contamination in the soil (Dražić et al., 2004). High concentrations of Cd may cause the reduction of Ca, Mg, and K in the tissues of tomatoes (Khan et al., 2016b) and cucumber plants (Burzynski, 1990). The increasing concentration of Cd in the external medium replaces Ca at the binding site by other heavy metal cations at the exterior surface of the plasma membrane, thereby increasing Ca requirement. Cadmium decreases Ca content because of the competition between Cd and Ca at both Ca channels (Nelson, 1986); but Ca applied exogenously can reverse the impact of Cd toxicity (Wan et al., 2011). Calcium reduces Cd toxicity, mainly by reducing its uptake, and competing with it at the transport site, and secondly, through influencing various physiological processes. It has been reported that liming or alkaline medium reduces Cd uptake, which is facilitated at low pH.

Magnesium shows an interaction with Cd toxicity. Mg-deficient shoots and roots of rice seedlings contain higher Cd concentrations (Chou et al., 2011). However, further results suggest that the protective effect of Mg against Cd toxicity was primarily due to enhanced antioxidant status, and not the inhibition of Cd uptake. In contrast, Kashem and Kawai (2007) have reported that 10 mM Mg alleviated the 0.25 μM Cd toxicity in Japanese mustard spinach plants, and exhibited two-fold higher shoot growth, with decreased shoot Cd concentration of 40%.

## 5.2 Effect of Cd Toxicity on Phytoavailability of Micronutrients

Cd has been shown to interfere with the uptake of micronutrients such as Cu, Fe, Zn, and Mo in medium and small plants, such as wild garlic (Street et al., 2010). In sunflowers, the uptake and translocation of essential nutrients has been imbalanced by Cd, causing depletion of Mn and Fe in their leaves. The decrease in Fe results in the decrease in ferredoxins necessary for the light-induced oxido-reduction process and reduction in chlorophyll content (Nada et al., 2007). Most of the literature regarding the impacts of Cd on nutrients' uptake by plants is based on artificial contamination of Cd at different application doses to plants either in hydroponic conditions, or soil contamination, which may not be fully representative of field conditions (Khan et al., 2016a; Dražić et al., 2004). Therefore, more extensive and long-term field studies need to be conducted to evaluate these effects under natural environmental conditions to give a more realistic view about the findings.

## 6 NUTRIENT TRANSPORT IN PLANTS UNDER Cd STRESS

Nutrients are important for the growth and development of plants. The transport of nutrients in plants through roots from the soil occurs by different ways, such as through the process of diffusion, root interception, and mass flow (Marschner, 2012). Different transporters are involved in the translocation of nutrients into the aerial part of the plant at different levels, and Cd can inhibit these transporters. The different nutrients such as Mg, Ca, and Fe are transported in the plants through membranes, and Cd also competes with them during their transportation (Wang et al., 2016). These nutrients are transported in plant by using cation transport systems and channels (Perfus-Barbeoch et al., 2002). The competition of toxic metals and nutrients occurs in different compartments of binding sites, such as cell membranes and cell walls, that may affect the spreading of the metal (Göthberg et al., 2004). Toxic metals such as Cd enter in the leaves of the plant and disturb the water relation of plants (Perfus-Barbeoch et al., 2002). This leads to the impact on the growth of plants, as well as it effects on the nutrients balancing in plants. Decreasing the concentration of different nutrients such as Mg, S, Fe, P, and Mn was observed in cabbage under Cd stress, which ultimately reduced the growth of the cabbage plant (Sun and Shen, 2007). Nutrients' uptake by the plants also change by the interfiling of Cd in the cell membrane, which results in the leakage of different nutrients from the cell membrane (Obata and Umebayashi, 1997) and also, the transport system of the membrane was effected (Rezvani et al., 2012).

## 7 CADMIUM TOXICITY-INDUCED WATER STRESS IN PLANTS

### 7.1 Less Water Uptake by Plant Roots Due to Cd Toxicity

The main functions of plant roots are the absorption of water and inorganic nutrients, supporting the plant body and anchoring it to the ground, storage of food and nutrients, and vegetative reproduction. These organs are also the first-contact sites for heavy metal ions, and significantly higher amounts of metal usually accumulate in them than in the above-ground plant parts (Feleafel and Mirdad, 2013). Cadmium (Cd) has been an environmental concern to the agricultural system for more than a thousand years (Di Toppi and Gabbrielli, 1999). The Cd accumulates in plants through the root system in the soil, and is responsible for different changes in plants, such as its physiology and biochemical activity (Khan et al., 2007; Feng et al., 2010). The stomata close after up taking of Cd, and as a result, there is less water uptake by the plants (Hossain et al., 2010). The accumulation of Cd in plants cause toxicity by disturbing the enzymes cycle, photosynthesis, and also changing the antioxidants' and carbohydrates' metabolism (Mobin and Khan, 2007; Khan et al., 2009; Shi et al., 2010).

## 7.2 Cell Expansion as Influenced by Cd Toxicity

The plant cell wall determines the cell size and shape through the mechanical control of cell expansion, which governs tissue and organ morphology. It can be considered a protective barrier, and is a complex structure composed of cellulose microfibrils and non-cellulosic neutral polysaccharides embedded in a physiologically active pectin matrix, cross-linked with structural proteins and, depending on the tissue/organ, with lignin (McCann and Roberts, 1991; Carpita and Gibeaut, 1993). (Wasteneys and Galway, 2003) stated that the microtubules play an important role in cell differentiation, expansion, division, and organization. While the polarization and depolarization of the microtubules are affected by the presence of heavy metals in living cells (Jiang et al., 2014).

## 7.3 Restricted Water Movement in Plant Parts Due to Cd Toxicity

Water absorbed from the ground by the roots is transported through the xylem network to the leaves. When transpiration occurs, water is not taken up actively by the roots, but instead moves passively through them in response to a water potential gradient established by transpiration. An excess of heavy metals influences the efficiency of the water flow by lowering the transpiration rate and/or through changes in stomatal resistance in leaves (Barceló and Poschenrieder, 1990). In the absence of transpiration, for example, when stomata are closed, water movement is driven by the active pumping of solutes in roots (Steudle, 2000; Javot and Maurel, 2002). Water transport through the roots is important because hydraulic conductivity (Lp) influences the water supply for the whole plant. It can also be supposed that extensive tissue damage was involved in the dehydration of roots evoked by Cd and Ni in *O. sativa* or *P. sativum* (Hernández et al., 1997; Llamas et al., 2000).

### 7.4 Interactions Between Drought and Cd Toxicity

Heavy metals toxicity and their effect on plants were studied under contaminated and pure conditions (Hall, 2002; Greger, 2004). The effect of heavy metals on plants also deals with another related environmental stress condition, drought, which gets less attention (Barceló and Poschenrieder, 1990). Drought stress is an important threat to the environment that mostly occurs where contamination of different metals are present, because the soils that contain metals have low microbe activity due to the presence of poor structure. This may lead to less presence of organic matter, and ultimately the water holding capacity of that soils are reduced (Derome and Nieminen, 1998). The heavy metals toxicity disturbs the water relationship of plants and reduces the concentration of ABA (Barcelö et al., 1986). The walls of the cell are less elastic (Barcelö et al., 1986), and negative impacts of the enlargement of roots (Kahle, 1993) also reduce the production and distribution of biomass to the roots (Ryser and Emerson, 2007), and ultimately, reduce the water uptake from the root (Przedpelska-Wasowicz and Wierzbicka, 2011). All stomata were also closed by the application of Cd; as a result, the stomatal conductance was affected, and transpiration losses were observed in Cd-treated plants (Barceló et al., 1986). The Cd uptake by the plant indicated that Cd-treated plants showed stomatal inhibitions (Perfus-Barbeoch et al., 2002).

## 8 MANAGEMENT STRATEGIES TO OFFSET NUTRIENT IMBALANCE UNDER Cd STRESS

The strategies used to offset the Cd stress-induced nutrient imbalance in plants have the main objective of lowering Cd uptake. The underlying mechanisms will involve lowering of bioavailable Cd in soil, or competitive inhibition of Cd uptake at the root surface. Based upon the higher ease of adaptability and degree of effectiveness, the strategies, including application of various organic and inorganic amendments, plant growth regulators, and microbial inoculants, are discussed in the following sections.

### 8.1 Use of Organic and Inorganic Amendments

Various organic and inorganic substance are used for phytostabilization of Cd in soil, and to reduce their uptake by plant roots (Putwattana et al., 2015). Among these are biochar manure compost, biosolid and municipal solid wastes, and farmyard manure, which are famous organic amendments.

Phosphate compounds, sepiolite, bentonite lignite, gypsum and mineral nutrients mentioning inorganic amendments. If organic amendments, such as biosolid and municipal solid wastes and manure compost can provide nutrients to crop, and if these have less Cd, they will act as a Cd sink and stimulate bioremediation of Cd (Park et al., 2011). A different kind of biochar can also help reduce bioavailable Cd in soil for a longer time (Li et al., 2016). Surface ligands of biochar have adsorptive effects for heavy metals, thus they can retain/immobilize metals such as Cd, Ni, Cu, and Pb (Uchimiya et al., 2011). Biochar, if modified like Fe-Biochar, can be significant in reducing Cd uptake by rice in rice systems (Yin et al., 2017). In stabilization of Cd, different biochar types, such as rice straw biochar (Lu et al., 2017; Jiang et al., 2012), corn straw and/or hardwood derived biochar (Li et al., 2016), and wood derived biochar (Oustriere et al., 2017) are effective. Biosolids can be effective in Alifisols and Entisols for Cd sorption (Shaheen et al., 2017). Manure can also help in sorption and stabilization of Cd in contaminated soils. Chicken manure compost helps reduce bioavailability of Cd to wheat (Liu et al., 2009). Manure compost increases soil's nutrient status and improves its physico-chemical properties, and helps adsorb Cd as well, such as poultry waste compost (Chen et al., 2010).

Inorganic amendments also have sportive effects for Cd in different soils. Application of bentonite, phosphate, and sepiolite convert available Cd to mostly carbonate-associated forms, thus reducing their toxicity (Sun et al., 2016). Amendments such as elemental sulphur, monoammonium phosphate, and gypsum help reduce negative impacts of Cd in plants (Qayyum et al., 2017). Silicon-rich amendments such as fly ash and steel slag can immobilize significant amounts of Cd in the form of silicates, and stop their uptake in plants (Gu et al., 2011). Si application in *Solanum nigrum* L. has proven to reduce Cd toxicity to plants (Liu et al., 2013a,b). Selenium is also reported to mitigate Cd toxicity in rice plants (Lin et al., 2012). Both organic and inorganic amendments can help crops withstand adverse effects of Cd toxicity, if applied in proper doses and time.

## 8.2 Exogenous Application of Plant Growth Regulators

Growth regulators such as auxins, abscisic acid, jasmonic acid, cytokinin, ethylene, NO, polyamines, and brassinosteroids are important in controlling plant growth in Cd toxicity (Asgher et al., 2015). Strigolactones, another class of plant growth promoter in abiotic stress, has shown to have a significant impact on decreasing Cd toxicity in *Panium virgatum* seedlings when applied exogenously. The auxins-to-cytokinin ratio influences root development of *Arabidopsis thaliana* in Cd-contaminated soils. Vitti et al. (2013) reported that auxins have a positive impact on root growth, while cytokinin has a negative influence in *Arabidopsis thaliana* under Cd stress. They also reported that auxins also have a positive response on the *Nitrilase (AtNIT) gene*. Cytokinin has been reported to positively influence growth and photosynthesis of *Pisum sativum* grown under Cd stress (Gangwar et al., 2010). Ethylene, another phytohormone, is reported to have a positive response to photosynthesis and S assimilation of *Brassica juncea* grown under Cd toxicity (Masood et al., 2012). On the other hand, ethylene suppresses production of $H_2O_2$ in tissues of *Lycopersicon esculentum* experiencing Cd toxicity (Yakimova et al., 2006).

Abscisic acid is a plant hormone involved in stress signaling (Tuteja, 2007). It has a tendency to slow down the uptake of Cd in Cd-tolerant rice grown in contaminated soils, as compared with the Cd-sensitive cultivar (Hsu and Kao, 2003). Salicylic acid is involved in antioxidant defense mechanisms, and can mitigate oxidative damage done by Cd in rice (Guo et al., 2007). It enhances photosynthetic activity in *Cannabis sativa* (Shi et al., 2009), and decreases oxidative stress in *Phaseolus aureus*, *Vicia sativa* (Zhang et al., 2011), and *Glycine* max (Noriega et al., 2012) under Cd toxicity. On the other hand, studies have shown that exogenous applied NO can show a protective role under Cd toxicity (He et al., 2014), and has a positive influence on chlorophyll contents of *Oryza sativa* (Hsu and Kao, 2003). NO is a signaling compound (Gill et al., 2013) that can mitigate Cd-induced oxidative stress (Laspina et al., 2005). Jamonic acid and jasmonates have multivariate effects on plants under stress conditions (Ahmad et al., 2016). Exogenously applied jasmonic acid has a suppressive effect on the accumulation of Cd, malondialdehyde, and $H_2O_2$ in *Vicia faba* L. under Cd stress (Ahmad et al., 2017). Brassinosteroids have supportive effects for plants experiencing abiotic stress (Fariduddin et al., 2014). These stress-suppressing phytohormones have shown an ameliorative influence against the inhibitory effects of Cd toxicity in radish seedlings (Anuradha and Rao, 2007). Foliar application of brassinosteroids enhances quality and yield of *Solanum lycopersicum* compared with the control group under Cd stress. Polyamines are organic compounds with multiple amino groups. They also act as growth regulators, and are involved in suppressing abiotic stress in plants (Gill and Tuteja, 2010; Ahmad et al., 2012). Polyamines can suppress oxidative stress in *Helianthus annuus*

(Groppa et al., 2001) and *Gracilaria dura* (Kumar et al., 2012) grown under Cd contamination. Exogenous application of phytohormones not only suppress Cd toxic effects, but also increases plant growth, quality, and nutritive values (Asgher et al., 2015).

## 8.3 Incorporation of Microbial Inoculants

In terrestrial ecosystems, soil is a rich consortium of microbes that play a role in nutrient dynamics (Van Der Heijden et al., 2008). Soil microbes have different metabolic mechanisms that can control bioavailability and ecotoxicology of metals in soils (Giller et al., 2009). Studies conducted by (Zhu et al., 2012) show that Cd and DDT co-contaminated soils can be remediated by using Cd hyper-accumulator plants *Sedum alfredii*, and DDT degrading microbes. Three strains of *Pseudomonas aeruginosa* species isolated from the active sludge of a food factory were mutated and used for accumulation/stabilization of Cd successfully (Kermani et al., 2010). Pseudomonas isolated from soil, water, and sediments of metal in a contaminated industrial area have also shown similar potential for Cd tolerance (Krishna et al., 2012). Bacterial strains resistant to Cd have been isolated form tannery effluents (Marzan et al., 2017), waste water (Chakraborty et al., 2014), metal contaminated sewage (Raja et al., 2009), heavy metal contaminated soils (Kanazawa and Mori, 1996), and phylloplane (Mohamed and Abo-Amer, 2012). *Rhodospirillum rubrum* inoculation has had significant effects on Cd bioaccumulation of absorbed Cd+2 (Smieljan et al., 2003). Bacillus species of soil microbes have a tendency for bioaccumulation of metals such as Cd, Zn, and copper (Costa and Duta, 2001). Soil bioaugmentation of *Bacillus* species have also proved effective in Cd phytoavailability in contaminated agricultural soils (Jézéquel et al., 2005). Bacterial Strains Cdrb1 isolated from contaminated soils of Gujrat have shown potential for Cd bioaccumulation (Dabhi et al., 2013).

Inoculum of plant-growth-promoting bacteria in soil may decrease metal toxicity by producing phytohormones, enzymes, or mineral solubilization (Ma et al., 2016a). Microbes produce organic compounds such as siderophore, which help them withstand Cd toxicity (Dimkpa et al., 2009). Metal-tolerant growth-promoting microbes such as *Bacillus subtilis*, *Alcaligenes feacalisand*, and *Pseudomonas aeruginosa* can be inoculated in soil to promote plant growth and improve soil health (Ndeddy Aka and Babalola, 2016). Inoculation of rhizosphere microorganisms in soil can produce an ideal plant-microbe interaction to mitigate toxic effects, and even remediate Cd toxicity to plants (Liang et al., 2014).

## 9 FUTURE PERSPECTIVES

Research needs to be done on a cumulative, multidisciplinary approach involving genetic engineering of Cd tolerance, phytohormone production, and exogenous amendments application for Cd-contaminated soils. Modern soil reclamation techniques need to be tested more frequently in alkaline soils as well, because these soils are also under threat of Cd toxicity, as seen in Pakistan. Interdisciplinary work needs to be done involving soil scientists, plant breeders for technical research on Cd decontamination of soils, and Cd-resistant plant development. While rural sociologists and agriculture extension workers in collaboration with dieticians can help disperse knowledge about sustainable and healthy crop production for animal and human use. A great deal of research needs to be done to understand cellular mechanisms involved in Cd inter-/intra-cellular translocation. Maybe in the future we will be able to check which proteins are responsible for Cd translocation into plant tissue. This can be done only with interdisciplinary modern research.

## 10 SUMMARY

Cadmium is one of the most toxic metals prevailing in agricultural soils, and it influences plant growth adversely. Different sources are elevating Cd concentrations in agriculture soils, which causes nutritional imbalance and water stress in a variety of crops, leading to growth reduction and contamination of food. It is an easy choice to reduce Cd input into the soil, but difficult to execute, especially in developing countries. The practical option is to manage the nutritional imbalance in soil to maintain the quantity and quality of food grown in contaminated soils. Different organic and inorganic amendments, along with microbial consortia and phytohormones, make a perfect recipes for the health of Cd-contaminated soils. The future research must be focused on integrated approaches for safe food production in the contaminated soils.

## REFERENCES

Abe, T., Fukami, M., Ogasawara, M., 2008. Cadmium accumulation in the shoots and roots of 93 weed species. Soil Sci. Plant Nutr. 54, 566–573. https://doi.org/10.1111/j.1747-0765.2008.00288.x.

Ahmad, P., Alyemeni, M.N., Wijaya, L., Alam, P., Ahanger, M.A., Alamri, S.A., 2017. Jasmonic acid alleviates negative impacts of cadmium stress by modifying osmolytes and antioxidants in faba bean (*Vicia faba* L.). Arch. Agron. Soil Sci. 63, 1889–1899. https://doi.org/10.1080/03650340.2017.1313406.

Ahmad, P., Kumar, A., Gupta, A., Hu, X., Ul Rehman Hakeem, K., Azooz, M.M., Sharma, S., 2012. Polyamines: role in plants under abiotic stress. In: Crop Production for Agricultural Improvement. Springer, Dordrecht, pp. 491–512. https://doi.org/10.1007/978-94-007-4116-4_19.

Ahmad, P., Rasool, S., Gul, A., Sheikh, S.A., Akram, N.A., Ashraf, M., Kazi, A.M., Gucel, S., 2016. Jasmonates: multifunctional roles in stress tolerance. Front. Plant Sci. 7. https://doi.org/10.3389/fpls.2016.00813.

Alahabadi, A., Ehrampoush, M.H., Miri, M., Ebrahimi Aval, H., Yousefzadeh, S., Ghaffari, H.R., Ahmadi, E., Talebi, P., Abaszadeh Fathabadi, Z., Babai, F., Nikoonahad, A., Sharafi, K., Hosseini-Bandegharaei, A., 2017. A comparative study on capability of different tree species in accumulating heavy metals from soil and ambient air. Chemosphere 172, 459–467. https://doi.org/10.1016/j.chemosphere.2017.01.045.

Anuradha, S., Rao, S.S.R., 2007. The effect of brassinosteroids on radish (*Raphanus sativus* L.) seedlings growing under cadmium stress. Plant Soil Environ. 53, 465–472.

Asgher, M., Khan, M.I.R., Anjum, N.A., Khan, N.A., 2015. Minimising toxicity of cadmium in plants—role of plant growth regulators. Protoplasma. https://doi.org/10.1007/s00709-014-0710-4.

Ashraf, M.A., Maah, M.J., Yusoff, I., 2012. Chemical speciation and potential mobility of heavy metals in the soil of former tin mining catchment. Sci. World J. 2012, 1–11. https://doi.org/10.1100/2012/125608.

Barcelö, J., Cabot, C., Poschenreider, C., 1986. Cadmium-induced decrease of water stress resistance in bush bean plants (*Phaseolus vulgaris* L. cv. Contender) II. Effects of cd on endogenous abscisic acid levels. J. Plant Physiol. 125, 27–34. https://doi.org/10.1016/S0176-1617(86)80240-1.

Barceló, J., Poschenrieder, C., 1990. Plant water relations as affected by heavy metal stress: a review. J. Plant Nutr. 13, 1–37. https://doi.org/10.1080/01904169009364057.

Benavides, M.P., Gallego, S.M., Tomaro, M.L., 2005. Cadmium toxicity in plants. Braz. J. Plant Physiol. https://doi.org/10.1590/S1677-04202005000100004.

Bi, X., Feng, X., Yang, Y., Qiu, G., Li, G., Li, F., Liu, T., Fu, Z., Jin, Z., 2006. Environmental contamination of heavy metals from zinc smelting areas in Hezhang County, western Guizhou, China. Environ. Int. 32, 883–890. https://doi.org/10.1016/j.envint.2006.05.010.

British Geological Survey, 2017. World mineral production 2011-2015. Brit. Geol. Surv.

Burzynski, M., 1990. Activity of some enzymes involved in $NO_3$ assimilation in cucumber seedlings treated with lead or cadmium. Acta Physiol. Plant. 12, 105–109.

Cao, R.L., Jia, X.K., Zhang, J.S., 1999. Prevention of treatment for cadmium contaminated farmland soils. Sci. Technol. Tianjin Agric. For. 12, 12–17 (in Mandarin).

Carpita, N.C., Gibeaut, D.M., 1993. Structural models of primary cell walls in flowering plants: consistency of molecular structure with the physical properties of the walls during growth. Plant J. 3, 1–30. https://doi.org/10.1111/j.1365-313X.1993.tb00007.x.

Chaffei, C., Pageau, K., Suzuki, A., Gouia, H., Ghorbel, M.H., Masclaux-Daubresse, C., 2004. Cadmium toxicity induced changes in nitrogen management in lycopersicon esculentum leading to a metabolic safeguard through an amino acid storage strategy. Plant Cell Physiol. 45, 1681–1693. https://doi.org/10.1093/pcp/pch192.

Chakraborty, S., Mukherjee, A., Khuda-Bukhsh, A.R., Das, T.K., 2014. Cadmium-induced oxidative stress tolerance in cadmium resistant *Aspergillus foetidus*: its possible role in cadmium bioremediation. Ecotoxicol. Environ. Saf. 106, 46–53. https://doi.org/10.1016/j.ecoenv.2014.04.007.

Chang, C.-Y., Xu, X.-H., Liu, C.-P., Li, S.-Y., Liao, X.-R., Dong, J., Li, F.-B., 2014. Heavy metal accumulation in balsam pear and cowpea related to the geochemical factors of variable-charge soils in the Pearl River Delta, South China. Environ. Sci. Process. Impacts 16, 1790–1798. https://doi.org/10.1039/c3em00637a.

Chen, H.S., Huang, Q.Y., Liu, L.N., Cai, P., Liang, W., Li, M., 2010. Poultry manure compost alleviates the phytotoxicity of soil cadmium: influence on growth of pakchoi (*Brassica chinensis* L.). Pedosphere 20, 63–70. https://doi.org/10.1016/S1002-0160(09)60283-6.

Chen, Z.J., Sheng, X.F., He, L.Y., Huang, Z., Zhang, W.H., 2013. Effects of root inoculation with bacteria on the growth, Cd uptake and bacterial communities associated with rape grown in Cd-contaminated soil. J. Hazard. Mater. 244–245, 709–717. https://doi.org/10.1016/j.jhazmat.2012.10.063.

Cheng, K., Tian, H.Z., Zhao, D., Lu, L., Wang, Y., Chen, J., Liu, X.G., Jia, W.X., Huang, Z., 2014. Atmospheric emission inventory of cadmium from anthropogenic sources. Int. J. Environ. Sci. Technol. 11, 605–616. https://doi.org/10.1007/s13762-013-0206-3.

Chou, T.-S., Chao, Y.-Y., Huang, W.-D., Hong, C.-Y., Kao, C.H., 2011. Effect of magnesium deficiency on antioxidant status and cadmium toxicity in rice seedlings. J. Plant Physiol. 168, 1021–1030. https://doi.org/10.1016/j.jplph.2010.12.004.

Clemens, S., 2006. Toxic metal accumulation, responses to exposure and mechanisms of tolerance in plants. Biochimie. https://doi.org/10.1016/j.biochi.2006.07.003.

Cloquet, C., Carignan, J., Libourel, G., Sterckeman, T., Perdrix, E., 2006. Tracing source pollution in soils using cadmium and lead isotopes. Environ. Sci. Technol. 40, 2525–2530. https://doi.org/10.1021/es052232.

Colacino, J.A., Arthur, A.E., Ferguson, K.K., Rozek, L.S., 2014. Dietary antioxidant and anti-inflammatory intake modifies the effect of cadmium exposure on markers of systemic inflammation and oxidative stress. Environ. Res. 131, 6–12. https://doi.org/10.1016/j.envres.2014.02.003.

Costa, A.C.A.D., Duta, F.P., 2001. Bioaccumulation of copper, zinc, cadmium and lead by *Bacillus* sp., *Bacillus cereus, Bacillus sphaericus* and *Bacillus subtilis*. Braz. J. Microbiol. 32, 1–5.

Dabhi, B., Mistry, K.N., Lal, S., Mehta, D., Thakar, R., 2013. Bioaccumulation of cadmium by bacterial strains Cdrb1 isolated from cadmium contaminated soil of Gujarat. JBB 4, 72–79.

Demenesku, J., Mirkov, I., Ninkov, M., Popov Aleksandrov, A., Zolotarevski, L., Kataranovski, D., Kataranovski, M., 2014. Acute cadmium administration to rats exerts both immunosuppressive and proinflammatory effects in spleen. Toxicology 326, 96–108. https://doi.org/10.1016/j.tox.2014.10.012.

Dermont, G., Bergeron, M., Mercier, G., Richer-Laflèche, M., 2008. Soil washing for metal removal: a review of physical/chemical technologies and field applications. J. Hazard. Mater. 152, 1–31. https://doi.org/10.1016/j.jhazmat.2007.10.043.

Derome, J., Nieminen, T., 1998. Metal and macronutrient fluxes in heavy-metal polluted scots pine ecosystems in SW Finland. Environ. Pollut. 103, 219–228. https://doi.org/10.1016/S0269-7491(98)00118-3.

Dimkpa, C.O., Merten, D., Svatoš, A., Büchel, G., Kothe, E., 2009. Siderophores mediate reduced and increased uptake of cadmium by Streptomyces tendae F4 and sunflower (*Helianthus annuus* L.), respectively. J. Appl. Microbiol. 107, 1687–1696. https://doi.org/10.1111/j.1365-2672.2009.04355.x.

Di Toppi, L.S., Gabbrielli, R., 1999. Response to cadmium in higher plants. Environ. Exp. Bot. 41, 105–130.

Dražić, G., Mihailović, N., Stojanović, Z., 2004. Cadmium toxicity: the effect on macro- and micro-nutrient contents in soybean seedlings. Biol. Plant. 48, 605–607. https://doi.org/10.1023/B:BIOP.0000047160.79306.b7.

Dunbar, K.R., McLaughlin, M.J., Reid, R.J., 2003. The uptake and partitioning of cadmium in two cultivars of potato (*Solanum tuberosum* L.). J. Exp. Bot. 54, 349–354. https://doi.org/10.1093/jxb/54.381.349.

Fariduddin, Q., Yusuf, M., Ahmad, I., Ahmad, A., 2014. Brassinosteroids and their role in response of plants to abiotic stresses. Biol. Plant. 58, 9–17. https://doi.org/10.1007/s10535-013-0374-5.

Feleafel, M.N., Mirdad, Z.M., 2013. Hazard and effects of pollution by lead on vegetable crops. J. Agric. Environ. Ethics. https://doi.org/10.1007/s10806-012-9403-1.

Feng, J., Shi, Q., Wang, X., Wei, M., Yang, F., Xu, H., 2010. Silicon supplementation ameliorated the inhibition of photosynthesis and nitrate metabolism by cadmium (Cd) toxicity in *Cucumis sativus* L. Sci. Hortic. 123, 521–530. https://doi.org/10.1016/j.scienta.2009.10.013.

Dwivedi, G.K., Upadhyay, S.K., Mishra, A.K., Singh, A.K., 2014. Hyper accumulation of cadmium in solanum nigrum l. and their effects on phyto-chemicals and antioxidant enzymatic activities. Int. J. Pharm. Sci. Res. 5, 1424–1430. https://doi.org/10.13040/IJPSR.0975-8232.5(4).1424-30.

Gamalero, E., Lingua, G., Berta, G., Glick, B.R., 2009. Beneficial role of plant growth promoting bacteria and arbuscular mycorrhizal fungi on plant responses to heavy metal stress. Can. J. Microbiol. 55, 501–514. https://doi.org/10.1139/W09-010.

Gangwar, S., Singh, V.P., Prasad, S.M., Maurya, J.N., 2010. Modulation of manganese toxicity in *Pisum sativum* L. seedlings by kinetin. Sci. Hortic. 126, 467–474. https://doi.org/10.1016/j.scienta.2010.08.013.

Ghani, A., Ali, Z., Islam, T., Sanaullah, S., Saeed, S., 2014. Nutrient evaluation and elemental analysis of four selected medicinal plants of soon valley Khushab, Punjab, Pakistan. Pak. J. Pharm. Sci. 27, 597–600.

Gill, S.S., Hasanuzzaman, M., Nahar, K., Macovei, A., Tuteja, N., 2013. Importance of nitric oxide in cadmium stress tolerance in crop plants. Plant Physiol. Biochem. https://doi.org/10.1016/j.plaphy.2012.12.001.

Gill, S.S., Khan, N.A., Tuteja, N., 2011. Differential cadmium stress tolerance in five indian mustard (*Brassica juncea* L.) cultivars: an evaluation of the role of antioxidant machinery. Plant Signal. Behav. 6, 293–300. https://doi.org/10.4161/psb.6.2.15049.

Gill, S.S., Tuteja, N., 2010. Polyamines and abiotic stress tolerance in plants. Plant Signal. Behav. 5, 26–33. https://doi.org/10.4161/psb.5.1.10291.

Giller, K.E., Witter, E., McGrath, S.P., 2009. Heavy metals and soil microbes. Soil Biol. Biochem. 41, 2031–2037. https://doi.org/10.1016/j.soilbio.2009.04.026.

Göthberg, A., Greger, M., Holm, K., Bengtsson, B.-E., 2004. Influence of nutrient levels on uptake and effects of mercury, cadmium, and lead in water spinach. J. Environ. Qual. 33, 1247–1255. https://doi.org/10.2134/jeq2004.1247.

Grant, C.A., Clarke, J.M., Duguid, S., Chaney, R.L., 2008. Selection and breeding of plant cultivars to minimize cadmium accumulation. Sci. Total Environ. 390, 301–310. https://doi.org/10.1016/j.scitotenv.2007.10.038.

Greger, M., 2004. Metal availability, uptake, transport and accumulation in plants. In: Heavy Metal Stress in Plants. Springer Verlag, Berlin/Heidelberg, p. 3218. https://doi.org/10.1007/978-3-662-07743-6.

Greenwood, N.N., Earnshaw, A., 2001. Chemistry of the Elements. Butterworth Heinemann, Oxford, UK, p. 1341.

Groppa, M.D., Tomaro, M.L., Benavides, M.P., 2001. Polyamines as protectors against cadmium or copper-induced oxidative damage in sunflower leaf discs. Plant Sci. 161, 481–488. https://doi.org/10.1016/S0168-9452(01)00432-0.

Gu, H.H., Qiu, H., Tian, T., Zhan, S.S., Deng, T.H.B., Chaney, R.L., Wang, S.Z., Tang, Y.T., Morel, J.L., Qiu, R.L., 2011. Mitigation effects of silicon rich amendments on heavy metal accumulation in rice (*Oryza sativa* L.) planted on multi-metal contaminated acidic soil. Chemosphere 83, 1234–1240. https://doi.org/10.1016/j.chemosphere.2011.03.014.

Guo, B., Liang, Y.C., Zhu, Y.G., Zhao, F.J., 2007. Role of salicylic acid in alleviating oxidative damage in rice roots (*Oryza sativa* L.) subjected to cadmium stress. Environ. Pollut. 147, 743–749. https://doi.org/10.1016/j.envpol.2006.09.007.

Gupta, N., Khan, D.K., Santra, S.C., 2008. An assessment of heavy metal contamination in vegetables grown in wastewater-irrigated areas of Titagarh, West Bengal, India. Bull. Environ. Contam. Toxicol. 80, 115–118. https://doi.org/10.1007/s00128-007-9327-z.

Hall, J.L., 2002. Cellular mechanisms for heavy metal detoxification and tolerance. J. Exp. Bot. 53, 1–11. https://doi.org/10.1093/jxb/53.366.1.

Hart, J.J., Welch, R.M., Norvell, W.A., Kochian, L.V., 2002. Transport interactions between cadmium and zinc in roots of bread and durum wheat seedlings. Physiol. Plant. 116, 73–78. https://doi.org/10.1034/j.1399-3054.2002.1160109.x.

Hasan, S.A., Fariduddin, Q., Ali, B., Hayat, S., Ahmad, A., 2009. Cadmium: toxicity and tolerance in plants. J. Environ. Biol. 30, 165–174.

Hasan, S.A., Hayat, S., Ali, B., Ahmad, A., 2008. 28-Homobrassinolide protects chickpea (*Cicer arietinum*) from cadmium toxicity by stimulating antioxidants. Environ. Pollut. 151, 60–66. https://doi.org/10.1016/j.envpol.2007.03.006.

He, J., Ren, Y., Chen, X., Chen, H., 2014. Protective roles of nitric oxide on seed germination and seedling growth of rice (*Oryza sativa* L.) under cadmium stress. Ecotoxicol. Environ. Saf. 108, 114–119. https://doi.org/10.1016/j.ecoenv.2014.05.021.

He, Z.L., Yang, X.E., Stoffella, P.J., 2005. Trace elements in agroecosystems and impacts on the environment. J. Trace Elem. Med. Biol. https://doi.org/10.1016/j.jtemb.2005.02.010.

Hentz, S., McComb, J., Miller, G., Begonia, M., Begonia, G., 2012. Cadmium uptake growth and phytochelatin contents of *Triticum aestivum* in response to various concentrations of cadmium. World Environ. 2, 44–50. https://doi.org/10.5923/j.env.20120203.05.

Hernández, L.E., Gárate, A., Carpena-Ruiz, R., 1997. Effects of cadmium on the uptake, distribution and assimilation of nitrate in *Pisum sativum*. Plant Soil 189, 97–106. https://doi.org/10.1023/A:1004252816355.

Hinojosa, M.B., Carreira, J.A., García-Ruíz, R., Dick, R.P., 2004. Soil moisture pretreatment effects on enzyme activities as indicators of heavy metal-contaminated and reclaimed soils. Soil Biol. Biochem. 36, 1559–1568. https://doi.org/10.1016/j.soilbio.2004.07.003.

Hossain, M.A., Hasanuzzaman, M., Fujita, M., 2010. Up-regulation of antioxidant and glyoxalase systems by exogenous glycinebetaine and proline in mung bean confer tolerance to cadmium stress. Physiol. Mol. Biol. Plants 16, 259–272. https://doi.org/10.1007/s12298-010-0028-4.

Hsu, Y.T., Kao, C.H., 2003. Role of abscisic acid in cadmium tolerance of rice (*Oryza sativa* L.) seedlings. Plant. Cell Environ. 26, 867–874. https://doi.org/10.1046/j.1365-3040.2003.01018.x.

Huang, J., Huang, R., Jiao, J.J., Chen, K., 2007. Speciation and mobility of heavy metals in mud in coastal reclamation areas in Shenzhen, China. Environ. Geol. 53, 221–228. https://doi.org/10.1007/s00254-007-0636-7.

Huang, S.H., Bing, P., Yang, Z.H., Chai, L.Y., Zhou, L.C., 2009. Chromium accumulation, microorganism population and enzyme activities in soils around chromium-containing slag heap of steel alloy factory. Trans. Nonferrous Met. Soc. China 19, 241–248. https://doi.org/10.1016/S1003-6326(08)60259-9.

Jamali, M.K., Kazi, T.G., Arain, M.B., Afridi, H.I., Jalbani, N., Kandhro, G.A., Shah, A.Q., Baig, J.A., 2009. Heavy metal accumulation in different varieties of wheat (*Triticum aestivum* L.) grown in soil amended with domestic sewage sludge. J. Hazard. Mater. 164, 1386–1391. https://doi.org/10.1016/j.jhazmat.2008.09.056.

Javot, H., Maurel, C., 2002. The role of aquaporins in root water uptake. Ann. Bot. https://doi.org/10.1093/aob/mcf199.

Jézéquel, K., Perrin, J., Lebeau, T., 2005. Bioaugmentation with a *Bacillus* sp. to reduce the phytoavailable Cd of an agricultural soil: comparison of free and immobilized microbial inocula. Chemosphere 59, 1323–1331. https://doi.org/10.1016/j.chemosphere.2004.11.050.

Ji, P., Song, Y., Jiang, Y., Tang, X., Tong, Y., Gao, P., Han, W., 2016. A two-year field study of phytoremediation using *Solanum nigrum* L. in China. Int. J. Phytoremediation 18, 924–928. https://doi.org/10.1080/15226514.2015.1058325.

Jiang, J., Xu, R., Jiang, T., Li, Z., 2012. Immobilization of Cu(II), Pb(II) and Cd(II) by the addition of rice straw derived biochar to a simulated polluted Ultisol. J. Hazard. Mater. 229–230, 145–150. https://doi.org/10.1016/j.jhazmat.2012.05.086.

Jiang, Z., Qin, R., Zhang, H., Zou, J., Shi, Q., Wang, J., Jiang, W., Liu, D., 2014. Determination of Pb genotoxic effects in *Allium cepa* root cells by fluorescent probe, microtubular immunofluorescence and comet assay. Plant Soil 383, 357–372. https://doi.org/10.1007/s11104-014-2183-9.

Kahle, H., 1993. Response of roots of trees to heavy metals. Environ. Exp. Bot. 33, 99–119. https://doi.org/10.1016/0098-8472(93)90059-O.

Kanazawa, S., Mori, K., 1996. Isolation of cadmium-resistant bacteria and their resistance mechanisms. Soil Sci. Plant Nutr. 42, 725–730. https://doi.org/10.1080/00380768.1996.10416619.

Kashem, M.A., Kawai, S., 2007. Alleviation of cadmium phytotoxicity by magnesium in Japanese mustard spinach. Soil Sci. Plant Nutr. 53, 246–251. https://doi.org/10.1111/j.1747-0765.2007.00129.x.

Kermani, A.J.N., Ghasemi, M.F., Khosravan, A., Farahmand, A., Shakibaie, M.R., 2010. Cadmium bioremediation by metal-resistant mutated bacteria isolated from active sludge of industrial effluent. Iran. J. Environ. Health Sci. Eng. 7, 279–286.

Khan, A., Khan, S., Alam, M., Khan, M.A., Aamir, M., Qamar, Z., Rehman, Z.U., Perveen, S., 2016a. Toxic metal interactions affect the bioaccumulation and dietary intake of macro- and micro-nutrients. Chemosphere 146, 121–128. https://doi.org/10.1016/j.chemosphere.2015.12.014.

Khan, A., Khan, S., Khan, M.A., Qamar, Z., Waqas, M., 2015. The uptake and bioaccumulation of heavy metals by food plants, their effects on plants nutrients, and associated health risk: a review. Environ. Sci. Pollut. Res. 22, 13772–13799. https://doi.org/10.1007/s11356-015-4881-0.

Khan, A.R., Ullah, I., Khan, A.L., Hong, S.J., Waqas, M., Park, G.S., Kwak, Y., Choi, J., Jung, B.K., Park, M., Lee, I.J., Shin, J.H., 2014. Phytostabilization and physicochemical responses of Korean ecotype *Solanum nigrum* L. to cadmium contamination. Water. Air. Soil Pollut. 225, 1–11. https://doi.org/10.1007/s11270-014-2147-y.

Khan, M.H., Kushwaha, K.P., Thomas, T., 2017. Soil physico-chemical properties, productivity and accumulation of heavy metals in fenugreek (*Trigonella foenum-graceum* L.), and their availability in soil as affected by Mussoorie rock phosphate. Int. J. Chem. Stud. 5, 24–28.

Khan, N.A., Anjum, N.A., Nazar, R., Iqbal, N., 2009. Increased activity of ATP-sulfurylase and increased contents of cysteine and glutathione reduce high cadmium-induced oxidative stress in mustard cultivar with high photosynthetic potential. Russ. J. Plant Physiol. 56, 670–677. https://doi.org/10.1134/S1021443709050136.

Khan, N.A., Samiullah, S., Nazar, R., 2007. Activities of antioxidative enzymes, sulphur assimilation, photosynthetic activity and growth of wheat (*Triticum aestivum* L.) cultivars differing in yield potential under cadmium stress. J. Agron. Crop Sci. 193, 435–444. https://doi.org/10.1111/j.1439-037X.2007.00272.x.

Khan, S., Munir, S., Sajjad, M., Li, G., 2016b. Urban park soil contamination by potentially harmful elements and human health risk in Peshawar City, Khyber Pakhtunkhwa, Pakistan. J. Geochem. Explor. 165, 102–110. https://doi.org/10.1016/j.gexplo.2016.03.007.

Khan, S., Rehman, S., Zeb Khan, A., Amjad Khan, M., Tahir Shah, M., 2010. Soil and vegetables enrichment with heavy metals from geological sources in Gilgit, northern Pakistan. Ecotoxicol. Environ. Saf. 73, 1820–1827. https://doi.org/10.1016/j.ecoenv.2010.08.016.

Krishna, M.P., Varghese, R., Babu, A.V., Mohamed Hatha, A.A., 2012. Bioaccumulation of cadmium by *Pseudomonas* Sp. isolated from metal polluted industrial region. Environ. Res. Eng. Manag. 3, 58–64. https://doi.org/10.5755/j01.erem.61.3.1268.

Kubo, K., Watanabe, Y., Matsunaka, H., Seki, M., Fujita, M., Kawada, N., Hatta, K., Nakajima, T., 2011. Differences in cadmium accumulation and root morphology in seedlings of japanese wheat varieties with distinctive grain cadmium concentration. Plant Prod. Sci. 14, 148–155. https://doi.org/10.1626/pps.14.148.

Kumar, M., Bijo, A.J., Baghel, R.S., Reddy, C.R.K., Jha, B., 2012. Selenium and spermine alleviate cadmium induced toxicity in the red seaweed *Gracilaria dura* by regulating antioxidants and DNA methylation. Plant Physiol. Biochem. 51, 129–138. https://doi.org/10.1016/j.plaphy.2011.10.016.

Laspina, N.V., Groppa, M.D., Tomaro, M.L., Benavides, M.P., 2005. Nitric oxide protects sunflower leaves against Cd-induced oxidative stress. Plant Sci. 169, 323–330. https://doi.org/10.1016/j.plantsci.2005.02.007.

Levy, G.J., Fine, P., Bar-Tal, A., 2010. Treated Wastewater in Agriculture: Use and Impacts on the Soil Environment and Crops. John Wiley & Sons, UK. https://doi.org/10.1002/9781444328561.

Li, H., Ye, X., Geng, Z., Zhou, H., Guo, X., Zhang, Y., Zhao, H., Wang, G., 2016. The influence of biochar type on long-term stabilization for Cd and Cu in contaminated paddy soils. J. Hazard. Mater. 304, 40–48. https://doi.org/10.1016/j.jhazmat.2015.10.048.

Liang, X., He, C.Q., Ni, G., Tang, G.E., Chen, X.P., Lei, Y.R., 2014. Growth and cd accumulation of orychophragmus violaceus as affected by inoculation of Cd-tolerant bacterial strains. Pedosphere 24, 322–329. https://doi.org/10.1016/S1002-0160(14)60018-7.

Lin, L., Zhou, W., Dai, H., Cao, F., Zhang, G., Wu, F., 2012. Selenium reduces cadmium uptake and mitigates cadmium toxicity in rice. J. Hazard. Mater. 235–236, 343–351. https://doi.org/10.1016/j.jhazmat.2012.08.012.

Liu, J., Li, K., Xu, J., Liang, J., Lu, X., Yang, J., Zhu, Q., 2003. Interaction of Cd and five mineral nutrients for uptake and accumulation in different rice cultivars and genotypes. Field Crop. Res. 83, 271–281. https://doi.org/10.1016/S0378-4290(03)00077-7.

Liu, J., Qian, M., Cai, G., Yang, J., Zhu, Q., 2007. Uptake and translocation of Cd in different rice cultivars and the relation with Cd accumulation in rice grain. J. Hazard. Mater. 143, 443–447. https://doi.org/10.1016/j.jhazmat.2006.09.057.

Liu, J., Zhang, H., Zhang, Y., Chai, T., 2013a. Silicon attenuates cadmium toxicity in *Solanum nigrum* L. by reducing cadmium uptake and oxidative stress. Plant Physiol. Biochemist 68, 1–7. https://doi.org/10.1016/j.plaphy.2013.03.018.

Liu, L., Chen, H., Cai, P., Liang, W., Huang, Q., 2009. Immobilization and phytotoxicity of Cd in contaminated soil amended with chicken manure compost. J. Hazard. Mater. 163, 563–567. https://doi.org/10.1016/j.jhazmat.2008.07.004.

Liu, Y., Du, Q., Wang, Q., Yu, H., Liu, J., Tian, Y., Chang, C., Lei, J., 2017. Causal inference between bioavailability of heavy metals and environmental factors in a large-scale region. Environ. Pollut. 226, 370–378. https://doi.org/10.1016/j.envpol.2017.03.019.

Liu, Y., Xiao, T., Ning, Z., Li, H., Tang, J., Zhou, G., 2013b. High cadmium concentration in soil in the Three Gorges region: geogenic source and potential bioavailability. Appl. Geochem. 37, 149–156. https://doi.org/10.1016/j.apgeochem.2013.07.022.

Llamas, A., Ullrich, C.I., Sanz, A., 2000. $Cd^{2+}$ effects on transmembrane electrical potential difference, respiration and membrane permeability of rice (*Oryza sativa* L.) roots. Plant Soil 219, 21–28. https://doi.org/10.1023/A:1004753521646.

Lu, K., Yang, X., Gielen, G., Bolan, N., Ok, Y.S., Niazi, N.K., Xu, S., Yuan, G., Chen, X., Zhang, X., Liu, D., Song, Z., Liu, X., Wang, H., 2017. Effect of bamboo and rice straw biochars on the mobility and redistribution of heavy metals (Cd, Cu, Pb and Zn) in contaminated soil. J. Environ. Manag. 186, 285–292. https://doi.org/10.1016/j.jenvman.2016.05.068.

Lux, A., Martinka, M., Vaculík, M., White, P.J., 2011. Root responses to cadmium in the rhizosphere: a review. J. Exp. Bot. 62, 21–37. https://doi.org/10.1093/jxb/erq281.

Ma, Y., Oliveira, R.S., Freitas, H., Zhang, C., 2016a. Biochemical and molecular mechanisms of plant-microbe-metal interactions: relevance for phytoremediation. Front. Plant Sci. 7. https://doi.org/10.3389/fpls.2016.00918.

Ma, Y., Rajkumar, M., Zhang, C., Freitas, H., 2016b. Beneficial role of bacterial endophytes in heavy metal phytoremediation. J. Environ. Manag. 174, 14–25. https://doi.org/10.1016/j.jenvman.2016.02.047.

Maghakyan, N., Tepanosyan, G., Belyaeva, O., Sahakyan, L., Saghatelyan, A., 2017. Assessment of pollution levels and human health risk of heavy metals in dust deposited on Yerevan's tree leaves (Armenia). Acta Geochim. 36, 16–26. https://doi.org/10.1007/s11631-016-0122-6.

Mahar, A., Wang, P., Li, R., Zhang, Z., 2015. Immobilization of lead and cadmium in contaminated soil using amendments: a review. Pedosphere 25, 555–568. https://doi.org/10.1016/S1002-0160(15)30036-9.

Marschner, P., 2012. Marschner's Mineral Nutrition of Higher Plants. Academic Press, USA. https://doi.org/10.1016/C2009-0-63043-9.

Marzan, L.W., Hossain, M., Mina, S.A., Akter, Y., Chowdhury, A.M.M.A., 2017. Isolation and biochemical characterization of heavy-metal resistant bacteria from tannery effluent in Chittagong city, Bangladesh: bioremediation viewpoint. Egypt. J. Aquat. Res. 43, 65–74. https://doi.org/10.1016/j.ejar.2016.11.002.

Masood, A., Iqbal, N., Khan, N.A., 2012. Role of ethylene in alleviation of cadmium-induced photosynthetic capacity inhibition by sulphur in mustard. Plant Cell Environ. 35, 524–533. https://doi.org/10.1111/j.1365-3040.2011.02432.x.

McCann, M., Roberts, K., 1991. Architecture of the primary cell wall. In: Lloyd, C.W. (Ed.), The Cytoskeletal Basis of Plant Growth and Form. Academic Press, London, UK, pp. 109–129.

Mobin, M., Khan, N.A., 2007. Photosynthetic activity, pigment composition and antioxidative response of two mustard (*Brassica juncea*) cultivars differing in photosynthetic capacity subjected to cadmium stress. J. Plant Physiol. 164, 601–610. https://doi.org/10.1016/j.jplph.2006.03.003.

Mohamed, R.M., Abo-Amer, A.E., 2012. Isolation and characterization of heavy-metal resistant microbes from roadside soil and phylloplane. J. Basic Microbiol. 52, 53–65. https://doi.org/10.1002/jobm.201100133.

Murtaza, G., Javed, W., Hussain, A., Wahid, A., Murtaza, B., Owens, G., 2015. Metal uptake via phosphate fertilizer and city sewage in cereal and legume crops in Pakistan. Environ. Sci. Pollut. Res. 22, 9136–9147. https://doi.org/10.1007/s11356-015-4073-y.

Mutlu, A., Lee, B.K., Park, G.H., Yu, B.G., Lee, C.H., 2012. Long-term concentrations of airborne cadmium in metropolitan cities in Korea and potential health risks. Atmos. Environ. 47, 164–173. https://doi.org/10.1016/j.atmosenv.2011.11.019.

Nada, E., Ferjani, B.A., Ali, R., Imed, B.R.B.M., Makki, B., 2007. Cadmium-induced growth inhibition and alteration of biochemical parameters in almond seedlings grown in solution culture. Acta Physiol. Plant. 29, 57–62.

Naeem, A., Saifullah, Rehman, M.Z.-U., Akhtar, T., Ok, Y.S., Rengel, Z., 2016. Genetic variation in cadmium accumulation and tolerance among wheat cultivars at the seedling stage. Commun. Soil Sci. Plant Anal. 47, 554–562. https://doi.org/10.1080/00103624.2016.1141918.

Nakanishi, H., Ogawa, I., Ishimaru, Y., Mori, S., Nishizawa, N.K., 2006. Iron deficiency enhances cadmium uptake and translocation mediated by the Fe2+ transporters OsIRT1 and OsIRT2 in rice. Soil Sci. Plant Nutr. 52, 464–469. https://doi.org/10.1111/j.1747-0765.2006.00055.x.

Nazar, R., Iqbal, N., Masood, A., Khan, M.I.R., Syeed, S., Khan, N.A., 2012. Cadmium toxicity in plants and role of mineral nutrients in its alleviation. Am. J. Plant Sci. 3, 1476–1489. https://doi.org/10.4236/ajps.2012.310178.

Ndeddy Aka, R.J., Babalola, O.O., 2016. Effect of bacterial inoculation of strains of *Pseudomonas aeruginosa*, *Alcaligenes feacalis* and *Bacillus subtilis* on germination, growth and heavy metal (Cd, Cr, and Ni) uptake of Brassica juncea. Int. J. Phytoremediation 18, 200–209. https://doi.org/10.1080/15226514.2015.1073671.

Nelson, M.T., 1986. Interactions of divalent cations with single calcium channels from rat brain synaptosomes. J. Gen. Physiol. 87, 201–222. https://doi.org/10.1085/jgp.87.2.201.

Nevo, Y., Nelson, N., 2006. The NRAMP family of metal-ion transporters. Biochim. Biophys. Acta: Mol. Cell Res. https://doi.org/10.1016/j.bbamcr.2006.05.007.

Nicholson, F.A., Smith, S.R., Alloway, B.J., Carlton-Smith, C., Chambers, B.J., 2003. An inventory of heavy metals inputs to agricultural soils in England and Wales. Sci. Total Environ. 311, 205–219. https://doi.org/10.1016/S0048-9697(03)00139-6.

Noriega, G., Caggiano, E., Lecube, M.L., Cruz, D.S., Batlle, A., Tomaro, M., Balestrasse, K.B., 2012. The role of salicylic acid in the prevention of oxidative stress elicited by cadmium in soybean plants. Biometals 25, 1155–1165. https://doi.org/10.1007/s10534-012-9577-z.

Nouha, K., Kumar, R.S., Tyagi, R.D., 2016. Heavy metals removal from wastewater using extracellular polymeric substances produced by *Cloacibacterium normanense* in wastewater sludge supplemented with crude glycerol and study of extracellular polymeric substances extraction by different methods. Bioresour. Technol. 212, 120–129. https://doi.org/10.1016/j.biortech.2016.04.021.

Nriagu, J.O., Pacyna, J.M., 1988. Quantitative assessment of worldwide contamination of air, water and soils by trace metals. Nature 333, 134–139. https://doi.org/10.1038/333134a0.

Nziguheba, G., Smolders, E., 2008. Inputs of trace elements in agricultural soils via phosphate fertilizers in European countries. Sci. Total Environ. 390, 53–57. https://doi.org/10.1016/j.scitotenv.2007.09.031.

Obata, H., Umebayashi, M., 1997. Effects of cadmium on mineral nutrient concentrations in plants differing in tolerance for cadmium. J. Plant Nutr. 20, 97–105. https://doi.org/10.1080/01904169709365236.

Oustriere, N., Marchand, L., Rosette, G., Friesl-Hanl, W., Mench, M., 2017. Wood-derived-biochar combined with compost or iron grit for in situ stabilization of Cd, Pb, and Zn in a contaminated soil. Environ. Sci. Pollut. Res. 24, 7468–7481. https://doi.org/10.1007/s11356-017-8361-6.

Paranychianakis, N.V., Angelakis, A.N., Leverenz, H., Tchobanoglous, G., 2006. Treatment of wastewater with slow rate systems: a review of treatment processes and plant functions. Crit. Rev. Environ. Sci. Technol. https://doi.org/10.1080/10643380500542756.

Park, J.H., Lamb, D., Paneerselvam, P., Choppala, G., Bolan, N., Chung, J.W., 2011. Role of organic amendments on enhanced bioremediation of heavy metal(loid) contaminated soils. J. Hazard. Mater. 185, 549–574. https://doi.org/10.1016/j.jhazmat.2010.09.082.

Perfus-Barbeoch, L., Leonhardt, N., Vavasseur, A., Forestier, C., 2002. Heavy metal toxicity: cadmium permeates through calcium channels and disturbs the plant water status. Plant J. 32, 539–548. https://doi.org/10.1046/j.1365-313X.2002.01442.x.

Perkins, R.B., Mason, C.E., 2015. The relative mobility of trace elements from short-term weathering of a black shale. Appl. Geochem. 56, 67–79. https://doi.org/10.1016/j.apgeochem.2015.01.014.

Przedpelska-Wasowicz, E.M., Wierzbicka, M., 2011. Gating of aquaporins by heavy metals in *Allium cepa* L. epidermal cells. Protoplasma 248, 663–671. https://doi.org/10.1007/s00709-010-0222-9.

Putwattana, N., Kruatrachue, M., Kumsopa, A., Pokethitiyook, P., 2015. Evaluation of organic and inorganic amendments on maize growth and uptake of Cd and Zn from contaminated paddy soils. Int. J. Phytoremediation 17, 165–174. https://doi.org/10.1080/15226514.2013.876962.

Qayyum, M.F., Rehman, M.Z.U., Ali, S., Rizwan, M., Naeem, A., Maqsood, M.A., Khalid, H., Rinklebe, J., Ok, Y.S., 2017. Residual effects of monoammonium phosphate, gypsum and elemental sulfur on cadmium phytoavailability and translocation from soil to wheat in an effluent irrigated field. Chemosphere 174, 515–523. https://doi.org/10.1016/j.chemosphere.2017.02.006.

Raja, E.C., Selvam, G.S., Omine, K., 2009. Isolation, identification and characterization of heavy metal resistant bacteria from sewage. International Joint Symposium on Geodisaster Prevention and Geoenvironment in Asia, pp. 205–211. https://doi.org/10.1007/s00128-012-0675-y.

Redjala, T., Zelko, I., Sterckeman, T., Legué, V., Lux, A., 2011. Relationship between root structure and root cadmium uptake in maize. Environ. Exp. Bot. 71, 241–248. https://doi.org/10.1016/j.envexpbot.2010.12.010.

Rehman, M.Z.U., Khalid, H., Akmal, F., Ali, S., Rizwan, M., Qayyum, M.F., Iqbal, M., Khalid, M.U., Azhar, M., 2017. Effect of limestone, lignite and biochar applied alone and combined on cadmium uptake in wheat and rice under rotation in an effluent irrigated field. Environ. Pollut. 227, 560–568. https://doi.org/10.1016/j.envpol.2017.05.003.

Rezvani, M., Zaefarian, F., Miransari, M., Nematzadeh, G.A., 2012. Uptake and translocation of cadmium and nutrients by *Aeluropus littoralis*. Arch. Agron. Soil Sci. 58, 1413–1425. https://doi.org/10.1080/03650340.2011.591385.

Rizwan, M., Ali, S., Abbas, T., Zia-ur-Rehman, M., Hannan, F., Keller, C., Al-Wabel, M.I., Ok, Y.S., 2016a. Cadmium minimization in wheat: a critical review. Ecotoxicol. Environ. Saf. https://doi.org/10.1016/j.ecoenv.2016.04.001.

Rizwan, M., Ali, S., Adrees, M., Rizvi, H., Zia-ur-Rehman, M., Hannan, F., Qayyum, M.F., Hafeez, F., Ok, Y.S., 2016b. Cadmium stress in rice: toxic effects, tolerance mechanisms, and management: a critical review. Environ. Sci. Pollut. Res. 23, 17859–17879. https://doi.org/10.1007/s11356-016-6436-4.

Roy, M., Mcdonald, L.M., 2015. Metal uptake in plants and health risk assessments in metal-contaminated smelter soils. Land Degrad. Dev. 26, 785–792. https://doi.org/10.1002/ldr.2237.

Ryser, P., Emerson, P., 2007. Growth, root and leaf structure, and biomass allocation in *Leucanthemum vulgare* Lam. (Asteraceae) as influenced by heavy-metal-containing slag. Plant Soil 301, 315–324. https://doi.org/10.1007/s11104-007-9451-x.

Satarug, S., 2012. Long-term exposure to cadmium in food and cigarette smoke, liver effects and hepatocellular carcinoma. Curr. Drug Metab. 13, 257–271. https://doi.org/10.2174/138920012799320446.

Saha, J.K., Selladurai, R., Coumar, M.V., Dotaniya, M.L., Kundu, S., Patra, A.K., 2017. Assessment of heavy metals contamination in soil. In: Soil Pollution—An Emerging Threat to Agriculture. Springer, Singapore, pp. 155–191.

Seregin, I.V., Shpigun, L.K., Ivanov, V.B., 2004. Distribution and toxic effects of cadmium and lead on maize roots. Russ. J. Plant Physiol. 51, 525–533. https://doi.org/10.1023/B:RUPP.0000035747.42399.84.

Shah, M.T., Begum, S., Khan, S., 2010. Pedo and biogeochemical studies of mafic and ultramfic rocks in the Mingora and Kabal areas, Swat, Pakistan. Environ. Earth Sci. 60, 1091–1102. https://doi.org/10.1007/s12665-009-0253-8.

Shaheen, S.M., Tsadilas, C.D., Niazi, N.K., Hseu, Z.Y., Ok, Y.S., Selim, M., Rinklebe, J., 2017. Impact of biosolid application rates on competitive sorption and distribution coefficients of Cd, Cu, Ni, Pb, and Zn in an Alfisol and an Entisol. Process. Saf. Environ. Prot. https://doi.org/10.1016/j.psep.2017.10.013.

Sharma, P., Dubey, R.S., 2007. Involvement of oxidative stress and role of antioxidative defense system in growing rice seedlings exposed to toxic concentrations of aluminum. Plant Cell Rep. 26, 2027–2038. https://doi.org/10.1007/s00299-007-0416-6.

Shi, G., Liu, C., Cai, Q., Liu, Q., Hou, C., 2010. Cadmium accumulation and tolerance of two safflower cultivars in relation to photosynthesis and antioxidantive enzymes.

Bull. Environ. Contam. Toxicol. 85, 256–263. https://doi.org/10.1007/s00128-010-0067-0.

Shi, G.R., Cai, Q.S., Liu, Q.Q., Wu, L., 2009. Salicylic acid-mediated alleviation of cadmium toxicity in hemp plants in relation to cadmium uptake, photosynthesis, and antioxidant enzymes. Acta Physiol. Plant. 31, 969–977. https://doi.org/10.1007/s11738-009-0312-5.

Smieljan, A., Wilkinson, K.J., Rossier, C., 2003. Cd bioaccumulation by a freshwater bacterium, *Rhodospirillum rubrum*. Environ. Sci. Technol. 37, 701–706. https://doi.org/10.1021/es025901h.

Smolders, E., Mertens, J., 2013. Cadmium. In: Alloway, B.J. (Ed.), Heavy Metals in Soils. In: vol. 22. Springer, The Netherlands, pp. 283–311.

Song, Y., Hudek, L., Freestone, D., Puhui, J., Michalczyk, A.A., Senlin, Z., Ackland, M.L., 2014. Comparative analyses of cadmium and zinc uptake correlated with changes in natural resistance-associated macrophage protein (NRAMP) expression in *Solanum nigrum* L. and *Brassica rapa*. Environ. Chem. 11, 653–660. https://doi.org/10.1071/EN14078.

Steudle, E., 2000. Water uptake by plant roots: an integration of views. Plant Soil 226, 45–56. https://doi.org/10.1023/A:1026439226716.

Street, R.A., Kulkarni, M.G., Stirk, W.A., Southway, C., Van Staden, J., 2010. Effect of cadmium on growth and micronutrient distribution in wild garlic (*Tulbaghia violacea*). S. Afr. J. Bot. 76, 332–336. https://doi.org/10.1016/j.sajb.2009.12.006.

Suèr, P., Gitye, K., Allard, B., 2003. Speciation and transport of heavy metals and macroelements during electroremediation. Environ. Sci. Technol. 37, 177–181. https://doi.org/10.1021/es010226h.

Sun, J.Y., Shen, Z.G., 2007. Effects of Cd stress on photosynthetic characteristics and nutrient uptake of cabbages with different Cd-tolerance. J. Appl. Ecol. 18, 2605–2610.

Sun, Y., Sun, G., Xu, Y., Liu, W., Liang, X., Wang, L., 2016. Evaluation of the effectiveness of sepiolite, bentonite, and phosphate amendments on the stabilization remediation of cadmium-contaminated soils. J. Environ. Manag. 166, 204–210. https://doi.org/10.1016/j.jenvman.2015.10.017.

Tamaddon, F., Hogland, W., 1993. Review of cadmium in plastic waste in Sweden. Waste Manag. Res. https://doi.org/10.1177/0734242X9301100403.

Tessier, A., Campbell, P.G.C., Bisson, M., 1979. Sequential extraction procedure for the speciation of particulate trace metals. Anal. Chem. 51, 844–851. https://doi.org/10.1021/ac50043a017.

Tudoreanu, L., Phillips, C.J.C., 2001. Modeling cadmium uptake and accumulation in plants. Adv. Agron. 84, 121–157. https://doi.org/10.1016/S0065-2113(04)84003-3.

Tuteja, N., 2007. Abscisic acid and abiotic stress signaling. Plant Signal. Behav. 2, 135–138. https://doi.org/10.4161/psb.2.3.4156.

U.S. Geological Survey, 2017. Mineral Commodity Summaries 2017. U.S. Geological Survey, USA. https://doi.org/10.3133/70140094.

Uchimiya, M., Chang, S., Klasson, K.T., 2011. Screening biochars for heavy metal retention in soil: role of oxygen functional groups. J. Hazard. Mater. 190, 432–441. https://doi.org/10.1016/j.jhazmat.2011.03.063.

Van Der Heijden, M.G.A., Bardgett, R.D., Van Straalen, N.M., 2008. The unseen majority: soil microbes as drivers of plant diversity and productivity in terrestrial ecosystems. Ecol. Lett. 11, 296–310. https://doi.org/10.1111/j.1461-0248.2007.01139.x.

Villiers, F., Ducruix, C., Hugouvieux, V., Jarno, N., Ezan, E., Garin, J., Junot, C., Bourguignon, J., 2011. Investigating the plant response to cadmium exposure by proteomic and metabolomic approaches. Proteomics 11, 1650–1663. https://doi.org/10.1002/pmic.201000645.

Vitti, A., Nuzzaci, M., Scopa, A., Tataranni, G., Remans, T., Vangronsveld, J., Sofo, A., 2013. Auxin and cytokinin metabolism and root morphological modifications in *Arabidopsis thaliana* seedlings infected with Cucumber mosaic virus (CMV) or exposed to cadmium. Int. J. Mol. Sci. 14, 6889–6902. https://doi.org/10.3390/ijms14046889.

Wan, G., Najeeb, U., Jilani, G., Naeem, M.S., Zhou, W., 2011. Calcium invigorates the cadmium-stressed *Brassica napus* L. plants by strengthening their photosynthetic system. Environ. Sci. Pollut. Res. 18, 1478–1486. https://doi.org/10.1007/s11356-011-0509-1.

Wang, M., Ding, L., Gao, L., Li, Y., Shen, Q., Guo, S., 2016. The interactions of aquaporins and mineral nutrients in higher plants. Int. J. Mol. Sci. 17, 1229. https://doi.org/10.3390/ijms17081229.

Wasteneys, G.O., Galway, M.E., 2003. Remodeling the cytoskeleton for growth and form: an overview with some new views. Annu. Rev. Plant Biol. 54, 691–722. https://doi.org/10.1146/annurev.arplant.54.031902.134818.

Wei, R., Guo, Q., Wen, H., Liu, C., Yang, J., Peters, M., Hu, J., Zhu, G., Zhang, H., Tian, L., Han, X., Ma, J., Zhu, C., Wan, Y., 2016. Fractionation of stable cadmium isotopes in the cadmium tolerant ricinus communis and hyperaccumulator *Solanum nigrum*. Sci. Rep. 6. https://doi.org/10.1038/srep24309.

Wei, S., Zeng, X., Wang, S., Zhu, J., Ji, D., Li, Y., Jiao, H., 2014. Hyperaccumulative property of *Solanum nigrum* L. to Cd explored from cell membrane permeability, subcellular distribution, and chemical form. J. Soils Sediments 14, 558–566. https://doi.org/10.1007/s11368-013-0810-3.

Yakimova, E.T., Kapchina-Toteva, V.M., Laarhoven, L.J., Harren, F.M., Woltering, E.J., 2006. Involvement of ethylene and lipid signalling in cadmium-induced programmed cell death in tomato suspension cells. Plant Physiol. Biochem. 44, 581–589. https://doi.org/10.1016/j.plaphy.2006.09.003.

Yang, X., Baligar, V.C., Martens, D.C., Clark, R.B., 1996. Cadmium effects on influx and transport of mineral nutrients in plant species. J. Plant Nutr. 19, 643–656. https://doi.org/10.1080/01904169609365148.

Yao, Z., Li, J., Xie, H., Yu, C., 2012. Review on remediation technologies of soil contaminated by heavy metals. Procedia Environ Sci 16, 722–729. https://doi.org/10.1016/j.proenv.2012.10.099.

Yin, D., Wang, X., Peng, B., Tan, C., Ma, L.Q., 2017. Effect of biochar and Fe-biochar on Cd and As mobility and transfer in soil-rice system. Chemosphere 186, 928–937. https://doi.org/10.1016/j.chemosphere.2017.07.126.

Yoneyama, T., Gosho, T., Kato, M., Goto, S., Hayashi, H., 2010. Xylem and phloem transport of Cd, Zn and Fe into the grains of rice plants (*Oryza sativa* L.) grown in continuously flooded Cd-contaminated soil. Soil Sci. Plant Nutr. 56, 445–453. https://doi.org/10.1111/j.1747-0765.2010.00481.x.

Zhang, C., Zeng, G., Huang, D., Lai, C., Huang, C., Li, N., Xu, P., Cheng, M., Zhou, Y., Tang, W., He, X., 2014. Combined removal of di(2-ethylhexyl)phthalate (DEHP) and Pb(II) by using a cutinase loaded nanoporous gold-polyethyleneimine adsorbent. RSC Adv. 4, 55511–55518. https://doi.org/10.1039/C4RA09563D.

Zhang, F., Zhang, H., Xia, Y., Wang, G., Xu, L., Shen, Z., 2011. Exogenous application of salicylic acid alleviates cadmium toxicity and reduces hydrogen peroxide accumulation in root apoplasts of *Phaseolus aureus* and *Vicia sativa*. Plant Cell Rep. 30, 1475–1483. https://doi.org/10.1007/s00299-011-1056-4.

Zhu, Z.Q., Yang, X.E., Wang, K., Huang, H.G., Zhang, X., Fang, H., Li, T.Q., Alva, A.K., He, Z.L., 2012. Bioremediation of Cd-DDT co-contaminated soil using the Cd-hyperaccumulator *Sedum alfredii* and DDT-degrading microbes. J. Hazard. Mater. 235, 144–151.

## FURTHER READING


Lopes Júnior, C.A., Mazzafera, P., Arruda, M.A.Z., 2014. A comparative ionomic approach focusing on cadmium effects in sunflowers (*Helianthus annuus* L.). Environ. Exp. Bot. 107, 180–186. https://doi.org/10.1016/j.envexpbot.2014.06.002.

Rizwan, M., Ali, S., Zia ur Rehman, M., Rinklebe, J., Tsang, D.C.W., Bashir, A., Ok, Y.S., 2018. Cadmium phytoremediation potential of *Brassica* crop species: a review. Sci. Total Environ. 631–632, 1175–1191. https://doi.org/10.1016/j.scitotenv.2018.03.104.

Zia-ur-Rehman, M., Sabir, M., Rizwan, M., Saifullah Ahmed, H.R., Nadeem, M., 2014. Remediating cadmium-contaminated soils by growing grain crops using inorganic amendments. In: Soil Remediation and Plants: Prospects and Challenges. Academic Press, Amsterdam, pp. 367–396. https://doi.org/10.1016/B978-0-12-799937-1.00013-9.

CHAPTER 13

# The Effect of Cd Stress in Mineral Nutrient Uptake in Plants

**Miguel Mourato, Filipa Pinto, Inês Moreira, Joana Sales, Inês Leitão, Luisa Louro Martins**
LEAF, School of Agriculture, University of Lisbon, Lisbon, Portugal

## Abstract

Plants need a whole range of mineral elements as nutrients in order to grow. Some of these essential elements are needed in higher amounts (macronutrients), and others in lower amounts (micronutrients). They are absorbed by the roots, and distributed through the different plant parts. The presence of heavy metals, such as cadmium, can disturb this equilibrium, and can affect both the uptake of several of these elements by the roots and/or its distribution through the other plant parts. This can be due to several effects, such as competition with the same transporters, a disturbance in water uptake, or affecting key enzymes in the transport process. Although there are many publications concerning the effect of Cd on nutrient uptake, the results are sometimes contradictory, due to the fact that the effects are highly dependent on plant species and experimental conditions, among other factors. In this chapter we present an update on the main effects caused by Cd in mineral nutrient uptake, describing the elements that are reported to be most affected, and the known mechanisms that can cause the nutrient imbalance.

**Keywords:** Cadmium, Mineral elements, Macronutrients, Micronutrients, Nutrient imbalance

## 1 INTRODUCTION

Plants need a wide range of mineral elements in order to survive and thrive, and they are usually classified according to the amounts present in the plant. Some of these are needed in large quantities, and are known as macronutrients, including nitrogen, phosphorus, potassium, calcium, magnesium, and sulfur. The other macronutrients are carbon, hydrogen, and oxygen, and they are not considered mineral elements, and are obtained from water or carbon dioxide. Elements needed in lower quantities are known as micronutrients, including iron, manganese, boron, zinc, copper, molybdenum, chlorine, and nickel (Marschner and Marschner, 2012). Other elements,

*Cadmium Toxicity and Tolerance in Plants*
https://doi.org/10.1016/B978-0-12-814864-8.00013-9

including sodium, silicon, selenium, vanadium, and cobalt have been shown to be beneficial to some, but not all plants, and so they are not usually considered micronutrients (Fageria et al., 2011; Pilon-Smits et al., 2009). Macronutrients are present in concentrations of 1000 mg/kg of dry matter (DM), and micronutrients occur in concentrations of 100 mg/kg DM or less. While the macronutrients are constituents of major physiological compounds, micronutrients are equally important, usually as components (cofactors) of enzymes (Hänsch and Mendel, 2009; Maathuis, 2009). It is thus essential for plants to keep an appropriate balance of these mineral nutrients, as both a deficiency, and an excess, will affect plant growth and development.

Numerous factors affect the uptake of these elements by the plants, most importantly soil parameters such as pH, redox potential, and organic matter content (Kirkham, 2006; Li et al., 2017). Different nutrients also interact with each other, and this interaction can have a positive, synergistic, effect; a negative, antagonistic, effect; or it can have no effect at all (Fageria, 2001). The presence of nonessential elements in the growth media can also affect the uptake and distribution of essential nutrients in the plant.

Cadmium is known to be toxic to plants at relatively low concentrations, and its effects have been extensively studied in several plant species (Adriano, 1986; Hasanuzzaman, 2013; Martins et al., 2013; Rizwan et al., 2017). Some of these effects might be due directly to the effect of Cd, but others might be caused by Cd displacing other essential elements, or reducing its uptake, and in this case, Cd can, in fact, induce deficiency symptoms of other nutrients (Alloway, 2008). However, these effects are highly variable, and depend on several factors, such as plant species, Cd concentration, growth media, duration of exposure, age of plant at time of contamination, and so forth (Andresen and Küpper, 2013; Grant et al., 2008).

The focus of this chapter will be on the effect Cd has on plant uptake of other essential nutrients. Although Cd's effect on the uptake of essential nutrients has been extensively studied over the years, a large number of studies just present the observed effects (increased or decreased uptake), and offer few mechanistic explanations for this behavior.

## 2 CADMIUM UPTAKE

Cadmium is not particularly abundant in the earth's crust, but anthropogenic activities can increase the amount of Cd present in the biosphere, mainly through mining, smelting, and industrial activities, and application of phosphate fertilizers and sewage sludge in agriculture (McLaughlin and Singh, 1999).

Once in the soil or water, a fraction of the total amount of cadmium may be available for plant uptake, depending on several factors, of which soil pH and organic matter content are perhaps the most important (Adriano, 1986; Clemens, 2006; DalCorso et al., 2008).

Cadmium can enter the roots using transporters/channels used for other essential elements, such as Ca, Fe, and Zn, and several transporters that mediate Cd uptake have been identified over the years (Clemens et al., 2013; Das et al., 1997; Gravot et al., 2004; Hussain et al., 2004; Leitenmaier et al., 2011; Li et al., 2017; Lombi et al., 2002; Mendoza-Cozatl et al., 2011; Milner et al., 2014; Pence et al., 2000; Song et al., 2017; Verbruggen et al., 2009; Wu et al., 2015; Xu et al., 2009). Once inside the plant, it can also use other metal transporters to enter different cell compartments, as many transmembrane transporters are able to transport a variety of elements (Assunção et al., 2008; Choppala et al., 2014). Thus, the uptake of Cd is frequently related to a decrease in the uptake of other essential elements, due to a competitive process (Choppala et al., 2014; DalCorso et al., 2008; Gonçalves et al., 2009). Of course, the toxicity of Cd could, by itself, induce damage to the metabolic mechanisms that regulate metal uptake, affecting the normal equilibrium established to control the necessary levels of essential elements. For example, it has been confirmed that Cd can affect $H^+$-ATPase proton pumps operating in plasma membranes that are important in regulating ion homeostasis (Janicka-Russak et al., 2008).

## 3 CALCIUM

Calcium is an essential element that plays a number of different roles in plant metabolism (Tang and Luan, 2017), and has long been known to be able to ameliorate the effect of the toxicity of different heavy metals, including Cd (Skórzyńska-Polit et al., 1998; Zhang and Xiao, 1998). Antosiewicz and Hennig (2004) studied the relationship between Ca uptake and Cd toxicity in tobacco plants, which is a fairly tolerant species, and notable for being able to accumulate relatively high concentrations of Cd in the leaves (Martins et al., 2011), and concluded that increased Ca in the growth medium increased LCT1 expression (a nonselective transmembrane transporter for Na, K, and Ca), leading to increased uptake of this essential element, and blocking constitutive Cd uptake, lowering its accumulation, and hence its toxicity, in roots. Thus, adding Ca to the growth media can lead to reduced uptake of Cd into the plant (Kim et al., 2002; Österås and Greger, 2006).

This could explain some results of induced Ca uptake by plants growing in Cd-containing media (Larbi et al., 2002; Mohamed et al., 2012), and generally, in studies regarding Cd toxicity, an increase both in Cd and Ca uptake is detected (Bah et al., 2011). In a study of Cd toxicity in tumbleweed, de la Rosa et al. (2005) reported an increase in Ca content in roots and leaves. The authors presented a similar explanation, claiming that the effect of Cd on Ca channels and transporters leads to an increase in its transcription and translation, allowing a larger Ca uptake, compensating for the blocking effect of Ca channels. Some authors have also attributed the increased Ca uptake as a protection mechanism against Cd toxicity, as Ca plays an important role in preserving several metabolic processes (Hédiji et al., 2015; Zorrig et al., 2010). According to Michalska and Asp (2001), the observed increase in Ca in both shoots and roots of lettuce could be due to the fact that Ca can be taken up passively via the plasma membranes, and Cd-induced disturbance of these would cause increased Ca influx. In this experiment, it should be noted that the Cd concentrations used in the nutrient solution (0.05 and 0.5 μM) were much lower than usual in most Cd toxicity studies.

A reduction of Ca levels in plants growing in Cd contaminated solutions have also been reported in different plants, possibly due to competition processes (Bertoli et al., 2012). Li et al. (2012) explained the Cd induced inhibition of Ca uptake by the depolarization of the membrane potential, reducing Ca (and K) uptake. In other plants under Cd stress, no significant differences were detected in Ca levels, although other micro and macro nutrients might be affected (Drazic and Mihailovic, 2005).

Douchiche et al. (2012) reported that in flax plants growing in Cd-contaminated sand, the concentration of Ca in the roots decreased, but in the stems, increased. As other micronutrients had the opposite behavior (increase in roots and decrease in shoots), the authors hypothesize that while Cd competes with Ca for the transporters in roots, it competes with several micronutrient transporters in the stems. The same behavior of decreased Ca in the roots and increased levels in the shoots of *Thlaspi arvense* plants under Cd stress were reported by Martin et al. (2012). On the other hand, in a short-term experiment (24 h) of *Arabidopsis thaliana* plants growing under Cd stress, it has been reported that Ca levels in the roots increased, while in the shoots it decreased, suggesting that xylem loading is being affected by Cd toxicity (Smeets et al., 2008). These results seem to show that in fact, there is a different effect of Cd in root and shoot nutrient uptake.

An important effect in relation to the levels of mineral elements in plants is related to a reduced biomass under Cd stress. This could lead to an apparent increase in the concentration of a certain element, but because of a smaller plant biomass, and not necessarily because of an increased uptake. This explanation was given by Ghnaya et al. (2005) for the observed behavior of Ca in *Sesuvium portulacastrum* and *Mesembryanthemum crystallinum* plants under Cd stress.

## 4 COPPER

Copper is an essential micronutrient that has numerous physiological roles in plants (Yruela, 2009). Most studies with Cd toxicity describe a reduction of Cu uptake with increasing Cd concentrations (Carpena et al., 2003; Redondo-Gómez et al., 2010). This reduced uptake of Cu is usually attributed to competition between Cd and Cu, as these metals are reported to use the same type of transporters (Burzynski et al., 2005).

In a study of Cd toxicity in the aquatic model plant *Ceratophyllum demers*, albeit with very low Cd concentrations, a decrease in Cu with increasing Cd concentrations was observed (Andresen et al., 2016). The authors admit that Cd may have prevented Cu uptake, but apparently the plants also actively lost Cu to the solution. A similar decrease in Cu content, both in roots and in leaves of tumbleweeds under different Cd concentrations, was reported, but no justification was given (de la Rosa et al., 2004); while in pea plants, a reduction in Cu was only observed in leaves (Sandalio et al., 2001).

In their study in tomato plants, Bertoli et al. (2012) reported an increase of Cu in roots, but a decrease in shoots. Although the authors claim that the absorption of Cu is stimulated, the observed increase in the roots could be due to the accumulation of Cu that is not able to be translocated to the shoots, probably because of competition with Cd or to the transport mechanism being affected by this heavy metal toxicity. The same behavior was reported in a study with white lupine plants, leading to the conclusion that Cd caused alterations to the vascular system, affecting element translocation (Zornoza et al., 2002). In *Matricaria chamomilla* plants, a similar effect was detected, although apparently Cd stress had no effect on other micronutrients (Kovacik et al., 2009); while in sunflower plants, this behavior was observed, not only for Cu, but also for Zn and Fe (Rivelli et al., 2014). López-Millán et al. (2009) also detected an increase in Cu concentration in roots, but not in leaves, of tomato plants growing under 100 μM Cd, with the same explanation given for the behavior of Fe (see the following).

Other studies have reported an increase in Cu content in several plant organs with Cd, such as the study of Chizzola and Mitteregger (2005) in chamomile flowers, or of Zhang et al. (2014) in king grass. In this last case, the observed increase not only of Cu, but also of other essential elements (such as Zn, Mg, and Ca) was seen as a possible defense mechanism against Cd toxicity.

## 5 IRON

Iron is an essential element for plants, and they have different strategies for Fe uptake that can be differently affected by Cd, either by increasing or by decreasing its uptake (Choppala et al., 2014; Moulis, 2010). The relationship between Fe and Cd uptake in different plant species using different strategies for Fe uptake have been well studied (Meda et al., 2007). Iron transporters have been reported to also be used by Cd (Lombi et al., 2002; Yoshihara et al., 2006). Both IRT1 (which belongs to the ZIP family of transporters) and the Nramp family of transporters are involved in Fe, and also in Cd transport into the plants (Takahashi et al., 2011; Thomine et al., 2000; Vert et al., 2002). Bao et al. (2012) results are in accordance with this, as they reported a decrease in Fe content in wheat and corn with Cd, both in shoots and in roots, and also an increase in Cd uptake in Fe-deficient conditions, and these results are consistent with a competition for a common transporter. Other studies reporting decreased levels of Fe in both roots and shoots with Cd are consistent with this explanation (de la Rosa et al., 2004).

López-Millán et al. (2009), in their study with tomato plants, reported an inhibited Fe uptake due to the direct effect of low concentrations of Cd (10 μM) on the root Fe reductase enzyme. However, the decrease in Fe levels was only observed in leaves. At higher Cd concentrations (100 μM), an opposite effect was detected, with an increase in Fe concentration in roots and stems, but not in leaves. The authors explain this as Cd-induced damage to the plasma membrane affecting its permeability, and this has also been reported as a mechanism for the reduced uptake of other elements (Hédiji et al., 2015). In a study with potato plants, Gonçalves et al. (2009) reported a different effect with an increase in Fe contents in roots, but not in shoots, at the lowest concentrations of Cd studied (however, the concentrations were still high, as this effect was detected up to 200 μM Cd), and a decrease at concentrations between 300 and 500 μM. The explanation given for the observed behavior was due to hyperpolarization of the plasma membranes of roots, increasing the transmembrane potential, and facilitating ion uptake. In a study of heavy metal toxicity in *Cistus ladanifer* plants, a similar accumulation of Fe in roots, and not in shoots, was also observed (Kidd et al., 2004).

A reduction of Fe levels in the roots, and a concomitant increase in the leaves has also been reported in soybean seedlings, but no explanation has been given for this effect, except for a general disturbance of the ion homeostasis (Drazic and Mihailovic, 2005); although these results reinforce the idea that Cd has a different effect in Fe root uptake and shoot translocation that is highly dependent on plant species and experimental conditions.

In several studies with different plant species such as *Brassica napus* (Mendoza-Cózatl et al., 2008), flax plants (Douchiche et al., 2012), lettuce (Monteiro et al., 2009), tomato plants (Bertoli et al., 2012), or *Thlaspi caerulescens* (Küpper and Kochian, 2010), Cd toxicity caused a decrease in Fe content in shoots, and an increase in roots, showing that although root uptake of Fe was not affected, Cd apparently affected vascular loading, and/or root-to-shoot iron translocation. The magnitude of this decrease in shoot Fe concentration was variable, and usually did not lead to chlorosis.

Bah et al. (2011) reported an increase in Fe content in Cd-stressed *Typha angustifolia*, a very tolerant species, and the authors claim that Fe (and also Ca) play a role in heavy metal detoxification, although it is not clear how they do it.

All the reported effects are highly dependent on several factors, including the applied Cd concentration. Pietrini et al. (2003) studied the effect of Cd in *Phragmites australis* plants, and found that the levels of Fe, Zn, and Ca increased for the lower concentration studied of 50 μM Cd, but showed no change for the higher concentration of 100 μM (except for Ca, which decreased). Although the authors failed to give an explanation for the observed results, they could be due to a defense mechanism, as with other plants, that is active for the lower concentration of Cd. At 100 μM, possibly the toxic effects overwhelm this mechanism.

Cadmium-induced deficiency in Fe could cause chlorosis so that, sometimes, the observed toxic effect of Cd might be indirectly caused by Fe-depletion, and not only to direct toxic effects of Cd (Choppala et al., 2014; Kashem and Kawai, 2007).

It has also been shown that, like with other nutrients, Fe can reduce the toxicity of Cd (Liu et al., 2017a). This has been explained by the role that Fe has in the formation of enzymes that are crucial in the plant antioxidative mechanisms, such as Fe-SOD and CAT. In this situation, the relative amounts of both Fe and Cd are probably crucial, as Cd can also replace iron in different macromolecules, leading on one hand to its inactivation, and on the other to the release of free Fe that can induce redox reactions, and thus oxidative stress (Cuypers et al., 2010).

Another factor that is decisive in the uptake behavior of different elements, such as Fe, under Cd contamination, is the duration of the applied stress, as short-term and long-term experiments can lead to different effects in plants (Hernandez et al., 1998).

## 6 POTASSIUM

Potassium is an essential element that has numerous physiological roles in plants, and is the most abundant inorganic cation in plant cells (Benito et al., 2014). Bertoli et al. (2012) reported that, in tomato plants, there was a negative correlation between Cd and K, and the authors attributed this to a competitive inhibition of the uptake, while no effect was detected in relation to other elements, such as Mg and P. Other authors have detected a reduction in K in different plants under Cd stress (Carpena et al., 2003; de la Rosa et al., 2004; Douchiche et al., 2012; Drazic and Mihailovic, 2005; Kovacik et al., 2009), and this can be due to decreased K uptake, K-leakage into the growing media, increased K transport to the shoots (Drazic and Mihailovic, 2005), or to a combination of these factors. Zornoza et al. (2002) detected a decrease in K concentrations both in roots and in shoots of white lupins, concluding that there was a Cd-induced net reduction in this macronutrient. In a study with *Sesuvium portulacastrum* and *Mesembryanthemum crystallinum* plants, the decrease in K root uptake was justified by the complexation of ATP with Cd, reducing the amount of energy available to the membrane transport systems (Ghnaya et al., 2005). The same authors attribute reduced shoot translocation to reduction of the number and diameter of xylems due to Cd's effect. Kidd et al. (2004) justified the observed decrease of K in roots under Cd concentrations as a result of K leakage due to disruption of the root in the plasma membrane. In fact, Li et al. (2012) proved that there is a double effect that acts to reduce K levels under Cd toxicity. Besides the reduction in K uptake as described above for Ca, Cd also induces K efflux through specific channels. In a study with *Genipa Americana*, the authors attributed the affected photosynthesis to the observed decrease in K translocation to the shoots (Souza et al., 2011).

As was observed for the micronutrients, the opposite effect has also been reported. de la Rosa et al. (2005) reported an increase in K levels with increased Cd concentration in a hydroponic solution. The explanation given is related to the fact that Cd can bind to the K channels, making them permanently open, and thus increasing K influx. However, in an earlier publication with the same species (de la Rosa et al., 2004), the same authors reported a decrease in K and Mg content, both in roots and in leaves, with the same Cd concentrations. The difference was in the growing medium, as in the earlier study the plants were grown in agar, while in the latter, the plants were grown in a hydroponic solution. This confirms that the experimental conditions are decisive in the observed response to the induced Cd toxicity. In a study of Cd toxicity in two potato cultivars in two different culture conditions (in vitro and hydroponics), the results reported for the uptake of different essential nutrients was completely different between the two experimental conditions (Gonçalves et al., 2009). In general, plants growing in vitro had a lower concentration of different macro and micronutrients, as compared with the plants growing in hydroponics, where few differences were observed. This effect was explained by the fact that in the in vitro experiment, the plants are enclosed in a nonventilated area with high humidity levels, while in the hydroponics; there is an environment with lower humidity, leading to increased plant transpiration that can enhance the uptake and translocation of other elements.

Another factor that is important to consider in studies regarding nutrient uptake is the presentation of the results, either on a dry matter or on a fresh matter basis. Most studies report results in the former mode, but changes in water uptake induced by Cd might affect the conclusions when the values are converted to fresh weight. In a study of Cd toxicity in lettuce, Zorrig et al. (2010) reported a decrease in K levels in roots and shoots on a dry matter basis, but when converted into fresh matter, K concentration was stable, or even increased. If the heavy metal stress affects the proportion of structural material in the plant, it can also have a variable effect in the nutrient concentration, when expressed on a dry matter basis (Michalska and Asp, 2001).

## 7 MANGANESE

Manganese is an essential element for plants, and is involved in a wide range of physiological processes (Campbell and Nable, 1988). However, Mn is only needed in relatively low amounts by plants, although they reportedly have higher tissue concentrations than needed (Shao et al., 2017). The mechanisms of Mn uptake by plants are now well established, and have been described in detail (Migocka et al., 2015; Millaleo et al., 2010;

Shao et al., 2017). In a field trial with rice, a significant correlation was found between Mn and Cd uptake from the soil, leading to the conclusion that Cd is transported into the grains via Mn transporters (Liu et al., 2017b).

Several studies have reported a decrease in Mn uptake in plants growing under Cd-contaminated media (Arduini et al., 1998; Bertoli et al., 2012; Hernandez et al., 1998; Larbi et al., 2002; López-Millán et al., 2009; Metwally et al., 2005; Wu et al., 2007; Zornoza et al., 2002). Both Arduini et al. (1998) and Hernandez et al. (1998) attributed the observed decrease in Mn content to changes in the plasma membrane integrity and permeability. This could also explain the decrease in Mn content in roots of soybean seedlings (Dražić et al., 2004), and of *Thlaspi arvense* (Martin et al., 2012); but in both cases, the plants were able to maintain the same levels of Mn in the leaves. Zhang et al. (2014) reported a decreased Mn content in stems and leaves of king grass, and concluded that Cd uses the same transporters as Mn due to this antagonistic effect. The effect of up to 50 μM Cd on the Cd/Zn hyperaccumulator *Thlaspi caerulescens* was also an inhibited uptake of Mn (and also of Zn) by competing with the transporters of these essential elements, or by interfering with transporter gene expression (Küpper and Kochian, 2010).

In a study with white lupin plants (Zornoza et al., 2002), although a reduction in Mn levels was detected with increased Cd concentrations, the authors report that the measured Mn concentrations were very high, and could contribute to a protective effect against the toxicity induced by Cd, namely in the photosynthetic processes.

On the other hand, increased Mn levels, as a function of applied Cd, although less frequent, have also been reported (de la Rosa et al., 2004). In a study with flax plants, it was reported that Mn levels increased in roots, but decreased in the shoots (Douchiche et al., 2012). In most plant species and experimental conditions studied, it does seem that Mn uptake restriction is a more common Cd effect.

Liu et al. (2013) made an extensive study of the protective effect of different ratios of Mn/Cd concentrations in *Phytolacca acinosa* plants (an Mn hyperaccumulator), and concluded that Mn induced a decrease in Cd uptake, and also a general protective effect against its toxicity. Another study in white lupins plants confirmed that Mn had a protective effect against Cd toxicity, even if the levels of cadmium were not decreased, mainly due to the protection of the photosynthetic apparatus, and to the increase in the antioxidant capacity of the roots (Zornoza et al., 2010).

## 8 ZINC

Zinc has several physiological roles in plants, being important in the maintenance of structural and functional integrity of biological membranes, and in protein synthesis and gene expression (Alloway, 2008). It is also an important component of several enzymes, and is the only metal that is present in all six enzyme classes (Broadley et al., 2007).

Zinc and Cd are chemically very similar, and so Cd uses the same Zn transporters in the plants (Wong and Cobbett, 2009; Wong et al., 2009), and also has been reported to replace it in several biological functions; for example, replacing Zn in enzymes containing the metal (Moulis, 2010). As the presence of Cd can lead to the induction of oxidative stress (Cuypers et al., 2010), this can lead to the increase in the expression and/or activity of enzymes involved in the response to this stress. Superoxide dismutase (SOD) exists in three different forms located in different cell compartments: FeSOD (chloroplasts), MnSOD (mitochondria and peroxisomes), and Cu/ZnSOD (chloroplast and cytosol) (Alscher et al., 2002). Thus, if Cd can replace Zn in these enzymes, it can inactivate this form of the enzyme, and reduce the capacity of the plant's antioxidative defense system (Clemens, 2006). Andresen et al. (2016) attributed to this effect the increase in superoxide production as a consequence of Cd stress in *Ceratophyllum demers* plants. This type of effect (the replacement/displacement of Zn in enzymes or other molecules by Cd) can be difficult to identify if the global levels of Zn remain constant, that is, if Zn uptake is not affected, as this is what is usually measured experimentally. In another study with the same plants, Andresen et al. (2013) reported a Cd-induced inhibition of Zn distribution in plant cells, leading to Zn deficiency of tissues outside the vein, and an increase in Zn content inside the vein, possibly leading to toxic effects due to Zn excess, confirming that even if the global state of Zn nutrition levels are maintained, there can be toxic or deficiency symptoms in different cellular locations induced by Cd toxicity. This type of effect could even lead to an increased uptake of Zn, as the plant tries to keep the homeostasis of this element in all plant parts, and this could be an explanation for the reported increase in Zn levels in plants in different studies (López-Millán et al., 2009). Thus, the displacement of Zn from some essential molecules could trick the regulation mechanisms into a state of Zn deficiency, leading to its increased uptake as an overcompensation mechanism (Assunção et al., 2008). In a study with several lettuce varieties, Zorrig et al. (2010) presented this same explanation for the results observed, as increased Cd concentrations led to increased Zn in all studied varieties, both in roots and in shoots.

In some situations it has been reported that the levels of Zn in the roots increase, but in the shoots, decrease (Hédiji et al., 2015), leading to the conclusion that the uptake by the roots is not affected, but the translocation to upper plant parts is restricted.

On the other hand, the similarity between Zn and Cd has been attributed to a decreased uptake of this essential element by Cd-stressed plants, as they can compete for the same transporters, both at the root level, and inside the plant (Hart et al., 2002). This justification can explain the results observed in several studies with different plant species where this reduction effect on the uptake of Zn has been detected (Arduini et al., 1998; Bertoli et al., 2012; de la Rosa et al., 2004; Hart et al., 2002; Metwally et al., 2005; Pinto et al., 2017). Gardea-Torresdey et al. (2004) attributed the competition of Cd to Zn in ligand compounds as the reason for a general decrease in the uptake of mineral elements in *Convolvulus arvensis*. In a study with four *Tillandsia* species under irrigation with 2 μM Cd over a period of 30 days, Zn was the only element that had a reduced uptake with no differences being detected in K, Ca, Mg, or Cu (Kovácik and Backor, 2008). This has been explained as being due to competition with Cd owing to a generally higher basal level of Zn.

Zinc levels in the growth media can also have a protective effect in Cd toxicity (Kumari et al., 2011), due to several factors other than decreased Cd uptake, such as the minimization of chlorophyll breakdown and lipid peroxidation, and even via the reparation of damaged DNA (Choppala et al., 2014). In a study with a freshwater macrophyte, *Ceratophyllum demersum*, it was proposed that Zn was effective in restoring the antioxidant defense system used by the plants, and putatively damaged by Cd (Aravind and Prasad, 2005). The authors proposed that Zn restored thiols, and inhibited the oxidation of ascorbate and glutathione, maintaining the levels of these important molecules used in the plants' stress defense mechanisms. In another study with the same plant, it was hypothesized that while Cd induced the oxidation of NADPH, producing reactive oxygen species (ROS), Zn would inhibit this oxidation, in effect reducing the toxic effect of Cd (Aravind et al., 2009). Papoyan et al. (2007) studied the interactions between Cd and Zn in *Thlaspi caerulescens*, a heavy metal hyperaccumulator, and also observed that plants growing in higher Zn concentrations had a higher Cd tolerance. If this higher tolerance is due to reduced Cd uptake, given the similar characteristics of Cd and Zn, this could be explained by competition for the negative charges in root cell walls. However, in the shoots, there was a positive correlation between Zn and Cd, and the authors postulate several hypotheses for this relationship, at least in hyperaccumulating plants, where translocation of the metals to the shoots is an important factor, as nonhyperaccumulating species will try to avoid transport of toxic metals to the shoots. These include the induction of other tolerance mechanisms that allow the uptake of higher Cd concentrations, or the upregulation of transporters that are common for Zn and Cd (such as HMA4, which participates in xylem loading of metals) as a function of the increase of either metal in the nutrient media (Papoyan et al., 2007). In a study with *Arabidopsis thaliana*, it has been proposed that AtHMA4 can have a role as an efflux pump under high Cd (or Zn) concentrations, and could thus lead to increased tolerance to Cd by actively exporting it to the soil solution (Mills et al., 2005). Anyway, apparently there are different mechanisms for hyperaccumulator and nonaccumulator species, as the former must have more efficient defense mechanisms in their shoots (Verbruggen et al., 2013).

The inverse relationship between Zn and Cd uptake observed in some plants has allowed using the application of Zn as a measure to reduce Cd contamination (Choppala et al., 2014; Rizwan et al., 2016; Wu et al., 2007), and this has been shown to be more or less effective in several plants under different experimental conditions and Zn/Cd ratios (Brennan and Bolland, 2014; Chizzola and Mitteregger, 2005; Luo and Rimmer, 1995; Sankaran and Ebbs, 2008). For example, in a study with rice, it has been reported that added Zn induced a reduced uptake of Cd in the roots, but an increase in the shoots (Liu et al., 2007).

## 9 OTHER ESSENTIAL ELEMENTS

The relationship between Cd and other macro and micronutrients, besides the ones described herein, have been less studied, although in most of the studies reported in this chapter, usually more than one essential element is quantified.

In relation to S, an increase in the uptake of this element was detected in tomato plants (Bertoli et al., 2012), and the authors attribute this to a defense mechanism based on the putative increased production of thiol groups as precursors of phytochelatins. Küpper and Kochian (2010) observed a similar effect in *Thlaspi caerulescens*, and also justified it as a defense mechanism based on the increased synthesis of S-containing compounds that are important in Cd detoxification mechanisms such as glutathione, metallothioneins, and phytochelatins. This effect has not been generally observed, as S content decrease in leaves under Cd stress has also been reported (Smeets et al., 2008). In fact, sulfur has been considered as having a very important role in Cd tolerance of plants (Gallego et al., 2012), and root sulphate availability could limit glutathione synthesis and, hence, Cd tolerance (Nocito et al., 2006). A recent study in rice reports a protective effect of applied sulfur as it reduced Cd uptake (Cao et al., 2018).

In a study regarding the toxic effect of Cd in soybean seedlings (Drazic and Mihailovic, 2005), a decrease in the uptake of Mg by the roots was detected, but with no change in the leaf content, showing that the observed decrease in the roots could be compensated for, or was not high enough to affect leaf uptake. As has been reported for other nutrients, it has been shown that added Mg to the growth media can alleviate Cd toxicity (Kashem and Kawai, 2007), in part due to a reduction in Cd uptake (Kim et al., 2002). In studying the effect of up to 50 μM Cd in *Thlaspi caerulescens* plants, an increase in Mg levels was measured, and the authors related this increase to a defense mechanism used by this hyperaccumulator plant, as increasing Mg levels in chloroplasts could protect chlorophyll from being affected by Cd (Küpper and Kochian, 2010; Kupper et al., 2000).

Other elements, such as Se, have also been reported to be able to alleviate Cd toxicity (Gao et al., 2018; Saidi et al., 2014), while P was described as having no effect in Cd uptake (Zhong et al., 2012).

## 10 CONCLUSIONS

Cadmium affects water uptake by plants, and this has an effect on the uptake of both macronutrients and micronutrients. The interactions between Cd and Zn have been extensively studied, and several mechanisms have been proposed to explain the synergistic or antagonistic effects observed. The uptake of other essential elements is usually reported to be affected by Cd contamination, although the results are highly dependent on plant species (and varieties, cultivars, genotypes, etc.), plant part (root, leaves, flowers, etc.), plant accumulating behavior (e.g., being or not being a hyperaccumulator), Cd concentration, plant age, and duration of the experiment, and on general experimental conditions (hydroponics, soils, etc.). Variations in all of these factors can lead to completely different results regarding the way micronutrient and macronutrient concentrations are affected, and this goes a long way in explaining some of the almost contradictory results reported in the literature. Most of the reported results describe a reduction in the uptake of other elements that could be due to competition with Cd for the same transporters. Other results indicate an increase in essential elements' uptake with increased Cd concentration. This could be derived from a release of these elements from different macromolecules, leading to a sensing of deficiency that would induce increased uptake of the same elements. Still, a number of studies report no difference in the uptake of essential elements. However, most studies report element totals in the different plant parts, and in most cases, it is difficult to identify exactly what is happening at a subcellular and molecular level, and to what extent there is a release or a replacement of essential elements by Cd in different biomolecules.

Another factor to take into account is that Cd toxicity can affect plant growth, and sometimes an apparent increase in element concentration, can be due, not to an increased uptake, but to the reduced growth of a certain plant part. This can affect the plant as a whole, or just some plant parts.

The antagonistic effect observed between Cd and different essential elements has served as a basis to propose several remediation techniques based on the application of excessive, but nontoxic, levels of these elements in order to reduce Cd uptake from contaminated media. As the results are very dependent on plant species (and even varieties), this cannot be regarded as a general solution.

## REFERENCES

Adriano, D.C., 1986. Cadmium. In: Trace Elements in the Terrestrial Environment. Springer-Verlag, New York, Berlin, Heidelberg, Tokyo.

Alloway, B., 2008. Zinc in Soils and Crop Nutrition. IZA and IFA, Brussels, Belgium and Paris, France.

Alscher, R.G., Erturk, N., Heath, L.S., 2002. Role of superoxide dismutases (SODs) in controlling oxidative stress in plants. J. Exp. Bot. 53, 1331–1341.

Andresen, E., Kappel, S., Stärk, H.J., Riegger, U., Borovec, J., Mattusch, J., Heinz, A., Schmelzer, C.E.H., Matoušková, Š., Dickinson, B., Küpper, H., 2016. Cadmium toxicity investigated at the physiological and biophysical levels under environmentally relevant conditions using the aquatic model plant *Ceratophyllum demersum*. New Phytol. 210, 1244–1258.

Andresen, E., Küpper, H., 2013. Cadmium toxicity in plants. In: Sigel, A., Sigel, H., Sigel, R.K.O. (Eds.), Cadmium: From Toxicity to Essentiality. In: vol. 11. Springer, The Netherlands, pp. 395–413.

Andresen, E., Mattusch, J., Wellenreuther, G., Thomas, G., Arroyo Abad, U., Kupper, H., 2013. Different strategies of cadmium detoxification in the submerged macrophyte *Ceratophyllum demersum* L. Metallomics 5, 1377–1386.

Antosiewicz, D.M., Hennig, J., 2004. Overexpression of LCT1 in tobacco enhances the protective action of calcium against cadmium toxicity. Environ. Pollut. 129, 237–245.

Aravind, P., Prasad, M.N.V., 2005. Modulation of cadmium-induced oxidative stress in *Ceratophyllum demersum* by zinc involves ascorbate–glutathione cycle and glutathione metabolism. Plant Physiol. Biochem. 43, 107–116.

Aravind, P., Prasad, M.N.V., Malec, P., Waloszek, A., Strzałka, K., 2009. Zinc protects *Ceratophyllum demersum* L. (free-floating hydrophyte) against reactive oxygen species induced by cadmium. J. Trace Elem. Med. Biol. 23, 50–60.

Arduini, I., Godbold, D.L., Onnis, A., Stefani, A., 1998. Heavy metals influence mineral nutrition of tree seedlings. Chemosphere 36, 739–744.

Assunção, A., Bleeker, P., ten Bookum, W., Vooijs, R., Schat, H., 2008. Intraspecific variation of metal preference patterns for hyperaccumulation in *Thlaspi caerulescens*: evidence from binary metal exposures. Plant Soil 303, 289–299.

Bah, A.M., Dai, H., Zhao, J., Sun, H., Cao, F., Zhang, G., Wu, F., 2011. Effects of cadmium, chromium and lead on growth, metal uptake and antioxidative capacity in *Typha angustifolia*. Biol. Trace Elem. Res. 142, 77–92.

Bao, T., Sun, T.H., Sun, L.N., 2012. Effect of cadmium on physiological responses of wheat and corn to iron deficiency. J. Plant Nutr. 35, 1937–1948.

Benito, B., Haro, R., Amtmann, A., Cuin, T.A., Dreyer, I., 2014. The twins K + and Na + in plants. J. Plant Physiol. 171, 723–731.

Bertoli, A.C., Cannata, M.G., Carvalho, R., Ribeiro Bastos, A.R., Freitas, M.P., Augusto, A.D.S., 2012. Lycopersicon esculentum submitted to Cd-stressful conditions in nutrition solution: nutrient contents and translocation. Ecotoxicol. Environ. Saf. 86, 176–181.

Brennan, R.F., Bolland, M.D.A., 2014. Application of increasing levels of zinc to soil reduced accumulation of cadmium in lupin grain. J. Plant Nutr. 37, 147–160.

Broadley, M.R., White, P.J., Hammond, J.P., Zelko, I., Lux, A., 2007. Zinc in plants. New Phytol. 173, 677–702.

Burzynski, M., Migocka, M., Klobus, K., 2005. Cu and Cd transport in cucumber (*Cucumis sativus* L.) root plasma membranes. Plant Sci. 168, 1609–1614.

Campbell, L.C., Nable, R.O., 1988. Physiological functions of manganese in plants. In: - Graham, R.D., Hannam, R.J., Uren, N.C. (Eds.), Manganese in Soils and Plants. In: - vol. 33. Springer, Dordrecht.

Cao, Z.-Z., Qin, M.-L., Lin, X.-Y., Zhu, Z.-W., Chen, M.-X., 2018. Sulfur supply reduces cadmium uptake and translocation in rice grains (*Oryza sativa* L.) by enhancing iron plaque formation, cadmium chelation and vacuolar sequestration. Environ. Pollut. 238, 76–84.

Carpena, R.O., Vazquez, S., Esteban, E., Fernandez-Pascual, M., de Felipe, M.R., Zornoza, P., 2003. Cadmium-stress in white lupin: effects on nodule structure and functioning. Plant Physiol. Biochem. 41, 911–919.

Chizzola, R., Mitteregger, U.S., 2005. Cadmium and zinc interactions in trace element accumulation in chamomile. J. Plant Nutr. 28, 1383–1396.

Choppala, G., Saifullah, N., Bolan, S., Bibi, M., Iqbal, Z., Rengel, A., Kunhikrishnan, N.A., Ok, Y.S., 2014. Cellular mechanisms in higher plants governing tolerance to cadmium toxicity. Crit. Rev. Plant Sci. 33, 374–391.

Clemens, S., 2006. Toxic metal accumulation, responses to exposure and mechanisms of tolerance in plants. Biochimie 88, 1707–1719.

Clemens, S., Aarts, M.G.M., Thomine, S., Verbruggen, N., 2013. Plant science: the key to preventing slow cadmium poisoning. Trends Plant Sci. 18, 92–99.

Cuypers, A., Plusquin, M., Remans, T., Jozefczak, M., Keunen, E., Gielen, H., Opdenakker, K., Nair, A., Munters, E., Artois, T., Nawrot, T., Vangronsveld, J., Smeets, K., 2010. Cadmium stress: an oxidative challenge. Biometals 23, 927–940.

DalCorso, G., Farinati, S., Maistri, S., Furini, A., 2008. How plants cope with cadmium: staking all on metabolism and gene expression. J. Integr. Plant Biol. 50, 1268–1280.

Das, P., Samantaray, S., Rout, G.R., 1997. Studies on cadmium toxicity in plants: a review. Environ. Pollut. 98, 29–36.

de la Rosa, G., Martinez-Martinez, A., Pelayo, H., Peralta-Videa, J.R., Sanchez-Salcido, B., Gardea-Torresdey, J.L., 2005. Production of low-molecular weight thiols as a response to cadmium uptake by tumbleweed (*Salsola kah*). Plant Physiol. Biochem. 43, 491–498.

de la Rosa, G., Peralta-Videa, J.R., Montes, M., Parsons, J.G., Cano-Aguilera, I., Gardea-Torresdey, J.L., 2004. Cadmium uptake and translocation in tumbleweed (*Salsola kali*), a potential Cd-hyperaccumulator desert plant species: ICP/OES and XAS studies. Chemosphere 55, 1159–1168.

Douchiche, O., Chaïbi, W., Morvan, C., 2012. Cadmium tolerance and accumulation characteristics of mature flax, cv Hermes: contribution of the basal stem compared to the root. J. Hazard. Mater. 235–236, 101–107.

Drazic, G., Mihailovic, N., 2005. Modification of cadmium toxicity in soybean seedlings by salicylic acid. Plant Sci. 168, 511–517.

Dražić, G., Mihailović, N., Stojanović, Z., 2004. Cadmium toxicity: the effect on macro- and micro-nutrient contents in soybean seedlings. Biol. Plant. 48, 605–607.

Fageria, N.K., Baligar, V.C., Jones, C.A., 2011. Growth and Mineral Nutrition Of Field Crops. vol. xxi. CRC Press, Boca Raton, FL, p. 560.

Fageria, V.D., 2001. Nutrient interactions in crop plants. J. Plant Nutr. 24, 1269–1290.

Gallego, S.M., Pena, L.B., Barcia, R.A., Azpilicueta, C.E., Lannone, M.F., Rosales, E.P., Zawoznik, M.S., Groppa, M.D., Benavides, M.P., 2012. Unravelling cadmium toxicity and tolerance in plants: insight into regulatory mechanisms. Environ. Exp. Bot. 83, 33–46.

Gao, M., Zhou, J., Liu, H., Zhang, W., Hu, Y., Liang, J., Zhou, J., 2018. Foliar spraying with silicon and selenium reduces cadmium uptake and mitigates cadmium toxicity in rice. Sci. Total Environ. 631–632, 1100–1108.

Gardea-Torresdey, J.L., Peralta-Videa, J.R., Montes, M., de la Rosa, G., Corral-Diaz, B., 2004. Bioaccumulation of cadmium, chromium and copper by *Convolvulus arvensis* L.: impact on plant growth and uptake of nutritional elements. Bioresour. Technol. 92, 229–235.

Ghnaya, T., Nouairi, I., Slama, I., Messedi, D., Grignon, C., Abdelly, C., Ghorbel, M.H., 2005. Cadmium effects on growth and mineral nutrition of two halophytes: *Sesuvium portulacastrum* and *Mesembryanthemum crystallinum*. J. Plant Physiol. 162, 1133–1140.

Gonçalves, J.F., Antes, F.G., Maldaner, J., Pereira, L.B., Tabaldi, L.A., Rauber, R., Rossato, L.V., Bisognin, D.A., Dressler, V.L., de Moraes Flores, É.M., Nicoloso, F.T., 2009. Cadmium and mineral nutrient accumulation in potato plantlets grown under cadmium stress in two different experimental culture conditions. Plant Physiol. Biochem. 47, 814–821.

Grant, C.A., Clarke, J.M., Duguid, S., Chaney, R.L., 2008. Selection and breeding of plant cultivars to minimize cadmium accumulation. Sci. Total Environ. 390, 301–310.

Gravot, A., Lieutaud, A., Verret, F., Auroy, P., Vavasseur, A., Richaud, P., 2004. AtHMA3, a plant P1B-ATPase, functions as a Cd/Pb transporter in yeast. FEBS Lett. 561, 22–28.

Hänsch, R., Mendel, R.R., 2009. Physiological functions of mineral micronutrients (Cu, Zn, Mn, Fe, Ni, Mo, B, Cl). Curr. Opin. Plant Biol. 12, 259–266.

Hart, J.J., Welch, R.M., Norvell, W.A., Kochian, L.V., 2002. Transport interactions between cadmium and zinc in roots of bread and durum wheat seedlings. Physiol. Plant. 116, 73–78.

Hasanuzzaman, M., 2013. Cadmium: Characteristics, Sources of Exposure, Health and Environmental Effects. vol. ix. Nova Science Publishers, Inc., New York, 369pp.

Hédiji, H., Djebali, W., Belkadhi, A., Cabasson, C., Moing, A., Rolin, D., Brouquisse, R., Gallusci, P., Chaïbi, W., 2015. Impact of long-term cadmium exposure on mineral content of *Solanum lycopersicum* plants: consequences on fruit production. S. Afr. J. Bot. 97, 176–181.

Hernandez, L.E., Lozano-Rodriguez, E., Garate, A., Carpena-Ruiz, R., 1998. Influence of cadmium on the uptake, tissue accumulation and subcellular distribution of manganese in pea seedlings. Plant Sci. 132, 139–151.

Hussain, D., Haydon, M.J., Wang, Y., Wong, E., Sherson, S.M., Young, J., Camakaris, J., Harper, J.F., Cobbett, C.S., 2004. P-type ATPase heavy metal transporters with roles in essential zinc homeostasis in Arabidopsis. Plant Cell 16, 1327–1339.

Janicka-Russak, M., Kabala, K., Burzynski, M., Klobus, G., 2008. Response of plasma membrane H+-ATPase to heavy metal stress in *Cucumis sativus* roots. J. Exp. Bot. 59, 3721–3728.

Kashem, M.A., Kawai, S., 2007. Alleviation of cadmium phytotoxicity by magnesium in Japanese mustard spinach. Soil Sci. Plant Nutr. 53, 246–251.

Kidd, P.S., Diez, J., Martinez, C.M., 2004. Tolerance and bioaccumulation of heavy metals in five populations of *Cistus ladanifer* L. subsp ladanifer. Plant Soil 258, 189–205.

Kim, Y.Y., Yang, Y.Y., Lee, Y., 2002. Pb and Cd uptake in rice roots. Physiol. Plant. 116, 368–372.

Kirkham, M.B., 2006. Cadmium in plants on polluted soils: effects of soil factors, hyperaccumulation, and amendments. Geoderma 137, 19–32.

Kovácik, J., Backor, M., 2008. Oxidative status of *Matricaria chamomilla* plants related to cadmium and copper uptake. Ecotoxicology 17, 471–479.

Kovacik, J., Gruz, J., Hedbavny, J., Klejdus, B., Strnad, M., 2009. Cadmium and nickel uptake are differentially modulated by salicylic acid in *Matricaria chamomilla* plants. J. Agric. Food Chem. 57, 9848–9855.

Kumari, M., Sinhal, V.K., Srivastava, A., Singh, V.P., 2011. Zinc alleviates cadmium induced toxicity in *Vigna radiata* (L.) Wilczek. J. Phytol. 3.

Küpper, H., Kochian, L.V., 2010. Transcriptional regulation of metal transport genes and mineral nutrition during acclimatization to cadmium and zinc in the Cd/Zn hyperaccumulator, *Thlaspi caerulescens* (Ganges population). New Phytol. 185, 114–129.

Kupper, H., Lombi, E., Zhao, F.J., McGrath, S.P., 2000. Cellular compartmentation of cadmium and zinc in relation to other elements in the hyperaccumulator *Arabidopsis halleri*. Planta 212, 75–84.

Larbi, A., Morales, F., Abadia, A., Gogorcena, Y., Lucena, J.J., Abadia, J., 2002. Effects of Cd and Pb in sugar beet plants grown in nutrient solution: induced Fe deficiency and growth inhibition. Funct. Plant Biol. 29, 1453–1464.

Leitenmaier, B., Witt, A., Witzke, A., Stemke, A., Meyer-Klaucke, W., Kroneck, P.M.H., Kupper, H., 2011. Biochemical and biophysical characterisation yields insights into the mechanism of a Cd/Zn transporting ATPase purified from the hyperaccumulator plant *Thlaspi caerulescens*. Biochim. Biophys. Acta Biomembr. 1808, 2591–2599.

Li, H., Luo, N., Li, Y.W., Cai, Q.Y., Li, H.Y., Mo, C.H., Wong, M.H., 2017. Cadmium in rice: transport mechanisms, influencing factors, and minimizing measures. Environ. Pollut. 224, 622–630.

Li, S., Yu, J., Zhu, M., Zhao, F., Luan, S., 2012. Cadmium impairs ion homeostasis by altering K+ and Ca2+ channel activities in rice root hair cells. Plant Cell Environ. 35, 1998–2013.

Liu, H., Zhang, C., Wang, J., Zhou, C., Feng, H., Mahajan, M.D., Han, X., 2017a. Influence and interaction of iron and cadmium on photosynthesis and antioxidative enzymes in two rice cultivars. Chemosphere 171, 240–247.

Liu, H., Zhang, J., Christie, P., Zhang, F., 2007. Influence of external zinc and phosphorus supply on Cd uptake by rice (*Oryza sativa* L.) seedlings with root surface iron plaque. Plant Soil 300, 105–115.

Liu, H., Zhang, Y., Chai, T., Tan, J., Wang, J., Feng, S., Liu, G., 2013. Manganese-mitigation of cadmium toxicity to seedling growth of *Phytolacca acinosa* Roxb. is controlled by the manganese/cadmium molar ratio under hydroponic conditions. Plant Physiol. Biochemist 73, 144–153.

Liu, Y., Zhang, C., Zhao, Y., Sun, S., Liu, Z., 2017b. Effects of growing seasons and genotypes on the accumulation of cadmium and mineral nutrients in rice grown in cadmium contaminated soil. Sci. Total Environ. 579, 1282–1288.

Lombi, E., Tearall, K.L., Howarth, J.R., Zhao, F.-J., Hawkesford, M.J., McGrath, S.P., 2002. Influence of iron status on cadmium and zinc uptake by different ecotypes of the hyperaccumulator *Thlaspi caerulescens*. Plant Physiol. 128, 1359–1367.

López-Millán, A.-F., Sagardoy, R., Solanas, M., Abadía, A., Abadía, J., 2009. Cadmium toxicity in tomato (*Lycopersicon esculentum*) plants grown in hydroponics. Environ. Exp. Bot. 65, 376–385.

Luo, Y., Rimmer, D.L., 1995. Zinc-copper interaction affecting plant growth on a metal-contaminated soil. Environ. Pollut. 88, 79–83.

Maathuis, F.J.M., 2009. Physiological functions of mineral macronutrients. Curr. Opin. Plant Biol. 12, 250–258.

Marschner, H., Marschner, P., 2012. Marschner's Mineral Nutrition of Higher Plants. vol. xv. Elsevier/Academic Press, London/Waltham, MA, p. 651.

Martin, S., Llugany, M., Barceló, J., Poschenrieder, C., 2012. Cadmium exclusion a key factor in differential Cd-resistance in *Thlaspi arvense* ecotypes. Biol. Plant. 56, 729–734.

Martins, L.L., Mourato, M.P., Cardoso, A.I., Pinto, A.P., Mota, A.M., Goncalves, M.D.L.S., de Varennes, A., 2011. Oxidative stress induced by cadmium in *Nicotiana tabacum* L.: effects on growth parameters, oxidative damage and antioxidant responses in different plant parts. Acta Physiol. Plant. 33, 1375–1383.

Martins, L.L., Reis, R., Moreira, I., Pinto, F., Sales, J., Mourato, M.P., 2013. Antioxidative response of plants to oxidative stress induced by cadmium. In: Hasanuzzaman, M., Fujita, M. (Eds.), Cadmium—Characteristics, Sources of Exposure, Health and Environmental Effects. Nova Publishers, New York, p. 369.

McLaughlin, M.J., Singh, B.R., 1999. Cadmium in Soils and Plants. Springer, Dordrecht.

Meda, A.R., Scheuermann, E.B., Prechsl, U.E., Erenoglu, B., Schaaf, G., Hayen, H., Weber, G., von Wiren, N., 2007. Iron acquisition by phytosiderophores contributes to cadmium tolerance. Plant Physiol. 143, 1761–1773.

Mendoza-Cózatl, D.G., Butko, E., Springer, F., Torpey, J.W., Komives, E.A., Kehr, J., Schroeder, J.I., 2008. Identification of high levels of phytochelatins, glutathione and cadmium in the phloem sap of *Brassica napus*: a role for thiol-peptides in the long-distance transport of cadmium and the effect of cadmium on iron translocation. Plant J. 54, 249–259.

Mendoza-Cozatl, D.G., Jobe, T.O., Hauser, F., Schroeder, J.I., 2011. Long-distance transport, vacuolar sequestration, tolerance, and transcriptional responses induced by cadmium and arsenic. Curr. Opin. Plant Biol. 14, 554–562.

Metwally, A., Safronova, V.I., Belimov, A.A., Dietz, K.J., 2005. Genotypic variation of the response to cadmium toxicity in *Pisum sativum* L. J. Exp. Bot. 56, 167–178.

Michalska, M., Asp, H., 2001. Influence of lead and cadmium on growth, heavy metal uptake and nutrient concentration of three lettuce cultivars grown in hydroponic culture. Commun. Soil Sci. Plant Anal. 32, 571–583.

Migocka, M., Papierniak, A., Kosieradzka, A., Posyniak, E., Maciaszczyk-Dziubinska, E., Biskup, R., Garbiec, A., Marchewka, T., 2015. Cucumber metal transport protein CsMTP9 is a plasma membrane H+-coupled antiporter involved in the Mn2+ and Cd2+ efflux from root cells. Plant J. 84, 1045–1058.

Millaleo, R., Reyes-Diaz, M., Ivanov, A.G., Mora, M.L., Alberdi, M., 2010. Manganese as essential and toxic element for plants: transport, accumulation and resistance mechanisms. J. Soil Sci. Plant Nutr. 10, 470–481.

Mills, R.F., Francini, A., da Rocha, P., Baccarini, P.J., Aylett, M., Krijger, G.C., Williams, L.E., 2005. The plant P-1B-type ATPase AtHMA4 transports Zn and Cd and plays a role in detoxification of transition metals supplied at elevated levels. FEBS Lett. 579, 783–791.

Milner, M.J., Mitani-Ueno, N., Yamaji, N., Yokosho, K., Craft, E., Fei, Z., Ebbs, S., Zambrano, M.C., Ma, J.F., Kochian, L.V., 2014. Root and shoot transcriptome analysis of two ecotypes of *Noccaea caerulescens* uncovers the role of NcNramp1 in Cd hyperaccumulation. Plant J. 78 (3), 398–410.

Mohamed, A.A., Castagna, A., Ranieri, A., Sanità di Toppi, L., 2012. Cadmium tolerance in *Brassica juncea* roots and shoots is affected by antioxidant status and phytochelatin biosynthesis. Plant Physiol. Biochem. 57, 15–22.

Monteiro, M.S., Santos, C., Soares, A.M.V.M., Mann, R.M., 2009. Assessment of biomarkers of cadmium stress in lettuce. Ecotoxicol. Environ. Saf. 72, 811–818.

Moulis, J.M., 2010. Cellular mechanisms of cadmium toxicity related to the homeostasis of essential metals. Biometals 23, 877–896.

Nocito, F.F., Lancilli, C., Crema, B., Fourcroy, P., Davidian, J.C., Sacchi, G.A., 2006. Heavy metal stress and sulfate uptake in maize roots. Plant Physiol. 141, 1138–1148.

Österås, A., Greger, M., 2006. Interactions between calcium and copper or cadmium in Norway spruce. Biol. Plant. 50, 647–652.

Papoyan, A., Pineros, M., Kochian, L.V., 2007. Plant Cd2+ and Zn2+ status effects on root and shoot heavy metal accumulation in *Thlaspi caerulescens*. New Phytol. 175, 51–58.

Pence, N.S., Larsen, P.B., Ebbs, S.D., Letham, D.L.D., Lasat, M.M., Garvin, D.F., Eide, D., Kochian, L.V., 2000. The molecular physiology of heavy metal transport in the Zn/Cd hyperaccumulator *Thlaspi caerulescens*. Proc. Natl. Acad. Sci. U. S. A. 97, 4956–4960.

Pietrini, F., Iannelli, M.A., Pasqualini, S., Massacci, A., 2003. Interaction of cadmium with glutathione and photosynthesis in developing leaves and chloroplasts of *Phragmites australis* (Cav.) Trin. ex Steudel. Plant Physiol. 133, 829.

Pilon-Smits, E.A.H., Quinn, C.F., Tapken, W., Malagoli, M., Schiavon, M., 2009. Physiological functions of beneficial elements. Curr. Opin. Plant Biol. 12, 267–274.

Pinto, F.R., Mourato, M.P., Sales, J.R., Moreira, I.N., Martins, L.L., 2017. Oxidative stress response in spinach plants induced by cadmium. J. Plant Nutr. 40, 268–276.

Redondo-Gómez, S., Mateos-Naranjo, E., Andrades-Moreno, L., 2010. Accumulation and tolerance characteristics of cadmium in a halophytic Cd-hyperaccumulator, *Arthrocnemum macrostachyum*. J. Hazard. Mater. 184, 299–307.

Rivelli, A.R., Puschenreiter, M., De Maria, S., 2014. Assessment of cadmium uptake and nutrient content in sunflower plants grown under Cd stress. Plant Soil Environ. 60, 80–86.

Rizwan, M., Ali, S., Abbas, T., Zia-ur-Rehman, M., Hannan, F., Keller, C., Al-Wabel, M.I., Ok, Y.S., 2016. Cadmium minimization in wheat: a critical review. Ecotoxicol. Environ. Saf. 130, 43–53.

Rizwan, M., Ali, S., Adrees, M., Ibrahim, M., Tsang, D.C.W., Zia-Ur-Rehman, M., Zahir, Z.A., Rinklebe, J., Tack, F.M.G., Ok, Y.S., 2017. A critical review on effects, tolerance mechanisms and management of cadmium in vegetables. Chemosphere 182, 90–105.

Saidi, I., Chtourou, Y., Djebali, W., 2014. Selenium alleviates cadmium toxicity by preventing oxidative stress in sunflower (*Helianthus annuus*) seedlings. J. Plant Physiol. 171, 85–91.

Sandalio, L.M., Dalurzo, H.C., Gomez, M., Romero-Puertas, M.C., del Rio, L.A., 2001. Cadmium-induced changes in the growth and oxidative metabolism of pea plants. J. Exp. Bot. 52, 2115–2126.

Sankaran, R.P., Ebbs, S.D., 2008. Transport of Cd and Zn to seeds of Indian mustard (*Brassica juncea*) during specific stages of plant growth and development. Physiol. Plant. 132, 69–78.

Shao, J.F., Yamaji, N., Shen, R.F., Ma, J.F., 2017. The key to Mn homeostasis in plants: regulation of Mn transporters. Trends Plant Sci. 22, 215–224.

Skórzyńska-Polit, E., Tukendorf, A., Selstam, E., Baszyński, T., 1998. Calcium modifies Cd effect on runner bean plants. Environ. Exp. Bot. 40, 275–286.

Smeets, K., Ruytinx, J., Semane, B., Van Belleghem, F., Remans, T., Van Sanden, S., Vangronsveld, J., Cuypers, A., 2008. Cadmium-induced transcriptional and enzymatic alterations related to oxidative stress. Environ. Exp. Bot. 63, 1–8.

Song, Y., Jin, L., Wang, X., 2017. Cadmium absorption and transportation pathways in plants. Int. J. Phytoremed. 19, 133–141.

Souza, V.L., de Almeida, A.-A.F., Lima, S.G., de M. Cascardo, J.C., da C. Silva, D., Mangabeira, P.A., Gomes, F.P., 2011. Morphophysiological responses and programmed cell death induced by cadmium in *Genipa americana* L. (Rubiaceae). Biometals 24, 59–71.

Takahashi, R., Ishimaru, Y., Nakanishi, H., Nishizawa, N.K., 2011. Role of the iron transporter OsNRAMP1 in cadmium uptake and accumulation in rice. Plant Signal. Behav. 6, 1813–1816.

Tang, R.-J., Luan, S., 2017. Regulation of calcium and magnesium homeostasis in plants: from transporters to signaling network. Curr. Opin. Plant Biol. 39, 97–105.

Thomine, S., Wang, R., Ward, J.M., Crawford, N.M., Schroeder, J.I., 2000. Cadmium and iron transport by members of a plant metal transporter family in Arabidopsis with homology to Nramp genes. Proc. Natl. Acad. Sci. U. S. A. 97, 4991–4996.

Verbruggen, N., Hermans, C., Schat, H., 2009. Mechanisms to cope with arsenic or cadmium excess in plants. Curr. Opin. Plant Biol. 12, 364–372.

Verbruggen, N., Juraniec, M., Baliardini, C., Meyer, C.-L., 2013. Tolerance to cadmium in plants: the special case of hyperaccumulators. Biometals 26, 633–638.

Vert, G., Grotz, N., Dedaldechamp, F., Gaymard, F., Guerinot, M.L., Briata, J.F., Curie, C., 2002. IRT1, an Arabidopsis transporter essential for iron uptake from the soil and for plant growth. Plant Cell 14, 1223–1233.

Wong, C.K., Cobbett, C.S., 2009. HMA P-type ATPases are the major mechanism for root-to-shoot Cd translocation in *Arabidopsis thaliana*. New Phytol. 181, 71–78.

Wong, C.K., Jarvis, R.S., Sherson, S.M., Cobbett, C.S., 2009. Functional analysis of the heavy metal binding domains of the Zn/Cd-transporting ATPase, HMA2, in *Arabidopsis thaliana*. New Phytol. 181, 79–88.

Wu, D., Sato, K., Ma, J.F., 2015. Genome-wide association mapping of cadmium accumulation in different organs of barley. New Phytol. 208, 817–829.

Wu, F., Zhang, G., Dominy, P., Wu, H., Bachir, D.M.L., 2007. Differences in yield components and kernel Cd accumulation in response to Cd toxicity in four barley genotypes. Chemosphere 70, 83–92.

Xu, J., Chai, T., Zhang, Y., Lang, M., Han, L., 2009. The cation-efflux transporter BjCET2 mediates zinc and cadmium accumulation in *Brassica juncea* L. leaves. Plant Cell Rep. 28, 1235–1242.

Yoshihara, T., Hodoshima, H., Miyano, Y., Shoji, K., Shimada, H., Goto, F., 2006. Cadmium inducible Fe deficiency responses observed from macro and molecular views in tobacco plants. Plant Cell Rep. 25, 365–373.

Yruela, I., 2009. Copper in plants: acquisition, transport and interactions. Funct. Plant Biol. 36, 409–430.

Zhang, X., Zhang, X., Gao, B., Li, Z., Xia, H., Li, H., Li, J., 2014. Effect of cadmium on growth, photosynthesis, mineral nutrition and metal accumulation of an energy crop, king grass (*Pennisetum americanum* × *P. purpureum*). Biomass Bioenergy 67, 179–187.

Zhang, Y., Xiao, H., 1998. Antagonistic effect of calcium, zinc and selenium against cadmium induced chromosomal aberrations and micronuclei in root cells of *Hordeum vulgare*. Mutat. Res. 420, 1–6.

Zhong, W.L., Li, J.T., Chen, Y.T., Shu, W.S., Liao, B., 2012. A study on the effects of lead, cadmium and phosphorus on the lead and cadmium uptake efficacy of *Viola baoshanensis* inoculated with arbuscular mycorrhizal fungi. J. Environ. Monit. 14, 2497–2504.

Zornoza, P., Sanchez-Pardo, B., Carpena, R.N.O., 2010. Interaction and accumulation of manganese and cadmium in the manganese accumulator *Lupinus albus*. J. Plant Physiol. 167, 1027–1032.

Zornoza, P., Vazquez, S., Esteban, E., Fernandez-Pascual, M., Carpena, R., 2002. Cadmium-stress in nodulated white lupin: strategies to avoid toxicity. Plant Physiol. Biochem. 40, 1003–1009.

Zorrig, W., Rouached, A., Shahzad, Z., Abdelly, C., Davidian, J.-C., Berthomieu, P., 2010. Identification of three relationships linking cadmium accumulation to cadmium tolerance and zinc and citrate accumulation in lettuce. J. Plant Physiol. 167, 1239–1247.

# CHAPTER 14

# Role of Salicylic Acid in Mitigating Cadmium Toxicity in Plants

**Bin Guo**
Institute of Environment, Resource, Soil and Fertilizer, Zhejiang Academy of Agricultural Sciences, Hangzhou, People's Republic of China

## Abstract

Salicylic acid (SA) has been reorganized as a signaling molecule in plants involved in biotic and abiotic stresses. This chapter covers the recent progress of SA-mediated signaling pathways in response to Cd stress. Preliminary treatment of plants with low concentrations of SA manifested alleviation effects on Cd toxicity. However, excess exogenous SA exposure or abnormal endogenous SA levels aggravated Cd-induced oxidative stress. Here, we summarized the possible mechanisms of the SA-mediated alleviation of Cd toxicity in plants in terms of plant growth, antioxidant systems, Cd immobilization, Cd and other elements' uptake, photosynthesis, GSH and PC synthesis, and cell death. Although the biochemical mechanisms are mainly focused on its signaling role by altering the levels of reactive oxygen species of plant cells, several questions remain unanswered at both the theoretical and practical levels.

Keywords: Antioxidant system, Cadmium, Cell wall, $H_2O_2$, Ion uptake, Reactive oxygen species, Phytochelatins, Salicylic acid, Signaling

## ABBREVIATIONS

| | |
|---|---|
| **AsA** | ascorbic acid |
| **ABC transporters** | ATP-binding cassette transporters |
| **APX** | ascorbate peroxidase |
| **Cd** | cadmium |
| **CAT** | catalase |
| **GR** | glutathione reductase |
| **GSH** | glutathione |
| **GSH1** | γ-glutamylcysteine synthetase |
| **GSH2** | glutathione synthetase |
| **GPX** | glutathione peroxidase |
| **MAPK** | mitogen-activated protein kinase |
| **NPR1** | nonexpressor of PR1 |
| **NPT** | nonprotein thiols |
| **PAL** | phenylalanine ammonia-lyase |

*Cadmium Toxicity and Tolerance in Plants*
https://doi.org/10.1016/B978-0-12-814864-8.00014-0

| | |
|---|---|
| **PCs** | phytochelatins |
| **POD** | peroxidase |
| **PPO** | polyphenol oxidase |
| **PSII** | photosystem II |
| **ROS** | reactive oxygen species |
| **SA** | salicylic acid |
| **SAT** | serine acetyltransferase |
| **SOD** | superoxide dismutase |

## 1 INTRODUCTION

Salicylic acid (SA) is a key phenolic signal in plant defense against biotic and abiotic stresses. Despite its broad distribution in plants, the basal level of SA in plants differs widely among species, with up to 100-fold differences having been recorded (Raskin et al., 1990). SA is synthesized through two distinct and compartmentalized pathways (Shah, 2003; Abreu and Munné-Bosch, 2009; Rivassan and Plasencia, 2011). It is well established that SA derives from the shikimate-phenylpropanoid pathway in the cytoplasm. Phenylalanine ammonia lyase catalyzes the first step in this pathway, which is the conversion of phenylalanine to trans-cinnamic acid. *Trans*-cinnamic acid is subsequently converted into benzoic acid. Benzoic-acid-2-hydroxylase catalyzes the final step, the conversion of benzoic acid to SA. A second pathway is proposed that is based on isochorismate, and occurs in chloroplasts. In this pathway, chorismite is converted into isochorismate through the action of an isochorismate synthase (ICS), and SA is generated from isochorismate by an isochorismate pyruvate lyase.

The first report on the role of SA signaling was in a study of thermogenesis in the inflorescences of *Arum lilies* (Raskin et al., 1987). Since then, over the past 30 years, SA as a phytohormone involved in plant defense has been extensively studied. SA induces viral, fungal, and pathogen resistance in plants by inducing the expression of pathogenesis-related proteins and establishing systemic acquired resistance. The development of acquired resistance often follows a localized tissue death at the site of pathogen penetration called the hypersensitive response, in which SA plays a central role. The proposed mode of SA action was related to the inhibition of catalase (CAT) and ascorbate peroxidase (APX), two major $H_2O_2$-scavenging enzymes. The inhibition might cause the cellular concentrations of $H_2O_2$ to rise. Subsequently, $H_2O_2$ may act as a second messenger, and activate defense-related genes (Chen et al., 1993). This signaling pathway was well described in many top journals, including Plant Cell (Nawrath and Métraux, 1999), Science (Delaney et al., 1994), Nature (Wildermuth et al., 2001), Proceedings of the National Academy of Sciences (Delaney et al., 1995), and Critical Reviews in Plant Sciences (Dempsey et al., 1997).

In addition to its function in biotic stress, a great deal of evidence suggests that SA modulates plant responses to various abiotic stresses, such as drought, cold, heat, salt, ultraviolet radiation, and heavy metals (Horváth et al., 2007). For example, inhibitors of SA biosynthesis reduced the tolerance of pea plants to heat stress. Spraying banana seedlings with SA (Kang et al., 2003) and exposing seedlings of maize, rice, and cucumber to SA significantly strengthened tolerance to subsequent chilling stress (Kang and Saltveit, 2002). SA is also involved in plants exposed to salt. SA added to the soil had an ameliorating effect on the survival of maize plants during salt stress, and decreased NaCl accumulation (Gunes et al., 2007).

Furthermore, SA plays an important role in the reduction of Cd accumulation and enhancement of tolerance to Cd stress in a wide range of plant species. However, when SA exposure exceeds the appropriate range, it may result in oxidative toxicity to plants. According to Rao et al. (1997), high levels of SA act as a prooxidant because SA is able to oxidize proteins in *Arabidopsis* leaves, and multiplies $H_2O_2$ generation in plants. Elsewhere, it has been shown that excessive SA pretreatment aggravates Cd-induced oxidative damage in perennial ryegrass and soybeans. Apparently, the role of SA in Cd-stressed plants depends on various factors, such as the species and developmental stage of the plant, the mode of application, and the concentration of SA and its endogenous level in the given plant.

## 2 MODES OF SA TREATMENT

Table 1 shows a survey of SA roles in plant responses to Cd stress. Most of the literature reports that SA has a mitigating effect on Cd toxicity, except for two studies with SA-deficient mutants. SA treatments are applied mainly in three modes, including spray, presoaking, and hydroponic application. In these data analyses, we excluded a study in which SA and Cd treatments were 3000 μM and 560 $mg L^{-1}$, respectively, which are extremely high and might be misleading. From 76 values of SA and Cd exposure concentrations drawn from 40 international publications, the following data can be inferred (Fig. 1):

**(1)** Mean Cd exposure concentration: 28 $mg L^{-1}$ (variable coefficient, 249%);

**Table 1** Salicylic acid alleviates Cd toxicity in plants

| | SA treatment | Cd treatment | Timeline | Plant species | Main response | Reference |
|---|---|---|---|---|---|---|
| Spraying | 1000 μM for 10 mL, 45 times in a day interval | 100 and 200 $mg L^{-1}$ for 14 days | Simultaneous | Indian mustard (*Brassica juncea*) | I, II, III, IV, V | Ahmad et al. (2011) |
| | 50 μM, 4 times in a 3 days interval | 75, 150, and 300 $mg kg^{-1}$ for 14 days | Simultaneous | Oilseed rape (*Brassica napus*) | II, III, V, VI | Ali et al. (2015) |
| | 500 μM, 1 time | 40 $mg kg^{-1}$ for 6 days | Pretreatment | Soybean (*Glycine* max L. *cv. Liaoxing 1*) | I, II, III | Li et al. (2014) |
| | 300 $mg L^{-1}$ (2170 μM), 1 time | 56 and 112 $mg kg^{-1}$ Field experiment | Simultaneous | Radish (*Raphanus sativus*) | I, IV[a] | Raza and Shafiq (2013) |
| | 10, 50, 100, and 200 μM Each day treated for 50 mL last 4 days | 44.8 $mg kg^{-1}$ for 10 days | Pretreatment | Melon (*Cucumis melo* L.) | I, II, III | Zhang et al. (2015) |
| Presoaking | 500 μM for 12 h | 56 and 112 $mg kg^{-1}$ for 56 days | Pretreatment | Wheat (*Triticum aestivum* L. *cv. Giza 168*) | I, II, III, IV[a] | Agami and Mohamed (2013) |
| | 250 and 1000 μM for 12 h | 5.6 and 11.2 $mg L^{-1}$ for 12 days | Pretreatment | Flax (*Linum usitatissimum* L.) | I, II, III, IV, V | Belkhadi et al. (2010) |
| | 100 μM for 3 h | 3, 5, and 7 $mg kg^{-1}$ for 3 days | Pretreatment | Soybean | II, III, IV[a] | Dražić and Mihailovic (2009) |
| | 500 μM for 12 h | 0.56, 1.12, and 5.60 $mg L^{-1}$ for 7 days | Pretreatment | Kentucky bluegrass | I, II, III, IV, V[a] | Guo et al. (2013) |

| | | | | | | |
|---|---|---|---|---|---|---|
| | 500 μM for 6 h | 1.12, 1.68, and 2.80 mg L $^{-1}$ for 14 days | Pretreatment | Maize (*Zea mays* L., hybrid Norma) | I, II, III, IV | Krantev et al. (2008) |
| | 250 and 500 μM for 12 h | 5.6 mg L $^{-1}$ for 12 days | Pretreatment | Bean (*R. communis* cv. Zibi 5) | I, III, IV | Liu et al. (2011) |
| | 500 μM for 6 h | 2.8 mg L $^{-1}$ for 12 days | Pretreatment | Barley (*Hordeum vulgare* cv Gerbel) | I, II, III, IV, V, VI, VII | Metwally et al. (2003) |
| | 100 μM for 16 h | 11.2 and 112 mg L $^{-1}$ for 1 day | Pretreatment | *Oryza sativa* L. (cv: Longai) | I, II | Panda and Patra (2007) |
| | 500 μM for 6 h | 25, 50, and 100 mg kg $^{-1}$ | Pretreatment | Hemp (*Cannabis sativa* L.) | I, II, III, IV[a] | Shi et al. (2009) |
| | 500 μM for 6 h | 1.12, 1.68, and 2.80 mg L $^{-1}$ for 14 days | Pretreatment | Maize (*Zea mays*) | VI | Szalai et al. (2013) |
| | 100 μM for 12 h | 5.6 and 11.2 mg L $^{-1}$ for 6 days | Pretreatment | Legume (*Phaseolus aureus* and *Vicia sativa*) | I, II | Zhang et al. (2011) |
| | 250 or 1000 μM for 8 h | 5.6 and 11.2 mg L $^{-1}$ for 10 days | Pretreatment | Flax (*Linum usitatissimum* L.) | I, IV[a], VI | Belkadhi et al. (2012) |
| | 250 or 1000 μM for 8 h | 5.6 and 11.2 mg L $^{-1}$ for 10 days | Pretreatment | Flax (cv. Viking) | I, II | Belkadhi et al. (2013) |
| | 250 or 1000 μM for 8 h | 5.6 and 11.2 mg L $^{-1}$ for 10 days | Pretreatment | Flax (*Linum usitatissimum* L.) | I, II | Belkadhi et al. (2014) |
| | 250 or 1000 μM for 8 h | 5.6 and 11.2 mg L $^{-1}$ for 10 days | Pretreatment | Flax (*Linum usitatissimum* L.) | Lipids | Belkadhi et al. (2015) |
| | 100 μM for 8 h | 1.12, 11.2, and 112 mg L $^{-1}$ for 1 day | Pretreatment | *Oryza sativa* L. (cv: Longai) | I, II, IV[a] | Choudhury and Panda (2004) |
| | 100 μM for 1, 3, 6 h | 3 and 5 mg L $^{-1}$ for 7 days | Pretreatment | Alfalfa (*Medicago sativa* L. cv. Evropa) | I, IV, V | Dražić et al. (2006) |
| | 500 μM for 24 h. | 112 mg L $^{-1}$ for 72 h. | Pretreatment | Mungbean (*Vigna radiata* L. Wilczek) | I, II | Roychoudhury et al. (2016) |

*Continued*

**Table 1** Salicylic acid alleviates Cd toxicity in plants—cont'd

| | SA treatment | Cd treatment | Timeline | Plant species | Main response | Reference |
|---|---|---|---|---|---|---|
| Nutrition | 100, 200, 300,400 μM | 11.2 mg $L^{-1}$ for 14 days | Simultaneous | Ryegrass | I, II, III, IV[a], V | Bai et al. (2014) |
| | 3000 μM for 3 h | 560 mg $L^{-1}$ for 1 day | Pretreatment | Rice (*Oryza sativa* L., cv. Taichung Native 1) | II, IV | Chao et al. (2010) |
| | 1, 10, 100 μM for 72 h | 5.6 mg $L^{-1}$ for 1 day | Pretreatment | Alfalfa (*Medicago sativa* L. cv Zhongmu No.1) | I, II, IV[a], VI, VII (HO-1) | Cui et al. (2012) |
| | 1, 10, 100 μM | 3, 6 mg $L^{-1}$ for 3 days | Simultaneous | Soybean (*Glycine max* L. cvSG1) | I[a], IV[a], IV | Drazic and Mihailovic (2005) |
| | 10 μM for 72 h | 5.6 mg $L^{-1}$ for 6 days | Pretreatment | Rice (*O. sativa* cv Jiahua 1) | I, II | Guo et al. (2007a,b) |
| | 10 μM for 24 h | 5.6 mg $L^{-1}$ for 6 days | Pretreatment | Rice (*O. sativa* cv Jiahua 1) | I, II, IV | Guo et al. (2007a,b) |
| | 10 μM for 72 h | 5.6 mg $L^{-1}$ for 6 days | Pretreatment | Rice (*O. sativa* cv Jiahua 1) | I, II, IV | Guo et al. (2009) |
| | 500 μM for 24 h | 2.8 mg $L^{-1}$ for 10 days | Pretreatment | Barley leaves (*Hordeum vulgare* cv Gerbel) | I, II, III, IV, V, VI, VII | Metwally et al. (2003) |
| | 60, 120, 250, 500 mM | 5.6 mg $L^{-1}$ for 5 days | Pretreatment | Soybean (*Glycine max* L., A6445RG) | II, III, IV[a], V, VI, VII (HO-1) | Noriega et al. (2012) |
| | 500 μM for 24 h | 56 mg $L^{-1}$ for 1 day | Pretreatment and simultaneously | Maize (*Zea mays* L., hybrid Norma) | I[a], II[a], III[a], VI[a] | Pál et al. (2002) |
| | 200 μM | 11.2 mg $L^{-1}$ for 14 days | Simultaneously | Ryegrass (*Lolium perenne* L.) | I, II, III, VI | Wang et al. (2013) |
| | 100 μM | 22.4 mg $L^{-1}$ for 14 days | Simultaneously | Peanut (*Arachis hypogaea* L.) | I, II, III, IV[a], V | Xu et al. (2015) |
| | 50 μM | 11.2 mg $L^{-1}$ for 8 h | Pretreatment | Wheat (*Triticum aestivum* L.) | I, IV[a], Hormones | Shakirova et al. (2016) |

|  |  |  |  |  |  |  |
|---|---|---|---|---|---|---|
|  | 250, 500 μM | 1.68 mg $L^{-1}$ for 3 and 6 h | Posttreatment | Barley (*Hordeum vulgare* L.) cv. Slaven | I, II, Auxin | Tamás et al. (2015) |
|  | 500 μM for 24 h | 56 mg $L^{-1}$ for 1 day | Pretreatment | Maize (*Zea mays* L., hybrid Norma) | II, III, IV[a], VI | Gondor et al. (2016) |
|  | 50 μM | 1.12 mg $L^{-1}$ for 7 days | Simultaneously | *Lemna minor* | II, III, IV, V | Lu et al. (2018) |
| SA mutants | SA accumulation |  |  | *Lycium chinense* | II, III, IV, VII (*LcGSHS*) | Guan et al. (2015) |
|  | Up and down-regulating endogenesis SA |  |  | Accumulating mutant *snc1*, *npr1-1*, reducing mutant *nahG*, *snc1/nahG* | I[a], II[a], III[a] | Tao et al. (2013) |
|  | Down-regulating endogenesis SA |  |  | *NahG* | II[a], III[a], VII[a] (CAT1) | Zawoznik et al. (2007) |
|  | Down-regulating endogenesis SA |  |  | *Sid2* | I, II, III, IV, V, VI, VII | Guo et al. (2016) |

[a]Negative or no effects.

I, Growth; II, antioxidant defense; III, photosynthesis; IV, Cd uptake; V, ion uptake; VI, phytochelatins; VII, SA or Cd-induced genes.

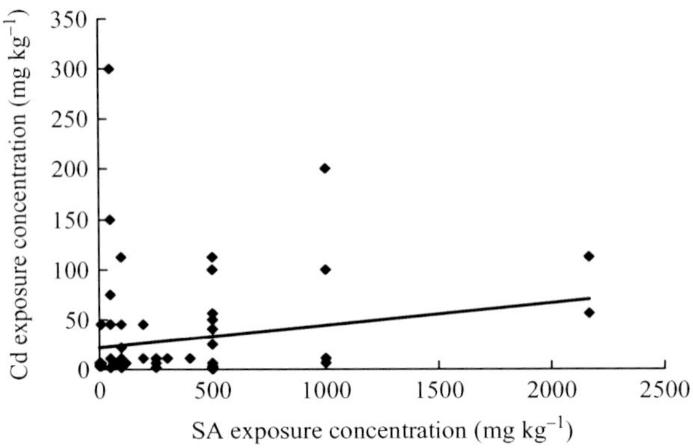


**Fig. 1** Representation of Cd exposure concentrations plotted against SA exposure concentrations, drawn from the data reported in Table 1.

**(2)** Mean SA spray concentration: 542 μM (variable coefficient, 147%);
**(3)** Mean SA presoaking concentration: 410 μM (variable coefficient, 72%); and
**(4)** Mean SA hydroponic concentration: 189 μM (variable coefficient 100%).

Generally, the concentrations used for SA spray and presoaking are higher than hydroponic concentrations, and have higher variable coefficients. Interestingly, most of the results show partial exposure of different organs (leaves, seeds, or roots) to SA can stimulate plants to build up a whole system to defend against Cd toxicity.

## 2.1 Spray

Until now, we found only five studies reporting the effect of SA spray on Cd stress in different plant species, including mustard, oilseed rape, soybeans, radishes, and melons. All of these studies showed that SA spray alleviated Cd toxicity to plants, mainly due to Cd-induced oxidative damage. For example, spraying 1000 μM SA onto mustard significantly mitigated Cd-induced oxidative stress and increased proline, lipid peroxidation, and electrolyte leakage. However, the exposure concentrations in the literature ranged from 10 to 2170 μM. Such a large gap in concentration is difficult to guide by SA spray.

## 2.2 Presoaking

Compared with the spray mode, there is more evidence in the literature (18) of the presoaking method. The presoaking concentrations ranged from

100 to 1000 μM, and presoaking time ranged from 3 to 12 h. Furthermore, the studied plants are mainly crops, i.e., wheat, beans, maize, barley, rice, and commercial crops, i.e., flax, hemp, and alfalfa, which are more meaningful for practice than the SA spray studies. Similar to the mode of spray, all of these studies showed that presoaking of SA in plant seeds alleviated the following Cd stress. In that study, SA exerted a significant negative effect on the growth and photosynthesis of Cd-exposed beans, in which a stomatal limitation was involved.

## 2.3 Hydroponical Application

There are 16 studies on SA, with Cd treatments via hydroponic application. The hydroponic presoaking concentrations ranged from 100 to 500 μM, and the hydroponic time ranged from 3 to 72 h. Many studies included pretreatment with SA followed by Cd stress, while other studies involved treatment with SA and Cd simultaneously. Furthermore, one experiment was conducted posttreatment for SA. Again, all of these studies showed that hydroponic application of SA to plant roots mitigated the Cd toxicity of the whole plant.

## 2.4 SA Mutants

It should be noted that the concentrations of SA and Cd used in these studies are high, or very high (i.e., 75% of the values are higher than 100 μM and $3\,mg\,kg^{-1}$, respectively), and the exposure time is short (hours, days, or weeks). Plants must rapidly manage this "acute" Cd stress by inducing an instant response regulated by the signaling of exogenous SA. Furthermore, various treatment modes and concentrations of SA and Cd, and different plant species, were included. All of the uncertain factors made these studies unpredictable. When SA application exceeds the appropriate range, it may result in oxidative toxicity to plants. Hence, comparison of wild-type plants with SA-deficient or SA-accumulating mutants is an effective way to evaluate the endogenous role of SA in response to Cd toxicity. Surprisingly, although *NahG* and *sid2* have SA deficiency in common, the two mutants showed conflicting results when exposed to Cd stress. With the *NahG* transgenic lines, encoding a bacterial SA-decomposing salicylate hydroxylase manifested higher tolerance than the wild-type under Cd toxicity (Zawoznik et al., 2007; Tao et al., 2013). However, depletion of SA in the *sid2* mutant, a mutation in the gene encoding ICS, decreased Cd damage (Guo et al., 2016).

## 3 POSSIBLE ROLE OF SALICYLIC ACID IN MITIGATING CADMIUM TOXICITY IN PLANTS

Studies have shown that SA, as an important phytohormone, plays important roles in alleviating Cd-induced damage in many plant species. This chapter summarizes recent advances in the understanding of the functions of the SA-mediated response to Cd toxicity. The possible physiological/molecular mechanisms are related to (1) growth, (2) reactive oxygen species (ROS) and the antioxidant defense system, (3) Cd immobilization in the cell wall, (4) Cd uptake and distribution, (5) element uptake, (6) photosynthesis, (7) synthesis of glutathione and phytochelatins (PCs), and (8) cell death (Fig. 2).

### 3.1 Growth

Increasing evidence has shown that SA as a phytohormone regulates plant growth throughout the entire lifespan, such as seed germination, vegetative and reproductive growth, and senescence (Rivassan and Plasencia, 2011). Furthermore, SA has been recognized as a regulatory signal in mitigating Cd-induced growth inhibition, as observed in many plant species (Table 1), such as mustard (Ahmad et al., 2011), rice (Panda and Patra, 2007; Guo et al., 2007a,b, 2009), barley (Metwally et al., 2003), and ryegrass

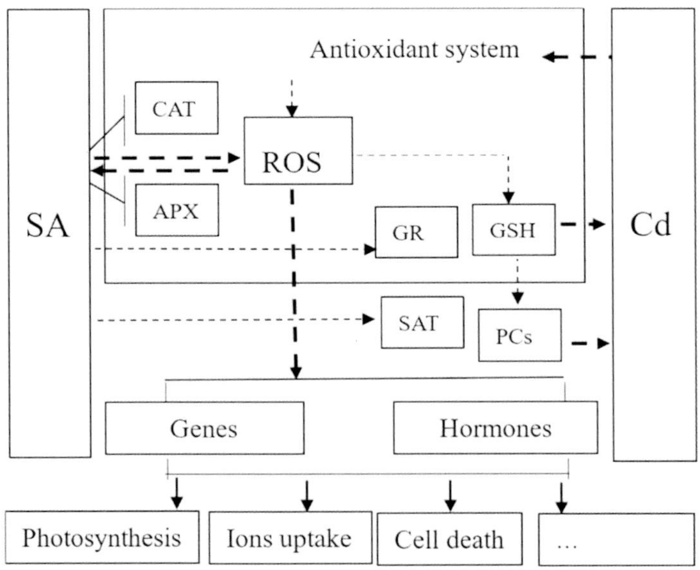


**Fig. 2** Possible mechanisms of SA alleviate the toxicity of Cd in plants.

(Bai et al., 2014). The growth-promoting effects of SA could be related to changes in the hormonal status (Shakirova et al., 2003; Abreu and Munné-Bosch, 2009), or by improvement of photosynthesis, transpiration, and stomatal conductance (Stevens et al., 2006).

Apparently, the effect of exogenous SA on plant growth is dependent on the plant species and the SA concentrations tested. Some plants are sensitive to SA, but are still protected by SA application when combined with Cd treatment. In the absence of Cd, the application of 250 and 500 μM SA significantly reduced the shoot and root biomass of castor plants. In the presence of Cd, a reduction in the shoot and root biomass was observed, only in the plants pretreated with 500 μM SA (Liu et al., 2011). Presoaking hemp seeds with 500 μM SA for 6 h decreased the biomass of shoots and roots by 8.6% and 8.1%, respectively. However, when plants were exposed to Cd stress, this SA pretreatment enhanced the biomass in comparison with the nonpretreated plants at the same Cd level (Shi et al., 2009).

Surprisingly, inconsistent conclusions were obtained when characterizing *Arabidopsis* mutants or transgenic plants. Based on the differential growth between wild-type plants and SA-altering mutants, Tao et al. (2013) found that Cd-induced growth inhibition in *Arabidopsis* was intensified by elevated endogenous SA (*snc1*) but ameliorated by reduced SA (*nahG*). It seems that endogenous SA plays a negative role in *Arabidopsis* tolerance to Cd. However, Guo et al. (2016) used another SA-deficiency *Arabidopsis* mutant, *sid2*, and found that depletion of SA caused negative effects on Cd tolerance by aggravating growth inhibition.

Furthermore, both Cd stress and the SA signaling pathway are involved in plant senescence. Exposure of 100 and 150 μM Cd to *Arabidopsis* accelerated senescence, as suggested by the expression of a marker senescence-associated gene (SAG12) (Michele et al., 2009). SA levels increased approximately fourfold at the midsenescent stage in *Arabidopsis* senescent leaves (Morris et al., 2000). It is still unclear whether SA and Cd manifest antagonistic or synergistic effects during senescence. Nevertheless, the balance of appropriate SA levels is vital for plants to respond to Cd toxicity. Insufficient SA may be insufficient to signal the defense system, whereas superfluous SA may aggravate Cd-induced growth inhibition.

## 3.2 Reactive Oxygen Species and Antioxidant Defense System

The majority of Cd toxicity to plants is interference with the electron transport chains or blocking of antioxidant enzyme structures, thereby altering the activity of antioxidative enzymes, such as superoxide dismutase (SOD), POD, CAT, APX, and glutathione reductase (GR), and disturbing the synthesis of antioxidant molecules, such as glutathione (GSH) and ascorbic acid (AsA). This leads to the accumulation of ROS, especially $H_2O_2$, and then to the oxidation of organic compounds and induction of membrane lipid peroxidation in the cellular environment.

In addition to oxidation toxicity, $H_2O_2$ is also an essential signal involved in plant defense against environmental stresses. SA is involved in the $H_2O_2$ signaling pathway. Upon pathogen attack, SA accumulates to high levels at the site of pathogenic infection and binds and inhibits tobacco CAT activity, both in vitro and in vivo, forming SA-binding proteins (SABPs), and thereby leading to an increase in the endogenous level of $H_2O_2$, which could then serve as a secondary signal to induce the expression of pathogenesis-related proteins and initiate the development of systemic acquired resistance (Chen et al., 1993). The SA-induced accumulation of $H_2O_2$ in rice leaves was visualized using the DAB staining method (Chao et al., 2010). Furthermore, the significant interplay between $H_2O_2$ and SA signaling was uncovered when applications of $H_2O_2$ and SA to tobacco and *Arabidopsis* plants induced each other, suggesting that they are involved in a selfamplifying feedback loop (Rao et al., 1997). In this process, SA may act as a one-electron donor that siphons the extremely fast CAT cycle into the relatively slow peroxidative cycle (approximately 1000 times slower) (Durner and Klessig, 1996). SA-deficient *Arabidopsis* plants, such as *sid2* mutants, have a higher CAT activity in leaves than wild-type plants (Guo et al., 2016). In addition to the inhibition of CAT, SA-blocked electron flow from the substrate dehydrogenizes to the ubiquinone pool in mitochondria might also be an important mechanism of triggering $H_2O_2$ generation (Norman et al., 2004).

Therefore, the mitigating role of SA in Cd-induced oxidative damage may be similar to that function in biotic stress. A body of studies was conducted by SA pretreatment by spray, presoaking, or hydroponic incubation, which aim to promote $H_2O_2$ status and "set up" the antioxidant system in advance. Guo et al. (2007a,b, 2009) showed that pretreatment with 10 μM SA initially caused $H_2O_2$ accumulation in vitro, which may act as a secondary signaling pathway in preparing the high level of antioxidant defense system of rice roots. During the subsequent Cd exposure period, the activities of SOD, CAT, and POD and the levels of GSH, NPT, and AsA in the SA pretreatment were higher than those in the non-SA treatment, which was accompanied by alleviation of Cd-induced oxidative damage. Similar conclusions were also reported by Chao et al. (2010). The effect of SA pretreatment was comparable to the exogenous $H_2O_2$ pretreatment, which both enhanced the antioxidant level in rice leaves and greatly improved tolerance of Cd-induced oxidant stress. In addition to rice, SA-signaled protection from oxidative injury caused by Cd has been found in many other plant species, such as mustard (Ahmad et al., 2011), wheat (Agami and Mohamed, 2013), oilseed rape (Ali et al., 2015), ryegrass (Wang et al., 2013; Bai et al., 2014), flax (Belkhadi et al., 2010), alfalfa (Cui et al., 2012), bluegrass (Guo et al., 2013), maize (Krantev et al., 2008; Noriega et al., 2012), barley (Metwally et al., 2003), peas (Popova et al., 2009), peanuts (Xu et al., 2015), beans (Zhang et al., 2011), melons (Zhang et al., 2015), and duckweed (Lu et al., 2018).

Although evidence has definitely shown the positive role of SA in Cd tolerance, there are still many gaps in the understanding of the mechanisms. Metwally et al. (2003) found that lower activities of antioxidant enzymes such as APX and CAT in SA-treated barley concurred with the stimulated antioxidant defense, thereby suggesting that SA alleviates Cd toxicity, not at the level of antioxidant defense. During the Cd stress period in peas, SA was reported to act as a direct scavenger of hydroxyl radicals and modulate redox balance to eliminate excessive $H_2O_2$ injury (Popova et al., 2009). SA could also be involved in the expression of specific proteins or defense-related enzymes (Zawoznik et al., 2007). Alleviation of Cd toxicity by heme oxygenase (HO-1)-mediated and SA-dependent signaling has been observed in alfalfa and soybeans, showing an upregulation in antioxidative behavior (Cui et al., 2012; Noriega et al., 2012). SA and NO synergistically promote the activities of antioxidant enzymes in peanut seedlings (Xu et al., 2015). High intrinsic levels of SA in *snc1* plants possess high POD activity, which can produce reactive oxygen species and lead to its sensitive phenotype under Cd stress (Tao et al., 2013). SA-pretreated seedlings are characterized by significantly lower Cd-induced accumulation of dehydrins, ABA-signaled water-soluble proteins that can bind and neutralize $H_2O_2$ (Shakirova et al., 2016).

Furthermore, the positive role of SA in Cd tolerance is still debatable. A few studies with NahG transgenic lines of *Arabidopsis* have shown that depletion of SA reduces Cd-induced oxidative stress (Zawoznik et al., 2007; Tao et al., 2013). However, Guo et al. (2016) suggested that the NahG genotype is not suitable for evaluating SA function because defense signaling results from catechol accumulation, which acts as an antioxidant that compromises the inhibitory effects of stresses. In contrast, another SA deficiency mutant, *sid2,* used in their study showed that the Cd-induced decrease in SOD and POD activities was more pronounced than in wild-type plants, thereby accentuating Cd-induced oxidative damage.

## 3.3 Cd Immobilization in the Cell Wall

For plants, the first barrier against Cd stress is immobilization by crosslinking Cd into the carboxyl groups of the cell wall, thereby limiting the Cd influx into the cytosol (Sanità di Toppi and Gabbrielli, 1999). Meanwhile, an important signaling role of SA is to highlight $H_2O_2$ levels to reinforce plant cell walls, thereby limiting the spread of invading pathogens (Durner and Klessig, 1996). Therefore, the beneficial effect of SA may be thickening the cell wall surrounding the xylem and pericycle of roots (Da Cunha and Do Nascimento, 2009), and improving Cd retention in this tissue. The results obtained from some studies concurred with this hypothesis. In peanuts, SA treatment increased Cd accumulation in the cell wall, and decreased Cd distribution to cell organelles (Xu et al., 2015). Similar results were also observed in ryegrass (Bai et al., 2014). These authors hypothesized that SA supply might protect the integrity of cell function and/or increase pectin and hemicellulose contents in the cell wall (Wang et al., 2013). However, some uncertainties remain. Guo et al. (2007a,b) reported that both Cd treatment alone and SA pretreatment alone enhanced the cell wall composition in root cells. However, there was no synergistic effect between SA and Cd treatments in the cell wall composition or in the activities related to lignin synthesis, such as phenylalanine ammonia-lyase (PAL), POD, and polyphenol oxidase (PPO). It was suggested that the strong effect of Cd stress might maximize the lignification of rice roots, which resulted in neglectable effects of SA on Cd immobilization.

## 3.4 Cd Uptake and Distribution

A survey of the literature indicates that exposure to SA can decrease the uptake of Cd and the root-to-shoot translocation of Cd, thus resulting in low Cd accumulation in shoots. Foliar application of SA substantially retarded Cd uptake in radish plants, thereby improving root vegetative characteristics and resulting in amelioration of Cd toxicity (Raza and Shafiq, 2013). Presoaking flax grains with 250 and 1000 μM SA for 8 h induced a significant decrease in internal Cd concentrations in roots, stems, and leaves (Belkhadi et al., 2010). Similar observations were also found in other plant species, such as ryegrass (Bai et al., 2014; Wang et al., 2013), oilseed rape

(Ali et al., 2015), Kentucky bluegrass (Guo et al., 2013), flax (Belkadhi et al., 2012), alfalfa (Cui et al., 2012), and wheat (Shakirova et al., 2016). The low Cd uptake in SA-treated plants might result from a decrease in the uptake area and/or inhibition of metal transporters, such as the ATP-binding cassette (ABC) (Shi et al., 2009); through PCs; or by activating Cd dislocation into vacuoles (Drazic and Mihailovic, 2005).

However, SA does not always show a preventative effect on Cd assimilation in plants. At 3 $mg kg^{-1}$ Cd, SA induced an increase in Cd uptake in soybean seedlings (Drazic and Mihailovic, 2005), which might be due to the activation of some divalent cation transporters capable of binding Cd, or changes in enzyme activities in S metabolism. In a split-root system, an appreciable amount of Cd was translocated from the Cd-exposed roots to the unexposed roots, while pretreatment of total roots with 10 μM SA enhanced this Cd translocation process (Guo et al., 2009). Depletion of SA in *sid2* and *NahG* was not related to Cd influx in either roots or shoots, suggesting that SA mediates a tolerance mechanism rather than an avoidance mechanism (Guo et al., 2016; Zawoznik et al., 2007).

## 3.5 Elements Uptake

One form of Cd toxicity to plants is disturbance of macronutrient and micronutrient homeostasis (Metwally et al., 2003) due to the inhibition of proton pumps in the cytomembrane, or competition for the same trans-membrane carriers of nutrient elements (Sanità di Toppi and Gabbrielli, 1999). An important beneficial effect of SA on plant growth is the maintenance of optimal mineral nutrition under Cd stress. It has been shown that Cd significantly disturbs ionic homeostasis of flax, ryegrass, peanuts, and *Lemna minor*, including Ca, Fe, Mg, and Mn. SA treatment was found to induce its transport from nutrient solution to plants (Bai et al., 2014; Belkhadi et al., 2010; Wang et al., 2013; Xu et al., 2015; Lu et al., 2018). In soybean seedlings, the influence of SA on the alleviation of toxic Cd effects is possibly mediated through the regulation of K in roots, promoting Fe absorption and increasing the Mg concentration in leaves, and the Ca content in roots (Drazic and Mihailovic, 2005). This positive effect of SA on K uptake was also reported by Guo et al. (2016) with an SA-deficient *Arabidopsis* mutant, *sid2*. Cd stress significantly enhanced Fe levels in roots, but significantly lowered them in shoots of wild-type plants. SA deficiency further aggravated the Cd-induced decrease in Fe and K assimilation observed in *sid2* roots. It is well known that $H^+$-ATPase in the plasma membrane plays an important role in the transport of multiple ions (Palmgren and Harper, 1999), and there are investigations indicating that SA mediation could induce $H^+$-ATPase activity (Gordon et al., 2004), which might be responsible for SA increasing the absorption of nutrient elements under Cd toxicity. Contrary to the results of Metwally et al. (2003); Guo et al. (2016) found that the S concentrations in *sid2* mutants were higher than those in wild-type plants under both Cd-free and Cd-stress conditions. It seems that *sid2* mutants resort to enhancing S supply and upregulating GSH biosynthesis genes to counteract the GSH depletion induced by Cd stress.

## 3.6 Photosynthesis

Generally, Cd toxicity causes the destruction of plant photosynthesis, resulting from severe alterations in the chlorophyll level, chloroplast structure, photochemistry, and carboxylating enzyme activities and the high extent of lipid peroxidation (Krantev et al., 2008). SA is an important regulator of photosynthesis because it affects leaf and chloroplast structure (Uzunova and Popova, 2000), stomatal closure (Melotto et al., 2006), chlorophyll and carotenoid contents (Fariduddin et al., 2003), and the activity of enzymes such as RuBisCO and carbonic anhydrase (Slaymaker et al., 2002). Therefore, many studies have resorted to treatment with SA to protect the photosynthesis of plants under Cd stress. Metwally et al. (2003) reported that application of SA either before or after Cd exposure significantly lowered the decline in the photosynthetic yield of PSII in barley leaf slices. It has been shown that pretreatment of flax, soybean, and oilseed rape with SA counteracted the chlorophyll destruction induced by Cd exposure (Belkhadi et al., 2010; Noriega et al., 2012; Ali et al., 2015). This protective role may be related to the regulation of elements that are associated with chlorophyll construction. For example, Fe is a key element in chlorophyll synthesis. Cd stress significantly enhanced Fe levels in roots, but significantly lowered them in shoots of *Arabidopsis* plants. Depletion of SA further aggravates this Cd-induced effect, which is consistent with the lower chlorophyll levels in the SA-deficient mutant *sid2*, compared with the wild-type plant (Guo et al., 2016).

The reversal of Cd-induced negative effects on photosynthesis by SA was accompanied by an increase in stomatal conductance, which could be due to a restored K concentration allowing better stomatal closure. Furthermore, the SA-dependent stomatal effect in Cd-stressed plants is not the sole factor. SA-induced stomatal opening synchronized with the capability for $CO_2$

fixation, transpiration, and net photosynthetic rates, and inhibition of chlorophyll-degrading enzymes, as shown in the studies conducted by Shi et al. (2009), Li et al. (2014). Again, the mechanism of the photosynthetic response to SA differs according to the dose and plant species tested. Liu et al. (2011) reported that pretreatment with 250 and 500 μM SA did not affect the chlorophyll contents, but induced aggravation of the negative effects of photosynthesis caused by Cd stress in castor bean leaves, which was highly associated with the decrease in stomatal conductance and the increase in stomatal limitation.

Fv/Fm is frequently used as an indicator of photoinhibition or stress damage to PSII. The SA-induced increase in Fv/Fm indicated that improved PSII efficiency by SA in Cd-stressed melon was associated with a decrease in the inhibition of electron flow at the oxidizing site of PSII (Zhang et al., 2015). SA application also changed the distribution of absorbed energy in both varieties under Cd stress, suggesting that SA can distribute absorbed energy by providing a greater fraction of light energy utilized by photochemistry.

## 3.7 Glutathione and Phytochelatins

In plants, the primary strategy in response to Cd toxicity is the sequestration of Cd into less sensitive organelles of cells (e.g., vacuoles) (Seth et al., 2012). Glutathione (γ-Glu-Cys-Gly, GSH) is a key molecular compound for Cd chelation by itself, or as a basic component for PCs. It is also a powerful reductant, acting to protect plant cells against Cd-induced oxidative damage. Due to the high demand for detoxification, Cd exposure rapidly depletes GSH, and subsequently induces transcripts of GSH biosynthesis enzymes, including serine acetyltransferase (SAT), γ-glutamylcysteine synthetase (GSH1), and glutathione synthetase (GSH2). Furthermore, GSH metabolic products are also involved in regeneration and degradation enzymes, such as GR and glutathione peroxidase (GPX), respectively (Noctor et al., 2012).

GSH biosynthesis, concentration, and redox status were recently identified as the link between SA accumulation. Under Cd stress, SA application enhanced Cd tolerance in flax seedlings by increasing NPT production (Belkadhi et al., 2012). The GSH levels in Cd-stressed *sid2* leaves (SA-deficient mutants) were always lower than in wild-type plants during the Cd exposure period, showing the direct interdependence of foliar SA and GSH (Guo et al., 2016). This phenomenon was observed in other *Arabidopsis* mutants, such as *mpk4-1*, *npr1-1*, and *cat2 atrbohF*

(Mateo et al., 2006; Yi et al., 2012). The physiologically coupled levels of SA and GSH may be related to the transcription of SAT, a precursor gene for GSH biosynthesis that catalyzes the formation of O-acetyl-Ser and leads to the synthesis of cysteine. Elevation of free SA concentrations in *Arabidopsis* leaves stimulated the specific activity of SAT, leading to elevated GSH contents and increased Ni tolerance (Freeman et al., 2005). After exposure to 5 μM Cd for 8 days, the contents of NPT paralleled by the transcription of SAT-p and SAT-c were significantly lower in *sid2* leaves than in wild-type plants, suggesting that SA may upregulate cysteine synthesis (Guo et al., 2016). Guan et al. (2015) showed that endogenous SA accumulation can regulate *LcGSHS* expression levels and GSH accumulation. The overexpression of *LcGSHS* in transgenic *Arabidopsis* results in a greater tolerance to Cd stress than in wild-type lines. Furthermore, GR1, the enzyme responsible for the regeneration and maintenance of GSH in its reduced state, is signaled through SA pathways (Mhamdi et al., 2010). SA deficiency in *sid2* mutants did not affect the transcript levels of GPXs, but significantly decreased GSH levels and GR activity, along with GR1 transcription, compared with wild-type plants under Cd stress. These results clearly indicate that SA signaling occurs through the GR/GSH system, rather than GPX/thioredoxin during the Cd exposure period (Guo et al., 2016).

## 3.8 Cell Death

Exposure of plants to Cd can induce oxidative damage, and eventually result in cell death. Studies with *Arabidopsis* (Michele et al., 2009) and pine (Schützendübel et al., 2001) have shown that Cd induced the morphology of cell death, which was related to the $H_2O_2$ burst. Similarly, SA signaling can also elevate $H_2O_2$ levels and induce programmed cell death at the infection site to create a physical barrier against pathogens. However, very few studies have attempted to test whether SA has negative roles in accelerating partial root death to avoid Cd uptake, in an analogue to the mode of action of SA-enhanced plant defense against pathogens through $H_2O_2$ bursts and consequent cell death. To test these hypotheses, Guo et al. (2009) performed a series of split-root hydroponic experiments with one-half of rice roots exposed to 50 μM Cd and the other half not exposed. In the split-root system, Cd exposure accelerated the cell death of the stressed part of roots, but this rigidification failed to lower Cd uptake. Interestingly, SA pretreatment in the Cd-stressed part of roots did not accelerate the process of cell death, but counteracted Cd toxicity by alleviating Cd-induced growth inhibition and increasing Cd transport and distribution from Cd-exposed parts to non-Cd-exposed parts. These obtained results may not be exactly the same as the original hypotheses, which might be due to the mode of SA pretreatment rather than the mode of SA with Cd treated simultaneously. In addition, it can be manifested more accurately by using *Arabidopsis* mutants that have constitutively high levels of SA, such as *snc-1*, as the assimilation of Cd in this phenotype is slightly lower than the wild-type (Tao et al., 2013).

## 4 CONCLUSION AND FUTURE PROSPECTS

The literature over the past 20 years has strongly indicated that SA could be a very promising compound for the reduction of Cd stress by mitigation of the Cd-caused damaging effects in plants. Here, we summarized the possible mechanisms of the SA-mediated alleviation of Cd toxicity in plants in terms of plant growth, antioxidant system, Cd immobilization, Cd and other elements' uptake, photosynthesis, GSH and PC synthesis, and cell death, although the biochemical mechanisms that mediate most of these responses remain largely unknown. Several questions remain unanswered at both the theoretical and practical levels.

### 4.1 SA Homeostasis

It is clear that SA has broad but divergent effects; hence, its endogenous concentration should maintain a proper balance between stress acclimation and damage development of plants. Most abiotic stresses, including Cd toxicity, increased the *in planta* concentration of SA. The free SA content of barley roots was significantly increased by Cd exposure, compared with the non-Cd treatment (Metwally et al., 2003). Similar results were observed in maize, mustard, and peas (Krantev et al., 2008; Horváth et al., 2007; Pál et al., 2005; Popova et al., 2009). Cd exposure induced an approximately twofold increase in SA in wild-type *Arabidopsis*. SA-deficient mutants, such as *sid2*, *nahG* and *snc1/nahG*, remained in a low steady state, indicating that Cd-induced SA increases are SID2- or NahG-dependent (Zawoznik et al., 2007; Tao et al., 2013; Guo et al., 2016).

It was supposed that the exogenous application of SA could increase the endogenous level of plants during the Cd exposure period. Chao et al. (2010) reported that SA pretreatment in rice leaves resulted in an increase in the contents of endogenous SA, as indicated by the expression of OsWRKY45 (a SA-responsive gene). Meanwhile, Lu et al. (2018) also found that endogenous SA increased by 251% compared to the control in plants treated with SA. However, few studies have shown that exogenous application of SA did not affect the endogenous concentration. Presoaking maize and barley seeds with 500 μM SA for 6 h failed to enhance the free SA level (Krantev et al., 2008; Metwally et al., 2003). SA spray did not influence the endogenous SA concentration of mustard leaves (Horváth et al., 2007). Niederl et al. (1998) reported that the plant cuticle hardly allows the entry of surface-applied SA in plants. However, they further explained that SA can pass through the tough cuticular layer in its methylated (MeSA) layer, or enter directly through the wound (infected) of tissues (Niederl et al., 1998; Ohashi et al., 2004).

## 4.2 SA-Regulated Gene Expression

*Nonexpressor of PR1*: The Cd tolerance induced by SA may have various origins of gene expression, but the direct effect on the regulation of the components of the DNA replication machinery is still largely unknown. In recent years, research has mainly focused on the interaction between SA and $H_2O_2$ and its metabolic genes. In addition to the SABPs mentioned herein, several other defense genes related to the SA pathway might also be influenced by Cd stress. SA is known to activate the pathogen response via a nonexpressor of PR1 (NPR1), a key SA signaling protein in inducing programmed cell death at the infection site to create a physical barrier against pathogen invasion (Vanacker et al., 2001). This subcellular localization of NPR1 is regulated by an $H_2O_2$-sensitive system through the SA signaling pathway. However, there is a lack of experimental studies about the SA signal transducer NPR1 involved in the alleviation of Cd toxicity in plants. Until now, only a negative conclusion was made by Freeman et al. (2005), who found that increases in GSH level and Ni tolerance were accompanied by SA accumulation in the SA-signaling mutant *npr1-1*, demonstrating that this regulation of GSH and Ni tolerance is independent of the NPR1 signaling pathway.

*Mitogen-activated protein kinase*: Another important SA-regulated gene expression that needs to be considered is mitogen-activated protein kinase (MAPK) cascades, although this has not yet been described for the Cd defense response. Arabidopsis MPK4 is closely related to the SA signaling pathway by acting as a negative regulator. The mpk4 mutants have a severely dwarfed phenotype that might be due to SA accumulation. In Arabidopsis, recognition of the bacterial flagellin activates an MAPK module, leading to expression of basal defense-related genes (Asai et al., 2002). MPK4 negatively regulates temperature-sensitive cell death and cell-specific cell death, highlighting the global role of SA in plant stress responses (Ichimura et al., 2006).

*ABC transporter*: It has been confirmed that plant ABC transporters are able to transport PC-Cd complexes into subcellular compartments, such as vacuoles or outside of the cell (Bovet et al., 2005). The ABC transporter belongs to a large family of similar proteins in plants. The genome project identified >130 different genes for this gene family in *Arabidopsis*. Eichhorn et al. (2006) reported that the ABC transporter, a member of the PDR-subfamily, was identified as an SA-induced gene from soybeans in a subtractive suppression hybridization approach. Thus, it would be interesting to determine whether the ABC transporter is involved in crosstalk with SA signaling, which could help to explain the mechanisms of SA-induced Cd tolerance.

## 4.3 Other Hormones

As a growth regulator, SA is likely to influence other hormones' synthesis, transport, and metabolism over a long range (e.g., auxins, gibberellins, cytokinins, abscisic acid, and ethylene), short range (NO, jasmonates, brassinosteroids, etc.), and very short range (ROS, especially for $H_2O_2$) in the resistance to Cd stress in plants. Recently, Shakirova et al. (2016) found that the Cd-stressed disturbance of abscisic acid, auxin, and cytokinins was recovered by SA pretreatment. Tamás et al. (2015) showed that SA alleviates Cd-induced stress responses through inhibition of Cd-induced auxin-mediated ROS production in barley root tips. However, much work is required to elucidate the exact crosstalk pathways. Furthermore, another important issue could also argue how the tissue-specific concentrations of SA transported systemically to play a role during broad-range Cd stresses.

The role of SA in plant Cd tolerance is still a controversial field in plant biology; however, various phenotypes are associated with deregulated SA levels, and new discoveries and mutant characterization should shed more light on this topic. Furthermore, SA has the characteristics of high efficiency, low cost, no toxicity, and no residue, and shows broad usefulness in agricultural production. Another important perspective is its plausible agricultural use for phytoremediation. Seed presoaking with SA is a pragmatic approach for this agricultural purpose. In brief, the future applications of this plant hormone hold great promise as a management tool for providing tolerance to our agricultural crops against Cd toxicity, consequently aiding to accelerate potential crop yield in the near future.

## REFERENCES

Abreu, M.E., Munné-Bosch, S., 2009. Salicylic acid deficiency in *NahG* transgenic lines and *sid2* mutants increases seed yield in the annual plant *Arabidopsis thaliana*. J. Exp. Bot. 60 (4), 1261–1271.

Agami, R.A., Mohamed, G.F., 2013. Exogenous treatment with indole-3-acetic acid and salicylic acid alleviates cadmium toxicity in wheat seedlings. Ecotoxicol. Environ. Saf. 94, 164–171.

Ahmad, P., Nabi, G., Ashraf, M., 2011. Cadmium-induced oxidative damage in mustard [*Brassica juncea* (L.) *czern.* & *coss.*] plants can be alleviated by salicylic acid. S. Afr. J. Bot. 77 (1), 36–44.

Ali, E., Maodzeka, A., Hussain, N., Shamsi, I.H., Jiang, L., 2015. The alleviation of cadmium toxicity in oilseed rape (*Brassica napus*), by the application of salicylic acid. Plant Growth Regul. 75 (3), 641–655.

Asai, T., Tena, G., Plotnikova, J., Willmann, M.R., Chiu, W.L., Gomez-Gomez, L., Boller, T., Ausubel, F.M., Sheen, J., 2002. Map kinase signalling cascade in Arabidopsis innate immunity. Nature 415 (6875), 977.

Bai, X., Dong, Y., Kong, J., Xu, L., Liu, S., 2014. Effects of application of salicylic acid alleviates cadmium toxicity in perennial ryegrass. Plant Growth Regul. 75 (3), 695–706.

Belkadhi, A., Hediji, H., Abbes, Z., Djebali, W., Chaibi, W., 2012. Influence of salicylic acid pre-treatment on cadmium tolerance and its relationship with non-protein thiol production in flax root. Afr. J. Biotechnol. 11 (41), 9788–9796.

Belkadhi, A., De Haro, A., Soengas, P., Obregón, S., Cartea, M.E., Djebali, W., Chaïbi, W., 2013. Salicylic acid improves root antioxidant defense system and total antioxidant capacities of flax subjected to cadmium. OMICS 17 (7), 398–406.

Belkadhi, A., De Haro, A., Soengas, P., Obregón, S., Cartea, M.E., Chaïbi, W., Djebali, W., 2014. Salicylic acid increases tolerance to oxidative stress induced by hydrogen peroxide accumulation in leaves of cadmium-exposed flax (*Linum usitatissimum* L.). J. Plant Interact. 9 (1), 647–654.

Belkadhi, A., De Haro, A., Obregon, S., Chaïbi, W., Djebali, W., 2015. Positive effects of salicylic acid pretreatment on the composition of flax plastidial membrane lipids under cadmium stress. Environ. Sci. Pollut. Res. 22, 1457–1467.

Belkhadi, A., Hediji, H., Abbes, Z., Nouairi, I., Barhoumi, Z., Zarrouk, M., Chaïbi, W., Djebali, W., 2010. Effects of exogenous salicylic acid pre-treatment on cadmium toxicity and leaf lipid content in *Linum usitatissimum*, L. Ecotoxicol. Environ. Saf. 73 (5), 1004–1011.

Bovet, L., Feller, U., Martinoia, E., 2005. Possible involvement of plant ABC transporters in cadmium detoxification: a cDNA sub-microarray approach. Environ. Int. 31 (2), 263–267.

Chao, Y.Y., Chen, C.Y., Huang, W.D., Chinghuei, K., 2010. Salicylic acid-mediated hydrogen peroxide accumulation and protection against Cd toxicity in rice leaves. Plant Soil 329 (1/2), 327–337.

Chen, Z., Ricigliano, W., Klessig, D.F., 1993. Purification and characterization of a soluble salicylic acid binding protein from tobacco. Proc. Natl. Acad. Sci. U. S. A. 90, 9533–9537.

Choudhury, S., Panda, S.K., 2004. Role of salicylic acid in regulating cadmium induced oxidative stress in Oryza sativa L. roots. Bulg. J. Plant Physiol. 30, 95–110.

Cui, W., Le, L., Gao, Z., Wu, H., Xie, Y., Shen, W., 2012. Haem oxygenase-1 is involved in salicylic acid-induced alleviation of oxidative stress due to cadmium stress in *Medicago* sativa. J. Exp. Bot. 63 (15), 5521–5534.

Da Cunha, K.P.V., Do Nascimento, C.W.A., 2009. Silicon effects on metal tolerance and structural changes in maize (*Zea mays* L.) grown on a cadmium and zinc enriched soil. Water Air Soil Pollut. 197, 323–330.

Delaney, T.P., Uknes, S., Vernooij, B., Friedrich, L., Weymann, K., Negrotto, D., Gaffney, T., Gut-Rella, M., Kessmann, H., Ward, E., Ryals, J., 1994. A central role of salicylic acid in plant disease resistance. Science 266, 1247–1250.

Delaney, T.P., Friedrich, L., Ryals, J.A., 1995. Arabidopsis signal transduction mutant defective in chemically and biologically induced disease resistance. Proc. Natl. Acad. Sci. U. S. A. 92, 6602–6606.

Dempsey, D.A., Shah, J., Klessig, D.F., 1997. Salicylic acid and disease resistance in plants. Crit. Rev. Plant Sci. 18, 547–575.

Drazic, G., Mihailovic, N., 2005. Modification of cadmium toxicity in soybean seedlings by salicylic acid. Plant Sci. 168 (2), 511–517.

Dražić, G., Mihailovic, N., 2009. Salicylic acid modulates accumulation of Cd in seedlings of Cd-tolerant and Cd-susceptible soybean genotypes. Arch. Biol. Sci. 61, 431–439.

Dražić, G., Mihailovic, N., Lojic, M., 2006. Cadmium accumulation in *Medicago sativa* seedlings treated with salicylic acid. Biol. Plant. 50, 239–244.

Durner, J., Klessig, D.F., 1996. Salicylic acid is a modulator of tobacco and mammalian catalases. J. Biol. Chem. 271 (45), 28492–28501.

Eichhorn, H., Klinghammer, M., Becht, P., Tenhaken, R., 2006. Isolation of a novel ABC-transporter gene from soybean induced by salicylic acid. J. Exp. Bot. 57 (10), 2193–2201.

Fariduddin, Q., Hayat, S., Ahmad, A., 2003. Salicylic acid influences net photosynthetic rate, carboxylation efficiency, nitrate reductase activity, and seed yield in *Brassica juncea*. Photosynthetica 41 (2), 281–284.

Freeman, J.L., Garcia, D., Kim, D., Hopf, A., Salt, D.E., 2005. Constitutively elevated salicylic acid signals glutathione-mediated nickel tolerance in *Thlaspi* nickel hyperaccumulators. Plant Physiol. 137 (3), 1082–1091.

Gondor, O.K., Pál, M., Darkó, É., Janda, T., Szalai, G., 2016. Salicylic acid and sodium salicylate alleviate cadmium toxicity to different extents in maize (*Zea mays* L.). PLoS ONE 11(8)e0160157.

Gordon, L.K., Minibayeva, F.V., Rakhmatullina, D.F., Alyabyev, A.J., Ogorodnikova, T.I., Loseva, N.L., Valitova, Y.N., 2004. Heat production of wheat roots induced by the disruption of proton gradient by salicylic acid. Thermochim. Acta 422 (1–2), 101–104.

Guan, C., Ji, J., Jia, C., Guan, W., Li, X., Jin, C., Wang, G., 2015. A GSHS-like gene from Lycium chinense may be regulated by cadmium induced endogenous salicylic acid and overexpression of this gene enhances tolerance to cadmium stress in Arabidopsis. Plant Cell Rep. 34, 871–884.

Gunes, A., Inal, A., Alpaslan, M., Eraslan, F., Bagci, E.G., Cicek, N., 2007. Salicylic acid induced changes on some physiological parameters symptomatic for oxidative stress and mineral nutrition in maize (*Zea mays* L.) grown under salinity. J. Plant Physiol. 164 (6), 728–736.

Guo, B., Liang, Y.C., Li, Z.J., Guo, W., 2007a. Role of salicylic acid in alleviating cadmium toxicity in rice roots. J. Plant Nutr. 30 (3), 427–439.

Guo, B., Liang, Y., Zhu, Y., Zhao, F., 2007b. Role of salicylic acid in alleviating oxidative damage in rice roots (*Oryza sativa*) subjected to cadmium stress. Environ. Pollut. 147, 743–749.

Guo, B., Liang, Y., Zhu, Y., 2009. Does salicylic acid regulate antioxidant defense system, cell death, cadmium uptake and partitioning to acquire cadmium tolerance in rice? J. Plant Physiol. 166, 20–31.

Guo, Q., Meng, L., Mao, P.C., Jia, Y.Q., Shi, Y.J., 2013. Role of exogenous salicylic acid in alleviating cadmium-induced toxicity in *kentucky* bluegrass. Biochem. Syst. Ecol. 50 (50), 269–276.

Guo, B., Liu, C., Li, H., Yi, K., Ding, N., Li, N., Lin, Y., Fu, Q., 2016. Endogenous salicylic acid is required for promoting cadmium tolerance of Arabidopsis by modulating glutathione metabolisms. J. Hazard. Mater. 316, 77–86.

Horváth, E., Szalai, G., Janda, T., 2007. Induction of abiotic stress tolerance by salicylic acid signaling. J. Plant Growth Regul. 26 (3), 290–300.

Ichimura, K., Casais, C., Peck, S.C., Shinozaki, K., Shirasu, K., 2006. Mekk1 is required for mpk4 activation and regulates tissue-specific and temperature-dependent cell death in *Arabidopsis*. J. Biol. Chem. 281 (48), 36969–36976.

Kang, H.M., Saltveit, M.E., 2002. Chilling tolerance of maize, cucumber and rice seedling leaves and roots are differentially affected by salicylic acid. Physiol. Plant. 115 (4), 571–576.

Kang, G., Wang, C., Sun, G., Wang, Z., 2003. Salicylic acid changes activities of $H_2O_2$—metabolizing enzymes and increases the chilling tolerance of banana seedlings. Environ. Exp. Bot. 50 (1), 9–15.

Krantev, A., Yordanova, R., Janda, T., Szalai, G., Popova, L., 2008. Treatment with salicylic acid decreases the effect of cadmium on photosynthesis in maize plants. J. Plant Physiol. 165 (9), 920–931.

Li, X.M., Ma, L.J., Bu, N., Li, Y.Y., Zhang, L.H., 2014. Effects of salicylic acid pre-treatment on cadmium and/or UV-B stress in soybean seedlings. Biol. Plant. 58 (1), 195–199.

Liu, C.F., Guo, J.L., Cui, Y.L., Lü, T.F., Zhang, X.H., Shi, G.R., 2011. Effects of cadmium and salicylic acid on growth, spectral reflectance and photosynthesis of castor bean seedlings. Plant Soil 344 (1–2), 131–141.

Lu, Q., Zhang, T., Zhang, W., Su, C., Yang, Y., Hu, D., Xu, Q., 2018. Alleviation of cadmium toxicity in *Lemna minor* by exogenous salicylic acid. Ecotoxicol. Environ. Saf. 147, 500–508.

Mateo, A., Funck, D., Muhlenbock, P., Kular, B., Mullineaux, P.M., Karpinski, S., 2006. Controlled levels of salicylic acid are required for optimal photosynthesis and redox homeostasis. J. Exp. Bot. 57, 1795–1807.

Melotto, M., Underwood, W., Koczan, J., Nomura, K., He, S.Y., 2006. Plant stomata function in innate immunity against bacterial invasion. Cell 126 (5), 969–980.

Metwally, A., Finkemeier, I., Georgi, M., Dietz, K.J., 2003. Salicylic acid alleviates the cadmium toxicity in barley seedlings. Plant Physiol. 132, 272–281.

Mhamdi, A., Hager, J., Chaouch, S., Queval, G., Yi, H., Taconnat, L., Saindrenan, P., Gouia, H., Issakidis-Bourguet, E., Renou, J.P., Noctor, G., 2010. *Arabidopsis glutathione reductase 1* plays a crucial role in leaf responses to intracellular hydrogen peroxide and in ensuring appropriate gene expression through both salicylic acid and jasmonic acid signaling pathways. Plant Physiol. 153 (3), 1144–1160.

Michele, R.D., Vurro, E., Rigo, C., Costa, A., Elviri, L., Valentin, M.D., Careri, M., Zottini, M., Sanità di Toppi, L., Lo Schiavo, F., 2009. Nitric oxide is involved in cadmium-induced programmed cell death in *Arabidopsis* suspension cultures. Plant Physiol. 150 (1), 217–228.

Morris, K., MacKerness, S.A., Page, T., John, C.F., Murphy, A.M., Carr, J.P., Buchanan-Wollaston, V., 2000. Salicylic acid has a role in regulating gene expression during leaf senescence. Plant J. 23 (5), 677–685.

Nawrath, C., Métraux, J.P., 1999. Salicylic acid induction-deficient mutants of *Arabidopsis* express PR-2 and PR-5 and accumulate high levels of camalexin after pathogen inoculation. Plant Cell 11, 1393–1404.

Niederl, S., Kirsch, T., Riederer, M., Schreiber, L., 1998. Co-permeability of $^3$H-labeled water and $^{14}$C-labeled organic acids across isolated plant cuticles. Plant Physiol. 116, 117–123.

Noctor, G., Mhamdi, A., Chaouch, S., Han, Y., Neukermans, J., Marquez-Garcia, B., Queval, G., Foyer, C.H., 2012. Glutathione in plants: an integrated overview. Plant Cell Environ. 35 (2), 454–484.

Noriega, G., Caggiano, E., Lecube, M.L., Santa-Cruz, D., Batlle, A., Tomaro, M., Balestrasse, K.B., 2012. The role of salicylic acid in the prevention of oxidative stress elicited by cadmium in soybean plants. Biometals 25, 1155–1165.

Norman, C., Howell, K.A., Millar, A.H., Whelan, J.M., Day, D.A., 2004. Salicylic acid is an uncoupler and inhibitor of mitochondrial electron transport. Plant Physiol. 134 (1), 492–501.

Ohashi, Y., Murakami, T., Mitsuhara, I., Seo, S., 2004. Rapid down and upward translocation of salicylic acid in tobacco plants. Plant Biotechnol. 21, 95–101.

Pál, M., Szalai, G., Horváth, E., Janda, T., Páldi, E., 2002. Effect of salicylic acid during heavy metal stress. Acta. Biol. Szeged. 46, 119–120.

Pál, M., Horváth, E., Janda, T., Páldi, E., Szalai, G., 2005. Cadmium stimulates the accumulation of salicylic acid and its putative precursors in maize (*Zea mays*) plants. Physiol. Plant. 125 (3), 356–364.

Palmgren, M.G., Harper, J.F., 1999. Pumping with plant P-type ATPases. J. Exp. Bot. 50 (Special), 883–893.

Panda, S.K., Patra, H.K., 2007. Effect of salicylic acid potentiates cadmium-induced oxidative damage in *Oryza sativa* L. leaves. Acta Physiol. Plant. 29 (6), 567–575.

Popova, L.P., Maslenkova, L.T., Yordanova, R.Y., Ivanova, A.P., Krantev, A.P., Szalai, G., Janda, T., 2009. Exogenous treatment with salicylic acid attenuates cadmium toxicity in pea seedlings. Plant Physiol. Biochem. 47 (3), 224–231.

Rao, M.V., Paliyath, G., Ormrod, D.P., 1997. Influence of salicylic acid on $H_2O_2$ production, oxidative stress, and $H_2O_2$-metabolizing enzymes salicylic acid-mediated oxidative damage requires $H_2O_2$. Plant Physiol. 115 (1), 137–149.

Raskin, I., Ehmann, A., Melander, W.R., Meusse, B.J.D., 1987. Salicylic acid: a natural inducer of hear production in Arum lilies. Science 237, 1601–1602.

Raskin, I., Skubatz, H., Tang, W., Meeuse, B.J.D., 1990. Salicylic acid levels in thermogenic and non-thermogenic plants. Ann. Bot. 66 (4), 369–373.

Raza, S.H., Shafiq, F., 2013. Exploring the role of salicylic acid to attenuate cadmium accumulation in radish (*Raphanus sativus*). Int. J. Agric. Biol. 15 (3), 547–552.

Rivassan, V.M., Plasencia, J., 2011. Salicylic acid beyond defence: its role in plant growth and development. J. Exp. Bot. 62 (10), 3321–3338.

Roychoudhury, A., Ghosh, S., Paul, S., Mazumdar, S., Das, G., Das, S., 2016. Pre-treatment of seeds with salicylic acid attenuates cadmium chloride-induced oxidative damages in the seedlings of mungbean (*Vigna radiata* L. *wilczek*). Acta Physiol. Plant. 38 (1), 1–18.

Sanità di Toppi, L., Gabbrielli, R., 1999. Response to cadmium in higher plants. Environ. Exp. Bot. 41, 105–130.

Schützendübel, A., Schwanz, P., Teichmann, T., Gross, K., Langenfeld-Heyser, R., Godbold, D.L., Polle, A., 2001. Cadmium-induced changes in antioxidative systems, hydrogen peroxide content, and differentiation in scots pine roots. Plant Physiol. 127 (3), 887–898.

Seth, C.S., Remans, T., Keunen, E., Jozefczak, M., Gielen, H., Opdenakker, K., Weyens, N., Vangronsveld, J., Cuypers, A., 2012. Phytoextraction of toxic metals: a central role for glutathione. Plant Cell Environ. 35 (2), 334–346.

Shah, J., 2003. The salicylic acid loop in plant defense. Curr. Opin. Plant Biol. 6 (4), 365–371.

Shakirova, F.M., Sakhabutdinova, A.R., Bezrukova, M.V., Fatkhutdinova, R.A., Fatkhutdinova, D.R., 2003. Changes in the hormonal status of wheat seedlings induced by salicylic acid and salinity. Plant Sci. 164 (3), 317–322.

Shakirova, F.M., Allagulova, C.R., Maslennikova, D.R., Klyuchnikova, E.O., Avalbaev, A.M., Bezrukova, M.V., 2016. Salicylic acid-induced protection against cadmium toxicity in wheat plants. Environ. Exp. Bot. 122, 19–28.

Shi, G.R., Cai, Q.S., Liu, Q.Q., Wu, L., 2009. Salicylic acid-mediated alleviation of cadmium toxicity in hemp plants in relation to cadmium uptake, photosynthesis, and antioxidant enzymes. Acta Physiol. Plant. 31 (5), 969–977.

Slaymaker, D.H., Navarre, D.A., Clark, D., del Pozo, O., Martin, G.B., Klessig, D.F., 2002. The tobacco salicylic acid-binding protein 3 (SABP3) is the chloroplast carbonic anhydrase, which exhibits antioxidant activity and plays a role in the hypersensitive defense response. Proc. Natl. Acad. Sci. U. S. A. 99 (18), 11640–11645.

Stevens, J., Senaratna, T., Sivasithamparam, K., 2006. Salicylic acid induces salinity tolerance in tomato (*Lycopersicon esculentum* cv. *roma*): associated changes in gas exchange, water relations and membrane stabilisation. Plant Growth Regul. 49 (1), 77–83.

Szalai, G., Krantev, A., Yordanova, R., Popova, L.P., Janda, T., 2013. Influence of salicylic acid on phytochelatin synthesis in *Zea mays* during Cd stress. Turk. J. Bot. 37, 708–714.

Tamás, L., Mistrík, I., Alemayehu, A., Zelinová, V., Bočová, B., Huttová, J., 2015. Salicylic acid alleviates cadmium-induced stress responses through the inhibition of Cd-induced auxin-mediated reactive oxygen species production in barley root tips. J. Plant Physiol. 173, 1–8.

Tao, S., Sun, L., Ma, C., Li, L., Li, G., Hao, L., 2013. Reducing basal salicylic acid enhances Arabidopsis tolerance to lead or cadmium. Plant Soil 372 (1–2), 309–318.

Uzunova, A.N., Popova, L.P., 2000. Effect of salicylic acid on leaf anatomy and chloroplast ultrastructure of barley plants. Photosynthetica 38 (2), 243–250.

Vanacker, H., Lu, H., Rate, D.N., Greenberg, J.T., 2001. A role for salicylic acid and NPR1 in regulating cell growth in Arabidopsis. Plant J. 28 (2), 209–216.

Wang, Q., Liang, X., Dong, Y., Xu, L., Zhang, X., Kong, J., Liu, S., 2013. Effects of exogenous salicylic acid and nitric oxide on physiological characteristics of perennial ryegrass under cadmium stress. J. Plant Growth Regul. 32 (4), 721–731.

Wildermuth, M.C., Dewdney, J., Wu, G., Ausubel, F.M., 2001. Isochorismate synthase is required to synthesize salicylic acid for plant defense. Nature 414, 562–565.

Xu, L.L., Fan, Z.Y., Dong, Y.J., Kong, J., Bai, X.Y., 2015. Effects of exogenous salicylic acid and nitric oxide on physiological characteristics of two peanut cultivars under cadmium stress. Biol. Plant. 59, 171–182.

Yi, H., Sejir, C., Amna, M., Guillaume, Q., Bernd, Z., Graham, N., 2012. Functional analysis of Arabidopsis mutants points to novel roles for glutathione in coupling $H_2O_2$ to activation of salicylic acid accumulation and signaling. Antioxid. Redox Signal. 18, 2106–2121.

Zawoznik, M.S., Groppa, M.D., Tomaro, M.L., Benavides, M.P., 2007. Endogenous salicylic acid potentiates cadmium-induced oxidative stress in *Arabidopsis thaliana*. Plant Sci. 173 (2), 190–197.

Zhang, F., Zhang, H., Xia, Y., Wang, G., Xu, L., Shen, Z., 2011. Exogenous application of salicylic acid alleviates cadmium toxicity and reduces hydrogen peroxide accumulation in root apoplasts of *Phaseolus aureus* and *Vicia sativa*. Plant Cell Rep. 30, 1475–1483.

Zhang, Y., Xu, S., Yang, S., Chen, Y., 2015. Salicylic acid alleviates cadmium-induced inhibition of growth and photosynthesis through upregulating antioxidant defense system in two melon cultivars (*Cucumis melo* L.). Protoplasma 252 (3), 911–924.

## FURTHER READING

Pan, Q., Zhan, J., Liu, H., Zhang, J., Chen, J., Wen, P., Huang, W.D., 2006. Salicylic acid synthesized by benzoic acid 2-hydroxylase participates in the development of thermotolerance in pea plants. Plant Sci. 171 (2), 226–233.

# CHAPTER 15

# Mitigating Cadmium Toxicity in Plants by Phytohormones

**Aditi Shreeya Bali*, Gagan Preet Singh Sidhu†, Vinod Kumar‡, Renu Bhardwaj§**

*Department of Botany, M.C.M. DAV College for Women, Chandigarh, India
†Department of Environment Education, Government College of Commerce and Business Administration, Chandigarh, India
‡Department of Botany, DAV University, Jalandhar, India
§Department of Botanical and Environmental Sciences, Guru Nanak Dev University, Amritsar, India

## Abstract

A wide range of technogenic activities stimulates the contamination of terrestrial and aquatic ecosystems with multifarious inorganic contaminants such as cadmium (Cd). Due to its pernicious influence in plants, it is one of the most extensively studied metals. Currently, various nonconventional strategies for mitigating Cd-toxicity in plants are on the rise. Phytoremediation is one such exclusively derived process for remediation of Cd-contaminated sites. However, it is a time consuming strategy. Exploiting phytohormones (salicylic acid, jasmonic acid, brassinosteroids, abscisic acid, ethylene, gibberellic acid, etc.) is a vital, sustainable approach in alleviating Cd-toxicity in plants. These phytohormones can ameliorate the plant-Cd interaction, and exploring this would definitely increase the importance of their role in plant growth under Cd-stress. Additionally, this chapter emphasizes: (i) the introduction of crucial phytohormones and outlines their signaling in cadmium-stressed plants, (ii) assesses the mechanisms potentially included in phytohormone-mediated increased Cd-tolerance in plants.

**Keywords:** Alleviating, Ameliorate, Cd-stress, Cd-tolerance, Phytoremediation, Phytohormones, Plant-Cd interactions, Signaling

## ABBREVIATIONS

**ABA** abscisic acid
**APX** ascorbate peroxidase
**CAT** catalase
**GA** gibberellic acid
**GPX** guaiacol peroxidase
**GR** glutathione reductase
**IAA** indole-3-acetic acid
**LPO** lipid peroxidation
**MDA** malondialdehyde
**POD** peroxidases
**SOD** superoxide dismutase

*Cadmium Toxicity and Tolerance in Plants*
https://doi.org/10.1016/B978-0-12-814864-8.00015-2

# 1 INTRODUCTION

Cadmium (Cd) is a nonessential, odorless, toxic heavy metal that is extensively perceived as a severe environmental pollutant that is noxious, even at lower concentrations (Sidhu et al., 2017a; Wagner, 1993). It has a long biological half-life and has been ranked seventh in the priority list of hazardous substances (Agency for Toxic Substances and Disease Registry (ATSDR), 2017). Due to increasing man-made activities, the range of Cd in agricultural soils ($\sim$100 mg Cd kg$^{-1}$ soil; Salt et al., 1995) is ascending day by day (Asgher et al., 2015). $Cd^{2+}$ is the principal form of Cd in soil (Tudoreanu and Phillips, 2004) that penetrates the root cells with the help of zinc-regulated transporters/iron-regulated transporters (Asgher et al., 2015). Cd is accrued at higher levels in leaves, and alters the photosynthetic process negatively by disturbing the electron transport chain, hampering the photosystems (Liu et al., 2011), and restricting the activity of different enzymes of the Calvin cycle (Mobin and Khan, 2007). Several reports suggesting the effects of Cd-toxicity on plant growth are related to the depletion in photochemical efficiency, increased deterioration of the net $CO_2$ assimilation rate, and reduced activity of enzymes, including ribulose-1,-5-bisphosphate carboxylase/oxygenase (Rubisco) (Asgher et al., 2014; Dias et al., 2013). Moreover, the reduction in the efficiency of the carbon assimilation pathway might be related to the Cd-induced damage to chlorophyll's structure, photosynthetic pigments, and chlorophyll-protein complexes (Parmar et al., 2013). Further, Cd-toxicity hampers the infusion of Mg in the chlorophyll ring, or may cause a substitution of Mg that ultimately leads to annihilation of the chlorophyll structure (Parmar et al., 2013). The activities of various mitochondrial enzymes, such as isocitrate dehydrogenase, succinate dehydrogenase, and malate dehydrogenase were reduced upon exposure to Cd-stress (Bansal et al., 2002). Moreover, the uptake, transport, and use of many essential nutrients were interrupted, which caused deficiency in the plants (Nedjimi and Daoud, 2009). Cd imposes toxic symptoms such as leaf chlorosis, growth retardation, generation of reactive oxygen species (ROS), and the influence of lipid peroxidation in plants (Li et al., 2012; Sidhu et al., 2017a).

Various defense strategies have been adopted by plants to impede cadmium toxicity. Phytoremediation is one such in situ, green, and clean technology that helps in the treatment of heavy metal contaminants (Sidhu, 2016; Sidhu et al., 2016, 2017b). However, because this technology is time consuming, plants have adopted other mechanisms to alleviate Cd-toxicity.

Phytohormones are natural chemical substances that are required in low concentrations, and can have stimulatory or retardatory effects on plants. They have been known to regulate various signaling pathways in plants. Phytohormones, such as auxins, ethylene, gibberellic acid (GA), abscisic acid (ABA), cytokinin, jasmonic acid (JA), brassinosteroids, and salicylic acid play an important part in different processes related to growth and metabolism of plants. Recently, plant biology studies have confirmed the role of plant hormones in alleviating harmful effects caused by abiotic stress (Khan et al., 2013). Different plant hormones such as ethylene (Shi et al., 2012), cytokinins (Peleg et al., 2011), jasmonates (Gonzalez-Aguilar et al., 2003), auxin (Ke et al., 2015), brassinosteroids (Ali et al., 2008), and ABA (Fujita et al., 2011) have been delineated to play an important role in stress signaling. Moreover, several reports suggest the significant role of plant hormones in mitigating Cd-stress in plants (Masood and Khan, 2013; Shi et al., 2009; Wang and Du, 2013). In light of the available literature, exploiting the role of plant hormones in alleviating Cd-toxicity in plants is an important sustainable approach. This chapter is divided into two sections that highlight the signaling of different phytohormones in Cd-stressed plants, and evaluate the processes involved in hormones in inducing enhanced Cd-tolerance in plants.

## 2 SIGNALING OF PHYTOHORMONES IN Cd-STRESSED PLANTS

The recognition of Cd may initiate various signaling pathways in plants. The endogenous and exogenous levels of plant hormones strictly regulate various mechanisms in plants that help in alleviating Cd-induced stress in plants (Fig. 1). Recently, an enormous development has been made toward understanding the signaling of phytohormones in Cd-stressed plants (Ahmad et al., 2017; Hasan et al., 2011; Tamás et al., 2015; Zhao et al., 2013). Indole-3-acetic acid (IAA) has been reported to play an important role in the signal transduction pathway, and in controlling plant growth and development under Cd-stressed conditions. Auxin increased the activities of peroxidase and catalase, and reduced $H_2O_2$ content, thereby providing tolerance toward Cd-toxicity (Ostrowski et al., 2016). Cytokinin treatment under Cd-exposed plants enhanced the activity of soluble sugars and proline, and also affected the ascorbate-glutathione cycle, as investigated by Al-Hakimi (2007) and Singh et al. (2018). Gibberellin signaling increased the expression of genes responsible for Cd uptake, and influenced NO

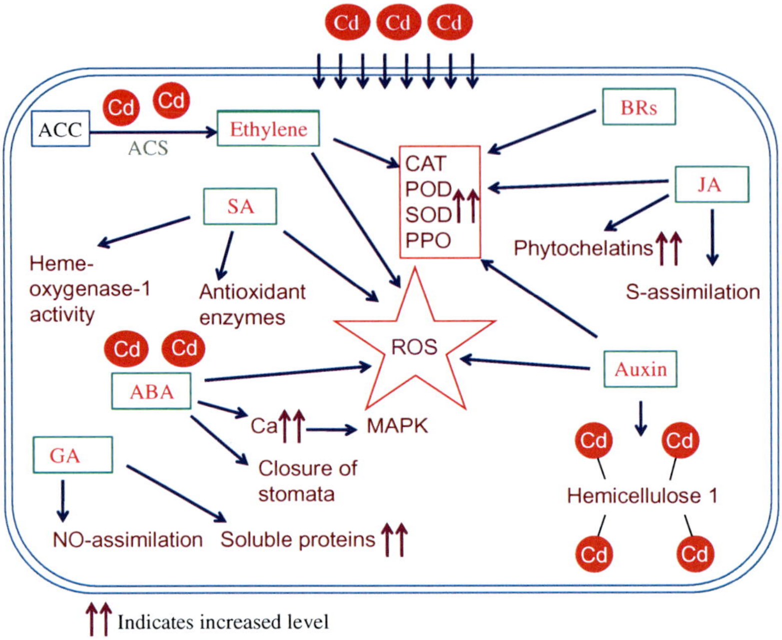


**Fig. 1** Diagrammatic representation of probable role of phytohormones in signaling mechanisms during Cd-stressed conditions. *ABA*, abscisic acid; *ACC:1*, aminocyclopropane-1-carboxylic acid; *ACS*, ACC synthase; *BRs*, brassinosteroids; *Ca*, calcium; *CAT*, catalase; *GA*, gibberellic aid; *JA*, jasmonic acid; *MAPK*, mitogen activated protein kinase; *NO*, nitric oxide; *POD*, peroxidases; *PPO*, polyphenol oxidases; *ROS*, reactive oxygen species; *S*, sulfur; *SA*, salicylic acid; *SOD*, superoxide dismutase.

accumulation, as revealed by Zhu et al. (2012). In addition to gibberellins, ethylene signaling is also a major aspect in Cd-signal transduction (Masood et al., 2012). The role of ABA under Cd-toxicity has also been reported, which leads to regulation of plant growth and development (Perfus-Barbeoch et al., 2002). Moreover, JA (Chen et al., 2014), salicylic acid (Shakirova et al., 2016), and brassinosteroids (Sharma et al., 2010) may also relay Cd-signals in plants (Fig. 1).

## 3 PLANT HORMONES CONFER Cd TOLERANCE IN PLANTS

Several reports have outlined the positive role of plant hormones in providing Cd-tolerance in plants (Table 1). This section highlights different plant hormones and the mechanisms instigated by them in providing tolerance in plants under Cd-stressed conditions.

**Table 1** Regulation of phytohormones in Cd-treated plants

| Growth hormone | Plants | Effect | Reference |
|---|---|---|---|
| Abscisic acid | *Oryza sativa* | Decreased Cd content, reduced transpiration rate | Hsu and Kao (2003) |
| | *Brassica napus* | Reduced intrinsic Cd accumulation | Meng et al. (2009) |
| | *Oryza sativa* | Decreased root-to-shoot Cd translocation | Uraguchi et al. (2009) |
| Auxin | *Hordeum vulgare* | Reduced Cd concentration | Bocova et al. (2013) |
| | *Arabidopsis thaliana* | Increased Cd accumulation in roots and decreased accumulation in shoots, increased cell wall hemicelluloses 1 content | Zhu et al. (2013) |
| | *Triticum aestivum* | Increased antioxidant activity and improved anatomical structure of wheat leaves | Agami and Mohamed (2013) |
| | *Solanum melongena* | Improved pigment content and photosynthetic activity | Singh and Prasad (2016) |
| Brassinosteroids | *Brassica napus* (winter rape) | Reduced toxicity on photochemical processes | Janeczko et al. (2005) |
| | *Lycopersicum esculentum* | Enhanced activity of antioxidant enzymes | Ahammed et al. (2013) |
| | *Raphanus* sp. | Improved photosynthesis | Anuradha and Rao (2009) |
| | *Solanum lycopersicum* | Proline content and activities of antioxidant enzymes increased | Hasan et al. (2011) |
| | *Solanum lycopersicum* | Increased growth, fresh weight and net photosynthetic rate | Hayat et al. (2012) |

*Continued*

**Table 1** Regulation of phytohormones in Cd-treated plants—cont'd

| Growth hormone | Plants | Effect | Reference |
|---|---|---|---|
| Cytokinin | *Bulbine natalensis* and *Rumex crispus* | Improved plant growth | Gemrotová et al. (2013) |
| | *Pisumsativum* | Increased proline content and soluble sugars | Al-Hakimi (2007) |
| | *Solanum melongena* | Enhanced antioxidant enzymes viz., SOD, POD, CAT | Singh and Prasad (2016) |
| Ethylene | *Brassica juncea* | Regulation of glutathione synthesis and modulation of antioxidant system | Masood et al. (2012) |
| | *Lycopersicon esculentum* | Improved Cd-tolerance | Liu et al. (2008) |
| | *Triticum aestivum* | Increased proline and glutathione production | Khan et al. (2015) |
| | *Brassica juncea* | Improved photosynthesis and growth | Asgher et al. (2014) |
| | *Catharanthus roseus* | Reduced Cd content, $H_2O_2$ and MDA production | Chen et al. (2017) |
| Gibberellic acid | *Arabidopsis thaliana* | Improved root growth, reduced Cd content and lipid peroxidation | Zhu et al. (2012) |
| | *Glycine max* | Improved plant growth rate and net assimilation rate and increased leaf area and stem length | Ghorbanli et al. (2000) |
| | *Chlorella vulgaris* | Increased cell number, photosynthetic pigment, protein content | Falkowska et al. (2011) |

| Jasmonic acid | *Avicennia marina* | Restored chlorophyll and MDA content and enhanced APX activity | Yan et al. (2015a) |
|---|---|---|---|
| | *Brassica napus* | Reduced Cd uptake by leaves, membrane damage and MDA content | Ali et al. (2018) |
| | *Vicia faba* | Restored growth, biomass, and reduced content of MDA, $H_2O_2$, and electrolyte leakage | Ahmad et al. (2017) |
| | *Solanum nigrum* | Regulated metal uptake and increased antioxidative capacity | Yan et al. (2015b) |
| Salicylic acid | *Phaseolus vulgaris* | Increased activity of SOD, APX, GPX, CAT, GR | Wael et al. (2015) |
| | *Triticum aestivum* | Reduced MDA content and electrolyte leakage | Shakirova et al. (2016) |
| | *Cucumis melo* | Reduced LPO and superoxide anion content | Zhang et al. (2015) |
| | *Nymphaea tetragona* | Decreased Cd concentration and increased content of glutathione | Gu et al. (2018) |
| | *Lemna minor* | Reduced Cd uptake and maintained ionic homeostasis | Lu et al. (2018) |

## 3.1 Auxin

Auxin is the most studied phytohormone, and is recognized to be concerned with growth and development, and cell division and elongation (Correa-Aragunde et al., 2016; Pacheco-Villalobos et al., 2016; Zažímalová et al., 2014). IAA is a known natural plant hormone synthesized from an amino acid precursor, tryptophan, that takes part in various physiological functions in plants (Enders and Strader, 2015). Many reports suggested the potential role of auxin in imparting tolerance to plants under various abiotic stresses (Bouzroud et al., 2018; Sharma et al., 2015). Hu et al. (2013) opined that Cd interferes in maintaining homeostasis of auxin in *Arabidopsis* seedlings. They also reported Cd-induced reduction in IAA levels, enhanced activity of IAA oxidase, and modified expression of different probable genes involved in biosynthesis and catabolism of auxin. Zhao et al. (2013) suggested the role of mitogen activated protein kinase (MAPKs) in promoting root growth in *Oryza sativa* plants, due to its effect on auxin signaling and expression of genes related to cell cycle, upon exposure to Cd (Zhao et al., 2013). Similarly, Farooq et al. (2015) observed increased growth and yield of rice plants upon treatment with different concentrations of L-TRP, a precursor of auxin, as compared with untreated seedlings in Cd-contaminated soil, without the application of L-TRP. Ostrowski et al. (2016) found that IAA-Asp, an auxin conjugate, affects pea responses under Cd-stress by regulating the activities of catalase and peroxidase, thus reducing the hydrogen peroxide content. Zhu et al. (2013) observed that hemicellulose 1 in the cell wall is an important component in detoxification of heavy metals. Further, naphthalene acetic acid (NAA) enhances metal retention in roots by directing it into hemicellulose under Cd-stress (Zhu et al., 2013). Exogenous application of NAA increases the level of hemicelluloses 1, thereby increasing Cd content (Zhu et al., 2013). Yu et al. (2015) studied that *OsAUX1* (auxin transporter) leads to the extension of primary roots and root hairs under Cd exposure in *O. sativa*. Fattorini et al. (2017) studied how Cd affects the development and continuity of the quiescent center in *Arabidopsis thaliana* postembryonic roots, and distorts the movement and biosynthesis of auxin (Fattorini et al., 2017). However, further research is required to study the mechanisms involved in auxin-mediated Cd-tolerance in plants.

## 3.2 Cytokinin

Cytokinin, a plant hormone derived from purine-adenine, plays an important role in cell division, differentiation, and seed germination (Ashikari et al., 2005; Yong et al., 2009). Cytokinins have been extensively studied for alleviating abiotic stress in plants (Kudoyarova et al., 2006; Stoll et al., 2000; Zwack and Rashotte, 2015). Piotrowska-Niczyporuk et al. (2012) documented the effect of exogenously applied cytokinin on growth and metabolism of heavy metal (Cd, Pb, and Cu) treated *Chlorella vulgaris*. The results indicated inhibition in heavy metal biosorption, and increased activities of antioxidant enzymes in response to stress. Earlier, Al-Hakimi (2007) reported that treatment of kinetin can regulate soluble sugars and proline content, and reverse Cd-induced retardatory effects on growth and photosynthesis in pea seedlings. Singh and Prasad (2016) pointed out the role of kinetin in alleviating Cd-induced toxicity in tomato seedlings by regulating the activity of different antioxidant enzymes. Recently, Singh et al. (2018) reported alleviation of Cd-toxicity upon exposure to kinetin in tomato seedlings due to its effect on the ascorbate-glutathione cycle.

## 3.3 Gibberellins

Gibberellins, or gibberellic acids (GAs), were discovered from a fungi, *Gibberella fujikori*, and play an important function in seed germination; and elongation of stems, leaves, and seeds; and fruit development and floral initiation (Davies, 2010; Fleet and Sun, 2005; Lambers et al., 2008). Meng et al. (2009) reported increased fresh weight, reduced MDA content, and antioxidant activities in leaves of *Brassica juncea* under Cd-induced stress. Sharaf et al. (2009) investigated the effect of GA in alleviating Cd-toxicity in broad beans and lupin plants. The results showed a marked increase in soluble proteins upon treatment of GA to Cd-exposed plants (Sharaf et al., 2009). The application of GA to *Parthenium hysterophorus* enhanced the growth and biomass of plants under Cd-stress (Hadi et al., 2014). The enhanced Cd-toxicity by GA was reported to be dependent upon expression of genes associated with Cd uptake and accumulation of NO in *A. thaliana* (Zhu et al., 2012). In Cd-exposed *B. juncea*, GA application incremented the relative growth rate and net assimilation rate, which might be due to GA-aroused oxidative stress (Masood and Khan, 2013). Further, GA also increased protein and RNA content, carbohydrate metabolism, mitotic activity, and Cd-tolerance in *Vicia faba* plants (Mohamed and Abdel-Razik, 2005).

## 3.4 Ethylene

Ethylene, a gaseous molecule with a simple two-carbon skeleton ($C_2H_4$), is treated as a crucial mediator of various stresses in plants. It is involved in several aspects of plant development, seed germination, seedling triple response, leaf and flower senescence, and fruit ripening, besides playing an important role in abiotic stress tolerance (Abeles et al., 1992). Heavy metal stress leads to excessive generation of ROS that induces oxidative stress in plants (Sharma and Chakraverty, 2013). An elevated quantity of these compounds damages various cellular (Nagajyoti et al., 2010), biochemical, and physiological mechanisms (Dugardeyn and Van Der Straeten, 2008) in plants. A high concentration of Cd in the leaves disrupts the electron transport chain (Liu et al., 2011) and inhibits activity of Calvin cycle enzymes (Mobin and Khan, 2007). Many studies have suggested increased ethylene production in response to Cd-stress in *B. juncea* (Masood et al., 2012), *A. thaliana* (Schellingen et al., 2014), and *Triticum aestivum* (Khan et al., 2015). According to Khan et al. (2015), Cd and Cu increased the activity of the ACC synthase enzyme that promoted ethylene biosynthesis. Masood et al. (2012) pointed out the participation of ethylene in sulfur-induced modulation of glutathione (GSH) synthesis and regulation of antioxidant systems under Cd application as a chief mechanism responsible for ethylene-induced defense of photosynthesis in *B. juncea* under Cd treatment. Earlier, Lakimova et al. (2008) reported the role of ethylene in controlling signal transduction during Cd-mediated programmed cell death and tolerance in *Lycopersicon esculentum*. Recently, Maksymiec (2011) observed a reduction in Cd-mediated hydrogen peroxide concentrations in *Allium cepa* upon treatment with silver thiosulfate, an ethylene inhibitor. Schellingen et al. (2014) documented that Cd-mediate increased formation of ethylene in *A. thaliana* was observed to be dependent on expression of ACS2 and ACS6. Moreover, ACS2 and ACS6 are recognized as the key isoforms that play a vital role in the Cd-mediated formation of ethylene in *A. thaliana* (Khan and Khan, 2014). According to Abozeid et al. (2017), ethylene stimulates the formation of root systems under Cd exposure by modulating superoxide anion content in *A. thaliana*. Recently, Chen et al. (2017) observed that exogenous application of ethylene provided Cd resistance and altered alkaloid biosynthesis in seedlings of *Catharanthus roseus*. However, prolonged Cd exposure decreased the production of ethylene in *A. thaliana* (Carrió-Seguí et al., 2015).

## 3.5 Abscisic Acid

ABA, an acidic, 15-C sesquiterpene, is an important phytohormone that imparts crucial function in several stages of plant life, such as dormancy and development of seeds (Baek et al., 2015). ABA has been associated with tolerance to various unfavorable environmental situations, and its signaling pathway is a key controller of various abiotic stress conditions in plants (Danquah et al., 2014; You and Chan, 2015). The Cd treatment enhanced the level of endogenous ABA in roots of *O. sativa*, *Phragmites australis*, and *Typha latifolia* (Fediuc et al., 2005; Kim et al., 2014; Stroinski et al., 2010). Hsu and Kao (2003) reported a change in ABA content in Cd-treated rice seedlings in two cultivars, viz. Cd-tolerant and Cd-sensitive. The results revealed increased ABA content in leaves and roots of Cd-tolerant cultivars upon application of $CdCl_2$ (Hsu and Kao, 2003). Meng et al. (2009) investigated the role of exogenous application of different phytohormones in alleviating Cd-induced stress on germination and seedling growth of *Brassica napus*. The authors found that among the various phytohormones studied, ABA resulted in a significant reduction in Cd accumulation in the plants (Meng et al., 2009). Moreover, closure of stomata by ABA under stress leads to reduction of transpiration flow, which results in reduction of root-to-shoot movement of metals (Perfus-Barbeoch et al., 2002). Further, Yeh et al. (2007) observed elevated ROS and calcium (Ca) levels in Cd-tolerant rice seedlings. These findings showed that under conditions of heavy metal stress, roots accumulated ROS and Ca, which stimulated MAP kinase activity; thus inducing Cd-tolerance (Yeh et al., 2007). In spite of improvements, more research is needed to understand the function of ABA in MAP kinase stimulation under heavy metal toxicity. The application of ABA declined root-to-shoot Cd translocation in *O. sativa*, thereby exhibiting high Cd-tolerance (Uraguchi et al., 2009). The exogenous application of ABA-induced Cd accumulation was studied by Wang et al. (2016) in two ecotypes of *Solanum photeinocarpum*. The findings confirmed that chlorophyll content and biomass of the two ecotypes were enhanced by incrementing the ABA level. Cd content was also mounted in two ecotypes of *S. photeinocarpum* with the exogenous application of ABA (Wang et al., 2016). Xu et al. (2018b) studied the role of the *JrVHAG1* gene in *Juglans regia* under exposure of Cd-stress. The results indicated that this gene functions under Cd-stress by taking part in the ABA-signaling pathway and myeloblastosis (MYB) transcription regulation network. Moreover, the gene is essential for tolerance of heavy metal stress in plant molecular breeding (Xu et al., 2018b).

### 3.6 Brassinosteroids

Brassinosteroids are ubiquitous steroidal plant hormones that were discovered in the pollen of *B. napus* (Rao et al., 2002). They regulate various developmental processes, which include growth, germination, flowering, senescence, and stress tolerance (Rao et al., 2002). Foliar treatment of homo-brassinolide to *B. juncea* enhances Cd-tolerance by enhancing the catalase, peroxidase, and superoxide dismutase activities (Hayat et al., 2007). The exogenous treatment of 24-epibrassinolide to *Phaseolus vulgaris* also improves Cd-tolerance (Rady, 2011). The treatment of brassinosteroids to *Solanum lycopersicum* increased the antioxidant activities and enhances the yield and quality of fruit (Hayat et al., 2012). Similar results were found by Hasan et al. (2008) in *Cicer arietinum* by the application of 28-homobrassinolide against Cd-stress. The treatment of Cd to *Raphanus sativus*, along with supplementation of 24-epibrassinolide, enhanced the activities of PPO and GST enzymes (Sharma et al., 2012). The application of 28-homobrassinolide to *R. sativus,* along with Cd, improves the activities of SOD, CAT, POD, APX, and GR, but when only Cd treatment was given, the activities of enzymes were reduced (Sharma et al., 2010). Hayat et al. (2012) studied that foliar treatment of brassinosteroids improves the production and quality of *S. lycopersicum* under Cd-stress. Under Cd-stress, plants showed reduced growth and production, whereas treatment with brassinosteroids neutralized the detrimental effect of Cd in *S. lycopersicum*. Kaur et al. (2017) studied that the coapplication of castasterone (a precursor of brassinosteroid) and citric acid increased Cd accumulation in the roots and shoots, and also enhanced morphological parameters, i.e., fresh weight, dry weight, root length, and shoot length in *B. juncea*. They also reported decreased malondialdehyde content and oxidative stress, and increased activities of antioxidant enzymes in response to the treatment. From their findings, it was confirmed that coapplication of castasterone and citric acid enhanced Cd accumulation, and alleviated Cd-toxicity in *B. juncea* seedlings (Kaur et al., 2018). Under Cd exposure to rice seedlings, Xu et al., 2018a studied that the combination of 24-epibrassinolide or 28-homobrassinolide reduced the accumulation of Cd in rice seedlings.

## 3.7 Jasmonic Acid

JA regulates growth and development in plants under both biotic and abiotic stress conditions (Creelman and Mullet, 1995; Kazan, 2015). Several studies outline the role of JA signaling and biosynthesis in providing resistance to plants in response to various abiotic stresses (Avramova, 2017; Dar et al., 2015). Maksymiec et al. (2007) demonstrated that application of methyl jasmonate in *A. thaliana* increased the content of phytochelatins that incremented the tolerance of plants upon exposure to Cd-stress. Similarly, methyl jasmonate decreased Cd-toxicity in *Capsicum frutescens* by enhancing the chlorophyll content and activities of antioxidant enzymes viz., catalase and ascorbate peroxidase (Yan et al., 2013). Recently, Chen et al. (2014) investigated the effect of methyl jasmonate on seedlings of *Kandelia obovata* upon exposure to Cd-toxicity. The results indicated reduced Cd accumulation and increased ascorbic acid content (Chen et al., 2014). The effect of methyl jasmonte (MeJA) and Cd on the structure and function of photosynthetic apparatus in *B. juncea* was studied by Per et al. (2016). Plants showed enhanced oxidative stress, and reduced photosynthetic activity when grown under Cd-stress, whereas a combination of MeJA and sulfur (S) mitigated the effect of Cd via enhanced S-assimilation, and formation of decreased GSH and enhanced photosynthetic functions. From their results, it was found that MeJA regulates S-assimilation and formation of GSH for preventing the structure and function of *B. juncea* under Cd-stress (Per et al., 2016). Ahmad et al. (2017) studied the effect of JA under Cd-stress in *V. faba*. The growth, biomass production, and pigment systems were decreased under Cd-stress, whereas application of JA with Cd enhanced the growth, biomass production, and pigment systems in *V. faba*. Further, JA reduces the harmful effects of Cd in *V. faba* by inhibiting the peaked Cd concentration, hydrogen peroxide and malondialdehyde content, and by increasing the osmolyte and antioxidant activities, which in turn decreases the oxidative stress (Ahmad et al., 2017). Ali et al. (2018) reported that JA decreased Cd uptake in leaves, and reduced malondialdehyde content, thereby preventing damage caused by Cd to the membrane.

## 3.8 Salicylic Acid

Salicylic acid, a phenolic plant hormone, plays a key role in regulating growth, development, and various immune responses in plants (An and Mou, 2011; Hayat et al., 2010; Rivas-San Vicente and Plasencia, 2011). Under unfavorable environmental conditions, salicylic acid scavenges hydroxyl radicals and provides resistance in plants (Khan et al., 2014). Salicylic acid is recognized as one of the main signaling molecules that participates in both biotic, as well as abiotic stress tolerance (Alavi et al., 2014; Zhang et al., 2011). Applications of salicylic acid improved the tolerance of plants to heavy metal stress by regulating the antioxidant enzyme activities, level of osmolytes, metal chelating compounds, and secondary metabolites (Kohli et al., 2017). Salicylic acid-induced alleviation of Cd-toxicity might be ascribed to decreased uptake of Cd, enhanced photosynthetic capacity, and improved antioxidant activities in *Cannabis sativa* (Shi et al., 2009). According to Chao et al. (2010), exogenous application of salicylic acid incremented expression of OsWRKY45 that led to the enhanced endogenous salicylic acid content, and averted membrane damage by reducing hydrogen peroxide content in response to Cd-toxicity in *O. sativa* (Chao et al., 2010). Popova et al. (2009) documented the role of salicylic acid in providing stability to the membrane, regulating the redox balance by triggering antioxidant enzyme activities, and also protecting photosynthetic carboxylation reactions in *Pisum sativum*, with respect to Cd exposure. In Cd-treated *Lolium perenne,* salicylic acid and sodium nitroprusside alone, and in combination, decreased the root-to-shoot uptake of Cd, and enhanced antioxidant enzyme activities, which lead to increased growth, photosynthesis, and reduced oxidative stress in plants (Wang et al., 2013). Salicylic acid reduced the accumulation of Cd, and increased heme-oxygenase-1 activity, which resulted in the reduction of Cd-toxicity in *Glycine max* (Noriega et al., 2012). Shakirova et al. (2016) studied the concept that salicylic acid provides a shield against Cd-toxicity in *T. aestivum* plants. The application of salicylic acid reduced the accumulation of Cd, MDA, and electrolyte leakage in wheat plants. Yotsova et al. (2018) found that application of salicylic acid to rice plants exposed to Cd considerably enhanced plant growth and photochemical activities of photosystems.

## 4 CONCLUSIONS AND FUTURE PERSPECTIVES

Cd-toxicity is a serious concern to the environment. Plants produce various physiological pathways, and mediate molecular mechanisms that alleviate Cd-induced toxic effects. Recently, plant biology studies have shown the role of plant hormones in reducing harmful effects of Cd-accrued toxicity. Exposure of plant hormones to Cd toxic plants plays an important role in protecting plant metabolism by regulating antioxidant systems, osmolyte production, reducing Cd uptake, and inducing up-regulation of genes responsible for providing stress tolerance. The regular plant breeding methods and phytoremediation technology are time consuming and costlier; hence, the treatment of plant hormones discussed herein can be an important approach to increasing plant metabolism under stressful conditions.

However, more elaborate research is needed to understand the interaction between plant hormones and abiotic stress tolerance, emphasizing Cd-toxicity. Efforts should be made to identify the expression of genes, as well as synthesis of certain plant-growth promoting substances involved in Cd-stress tolerance.

## REFERENCES

Abeles, F.B., Morgan, P.W., Saltveit Jr., M.E., 1992. Ethylene in Plant Biology, second ed. Academic Press, New York.

Abozeid, A., Ying, Z., Lin, Y., Liu, J., Zhang, Z., Tang, Z., 2017. Ethylene improves root system development under cadmium stress by modulating superoxide anion concentration in *Arabidopsis thaliana*. Front. Plant Sci. 8, 1–15. https://doi.org/10.3389/fpls.2017.00253.

Agami, R.A., Mohamed, G.F., 2013. Exogenous treatment with indole-3-acetic acid and salicylic acid alleviates cadmium toxicity in wheat seedlings. Ecotoxicol. Environ. Saf. 94, 164–171.

Agency for Toxic Substances and Disease Registry (ATSDR), 2017. CERCLA Priority List of Hazardous Substances. http://www.atsdr.cdc.gov/spl/index.html. [(Accessed 28 March 2018)].

Ahammed, G.J., Ruan, Y.P., Zhou, J., Xia, X.J., Shi, K., Zhou, Y.H., Yu, J.Q., 2013. Brassinosteroid alleviates polychlorinated biphenyls-induced oxidative stress by enhancing antioxidant enzymes activity in tomato. Chemosphere 90, 2645–2653.

Ahmad, P., Alyemeni, M.N., Wijaya, L., Alam, P., Ahanger, M.A., Alamri, S.A., 2017. Jasmonic acid alleviates negative impacts of cadmium stress by modifying osmolytes and antioxidants in faba bean (*Vicia faba* L.). Arch. Agron. Soil Sci. 13, 1889–1899.

Alavi, S.M.N., Arvin, M.J., Kalantari, K.M., 2014. Salicylic acid and nitric oxide alleviate osmotic stress in wheat (*Triticum aestivum* L.) seedlings. J. Plant Interact. 9, 683–688.

Al-Hakimi, A.M.A., 2007. Modification of cadmium toxicity in pea seedlings by kinetin. Plant Soil Environ. 53, 129–135.

Ali, B., Hasan, S.A., Hayat, S., Hayat, Q., Yadav, S., Fariduddin, Q., Ahmad, A., 2008. A role for brassinosteroids in the amelioration of aluminium stress through antioxidant system in mung bean (*Vigna radiata* L. Wilczek). Environ. Exp. Bot. 62, 153–159.

Ali, E., Hussain, N., Shamsi, I.H., Jabeen, Z., Siddiqui, M.H., Jiang, L.X., 2018. Role of jasmonic acid in improving tolerance of rapeseed (*Brassica napus* L.) to Cd toxicity. J. Zhejiang Univ. Sci. B 19, 130–146.

An, C., Mou, Z., 2011. Salicylic acid and its function in plant immunity. J. Integr. Plant Biol. 53, 412–428.

Anuradha, S., Rao, S.S.R., 2009. Effect of 24-epibrassinolide on the photosynthetic activity of radish plants under cadmium stress. Photosynthetica 47, 317–320.

Asgher, M., Khan, N.A., Khan, M.I.R., Fatma, M., Masood, A., 2014. Ethylene production is associated with alleviation of cadmium-induced oxidative stress by sulfur in mustard types differing in ethylene sensitivity. Ecotoxicol. Environ. Saf. 106, 54–61.

Asgher, M., Khan, M.I.R., Anjum, N.A., Khan, N.A., 2015. Minimising toxicity of cadmium in plants—role of plant growth regulators. Protoplasma 252, 399–413.

Ashikari, M., Sakakibara, H., Lin, S., Yamamoto, T., Takashi, T., Nishimura, A., Angeles, E.R., Qian, Q., Kitano, H., Matsuoka, M., 2005. Cytokinin oxidase regulates rice grain production. Science 309, 741–745.

Avramova, Z., 2017. The jasmonic acid-signaling and abscisic acid-signaling pathways cross talk during one, but not repeated, dehydration stress: a non-specific 'panicky' or a meaningful response? Plant Cell Environ. 40, 1704–1710.

Baek, D., Cha, J.Y., Kang, S., Park, B., Lee, H.J., Hong, H., Chun, H.J., Kim, D.H., Kim, M.C., Lee, S.Y., Yun, D.J., 2015. The *Arabidopsis* a zinc finger domain protein ARS1 is essential for seed germination and ROS homeostasis in response to ABA and oxidative stress. Front. Plant Sci. 6, 1–10.

Bansal, P., Sharma, P., Goyal, V., 2002. Impact of lead and cadmium on enzyme of citric acid cycle in germinating pea seeds. Biol. Plant. 45, 125–127.

Bocova, B., Huttova, J., Mistrık, I., Tamas, L., 2013. Auxin signaling is involved in cadmium-induced glutathione-S-transferase activity in barley root. Acta Physiol. Plant. 35, 2685–2690.

Bouzroud, S., Gouiaa, S., Hu, N., Bernadac, A., Mila, I., Bendaou, N., Smouni, A., Bouzayen, M., Zouine, M., 2018. Auxin response factors (ARFs) are potential mediators of auxin action in tomato response to biotic and abiotic stress (*Solanum lycopersicum*). PLoS ONE 13. e0193517.

Carrió-Seguí, A., Garcia-Molina, A., Sanz, A., Peñarrubia, L., 2015. Defective copper transport in the copt5 mutant affects cadmium tolerance. Plant Cell Physiol. 56, 442–454.

Chao, Y.Y., Chen, C.Y., Huang, W.D., Ching, C.H., 2010. Salicylic acid mediated hydrogen peroxide accumulation and protection against Cd toxicity in rice leaves. Plant Soil 329, 327–337.

Chen, J., Yan, Z., Li, X., 2014. Effect of methyl jasmonate on cadmium uptake and antioxidative capacity in *Kandelia obovata* seedlings under cadmium stress. Ecotoxicol. Environ. Saf. 104, 349–356.

Chen, Q., Wu, K., Tang, Z., Guo, Q., Guo, X., Wang, H., 2017. Exogenous ethylene enhanced the cadmium resistance and changed the alkaloid biosynthesis in *Catharanthus roseus* seedlings. Acta Physiol. Plant. 39, 1–12.

Correa-Aragunde, N., París, R., Foresi, N., Terrile, C., Casalongué, C., Lamattina, L., 2016. The auxin-nitric oxide highway: a right direction in determining the plant root system. In: Gasotransmitters in Plants. Springer, Cham, pp. 117–136.

Creelman, R.A., Mullet, J.E., 1995. Jasmonic acid distribution and action in plants: regulation during development and response to biotic and abiotic stress. Proc. Natl. Acad. Sci. U. S. A. 92, 4114–4119.

Danquah, A., de Zelicourt, A., Colcombet, J., Hirt, H., 2014. The role of ABA and MAPK signaling pathways in plant abiotic stress responses. Biotechnol. Adv. 32, 40–52.

Dar, T.A., Uddin, M., Khan, M.M.A., Hakeem, K.R., Jaleel, H., 2015. Jasmonates counter plant stress: a review. Environ. Exp. Bot. 115, 49–57.

Davies, P.J., 2010. Plant Hormones: Biosynthesis, Signal Transduction, Action, revised third ed. Springer, Dordrecht, pp. 1–15.

Dias, M.C., Monteiro, C., Moutinho-Pereira, J., Correia, C., Gonçalves, B., Conceiçã o, S., 2013. Cadmium toxicity affects photosynthesis and plant growth at different levels. Acta Physiol. Plant. 35, 1281–1289.

Dugardeyn, J., Van Der Straeten, D., 2008. Ethylene: fine-tuning plant growth and development by stimulation and inhibition of elongation. Plant Sci. 175, 59–70.

Enders, T.A., Strader, L.C., 2015. Auxin activity: past, present, and future. Am. J. Bot. 102, 180–196.

Falkowska, M., Pietryczuk, A., Piotrowska, A., Bajguz, A., Grygoruk, A., Czerpak, R., 2011. The effect of gibberellic acid (GA3) on growth, metal biosorption and metabolism of the green algae *Chlorella vulgaris* (Chlorophyceae) Beijerinck exposed to cadmium and lead stress. Pol. J. Environ. Stud. 20, 53–59.

Farooq, H., Asghar, H.N., Khan, M.Y., Saleem, M., Zahir, Z.A., 2015. Auxin-mediated growth of rice in cadmium-contaminated soil. Turk. J. Agric. For. 39, 272–276.

Fattorini, L., Ronzan, M., Piacentini, D., Della Rovere, F., De Virgilio, C., Sofo, A., Altamura, M.M., Falasca, G., 2017. Cadmium and arsenic affect quiescent centre formation and maintenance in *Arabidopsis thaliana* post-embryonic roots disrupting auxin biosynthesis and transport. Environ. Exp. Bot. 144, 37–48.

Fediuc, E., Lips, S.H., Erdei, L., 2005. O-acetylserine (thiol) lyase activity in *Phragmites* and *Typha* plants under cadmium and NaCl stress conditions and the involvement of ABA in the stress response. J. Plant Physiol. 162, 865–872.

Fleet, C.M., Sun, T.P., 2005. A DELLAcate balance: the role of gibberellin in plant morphogenesis. Curr. Opin. Plant Biol. 8, 77–85.

Fujita, Y., Fujita, M., Shinozaki, K., Yamaguchi-Shinozaki, K., 2011. ABA-mediated transcriptional regulation in response to osmotic stress in plants. J. Plant Res. 124, 509–525.

Gemrotová, M., Kulkarni, M.G., Stirk, W.A., Strnad, M., Van Staden, J., Spíchal, L., 2013. Seedlings of medicinal plants treated with either a cytokinin antagonist (PI-55) or an inhibitor of cytokinin degradation (INCYDE) are protected against the negative effects of cadmium. Plant Growth Regul. 71, 137–145.

Ghorbanli, M., Kaveh, S.H., Sepehr, M.F., 2000. Effects of cadmium and gibberellin on growth and photosynthesis of *Glycine max*. Photosynthetica 37, 627–631.

Gonzalez-Aguilar, G.A., Buta, J.G., Wang, C.Y., 2003. Methyl jasmonate and modified atmosphere packaging (MAP) reduce decay and maintain postharvest quality of papaya 'sunrise'. Postharvest Biol. Technol. 28, 361–370.

Gu, C.S., Yang, Y.H., Shao, Y.F., Wu, K.W., Liu, Z.L., 2018. The effects of exogenous salicylic acid on alleviating cadmium toxicity in *Nymphaea tetragona* Georgi. S. Afr. J. Bot. 114, 267–271.

Hadi, F., Ali, N., Ahmad, A., 2014. Enhanced phytoremediation of cadmium-contaminated soil by *Parthenium hysterophorus* plant: effect of gibberellic acid (GA3) and synthetic chelator, alone and in combinations. Biorem. J. 18, 46–55.

Hasan, S.A., Hayat, S., Ali, B., Ahmad, A., 2008. 28-Homobrassinolide protects chickpea (*Cicer arietinum*) from cadmium toxicity by stimulating antioxidants. Environ. Pollut. 151, 60–66.

Hasan, S.A., Hayat, S., Ahmad, A., 2011. Brassinosteroids protect photosynthetic machinery against the cadmium induced oxidative stress in two tomato cultivars. Chemosphere 84, 1446–1451.

Hayat, S., Ali, B., Hasan, S.A., Ahmad, A., 2007. Brassinosteroid enhanced the level of antioxidants under cadmium stress in *Brassica juncea*. Environ. Exp. Bot. 60, 33–41.

Hayat, Q., Hayat, S., Irfan, M., Ahmad, A., 2010. Effect of exogenous salicylic acid under changing environment: a review. Environ. Exp. Bot. 68, 14–25.

Hayat, S., Alyemeni, M.N., Hasan, S.A., 2012. Foliar spray of brassinosteroid enhances yield and quality of *Solanum lycopersicum* under cadmium stress. Saudi J. Biol. Sci. 19, 325–335.

Hsu, Y.T., Kao, C.H., 2003. Role of abscisic acid in cadmium tolerance of rice (*Oryza sativa* L.) seedlings. Plant Cell Environ. 26, 867–874.

Hu, Y.F., Zhou, G., Na, X.F., Yang, L., Nan, W.B., Liu, X., Zhang, Y.Q., Li, J.L., Bi, Y.R., 2013. Cadmium interferes with maintenance of auxin homeostasis in *Arabidopsis* seedlings. J. Plant Physiol. 170, 965–975.

Janeczko, A., Koscielniak, J., Pilipowicz, M., Szarek-Lukaszewska, G., Skoczowski, A., 2005. Protection of winter rape photosystem 2 by 24-epibrassinolide under cadmium stress. Photosynthetica 43, 293–298.

Kaur, R., Yadav, P., Sharma, A., Thukral, A.K., Kumar, V., Kohli, S.K., Bhardwaj, R., 2017. Castasterone and citric acid treatment restores photosynthetic attributes in *Brassica juncea* L. under Cd (II) toxicity. Ecotoxicol. Environ. Saf. 145, 466–475.

Kaur, R., Yadav, P., Thukral, A.K., Sharma, A., Bhardwaj, R., Alyemeni, M.N., Ahmad, P., 2018. Castasterone and citric acid supplementation alleviates cadmium toxicity by modifying antioxidants and organic acids in *Brassica juncea*. J. Plant Growth Regul. 37, 286–299.

Kazan, K., 2015. Diverse roles of jasmonates and ethylene in abiotic stress tolerance. Trends Plant Sci. 20, 219–229.

Ke, Q., Wang, Z., Ji, C.Y., Jeong, J.C., Lee, H.S., Li, H., Xu, B., Deng, X., Kwak, S.S., 2015. Transgenic poplar expressing *Arabidopsis* YUCCA6 exhibits auxin-overproduction phenotypes and increased tolerance to abiotic stress. Plant Physiol. Biochem. 94, 19–27.

Khan, M.I.R., Khan, N.A., 2014. Ethylene reverses photosynthetic inhibition by nickel and zinc in mustard through changes in PS II activity, photosynthetic nitrogen use efficiency, and antioxidant metabolism. Protoplasma 251, 1007–1019.

Khan, M.I.R., Iqbal, N., Masood, A., Per, T.S., Khan, N.A., 2013. Salicylic acid alleviates adverse effects of heat stress on photosynthesis through changes in proline production and ethylene formation. Plant Signal. Behav. 8. e26374.

Khan, M.I.R., Asgher, M., Khan, N.A., 2014. Alleviation of salt-induced photosynthesis and growth inhibition by salicylic acid involves glycinebetaine and ethylene in mungbean (*Vigna radiata* L.). Plant Physiol. Biochem. 80, 67–74.

Khan, M.I.R., Nazir, F., Asgher, M., Per, T.S., Khan, N.A., 2015. Selenium and sulfur influence ethylene formation and alleviate cadmium-induced oxidative stress by improving proline and glutathione production in wheat. J. Plant Physiol. 173, 9–18.

Kim, Y.H., Khan, A.L., Kim, D.H., Lee, S.Y., Kim, K.M., Waqas, M., Jung, H.Y., Shin, J.H., Kim, J.G., Lee, I.J., 2014. Silicon mitigates heavy metal stress by regulating P-type heavy metal ATPases, *Oryza sativa* low silicon genes, and endogenous phytohormones. BMC Plant Biol. 14, 1–13.

Kohli, S.K., Handa, N., Kaur, R., Kumar, V., Khanna, K., Bakshi, P., Bhardwaj, R., 2017. Role of salicylic acid in heavy metal stress tolerance: insight into underlying mechanism. In: Salicylic Acid: A Multifaceted Hormone. Springer, Singapore, pp. 123–144.

Kudoyarova, G.R., Vysotskaya, L.B., Cherkozyanova, A., Dodd, I.C., 2006. Effect of partial rootzone drying on the concentration of zeatin-type cytokinins in tomato (*Solanum lycopersicum* L.) xylem sap and leaves. J. Exp. Bot. 58, 161–168.

Lakimova, E.T., Woltering, E.J., Kapchina-Toteva, V.M., Harren, F.J.M., Cristescu, S.M., 2008. Cadmium toxicity in cultured tomato cells-role of ethylene, proteases and oxidative stress in cell death signaling. Cell Biol. Int. 32, 1521–1529.

Lambers, H., Chapin, F.S., Pons, T.L., 2008. Growth and allocation. In: Plant Physiological Ecology. Springer, New York, pp. 321–374.

Li, L., Wang, Y., Shen, W., 2012. Roles of hydrogen sulfide and nitric oxide in the alleviation of cadmium-induced oxidative damage in alfalfa seedling roots. Biometals 25, 617–631.

Liu, K., Shen, L., Sheng, J., 2008. Improvement in cadmium tolerance of tomato seedlings with an antisense DNA for 1-aminocyclopropane-1-carboxylate synthase. J. Plant Nutr. 31, 809–827.

Liu, C., Guo, J., Cui, Y., Lu, T., Zhang, X., Shi, G., 2011. Effects of cadmium and salicylic acid on growth, spectral reflectance and photosynthesis of castor bean seedlings. Plant Soil 344, 131–141.

Lu, Q., Zhang, T., Zhang, W., Su, C., Yang, Y., Hu, D., Xu, Q., 2018. Alleviation of cadmium toxicity in Lemna minor by exogenous salicylic acid. Ecotoxicol. Environ. Saf. 147, 500–508.

Maksymiec, W., 2011. Effects of jasmonate and some other signaling factors on bean and onion growth during the initial phase of cadmium action. Biol. Plant. 55, 112–118.

Maksymiec, W., Wojcik, M., Krupa, Z., 2007. Variation in oxidative stress and photochemical activity in *Arabidopsis thaliana* leaves subjected to cadmium and excess copper in the presence or absence of jasmonate and ascorbate. Chemosphere 66, 421–427.

Masood, A., Khan, N.A., 2013. Ethylene and gibberellic acid interplay in regulation of photosynthetic capacity inhibition by cadmium. J. Plant Biochem. Physiol. 1, 1–3.
Masood, A., Iqbal, N., Khan, N.A., 2012. Role of ethylene in alleviation of cadmium-induced photosynthetic capacity inhibition by sulphur in mustard. Plant Cell Environ. 35, 524–533.
Meng, H., Hua, S., Shamsi, I.H., Jilani, G., Li, Y., Jiang, L., 2009. Cadmium induced stress on the seed germination and seedling growth of *Brassica napus* L, and its alleviation through exogenous plant growth regulators. Plant Growth Regul. 58, 47–59.
Mobin, M., Khan, N.A., 2007. Photosynthetic activity, pigment composition and antioxidative response of two mustard (*Brassica juncea*) cultivars differing in photosynthetic capacity subjected to cadmium stress. J. Plant Physiol. 164, 601–610.
Mohamed, M.M., Abdel-Razik, K.E., 2005. Interactive effect of heavy metals and gibberellic acid on mitotic activity and some metabolic changes of *Vicia faba* L. plants. Cytologia 70, 275–282.
Nagajyoti, P., Lee, K., Sreekanth, T., 2010. Heavy metals, occurrence and toxicity for plants: a review. Environ. Chem. Lett. 8, 199–216.
Nedjimi, B., Daoud, Y., 2009. Cadmium accumulation in *Atriplex halimus* subsp. *schweinfurthii* and its influence on growth, proline, root hydraulic conductivity and nutrient uptake. Flora 204, 316–324.
Noriega, G., Caggiano, E., Lecube, M.P., Santa Cruz, D., Batlle, A., Tomaro, M., Balestrasse, K.B., 2012. The role of salicylic acid in the prevention of oxidative stress elicited by cadmium in soybean plants. Biometals 25, 1155–1165.
Ostrowski, M., Ciarkowska, A., Jakubowska, A., 2016. The auxin conjugate indole-3-acetyl-aspartate affects responses to cadmium and salt stress in *Pisum sativum* L. J. Plant Physiol. 191, 63–72.
Pacheco-Villalobos, D., Díaz-Moreno, S.M., van der Schuren, A., Tamaki, T., Kang, Y.H., Gujas, B., Novak, O., Jaspert, N., Li, Z., Wolf, S., Oecking, C., Ljung, K., Bulone, V., Hardtke, C.S., 2016. The effects of high steady state auxin levels on root cell elongation in *Brachypodium*. Plant Cell 28, 1009–1024.
Parmar, P., Kumari, N., Sharma, V., 2013. Structural and functional alterations in photosynthetic apparatus of plants under cadmium stress. Bot. Stud. 54, 45.
Peleg, Z., Reguera, M., Tumimbang, E., Walia, H., Blumwald, E., 2011. Cytokinin-mediated source/sink modifications improve drought tolerance and increase grain yield in rice under water-stress. Plant Biotechnol. J. 9, 747–758.
Per, T.S., Khan, N.A., Masood, A., Fatma, M., 2016. Methyl jasmonate alleviates cadmium-induced photosynthetic damages through increased S-assimilation and glutathione production in mustard. Front. Plant Sci. 7, 1–13.
Perfus-Barbeoch, L., Leonhardt, N., Vavasseur, A., Forestier, C., 2002. Heavy metal toxicity: cadmium permeates through calcium channels and disturbs the plant water status. Plant J. 32, 539–548.
Piotrowska-Niczyporuk, A., Bajguz, A., Zambrzycka, E., Godlewska-Żyłkiewicz, B., 2012. Phytohormones as regulators of heavy metal biosorption and toxicity in green alga *Chlorella vulgaris* (Chlorophyceae). Plant Physiol. Biochem. 52, 52–65.
Popova, L.P., Maslenkova, L.T., Yordanova, R.Y., Ivanova, A.P., Krantev, A.P., Szalai, G., Janda, T., 2009. Exogenous treatment with salicylic acid attenuates cadmium toxicity in pea seedlings. Plant Physiol. Biochem. 47, 224–231.
Rady, M.M., 2011. Effect of 24-epibrassinolide on growth, yield, antioxidant system and cadmium content of bean (*Phaseolus vulgaris* L.) plants under salinity and cadmium stress. Sci. Hortic. (Amsterdam) 129, 232–237.
Rao, S.S.R., Vardhini, B.V., Sujatha, E., Anuradha, S., 2002. Brassinosteroids—a new class of phytohormones. Curr. Sci. India 82, 1239–1245.
Rivas-San Vicente, M., Plasencia, J., 2011. Salicylic acid beyond defence: its role in plant growth and development. J. Exp. Bot. 62, 3321–3338.

Salt, D.E., Prince, R.C., Pickering, I.J., Raskin, I., 1995. Mechanisms of cadmium mobility and accumulation in Indian mustard. Plant Physiol. 109, 1427–1433.

Schellingen, K., Der Straeten, D.V., Vandenbussche, F., Prinsen, E., Remans, T., Vangronsveld, J., Cuypers, A., 2014. Cadmium-induced ethylene production and responses in *Arabidopsis thaliana* rely on ACS2 and ACS6 gene expression. BMC Plant Biol. 14, 1–14.

Shakirova, F.M., Allagulova, C.R., Maslennikova, D.R., Klyuchnikova, E.O., Avalbaev, A.M., Bezrukova, M.V., 2016. Salicylic acid-induced protection against cadmium toxicity in wheat plants. Environ. Exp. Bot. 122, 19–28.

Sharaf, A.E.M.M., Farghal, I.I., Sofy, M.R., 2009. Role of gibberellic acid in abolishing the detrimental effects of Cd and Pb on broad bean and lupin plants. Res. J. Agric. Biol. Sci. 5, 668–673.

Sharma, J., Chakraverty, N., 2013. Mechanism of plant tolerance in response to heavy metals. In: Rout, G.R., Das, A.B. (Eds.), Molecular Stress Physiology of Plants. Springer India, New Delhi, pp. 289–308.

Sharma, I., Pati, P.K., Bhardwaj, R., 2010. Regulation of growth and antioxidant enzyme activities by 28-homo brassinolide in seedlings of *Raphanus sativus* L. under cadmium stress. Indian J. Biochem. Biophys. 47, 172–177.

Sharma, N., Hundal, G.S., Sharma, I., Bharadwaj, R., 2012. Effect of 24-epibrassinolide on protein content and activities of glutathione-S-transferase and polyphenoloxidase in *Raphanus sativus* L. plants under cadmium and mercury metal stress. Terr. Aquat. Environ. Toxicol. 6, 1–7.

Sharma, E., Sharma, R., Borah, P., Jain, M., Khurana, J.P., 2015. Emerging roles of auxin in abiotic stress responses. In: Elucidation of Abiotic Stress Signaling in Plants. Springer, New York, pp. 299–328.

Shi, G.R., Cai, Q.S., Liu, Q.Q., Wu, L., 2009. Salicylic acid-mediated alleviation of cadmium toxicity in hemp plants in relation to cadmium uptake, photosynthesis, and antioxidant enzymes. Acta Physiol. Plant. 31, 969–977.

Shi, Y., Tian, S., Hou, L., Huang, X., Zhang, X., Guo, H., Yang, S., 2012. Ethylene signaling negatively regulates freezing tolerance by repressing expression of CBF and type-A ARR genes in *Arabidopsis*. Plant Cell 24, 2578–2595.

Sidhu, G.P.S., 2016. Heavy metal toxicity in soils: Sources, remediation technologies and challenges. Adv. Plants Agric. Res. 500166.

Sidhu, G.P.S., Singh, H.P., Batish, D.R., Kohli, R.K., 2016. Effect of lead on oxidative status, antioxidative response and metal accumulation in *Coronopus didymus*. Plant Physiol. Biochem. 105, 290–296.

Sidhu, G.P.S., Singh, H.P., Batish, D.R., Kohli, R.K., 2017a. Tolerance and hyperaccumulation of cadmium by a wild, unpalatable herb *Coronopus didymus* (L.) Sm. (Brassicaceae). Ecotoxicol. Environ. Saf. 135, 209–215.

Sidhu, G.P.S., Singh, H.P., Batish, D.R., Kohli, R.K., 2017b. Appraising the role of environment friendly chelants in alleviating lead by *Coronopus didymus* from Pb-contaminated soils. Chemosphere 182, 129–136.

Singh, S., Prasad, S.M., 2016. Kinetin ameliorates cadmium induced toxicity on growth, pigments and photosynthesis by regulating antioxidant potential in tomato seedlings. Int. J. Sci. Eng. Appl. Sci. 2, 1–10.

Singh, S., Singh, A., Srivastava, P.K., Prasad, S.M., 2018. Cadmium toxicity and its amelioration by kinetin in tomato seedlings vis-à-vis ascorbate-glutathione cycle. J. Photochem. Photobiol. B Biol. 178, 76–84.

Stoll, M., Loveys, B., Dry, P., 2000. Hormonal changes induced by partial rootzone drying of irrigated grapevine. J. Exp. Bot. 51, 1627–1634.

Stroinski, A., Chadzinikolau, T., Gizewska, K., Zielezinska, M., 2010. ABA or cadmium induced phytochelatin synthesis in potato tubers. Biol. Plant. 54, 117–120.

Tamás, L., Mistrík, I., Alemayehu, A., Zelinová, V., Bočová, B., Huttová, J., 2015. Salicylic acid alleviates cadmium-induced stress responses through the inhibition of Cd-induced auxin-mediated reactive oxygen species production in barley root tips. J. Plant Physiol. 173, 1–8.

Tudoreanu, L., Phillips, C.J.C., 2004. Modelling cadmium uptake and accumulation in plants. Adv. Agron. 84, 121–157.

Uraguchi, S., Mori, S., Kuramata, M., Kawasaki, A., Arao, T., Ishikawa, S., 2009. Root-to-shoot Cd translocation via the xylem is the major process determining shoot and grain cadmium accumulation in rice. J. Exp. Bot. 60, 2677–2688.

Wael, M.S., Mostafa, M.R., Taia, A.A.E.M., Saad, M.H., Magdi, T.A., 2015. Alleviation of cadmium toxicity in common bean (*Phaseolus vulgaris* L.) plants by the exogenous application of salicylic acid. J. Hortic. Sci. Biotechnol. 90, 83–91.

Wagner, G.J., 1993. Accumulation of cadmium in crop plants and its consequences to human health. Adv. Agron. 51, 173–212.

Wang, B., Du, Y., 2013. Cadmium and its neurotoxic effects. Oxidative Med. Cell. Longev. 2013, 1–12.

Wang, Q., Liang, X., Dong, Y., Xu, L., Zhang, X., Kong, J., Liu, S., 2013. Effects of exogenous salicylic acid and nitric oxide on physiological characteristics of perennial ryegrass under cadmium stress. J. Plant Growth Regul. 32, 721–731.

Wang, J., Lin, L., Luo, L., Liao, M.A., Lv, X., Wang, Z., Liang, D., Xia, H., Wang, X., Lai, Y., Tang, Y., 2016. The effects of abscisic acid (ABA) addition on cadmium accumulation of two ecotypes of *Solanum photeinocarpum*. Environ. Monit. Assess. 188, 1–8.

Xu, B., Yu, J.Y., Xie, T., Li, Y.L., Liu, M.J., Guo, J.X., Li, H.L., Yu, Y., Zheng, C.Y., Chen, Y.H., Wang, G., 2018a. Brassinosteroids and iron plaque affect arsenic and cadmium uptake by rice seedlings grown in hydroponic solution. Biol. Plant., 1–7. https://doi.org/10.1007/s10535-018-0784-5.

Xu, Z., Ge, Y., Zhang, W., Zhao, Y., Yang, G., 2018b. The walnut JrVHAG1 gene is involved in cadmium stress response through ABA-signal pathway and MYB transcription regulation. BMC Plant Biol. 18, 1–19.

Yan, Z., Chen, J., Li, X., 2013. Methyl jasmonate as modulator of Cd toxicity in *Capsicum frutescens* var. *fasciculatum* seedlings. Ecotoxicol. Environ. Saf. 98, 203–209.

Yan, Z., Li, X., Chen, J., Tam, N.F.Y., 2015a. Combined toxicity of cadmium and copper in *Avicennia marina* seedlings and the regulation of exogenous jasmonic acid. Ecotoxicol. Environ. Saf. 113, 124–132.

Yan, Z., Zhang, W., Chen, J., Li, X., 2015b. Methyl jasmonate alleviates cadmium toxicity in *Solanum nigrum* by regulating metal uptake and antioxidative capacity. Biol. Plant. 59, 373–381.

Yeh, C.M., Chien, P.S., Huang, H.J., 2007. Distinct signaling pathways for induction of MAP kinase activities by cadmium and copper in rice roots. J. Exp. Bot. 58, 659–671.

Yong, J.W., Ge, L., Ng, Y.F., Tan, S.N., 2009. The chemical composition and biological properties of coconut (*Cocos nucifera* L.) water. Molecules 14, 5144–5164.

Yotsova, E.K., Dobrikova, A.G., Stefanov, M.A., Kouzmanova, M., Apostolova, E.L., 2018. Improvement of the rice photosynthetic apparatus defence under cadmium stress modulated by salicylic acid supply to roots. Theor. Exp. Plant Phys. 30, 57–70.

You, J., Chan, Z., 2015. ROS regulation during abiotic stress responses in crop plants. Front. Plant Sci. 6, 1092.

Yu, C., Sun, C., Shen, C., Wang, S., Liu, F., Liu, Y., Chen, Y.L., Li, C., Qian, Q., Aryal, B., Geisler, M., Jiang, D.A., Qi, Y.H., 2015. The auxin transporter, OsAUX1, is involved in primary root and root hair elongation and in Cd stress responses in rice (*Oryza sativa* L.). Plant J. 83, 818–830.

Zažímalová, E., Petrášek, J., Benková, E., 2014. Auxin and Its Role in Plant Development. Springer, Heidelberg.

Zhang, F., Zhang, H., Xia, Y., Wang, G., Xu, L., Shen, Z., 2011. Exogenous application of salicylic acid alleviates cadmium toxicity and reduces hydrogen peroxide accumulation in root apoplasts of *Phaseolus aureus* and *Vicia sativa*. Plant Cell Rep. 30, 1475–1483.

Zhang, Y., Xu, S., Yang, S., Chen, Y., 2015. Salicylic acid alleviates cadmium-induced inhibition of growth and photosynthesis through upregulating antioxidant defense system in two melon cultivars (*Cucumis melo* L.). Protoplasma 252, 911–924.

Zhao, F.Y., Hu, F., Zhang, S.Y., Wang, K., Zhang, C.R., Liu, T., 2013. MAPKs regulate root growth by influencing auxin signaling and cell cycle-related gene expression in cadmium-stressed rice. Environ. Sci. Pollut. Res. Int. 20, 5449–5460.

Zhu, X.F., Jiang, T., Wang, Z.W., Lei, G.J., Shi, Y.Z., Li, G.X., Zheng, S.J., 2012. Gibberellic acid alleviates cadmium toxicity by reducing nitric oxide accumulation and expression of IRT1 in *Arabidopsis thaliana*. J. Hazard. Mater. 239, 302–307.

Zhu, X.F., Wang, Z.W., Dong, F., Lei, G.J., Shi, Y.Z., Li, G.X., Zheng, S.J., 2013. Exogenous auxin alleviates cadmium toxicity in *Arabidopsis thaliana* by stimulating synthesis of hemicelluloses 1 and increasing the cadmium fixation capacity of root cell walls. J. Hazard. Mater. 263, 98–403.

Zwack, P.J., Rashotte, A.M., 2015. Interactions between cytokinin signaling and abiotic stress responses. J. Exp. Bot. 66, 4863–4871.

# CHAPTER 16

# Use of Fungi in Mitigating Cadmium Toxicity in Plants

**Gagan Preet Singh Sidhu*, Aditi Shreeya Bali†, Renu Bhardwaj‡**

*Department of Environment Education, Government College of Commerce and Business Administration, Chandigarh, India

†Department of Botany, M.C.M. DAV College for Women, Chandigarh, India

‡Department of Botanical and Environmental Sciences, Guru Nanak Dev University, Amritsar, India

## Abstract

Environmental contamination due to Cd is a global concern. Development of phytoremediation, a plant-based strategy for the clean-up of Cd-contaminated sites, is therefore a great area of interest. The screening of Cd-tolerant plant species to mitigate Cd toxicity has become a major issue in the scientific world. One promising technique that helps ameliorate Cd accumulation and its translocation to upper-aerial plant parts is the use of fungal symbionts (endophytes, arbuscular mycorrhizal fungi, etc.). This chapter discusses: (i) fungi-induced modifications in the physiology of plants exposed to Cd, (ii) the role of fungal symbionts to ameliorate the Cd phytoextraction and its alleviation from the contaminated sites, (iii) development of biotechnological applications of fungal symbionts for enhancing plant stress tolerance to Cd. The association of fungal symbionts in Cd-stressed plants offers an interesting perspective on tolerating Cd excess and improving Cd exclusion from contaminated environments.

Keywords: Alleviation, Cd toxicity, Fungal symbionts, Phytoremediation, Sustainable, Tolerance, Translocation

## ABBREVIATIONS

**AMF** arbuscular mycorrhizal fungi
**APX** ascorbate peroxidase
**CAT** catalase
**GR** glutathione reductase
**POD** peroxidases
**ROS** reactive oxygen species
**SOD** superoxide dismutase


https://doi.org/10.1016/B978-0-12-814864-8.00016-4

# 1 INTRODUCTION

Heavy metal contamination of soil and water ecosystems is increasing at an alarming rate, and is a major matter of concern globally. Technogenic advancements such as increasing urbanization and developing industrialization particularly, in the second half of the 20th century, have increased the heavy metal concentration in arable lands. These ludicrous practices created a holocaust due to the nonbiodegradable, tenacious, and recalcitrant traits of heavy metals in the soil ecosystems. Among the heavy metals, cadmium (Cd) is a redundant, pernicious, highly mobile metal found in soil. Throughout the world, Cd is released at the rate of 22,000 tons annually into the soil (Anjum et al., 2008). High Cd mobility at the soil-plant interface easily mediates its entry into the food chain (United Nations Environment Programme, 2008). On the basis of its toxicity and potential for human exposure, the Agency for Toxic Substances and Disease Registry (ATSDR) has listed Cd as the seventh most toxic substance (Agency for Toxic Substances and Disease Registry, 2017). Cd has no physiological importance in plants, and is easily extracted by the plant roots due to its high mobility (Sidhu et al., 2017). High Cd levels in plant tissues induce impeded growth, promote chlorosis by restricting chlorophyll biosynthesis, and alter uptake of vital micronutrients (Sidhu et al., 2017). Additionally, Cd toxicity imparts oxidative stress (Kapoor et al., 2014; Sidhu et al., 2017) by altering carbohydrate metabolism; antioxidative response; and the oxidation of lipids, proteins, and nucleic acids in plants (Ahmad et al., 2016).

In recent decades, various conventional strategies (physical, chemical, and biological) have been adopted to remediate Cd-contaminated soils. These physio-chemical treatment strategies include soil acidification, washing, vitrification, excavation, and land filling. These conventional approaches are laborious, costlier, and time consuming. Furthermore, these physio-chemical treatment methods may impart negative effects on soil quality and the rhizospheric microbial community. Therefore, exploitation of vascular plants for remediation of Cd-contaminated soils has acquired impetus. In contrast to the physio-chemical strategies, growing plants in Cd-contaminated soils does not require soil excavation. Moreover, the inception of green cover on metal-contaminated areas looks aesthetically pleasing. Therefore, phytoremediation is a profitable, ecofriendly surrogate to traditional conventional strategies such as excavation followed by chemical processing. Additionally, biomass derived after the remediation of metal-contaminated soils can be utilized as biofuel. In addition, plants that continue to grow in contaminated soils can reconstruct their structure by potentially acting as pioneers for ecosystem refabrication, mounting ecological biodiversity, and effectively contribute to $CO_2$ fixation.

The variable effectiveness of plant species in warding off different metal contaminants poses hurdles to the viability of phytoremediation technology. To overcome such implications, advancements in biotechnological approaches have introduced microorganisms, especially certain fungi-like fungal endophytes, arbuscular mycorrhizal fungi (AMF), and so forth, that aid synergistic symbiotic interactions with plants to facilitate tolerance and remediation of metal-contaminated soils (Miransari, 2011). Fungal endophytic association with the roots of higher plants assists the plants in conferring enhanced growth and biomass with high phytoremediation capacity (Mei and Flinn, 2010). Heavy metal-tolerant endophytic fungi exhibited high efficacy in promoting plant growth and remediation of heavy metal-contaminated soils in an eco-friendly manner, thus regulating metal toxicity in plants (Aishwarya et al., 2014). Different researchers have documented the role of endophytic fungi in enhancing plant tolerance against Cd stress, alleviating its toxicity from contaminated soils (Khan et al., 2017b; Xu et al., 2015). On the parallel, AMF are abundant, widely distributed, and are commonly found in highly contaminated sites (Schneider et al., 2013). Studies have documented the substantial prospectives of AMF to advance the tolerance of plants to Cd, thus ameliorating their ability to remediate Cd-contaminated soils (Garg and Bhandari, 2014). In Cd-contaminated soils, plant species form mycorrhizal symbiosis with AMF (Gao et al., 2010; Luo et al., 2011), thus marking their potential to alleviate Cd from the contaminated soils. In this chapter, we emphasize the role of fungal endophytes and AMF in various physiological and biochemical mechanisms involved in Cd tolerance and their detoxification in plants. Additionally, the contribution of fungal symbionts in enhancing phytoremediation of Cd-contaminated soils is discussed.

## 2 Cd MOBILITY, UPTAKE, AND TRANSLOCATION IN PLANTS

Cd is a nonessential trace element, and is often very noxious for plants (Sidhu et al., 2017). According to the findings of Nagajyoti et al. (2010), $100\,mg\,kg^{-1}$ is the critical limit for Cd in agricultural soils. The uptake of Cd by plant roots depends on the availability of $Cd^{2+}$ in the soil solution. The root epidermal layer is the first site of contact with free $Cd^{2+}$ in the soil,

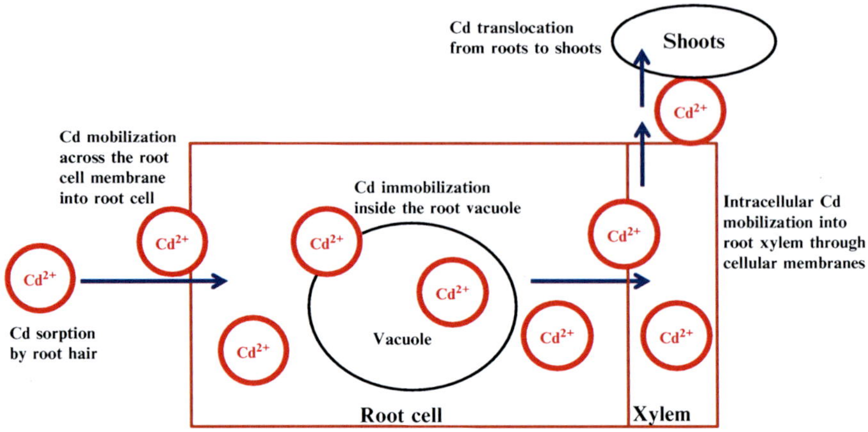


**Fig. 1** Cadmium mobilization, uptake, and translocation in the plants.

and it imparts a pivotal role in plant resistance to Cd stress. The absorption of $Cd^{2+}$ in plant roots is aided by the root tip (Fig. 1). Root hairs, by enhancing the soil surface area, play a crucial role for $Cd^{2+}$ absorption. Root hairs provide a large surface area for Cd absorption, and through diffusion, $Cd^{2+}$ is taken up by the plant roots (Seregin and Ivanov, 1997). Three pathways are involved during Cd entry in plant roots. (i) During plant respiration, carbonic acid ($H_2CO_3$) at the plasma membrane of the root epidermal cells breaks down into $H^+$ and $HCO_3^-$. The $H^+$ is instantly adsorbed and replaced by $Cd^{2+}$. This process needs no energy, and $Cd^{2+}$ enters the root epidermis layer through apoplastic movement. (ii) Through symplastic movement, $Cd^{2+}$ competes with other ions such as $Fe^{2+}$, $Ca^{2+}$, and $Zn^{2+}$ and enters the root epidermal cells through ion channels. Instead of $Ca^{2+}$ and $Zn^{2+}$, $Cd^{2+}$ combines with transporter proteins, thus making its entry in the root epidermis layer through ion channels. (iii) Plant roots secrete organic compounds as chelates in soil that form ligands with free $Cd^{2+}$, thus permitting their entry into the root epidermis layer as metal-ligand complexes.

After passing through all the barriers from the root epidermis to the cortex, and especially via casparian strips of the endodermis, $Cd^{2+}$ is transported to the stele and xylem elements through symplastic movement. The loading of $Cd^{2+}$ into the root xylem is a daunting task, and is mediated by both apoplastic and symplastic pathways. The root cortex plays a crucial role where various processes have been controlled at specific sites, and the cells are involved in Cd loading into the root xylem. According to Hu et al. (2009), before loading into the xylem vessels, Cd undergoes the apoplastic movement after its entry into the stele. The activity of various transporter proteins is involved in $Cd^{2+}$ loading from the xylem parenchyma to the xylem vessels (Mori et al., 2009). Yoneyama et al. (2015) documented that Cd uptake in rice plants is mediated by a xylem-to-phloem transport system that is operational at the stem nodes. They further pointed out that removal of the OsNRAMP5 transporter inhibits Cd uptake and root-to-shoot transport in rice, resulting in a reduction of Cd accumulation in rice grains. This means that after Cd uptake by plant roots, xylem loading, and its acclimatization at the shoot base, Cd enters the developing grain through the phloem, and the condition that facilitates Cd accumulation in leaves and grains is xylem-to-phloem transport (Rodda et al., 2011). Studies have suggested that synthesis of phytochelatins and glutathione in the phloem sap can function as a long-distance transport of Cd in plants (Yadav, 2010). In *Brassica napus*, Mendoza-Cózatl et al. (2008) stated that Cd- phytochelatins and Cd-glutathione complexes are actively transported through the phloem channel. In Cd-hyperaccumulator plants, complexation of Cd with organic ligands provides tolerance and induces detoxification. In upper aerial leaf tissues, Cd gets accumulated and quenched into extracellular and subcellular compartments. Additionally, transpiration in hyperaccumulator plants plays a crucial role in metal transport, which is facilitated by the formation of a metal, low-molecular-weight chelant complex.

## 3 ROLE OF MICROBES IN ENHANCING PHYTOREMEDIATION AND ALLEVIATING Cd FROM CONTAMINATED SOILS

In various biochemical processes such as soil formation, nutrient cycling, energy transfer, and so forth, soil microbes are actively implicated. These processes are vital to rejuvenate and to mount revegetation in and around the metal-contaminated areas. Rhizospheric soil succor large number of microorganisms having the inherent potential to alter the mobility of soil contaminants such as Cd (Aghababaei et al., 2014; Sharma and Archana, 2016). Microorganisms in the soil create an acidic environment by releasing hydronium ions ($H^+$), which in turn increment Cd bioavailability (Lamelas et al., 2006). Nevertheless, high Cd levels (alone or in combination with other metals) in soils often pose pernicious effects to the microbial community structure, and activities in soil (Khan et al., 2010; Pan and Yu, 2011). The rhizospheric microbial population on prolonged plant exposure to high Cd concentrations instigates tolerance, ameliorates biomass, and mitigates

Cd toxicity (Guo and Chi, 2014). However, the success of phytoremediation is dependent upon the tolerance ability of the plant species to Cd toxicity. Thus, enhancing the tolerance of a plant species to Cd can advance the phytoremediation efficiency. The contribution of fungal symbionts to confer plant tolerance to Cd stress, and its alleviation from contaminated soils, is noteworthy. Soil fungi play a crucial role in detoxification and rectification of Cd-contaminated soils. The fungal cell wall is made up of polysaccharides and chitin (Ruiz-Herrera, 2016), and is efficaciously involved in controlling Cd tolerance and its uptake, and serve as a barrier for the entry of $Cd^{2+}$ into the plant cell (Krznaric et al., 2009). Various functional groups, such as the carboxylic group, hydroxylic group, and so forth, along with amino acids in the fungal cell wall, impart a negative charge, promoting the negatively charged cell wall to bind to positively charged metallic ions (Meier et al., 2012). Studies have demonstrated that species such as *Absidia cylindrospora*, *Suillus luteus* (ectomycorrhizal fungi), and the group of *Neotyphodium* endophytes exhibited great potential for Cd-tolerance, and can be exploited effectively in the remediation of Cd-contaminated soils (Albert et al., 2018; Krznaric et al., 2009; Soleimani et al., 2010). Besides, mycorrhizoremediation, in which AMF are employed, can be exploited to enhance phytoremediation of Cd-contaminated soils. AMF are ubiquitously found in the soils, and form mycorrhizal symbiosis with the roots of higher vascular plants. AMF belong to the order glomales, and are austere symbionts, as they complete their life cycle in association with the plant roots. The fungi develop dense mycelium around the root cortical cells, thus serving as an intermediate between the soil and roots of higher plants. AMF provides nutrients to the plants by absorbing them from soil, while metabolic activities of the fungi are carried out with the help of the photosynthates obtained from the plants.

## 4 PARTICIPATION OF FUNGAL SYMBIONTS TO CONFER Cd TOLERANCE AND DETOXIFICATION IN PLANTS

High mobility of Cd in the tissue of plants induces Cd toxicity in them. Association of fungal symbionts (endophytic fungi, AMF, etc.) with the roots of the plants induces high plant tolerance to Cd toxicity. The effects of fungal symbionts on plant potential in Cd-contaminated soils depend upon fungal isolates, quantification of soil Cd concentration, and plant speciation (Garg and Bhandari, 2014). Endophytic fungi are crucial organisms due to their asymptomatic and mutualistic role within plants (Khan et al., 2011).

They form symbiotic associations, and colonize in the inner tissues of healthy plants to exhibit a balanced relationship with host plants without exerting any disease symptoms. Endophytic fungi are ubiquitously present in all higher plants, and can be easily isolated from plants, ranging from herbaceous crop plants to large forest trees (Luo et al., 2011). Endophytic fungi induce morphological and chemical alterations inside the host plants by changing the nutrient composition, thus enhancing plant protection against heavy metal stresses (Singh et al., 2011). Plants growing in Cd-contaminated soils may harbor fungal endophytes with high bioremediation potential, ensuring the survival of endophytes in contaminated soils (Xiao et al., 2010). The isolation of fungal endophytes from metal-resistant hyperaccumulators is a potential application for the alleviation of Cd from contaminated soils. Deng et al. (2011) isolated a Cd- and Pb-tolerant fungal strain *Mucor* sp. CBRF59 from *Brassica chinensis* growing in metal-contaminated soils. They revealed that the maximum biosorption capacity of Cd by the dead and active biomass of *Mucor* sp. CBRF59 was 108 and 173 $mg\,g^{-1}$. Furthermore, the addition of active mycelia of CBRF59 dramatically advances the soil's Cd availability by 11.5-fold, exhibiting strong potential for endophytic fungus to detoxify Cd-contaminated soils. The probable mechanism involves the contribution of certain enzymes, or compounds such as metal binding peptides synthesized by fungus to detoxify Cd inside the cells, hence maximizing the percentage of Cd bioaccumulation (Malik, 2004). Other mechanisms that contribute to high Cd bioaccumulation by a fungal strain are the increased surface area and exposure of intracellular binding sites, and the biosynthesis of siderophores in the active strain that increase Cd bioavailability (Gadd, 2004). Likewise, *Paxillus involutus*, an ectomycorrhizal fungus, enhances the bioavailability, as well as phytoextraction of Zn and Cd by willows (Baum et al., 2006).

The potential of endophytes to synthesize plant hormones such as indole-3-acetic acid (IAA), gibberellins, and so forth as secondary metabolites enhances plant growth by alleviating the heavy metal uptake by plants (Khan and Lee, 2013). A study conducted by Khan et al. (2014) revealed the active role of gibberellins producing endophyte *Pencillium janthinellum* LK5 to enhance the responses in ABA-deficient mutants and wild-type mutants of *Solanum lycopersicum* to Cd toxicity by exhibiting higher plant growth and nutrient content in the form of sulfur and calcium. They found that the concentration of salicylic acid was approximately 1.7-fold higher in colonized plants than the control plants. Besides, the concentration of solubilized phosphates was more in the inoculated plants than the control group plants. The findings of these authors suggest the active role of endophytic fungi in mitigating Cd toxicity. On the parallel, Shahabivand et al. (2012) documented that the colonization of endophyte *Piriformospora indica* with the roots of *Triticum aestivum* under Cd stress yielded higher plant growth, chlorophyll content, phosphorus concentration, and reduced Cd translocation in shoots of wheat plants. These findings illustrated the contribution of endophytes in conferring plant tolerance to withstand Cd stress.

Additionally, the contributions of endophytic fungi in enhancing Cd phytoremediation are noteworthy. A Cd-tolerant endophytic fungal strain, RSF-6L, applied in vivo to *Solanum nigrum* under different Cd concentrations significantly decreased the activities of POD and PPO, while CAT activity was dramatically enhanced in fungal inoculated plants grown under 30 mg $kg^{-1}$ Cd treatment (Khan et al., 2017a). They further documented that the inoculation of the appropriate fungal strain in *S. nigrum* can advance its potential usage in phytostabilization of Cd-contaminated soils. The colonization of metal-tolerant dark septate endophytes (DSEs) with the roots of *Alnus nepalensis* revealed its potential application in the phytoremediation of Pb, Zn, and Cd mining tailings (Xu et al., 2015). Furthermore, they opined that more than 50% of the tested isolates of the fungi were tolerant to high Cd concentrations ranging up to 100 mg $L^{-1}$, and even more than 600 mg $L^{-1}$. This tolerance of harsh environmental conditions is correlated with the presence of chitin and melanin in DSE fungal walls that can enhance the cell wall rigidity by decrementing the membrane permeability (Zhan et al., 2011). Likewise, *S. nigrum* plants inoculated with endophytic isolates of *Glomerella truncata* PDL-1 and *Phomopsis fukushii* PDL-10 under 5, 10, and 25 mg $kg^{-1}$ Cd concentrations exhibited high tolerance to increased Cd concentrations (Khan et al., 2017a). It is noteworthy that Cd concentration in the shoots of inoculated plants was 40%–100% higher than noninoculated plants. Additionally, Cd concentration in roots and shoots of PDL-10 inoculated plants was higher than PDL-1 inoculated plants, suggesting the active participation of fungal strains in ameliorating phytostabilization and the phytoextraction potential of *S. nigrum* plants to Cd-contaminated soils.

## 5 ROLE OF AMF IN Cd TOLERANCE AND ITS DETOXIFICATION IN PLANTS

In association with AMF, a number of Cd tolerance and its detoxification mechanisms are triggered simultaneously in plants. These mechanisms include (i) Cd retention in fungal hyphae and mycorrhizal roots

(Carvalho et al., 2006); (ii) Cd sequestration and acquisition in plant and fungal vacuoles (Li et al., 2016); (iii) Cd exclusion from uptake by plants through precipitation, binding, or chelation by metallothioneins, glutathione, and so forth present in the cytosol of plants and fungi (Khullar and Reddy, 2016); (iv) regulation of antioxidative responses in plants and fungi; and (v) glomalin production. A description of the mechanisms related to the contribution of AMF is described as follows, and is presented diagrammatically in Fig. 2.

## 5.1 Cd Retention in Fungal Hyphae and Mycorrhizal Roots

The colonization of AMF with the roots of higher plants enables the plants to retain Cd in the cell walls of hyphae and roots of plants. The immobilization of Cd in aerial plant tissues, and its retention in fungal mycelium, is the possible tolerance mechanism adapted to plants conferred by AMF to overcome Cd phytotoxicity. The specific retention mechanisms imparted by AMF may cause a reduction in Cd translocation in aerial plant tissues by retaining the metal in roots, therefore contributing to Cd-phytostabilization. It has been previously reported that AMF can improve the soil structure by incrementing the number, quantity, and activity of soil enzymes, especially acid phosphatase activity (Hu et al., 2013; Liu et al., 2015), and enhance the nutritional status and growth of the host plants through increasing the surface area by expansion of their hyphae, thus allowing the host plants to extract and accumulate metals in their roots (Rajkumar et al., 2012). Cd tolerance potential in mycorrhizal plants is correlated with increased fungal growth, and greater yield of mycelial biomass as the toxic metals are acclimatized in the fungal mycelium. The extraradicular mycelium of AMF has shown to be effective in Cd uptake as demonstrated by Janousková et al. (2006), who found 10–20 times the enhancement of Cd accumulation per unit root biomass of *Nicotiana tabacum* plants than that of nonmycorrhizal plants. The influence of AMF on Cd retention in *Aster tripolium* plants was demonstrated by Carvalho et al. (2006), suggesting that AMF regulates a dual behavior in Cd tolerance and accumulation in plant tissues. At high Cd concentrations, AMF impeded Cd translocation, retained maximum Cd concentration in roots, and maintained a high root to shoot Cd concentration in mycorrhizal plants. These authors further suggested that AMF colonization with the roots of *A. tripolium* enable the fungi to retain maximum Cd in their mycelium, thus promoting high plant tolerance to Cd. In the same way, Hassan et al. (2013) depicted that

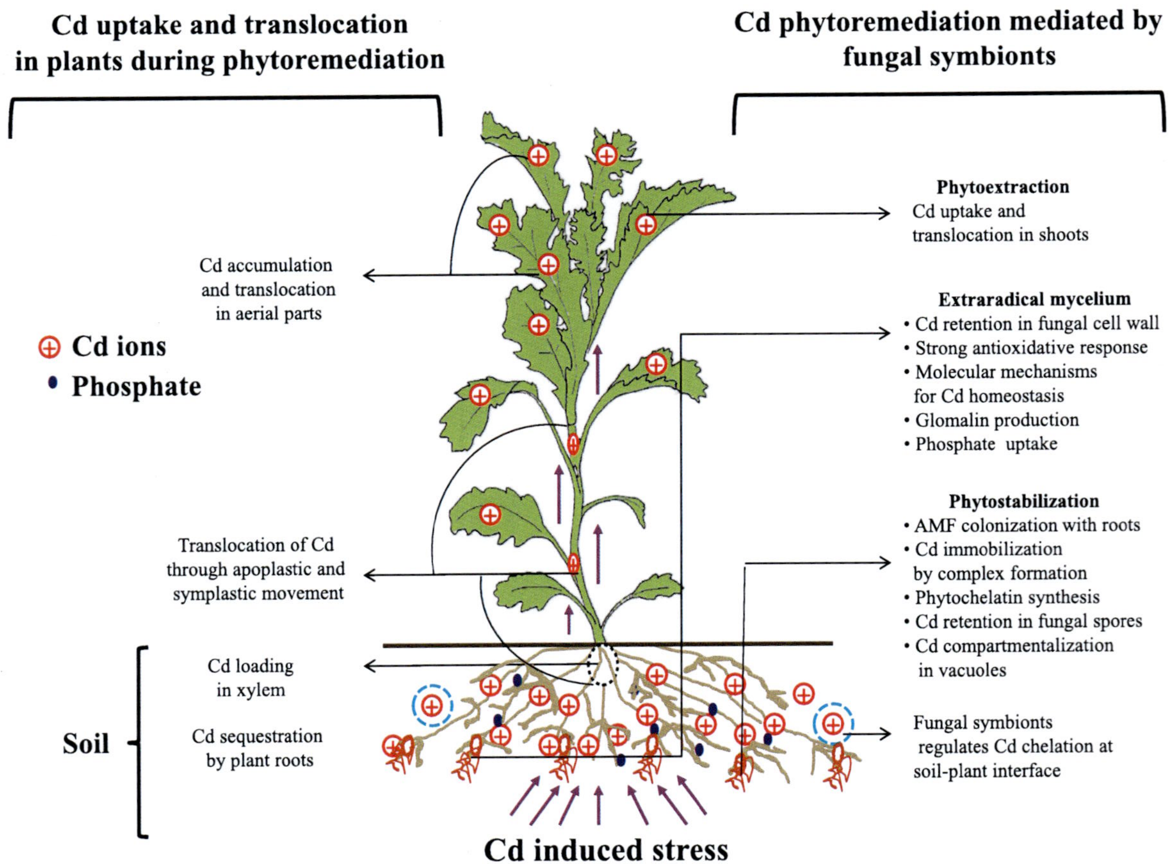


**Fig. 2** Symbiotic fungi confer Cd tolerance and its phytoremediation in plants.

*Funneliformis mosseae* can significantly reduce Cd and Zn translocation in shoots of sunflower plants grown on Cd-contaminated soils. On the parallel, Li et al. (2009) opined that inoculation of *F. mosseae* with *Astragalus sinicus* plants induced lower Cd translocation to plant shoots than nonmycorrhizal plants. On the contrary, studies have demonstrated the effectiveness of AMF in ameliorating Cd-phytoextraction and its translocation to the shoots of the plants. In AMF-assisted remediation studies, plants exhibited enormous growth and high biomass yield, a vital trait for phytoremediation-based studies. In other AMF-related soil remediation studies, Liu et al. (2015) demonstrated that inoculation of *Glomus versiforme* with 25 and 50 $mg\,kg^{-1}$ soil Cd treatments enhanced Cd phytoavailability in *S. nigrum* plants. However, Cd phytoavailability was lowered at 100 $mg\,kg^{-1}$ soil Cd treatment, probably due to the biological dilution effect (Hu et al., 2013) in which *G. versiforme* colonization significantly increased the plant biomass, but decreased Cd phytoavailability. Similar findings were reported by Chen et al. (2004) in maize plants that exhibited peaked Cd levels in maize shoots, when *Glomus mosseae* was amended with 25 $mg\,kg^{-1}$ soil Cd, while it decreased in soils with 100 $mg\,kg^{-1}$ Cd, respectively.

## 5.2 AMF Mediates Cd Compartmentalization in Vacuoles

The sequestration and acclimatization of Cd in plant and fungal vacuoles is an important mechanism conferred to plants by AMF against Cd toxicity (Li et al., 2016). The acquisition of $Cd^{2+}$ in the fungal storage structures circumvent its translocation to aerial plant parts, thereby providing plants an additional protection mechanism to mitigate Cd toxicity. Energy-dispersive X-ray spectroscopic (EDXS) studies conducted by Gonzalez-Guerrero et al. (2008) revealed an accumulation of Zn, Cu, and Cd in vacuoles (spores and hyphae) of extraradical mycelium of *Glomus intraradices* grown in association with carrot root organ cultures when exposed to these metals. The authors further documented that the fungi act as a first line of defense against the increasing metal concentrations, as most of the metals were accumulated in walls and vacuoles, while very low cytoplasmic metal concentrations were maintained. Moreover, Wang et al. (2012) found that inoculation of *G. intraradices* in Cd-contaminated soils significantly enhanced the growth of *Medicago sativa* plants by restricting high Cd translocation to the aerial parts. Additionally, the plants exhibited very little Cd acclimatization in subcellular fragments of shoots. However, Cd accumulated more in the plant cell wall compared with the membranes and organelles, suggesting the efficacy of AMF in providing resistance and tolerance of plants to Cd stress (Wang et al., 2012). Yao et al. (2014) documented that symbiotic association of *Rhizophagus irregularis* with clove plants involved the colocalization of Cd with polyP that assisted the vacuolar accumulation of Cd in the extraradical mycelium of *R. irregularis*. Concomitantly, the capacity to accumulate Cd in the vacuole of extraradical mycelium of *Gigaspora margarita* in association with *Lotus japonicas* was revealed by Nayuki et al. (2014), suggesting the colocalization of Cd with polyP that served as a counter ion to stabilize Cd in the fungal vacuoles. Interestingly, Li et al. (1996) found the relationship between vacuolar Cd compartmentalization and ABC transporter proteins encoded by a homologue of a yeast Cd factor gene that transports bis-GSH-Cd complexes across the tonoplast to confer Cd resistance. Likewise, González-Guerrero et al. (2010) verified that *GintABC1* encodes a putative ABC transporter instigated by Cd and Cu stress in *G. intraradices*, which has also been involved in compartmentalization of these metals in the vacuoles.

## 5.3 Molecular Mechanisms of AMF-Assisted Cd Chelation and Its Homeostasis

Several homeostatic mechanisms operate in mycorrhizal roots to regulate Cd tolerance and its homeostasis. AMF has developed a variety of intracellular mechanisms to chelate Cd. Although the cell wall acts as a first-line barrier against metal extraction, in multi-metal-contaminated soils, metals can cross this barrier and enter the cytoplasm of fungus (Meier et al., 2012). Chelation of heavy metals such as Cd with high-affinity ligands in the cytosol is a vital detoxification strategy adopted by AMF and plants. Various chelating agents such as phytochelatins, metallothioneins, organic acids, and so forth present in AMF, and plants are actively involved in chelating metals such as Cd in cytosol, and mediate its tolerance and detoxification (Anjum et al., 2015; Cobbett and Goldsbrough, 2002). In the mycorrhiza, polyphosphate granules, in association with high nitrogen and sulfur content, are believed to be involved in metal-thiol binding (Galli et al., 1994). According to Hall (2002), the metal-chelator complex travels across the cytoplasm, then the tonoplast, and is later quenched in the vacuoles. Glutathione derivatives such as phytochelatins are concerned with Cd detoxification in AMF and plants. Enhanced synthesis of phytochelatins in AMF and plants enable intracellular Cd tolerance and detoxification. Garg and Aggarwal (2011) analyzed the correlation of higher Cd accumulation with greater phytochelatin synthesis in nodules of *Cajanus cajan* formed in conjunction with *G. mosseae*. The authors further revealed the participation of *G. mosseae* in enhancing the number of nodules, plant dry weights, and nitrogenase activity. Furthermore, the results of their study clearly illustrated the greater phytochelatin synthesis in root nodules with increased Cd concentrations in them. In a similar study, Garg and Chandel (2012) determined the combined effect of Cd and NaCl concentrations on *C. cajan* in the presence of *G. mosseae* and found that mycorrhizal plants synthesize more phytochelatins, nonprotein thiols, and glutathione in the roots, which can be attributed to high Cd levels in the roots of the plants. It was further suggested that rapid phytochelatins' biosynthesis in roots contributed to the plant tolerance response to Cd contamination by restricting its translocation to the shoots. Concomitantly, the P792 genotype of *C. cajan*, in the presence of combined interactions of Zn, Cd, and *G. mosseae*, exhibited greater biosynthesis of phytochelatins, especially in the roots, which can be correlated with increased Cd accumulation in roots. Besides, studies have revealed the evidence related to the mycorrhizal association of plants involved in Cd phytoextraction (Wang et al., 2005, 2007). At the lowest and highest soil Cd concentrations (0.75 and 40 $mg\,kg^{-1}$), inoculation of *R. irregularis* induced a drastic increase in shoot Cd concentrations in sunflower plants (Hassan et al., 2013). Likewise, a field study conducted by Wang et al. (2005) revealed that combined inoculation of AMF with chitosan enhanced the phytoextraction efficiency of *Elsholtzia splendens* grown under Cd-contaminated soil.

Metallothioneins are another group of proteins involved in Cd accumulation and sequestration in plants. These proteins are low molecular weight, cysteine rich, long polypeptide chains that contain large numbers of amino acids. They are biosynthesized in AMF and mycorrhizal plants by certain hormones, and metalloid stress such as Cu, Cd, and so forth (Haq et al., 2003) play a crucial role in both Cd homeostasis (Janoušková et al., 2005a), as well as conferring effective protection against biochemical stress induced by metalloids (Maret, 2000, 2003). Some studies correlated the involvement and contribution of AMF in Cd-phytostabilization through metallothioneins synthesis. Hasegawa et al. (1997) documented an enhanced Cd tolerance in cauliflower on the insertion of *CUP1* genes from yeast that in turn increased its accumulation by 16-fold in plant tissues. In another study conducted by Janoušková et al. (2005b), transgenic tobacco plants with *CUP1* genes, in combination with AMF inoculation exposed to varied Cd levels, exhibited a very low Cd concentration in leaves and shoots of the plants of the same variety, compared with that of the nontransgenic and uninoculated tobacco plants. However, not much has been mentioned in the literature regarding the role of metallothioneins in the identification of AMF genes in maintaining cellular homeostasis against Cd. Studies conducted by González-Guerrero et al. (2006, 2007) have recognized metallothioneins such as GmarMT1 in *G. margarita* that regulate fungal redox potential and protect it from the oxidative stress induced by Zn and Cd.

## 5.4 AMF-Conferred Induction of Strong Antioxidative Responses in Plants

Cd is a redox-inactive metal, and it induces oxidative damage to plant cells by generating reactive oxygen species (ROS) (Sidhu et al., 2017). The positive effects of AMF on the survival, growth, and physiology of plants against Cd toxicity have been widely recognized. Bhaduri and Fulekar (2012) documented that inoculation of AMF with increased Cd levels (5–100 ppm) in the soil triggered the activities of antioxidant enzymes in shoots of *Ipomoea aquatica* plants. Strong antioxidant activity in AMF-inoculated *I. aquatica* is a tolerance strategy acquired by plants against Cd toxicity. Likewise, in Cd-stressed *Bassia indica* seedlings, Hashem et al. (2016b) found that calcium application, in combination with AMF, enhanced the activities of SOD, CAT, APX, GR, POX, and PPO by 133.5%, 201.4%, 536.8%, 269.3%, 147.5%, and 93.4%, respectively. They further reported that the combination of both calcium and AMF has a stimulatory effect on the activities of antioxidant enzymes that triggers the effective tolerance mechanisms in *B. indica* against Cd toxicity. On the parallel, Hashem et al. (2016a) demonstrated an increment in SOD, CAT, APX, POD, and GR activities by 82%, 16.7%, 65.3%, 94.9%, and 76.2%, respectively, in *S. lycopersicum* plants when raised in colonization with AMF under Cd excess. Recently, a study conducted by Chen et al. (2016) revealed that colonization of *G. intraradices* with 10 $mg\,kg^{-1}$ Cd-stressed male and female plants of *Populus deltoides* up-regulated the antioxidative response, especially in the female plants. Cd stress increases the activities of SOD, APX, and POD in female *P. deltoides* plants when preinoculated with *G. intraradices*. The findings of the study illustrated that Cd tolerance in the presence of AMF is strengthened in female *P. deltoides* plants. According to Liu et al. (2011), the inoculation of *Glomus constrictum* and *G. mosseae* in Cd-contaminated soils increased the SOD and CAT activities in sunflower plants. Plant colonization with

AMF up regulates the antioxidative response, thus influencing the plant biomass by quenching ROS, and hence mitigating Cd stress in *Tagetes erecta* by restricting Cd translocation in aerial tissues.

## 5.5 Induction of Glomalins by AMF

Glomalins are the glycoproteins generated in the hyphae of AMF (Wright and Upadhyaya, 1998). The proteins have variable functions in the soil. Glomalins play a key role as active agents in providing tolerance to Cd-stressed plants, and assist the plants in reducing phytotoxicity induced by Cd excess (Gonzalez-Chavez et al., 2009). The concentration and quantity of glomalin production depend upon the level of metal contamination in soils (Cornejo et al., 2008). However, the type of soil, its physiochemical properties such as pH, and redox potential determine the metal-binding potential of glomalin proteins in AMF (Chern et al., 2007; Wright and Upadhyaya, 1998). According to Purin and Rillig (2007), these proteins induce a protective role by preventing the decomposition of AMF by soil microbes. They pointed out that the localization of glomalins primarily happens more frequently on the inner layers of the cell wall, rather than the outer cell wall layer. These authors further appraised the functions of glomalins in AMF by forming protein-metal complexes. The findings suggest that the glomalin proteins' assistance in wall binding has additional functions other than the cytoplasmic heat shock proteins, which further influence the AMF and plants to withstand metal stress. Gonzalez-Chavez et al. (2009) documented that glomalins have certain functional groups in their structure that assist AMF in sequestering Cd. Studies conducted by Chern et al. (2007) and Gonzalez-Chavez et al. (2009) revealed the contribution of glomalins in decrementing the phytotoxicity induced by Cd, Cu, and Zn. In the soils contaminated with heavy metals, Gonzalez-Chavez et al. (2004) found that the glomalin proteins can quench up to 80 mg of Cd, 1120 mg of Pb, and 4300 $mg\,kg^{-1}$ of Cu. Investigations on a Cd-containing slag heap showed glomalin abundance in the rhizosphere of 22 AMF colonized plants Gonzalez-Chavez et al. (2009). The study suggested that the mycelium mass ranged from 0.11 to 26.3 $mg\,kg^{-1}$, and the glomalin ranged between 13 and 75 $mg\,g^{-1}$. They further suggested that Cd quenched by glomalins extracted from mycelium was 0.028 $mg\,g^{-1}$, thus contributing to maintenance of the soil structure of previously contaminated soils. Hence, the findings proclaimed the active involvement of glomalin proteins in Cd sequestration, and its alleviation from contaminated soils.

## 6 AMF MEDIATE PHYTOREMEDIATION OF Cd-CONTAMINATED SOILS

The advantage of AMF in ameliorating the plant tolerance and phytoremediation of Cd-contaminated soils is well elucidated, and is diagrammatically presented in Fig. 2. In Cd-contaminated soils, the contribution of AMF in the phytoremediation process is correlated with phytostabilization and phytoextraction (Meier et al., 2012). Mycorrhizal colonization of AMF in the roots of plants enables the retention and stabilization of Cd in the rhizospheric regions of plant roots. Studies have revealed that AMF influences the mechanisms for retention and accumulation of Cd in plant roots, and mitigates Cd toxicity by restricting its translocation to the aerial parts; hence, AMF mediates phytostabilization of Cd-contaminated soils (Arora et al., 2016; Hassan et al., 2013; Sun et al., 2018). On the contrary, hyperaccumulator plants, in association with mycorrhizal colonization, can extract and translocate Cd to the above-ground plant tissues, thus promoting Cd-phytoextraction in plants. Cd-phytoextraction mechanisms have been reported for *Sedum alfredii* (Hu et al., 2013); and sunflowers (Hassan et al., 2013). However, the success of phytostabilization and phytoextraction in AMF colonized plants depends on the growth, biomass yield, and ability of Cd-tolerant plants and AMF species to withstand high toxic Cd levels in the soils. In addition to this, Cd bioavailability is a vital factor that determines the effectiveness of phytostabilization and phytoextraction processes (Ma et al., 2011).

AMF fungal hyphae impart a significant role in Cd sequestration and confer plant tolerance to Cd stress Table 1. Many studies have illustrated the contribution of extraradical mycelium of AMF in ameliorating the remediation of Cd-contaminated soils (Table 1). Colonization of plant roots with AMF transfers the additional Cd resistance to roots. Nayuki et al. (2014) reported that the outer surface extraradical mycelium of *G. margarita* influence Cd sorption in the cell with the highest signaling in the cell wall, selective accumulation in vacuoles, and less Cd signaling in the cytoplasm. They further found that colocalization of charged ions polyP with Cd in the vacuoles of AMF provide an efficient mechanism in detoxification and translocation of Cd in the fungal hyphae. de Andrade et al. (2008) opined that extraradical mycelium of *G. intraradices* absorbed Cd as high as 728 $\mu g g^{-1}$. Likewise, Janoušková and Pavlíková (2010) revealed that Cd-tolerance and its stabilization mechanism are also influenced outside the fungal hyphae as the extraradical mycelium instigated alkalinization of

**Table 1** Colonization of AMF with host plant increment the tolerance and phytoremediation potential of plants in Cd-contaminated soils

| AMF | Colonized plant | AMF role in enhancing phytoremediation and mitigating Cd toxicity | References |
|---|---|---|---|
| *Glomus intraradices* | *Thlaspi praecox* | Mycorrhizal plants translocate Cd in shoots up to the level of $5960\,mg\,kg^{-1}$ | Vogel-Mikuš et al. (2005) |
| *Glomus mosseae* | *Tagetes erecta* | Lemmatize Cd concentration in plant tissues to overcome Cd stress | Liu et al. (2011) |
| *Glomus versiforme* | *Solanum nigrum* | Fungal colonization enhanced soil acid phosphatase activity, ameliorated the plant growth at 25, $50\,mg\,Cd\,kg^{-1}$ soil, and Cd concentration in plant tissues at 25, 50, and $100\,mg\,Cd\,kg^{-1}$ soil | Liu et al. (2015) |
| *Rhizophagus irregularis* | Sunflower | Colonized plants promote Cd-phytoextraction by enhancing Cd translocation in shoots | Hassan et al. (2013) |
| *Funneliformis mosseae* | Sunflower | Mycorrhizal roots promote Cd-phytostabilization | Hassan et al. (2013) |
| *Glomus intraradices* | *Salix viminalis* | AMF inoculation reduced Cd and Cu translocation to shoots and retain maximum metal concentration in roots exhibiting metal exclusion mechanisms | Bissonnette et al. (2010) |

*Continued*

**Table 1** Colonization of AMF with host plant increment the tolerance and phytoremediation potential of plants in Cd-contaminated soils—cont'd

| AMF | Colonized plant | AMF role in enhancing phytoremediation and mitigating Cd toxicity | References |
|---|---|---|---|
| *Glomus etunicatum* | *Zea mays* | Inoculated plant roots contain more Cd concentration. Coexistence of atrazine and AMF in contaminated soil increased Cd level in roots by 25.6% and 51.3% at 0.5 and 5.0 mg $kg^{-1}$ atrazine concentrations, respectively | Huang et al. (2006) |
| *Glomus geosporum* | *Aster tripolium* | Inoculated roots accumulate maximum Cd concentration. Root and shoot Cd concentration in plants reached up to 534.4 and 710.2 mg $kg^{-1}$ | Carvalho et al. (2006) |
| *Rhizophagus intraradices* | *Lonicera japonica* | Colonization increment phosphorus acquisition and growth of the plants at all Cd levels. Though Cd concentration in shoots decreased but a mount in Cd levels by 62% and 28% was noticed hence favoring phytostabilization of Cd at 10 and 20 μg Cd $g^{-1}$ soil, respectively | Jiang et al. (2016) |

**Table 1** Colonization of AMF with host plant increment the tolerance and phytoremediation potential of plants in Cd-contaminated soils—cont'd

| AMF | Colonized plant | AMF role in enhancing phytoremediation and mitigating Cd toxicity | References |
|---|---|---|---|
| *Glomus constrictum* | *Zea mays* | Mycorrhizal roots lemmatize Cd translocation to the shoots, promoted Cd trapping and conferred plant to exhibit its potential in phytostabilization applications | Liu et al. (2014) |
| *Glomus intraradices* | *Linum usitatissimum* | Mycorrhizal association enhanced proline content, Cd and Cr uptake incremented by 23% and 33% in plant roots | Ali et al. (2015) |
| *Glomus intraradices* | *Helianthus annuus* | Mycorrhizal plants exhibited 24% higher phosphorus content with 23% and 34% higher Cd levels in roots and shoots when compared with noninoculated plants | de Andrade et al. (2008) |
| *Glomus etunicatum* | *Canavalia ensiformis* | Mycorrhizal plants exerted high foliar phosphorus concentration and accumulated more Cd in roots and shoots compared with the noninoculated plants | de Andrade et al. (2005) |

*Continued*

**Table 1** Colonization of AMF with host plant increment the tolerance and phytoremediation potential of plants in Cd-contaminated soils—cont'd

| AMF | Colonized plant | AMF role in enhancing phytoremediation and mitigating Cd toxicity | References |
|---|---|---|---|
| *Glomus* species (AMF) inocula (IN2) as fungal isolates obtained from mine-contaminated soils | *Canna indica* | AMF IN2 inoculated plants growing in mine contaminated soils accumulated twofold more Cu, Zn, and Cd concentrations compared with noninoculated plants | El Faiz et al. (2015) |
| *Glomus lamellosum* | *Medicago sativa* | Inoculated plants accumulated more Cd and Zn in roots, impeded their translocation to the shoots, hence favoring efficient plant resistance against metal contaminated soils | Huang et al. (2012) |
| *Funneliformis mosseae* | *Oryza sativa* | Colonization reduced Cd accumulation both in roots and shoots of rice plants and convert active toxic Cd forms into inactive Cd forms | Li et al. (2016) |
| *Glomus mosseae* | *Sesbania rostrata*, *Sesbania cannabina*, *Medicago sativa* | In AMF inoculated plants, concentration of Cu and Cd in the shoots of *Sesbaniarostrata* were relatively higher than the other two plants. However, concentration of Cu, Zn, Pb, and Cd in the roots increased progressively in all | Lin et al. (2007) |

**Table 1** Colonization of AMF with host plant increment the tolerance and phytoremediation potential of plants in Cd-contaminated soils—cont'd

| AMF | Colonized plant | AMF role in enhancing phytoremediation and mitigating Cd toxicity | References |
|---|---|---|---|
| | | the three plants, thus AMF conferred strong plant tolerance against metals | |
| *Glomus intraradices* | *Nicotiana tobacum* | Symbiotic association increased the plant growth and tolerance to Cd as Cd uptake per unit biomass of the plant decreased | Janoušková et al. (2005b) |
| *Acaulospora* sp. | *Jatropha curcas* | Coexistence of biochar and fungal inoculation increased the BCF values of roots and shoots by more than 1 under Cd stress | González-Chávez et al. (2017) |
| *Glomus intraradices* | *Linum usitatissimum* | Mycorrhizal fungal association assist to remediate Cd and Cr contaminated soils upto 500 ppm. High proline accumulation in inoculated plants impart tolerance mechanisms in plants toward heavy metal stress | Ali et al. (2015) |
| *Glomus caledonium* | *Sedum alfredii* | Mycorrhizal colonization peaked the phytoextraction efficiency of plants by 78%, thus marked *Sedum alfredii* for both | Hu et al. (2013) |

*Continued*

**Table 1** Colonization of AMF with host plant increment the tolerance and phytoremediation potential of plants in Cd-contaminated soils—cont'd

| AMF | Colonized plant | AMF role in enhancing phytoremediation and mitigating Cd toxicity | References |
|---|---|---|---|
| | | phytostabilization and phytoextraction of Cd-contaminated soils | |
| Indigenous AMF inoculum of the undisturbed wetland in dryzone environment | *Eichhornia crassipes* | Colonization with plant roots ameliorated growth, Cd accumulation both in roots and shoots and tolerance potential in plants to cope Cd stress, hence inoculation promote phytoremediation of Cd-contaminated soil and water ecosystems | Gunathilakae et al. (2018) |
| *Glomus intraradices* | *Medicago sativa* | Fungal inoculation increased the plant growth and Cd concentration more in the roots than shoots | Wang et al. (2012) |
| Indigenous mycorrhizal inoculum from contaminated soils | *Solanum melongena* | AMF colonization enhanced uptake in plant roots by 1.11 times for Pb, 1.09 for Cd and 1.36 for As but the translocation factor was more than 2 only for Cd and As | Chaturvedi et al. (2018) |
| *Glomus mosseae* | *Cajanus cajan* | Fungal inoculation accelerated the biosynthesis of phytochelatins and glutathione as biomarkers in roots, enabling the plants to counteract Cd toxicity | Garg and Aggarwal (2011) |

the rhizospheric substrates, hence promoting Cd immobilization. Nevertheless, AMF-mediated Cd accumulation and its stabilization mechanisms mostly occur inside the fungal hyphae. In a study conducted by Yao et al. (2014), polyP formed inside the AMF hyphae forms chelation with Cd and phytostabilize it around the root zone of clove plants, thus influencing Cd phytostabilization by lemmatizing its translocation to upper-aerial plant parts. Besides, AMF mediates the phytoavailability and bioaccumulation of Cd in plants tissues by regulating the soil pH, enhancing the phosphorus availability, and ameliorating its fertilization rate. Hu et al. (2013) reported that AMF *Glomus caledonium* colonization with roots of *S. alfredii* increase the phosphorus availability in soil, decrease the soil pH, and hence enhance the phytoextraction potential of the plants to Cd toxicity. Likewise, *G. caledonium* and *G. mosseae* colonization ameliorate Cd accumulation, its absorption, and extraction in *Medicago truncatula* by increasing the phosphorus availability in soil when the plants were exposed to 108–148 $mg\,kg^{-1}$ Cd concentrations (Redon et al., 2008). This clearly illustrates that Cd accumulation and its extraction in plant tissues depends upon soil composition, nutrient availability, metal speciation, type of plant, and fungal species (Wang et al., 2006). Additionally, in a pot experiment, Gunathilakae et al. (2018) revealed that AMF inoculation, alongside water hyacinths in a Cd-contaminated medium, increased the Cd concentration in the roots relative to the shoots of *Eichhornia crassipes*. The prevention of Cd translocation, its immobilization, and its compartmentalization by AMF hyphae in the roots by binding and adsorption is the protective mechanism adopted by *E. crassipes* in phytostabilization of Cd-contaminated soils. Similar findings were documented in the studies of Liu et al. (2015) in *S. nigrum*; Bissonnette et al. (2010) in *Salix viminalis* and *Populus generosa*; and Garg and Chandel (2012) in *C. cajan* plants.

## 7 CONCLUSIONS AND FUTURE PROSPECTIVE

High Cd mobility induces toxicity in plants. Colonization of fungal symbionts with the roots of plants ensures high plant tolerance and their survival in Cd-contaminated soils. Within the root cortical cells of vascular plants, there exist specialized branched structures known as vesicles, or arbuscules, that act as a bilateral growth factor for both fungal symbionts and plants. Mycorrhizal fungi, especially AMF, have developed specific intrinsic mechanisms that may contribute effectively to phytoremediation of Cd-contaminated soils.

In the literature, the mechanisms related to AMF in ameliorating plant nutrition are well established. However, the participation of endophytic fungi and AMF in Cd homeostasis and its tolerance mechanisms against Cd toxicity needs to be pondered. Various mechanisms related to Cd uptake and its long distance transport in fungal symbionts and plants would be of significant importance to appraise the profitability of these organisms in phytoremediation-based studies.

In the future, focus should be placed on the isolation of indigenous and presumably adapted fungal symbionts from Cd-contaminated soils; their number, and functional compatibility with Cd-tolerant plants can work as a crucial biotechnological approach for reclamation of Cd-contaminated soils. An understanding of molecular mechanisms related to Cd tolerance in fungal symbionts will enlighten researchers about the future design of phytoremediation strategies for restoring degraded soils and promoting sustainable agricultural systems.

## REFERENCES


Agency for Toxic Substances and Disease Registry (ATSDR), 2017. CERCLA Priority List of Hazardous Substances. http://www.atsdr.cdc.gov/spl/index.html. [(Accessed 2 April 2018)].

Aghababaei, F., Raiesi, F., Hosseinpur, A., 2014. The combined effects of earthworms and arbuscular mycorrhizal fungi on microbial biomass and enzyme activities in a calcareous soil spiked with cadmium. Appl. Soil Ecol. 75, 33–42.

Ahmad, P., Abd Allah, E.F., Hashem, A., Sarwat, M., Gucel, S., 2016. Exogenous application of selenium mitigates cadmium toxicity in *Brassica juncea* L. (Czern& Cross) by up-regulating antioxidative system and secondary metabolites. J. Plant Growth Regul. 35, 936–950.

Aishwarya, S., Venkateswarlu, N., Chandramouli, K., Vijaya, T., 2014. Role of endophytic fungi in restoration of heavy metal contaminated soils. Indo Am. J. Pharm. Res. 4, 5427–5436.

Albert, Q., Leleyter, L., Lemoine, M., Heutte, N., Rioult, J.P., Sage, L., Baraud, F., Garon, D., 2018. Comparison of tolerance and biosorption of three trace metals (Cd, Cu, Pb) by the soil fungus *Absidia cylindrospora*. Chemosphere 196, 386–392.

Ali, N., Masood, S., Mukhtar, T., Kamran, M.A., Rafique, M., Munis, M.F.H., Chaudhary, H.J., 2015. Differential effects of cadmium and chromium on growth, photosynthetic activity, and metal uptake of *Linum usitatissimum* in association with *Glomus intraradices*. Environ. Monit. Assess. 187, 311.

Anjum, N.A., Umar, S., Ahmad, A., Iqbal, M., 2008. Responses of components of antioxidant system in moongbean genotypes to cadmium stress. Commun. Soil Sci. Plant Anal. 39, 2469–2483.

Anjum, N.A., Hasanuzzaman, M., Hossain, M.A., Thangavel, P., Roychoudhury, A., Gill, S.S., Rodrigo Merlos, M.A., Adam, V., Fujita, M., Kizek, R., Duarte, A.C., Pereira, E., Ahmad, I., 2015. Jacks of metal/metalloid chelation trade in plants—an overview. Front. Plant Sci. 6, 1–17.

Arora, K., Sharma, S., Monti, A., 2016. Bio-remediation of Pb and Cd polluted soils by switchgrass: a case study in India. Int. J. Phytoremediation 18, 704–709.

Baum, C., Hrynkiewicz, K., Leinweber, P., Meißner, R., 2006. Heavy-metal mobilization and uptake by mycorrhizal and nonmycorrhizal willows (*Salix* × *dasyclados*). J. Plant Nutr. Soil Sci. 169, 516–522.

Bhaduri, A.M., Fulekar, M.H., 2012. Assessment of arbuscular mycorrhizal fungi on the phytoremediation potential of *Ipomoea aquatica* on cadmium uptake. 3 Biotech 2, 193–198.

Bissonnette, L., St-Arnaud, M., Labrecque, M., 2010. Phytoextraction of heavy metals by two Salicaceae clones in symbiosis with arbuscular mycorrhizal fungi during the second year of a field trial. Plant Soil 332, 55–67.

Carvalho, L.M., Caçador, I., Martins-Loução, M.A., 2006. Arbuscular mycorrhizal fungi enhance root cadmium and copper accumulation in the roots of the salt marsh plant *Aster tripolium* L. Plant Soil 285, 161–169.

Chaturvedi, R., Favas, P., Pratas, J., Varun, M., Paul, M.S., 2018. Assessment of edibility and effect of arbuscular mycorrhizal fungi on *Solanum melongena* L. grown under heavy metal (loid) contaminated soil. Ecotoxicol. Environ. Saf. 148, 318–326.

Chen, B.D., Liu, Y., Shen, H., Li, X.L., Christie, P., 2004. Uptake of cadmium from an experimentally contaminated calcareous soil by arbuscular mycorrhizal maize (*Zea mays* L.). Mycorrhiza 14, 347–354.

Chen, L., Zhang, D., Yang, W., Liu, Y., Zhang, L., Gao, S., 2016. Sex-specific responses of Populus deltoides to *Glomus intraradices* colonization and Cd pollution. Chemosphere 155, 196–206.

Chern, E.C., Tsai, D.W., Ogunseitan, O.A., 2007. Deposition of glomalin-related soil protein and sequestered toxic metals into watersheds. Environ. Sci. Technol. 41, 3566–3572.

Cobbett, C., Goldsbrough, P., 2002. Phytochelatins and metallothioneins: roles in heavy metal detoxification and homeostasis. Annu. Rev. Plant Biol. 53, 159–182.

Cornejo, P., Meier, S., Borie, G., Rillig, M., Borie, F., 2008. Glomalin-related soil protein in a Mediterranean ecosystem affected by a copper smelter and its contribution to Cu and Zn sequestration. Sci. Total Environ. 406, 154–160.

de Andrade, S.A.L., Jorge, R.A., da Silveira, A.P.D., 2005. Cadmium effect on the association of jackbean (*Canavalia ensiformis*) and arbuscular mycorrhizal fungi. Sci. Agric. 62, 389–394.

de Andrade, S.A.L., da Silveira, A.P.D., Jorge, R.A., de Abreu, M.F., 2008. Cadmium accumulation in sunflower plants influenced by arbuscular mycorrhiza. Int. J. Phytoremediation 10, 1–13.

Deng, Z., Cao, L., Huang, H., Jiang, X., Wang, W., Shi, Y., Zhang, R., 2011. Characterization of Cd-and Pb-resistant fungal endophyte Mucor sp. CBRF59 isolated from rapes (*Brassica chinensis*) in a metal-contaminated soil. J. Hazard. Mater. 185, 717–724.

El Faiz, A., Duponnois, R., Winterton, P., Ouhammou, A., Meddich, A., Boularbah, A., Hafidi, M., 2015. Effect of different amendments on growing of *Canna indica* L. inoculated with AMF on mining substrate. Int. J. Phytoremediation 17, 503–513.

Gadd, G.M., 2004. Microbial influence on metal mobility and application for bioremediation. Geoderma 122, 109–119.

Galli, U., Schüepp, H., Brunold, C., 1994. Heavy metal binding by mycorrhizal fungi. Physiol. Plant. 92, 364–368.

Gao, Y., Miao, C., Mao, L., Zhou, P., Jin, Z., Shi, W., 2010. Improvement of phytoextraction and antioxidative defense in *Solanum nigrum* L. under cadmium stress by application of cadmium-resistant strain and citric acid. J. Hazard. Mater. 181, 771–777.

Garg, N., Aggarwal, N., 2011. Effects of interactions between cadmium and lead on growth, nitrogen fixation, phytochelatin, and glutathione production in mycorrhizal *Cajanus cajan* (L.) Millsp. J. Plant Growth Regul. 30, 286–300.

Garg, N., Bhandari, P., 2014. Cadmium toxicity in crop plants and its alleviation by arbuscular mycorrhizal (AM) fungi: an overview. Plant Biosyst. 148, 609–621.

Garg, N., Chandel, S., 2012. Role of arbuscular mycorrhizal (AM) fungi on growth, cadmium uptake, osmolyte, and phytochelatin synthesis in *Cajanus cajan* (L.) Millsp. under NaCl and Cd stresses. J. Plant Growth Regul. 31, 292–308.

Gonzalez-Chavez, M.C., Carrillo-Gonzalez, R., Wright, S.F., Nichols, K.A., 2004. The role of glomalin, a protein produced by arbuscular mycorrhizal fungi, in sequestering potentially toxic elements. Environ. Pollut. 130, 317–323.

Gonzalez-Chavez, M.C., Carrillo-Gonzalez, R., Gutierrez-Castorena, M.C., 2009. Natural attenuation in a slag heap contaminated with cadmium: the role of plants and arbuscular mycorrhizal fungi. J. Hazard. Mater. 161, 1288–1298.

González-Chávez, M.D.C.A., Carrillo-González, R., Hernández Godínez, M.I., Evangelista Lozano, S., 2017. *Jatropha curcas* and assisted phytoremediation of a mine tailing with biochar and a mycorrhizal fungus. Int. J. Phytoremediation 19, 174–182.

González-Guerrero, M., Azcón-Aguilar, C., Ferrol, N., 2006. GintABC1 and GintMT1 are involved in Cu and Cd homeostasis in *Glomus intraradices*. Paper Presented at the 5th International Conference on Mycorrhiza. Mycorrhiza Forscience and Society, Granada, Spain.

González-Guerrero, M., Cano, C., Azcón-Aguilar, C., Ferrol, N., 2007. GinMT1 encodes a functional metallothionein in *Glomus intraradices* that responds to oxidative stress. Mycorrhiza 17, 327–335.

Gonzalez-Guerrero, M., Melville, L.H., Ferrol, N., Lott, J.N., Azcón-Aguilar, C., Peterson, R.L., 2008. Ultrastructural localization of heavy metals in the extraradical mycelium and spores of the arbuscular mycorrhizal fungus *Glomus intraradices*. Can. J. Microbiol. 54, 103–110.

González-Guerrero, M., Benabdellah, K., Valderas, A., Azcón-Aguilar, C., Ferrol, N., 2010. GintABC1 encodes a putative ABC transporter of the MRP subfamily induced by Cu, Cd, and oxidative stress in *Glomus intraradices*. Mycorrhiza 20, 137–146.

Gunathilakae, N., Yapa, N., Hettiarachchi, R., 2018. Effect of arbuscular mycorrhizal fungi on the cadmium phytoremediation potential of *Eichhornia crassipes* (Mart.) Solms. Groundwater Sustain. Dev. https://doi.org/10.1016/j.gsd.2018.03.008.

Guo, J., Chi, J., 2014. Effect of Cd-tolerant plant growth-promoting rhizobium on plant growth and Cd uptake by *Lolium multiflorum* Lam. and *Glycine max* (L.) Merr. in Cd-contaminated soil. Plant Soil 375, 205–214.

Hall, J.L., 2002. Cellular mechanisms for heavy metal detoxification and tolerance. J. Exp. Bot. 53, 1–11.

Haq, F., Mahoney, M., Koropatnick, J., 2003. Signaling events for metallothionein induction. Mutat. Res. 533, 211–226.

Hasegawa, I., Terada, E., Sunairi, M., Wakita, H., Shinmachi, F., Noguchi, A., Nakajima, M., Yazaki, J., 1997. Genetic improvement of heavy metal tolerance in plants by transfer of the yeast metallothionein gene (CUP1). In: Plant Nutrition for Sustainable Food Production and Environment. Springer, Dordrecht, pp. 391–395.

Hashem, A., Abd_Allah, E.F., Alqarawi, A.A., Al Huqail, A.A., Egamberdieva, D., Wirth, S., 2016a. Alleviation of cadmium stress in *Solanum lycopersicum* L. by arbuscular mycorrhizal fungi via induction of acquired systemic tolerance. Saudi J. Biol. Sci. 23, 272–281.

Hashem, A., Abd_Allah, E.F., Alqarawi, A.A., Malik, J.A., Wirth, S., Egamberdieva, D., 2016b. Role of calcium in AMF-mediated alleviation of the adverse impacts of cadmium stress in *Bassia indica* [Wight] AJ Scott. Saudi J. Biol. Sci. https://doi.org/10.1016/j.sjbs.2016.11.003.

Hassan, S.E., Hijri, M., St-Arnaud, M., 2013. Effect of arbuscular mycorrhizal fungi on trace metal uptake by sunflower plants grown on cadmium contaminated soil. New Biotechnol. 30, 780–787.

Hu, P.J., Qiu, R.L., Senthilkumar, P., Jiang, D., Chen, Z.W., Tang, Y.T., Liu, F.J., 2009. Tolerance accumulation and distribution of zinc and cadmium in hyperaccumulator *Potentilla griffithii*. Environ. Exp. Bot. 66, 317–325.

Hu, J., Wu, S., Wu, F., Leung, H.M., Lin, X., Wong, M.H., 2013. Arbuscular mycorrhizal fungi enhance both absorption and stabilization of Cd by Alfred stonecrop (*Sedum alfredii* Hance) and perennial ryegrass (*Lolium perenne* L.) in a Cd-contaminated acidic soil. Chemosphere 93, 1359–1365.

Huang, H., Zhang, S., Chen, B.D., Wu, N., Shan, X.Q., Christy, P., 2006. Uptake of atrazine and cadmium from soil by maize (*Zea mays* L.) in association with the arbuscular mycorrhizal fungus *Glomus etunicatum*. J. Agric. Food Chem. 54, 9377–9382.

Huang, J., Ling, W.T., Sun, Y.D., Liu, J., 2012. Impacts of arbuscular mycorrhizal fungi inoculation on the uptake of cadmium and zinc by alfalfa in contaminated soil. J. Agro-Environ. Sci. 31, 99–105.

Janoušková, M., Pavlíková, D., 2010. Cadmium immobilization in the rhizosphere of arbuscular mycorrhizal plants by the fungal extraradical mycelium. Plant Soil 332, 511–520.

Janoušková, M., Pavlikova, D., Macek, T., Vosátka, M., 2005a. Arbuscular mycorrhiza decreases cadmium phytoextraction by transgenic tobacco with inserted metallothionein. Plant Soil 272, 29–40.

Janoušková, M., Pavlikova, D., Macek, T., Vosátka, M., 2005b. Influence of arbuscular mycorrhiza on the growth and cadmium uptake of tobacco with inserted metallothionein gene. Appl. Soil Ecol. 29, 209–214.

Janousková, M., Pavlíková, D., Vosátka, M., 2006. Potential contribution of arbuscular mycorrhiza to cadmium immobilisation in soil. Chemosphere 65, 1959–1965.

Jiang, Q.Y., Zhuo, F., Long, S.H., Zhao, H.D., Yang, D.J., Ye, Z.H., Li, S.S., Jing, Y.X., 2016. Can arbuscular mycorrhizal fungi reduce Cd uptake and alleviate Cd toxicity of *Lonicera japonica* grown in Cd-added soils? Sci. Rep. 6, 1–9.

Kapoor, D., Kaur, S., Bhardwaj, R., 2014. Physiological and biochemical changes in *Brassica juncea* plants under Cd-induced stress. Biomed. Res. Int. 2014, 1–13.

Khan, A.L., Lee, I.J., 2013. Endophytic *Penicillium funiculosum* LHL06 secretes gibberellin that reprograms *Glycine max* L. growth during copper stress. BMC Plant Biol. 13, 86.

Khan, S., Hesham, A.E.L., Qiao, M., Rehman, S., He, J.Z., 2010. Effects of Cd and Pb on soil microbial community structure and activities. Environ. Sci. Pollut. Res. 17, 288–296.

Khan, A.L., Hamayun, M., Ahmad, N., Waqas, M., Kang, S.M., Kim, Y.H., Lee, I.J., 2011. *Exophiala* sp. LHL08 reprograms *Cucumis sativus* to higher growth under abiotic stresses. Physiol. Plant. 143, 329–343.

Khan, A.L., Waqas, M., Hussain, J., Al-Harrasi, A., Lee, I.J., 2014. Fungal endophyte *Penicillium janthinellum* LK5 can reduce cadmium toxicity in *Solanum lycopersicum* (Sitiens and Rhe). Biol. Fertil. Soils 50, 75–85.

Khan, A.R., Ullah, I., Waqas, M., Park, G.S., Khan, A.L., Hong, S.J., Ullah, R., Jung, B.K., Park, C.E., Ur-Rehman, S., Lee, I.J., Shin, J.H., 2017a. Host plant growth promotion and cadmium detoxification in *Solanum nigrum*, mediated by endophytic fungi. Ecotoxicol. Environ. Saf. 136, 180–188.

Khan, A.R., Waqas, M., Ullah, I., Khan, A.L., Khan, M.A., Lee, I.J., Shin, J.H., 2017b. Culturable endophytic fungal diversity in the cadmium hyperaccumulator *Solanum nigrum* L. and their role in enhancing phytoremediation. Environ. Exp. Bot. 135, 126–135.

Khullar, S., Reddy, M.S., 2016. Ectomycorrhizal fungi and its role in metal homeostasis through metallothionein and glutathione mechanisms. Curr. Biotechnol. 5, 1–11.

Krznaric, E., Verbruggen, N., Wevers, J.H., Carleer, R., Vangronsveld, J., Colpaert, J.V., 2009. Cd-tolerant *Suillus luteus*: a fungal insurance for pines exposed to Cd. Environ. Pollut. 157, 1581–1588.

Lamelas, C., Benedetti, M., Wilkinson, K.J., Slaveykova, V.I., 2006. Characterization of $H^+$ and $Cd^{2+}$ binding properties of the bacterial exopolysaccharides. Chemosphere 65, 1362–1370.

Li, Z.S., Szczypka, M., Lu, Y.P., Thiele, D.J., Rea, P.A., 1996. The yeast cadmium factor protein (YCF1) is a vacuolar glutathione S-conjugate pump. J. Biol. Chem. 271, 6509–6517.

Li, Y., Peng, J., Shi, P., Zhao, B., 2009. The effect of Cd on mycorrhizal development and enzyme activity of *Glomus mosseae* and *Glomus intraradices* in *Astragalus sinicus* L. Chemosphere 75, 894–899.

Li, H., Luo, N., Zhang, L.J., Zhao, H.M., Li, Y.W., Cai, Q.Y., Wong, M.H., Mo, C.H., 2016. Do arbuscular mycorrhizal fungi affect cadmium uptake kinetics, subcellular distribution and chemical forms in rice. Sci. Total Environ. 571, 1183–1190.

Lin, A.J., Zhang, X.H., Wong, M.H., Ye, Z.H., Lou, L.Q., Wang, Y.S., Zhu, Y.G., 2007. Increase of multi-metal tolerance of three leguminous plants by arbuscular mycorrhizal fungi colonization. Environ. Geochem. Health 29, 473–481.

Liu, L.-Z., Gong, Z.-Q., Zhang, Y.L., Li, P.-J., 2011. Growth, cadmium accumulation and physiology of marigold (*Tagetes erecta* L.) as affected by arbuscular mycorrhizal fungi. Pedosphere 21, 319–327.

Liu, L., Gong, Z., Zhang, Y., Li, P., 2014. Growth, cadmium uptake and accumulation of maize (*Zea mays* L.) under the effects of arbuscular mycorrhizal fungi. Ecotoxicology 23, 1979–1986.

Liu, H., Yuan, M., Tan, S., Yang, X., Lan, Z., Jiang, Q., Ye, Z., Jing, Y., 2015. Enhancement of arbuscular mycorrhizal fungus (*Glomus versiforme*) on the growth and Cd uptake by Cd-hyperaccumulator *Solanum nigrum*. Appl. Soil Ecol. 89, 44–49.

Luo, S.L., Chen, L., Chen, J.L., Xiao, X., Xu, T.Y., Wan, Y., Rao, C., Liu, C.B., Liu, Y.T., Lai, C., Zeng, G.M., 2011. Analysis and characterization of cultivable heavy metal-resistant bacterial endophytes isolated from Cd-hyperaccumulator *Solanum nigrum* L. and their potential use for phytoremediation. Chemosphere 85, 1130–1138.

Ma, Y., Prasad, M.N.V., Rajkumar, M., Freitas, H., 2011. Plant growth promoting rhizobacteria and endophytes accelerate phytoremediation of metalliferous soils. Biotechnol. Adv. 29, 248–258.

Malik, A., 2004. Metal bioremediation through growing cells. Environ. Int. 30, 261–278.

Maret, W., 2000. The function of zinc metallothionein: a link between cellular zinc and redox state. J. Nutr. 130, 1455–1458.

Maret, W., 2003. Cellular zinc and redox states converge in the metallothionein/thionein pair. J. Nutr. 133, 1460–1462.

Mei, C., Flinn, B.S., 2010. The use of beneficial microbial endophytes for plant biomass and stress tolerance improvement. Recent Pat. Biotechnol. 4, 81–95.

Meier, S., Borie, F., Bolan, N., Cornejo, P., 2012. Phytoremediation of metal-polluted soils by arbuscular mycorrhizal fungi. Crit. Rev. Environ. Sci. Technol. 42, 741–775.

Mendoza-Cózatl, D.G., Butko, E., Springer, F., Torpey, J.W., Komives, E.A., Kehr, J., Schroeder, J.I., 2008. Identification of high levels of phytochelatins, glutathione and cadmium in the phloem sap of *Brassica napus*. A role for thiol-peptides in the long-distance transport of cadmium and the effect of cadmium on iron translocation. Plant J. 54, 249–259.

Miransari, M., 2011. Hyperaccumulators, arbuscular mycorrhizal fungi and stress of heavy metals. Biotechnol. Adv. 29, 645–653.

Mori, S., Uraguchi, S., Ishikawa, S., Arao, T., 2009. Xylem loading process is a critical factor in determining Cd accumulation in the shoots of *Solanum melongena* and *Solanum torvum*. Environ. Exp. Bot. 67, 127–132.

Nagajyoti, P.C., Lee, K.D., Sreekanth, T.V.M., 2010. Heavy metals, occurrence and toxicity for plants: a review. Environ. Chem. Lett. 8, 199–216.

Nayuki, K., Chen, B., Ohtomo, R., Kuga, Y., 2014. Cellular imaging of cadmium in resin sections of arbuscular mycorrhizas using synchrotron micro X-ray fluorescence. Microbes Environ. 29, 60–66.

Pan, J., Yu, L., 2011. Effects of Cd or/and Pb on soil enzyme activities and microbial community structure. Ecol. Eng. 37, 1889–1894.

Purin, S., Rillig, M., 2007. The arbuscular mycorrhizal fungal protein glomalin: limitations, progress, and a new hypothesis for its function. Pedobiologia 51, 123–130.

Rajkumar, M., Sandhya, S., Prasad, M.N.V., Freitas, H., 2012. Perspectives of plant associated microbes in heavy metal phytoremediation. Biotechnol. Adv. 30, 1562–1574.

Redon, P.O., Béguiristain, T., Leyval, C., 2008. Influence of *Glomus intraradices* on Cd partitioning in a pot experiment with *Medicago truncatula* in four contaminated soils. Soil Biol. Biochem. 40, 2710–2712.

Rodda, M.S., Li, G., Reid, R.J., 2011. The timing of grain Cd accumulation in rice plants: the relative importance of remobilisation within the plant and root Cd uptake post-flowering. Plant Soil 347, 105–114.

Ruiz-Herrera, J., 2016. Fungal Cell Wall: Structure, Synthesis, and Assembly. CRC Press.

Schneider, J., Stürmer, S.L., Guilherme, L.R., de Souza Moreira, F.M., Soares, C.R., 2013. Arbuscular mycorrhizal fungi in arsenic contaminated areas in Brazil. J. Hazard. Mater. 262, 1105–1115.

Seregin, I.V., Ivanov, V.B., 1997. Histochemical investigation of cadmium and lead distribution in plants. Russ. J. Plant Physiol. 44, 791–796.

Shahabivand, S., Maivan, H.Z., Goltapeh, E.M., Sharifi, M., Aliloo, A.A., 2012. The effects of root endophyte and arbuscular mycorrhizal fungi on growth and cadmium accumulation in wheat under cadmium toxicity. Plant Physiol. Biochem. 60, 53–58.

Sharma, R.K., Archana, G., 2016. Cadmium minimization in food crops by cadmium resistant plant growth promoting rhizobacteria. Appl. Soil Ecol. 107, 66–78.

Sidhu, G.P.S., Singh, H.P., Batish, D.R., Kohli, R.K., 2017. Tolerance and hyperaccumulation of cadmium by a wild, unpalatable herb *Coronopus didymus* (L.) Sm.(Brassicaceae). Ecotoxicol. Environ. Saf. 135, 209–215.

Singh, L.P., Gill, S.S., Tuteja, N., 2011. Unraveling the role of fungal symbionts in plant abiotic stress tolerance. Plant Signal. Behav. 6, 175–191.

Soleimani, M., Hajabbasi, M.A., Afyuni, M., Mirlohi, A., Borggaard, O.K., Holm, P.E., 2010. Effect of endophytic fungi on cadmium tolerance and bioaccumulation by *Festuca arundinacea* and *Festuca pratensis*. Int. J. Phytoremediation 12, 535–549.

Sun, H., Xie, Y., Zheng, Y., Lin, Y., Yang, F., 2018. The enhancement by arbuscular mycorrhizal fungi of the Cd remediation ability and bioenergy quality-related factors of five switchgrass cultivars in Cd-contaminated soil. PeerJ 6, e4425.

United Nations Environment Programme (UNEP), 2008. Draft Final Review of Scientific Information on Lead and Cadmium, Version of November 2008. http://www.chem.unep.ch/pb_and_cd/SR/Draft_final_reviews_Nov2008.htm. [(Accessed 2 April 2018)].

Vogel-Mikuš, K., Drobne, D., Regvar, M., 2005. Zn, Cd and Pb accumulation and arbuscular mycorrhizal colonisation of pennycress *Thlaspi praecox* Wulf. (Brassicaceae) from the vicinity of a lead mine and smelter in Slovenia. Environ. Pollut. 133, 233–242.

Wang, F., Lin, X., Yin, R., 2005. Heavy metal uptake by arbuscular mycorrhizas of *Elsholtzia splendens* and the potential for phytoremediation of contaminated soil. Plant Soil 269, 225–232.

Wang, F.Y., Lin, X.G., Yin, R., Wu, L.H., 2006. Effects of arbuscular mycorrhizal inoculation on the growth of *Elsholtzia splendens* and *Zea mays* and the activities of fungal characteristics and soil enzymatic activities in Cd polluted soil. Environ. Pollut. 134, 257–266.

Wang, F.Y., Lin, X.G., Yin, R., 2007. Role of microbial inoculation and chitosan in phytoextraction of Cu, Zn, Pb and Cd by *Elsholtzia splendens*–a field case. Environ. Pollut. 147, 248–255.

Wang, Y., Huang, J., Gao, Y., 2012. Arbuscular mycorrhizal colonization alters subcellular distribution and chemical forms of cadmium in *Medicago sativa* L. and resists cadmium toxicity. PLoS ONE 7, e48669.

Wright, S.F., Upadhyaya, A., 1998. A survey of soils for aggregate stability and glomalin, a glycoprotein produced by hyphae of arbuscular mycorrhizal fungi. Plant Soil 198, 97–107.

Xiao, X., Luo, S., Zeng, G., Wei, W., Wan, Y., Chen, L., Guo, H., Cao, Z., Yang, L., Chen, J., Xi, Q., 2010. Biosorption of cadmium by endophytic fungus (EF) Microsphaeropsis sp. LSE10 isolated from cadmium hyperaccumulator *Solanum nigrum* L. Bioresour. Technol. 101, 1668–1674.

Xu, R., Li, T., Cui, H., Wang, J., Yu, X., Ding, Y., Wang, C., Yang, Z., Zhao, Z., 2015. Diversity and characterization of Cd-tolerant dark septate endophytes (DSEs) associated with the roots of Nepal alder (*Alnus nepalensis*) in a metal mine tailing of southwest China. Appl. Soil Ecol. 93, 11–18.

Yadav, S.K., 2010. Heavy metals toxicity in plants: an overview on the role of glutathione and phytochelatins in heavy metal stress tolerance of plants. S. Afr. J. Bot. 76, 167–179.

Yao, Q., Yang, R., Long, L., Zhu, H., 2014. Phosphate application enhances the resistance of arbuscular mycorrhizae in clover plants to cadmium via polyphosphate accumulation in fungal hyphae. Environ. Exp. Bot. 108, 63–70.

Yoneyama, T., Ishikawa, S., Fujimaki, S., 2015. Route and regulation of zinc, cadmium, and iron transport in rice plants (*Oryza sativa* L.) during vegetative growth and grain filling: metal transporters, metal speciation, grain Cd reduction and Zn and Fe biofortification. Int. J. Mol. Sci. 16, 19111–19129.

Zhan, F., He, Y., Zu, Y., Li, T., Zhao, Z., 2011. Characterization of melanin isolated from a dark septateendophyte (DSE), *Exophiala pisciphila*. World J. Microbiol. Biotechnol. 27, 2483–2489.

# CHAPTER 17

# Microbe-Mediated Mitigation of Cadmium Toxicity in Plants

**Muhammad Shahid*, Muhammad Tariq Javed†, Aqsa Mushtaq*, Muhammad Sohail Akram†, Faisal Mahmood‡, Temoor Ahmed*, Muhammad Noman*, Muhammad Azeem†**

*Department of Bioinformatics and Biotechnology, Government College University, Faisalabad, Pakistan
†Department of Botany, Faculty of Life Sciences, Government College University, Faisalabad, Pakistan
‡Department of Environmental Sciences and Engineering, Government College University, Faisalabad, Pakistan

## Abstract

There has been increasing evidence of cadmium (Cd) toxicity and accumulation in various plant parts. Due to its mobile nature, Cd becomes an integral part of the food chain, causing a threat to humans in terms of kidney, liver, and lung diseases. In Cd-prone areas, leaf chlorosis, stunted growth, and inhibition of photosynthesis are the symptoms displayed by many plants. Microbe-mediated induction of Cd-stress tolerance is associated with various plant-based mechanisms, including alteration of phytohormone levels, modulation of the expression of defense-related proteins, upregulation of antioxidant enzymes, and expression modulation of metal transporters; and microbe-based mechanisms, including growth hormone production, better nutrient assimilation, siderophore production, 1-amino cyclopropane-1-carboxylate (ACC) deaminase activity, exopolysaccharide production, and so forth. Moreover, the rhizosphere microbes, especially bacteria and fungi, alter the state of toxic metal compounds and trigger the process of phytoremediation. Nevertheless, scientists need to explore new mechanisms and microbial signals for Cd-stress alleviation, coupled with increased plant productivity.

**Keywords:** Mechanisms, Microbes, Phytoremediation, Bioremediation, Stress alleviation, Antioxidants

## 1 INTRODUCTION

Because it is a toxic heavy metal, plants exposed to cadmium (Cd) may experience hazardous chemical reactions. In plants, Cd is a mobile element, and is not valuable for plant growth (Badri et al., 2009; Gupta et al., 2017). Accumulation and subsequent mobility of Cd in soil systems make it toxic for plants (Dugar et al., 2013). Although the pure form of Cd does not exist in nature, it oxidizes to cadmium oxide in the air, and readily reacts with water vapors,


https://doi.org/10.1016/B978-0-12-814864-8.00017-6

carbon dioxide, sulfur trioxide, and hydrogen chloride to produce various molecules such as $CdCO_3$, CdS, $CdCl_2$, and so forth (Dugar et al., 2013).

Agricultural soils contaminated with Cd are far less productive, especially in urban areas (de Araújo et al., 2017; Mahar et al., 2016; Rizwan et al., 2016). Plants exposed to Cd contamination show symptoms such as browning, root mucilage, biomass reduction, curling of leaves, and chlorosis. Cd has been found to impede lateral root formation due to browning and twisting of the main roots (Krantev et al., 2008; Rascio and Navari-Izzo, 2011; Yadav, 2010). Farm yield reductions occur in response to Cd toxicity. Cd accumulates in various organs of humans and animals due to contamination of edible portions of food crops. However, a rational approach can be adopted to overcome Cd toxicity in plants (Nymark et al., 2016).

A sustainable and cheaper method to reduce heavy metal contamination by transformation and mineralization processes is bioremediation (Vidali, 2001). Bioremediation is an effective technique that ensures the efficient degradation of pollutants. Various microorganisms, such as most microbes, have the potential to enzymatically degrade organic pollutants, and/or transform inorganic toxic chemicals into less toxic forms (Cristaldi et al., 2017; Vidali, 2001). Various metabolic functions of microbes depend upon the energy they derive from carbon sources utilized in the environment. Heavy metals, including Cd, cannot be degraded biologically, and thus, are transformed from toxic to less toxic states to guarantee soil and plant health (Dzionek et al., 2016). Bioremediation is an environmentally-friendly and cheaper approach than many other conventional methods (Garbisu and Alkorta, 2003; Perelo, 2010). The mechanisms by which microorganisms confer plants' tolerance to heavy metals, including Cd, include alteration of phytohormone levels, modulation of expression of defense-related proteins, upregulation of antioxidant enzymes, and expression modulation of metal transporters. Various microbe-based mechanisms, such as growth hormone production, better nutrient assimilation, siderophore production, 1-amino cyclopropane-1-carboxylate (ACC) deaminase activity, and exopolysaccharide production also contribute to mitigating Cd toxicity in plants.

Furthermore, the rhizosphere microbes, particularly bacteria and fungi, also support plant accumulation and transformation of metals from soil, which is a process known as phytoremediation (Gerhardt et al., 2016). In this chapter, Cd toxicity in plants, and its mitigation by microbe-based, as well as microbially triggered, mechanisms in plants, has been discussed. Detailed knowledge of these mechanisms is helpful for adopting microbe-mediated remediation technology for Cd-contaminated soils.

## 2 CADMIUM AS A HEAVY METAL

Cd as a heavy metal is extremely lethal for plant growth. Toxicity of Cd in agricultural soils leads to many environmental-related problems, such as a shift in microbial community, soil structure deterioration, and plant growth reduction (Nymark et al., 2016). Cd plays an insignificant physiological role in plants' development (Gallego et al., 2012). It is hazardous for plant and human health in terms of its presence in soil and uptake by plants. It is regarded as a very toxic metal (ATSDR, 1992; Nordberg, 2009). It interacts with DNA and many proteins to cause oxidative damage inside the plant cells. The transport of Cd into the bacterial cells is similar in Cd to that of other divalent heavy metals (Nymark et al., 2016).

## 3 GLOBAL POLLUTION STATUS OF Cd

Earth's crust has various concentrations of Cd, depending upon many geological and environmental factors, but an average concentration ranges between 0.1 and 0.5 ppm. The Cd abundance ranges from 0.02 to 0.2 ppm in igneous and metamorphic rocks, whereas sedimentary rocks contain 0.1–25 ppm, which is a much higher concentration. Fossil fuels have 0.5–1.5 ppm Cd, but the amount of Cd in phosphate fertilizers ranges from 10 to 200 ppm (Tran and Popova, 2013). Soil solutions containing a Cd level ranging from 0.32 to about 1 mM can be referred to as polluted soil. The Cd is naturally present in the soil, and enters the food chain as a component of food stuffs (Mench et al., 1998; Wagner, 1993).

## 4 SOURCES OF Cd IN SOIL

Environmental and public health concerns are associated with Cd contamination in soil and food crops (Sobariu et al., 2016). Out of the 20 toxins in plants, the toxicity ranking of Cd is seven (Gill and Tuteja, 2011; Zhou et al., 2017). The Cd contamination in soil is originated from both geogenic (e.g., natural) and anthropogenic (e.g., human originated) sources (Tran and Popova, 2013). The natural reservoirs of Cd are the ores of zinc and lead. It is also found in high concentrations in phosphate fertilizers. Different agricultural and industrial activities have been recognized as major contributors of the metal dispersion, at low quantities, into the general environment, and human food stuffs (Järup and Åkesson, 2009; Satarug et al., 2003). Moreover, leafy vegetables and potato tubers are hyperaccumulators of Cd, more so than other fruits and cereals (Mench et al., 1998). The major natural sources of Cd mobilizations from earth's crust include volcanoes and weathering of rocks (Tran and Popova, 2013). Naturally, a huge amount of Cd (almost 25,000 tons/annum) is released into the environment, a major part of which comes from weathering of rocks, volcanoes, and forest fires (Fig. 1).

Naturally, approximately 25,000 tons per year is released in the environment. A great portion of this is released into the rivers via disintegration of rocks, and some quantity enters in the air through the process of volcanoes and forest fire (Dugar et al., 2013). The main anthropogenic source of this heavy metal is industrial waste of various processes such as electroplating, the plastics industry, pigments, alloy formation, mining, Cd-containing batteries, and composts or fertilizers. Soils flooded with domestic or industrial wastewater also contain Cd in high concentrations. The Cd enters into streams and rivers from the sewage systems and industrial waste, and thus, becomes the source of aquatic pollution. The Cd accumulation in water bodies may also cause groundwater contamination (Moradi et al., 2005).

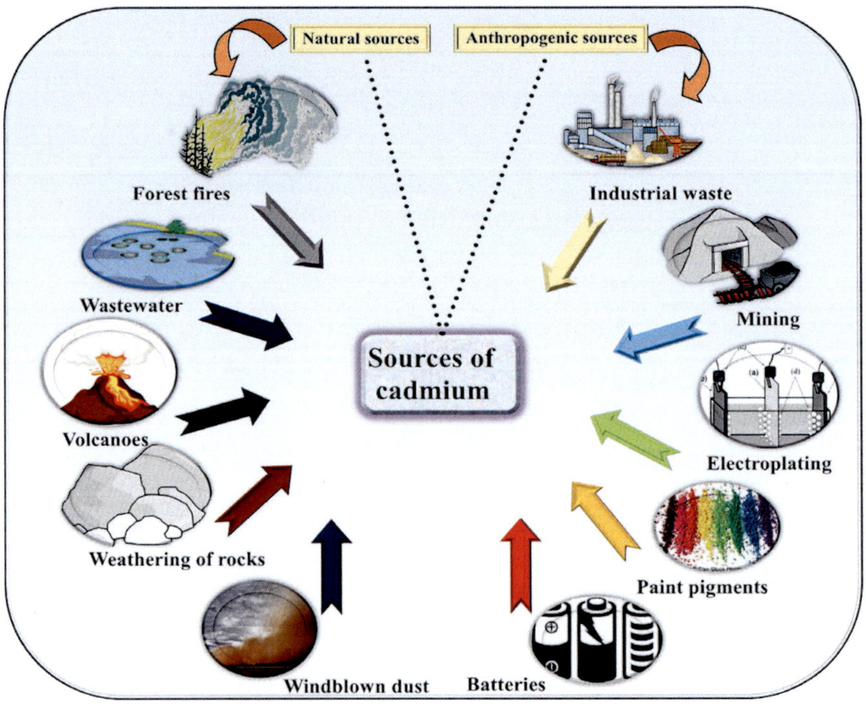


**Fig. 1** Schematic representation of the sources of Cd in agricultural soils.

Other industry-based sources of Cd are manufacturing of alloys, motor oils, solders, rubber, fungicide, and textiles (Gupta et al., 2017). The total production of Cd was almost 24,900 metric tons worldwide in 2015 (Brown et al., 2017). Furthermore, farm activities also have a major effect on the Cd contents of food (Dugar et al., 2013).

## 5 Cd ACCUMULATION IN PLANTS

Plants' Cd concentration depends upon its biomass, organic matter, and pH of soil (Zeng et al., 2011). The Cd uptake by plants is encouraged by lowering the pH of soil (Kirkham, 2006). At high pH, adsorption of Cd by soil is favored, which consequently reduces the Cd extractability (Dugar et al., 2013). Adams et al. (2004) studied the relationship among the soil properties and Cd concentration in wheat and barley grains by examining wheat and barley grain samples that were collected from consecutive soil and crop surveys. They concluded that the soil pH was the key factor that affected the Cd concentration in the grain.

Several studies have revealed the transfer of Cd from contaminated soil to plants. On the other hand, the mechanisms that regulate the mobility of Cd in soil and plant systems have been studied less, leading to biases in the data (Wehner et al., 2010; Yang et al., 2011). The opinions of researchers about the Cd uptake by plants are different in terms of active and passive transport. Metabolic energy is needed to carry out the active transport for the transfer of the ions through the carriers. Similarly, the carriers have attraction toward the ions that regulate the elements within plant species (Stein and Litman, 2014). Conversely, the passive uptake is independent of energy consumption in the form of ATP. After Cd entry into the plants, the hypothesis of Cd motility through xylem, along with water, is mostly accepted (Dugar et al., 2013). Although the processes underlying Cd uptake and transfer require further investigation for a full understanding, we know that Cd is absorbed by the roots, and then translocated to shoots, primarily through the xylem, with subsequent translocation to the grains or fruits, e.g., tomatoes and other vegetables.

## 6 Cd TOXICITY IN PLANTS

Cd is not in the category of essential nutrients for plant growth (Benavides et al., 2005). The Cd ions' persistence in the soil can affect the water and mineral uptake, transport, and consequent distribution by plants. Thus, soil

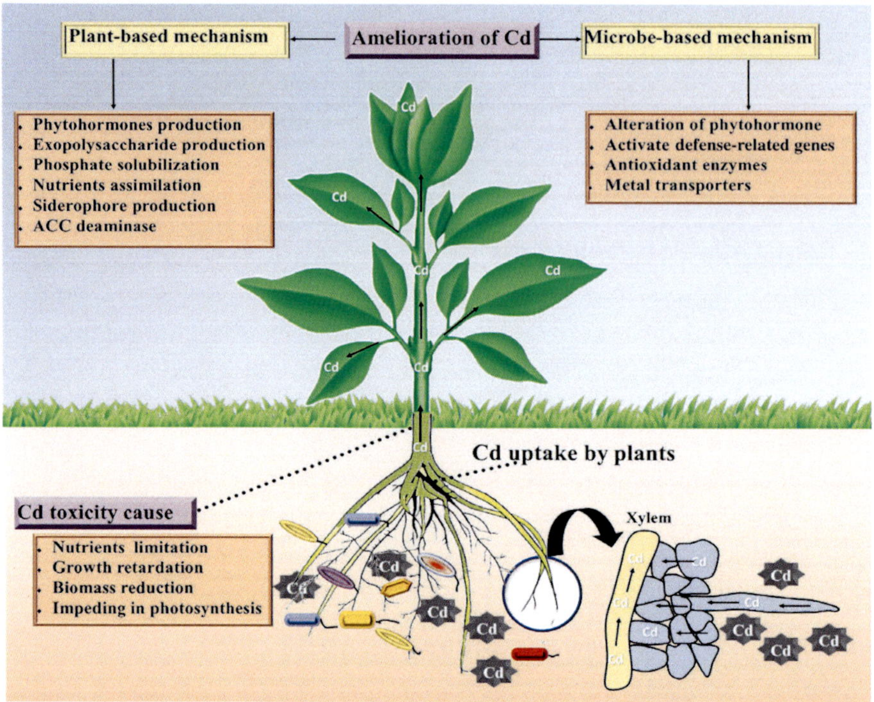


**Fig. 2** Uptake of Cd by plants from soil, and different mechanisms of Cd toxicity mitigation in plants.

contaminated with Cd also destroys beneficial soil microbial communities (Tran and Popova, 2013). Long- and short-term Cd exposure has also been reported to affect the stomatal opening, transpiration, and photosynthesis. The symptoms of Cd toxicity in plants include leaf rolls, chlorophyll deficiency, and stunted growth (Fig. 2). Chlorosis in Cd-prone areas may result from iron and phosphorus deficiencies, or decreased manganese transport (Benavides et al., 2005). Cd also reduces the Fe(II) uptake and transport, due to the inhibition of Fe (III) reductase production in roots, thus affecting photosynthesis (Hasan et al., 2009). Cd interferes with nitrate absorption and its distribution within plants by retarding the nitrate reductase activity within shoots. Due to Cd toxicity, remarkable suppression of nitrate reductase activity was observed in *Silene cucubalus* plants (Tran and Popova, 2013). Nitrogen fixation and ammonia assimilation within root nodules of soybean plants were also reduced under Cd stress (Balestrasse et al., 2003). Heavy metal stress, particularly Cd stress, has been reported to alter the plasma membrane permeability, and interfere with the water balance (Benavides et al., 2005). The Cd accumulation in plants also induces alterations in ATPase activity of plasma membrane components of plants, such as wheat (Fodor et al., 1995), oats (Astolfi et al., 2003), and maize (Astolfi et al., 2005). The lipid peroxidation and chloroplast metabolism disturbances induced by Cd may lead to changes in the membranes' functionality (Fodor et al., 1995; Tran and Popova, 2013). Cd toxicity creates oxidative stress in plants, either by triggering the generation of oxygen-free radicals, or by dropping the enzymatic and nonenzymatic antioxidant level in plants (Benavides et al., 2005; Cho and Seo, 2005; Gratão et al., 2005). For example, under Cd stress, soybean plants showed rapid senescence due to oxidative stress (Balestrasse et al., 2004).

# 7 MECHANISMS OF MICROBE-ASSISTED AMELIORATION OF Cd TOXICITY IN PLANTS

## 7.1 Plant-Based Mechanisms

In the middle of the 20th century, the term "phytoremediation" was developed to describe the application of using plants to remediate the contaminants from soil (Cristaldi et al., 2017; Visoottiviseth et al., 2002). This plant-based technology is advantageous in terms of its low cost, efficient remediation, and environmentally-friendly nature (McCutcheon and Schnoor, 2004; Vamerali et al., 2010). The phytoremediation process is primarily based on the use of green plants, herbs, and woody species (e.g., *Salix* spp., *Populus* spp.), due to their promising characteristics of accumulating and transporting harmful ecological pollutants such as heavy metals, organic compounds, and radioactive compounds from soil to plants, and their subsequent transformation into less-toxic compounds (Tahir et al., 2016).

Phytoremediation techniques can be further stimulated by using microbes such as bacteria and fungi to tolerate the stress of heavy metals, including Cd, by altering the phytohormone level, upregulation of antioxidant enzymes, modulation of defense-related proteins, and expression modulation of metal transporters (Hall, 2002; Kim et al., 2007; Jadia and Fulekar, 2009; Gallego et al., 2012;). The success of the microbe-assisted phytoremediation process relies on several factors, such as the nature of the contaminant, bioavailability, soil properties, and, most importantly, plant species (DalCorso et al., 2008; Tripathi et al., 2016).

### 7.1.1 Alteration of Phytohormones Level in Plants

Plant hormones act as a vital component of regulating growth and development in plants. Different phytohormones were found effective in relieving the plant from Cd stress under observed abiotic stresses. By exposing them to Cd stress, plants exhibit an induced mechanism that produces elevated levels of hormones such as jasmonic acid (JA), abscisic acid (ABA), ethylene (ET), and salicylic acid (SA) (Gallego et al., 2012). These phytohormones cause an increase in plant resistance to ecological stresses, and the expression of various genes, enzymes, metabolites, and pigments (Kaur et al., 2017; Tsavkelova et al., 2006). The auxins are involved in plant cell division and root elongation, and are also responsible for the biosynthesis of various metabolites and pigments to make the plants immune to stresses (Tsavkelova et al., 2006). Pea plants produced an immense amount of ET and ammonic acid when exposed to Cd (Rodríguez-Serrano et al., 2009). In *Arabidopsis*, SA was shown to potentiate oxidative reactions induced by Cd stress, while after 5 days of Cd treatment, lipid peroxidation significantly increased, and chlorophyll contents were significantly reduced, in wild type seedlings, as compared with SA-deficient lines (Zawoznik et al., 2007). In barley (*Hordeum vulgare*) roots, SA content was substantially increased by the exposure of Cd stress (Metwally et al., 2003). The research concluded that the alleviated level of hormones not only act as an antioxidant defense against Cd toxicity, but also affect the mechanism of Cd detoxification.

### 7.1.2 Modulation of Defense-Related Genes/Proteins

Interaction of genotypes with the environment is vital to minimize Cd toxicity in plants. A coordinated modulation of many metabolic and genetic mechanisms is activated in terms of gene expression, protein actions, and altered levels of many primary and secondary metabolites (Gallego et al., 2012; Hall, 2002). Upon the onset of stress conditions, many signal transduction mechanisms are also triggered in plant cells, making cells able to perceive signals and act in response to nature, type, and intensity of stress conditions. Such signals confer plants' resistance to unfavorable stresses, including Cd. Many efforts have already been made to unravel such signal transduction mechanisms through functional genomics, proteomics, and transcriptomics approaches (Thakur et al., 2016; Vamerali et al., 2010). For example, AtOS1 is a well-known and putative gene for perceiving signals under Cd stress conditions. Being a member of the Acb1 family, this gene is reported to induce Cd tolerance in plants through enzymatic actions (Jasinski et al., 2008). The transcription factor OBF5, which belongs to basic leucine zipper (B-ZIP), is known to bind to the promoter region of glutathione S-transferase and induce Cd tolerance in plants (Onodera et al., 2001). Similarly, the TGA3 protein has also been reported to play a vital role under Cd stress by modulating gene expression (Johnson et al., 2003). Recently, the transcription factor BjCdR15 has been shown to act as an orthologue of TGA3 in *Brassica juncea* under induced Cd stress (Fusco et al., 2005). This transcription factor has a defined role in Cd transportation from roots to shoots by several metal transporters and their overexpression in *Arabidopsis thaliana* and tobacco plants (Farinati et al., 2010). Similarly, transcription factors from the WRKY family have been found to be differently expressed in Cd-treated *Thlaspi caerulescens* plants (Wang et al., 2008).

### *7.1.3 Upregulation of Antioxidant Enzymes*

Modulation of several enzymatic activities occurs in plants under Cd stress. This enzyme modulation takes place via metal binding with the sulfyl-hydrl group, or the replacement of essential metals with toxic metals. Mostly, the metal uptake also activates the enzyme activities (Jadia and Fulekar, 2009). Catalase, a metallo-enzyme, is affected by the high concentration of heavy metals (Fayiga et al., 2004). Conversely, when *A. thaliana* was induced with Cd stress, the catalase activity in leaves was reduced (Leterrier et al., 2012; Munasinghe and Khanal, 2010). The nature and number of antioxidant enzymes depend on the level of Cd stress and plant species. For example, the increased activity of superoxide dismutase, glutathione reductase, and catalase was measured in *Avena strigose* (Uraguchi et al., 2006). Antioxidant enzymes are modulated in response to varying concentrations of Cd pollution of soil.

### *7.1.4 Modulation of Metal Transporters*

As the Cd gains entry into the cells, plants activate several mechanisms to fight against Cd toxicity. One such mechanism is the Cd efflux, or compartmentalization within the cell vacuole via activation of specific transport processes (DalCorso et al., 2010). The Cd entry into plant cells is mediated by transport systems specifically for micronutrients' (such as $Ca^{2+}$, $Fe^{2+}$, $Zn^{2+}$, $Mg^{2+}$, $Cu^{2+}$, etc.) uptake (Clemens, 2006; Roth et al., 2006). The members of various transporter families confer resistance to plants against Cd toxicity. For example, the AtPDR8 belongs to the ABC transporter family, and is engaged in the transportation of Cd out of the epidermal cells of the roots (Kim et al., 2007). The members of zinc and iron transporter-like proteins, e.g., ZIP (zinc regulated transporter and iron regulated transporter-like proteins), are best known examples of their Cd detoxification potential. These transporters are plasma membrane transport proteins that were characterized for the first time in *Arabidopsis*. They are triggered in both roots and shoots of *Arabidopsis* under Zn scarcity (DalCorso et al., 2010). More recently, ZIP transporters have been identified in various plant species, and also in fungi, bacteria, and animals. They are specific for the transportation of divalent cations across the plasma membrane (Lopez-Millan et al., 2004). The IRT1 is involved in root $Fe^{2+}$ uptake under Fe-limiting conditions, but under specific circumstances, it also acts as a channel for Cd transportation across the plasma membrane (Rogers et al., 2000). The P-type metal ATPase transporter HMA4 is involved in Zn/Cd transportation. It is involved in mobilization of Cd toward the shoot through the xylem, where Cd has less toxic effects (Hanikenne et al., 2008; Verret et al., 2004). The natural resistance–associated macrophage protein (NRAMP) is another family of metal transporter involved in the translocation of Cd in plants. The AtNRAMP6 works intracellularly either by translocating Cd from its storage sites, or by compartmentalization of Cd within cells (Cailliatte et al., 2009). The divalent metal cations' extrusion from the cell, or storage within the vacuoles, can also be accomplished by the cation diffusion facilitator (CDF) family of metal transporters (Krämer et al., 2007).

## 7.2 Microbe-Based Mechanisms

Plants inhabiting heavy-metal contaminated soil have a diverse variety of heavy metals immune microorganisms in their rhizosphere (Idris et al., 2004; Zarei et al., 2010, 2008). These plant-associated microbes are beneficial for both soil and plants. The rhizospheric bacteria have gained much attention among all microorganisms as they have the potential to influence the phytoremediation process directly through the fluctuation of metal bioavailability via changing the pH of soil, and production of metal chelating compounds (such as organic acids and siderophores) (Table 1) (Rajkumar et al., 2012).

Likewise, various species of mycorrhizal fungi have also been reported to be associated with plants growing in the soil polluted with heavy metals, indicating that these fungi have become metal resistant during their evolutionary course. Such fungal species may play a beneficial role in phytoremediation of metal-contaminated soil (Miransari, 2011; Orłowska et al., 2011; Zarei et al., 2010).

### *7.2.1 Phytohormones Production*

The production of phytohormones by microbes boosts plants' growth and nutrition. The phytohormones trigger plant growth under both normal and

**Table 1** Microbial species and their mechanisms to reduce Cd toxicity in plants

| Microbes | Plants | Resistance traits | References |
|---|---|---|---|
| *Flavobacterium* | *Hordeum vulgare* (barley) | Cd accumulation | Pishchik et al. (2002) |
| *Pseudomonas brassicacearum and Pseudomonas marginalis* | *Pisum sativum* (Pea) | ACC deaminase activity Enhancement of nutrient uptake processes | Safronova et al. (2006) |
| *Methylobacterium oryzae* | *Lycopersicon esculentum* (tomato) | ACC deaminase activity | Madhaiyan et al. (2007) |
| *Pseudomonas aeruginosa* | *Brassica* (mustard) | Siderophore production | Sinha and Mukherjee (2008) |
| *Pseudomonas aeruginosa* | *Cucurbita pepo* (pumpkin) | Siderophore production | Sinha and Mukherjee (2008) |
| *Enterobacter asburiae* | *Vigna radiata* (mung bean) | Gluconic acid production Phosphate solubilization | Kavita et al. (2008) |
| *Burkholderia* sp. RX232 | *Salix caprea* (goat willow) | ACC deaminase activity Siderophore production | (Kuffner et al. (2010) |
| *Enterobacter aerogenes Solanum nigrum* | *Solanum nigrum* (nightshade) | IAA production ACC deaminase activity Siderophore production and phosphate solubilization | Creyghton et al. (2010) |
| *Ralstonia Chryseobacterium humi* | *Zea mays* (Maize) | Eutropha Cd immobilization and low Cd translocation to the shoots | (Moreira et al. (2014) |
| *Bradyrhizobium* sp. | *Glycine max* (soybean) | Increasing mineral nutrient (Fe) uptake | Guo and Chi (2014) |
| *Pseudomonas putida* | *Vigna radiata* (mung bean) | P-type ATPase's, Cd bioaccumulation | Saluja and Sharma (2014) |
| *Klebsiella* sp. *Bacillus* sp. *Stenotrophomonas* sp. | *Triticum aestivum* (wheat) | Not reported | Ahmad et al. (2014) |

*Continued*

**Table 1** Microbial species and their mechanisms to reduce Cd toxicity in plants—cont'd

| Microbes | Plants | Resistance traits | References |
|---|---|---|---|
| *Glomus intraradices*, *Glomus constrictum*, and *Glomus mosseae* | *Tagetes erecta* L. (marigold) | Cd accumulation | Ling-Zhi et al. (2011) |
| *Piriformospora indica* and *Glomus mosseae* | *Triticum aestivum* (wheat) | Enhanced the chlorophyll content | Shahabivand et al. (2012) |
| *Aspergillus aculeatus* | *Cynodondactylon* L. *Pers* (Bermudagrass) | Improve Cd tolerance | Xie et al. (2014) |
| *Chromosporium*, *Fusarium*, and *Gonatobotrys* | *Dracaena fragrans* (corn) | Significantly enhanced the Cd tolerance | An et al. (2015) |
| *Piriformospora indicas* | *Helianthus annuus* (sunflower) | Significantly enhanced growth, Chl a, Chl b, and proline contents | Shahabivand et al. (2017) |
| *Alternaria alternata* | *Solanum nigrum* (nightshade) | Low peroxidase (POD), polyphenol peroxidase (PPO) activities or high catalase (CAT) activity | Li et al. (2017) |
| *Aspergillus aculeatus* | *Cynodondactylon* (L.) *Pers* (Bermudagrass) | IAA production, Cd accumulation | Li et al. (2017) |

stress conditions. In the presence of Cd, the reactive oxygen species (ROS) is generated by plant cells, which results in stimulation of the mitogen activated protein (MAP) kinase (MAPK) signaling pathway (Sharma and Archana, 2016). The Cd-ROS-MAPK signal induces damage in metal-sensitive plants through various mechanisms. In metal-resistant plants, the MAPK cascades trigger metal-resistant pathways, rather than aging and apoptosis (Lin and Aarts, 2012). It has been reported that Cd toxicity affects the auxin physiology in *Arabidopsis* (Hu et al., 2013) and poplars (Elobeid et al., 2011) by reducing auxin concentrations, altering its distribution, and increasing degradation. The auxin and other plant hormones produced by rhizobacteria can alleviate the harmful effects of Cd toxicity on the growth and development of the plant. Bhattacharyya and Jha (2012) reported that phytohormones induce the growth of Cd-treated rice seedlings. Indole-3-acetic acid (IAA) production is among the most important characteristics of numerous PGPR. Under Cd stress, rhizobacteria spp. such as *Arthrobacte rmysorens*, *Agrobacterium radiobacter*, *Azospirillum lipoferum*, and *Flavobacterium* produce auxins and enhance the growth of barley (Belimov et al., 2004; Belimov and Dietz, 2000). Auxins support plant growth by the low sorption of Cd from soil, and enhanced activity of antioxidant enzymes (Chmielowska-Bąk et al., 2014). Better root development was observed in barley seedlings treated with auxin produced by bacteria, thus resulting in enhanced nutrient acquisition (Belimov and Dietz, 2000).

### 7.2.2 Better Nutrient Assimilation

High Cd concentration in soil leads to the deficiency and imbalance of nutrients in plants. There is a competition between $Cd^{2+}$ and other divalent cations $Ca^{2+}$, $Mg^{2+}$, $Zn^{2+}$, and $Fe^{2+}$, that causes mineral deficiencies in plants (Sharma and Archana, 2016). The Cd and various essential metals act as rivals due to the presence of similar transport mechanisms in plants (Nazar et al., 2012). In rice, for Cd uptake, translocation, and storage, more than 30 genes have been identified (Sebastian and Prasad, 2014). Cd toxicity and tolerance is linked with the homeostasis of various metals such as Zn, Ca, and Fe (Tkalec et al., 2014; Wu et al., 2012; Zhai et al., 2014). It has been observed that barley root-associated microbes improve the nutrient content, including P, Mg, Ca, S, Fe, and Mn in roots and shoots when cultivated in Cd-contaminated soils (Sharma and Archana, 2016). These microbes improve the uptake of nutrient elements, thus reducing the Cd toxicity to barley plants. Under Cd stress, Cd-tolerant microbes induce the various minerals' uptake and increase the general nutrient profile of the plants, thus helping to reduce Cd toxicity in the plant (Gomes et al., 2012).

### 7.2.3 Siderophore Production

Under aerobic environments, iron is present in insoluble Fe(III) form in the soil. Thus, Fe bioavailability is halted, affecting the growth of plants and microbes. Under such conditions, certain chemical compounds are produced by microorganisms associated with plant roots that have an affinity with iron (Rajkumar et al., 2012). These compounds have a very high affinity towards $Fe^{3+}$ ions, and help plants and microbes in the utilization of iron for metabolic needs (Sharma and Archana, 2016). Such compounds have been produced both by plants, as well as microbes. Siderophore-mediated iron uptake has been reported in graminaceous plants by the synthesis of $Fe^{3+}$ chelating phytosiderophores, whereas in nongraminaceous plants, a root-surface localized enzyme ferric reductase transform Fe(III) to Fe(II) is followed by $Fe^{2+}$ uptake by roots in epidermal cells (Kobayashi and Nishizawa, 2012). Thus, improved iron uptake makes the plant strong enough to fight the Cd stress. The iron uptake also reduces the transportation of heavy metals, including Cd to plant shoots, and thus, mitigating the Cd toxicity (Wu et al., 2012). The accumulation of Cd in roots might be due to the high concentration of iron.

Siderophore are produced in a variety of microorganisms to provide soluble iron to plants (Desai and Archana, 2011; Rajkumar et al., 2012). Cd stress stimulates the production of siderophores (e.g., pyoverdin and pyochelin) in root-associated *Pseudomonas aeruginosa*. High iron content was found in the leaves of *P. aeruginosa*-treated mustard and pumpkin plants when grown in Cd-amended soils (Braud et al., 2009; Sinha and Mukherjee, 2008). Cd-induced siderophore production was also found in the *Streptomyces tendae* strain F4, which improved plant growth, bioavailability of soil metals, and increased the Fe and Cd uptake in sunflowers (Dimkpa et al., 2009). Some mycorrhizal fungi species (*Scleroderma verrucosum*, *Rhizopogon luteolus*, and *Suillus luteus*) also produce a variety of siderophores (Rajkumar et al., 2012).

### 7.2.4 ACC Deaminase Activity

The ACC deaminase is an ET precursor in plants. In Cd-prone areas, the ET level in plants automatically increases due to oversynthesis of ACC by ACC synthase in plants, thus reducing the growth of many plants (Chmielowska-Bąk et al., 2014; Ovečka and Takáč, 2014). Due to an elevated level of ET in plants, plant roots do not grow normally in Cd-stress conditions (Sun and Guo, 2013). Plant-associated microbes produce an enzyme called ACC deaminase, which breaks ACC into ammonia and α-ketobutyrate (Glick, 2005; Gamalero et al., 2009), and thus, disrupts the ET biosynthesis pathway. In this way, microbes help plants to continue their normal growth in Cd polluted soils. In the presence of Cd, ACC deaminase activity is reported to increase seed germination and improve root elongation in tomato plants (Grichko et al., 2000). The ACC deaminase enzyme was discovered and isolated for the first time in the *Pseudomonas* sp. ACP strain (Honma and Shimomura, 1978). Later, the enzyme activity was reported in various other soil-borne microbes, such as the fungal strain *Penicillium citrinum,* the yeast strain *Hansenula saturnus*, and in numerous bacterial strains (Gamalero et al., 2009). ACC deaminase synthesizing metal-tolerant microbes promoted the growth of *Brassica* when grown in Cd-amended soils (Belimov et al., 2001). Under Cd stress, pea plants inoculated with ACC-producing bacterial strains *Pseudomoas brassicacearum* and *Pseudomonas marginalis* evidenced better growth and root development (Safronova et al., 2006). Thus, the characterization of ACC deaminase-producing microbes with metal tolerance ability in order to make new bioinoculant formulations for heavy metal-contaminated soils should be examined.

### 7.2.5 Extracellular Polymeric Substances and Glycoprotein Production

Plant-associated microorganisms also produce extra-cellular polymeric substances (EPS), proteins, and mucopolysaccharides that immobilize toxic heavy metals within the soils and prevent their translocation (Rajkumar et al., 2012). High Cd immobilizing ability has been evidenced in EPS-producing *Azotobacter* sp. Furthermore, it has been demonstrated that *Azotobacter* sp. inoculation in Cd-contaminated soils reduced the Cd uptake by wheat (Joshi and Juwarkar, 2009). The ability of AMF fungi-synthesized insoluble glycoprotein to bind Cd was also investigated, and it was found that 0.1 mg/g of Cd was chelated (Gonzalez-Chavez et al., 2004; Rajkumar et al., 2012). The plant inoculation with such EPS and glycoprotein producing microbes would help plants tolerate Cd toxicity with a subsequent increase in plant growth.

### 7.2.6 Production of Organic Acids

Microbes (Kavita et al., 2008; Archana et al., 2012; Rajkumar et al., 2012) and plants (Majewska and Kurek, 2011) are able to synthesize various organic acids (malic acid, gluconic acid, acetic acid, citric acid, etc.) in soils in Cd-rich conditions. The organic acids reduce the soil pH and detoxify the metal ions by binding them with themselves. Rhizobacteria are able to utilize rhizospheric carbohydrates as a substrate with the subsequent secretion of organic acids in soil. Moreover, an enzyme glucose dehydrogenase produces gluconic acid, which is an important organic acid involved in nutrient uptake from soil (Sharma and Archana, 2016). Under Cd-prone conditions, the inoculation of gluconic acid producing *Enterobacter asburiae* with *Vigna radiata* seedlings promoted the growth and altered the physiology of plants (Kavita et al., 2008). Organic acids were also produced by some fungal spp., e.g., *Oidiodendron maius* (Martino et al., 2003), *Beauveria caledonica* (Fomina et al., 2005), and *Aspergillus niger* (Rajkumar et al., 2012). In addition, organic acids synthesis is strongly correlated with mineral phosphate solubilization and mineralization (Patel et al., 2010), and zinc solubilization (Wani et al., 2007), which subsequently confer plants resistance against under unfavorable environmental conditions, including Cd stress.

## 8 CONCLUSIONS

Microbes play a vital role in ameliorating Cd toxicity in plants by triggering various defense-related mechanisms, such as signal transduction to perceive stress, and activation of stress-responsive mechanisms in plants, such as metal transporters, synthesis of primary and secondary metabolites, and modulation of antioxidant enzymes. In addition, many microbes, such as bacteria and fungi, directly support plants in mitigating Cd toxicity by making essential nutrients available to plants, producing phytohormones and enzymes, secreting nutrient chelating siderophores, synthesizing organic acids, and so forth. Thus, many potential microbes might be used as bioinoculants in Cd-prone soils to mitigate the toxic effects and increase plant productivity. Moreover, future research should focus on the exploration of the hidden potential of these microbes to mitigate the toxic effects of Cd.

## ACKNOWLEDGMENTS

The authors are thankful to all their colleagues for the encouragement to write this manuscript.

## REFERENCES

Adams, M., Zhao, F., McGrath, S., Nicholson, F., Chambers, B., 2004. Predicting cadmium concentrations in wheat and barley grain using soil properties. J. Environ. Qual. 33 (2), 532–541.

Ahmad, I., Akhtar, M.J., Zahir, Z.A., Naveed, M., Mitter, B., Sessitsch, A., 2014. Cadmium-tolerant bacteria induce metal stress tolerance in cereals. Environ. Sci. Pollut. Res. 21 (18), 11054–11065.

An, H., Liu, Y., Zhao, X., Huang, Q., Yuan, S., Yang, X., Dong, J., 2015. Characterization of cadmium-resistant endophytic fungi from *Salix variegata* Franch. in three gorges reservoir region, China. Microbiol. Res. 176, 29–37.

Archana, G., Buch, A., Kumar, G.N., 2012. Pivotal role of organic acid secretion by rhizobacteria in plant growth promotion. In: Satyanarayana, T., Johri, B.N. (Eds.), Microorganisms in Sustainable Agriculture and Biotechnology. Springer, Netherlands, pp. 35–53.

Astolfi, S., Zuchi, S., Chiani, A., Passera, C., 2003. In vivo andin vitro effects of cadmium on $H^+$ ATPase activity of plasma membrane vesicles from oat (*Avena sativa* L.) roots. J. Plant Physiol. 160 (4), 387–393.

Astolfi, S., Zuchi, S., Passera, C., 2005. Effect of cadmium on $H^+$ ATPase activity of plasma membrane vesicles isolated from roots of different S-supplied maize (*Zea mays* L.) plants. Plant Sci. 169 (2), 361–368.

ATSDR (Agency for Toxic Substances and Disease Registry), 1992. Prepared by Clement International Corp., under contract, 205, 88-0608.

Badri, D.V., Quintana, N., El Kassis, E.G., Kim, H.K., Choi, Y.H., Sugiyama, A., Verpoorte, R., Martinoia, E., Manter, D.K., Vivanco, J.M., 2009. An ABC transporter mutation alters root exudation of phytochemicals that provoke an overhaul of natural soil microbiota. Plant Physiol. 151 (4), 2006–2017.

Balestrasse, K.B., Benavides, M.P., Gallego, S.M., Tomaro, M.L., 2003. Effect of cadmium stress on nitrogen metabolism in nodules and roots of soybean plants. Funct. Plant Biol. 30 (1), 57–64.

Balestrasse, K., Gallego, S., Tomaro, M., 2004. Cadmium-induced senescence in nodules of soybean (Glycine max L.) plants. Plant Soil 262 (1), 373–381.

Belimov, A.A., Dietz, K.J., 2000. Effect of associative bacteria on element composition of barley seedlings grown in solution culture at toxic cadmium concentrations. Microbiol. Res. 155 (2), 113–121.

Belimov, A.A., Safronova, V.I., Sergeyeva, T.A., Egorova, T.N., Matveyeva, V.A., Tsyganov, V.E., et al., 2001. Characterization of plant growth promoting rhizobacteria isolated from polluted soils and containing 1-aminocyclopropane-1-carboxylate deaminase. Can. J. Microbiol. 47 (7), 642–652.

Belimov, A., Kunakova, A., Safronova, V., Stepanok, V., Yudkin, L.Y., Alekseev, Y.V., et al., 2004. Employment of rhizobacteria for the inoculation of barley plants cultivated in soil contaminated with lead and cadmium. Microbiology 73 (1), 99–106.

Benavides, M.P., Gallego, S.M., Tomaro, M.L., 2005. Cadmium toxicity in plants. Braz. J. Plant Physiol. 17 (1), 21–34.

Bhattacharyya, P., Jha, D., 2012. Plant growth-promoting rhizobacteria (PGPR): emergence in agriculture. World J. Microbiol. Biotechnol. 28 (4), 1327–1350.

Braud, A., Jézéquel, K., Bazot, S., Lebeau, T., 2009. Enhanced phytoextraction of an agricultural Cr-and Pb-contaminated soil by bioaugmentation with siderophore-producing bacteria. Chemosphere 74 (2), 280–286.

Brown, T., Idoine, N., Raycraft, E., Shaw, R., Deady, E., Hobbs, S., Bide, T., 2017. World Mineral Production 2011-15. British Geological Survey.

Cailliatte, R., Lapeyre, B., Briat, J.F., Mari, S., Curie, C., 2009. The NRAMP6 metal transporter contributes to cadmium toxicity. Biochem. J. 422 (2), 217–228.

Chmielowska-Bąk, J., Gzyl, J., Rucińska-Sobkowiak, R., Arasimowicz-Jelonek, M., Deckert, J., 2014. The new insights into cadmium sensing. Front. Plant Sci. 5, 245. https://doi.org/10.3389/fpls.2014.00245.

Cho, U.H., Seo, N.H., 2005. Oxidative stress in *Arabidopsis thaliana* exposed to cadmium is due to hydrogen peroxide accumulation. Plant Sci. 168 (1), 113–120.

Clemens, S., 2006. Toxic metal accumulation, responses to exposure and mechanisms of tolerance in plants. Biochimie 88 (11), 1707–1719.

Creyghton, M.P., Cheng, A.W., Welstead, G.G., Kooistra, T., Carey, B.W., Steine, E.J., et al., 2010. Histone H3K27ac separates active from poised enhancers and predicts developmental state. Proc. Natl. Acad. Sci. U. S. A. 107 (50), 21931–21936.

Cristaldi, A., Conti, G.O., Jho, E.H., Zuccarello, P., Grasso, A., Copat, C., et al., 2017. Phytoremediation of contaminated soils by heavy metals and PAHs. A brief review. Environ. Technol. Innov. 8, 309–326.

DalCorso, G., Farinati, S., Maistri, S., Furini, A., 2008. How plants cope with cadmium: staking all on metabolism and gene expression. J. Integr. Plant Biol. 50 (10), 1268–1280.

DalCorso, G., Farinati, S., Furini, A., 2010. Regulatory networks of cadmium stress in plants. Plant Signal. Behav. 5 (6), 663–667.

de Araújo, R.P., de Almeida, A.A.F., Pereira, L.S., Mangabeira, P.A., Souza, J.O., Pirovani, C.P., et al., 2017. Photosynthetic, antioxidative, molecular and ultrastructural responses of young cacao plants to Cd toxicity in the soil. Ecotoxicol. Environ. Saf. 144, 148–157.

Desai, A., Archana, G., 2011. Role of siderophores in crop improvement. In: Bacteria in Agrobiology: Plant Nutrient Management. Springer, pp. 109–139.

Dimkpa, C., Merten, D., Svatoš, A., Büchel, G., Kothe, E., 2009. Siderophores mediate reduced and increased uptake of cadmium by *Streptomyces tendae* F4 and sunflower (*Helianthus annuus*), respectively. Appl. Microbiol. Biotechnol. 107 (5), 1687–1696.

Dugar, G., Herbig, A., Förstner, K.U., Heidrich, N., Reinhardt, R., Nieselt, K., et al., 2013. High-resolution transcriptome maps reveal strain-specific regulatory features of multiple *Campylobacter jejuni* isolates. PLoS Genet. 9(5)e1003495.

Dzionek, A., Wojcieszyńska, D., Guzik, U., 2016. Natural carriers in bioremediation: a review. Electron. J. Biotechnol. 23, 28–36.

Elobeid, M., Göbel, C., Feussner, I., Polle, A., 2011. Cadmium interferes with auxin physiology and lignification in poplar. J. Exp. Bot. 63 (3), 1413–1421.

Farinati, S., DalCorso, G., Varotto, S., Furini, A., 2010. The *Brassica juncea* BjCdR15, an ortholog of *Arabidopsis* TGA3, is a regulator of cadmium uptake, transport and accumulation in shoots and confers cadmium tolerance in transgenic plants. New Phytol. 185 (4), 964–978.

Fayiga, A.O., Ma, L.Q., Cao, X., Rathinasabapathi, B., 2004. Effects of heavy metals on growth and arsenic accumulation in the arsenic hyperaccumulator *Pteris vittata* L. Environ. Pollut. 132 (2), 289–296.

Fodor, E., Szabó-Nagy, A., Erdei, L., 1995. The effects of cadmium on the fluidity and $H^+$-ATPase activity of plasma membrane from sunflower and wheat roots. J. Plant Physiol. 147 (1), 87–92.

Fomina, M., Alexander, I., Colpaert, J., Gadd, G., 2005. Solubilization of toxic metal minerals and metal tolerance of mycorrhizal fungi. Soil Biol. Biochem. 37 (5), 851–866.

Fusco, N., Micheletto, L., Dal Corso, G., Borgato, L., Furini, A., 2005. Identification of cadmium-regulated genes by cDNA-AFLP in the heavy metal accumulator *Brassica juncea* L. J. Exp. Bot. 56 (421), 3017–3027.

Gallego, S.M., Pena, L.B., Barcia, R.A., Azpilicueta, C.E., Iannone, M.F., Rosales, E.P., et al., 2012. Unravelling cadmium toxicity and tolerance in plants: insight into regulatory mechanisms. Environ. Exp. Bot. 83, 33–46.

Gamalero, E., Lingua, G., Berta, G., Glick, B.R., 2009. Beneficial role of plant growth promoting bacteria and arbuscular mycorrhizal fungi on plant responses to heavy metal stress. Can. J. Microbiol. 55 (5), 501–514.

Garbisu, C., Alkorta, I., 2003. Basic concepts on heavy metal soil bioremediation. Eur. J. Miner. Process. Environ. Prot. 3 (1), 58–66.

Gerhardt, K.E., Gerwing, P.D., Greenberg, B.M., 2016. Opinion: taking phytoremediation from proven technology to accepted practice. Plant Sci. 256, 170–185.

Gill, S.S., Tuteja, N., 2011. Cadmium stress tolerance in crop plants: probing the role of sulfur. Plant Signal. Behav. 6 (2), 215–222.

Glick, B.R., 2005. Modulation of plant ethylene levels by the bacterial enzyme ACC deaminase. FEMS Microbiol. Lett. 251 (1), 1–7.

Gomes, M., Marques, T., Carneiro, M., Soares, Â.M., 2012. Anatomical characteristics and nutrient uptake and distribution associated with the Cd-phytoremediation capacity of *Eucalyptus camaldulenses* Dehnh. J. Soil Sci. Plant Nutr. 12 (3), 481–496.

Gonzalez-Chavez, M., Carrillo-Gonzalez, R., Wright, S., Nichols, K., 2004. The role of glomalin, a protein produced by arbuscular mycorrhizal fungi, in sequestering potentially toxic elements. Environ. Pollut. 130 (3), 317–323.

Gratão, P.L., Polle, A., Lea, P.J., Azevedo, R.A., 2005. Making the life of heavy metal-stressed plants a little easier. Funct. Plant Biol. 32 (6), 481–494.

Grichko, V.P., Filby, B., Glick, B.R., 2000. Increased ability of transgenic plants expressing the bacterial enzyme ACC deaminase to accumulate Cd, Co, Cu, Ni, Pb, and Zn. J. Biotechnol. 81 (1), 45–53.

Guo, J., Chi, J., 2014. Effect of Cd-tolerant plant growth-promoting rhizobium on plant growth and Cd uptake by *Lolium multiflorum* Lam. and *Glycine max* (L.) Merr. in Cd-contaminated soil. Plant Soil 375 (1–2), 205–214.

Gupta, A., Gupta, R., Singh, R.L., 2017. Microbes and environment. In: Principles and Applications of Environmental Biotechnology for a Sustainable Future. Springer, pp. 43–84.

Hall, J., 2002. Cellular mechanisms for heavy metal detoxification and tolerance. J. Exp. Bot. 53 (366), 1–11.

Hanikenne, M., Talke, I.N., Haydon, M.J., Lanz, C., Nolte, A., Motte, P., et al., 2008. Evolution of metal hyperaccumulation required cis-regulatory changes and triplication of HMA4. Nature 453 (7193), 391–395.

Hasan, S.A., Fariduddin, Q., Ali, B., Hayat, S., Ahmad, A., 2009. Cadmium: toxicity and tolerance in plants. J. Environ. Biol. 30 (2), 165–174.

Honma, M., Shimomura, T., 1978. Metabolism of 1-aminocyclopropane-1-carboxylic acid. Agric. Biol. Chem. 42 (10), 1825–1831.

Hu, Y.F., Zhou, G., Na, X.F., Yang, L., Nan, W.B., Liu, X., et al., 2013. Cadmium interferes with maintenance of auxin homeostasis in *Arabidopsis* seedlings. J. Plant Physiol. 170 (11), 965–975.

Idris, R., Trifonova, R., Puschenreiter, M., Wenzel, W.W., Sessitsch, A., 2004. Bacterial communities associated with flowering plants of the Ni hyperaccumulator *Thlaspi goesingense*. Appl. Environ. Microbiol. 70 (5), 2667–2677.

Jadia, C.D., Fulekar, M., 2009. Phytoremediation of heavy metals: recent techniques. Afr. J. Biotechnol. 8 (6), 921–928.

Järup, L., Åkesson, A., 2009. Current status of cadmium as an environmental health problem. Toxicol. Appl. Pharmacol. 238 (3), 201–208.

Jasinski, M., Sudre, D., Schansker, G., Schellenberg, M., Constant, S., Martinoia, E., et al., 2008. AtOSA1, a member of the Abc1-like family, as a new factor in cadmium and oxidative stress response. Plant Physiol. 147 (2), 719–731.

Johnson, C., Boden, E., Arias, J., 2003. Salicylic acid and NPR1 induce the recruitment of trans-activating TGA factors to a defense gene promoter in *Arabidopsis*. Plant Cell 15 (8), 1846–1858.

Joshi, P.M., Juwarkar, A.A., 2009. In vivo studies to elucidate the role of extracellular polymeric substances from *Azotobacter* in immobilization of heavy metals. Environ. Sci. Technol. 43 (15), 5884–5889.

Kaur, G., Prabhavathi, V., Bamel, K., Sarwat, M., 2017. Phosphate signaling in plants: biochemical and molecular approach. In: Stress Signaling in Plants: Genomics and Proteomics Perspective. vol. 2. Springer, pp. 83–110.

Kavita, B., Shukla, S., Kumar, G.N., Archana, G., 2008. Amelioration of phytotoxic effects of Cd on mung bean seedlings by gluconic acid secreting rhizobacterium *Enterobacter asburiae* PSI3 and implication of role of organic acid. World J. Microbiol. Biotechnol. 24 (12), 2965–2972.

Kim, D.Y., Bovet, L., Maeshima, M., Martinoia, E., Lee, Y., 2007. The ABC transporter AtPDR8 is a cadmium extrusion pump conferring heavy metal resistance. Plant J. 50 (2), 207–218.

Kirkham, M., 2006. Cadmium in plants on polluted soils: effects of soil factors, hyperaccumulation, and amendments. Geoderma 137 (1), 19–32.

Kobayashi, T., Nishizawa, N.K., 2012. Iron uptake, translocation, and regulation in higher plants. Annu. Rev. Plant Biol. 63, 131–152.

Krämer, U., Talke, I.N., Hanikenne, M., 2007. Transition metal transport. FEBS Lett. 581 (12), 2263–2272.

Krantev, A., Yordanova, R., Janda, T., Szalai, G., Popova, L., 2008. Treatment with salicylic acid decreases the effect of cadmium on photosynthesis in maize plants. J. Plant Physiol. 165 (9), 920–931.

Kuffner, M., De Maria, S., Puschenreiter, M., Fallmann, K., Wieshammer, G., Gorfer, M., et al., 2010. Culturable bacteria from Zn-and Cd-accumulating *Salix caprea* with differential effects on plant growth and heavy metal availability. J. Appl. Microbiol. 108 (4), 1471–1484.

Leterrier, M., Airaki, M., Palma, J.M., Chaki, M., Barroso, J.B., Corpas, F.J., 2012. Arsenic triggers the nitric oxide (NO) and S-nitrosoglutathione (GSNO) metabolism in *Arabidopsis*. Environ. Pollut. 166, 136–143.

Li, X., Gitau, M.M., Han, S., Fu, J., Xie, Y., 2017. Effects of cadmium-resistant fungi *Aspergillus aculeatus* on metabolic profiles of bermudagrass [*Cynodondactylon* (L.) Pers.] under Cd stress. Plant Physiol. Biochem. 114, 38–50.

Lin, Y.F., Aarts, M.G., 2012. The molecular mechanism of zinc and cadmium stress response in plants. Cell. Mol. Life Sci. 69 (19), 3187–3206.

Ling-Zhi, L., Zong-Qiang, G., Zhang, Y.L., Pei-Jun, L., 2011. Growth, cadmium accumulation and physiology of marigold (*Tagetes erecta* L.) as affected by arbuscular mycorrhizal fungi. Pedosphere 21 (3), 319–327.

Lopez-Millan, A.F., Ellis, D.R., Grusak, M.A., 2004. Identification and characterization of several new members of the ZIP family of metal ion transporters in *Medicago truncatula*. Plant Mol. Biol. 54 (4), 583–596.

Madhaiyan, M., Poonguzhali, S., Sa, T., 2007. Metal tolerating methylotrophic bacteria reduces nickel and cadmium toxicity and promotes plant growth of tomato (*Lycopersicon esculentum* L.). Chemosphere 69 (2), 220–228.

Mahar, A., Wang, P., Ali, A., Awasthi, M.K., Lahori, A.H., Wang, Q., et al., 2016. Challenges and opportunities in the phytoremediation of heavy metals contaminated soils: a review. Ecotoxicol. Environ. Saf. 126, 111–121.

Majewska, M., Kurek, E., 2011. Effect of Cd concentration in growth media on Secale cereale roots and Cd interaction with rhizosphere microorganisms originating from different parts of the grain. Eur. J. Soil Biol. 47 (2), 95–101.

Martino, E., Perotto, S., Parsons, R., Gadd, G.M., 2003. Solubilization of insoluble inorganic zinc compounds by ericoid mycorrhizal fungi derived from heavy metal polluted sites. Soil Biol. Biochem. 35 (1), 133–141.

McCutcheon, S.C., Schnoor, J.L., 2004. Phytoremediation: Transformation and Control of Contaminants. vol. 121. John Wiley and Sons.

Mench, M., Vangronsveld, J., Lepp, N., Edwards, R., 1998. Physicochemical Aspects and Efficienty of Trace Element Immobilization by Soil Amendments. Springer, pp. 151–182.

Metwally, A., Finkemeier, I., Georgi, M., Dietz, K.J., 2003. Salicylic acid alleviates the cadmium toxicity in barley seedlings. Plant Physiol. 132 (1), 272–281.

Miransari, M., 2011. Hyperaccumulators, arbuscular mycorrhizal fungi and stress of heavy metals. Biotechnol. Adv. 29 (6), 645–653.

Moradi, A., Abbaspour, K., Afyuni, M., 2005. Modelling field-scale cadmium transport below the root zone of a sewage sludge amended soil in an arid region in Central Iran. J. Contam. Hydrol. 79 (3), 187–206.

Moreira, H., Marques, A.P., Franco, A.R., Rangel, A.O., Castro, P.M., 2014. Phytomanagement of Cd-contaminated soils using maize (*Zea mays* L.) assisted by plant growth-promoting rhizobacteria. Environ. Sci. Pollut. Res. 21 (16), 9742–9753.

Munasinghe, P.C., Khanal, S.K., 2010. Biomass-derived syngas fermentation into biofuels: opportunities and challenges. Bioresour. Technol. 101 (13), 5013–5022.

Nazar, R., Iqbal, N., Masood, A., Khan, M.I.R., Syeed, S., Khan, N.A., 2012. Cadmium toxicity in plants and role of mineral nutrients in its alleviation. Am. J. Plant Sci. 3 (10), 1476.

Nordberg, G.F., 2009. Historical perspectives on cadmium toxicology. Toxicol. Appl. Pharmacol. 238 (3), 192–200.

Nymark, M., Sharma, A.K., Sparstad, T., Bones, A.M., Winge, P., 2016. A CRISPR/Cas9 system adapted for gene editing in marine algae. Sci. Rep. 624951.

Onodera, Y., Suzuki, A., Wu, C.Y., Washida, H., Takaiwa, F., 2001. A rice functional transcriptional activator, RISBZ1, responsible for endosperm-specific expression of storage protein genes through GCN4 motif. J. Biol. Chem. 276 (17), 14139–14152.

Orłowska, E., Przybyłowicz, W., Orlowski, D., Turnau, K., Mesjasz-Przybyłowicz, J., 2011. The effect of mycorrhiza on the growth and elemental composition of Ni-hyperaccumulating plant *Berkheya coddii* Roessler. Environ. Pollut. 159 (12), 3730–3738.

Ovečka, M., Takáč, T., 2014. Managing heavy metal toxicity stress in plants: biological and biotechnological tools. Biotechnol. Adv. 32 (1), 73–86.

Patel, K.J., Singh, A.K., Nareshkumar, G., Archana, G., 2010. Organic-acid-producing, phytate-mineralizing rhizobacteria and their effect on growth of pigeon pea (*Cajanus cajan*). Appl. Soil Ecol. 44 (3), 252–261.

Perelo, L.W., 2010. In situ and bioremediation of organic pollutants in aquatic sediments. J. Hazard. Mater. 177 (1), 81–89.

Pishchik, V., Vorobyev, N., Chernyaeva, I., Timofeeva, S., Kozhemyakov, A., Alexeev, Y., et al., 2002. Experimental and mathematical simulation of plant growth promoting rhizobacteria and plant interaction under cadmium stress. Plant Soil 243 (2), 173–186.

Rajkumar, M., Sandhya, S., Prasad, M., Freitas, H., 2012. Perspectives of plant-associated microbes in heavy metal phytoremediation. Biotechnol. Adv. 30 (6), 1562–1574.

Rascio, N., Navari-Izzo, F., 2011. Heavy metal hyperaccumulating plants: how and why do they do it? And what makes them so interesting? Plant Sci. 180 (2), 169–181.

Rizwan, M., Ali, S., Adrees, M., Rizvi, H., Zia-ur-Rehman, M., Hannan, F., et al., 2016. Cadmium stress in rice: toxic effects, tolerance mechanisms, and management: a critical review. Environ. Sci. Pollut. Res. 23 (18), 17859–17879.

Rodríguez-Serrano, M., Romero-Puertas, M.C., Pazmino, D.M., Testillano, P.S., Risueño, M.C., Luis, A., et al., 2009. Cellular response of pea plants to cadmium toxicity: cross talk between reactive oxygen species, nitric oxide, and calcium. Plant Physiol. 150 (1), 229–243.

Rogers, E.E., Eide, D.J., Guerinot, M.L., 2000. Altered selectivity in an *Arabidopsis* metal transporter. Proc. Natl. Acad. Sci. U. S. A. 97 (22), 12356–12360.

Roth, U., von Roepenack-Lahaye, E., Clemens, S., 2006. Proteome changes in *Arabidopsis thaliana* roots upon exposure to $Cd^{2+}$. J. Exp. Bot. 57 (15), 4003–4013.

Safronova, V.I., Stepanok, V.V., Engqvist, G.L., Alekseyev, Y.V., Belimov, A.A., 2006. Root-associated bacteria containing 1-aminocyclopropane-1-carboxylate deaminase improve growth and nutrient uptake by pea genotypes cultivated in cadmium supplemented soil. Biol. Fertil. Soils 42 (3), 267–272.

Saluja, B., Sharma, V., 2014. Cadmium resistance mechanism in acidophilic and alkalophilic bacterial isolates and their application in bioremediation of metal-contaminated soil. Soil Sediment Contam. Int. J. 23 (1), 1–17.

Satarug, S., Baker, J.R., Urbenjapol, S., Haswell-Elkins, M., Reilly, P.E., Williams, D.J., et al., 2003. A global perspective on cadmium pollution and toxicity in non-occupationally exposed population. Toxicol. Lett. 137 (1), 65–83.

Sebastian, A., Prasad, M.N.V., 2014. Cadmium minimization in rice. A review. Agron. Sustain. Dev. 34 (1), 155–173.

Shahabivand, S., Maivan, H.Z., Goltapeh, E.M., Sharifi, M., Aliloo, A.A., 2012. The effects of root endophyte and arbuscular mycorrhizal fungi on growth and cadmium accumulation in wheat under cadmium toxicity. Plant Physiol. Biochem. 60, 53–58.

Shahabivand, S., Parvaneh, A., Aliloo, A.A., 2017. Root endophytic fungus *Piriformospora indica* affected growth, cadmium partitioning and chlorophyll fluorescence of sunflower under cadmium toxicity. Ecotoxicol. Environ. Saf. 145, 496–502.

Sharma, R.K., Archana, G., 2016. Cadmium minimization in food crops by cadmium resistant plant growth promoting rhizobacteria. Appl. Soil Ecol. 107, 66–78.

Sinha, S., Mukherjee, S.K., 2008. Cadmium–induced siderophore production by a high Cd-resistant bacterial strain relieved Cd toxicity in plants through root colonization. Curr. Microbiol. 56 (1), 55–60.

Sobariu, D.L., Fertu, D.I.T., Diaconu, M., Pavel, L.V., Hlihor, R.M., Drăgoi, E.N., et al., 2016. Rhizobacteria and plant symbiosis in heavy metal uptake and its implications for soil bioremediation. New Biotechnol. 39, 125–134.

Stein, W.D., Litman, T., 2014. Channels, Carriers, and Pumps: An Introduction to Membrane Transport. Elsevier.

Sun, X., Guo, L., 2013. Relationship between cadmium-induced root subapical hair development and ethylene biosynthesis in oilseed rape seedlings. Acta Biol. Cracov. Ser. Bot. 55 (2), 68–75.

Tahir, U., Yasmin, A., Khan, U.H., 2016. Phytoremediation: potential flora for synthetic dyestuff metabolism. J. King Saud Univ. Sci. 28 (2), 119–130.

Thakur, S., Singh, L., Ab Wahid, Z., Siddiqui, M.F., Atnaw, S.M., Din, M.F.M., 2016. Plant-driven removal of heavy metals from soil: uptake, translocation, tolerance mechanism, challenges, and future perspectives. Environ. Monit. Assess. 188 (4), 206.

Tkalec, M., Štefanić, P.P., Cvjetko, P., Šikić, S., Pavlica, M., Balen, B., 2014. The effects of cadmium-zinc interactions on biochemical responses in tobacco seedlings and adult plants. PLoS ONE 9(1)e87582.

Tran, T.A., Popova, L.P., 2013. Functions and toxicity of cadmium in plants: recent advances and future prospects. Turk. J. Bot. 37 (1), 1–13.

Tripathi, A., Singh, Y., Verma, D.K., Ranjan, M.R., Srivastava, S., 2016. Bioremediation of hazardous azo dye methyl red by a newly isolated Bacillus megaterium ITBHU01: process improvement through ANN-GA based synergistic approach. Indian J. Biochem. Biophys. 53, 112–125.

Tsavkelova, E., Klimova, S.Y., Cherdyntseva, T., Netrusov, A., 2006. Microbial producers of plant growth stimulators and their practical use: a review. Appl. Biochem. Microbiol. 42 (2), 117–126.

Uraguchi, S., Watanabe, I., Yoshitomi, A., Kiyono, M., Kuno, K., 2006. Characteristics of cadmium accumulation and tolerance in novel Cd-accumulating crops, *Avena strigosa* and *Crotalaria juncea*. J. Exp. Bot. 57 (12), 2955–2965.

Vamerali, T., Bandiera, M., Mosca, G., 2010. Field crops for phytoremediation of metal-contaminated land: a review. Environ. Chem. Lett. 8 (1), 1–17.

Verret, F., Gravot, A., Auroy, P., Leonhardt, N., David, P., Nussaume, L., et al., 2004. Overexpression of AtHMA4 enhances root-to-shoot translocation of zinc and cadmium and plant metal tolerance. FEBS Lett. 576 (3), 306–312.

Vidali, M., 2001. Bioremediation. An overview. Pure Appl. Chem. 73 (7), 1163–1172.

Visoottiviseth, P., Francesconi, K., Sridokchan, W., 2002. The potential of Thai indigenous plant species for the phytoremediation of arsenic contaminated land. Environ. Pollut. 118 (3), 453–461.

Wagner, G.J., 1993. Accumulation of cadmium in crop plants and its consequences to human health. Adv. Agron. 51, 173–212.

Wang, Z., Zang, C., Rosenfeld, J.A., Schones, D.E., Barski, A., Cuddapah, S., et al., 2008. Combinatorial patterns of histone acetylations and methylations in the human genome. Nat. Genet. 40 (7), 897–903.

Wani, P.A., Khan, M.S., Zaidi, A., 2007. Chromium reduction, plant growth–promoting potentials, and metal solubilizatrion by *Bacillus* sp. isolated from alluvial soil. Curr. Microbiol. 54 (3), 237–243.

Wehner, J., Antunes, P.M., Powell, J.R., Mazukatow, J., Rillig, M.C., 2010. Plant pathogen protection by arbuscular mycorrhizas: a role for fungal diversity? Pedobiologia 53 (3), 197–201.

Wu, H., Chen, C., Du, J., Liu, H., Cui, Y., Zhang, Y., et al., 2012. Co-overexpression FIT with AtbHLH38 or AtbHLH39 in *Arabidopsis*-enhanced cadmium tolerance via increased cadmium sequestration in roots and improved iron homeostasis of shoots. Plant Physiol. 158 (2), 790–800.

Xie, Y., Luo, H., Du, Z., Hu, L., Fu, J., 2014. Identification of cadmium-resistant fungi related to Cd transportation in bermudagrass [*Cynodon dactylon* (L.) Pers.]. Chemosphere 117, 786–792.

Yadav, S., 2010. Heavy metals toxicity in plants: an overview on the role of glutathione and phytochelatins in heavy metal stress tolerance of plants. S. Afr. J. Bot. 76 (2), 167–179.

Yang, Y., Singer, S.D., Liu, Z., 2011. Evaluation and comparison of the insulation efficiency of three enhancer-blocking insulators in plants. Plant Cell Tissue Organ Cult. 105 (3), 405–414.

Zarei, M., König, S., Hempel, S., Nekouei, M.K., Savaghebi, G., Buscot, F., 2008. Community structure of arbuscular mycorrhizal fungi associated to *Veronica rechingeri* at the Anguran zinc and lead mining region. Environ. Pollut. 156 (3), 1277–1283.

Zarei, M., Hempel, S., Wubet, T., Schäfer, T., Savaghebi, G., Jouzani, G.S., et al., 2010. Molecular diversity of arbuscular mycorrhizal fungi in relation to soil chemical properties and heavy metal contamination. Environ. Pollut. 158 (8), 2757–2765.

Zawoznik, M.S., Groppa, M.D., Tomaro, M.L., Benavides, M.P., 2007. Endogenous salicylic acid potentiates cadmium-induced oxidative stress in *Arabidopsis thaliana*. Plant Sci. 173 (2), 190–197.

Zeng, F., Ali, S., Zhang, H., Ouyang, Y., Qiu, B., Wu, F., et al., 2011. The influence of pH and organic matter content in paddy soil on heavy metal availability and their uptake by rice plants. Environ. Pollut. 159 (1), 84–91.

Zhai, Z., Gayomba, S.R., Jung, H.I., Vimalakumari, N.K., Piñeros, M., Craft, E., et al., 2014. OPT3 is a phloem-specific iron transporter that is essential for systemic iron signaling and redistribution of iron and cadmium in *Arabidopsis*. Plant Cell 26 (5), 2249–2264.

Zhou, C., Zhu, L., Ma, Z., Wang, J., 2017. *Bacillus amyloliquefaciens* SAY09 increases cadmium resistance in plants by activation of auxin-mediated signaling pathways. Genes 8 (7), 173.

## FURTHER READING

Jin, C.W., Ye, Y.Q., Zheng, S.J., 2013. An underground tale: contribution of microbial activity to plant iron acquisition via ecological processes. Ann. Bot. 113 (1), 7–18.

Lombi, E., Tearall, K.L., Howarth, J.R., Zhao, F.J., Hawkesford, M.J., McGrath, S.P., 2002. Influence of iron status on cadmium and zinc uptake by different ecotypes of the hyperaccumulator *Thlaspi caerulescens*. Plant Physiol. 128 (4), 1359–1367.

Ma, Y., Prasad, M., Rajkumar, M., Freitas, H., 2011. Plant growth promoting rhizobacteria and endophytes accelerate phytoremediation of metalliferous soils. Biotechnol. Adv. 29 (2), 248–258.

Moe, L.A., 2013. Amino acids in the rhizosphere: from plants to microbes. Am. J. Bot. 100 (9), 1692–1705.

Pinto, A., Mota, A., De Varennes, A., Pinto, F., 2004. Influence of organic matter on the uptake of cadmium, zinc, copper and iron by sorghum plants. Sci. Total Environ. 326 (1), 239–247.

Yoshihara, T., Hodoshima, H., Miyano, Y., Shoji, K., Shimada, H., Goto, F., 2006. Cadmium inducible Fe deficiency responses observed from macro and molecular views in tobacco plants. Plant Cell Rep. 25 (4), 365–373.

# CHAPTER 18

# *Solanum nigrum* L.: A Novel Hyperaccumulator for the Phyto-Management of Cadmium Contaminated Soils


**Hinnan Khalid*[,†], Muhammad Zia-ur-Rehman*, Asif Naeem[‡,§], Muhammad Usman Khalid[¶], Muhammad Rizwan[‖], Shafaqat Ali[‖], Muhammad Umair*, Muhammad Irfan Sohail*[,#]**

*Faculty of Agriculture, Institute of Soil and Environmental Sciences, University of Agriculture, Faisalabad, Pakistan

[†]Punjab Agriculture Department, Shahkot, Pakistan

[‡]Soil and Environmental Sciences Division, Nuclear Institute for Agriculture and Biology (NIAB), Faisalabad, Pakistan

[§]Institute of Soil and Environmental Sciences, University of Agriculture, Faisalabad, Pakistan

[¶]Department of Fisheries and Aquaculture, University of Veterinary and Animal Sciences, Lahore, Pakistan

[‖]Department of Environmental Sciences and Engineering, Government College University, Faisalabad, Pakistan

[#]Indian River Research and Education Center, Institute of Food and Agricultural Sciences, University of Florida, Fort Pierce, FL, United States



## Abstract

Cadmium (Cd) toxicity is a major threat for food security all over the world. High mobility, persistent bio-availability, and unavoidable toxicity risks, even at low concentrations, have highlighted Cd as one of the most dangerous heavy metals (HMs). However, use of *Solanum nigrum* (*S. nigrum*) for phyto-management of Cd-contaminated soil is well documented. This chapter will explain the negative impacts of Cd toxicity and the role of *S. nigrum* in phyto-management of Cd-contaminated soils. *S. nigrum,* commonly known as black nightshade, is a perennial weed with a fast-growing nature and high biomass production, and bulk Cd accumulation in its various tissues characterizes it as a Cd hyperaccumulator plant. *S. nigrum* possesses some unique characteristics, such as release of organic acids, activation of natural resistance-associated macrophage proteins (NRAMPs), and use of the antioxidant defense mechanism for phyto-management of Cd. Therefore, *S. nigrum* restricts Cd entry into the food chain by binding Cd into its nonactive parts, coupled with Cd complexation and/or stabilization into the rhizosphere with the release of root exudates. However, use of some additional strategies, such as farming practices, growth activators, soil amendments, and transgenic aid accelerates the *S. nigrum*-based Cd phyto-management technique significantly. At the moment,



*Cadmium Toxicity and Tolerance in Plants*
https://doi.org/10.1016/B978-0-12-814864-8.00018-8

long-term field trials are required for testing the effectiveness and economic viability of this proximal technique for decontamination of Cd-contaminated soils on a sustainable basis.

**Keywords:** Soil contamination, Cadmium, Phyto-extraction, *Solanum nigrum* L., Management

## 1 INTRODUCTION

Although this revolutionized century has provided leisure to humanity, however, this advancement has been made at the cost of natural resources. The environment has become polluted, and soil resources have been increasingly damaged through anthropogenic contamination with a wide range of toxic contaminants. Among these contaminations, Cd contamination into agri-lands due to aggregated human activities has become the most threatening and substantial global concern for both society and the scientific community (Wesam and Hajer, 2014; Yu et al., 2015; Rizwan et al., 2016b; Rehman et al., 2017a) because of its toxicity to almost all biota, even at low intakes, enhanced solubility in all environmental components, persistence over a longer time, and particularly due to its manifold risk of transfer into the human food chain. Major sources of Cd contamination include dumping of sewage and/or raw city effluents into agri-lands (Farahat and Linderholm, 2015; Khan et al., 2016; Qayyum et al., 2017), burning of fossil fuels (Muradoglu et al., 2015), wastes of mining and smelting industries (Chen et al., 2016), and extensive use of pesticides and fertilizers (Hamzah et al., 2016; Iqbal et al., 2016; Rafique and Tariq, 2016). Cd, being highly mobile and persistent, has increased its accumulation level into the food chain, and thus, imparts unavoidable health issues (Kim and Lee, 2010; Yu et al., 2015; Rehman et al., 2017a; Qayyum et al., 2017). International environmental and food agencies have declared Cd a food chain contaminant and carcinogenic element for human consumption. Moreover, Cd toxicity impairs plant growth and development by perturbing the plant's physiological, enzymatic, and photosynthetic processes, coupled with a disturbance in gas exchange attributes and the nutritional status of plants (Adrees et al., 2015; Yang et al., 2015; Yu et al., 2015). Thus, it has become a threat to global food productivity. Therefore, it is urgently necessary to adopt eco-friendly and economically feasible strategies to combat the prevailing challenges of remediation of Cd-contaminated soils. Among countless promising approaches, phytoremediation has gained the spotlight, being socially and economically acceptable green strategy to remediate Cd-contaminated soils, with the use of hyperaccumulators (Luo et al., 2011; Liu et al., 2015; Rehman et al., 2017b). Cd hyperaccumulators are of particular interest because of their ability to tolerate and take up significant amounts of Cd from growing media. The Cd hyperaccumulators can efficiently grow under a stressed environment, and can translocate >100 mg Cd $kg^{-1}$ dry weight in above-ground tissues. This green technology may suffer a few limitations, such as low Cd availability in soil, a decreased translocation rate, plant biomass, and so forth.

Among the plants that are Cd hyperaccumulators, most belong to the *Brassicaceae* family. However *Solanum nigrum*, which belongs to the *Solanaceae* family, is acknowledged as a medicinal plant. *S. nigrum* has been distinguished as the most efficient potential hyperaccumulator for the remediation of Cd-contaminated soils (Song et al., 2014; Wei et al., 2014b; Liu et al., 2015; Yu et al., 2015; Rehman et al., 2017b). Many Cd hyperaccumulators were identified from contaminated sites, but *S. nigrum* was identified in Cd-free soil. *S. nigrum*, being a perennial weed, has been preferred over other Cd hyperaccumulators due to its extensive biomass production and its fast-growing systems, along with its potential for stress resistance (Chen et al., 2014; Saad-Allah and Elhaak, 2015; Cheng et al., 2016; Hammami et al., 2016; Chandra et al., 2017). *S. nigrum* has an advantage over other hyperaccumulators because it can be transplanted/cropped twice a year, thus, its Cd removal efficiency is far better than other Cd hyperaccumulators (Ji et al., 2011b). Many researchers have found *S. nigrum* good for the remediation of Cd-contaminated soils, and their pot experiments have concluded that they are in favor of this annual herb as the best potential Cd remediation plant, which can remediate Cd up to 125 mg $kg^{-1}$ soil (Wei and Zhou, 2004). Due to these facts, *S. nigrum* has become the preferred hyperaccumulator for Cd remediation during the phyto-remediation technique, and therefore, many field trials are still underway to determine additional beneficial aspects of *S. nigrum* for the remediation of Cd-contaminated soils under natural conditions (Peng et al., 2009; Ji et al., 2011a,b; Gao et al., 2012; Chen et al., 2014; Song et al., 2014; Wesam and Hajer, 2014; Yu et al., 2015; Jiang et al., 2016; Rehman et al., 2017b).

Various studies reveal that *S. nigrum* adopts many processes, such as the use of antioxidants, amino acids, organic acids, proline, and phyto-chelatins that support it for tolerance under unfavorable conditions, and higher uptake of Cd from Cd-contaminated soils (Xiao et al., 2010; Chen et al., 2012; Freitas et al., 2013; Khan et al., 2014; Wei et al., 2014a,b; Liu et al., 2015). However, researchers are also of the view that under elevated Cd contamination, *S. nigrum* suffers from severe oxidative stress that reduces plant biomass and slows down the growth rate. Therefore, to overcome such concerns, it has been recommended to use phyto-management enhancing strategies, such as farming practices, growth regulators, soil amendments, and transgenic aids that improve the phyto-remediation potential of *S. nigrum* (Sun et al., 2007; Chen et al., 2014; Khan et al., 2015a,b; Mao et al., 2015; Zou, 2015; Rehman et al., 2017b). Thus, at the moment, it is beneficial to focus on the physiological and biochemical mechanisms adopted by *S. nigrum* for the remediation of Cd-contaminated soils, and to identify of the best strategies for increasing its Cd accumulation potential.

## 2 Cd IN SOIL-PLANT SYSTEMS

To expound on the dynamics of Cd in soil-plant systems, it is necessary to take a comprehensive look at the sources of Cd, distribution of various forms of Cd, and its potential for toxicity in biological systems.

### 2.1 Sources of Soil Cd Contamination

Major sources of soil Cd contamination include Cd-enriched parent material, irrigation with sewage water and/or raw city effluents, extensive use of agricultural chemicals such as pesticides and especially phosphatic fertilizers, use of soil as a sink for industrial waste and/or by-products, and aerial deposition of Cd into the soil (Wuana and Okieimen, 2011; Rehman et al., 2015). Fig. 1 depicts a clear picture of sources of soil Cd contamination.

Due to variations in all these processes, Cd concentration varies from 0.02 to 6.9 $mg\,kg^{-1}$ soil (Baize, 1997; Mench et al., 1997). However, it has been reported that soil solution Cd contamination varies from 0.2 $\mu g\,L^{-1}$ to 300 $\mu g\,L^{-1}$ (Itoh and Yumura, 1979). Rehman et al. (2015) stated that Cd is present in higher concentrations in household sewage, as compared with other heavy metals (HMs), due to flushing of cigarette butts into domestic toilets, along with frictional particles of car tires on roads, and dumping of deteriorated rubber tires into open fields. Moreover, composts and sludge incorporated into agricultural fields contain high levels of Cd contamination. Liphadzi and Kirkham (2006) revealed the presence of 4.2 $mg\,Cd\,kg^{-1}$ soil in Topeka, Kansas, where composted sludge was used to enhance the agricultural productivity. Moreover, higher levels of Cd contamination occur due to the extensive use of phosphatic fertilizers, pesticides, manure, and raw city effluents for irrigational purposes (Nagajyoti et al., 2010; Qayyum et al., 2017; Rehman et al., 2017a). Besides these sources, aerial deposition of airborne Cd into agri-lands is also an important input of Cd

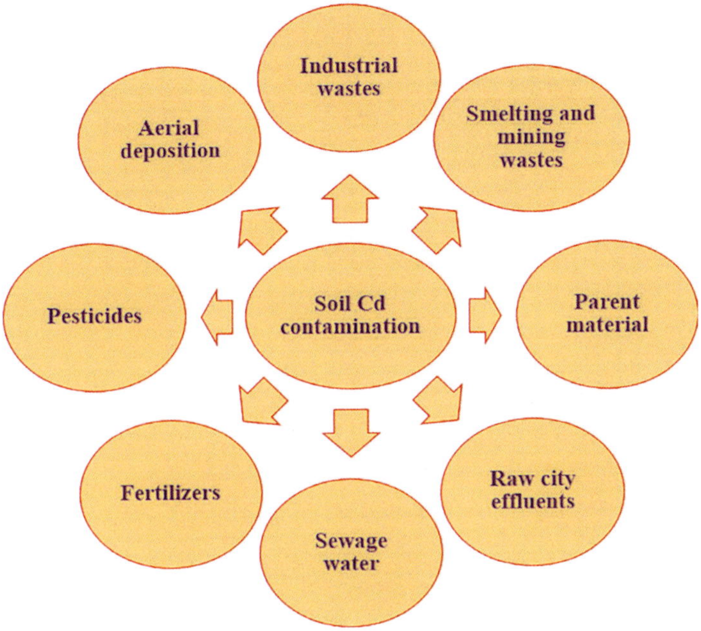


**Fig. 1** Major sources of soil Cd contamination.

concentration. Currently, the nonstop burning of fossil fuels, wild forest fires, and emissions from vehicles have increased environmental pollution, as well as soil Cd contamination, up to many-folds. Therefore, in many parts of the world, a raised level of Cd, beyond the threshold level of soil Cd contamination ($3\,mg\,kg^{-1}$), has created a serious challenge for the whole world.

## 2.2 Bio-Available Cd in Soil

Cd has been declared a relatively rare element. However, it is widely distributed in the earth's crust, ranging from 0.1 to $0.2\,mg\,kg^{-1}$. According to the report of UNEP (2010) Cd is available in various complex forms such as Zn, Pb, and Cu ores. Naturally, Cd is present 0.1–0.5 ppm in the earth's crust, which varies based on differing factors. Cook and Morrow (1995) stated that sedimentary rocks contain a relatively higher Cd concentration (0.1–25 ppm) as compared with igneous and metamorphic rocks (0.02–0.2 ppm). Cd is also present in various forms in the soil, such as free Cd ions, exchangeable Cd, surface adsorption of Cd on clay particles, iron oxides, and organic matter (OM) (Rehman et al., 2015). Moreover, Cd also forms various complexes with chlorides, sulfates, amino acids, and carboxylic acids (Sammut et al., 2010; Vega et al., 2010). Adriano et al. (2004) believe that Cd behavior is directly related to the forms of Cd, and interaction of these forms with each other. Because plants uptake only bio-available Cd concentrations, this is more important as compared with the total Cd concentration in the soil (Kirkham, 2006). Therefore, for the evaluation of the bio-available Cd concentration, it is necessary to extract the specific soil part, and then analyze the Cd concentration that plants uptake from that portion of soil. Many protocols are in use to determine the bio-available metal concentration. For the proper completion of any procedure of determining the bio-available metal concentration, use of an efficient extractant is of high importance. The concerned extractant should clearly depict the Cd concentration that plants uptake from the soil, or in other words, as Cd is present in various forms in soil, it should only remove the bio-available Cd concentration. However, until now there has been no uniform extractant used for this purpose; rather, the use of certain chemicals has been recommended as the most efficient technique (Gupta and Aten, 1993). Kennedy et al. (1997) have declared that the ideal extractant sustains the soil solution equilibrium, and the soil solution pH, and forms of Cd ions present at the exchange sites. Among the most efficient protocols used for the extraction of bio-available Cd is the use of DTPA-TEA (di-ethylene tri-amine penta acetic-acid and tri-ethanol amine) (Borges, 2002; Keller et al., 2005). Many reputable researchers have used the DTPA-TEA method for the determination of bio-available Cd concentrations (Rehman et al., 2015; Rizwan et al., 2016a,b; Rehman et al., 2017a; Qayyum et al., 2017). Therefore, it can be concluded that use of the DTPA-TEA procedure for the determination of bio-available Cd concentrations is the most reliable method.

## 2.3 Cd Phytotoxicity

Cd phytotoxicity is a serious threat for the agriculture sector, and this challenge must be addressed to ensure existence of healthy life on this planet. It has been reported that Cd concentration $>110.6\,mg\,kg^{-1}$ dry matter (DM) in plants is responsible for reducing the plant growth up to 50%. However, this value varies from plant to plant, and is $>10\,mg\,kg^{-1}$ DM, $>500\,mg\,kg^{-1}$ DM and $>300\,mg\,kg^{-1}$ DM for soybeans, radish tops, and radish roots respectively; when grown under Cd contamination of 10–190 $mg\,kg^{-1}$ soil with the use of $Cd(NO_3)_2$ salt (Miller et al., 1976; John et al., 1972). Turner (1973) proposed this value as the $2.5\,mg\,kg^{-1}$ DM for beet root, $2.0\,mg\,kg^{-1}$

DM for carrots, $150\,mg\,kg^{-1}$ DM for swiss chard, and $158\,mg\,kg^{-1}$ DM for tomatoes. However, a higher concentration of organic matter (OM) in agricultural soil creates Cd complexes (Cd-OM) and lowers the bio-available Cd concentration (Street et al., 1978; Sauve et al., 2000). Similarly, Street et al. (1978) also assert that lime-rich soils reduce the plant-available Cd concentration due to the formation of inorganic complexes of Cd, such as cadmium carbonate ($CdCO_3^{\circ}$) and cadmium hydro-oxide ($CdOH^{+}$). Moreover, accumulation of Cd in edible plant parts poses serious health issues for humans and animals (Yu et al., 2015; Rizwan et al., 2016a; Rehman et al., 2017a; Qayyum et al., 2017). Coupled with all this, Cd toxicity also lowers the market value of agricultural products, such as cereal crops (Nogawa and Kido, 1996). In addition, Li et al. (2005) stated that rice crops face serious market issues in China due to Cd impurity in the paddies. Schnoor (2004) revealed that rice grown in Cd-contaminated soil accumulates Cd $>3\,mg\,kg^{-1}$ DM, while the permissible limit explained by the World Health Organization (WHO) is $<0.2\,mg\,kg^{-1}$ DM. Therefore, Cd toxicity has hazardous effects on plants, animals, and human beings.

## 3 TRANSFORMATION BEHAVIOR OF Cd IN SOIL-PLANT SYSTEMS

Plants take up Cd from soil solutions through plant root systems, along with the absorption of plant nutrients. Moreover, various ion transporters, such as magnesium ($Mg^{2+}$), iron ($Fe^{2+}$), zinc ($Zn^{2+}$), and calcium ($Ca^{2+}$) also facilitate Cd uptake and accumulation into various plant parts (Clemens, 2006). On the other hand, Cd uptake and accumulation varies with plant type and species (Liu et al., 2003; Adriano et al., 2004; Grant et al., 2008; Vega et al., 2010). Mainly, plants uptake Cd through roots, and then, after passing through xylem, accumulate into the grain and/or fruiting parts. Besides Cd uptake by roots, some portion from the soil solution also becomes adsorbed on the surface of the plant roots. Moreover, Hentz et al. (2012) has explained the direct relation between the Cd uptake and exposure time, along with the Cd concentration into the soil solution.

Gill et al. (2011) stated that after Cd uptake by the roots, some portion becomes accumulated into the roots; while some part is transferred to the shoots, and becomes accumulated in above-ground tissues. However, Abe et al. (2008) are of the view that Cd transfer and accumulation varies from plant to plant, and generally, a small portion get transferred from roots to shoots. Cd uses the apoplastic and/or symplastic pathways for transferring from roots to shoots via xylem (Benavides et al., 2005). Hart et al. (2006) and Lux et al. (2011) have revealed that most of the Cd becomes accumulated in the vacuoles and/or nuclei of the roots in the forms of Cd-organic acid, and/or phyto-chelatin complexes. In addition, for transferring Cd from the lower parts of the plants to the upper parts, plasmalemma and/or casparian strips play a supportive role for Cd movement through the xylem and phloem (Clemens et al., 2002; Seregin et al., 2004). Thus, plants use either passive transport (with the help of transpiration pull), and/or active transport (with the use of ion transporters) for Cd transfer from the roots to the shoots (Nakanishi et al., 2006; Van der Vliet et al., 2007; Zorrig et al., 2010). However, Cd transfer and/or accumulation into grains of cereal crops takes place with the help of phloem sap, while the existence of other ions creates competition for Cd accumulation/storage (Liu et al., 2007; Van der Vliet et al., 2007; Yoneyama et al., 2010).

## 4 PHYTO-ACCUMULATION OF Cd BY *S. nigrum*

*S. nigrum* has a greater ability for Cd accumulation than most other plant species (Ghani et al., 2014; Khan et al., 2015b; Ji et al., 2016). Cd bioavailability is directly related to the pH of the rhizosphere, as pH plays an important role for determining the mobility and/or solubility of the Cd ions (Rehman et al., 2015, 2017b). Bao et al. (2011) has explained that the *S. nigrum* releases various types of organic acids through root hairs that enhance the mobility and/or solubility of the bound Cd. High availability of the Cd stimulates the Cd uptake and accumulation phenomenon of *S. nigrum* within the extensive root system, and then, transfers it to the aerial parts (Wei et al., 2014b; Yu et al., 2015; Rizwan et al., 2016b). Song et al. (2014) has declared the supportive role of NRAMPs for Cd uptake and translocation within *S. nigrum*. During Cd toxicity, an excess release of NRAMPs occurs into the *S. nigrum* that enhance the Cd accumulation ability and make it an efficient hyperaccumulator for remediation of Cd-contaminated soils. However, concentration of NRAMPs varies from roots to shoots, and thus, variation occurs for Cd accumulation in different parts of *S. nigrum*. Dwivedi et al. (2014) has explained the Cd accumulation trend in various parts of *S. nigrum* with the maximum accumulation in the leaf, followed by the shoots and roots. However, contrary to this, many researchers are in favor of maximum Cd accumulation in the roots of the *S. nigrum*, as compared with the other plant parts (Fidalgo et al., 2011; Khan et al., 2014). Similarly, among 12 different plant species, Kumar et al. (2013) and Liu et al. (2014) have suggested that *S. nigrum* is the most efficient potential hyperaccumulator for the phyto-remediation of Cd-contaminated soils. In addition, Peng et al. (2009) has described that the Cd concentration can be accumulated up to 262 mg kg$^{-1}$ in leaves, and 142 mg kg$^{-1}$ in shoots of *S. nigrum* during the phyto-remediation process. In a study conducted by Sun et al. (2009), the bio-accumulation factor for Cd was found to be 38 for *S. nigrum* grown under stress of 50 mg Cd kg$^{-1}$ soil. Thus, many reputed researchers are in favor of *S. nigrum* because of its high bioaccumulation and translocation factors for Cd, and the fact that it is the most efficient Cd hyperaccumulator, compared with other plants, such as *Brassica rapa*, *Ricinus communis*, *Brassica juncea*, and *Nicotiana tabacum* (Pinto et al., 2009; Song et al., 2014; Singh et al., 2016).

## 5 MECHANISMS OF Cd TOLERANCE IN *S. nigrum*

*S. nigrum* uses various physiological and biochemical processes for Cd tolerance under Cd toxicity stress. For this purpose, *S. nigrum* involves the phyto-chelation and antioxidant defense mechanism that has been described as follows.

### 5.1 Phyto-Chelation

Phyto-chelations act as the key process for improving the tolerance in *S. nigrum* against Cd stress (Yadav, 2010; Rehman et al., 2017b). During stress conditions, an excess release of phyto-chelatins enhances the ability of *S. nigrum* to lower the damaging effects of Cd toxicity, and thus, is a major reason behind the declaration of *S. nigrum* as the potential Cd hyperaccumulator (Gao et al., 2010). Sun et al. (2007) reported that with increasing the Cd exposure to *S. nigrum*, the increased production of phyto-chelatins was found more in the root parts, as compared with the shoots of the plant. Moreover, Ferraz et al. (2012) and Wei et al. (2016) stated that production of metallothioneins (metal chelator) in the *S. nigrum* stimulates Cd homeostasis and detoxification during Cd toxicity stress. Moreover, proline production in *S. nigrum* also plays a supportive role against Cd stress conditions, and enhances the Cd tolerance capacity of *S. nigrum* (Sun et al., 2007) by detoxifying the reactive oxygen species (ROS) generated during the stressed period. Siripornadulsil et al. (2002) and Sun et al. (2007) have reported that proline form the proline-Cd complex by following the chelation process, and support *S. nigrum* to tolerate Cd toxicity stress. Therefore, chelation is an important mechanism for improving tolerance against Cd toxicity stress in *S. nigrum*.

### 5.2 Antioxidant Defense System

Among various tolerance mechanisms of *S. nigrum*, activation of the antioxidant defense system is prominent (Gao et al., 2010; Khan et al., 2014). Major antioxidants, including glutathione (GSH), catalase (CAT), guaiacol peroxidase (POD), ascorbate peroxidase (APX), and superoxide dismutase (SOD) increase their activity in different parts of the *S. nigrum,* and improve the Cd tolerance mechanism under Cd toxicity (Yadav, 2010). However, the Cd tolerance ability of *S. nigrum* depends on the type of the antioxidant involved, Cd toxicity level, Cd exposure duration, and plant tissue. Khan et al. (2014) has reported that with the increase in Cd toxicity stress, activities of POD and GSH enzymes increase up to many-fold for enhancing the Cd tolerance ability of *S. nigrum*. Contrary to this, Bao et al. (2011) has stated that under Cd stress conditions, activities of POD and SOD decrease, while activity of CAT increase in various parts of the *S. nigrum*. Moreover, they have also reported that antioxidant enzyme activities were much higher in *S. nigrum,* as compared with the tomato plant. In addition, Gao et al. (2010) explained that during their experimentation, they noted the increased activities of SOD and POD, while a significant reduction was also observed for GSH with the increase in Cd exposure duration to the *S. nigrum*. Overall, Cd tolerance ability of *S. nigrum* increased with the variation in the activities of the antioxidant enzymes (either decrease or increase), depending on the Cd exposure duration and type of the plant tissue.

## 6 PHYTO-MANAGEMENT ENHANCING STRATEGIES OF *S. nigrum*

Possible techniques of bio-remediation include phyto-extraction, bio-stimulation, phyto-degradation, bio-augmentation, rhizo-remediation, rhizo-degradation, and geno-remediation. Phyto-management of Cd-contaminated soils using *S. nigrum* can be improved through resistance and tolerance, or creating metal detoxification by improving chelation and antioxidant systems, which may help in higher Cd uptake, in addition to mineral nutrients under integrated management practices.

### 6.1 Farming Practices

Biomass of the remediation plant, the most important parameter of phyto-remediation, can be increased by using certain farming practices, such as cocropping, appropriate planting density, preharvesting, and so forth (Ji et al., 2011b; Niu et al., 2015; Wang et al., 2015; Zou, 2015). By following these protocols, phyto-remediation activity of *S. nigrum* in Cd-polluted soils can be increased up to many-folds (Liu et al., 2012; Kidd et al., 2015; Ogbazghi et al., 2015; Xu et al., 2015).

### 6.1.1 Double Cropping

The double cropping pattern (growing hyperaccumulator plants along with low accumulating crops) was developed to boost the phyto-extraction potency of HMs from contaminated soil, along with the cultivation of agriculture crops (Whiting et al., 2001; Wu et al., 2007). However, output of this planting strongly depends on the interaction between the coinciding plant species and the type of metals on site (Zou, 2015). In this system, plants interact with each other for the distribution of the nutrient, and show symbiotic association; hence, supporting each other (Lu et al., 2012). Normally, the microbes and root exudates present among the co-cropped plant species support the growth and metal uptake (Zhang and Li, 2003). So, the phyto-extraction capability of *S. nigrum* can be beneficial, in ways other than enhancement by co-croppnig with certain plants such as rye grass (Zou, 2015). *S. nigrum* has more potential (hyperaccumulator) to extract available Cd and/or other HMs from soil than (low accumulator) rye grass (Wei et al., 2006). Similarly, *S. nigrum* showed significant results when intercropped with Welsh onions and Chinese cabbage (Niu et al., 2015; Wang et al., 2015). Thus, it is economical, and helpful, to use this pattern in phyto-remediation of Cd-containing soils by accumulation of bio-available Cd in *S. nigrum* biomass, which makes the soil-free of it, and allows farmers to proceed by growing crops within the remediation period, without waiting for the decontamination stage (Zou, 2015).

### 6.1.2 Premature Harvesting

The major restriction in the use of phyto-remediation technology is the long growing period of hyperaccumulator plants to obtain high biomass (Zhou and Song, 2004). Phyto-remediation efficacy of *S. nigrum* can be increased using a newly suggested premature harvesting technique, as it decreases the time for remediation by improving the rate of extraction of Cd up to many times (Wei et al., 2006). It has been recorded in a study by Wei et al. (2006) that harvesting at the inflorescence stage twice in a year, rather than waiting for the maturity stage, not only increases the Cd extraction efficiency of *S. nigrum* up to 75%, but also shortens the remediation period, as it takes

35 days to reach it at the flowering stage, and a further 60 days to reach it at maturity. Comparable results were disclosed by Ji et al. (2011a,b) and Ningyu et al. (2016) related to Cd extraction by *S. nigrum*, sweet sorghum, amaranth, and Sudan grass using multiple harvesting techniques.

### 6.1.3 Plant Density

Plant density has a significant role in increasing the biomass of the plant, and hence, in the accumulation of HMs. Liphadzi et al. (2003) have reported that higher plant density can increase the phyto-extraction ability of Cd by 1.41 times in the case of sunflowers, relative to lower plant density. Plant density is affected by many factors, including both external and internal, which directly or indirectly affect the seedling emergence and growth (Grundy et al., 2003; Zhou et al., 2005). The optimum burial depth reported for *S. nigrum* is 1–2 cm; a further increase causes a decrease in the emergence of seedlings, and at or above 8 cm, there is no emergence at all (Zhou et al., 2005). There is a substantial influence of seed density on the emergence and growth of seedlings of many crops and weed species (Grundy et al., 2003; Maddonni and Otegui, 2004; Chauvel et al., 2005), including *S. nigrum* (Lati et al., 2012). There is always an intra-specific competition for space, nutrients, and light when in the per unit area, plants exceed the optimum limit, which ultimately affects the plant growth and development, leading to a decrease in plant biomass (Evans et al., 2003). Hence, an ideal seed density is always required for healthy plants. In the case of *S. nigrum,* one plant/30 $cm^2$ or 11 plants per $m^2$ is ideal for the highest biomass.

## 6.2 Growth Activators

Bioavailability of metals in soil systems, and good biomass of plants, can reinforce the phyto-remediation activity, while their limitation can decrease the phyto-extraction (Gerhardt et al., 2009). To rectify this, auxiliary substrates can be used, such as soil conditioners, chelating agents, microbes, and different chemical coordinators, as they promote the metal mobility, both in soil and plants, and plant growth, by enhancing different regulatory mechanisms, including *S. nigrum* (Vaxevanidou et al., 2008; Schoebitz et al., 2013; Ji et al., 2015a,b). Pretreatment of seeds/seedlings with some growth regulators also help the plant to sustain under a stressed environment. A study by Xu et al. (2009) proves that pretreatment of *S. nigrum* seedlings with proline significantly improves the callus and plantlet's growth of *S. nigrum* under Cd stress, helps in regulating biochemical and intracellular responses to Cd toxicity, and enhances the Cd absorption.

### 6.2.1 Soil Conditions

All of the plant activities that are associated with its growth and development (i.e., cell division, seed dormancy, seedling emergence, leaf senescence, cell growth, and differentiation) are related to defense toward biotic and abiotic stresses, including tolerance or resistance to HMs, which are controlled by the chemical coordinators (Chen et al., 2011; Ma et al., 2013; Yan et al., 2015; Liu et al., 2016; Soares et al., 2016; Wei et al., 2016). Similarly, it can be helpful to cope with Cd stress through exogenous application of assisting conditioners. These can be made available to the plant exogenously, or by inducing the in-built production of these in the plant itself. In this respect, low molecular weight acids, or some plant growth regulators, including gibberellins, jasmonates, indole acetic acid, *N*-acetyl-L-cysteine, citric acid, and so forth can be used, as they enhance the phytoextraction ability by making the HMs available to the *S. nigrum* and transporting them to the biomass of the plant by increasing translocation from soil to plant (Braud et al., 2009; Deng et al., 2010; Gao et al., 2012; Ji et al., 2015a,b; Yan et al., 2015).

Citric acid can reduce the Cd oxidative stress by activating antioxidant enzymes, formation of complexes with ligands, improving carbon dioxide fixation in plants, and increasing metal solubility (Gao et al., 2010, 2012). In addition, it strengthens the roots of the *S. nigrum* (Mao et al., 2015). Likewise, Xu et al. (2012) has reported about amino acid and malic acid pretreatment, that it can increase the bioaccumulation factor of Cd in *S. nigrum*. Gibberellins promote *S. nigrum* leaf biomass, as well as that of shoots, so, the Cd concentration in shoots increases (Ji et al., 2015b). Indole acetic acid is also effective for promoting Cd concentration in the upper portion of *S. nigrum* (Ji et al., 2015a).

### 6.2.2 Microbes

Hyper-accumulation is one of the eco-friendly and economically feasible treatments for the remediation of heavy metal polluted soils, but the technique has limitations, due to sensitivity of roots to HMs and low biomass production. This effect can be compensated by the inoculation of microbes. Rhizo-bacteria, arbuscular mycorrhizal fungi (AMF), and endophytic bacteria enhance the heavy metal uptake by supporting the plant, and improving the nutrient uptake and tolerance against HMs. AMF support the plant by making hyphal networks (Garg and Chandel, 2010), which increase the contact of roots with soil, which ultimately increases the metal uptake rate.

It not only improves the tolerance to metal stress (Wan et al., 2012), it also promotes growth and enhances the mineral nutrients uptake.

Many studies have revealed the phyto-remediation activity of *S. nigrum* in Cd-contaminated soils by inoculation of microbes. All the results showed amelioration in Cd tolerance, uptake, and plant biomass. This behavior of *S. nigrum* toward HMs may be due to the presence of a large diversity of microbes in it (Guo et al., 2010; Chen et al., 2012, 2014). Inoculation of endophytic bacteria, such as pseudomonas specie "Lk9," enhanced the uptake of mineral nutrients, as well as the availability of Cd, and the growth of *S. nigrum,* as this strain has the ability to improve plant health, even under heavy metal stress (Chen et al., 2014). Exogenously applied AMF increase Cd uptake in *S. nigrum,* especially the aerial parts, by increasing the antioxidant enzymes system (Jiang et al., 2016). All of these studies show that the hyper-accumulation capacity and growth of *S. nigrum* in Cd-contaminated soils is enhanced. Similarly, Khan et al. (2015a,b) has reported that microbes can modify the antioxidant system of *S. nigrum* under Cd-induced oxidative stress, and thus, increase the Cd bioconcentration and translocation factor, compared with the control group. Endophytic strains, including *Glomerella truncate*, and *Phomopsis fukushii,* boost the phyto-remediation activity of *S. nigrum* in Cd-contaminated soil by increasing the Cd uptake through increased plant growth and chlorophyll content. It boosts the root's Cd uptake. Similar results have been reported for fungus *Glomus claroideum* and *Glomus intraradices* by Marques et al. (2007). Application of microbes (*Hypocrea virens* and *Paecilomyces lilacinus*) in combination with some soil conditioners (citric acid) can impart more beneficial effects on phyto-extraction of Cd, rather than a single treatment (Gao et al., 2012; Mao et al., 2015). It can enhance *S. nigrum* biomass up to 50%, and increase Cd accumulation ability up to 35% (Gao et al., 2012), which shows the synergistic effect of these microbes. Similarly, fungi, in combination with chelating agents, also imparts beneficial effects on growth and phyto-extraction of *S. nigrum* on Cd soils up to 47%, relative to separate application (Mao et al., 2015). All of these studies show that microbes can play an effective role in growth and uptake, thus, may be useful in the phyto-remediation activity of *S. nigrum* (Gao et al., 2012; Sun et al., 2017).

## 6.3 Soil Amendments

### *6.3.1 Use of Organic Substances*

Organic amendments (OAs) are widely used to accelerate phyto-remediation process in HMs contaminated soils (Rehman et al., 2016;

Younis et al., 2016; Abbas et al., 2017). Contrasting theories exist on the effect of OM interactions with HMs. The soil's OM complexes the metals, and decreases the metal mobility, while after decomposition, the dissolved OM is thought to enhance the mobility of metal due to formation of the HM-DOM complex. But the direction of this process is not independent of biochemical properties of added organic matter, physiochemical properties of soil, total and available pools of metals, and environmental factors. It is well known by various researchers that application of OAs greatly reduces HMs uptake by reducing the bio-available fraction. Bio-availability of HMs mainly depends on the nature of OAs, soil properties, and concerned metal; usually, OAs involves in the formation of insoluble metal species, which is less prone to leach into the deeper soil profile. Phyto-remediation efficiency can be increased either by phyto-extraction (Park et al., 2011; Houben et al., 2013) or by phyto-stabilizing of HMs in soil (Hattab et al., 2014; Wang et al., 2014). Dissolved OM is enhanced with OAs, which in turn increases HMs mobility (Smebye et al., 2016; Tang et al., 2016). Chicken manure increased the biomass, which significantly increases Cd uptake and accumulation in plant parts (Wei et al., 2010). Growth of *S. nigrum* was significantly enhanced with compost, vegetable waste, and cattle dung (Mary and Nithiya, 2015). Injudicious use of OAs can cause secondary HMs toxicity, so these OAs must be used carefully during the phyto-remediation process (Rizwan et al., 2016b).

### 6.3.2 Use of Inorganic Materials

Phyto-remediation potential of *S. nigrum* can be encouraged by the use of inorganic substances. The addition of inorganic materials can enhance Cd mobility, and mineral uptake, which in turn can support plant biomass and enhance the metal tolerance ability of *S. nigrum* (Liu et al., 2013; Cheng et al., 2016; and Rizwan et al., 2016a). A number of inorganic amendments have been proven beneficial in this respect, and some are discussed as follows.

Nitrogen application boosts the plant's biomass, and thus, supports the phyto-management of Cd contamination with the use of efficient hyperaccumulators. Moreover, different forms of nitrogen application have different behaviors, such as $NH_4^+$, which has more potential to increase uptake than $NO_3^-$. Likewise, application of urea also increases the accumulation rate of Cd in plants (Wei et al., 2010; Cheng et al., 2016). Elemental sulfur is also an economical amendment for increasing the phyto-remediation activity of plants (Rehman et al., 2015; Saifullah et al., 2016). Similarly, Rehman et al. (2015) has reported that sulfur increases the Cd concentration by decreasing the soil pH, and enhancing the biomass production. Phyto-remediation enhanced by chemical chelators is advocated for the restoration of soils disturbed by HMs. Many chelators can be used to manage the phyto-remediation activity of *S. nigrum* in heavy metal contaminated soils, such as ethylene diamine tetra-acetic acid (EDTA), S, and S-ethylene diamine dissucinic acid (EDDS), as they enhance the Cd uptake and promote biomass production. Sun et al. (2009) has reported that EDTA application showed an increase in Cd concentration in *S. nigrum* compared with a control group at the flowering stage, as well as other growth stages. This study favors the exogenous application of chelators, but this might also be a risk to the environment, as excessive application can lead to high metal solubilization, so care must be taken during handling this approach (Rizwan et al., 2017). Many studies showed that exogenous application of chelating agents might be helpful in metal extraction by *S. nigrum,* but these chemicals should be applied with great care in the soil to avoid higher metal solubilization, and thus, contamination of the surrounding environment (Rizwan et al., 2017).

The presence of iron (Fe) decreases the Cd uptake rate of *S. nigrum,* and while Fe-insufficient plants show a higher uptake, this might be due to the release of low-molecular-weight organic acids (LMWOAs) and a decrease in pH (Bao et al., 2009, 2011). Similarly, in *S. nigrum,* exogenous application of silicon increased the plant biomass and growth, but reduced the Cd concentration, thus, also decreasing the oxidative stress due to Cd (Li et al., 2011; Liu et al., 2013). Plant tolerance for many abiotic stresses can be increased by a silicon supply, as it takes part in many soil–plant system mechanisms (Rizwan et al., 2015; Meunier et al., 2017).

## 6.4 Transgenic Aids

The phyto-remediation technique for the management of heavy-metal contaminated soils is gaining importance day by day to remove HMs. In this respect, genetic engineering can also play a significant role by modifying the hyperaccumulator plants for enhancing metal uptake efficiency, and their sequestration. For this purpose, several metal chelator and metal transporter-producing genes in the plant system can be inserted. Transgenic plants, which remediate the heavy-metal-contaminated soils, have been introduced. A better understanding of this concept can also open up new avenues for the better management of Cd-contaminated soils. Phillips (2004) has reported that plants of *S. nigrum* showed an increase in internode length, and also in plant stature, when transformed with the AtGA20ox1 gibberellins synthesis gene. It also increased the canopy diameter, which might be due to the healthy growth of branches.

## 7 FUTURE CHALLENGES

*S. nigrum* has great potential for Cd remediation using the principles of phyto-remediation. However, there is still a need for further research for the confirmation of the beneficial aspects of *S. nigrum* during phyto-remediation, and exploration of the other metabolic processes involved for higher Cd uptake and accumulation. Moreover, there is the need to use genetic engineering for improving the Cd phyto-remediation potential of *S. nigrum* with the use of transgenic aids. In addition, a detailed genomic study of *S. nigrum* could be helpful for determining Cd translocation and accumulation behavior within different parts of *S. nigrum*. Further research is needed to elucidate the molecular details in response to Cd toxicity. Long-term field experimentation is required for the evaluation of the results of the phyto-management enhancing strategies of *S. nigrum*, such as farming practices, transgenic aid, soil amendments, and growth regulators. Moreover, health concerns for the consumption of this medicinal plant used for Cd remediation are still unknown, but a lot of research findings confirm the suitability of *S. nigrum* in different agro-regions, thus its commercial and large-scale application should be practiced for restoration of soil resources from Cd contaminations.

## 8 SUMMARY

With the increase in anthropogenic activities, and disturbances in the natural ecosystem, environmental threats such as Cd contamination have become a global concern. Cd toxicity has imposed serious hazards for the survival of plants, animals, and human beings. Under such despondent conditions, remediation of Cd-contaminated soils is a significant challenge. Among various strategies that are in practice for the remediation of Cd-contaminated soils, use of the phyto-remediation technique is of great importance, due to its social and economic acceptability. However, use of *S. nigrum* as a potential Cd hyperaccumulator is the most promising element of the phyto-remediation strategy. *S. nigrum,* being a perennial weed, is equipped with a fast-growing system and high biomass production that distinguishes the potential of *S. nigrum* for Cd de-contamination. After uptake from the soil solution, *S. nigrum* uses various mechanisms for Cd translocation and accumulation into various plant parts. Similarly, along with high Cd accumulation ability, *S. nigrum* also adopts various tolerance mechanisms (such as use of antioxidant defense mechanisms, stimulation of the chelation process, and production of organic acids) under Cd toxicity stress to avoid the hazardous effects of Cd toxicity. However, to improve the efficacy of the *S. nigrum* plant for Cd decontamination, some phyto-remediation-assisting strategies (such as farming practices, use of soil amendments, transgenic aids, and growth regulators) must be used. In addition, there is still the need for genomic research to determine the Cd translocation and accumulation mechanism in various parts of *S. nigrum*, along with the effects of this medicinal plant's consumption after being grown in Cd-contaminated soils.

## REFERENCES

Abbas, T., Rizwan, M., Ali, S., Rehman, M.Z., Qayyum, M.F., Abbas, F., Hannan, F., Rinklebe, J., Ok, Y.S., 2017. Effect of biochar on cadmium bioavailability and uptake in wheat (*Triticum aestivum* L.) grown in a soil with aged contamination. Ecotoxicol. Environ. Saf. 140, 37–47.

Abe, T., Fukami, M., Ogasawara, M., 2008. Cadmium accumulation in the shoots and roots of 93 weed species. Soil Sci. Plant Nutr. 54, 566–573.

Adrees, M., Ali, S., Rizwan, M., Ibrahim, M., Abbas, F., Farid, M., Rehman, M.Z., Irshad, M.K., Bharwana, S.A., 2015. The effect of excess copper on growth and physiology of important food crops: a review. Environ. Sci. Pollut. Res. 22, 8148–8162.

Adriano, D.C., Wenzel, W.W., Vangronsvel, J., Bolan, N.S., 2004. Role of assisted natural remediation in environmental cleanup. Geoderma 122, 121–142.

Baize, D., 1997. Teneurs Totales en Eléments Traces Métalliques dans les Sols Français. Références et Stratégies d'Interprétation. INRA Editions, Paris.

Bao, T., Sun, L., Sun, T., Zhang, P., Niu, Z., 2009. Iron-deficiency induces cadmium uptake and accumulation in *Solanum nigrum* L. Bull. Environ. Contam. Toxicol. 82, 338–342.

Bao, T., Sun, T., Sun, L., 2011. Lows molecular weight organic acids in root exudates and cadmium accumulation in cadmium hyperaccumulator *Solanum nigrum* L. and non-hyperaccumulator Solanum lycopersicum L. Afr. J. Biotechnol. 10, 17180–17185.

Benavides, M.P., Gallego, S.M., Tomaro, M.L., 2005. Cadmium toxicity in plants. Braz. J. Plant Physiol. 17, 21–34.

Borges, M., 2002. Extratabilidade do cádmio: influência de atributos de solos muito intemperizados em extratores convencionais e potencialidade de ácidos orgânicos de baixo peso molecular. 76 f. (Dissertaçao em Agronomia)Escola superior de Agricultura Luiz de Queiroz, USP, Piracicaba.

Braud, A., Jézéquel, K., Bazot, S., Lebeau, T., 2009. Enhanced phytoextraction of an agricultural Cr- and Pb-contaminated soil by bioaugmentation with siderophore-producing bacteria. Chemosphere 74, 280–286.

Chandra, R., Yadav, S., Yadav, S., 2017. Phytoextraction potential of heavy metals by native wetland plants growing on chlorolignin containing sludge of pulp and paper industry. Ecol. Eng. 98, 134–145.

Chauvel, B., Guillemin, J.P., Letouzé, A., 2005. Effect of intra-specific competition on development and growth of *Alopecurus myosuroides* Hudson. Eur. J. Agron. 22, 301–308.

Chen, J., Huang, B., Li, Y., Du, H., Gu, Y., Liu, H., Zhang, J., Huang, Y., 2011. Synergistic influence of sucrose and abscisic acid on the genes involved in starch synthesis in maize endosperm. Carbohydr. Res. 346, 1684–1691.

Chen, L., Luo, S., Chen, J., Wan, Y., Liu, C., Liu, Y., Pang, X., Lai, C., Zeng, G., 2012. Diversity of endophytic bacterial populations associated with Cd-hyperaccumulator plant *Solanum nigrum* L. grown in mine tailings. Appl. Soil Ecol. 62, 24–30.

Chen, L., Luo, S., Li, X., Wan, Y., Chen, J., Liu, C., 2014. Interaction of Cd hyperaccumulator *Solanum nigrum* L. and functional endophyte Pseudomonas sp. Lk9 on soil heavy metals uptake. Soil Biol. Biochem. 68, 300–308.

Chen, B., Stein, A.F., Castell, N., Gonzalez-Castanedo, Y., de la Campa, A.S., de la Rosa, J.D., 2016. Modeling and evaluation of urban pollution events of atmospheric heavy metals from a large Cu-smelter. Sci. Total Environ. 539, 17–25.

Cheng, M., Wang, P., Kopittke, P.M., Wang, A., Sale, P.W., Tang, C., 2016. Cadmium accumulation is enhanced by ammonium compared to nitrate in two hyperaccumulators, without affecting speciation. J. Exp. Bot. 67, 5041–5050.

Clemens, S., 2006. Toxic metal accumulation, responses to exposure and mechanisms of tolerance in plants. Biochimie 88, 1707–1719.

Clemens, S., Palmgren, M.G., Kramer, U., 2002. A long way ahead: understanding and engineering plant metal accumulation. Trends Plant Sci. 7, 309–315.

Cook, M., Morrow, H., 1995. Anthropogenic Sources of Cadmium to Canadian Environment. In: Workshop Proceeding of Workshop on Cadmium Transport Into Plants. Canadian Network of Toxicology Centre, Ottawa, ON, Canada, pp. 19–21.

Deng, X., Xia, Y., Hu, W., Zhang, H., Shen, Z., 2010. Cadmium-induced oxidative damage and protective effects of N-acetyl-L-cysteine against cadmium toxicity in *Solanum nigrum* L. J. Hazard. Mater. 180, 722–729.

Dwivedi, G.K., Upadhyay, S.K., Mishra, A.K., Singh, A.K., 2014. Hyper accumulation of cadmium in *Solanum nigrum* L. and their effects on phyto-chemicals and antioxidant enzymatic activities. Int. J. Pharm. Sci. Res. 5, 1424–1430.

Evans, S.P., Knezevic, S.Z., Lindquist, J.L., Shapiro, C.A., Blankenship, E.E., 2003. Nitrogen application influences the critical period of weed control in corn. Weed Sci. 51, 408–417.

Farahat, E., Linderholm, H.W., 2015. The effect of long-term wastewater irrigation on accumulation and transfer of heavy metals in Cupressus sempervirens leaves and adjacent soils. Sci. Total Environ. 512, 1–7.

Ferraz, P., Fidalgo, F., Almeida, A., Teixeira, J., 2012. Phytostabilization of nickel by the zinc and cadmium hyperaccumulator *Solanum nigrum* L. Are metallothioneins involved? Plant Physiol. Biochem. 57, 254–260.

Fidalgo, F., Freitas, R., Ferreira, R., Pessoa, A.M., Teixeira, J., 2011. Solanum nigrum L. antioxidant defence system isozymes are regulated transcriptionally and post-translationally in Cd-induced stress. Environ. Exp. Bot. 72 (2), 312–319.

Freitas, E.V., Nascimento, C.W., Souza, A., Silva, F.B., 2013. Citric acid-assisted phytoextraction of lead: a field experiment. Chemosphere 92, 213–217.

Gao, Y., Zhou, P., Mao, L., Shi, W.J., Zhi, Y.E., 2010. Phytoextraction of cadmium and physiological changes in *Solanum nigrum* as a novel cadmium hyperaccumulator. Russ. J. Plant Physiol. 57, 501–508.

Gao, Y., Miao, C., Xia, J., Luo, C., Mao, L., Zhou, P., Shi, W., 2012. Effect of citric acid on phytoextraction and antioxidative defence in *Solanum nigrum* L. as a hyperaccumulator under Cd and Pb combined pollution. Environ. Earth Sci. 65, 1923–1932.

Garg, N., Chandel, S., 2010. Arbuscular mycorrhizal networks: process and functions. A review. Agron. Sustain. Dev. 30 (3), 581–599.

Gerhardt, K.E., Huang, X.D., Glick, B.R., Greenberg, B.M., 2009. Phytoremediation and rhizoremediation of organic soil contaminants: potential and challenges. Plant Sci. 176 (1), 20–30.

Ghani, A., Ali, Z., Islam, T., Sanaullah, S., Saeed, S., 2014. Nutrient evaluation and elemental analysis of four selected medicinal plants of soon valley Khushab, Punjab, Pakistan. Pak. J. Pharm. Sci. 27, 597–600.

Gill, S.S., Khan, N.A., Tuteja, N., 2011. Differential cadmium stress tolerance in five Indian mustard (*Brassica juncea* L.) cultivars. An evaluation of the role of antioxidant machinery. Plant Signal. Behav. 6, 293–300.

Grant, C.A., Clarke, J.M., Duguid, S., Chaney, R.L., 2008. Selection and breeding of plant cultivars to minimize cadmium accumulation. Sci. Total Environ. 390, 301–310.

Grundy, A.C., Mead, A., Burston, S., 2003. Modeling the emergence response of weed seeds to burial depth: interactions with seed density, weight and shape. J. Appl. Ecol. 40, 757–770.

Guo, H., Luo, S., Chen, L., Xiao, X., Xi, Q., Wei, W., Zeng, G., Liu, C., Wan, Y., Chen, J., He, Y., 2010. Bioremediation of heavy metals by growing hyperaccumulator endophytic bacterium *Bacillus* sp. L14. Bioresour. Technol. 101, 8599–8605.

Gupta, S.K., Aten, C., 1993. Comparison and evaluation of extraction media and their suitability in a simple model to predict the biological relevance of heavy metal concentrations in contaminated soils. Int. J. Environ. Anal. Chem. 51, 25–46.

Hammami, H., Parsa, M., Mohassel, M.H.R., Rahimi, S., Mijani, S., 2016. Weeds ability to phytoremediate cadmium-contaminated soil. Int. J. Phytoremediation 18, 48–53.

Hamzah, A., Hapsari, R.I., Wisnubroto, E.I., 2016. Phytoremediation of cadmium contaminated agricultural land using indigenous plants. Int. J. Environ. Agric. Res. 2, 8–14.

Hart, J.J., Welch, R.M., Norvell, W.A., Kochian, L.V., 2006. Characterization of cadmium uptake, translocation and storage in near-isogenic lines of durum wheat that differ in grain cadmium concentration. New Phytol. 172, 261–271.

Hattab, N., Soubrand, M., Guégan, R., Motelica-Heino, M., Bourrat, X., Faure, O., Bouchardon, J.L., 2014. Effect of organic amendments on the mobility of trace elements in phytoremediated techno-soils: role of the humic substances. Environ. Sci. Pollut. Res. 21, 10470–10480.

Hentz, S., McComb, J., Miller, G., Begonia, M., Begonia, G., 2012. Cadmium uptake, growth and phytochelatin contents of *Triticum aestivum* L. in response to various concentrations of cadmium. World Environ. 2, 44–50.

Houben, D., Evrard, L., Sonnet, P., 2013. Beneficial effects of biochar application to contaminated soils on the bioavailability of Cd, Pb and Zn and the biomass production of rapeseed (*Brassica napus* L.). Biomass Bioenergy 57, 196–204.

Iqbal, M., Iqbal, N., Bhatti, I.A., Ahmad, N., Zahid, M., 2016. Response surface methodology application in optimization of cadmium adsorption by shoe waste: a good option of waste mitigation by waste. Ecol. Eng. 88, 265–275.

Itoh, S., Yumura, Y., 1979. Studies on the contamination of vegetables crops by excessive absorption of heavy metals. Bull. Veg. Ornamental Crops Res. Stn. 6a, 123 (Ja).

Ji, P., Song, Y., Sun, T., Liu, Y., Cao, X., Xu, D., Yang, X., McRae, T., 2011a. In-situ cadmium phytoremediation using *Solanum nigrum* L.: the bio-accumulation characteristics trail. Int. J. Phytoremediation 13, 1014–1023.

Ji, P., Sun, T., Song, Y., Ackland, M.L., Liu, Y., 2011b. Strategies for enhancing the phytoremediation of cadmium-contaminated agricultural soils by *Solanum nigrum* L. Environ. Pollut. 159, 762–768.

Ji, P., Jiang, Y., Tang, X., Nguyen, T.H., Tong, Y.A., Gao, P., Han, W., 2015a. Enhancing of phytoremediation efficiency using indole-3-acetic acid (IAA). Soil Sediment Contam. 24, 909–916.

Ji, P., Tang, X., Jiang, Y., Tong, Y.A., Gao, P., Han, W., 2015b. Potential of gibberellic acid 3 (GA3) for enhancing the phytoremediation efficiency of *Solanum nigrum* L. Bull. Environ. Contam. Toxicol. 95, 810–814.

Ji, P., Song, Y., Jiang, Y., Tang, X., Tong, Y.A., Gao, P., Han, W., 2016. A two-year field study of phytoremediation using *Solanum nigrum* L. in China. Int. J. Phytoremediation 18, 924–928.

Jiang, Q.Y., Tan, S.Y., Zhuo, F., Yang, D.J., Ye, Z.H., Jing, Y.X., 2016. Effect of *Funneliformis mosseae* on the growth, cadmium accumulation and antioxidant activities of Solanum nigrum. Appl. Soil Ecol. 98, 112–120.

John, M.K., van Laerhoven, C.J., Chuah, H., 1972. Factors affecting plant uptake and phytoavailability of cadmium added to soils. Environ. Sci. Technol. 6, 1005–1009.

Keller, C., Marchetti, M., Rossi, L., Lugon-Moulin, N., 2005. Reduction of cadmium availability to tobacco (*Nicotiana tabacum*) plants using soil amendments in low cadmium-contaminated agricultural soils: a pot experiment. Plant Soil 276, 69–84.

Kennedy, V.H., Sanchez, A.L., Oughton, D.H., Rowland, A.P., 1997. Use of single and sequential chemical extractants to assess radionuclide and heavy metal availability from soils for root uptake. Analyst 122, 89–100.

Khan, A.R., Ullah, I., Khan, A.L., Hong, S.J., Waqas, M., Park, G.S., Kwak, Y., Choi, J., Jung, B.K., Park, M., Lee, I.J., 2014. Phytostabilization and physicochemical responses of Korean ecotype *Solanum nigrum* L. to cadmium contamination. Water Air Soil Pollut. 225, 1–11.

Khan, A.R., Ullah, I., Khan, A.L., Park, G.S., Waqas, M., Hong, S.J., Jung, B.K., Kwak, Y., Lee, I.J., Shin, J.H., 2015a. Improvement in phytoremediation potential of *Solanum nigrum* under cadmium contamination through endophytic-assisted Serratia sp. RSC14 inoculation. Environ. Sci. Pollut. Res. 22, 14032–14042.

Khan, M.U., Sessitsch, A., Harris, M., Fatima, K., Imran, A., Arslan, M., Shabir, G., Khan, Q.M., Afzal, M., 2015b. Cr-resistant rhizo-and endophytic bacteria associated with Prosopis juliflora and their potential as phytoremediation enhancing agents in metal degraded soils. Front. Plant Sci. 5, 1–10.

Khan, A.R., Waqas, M., Ullah, I., Khan, A.L., Khan, M.A., Lee, I.J., Shin, J.H., 2016. Culturable endophytic fungal diversity in the cadmium hyperaccumulator *Solanum nigrum* L. and their role in enhancing phytoremediation. Environ. Exp. Bot. 135, 126–135.

Kidd, P., Mench, M., Álvarez-López, V., Bert, V., Dimitriou, I., Friesl-Hanl, W., Herzig, R., Olga Janssen, J., Kolbas, A., Müller, I., Neu, S., 2015. Agronomic practices for improving gentle remediation of trace element-contaminated soils. Int. J. Phytoremediation 17, 1005–1037.

Kim, S.H., Lee, I.S., 2010. Comparison of the ability of organic acids and EDTA to enhance the phytoextraction of metals from a multi metal contaminated soil. Bull. Environ. Contam. Toxicol. 84 (2), 255–259.

Kirkham, M.B., 2006. Cadmium in plants on polluted soils: effects of soil factors, hyperaccumulation, and amendments. Geoderma 137, 19–32.

Kumar, N., Bauddh, K., Kumar, S., Dwivedi, N., Singh, D.P., Barman, S.C., 2013. Accumulation of metals in weed species grown on the soil contaminated with industrial waste and their phytoremediation potential. Ecol. Eng. 61, 491–495.

Lati, R.N., Filin, S., Eizenberg, H., 2012. Black nightshade (*Solanum nigrum*) emergence and development is affected by seed density and burial depth. Phytoparasitica 40 (2), 195–203.

Li, S., Lin, B., Zhou, W., 2005. Effects of previous elemental sulfur applications on oxidation of additional applied elemental sulfur in soils. Biol. Fertil. Soils 42, 146–152.

Li, L., Zhang, Y., Liu, J., Hu, Z., Chai, T., 2011. Silicon-induced alleviation of cadmium toxicity in hyperaccumulator Solanum nigrum L. In: Water Resource and Environmental Protection (ISWREP), International Symposium on IEEE. vol. 3. pp. 1908–1911.

Liphadzi, M.S., Kirkham, M.B., 2006. Availability and plant uptake of heavy metals in EDTA assisted phytoremediation of soil and composted biosolids. S. Afr. J. Bot. 72, 391–397.

Liphadzi, M.S., Kirkham, M.B., Mankin, K.R., Paulsen, G.M., 2003. EDTA assisted heavy-metal up take by poplar and sunflower grown at a long-term sewage sludge farm. Plant Soil 257, 171–182.

Liu, J.G., Li, K., Xu, J.K., Liang, J.S., Lu, X.L., Yang, J.C., Zhu, Q.S., 2003. Interaction of Cd and five mineral nutrients for uptake and accumulation in different rice cultivars and genotypes. Field Crop Res. 83, 271–281.

Liu, J.G., Qian, M., Cai, G.L., Yang, J.C., Zhu, Q.S., 2007. Uptake and translocation of Cd in different rice cultivars and the relation with Cd accumulation in rice grain. J. Hazard. Mater. 143, 443–447.

Liu, Z., Gu, C., Chen, F., Yang, D., Wu, K., Chen, S., Jiang, J., Zhang, Z., 2012. Heterologous expression of a Nelumbo nucifera phytochelatin synthase gene enhances cadmium tolerance in *Arabidopsis thaliana*. Appl. Biochem. Biotechnol. 166, 722–734.

Liu, J., Zhang, H., Zhang, Y., Chai, T., 2013. Silicon attenuates cadmium toxicity in *Solanum nigrum* L. by reducing cadmium uptake and oxidative stress. Plant Physiol. Biochem. 68, 1–7.

Liu, J., Zhang, X.H., Li, T.Y., Wu, Q.X., Jin, Z.J., 2014. Soil characteristics and heavy metal accumulation by native plants in a Mn mining area of Guangxi, South China. Environ. Monit. Assess. 186, 2269–2279.

Liu, H., Yuan, M., Tan, S., Yang, X., Lan, Z., Jiang, Q., Ye, Z., Jing, Y., 2015. Enhancement of arbuscular mycorrhizal fungus (*Glomus versiforme*) on the growth and Cd uptake by Cd-hyperaccumulator *Solanum nigrum*. Appl. Soil Ecol. 89, 44–49.

Liu, X., Chen, J., Wang, G.H., Wang, W.H., Shen, Z.J., Luo, M.R., Gao, G.F., Simon, M., Ghoto, K., Zheng, H.L., 2016. Hydrogen sulfide alleviates zinc toxicity by reducing zinc uptake and regulating genes expression of antioxidative enzymes and metallothioneins in roots of the cadmium/zinc hyperaccumulator *Solanum nigrum* L. Plant Soil 400, 177–192.

Lu, N., Maruo, T., Johkan, M., Hohjo, M., Tsukagoshi, S., Ito, Y., Chimura, T., Shinohara, Y., 2012. Effects of supplemental lighting within the canopy at different developing stages on tomato yield and quality of single-truss tomato plants grown at high density. Environ. Control. Biol. 50, 1–11.

Luo, S.L., Chen, L., Chen, J.L., Xiao, X., Xu, T.Y., Wan, Y., Rao, C., Liu, C.B., Liu, Y.T., Lai, C., Zeng, G.M., 2011. Analysis and characterization of cultivable heavy metal resistant bacterial endophytes isolated from Cd-hyperaccumulator *Solanum nigrum* L. and their potential use for phytoremediation. Chemosphere 85, 1130–1138.

Lux, A., Martinka, M., Vaculik, M., White, P.J., 2011. Root responses to cadmium in the rhizosphere: a review. J. Exp. Bot. 62, 21–37.

Ma, L., Zhang, H., Xu, W., He, X., Yang, L., Luo, Y., Huang, K., 2013. Simultaneous determination of 15 plant growth regulators in bean sprout and tomato with liquid chromatography–triple quadrupole tandem mass spectrometry. Food Anal. Methods 6, 941–951.

Maddonni, G.A., Otegui, M.E., 2004. Intra-specific competition in maize: early establishment of hierarchies among plants affects final kernel set. Field Crop Res. 85, 1–13.

Mao, L., Tang, D., Feng, H., Gao, Y., Zhou, P., Xu, L., Wang, L., 2015. Determining soil enzyme activities for the assessment of fungi and citric acid-assisted phytoextraction under cadmium and lead contamination. Environ. Sci. Pollut. Res. 22, 19860–19869.

Marques, A.P., Oliveira, R.S., Samardjieva, K.A., Pissarra, J., Rangel, A.O., Castro, P.M., 2007. *Solanum nigrum* grown in contaminated soil: effect of arbuscular mycorrhizal fungi on zinc accumulation and histolocalisation. Environ. Pollut. 145, 691–699.

Mary, J.A.L., Nithiya, T., 2015. Effect of organic and inorganic fertilizer on growth, phenolic compounds and antioxidant activity of *Solanum Nigrum* L. World J. Pharm. Pharm. Sci. 4 (5), 808–822.

Mench, M., Baize, D., Mocquot, B., 1997. Cadmium availability to wheat in five soil series from the Yonne district, Burgundy, France. Environ. Pollut. 95, 93–103.

Meunier, J.D., Barboni, D., Anwar-ul-Haq, M., Levard, C., Chaurand, P., Vidal, V., Grauby, O., Huc, R., Laffont-Schwob, I., Rabier, J., Keller, C., 2017. Effect of phytoliths for mitigating water stress in durum wheat. New Phytol. 215 (1), 229–239.

Miller, J.E., Hassett, J.J., Koeppe, D.E., 1976. Uptake of cadmium by soybeans as influenced by soil cation exchange capacity, pH, and available phosphorus. J. Environ. Qual. 6, 157–160.

Muradoglu, F., Gundogdu, M., Ercisli, S., Encu, T., Balta, F., Jaafar, H.Z., Zia-ul-Haq, M., 2015. Cadmium toxicity affects chlorophyll a and b content, antioxidant enzyme activities and mineral nutrient accumulation in strawberry. Biol. Res. 48, 1–7.

Nagajyoti, P.C., Lee, K.D., Sreekanth, T.V.M., 2010. Heavy metals, occurrence and toxicity for plants: a review. Environ. Chem. Lett. 8, 199–216.

Nakanishi, H., Ogawa, I., Ishimaru, Y., Mori, S., Nishizawa, N.K., 2006. Iron deficiency enhances cadmium uptake and translocation mediated by the $Fe^{2+}$ transporters OsIRT1 and OsIRT2 in rice. Soil Sci. Plant Nutr. 52, 464–469.

Ningyu, L.I., Guo, B., Hua, L.I., Qinglin, F.U., Ren-wei, F., Yongzhen, D., 2016. Effects of double harvesting on heavy metal uptake by six forage species and the potential for phytoextraction in field. Pedosphere 26, 717–724.

Niu, M., Wei, S., Bai, J., Wang, S., Ji, D., 2015. Remediation and safe production of Cd contaminated soil via multiple cropping hyperaccumulator *Solanum nigrum* L. and low accumulation Chinese cabbage. Int. J. Phytoremediation 17, 657–661.

Nogawa, K., Kido, M., 1996. Itai-Itai disease and health effects of cadmium. In: Chang, L.W. (Ed.), Toxicology of Metals. CRC Lewis Publishers, New York, pp. 353–370.

Ogbazghi, Z.M., Tesfamariam, E.H., Annandale, J.G., De Jager, P.C., 2015. Mobility and uptake of zinc, cadmium, nickel, and lead in sludge-amended soils planted to dryland maize and irrigated maize–oat rotation. J. Environ. Qual. 44, 655–667.

Park, J.H., Lamb, D., Paneerselvam, P., Choppala, G., Bolan, N., Chung, J.-W., 2011. Role of organic amendments on enhanced bioremediation of heavy metal(loid) contaminated soils. J. Hazard. Mater. 185, 549–574.

Peng, K.J., Luo, C.L., Chen, Y.H., Wang, G.P., Li, X.D., Shen, Z.G., 2009. Cadmium and other metal uptake by *Lobelia chinensis* and *Solanum nigrum* from contaminated soils. Bull. Environ. Contam. Toxicol. 83, 260–264.

Phillips, A.L., 2004. Plant hormones: biosynthesis, signal transduction, action. In: Davies, P.J. (Ed.), Genetic and Transgenic Approaches to Improving Crop Performance. Kluwer Academic Publishers, Dordrecht, pp. 582–609.

Pinto, A.P., Alves, A.S., Candeias, A.J., Cardoso, A.I., de Varennes, A., Martins, L.L., Mourato, M.P., Gonçalves, M.L., Mota, A.M., 2009. Cadmium accumulation and antioxidative defences in Brassica juncea L. Czern, Nicotiana tabacum L. and *Solanum nigrum* L. Int. J. Environ. Anal. Chem. 89, 661–676.

Qayyum, M.F., Rehman, M.Z., Ali, S., Rizwan, M., Naeem, A., Maqsood, M.A., Khalid, H., Rinklebe, J., Ok, Y.S., 2017. Residual effects of monoammonium phosphate, gypsum and elemental sulfur on cadmium phytoavailability and translocation from soil to wheat in an effluent irrigated field. Chemosphere 174, 515–523.

Rafique, N., Tariq, S.R., 2016. Distribution and source apportionment studies of heavy metals in soil of cotton/wheat fields. Environ. Monit. Assess. 188, 1–10.

Rehman, M.Z., Rizwan, M., Ghafoor, A., Naeem, A., Ali, S., Sabir, M., Qayyum, M.F., 2015. Effect of inorganic amendments for in situ stabilization of cadmium in contaminated soils and its phyto-availability to wheat and rice under rotation. Environ. Sci. Pollut. Res. 22, 16897–16906.

Rehman, M.Z., Rizwan, M., Ali, S., Fatima, N., Yousaf, B., Naeem, A., Sabir, M., Ahmad, H.R., Ok, Y.S., 2016. Contrasting effects of biochar, compost and farm manure on alleviation of nickel toxicity in maize (*Zea mays* L.) in relation to plant growth, photosynthesis and metal uptake. Ecotoxicol. Environ. Saf. 133, 218–225.

Rehman, M.Z., Khalid, H., Akmal, F., Ali, S., Rizwan, M., Qayyum, M.F., Iqbal, M., Khalid, M.U., Azhar, M., 2017a. Effect of limestone, lignite and biochar applied alone and combined on cadmium uptake in wheat and rice under rotation in an effluent irrigated field. Environ. Pollut.. https://doi.org/10.1016/j.envpol.2017.05.003.

Rehman, M.Z., Rizwan, M., Ali, S., Ok, Y.S., Ishaque, W., Nawaz, M.F., Akmal, F., Waqar, M., 2017b. Remediation of heavy metal contaminated soils by using *Solanum nigrum*: a review. Ecotoxicol. Environ. Saf. 143, 236–248.
Rizwan, M., Ali, S., Ibrahim, M., Farid, M., Adrees, M., Bharwana, S.A., Rehman, M.Z., Qayyum, M.F., Abbas, F., 2015. Mechanisms of silicon-mediated alleviation of drought and salt stress in plants: a review. Environ. Sci. Pollut. Res. 22, 15416–15431.
Rizwan, M., Ali, S., Abbas, T., Rehman, M.Z., Hannan, F., Keller, C., Al-Wabel, M.I., Ok, Y.S., 2016a. Cadmium minimization in wheat: a critical review. Ecotoxicol. Environ. Saf. 130, 43–53.
Rizwan, M., Ali, S., Rizvi, H., Rinklebe, J., Tsang, D.C., Meers, E., Ok, Y.S., Ishaque, W., 2016b. Phytomanagement of heavy metals in contaminated soils using sunflower: a review. Crit. Rev. Environ. Sci. Technol. 46, 1498–1528.
Rizwan, M., Ali, S., Adrees, M., Ibrahim, M., Tsang, D.C.W., Rehman, M.Z., Zahir, Z.A., Rinklebe, J., Tack, F.M.G., Ok, Y.S., 2017. A critical review on effects, tolerance mechanisms and management of cadmium in vegetables. Chemosphere. https://doi.org/10.1016/j.chemosphere.2017.05.013.
Saad-Allah, K.M., Elhaak, M.A., 2015. Hyperaccumulation activity and metabolic responses of *Solanum nigrum* in two differentially polluted growth habitats. J. Saudi Soc. Agric. Sci.. https://doi.org/10.1016/j.jssas.2015.08.001.
Saifullah, M.N.K., Iqbal, M., Naeem, A., Bibi, S., Waraich, E.A., 2016. Elemental sulfur improves growth and phytoremediative ability of wheat grown in lead (Pb) contaminated calcareous soil. Int. J. Phytoremediation 18, 1022–1028.
Sammut, M.L., Noack, Y., Rose, J., Hazemann, J.L., Proux, O., Depoux, M., Ziebel, A., Fiani, E., 2010. Speciation of Cd and Pb in dust emitted from sinter plant. Chemosphere 78, 445–450.
Sauve, S., Norvell, W.A., McBride, M., Hendershot, W., 2000. Speciation and complexation of cadmium in extracted soil solutions. Environ. Sci. Technol. 34, 291–296.
Schnoor, J., 2004. Australasian soil contamination gets attention. Environ. Sci. Technol. 38, 53A.
Schoebitz, M., Ceballos, C., Ciamp, L., 2013. Effect of immobilized phosphate solubilizing bacteria on wheat growth and phosphate uptake. J. Soil Sci. Plant Nutr. 13 (1), 1–10.
Seregin, I.V., Shpigun, L.K., Ivanov, V.B., 2004. Distribution and toxic effects of cadmium and lead on maize roots. Russ. J. Plant Physiol. 51, 525–533.
Singh, A., Prasad, S.M., Singh, S., Singh, M., 2016. Phytoremediation potential of weed plants' oxidative biomarker and antioxidant responses. Chem. Ecol. 32, 684–706.
Siripornadulsil, S., Traina, S., Verma, D.P.S., Sayre, R.T., 2002. Molecular mechanisms of proline-mediated tolerance to toxic heavy metals in transgenic microalgae. Plant Cell 14, 2837–2847.
Smebye, A., Alling, V., Vogt, R.D., Gadmar, T.C., Mulder, J., Cornelissen, G., Hale, S.E., 2016. Biochar amendment to soil changes dissolved organic matter content and composition. Chemosphere 142, 100–105.
Soares, C., de Sousa, A., Pinto, A., Azenha, M., Teixeira, J., Azevedo, R.A., Fidalgo, F., 2016. Effect of 24-epibrassinolide on ROS content, antioxidant system, lipid peroxidation and Ni uptake in *Solanum nigrum* L. under Ni stress. Environ. Exp. Bot. 122, 115–125.
Song, Y., Hudek, L., Freestone, D., Puhui, J., Michalczyk, A.A., Senlin, Z., Ackland, M.L., 2014. Comparative analyses of cadmium and zinc uptake correlated with changes in natural resistance-associated macrophage protein (NRAMP) expression in *Solanum nigrum* L. and Brassica rape. Environ. Chem. 11, 653–660.
Street, J.J., Sabey, B.R., Lindsay, W.L., 1978. Influence of pH, phosphorus, cadmium, sewage sludge, and incubation time on the solubility and plant uptake of cadmium. J. Environ. Qual. 7, 286–290.

Sun, R.L., Zhou, Q.X., Sun, F.H., Jin, C.X., 2007. Antioxidative defence and proline/phytochelatin accumulation in a newly discovered Cd-hyperaccumulator, *Solanum nigrum* L. Environ. Exp. Bot. 60, 468–476.

Sun, Y., Zhou, Q., Wang, L., Liu, W., 2009. The influence of different growth stages and dosage of EDTA on Cd uptake and accumulation in Cd-hyperaccumulator (*Solanum nigrum* L.). Bull. Environ. Contam. Toxicol. 82, 348–353.

Sun, L., Cao, X., Li, M., Zhang, X., Li, X., Cui, Z., 2017. Enhanced bioremediation of lead contaminated soil by *Solanum nigrum* L. with Mucor circinelloides. Environ. Sci. Pollut. Res.. https://doi.org/10.1007/s11356-017-8637-x.

Tang, J., Li, X., Luo, Y., Li, G., Khan, S., 2016. Spectroscopic characterization of dissolved organic matter derived from different biochars and their polycylic aromatic hydrocarbons (PAHs) binding affinity. Chemosphere 152, 399–406.

Turner, M.A., 1973. Effects of cadmium treatment on cadmium and zinc uptake by selected vegetable species. J. Environ. Qual. 2, 118–119.

UNEP (United Nations Environment Programme), 2010. Final Review of Scientific Information on Cadmium—Version of December 2010.

Van der Vliet, L., Peterson, C., Hale, B., 2007. Cd accumulation in roots and shoots of durum wheat: the roles of transpiration rate and apoplastic bypass. J. Exp. Bot. 58, 2939–2947.

Vaxevanidou, K., Papassiopi, N., Paspaliaris, I., 2008. Removal of heavy metals and arsenic from contaminated soils using bioremediation and chelant extraction techniques. Chemosphere 70 (8), 1329–1337.

Vega, F., Andrade, M., Covelo, E., 2010. Influence of soil properties on the sorption and retention of cadmium, copper and Pb, separately and together, by 20 soil horizons: comparison of linear regression and tree regression analyses. J. Hazard. Mater. 174 (1–3), 522–533.

Wan, Y., Luo, S., Chen, J., Xiao, X., Chen, L., Zeng, G., Liu, C., He, Y., 2012. Effect of endophyte-infection on growth parameters and Cd induced phytotoxicity of Cd-hyperaccumulator *Solanum nigrum* L. Chemosphere 89, 743–750.

Wang, F.L., Ouyang, W., Hao, F.H., Lin, C.Y., Song, N.N., 2014. In situ remediation of cadmium-polluted soil reusing four byproducts individually and in combination. J. Soils Sediments 14, 451–461.

Wang, S., Wei, S., Ji, D., Bai, J., 2015. Co-planting Cd contaminated field using hyperaccumulator *Solanum nigrum* L. through interplant with low accumulation welsh onion. Int. J. Phytoremediation 17, 879–884.

Wei, S.H., Zhou, Q.X., 2004. Discussion on basic principles and strengthening measures for phytoremediation of soils contaminated by heavy metals. Chin. J. Ecol. 23 (1), 65–72.

Wei, S., Zhou, Q., Koval, P.V., 2006. Flowering stage characteristics of cadmium hyperaccumulator *Solanum nigrum* L. and their significance to phytoremediation. Sci. Total Environ. 369, 441–446.

Wei, S., Li, Y., Zhou, Q., Srivastava, M., Chiu, S., Zhan, J., Wu, Z., Sun, T., 2010. Effect of fertilizer amendments on phytoremediation of Cd-contaminated soil by a newly discovered hyperaccumulator *Solanum nigrum* L. J. Hazard. Mater. 176, 269–273.

Wei, S., Anders, I., Feller, U., 2014a. Selective uptake, distribution, and redistribution of 109Cd, 57Co, 65Zn, 63Ni, and 134Cs via xylem and phloem in the heavy metal hyperaccumulator *Solanum nigrum* L. Environ. Sci. Pollut. Res. 21, 7624–7630.

Wei, S., Zeng, X., Wang, S., Zhu, J., Ji, D., Li, Y., Jiao, H., 2014b. Hyperaccumulative property of *Solanum nigrum* L. to Cd explored from cell membrane permeability, subcellular distribution, and chemical form. J. Soils Sediments 14, 558–566.

Wei, S., Bai, J., Yang, C., Zhang, Q., Knorr, K.H., Zhan, J., Gao, Q., 2016. Compound amino acids added in media improved *Solanum nigrum* L. phytoremediating CD-PAHS contaminated soil. Int. J. Phytoremediation 18, 358–363.

Wesam, A.K., Hajer, A.-Q., 2014. Cadmium, copper and zinc toxicity effects on growth, proline content and genetic stability of *Solanum nigrum* L., a crop wild relative for tomato; comparative study. Physiol. Mol. Biol. Plants 20 (1), 31–39.

Whiting, S.N., Leake, J.R., McGrath, S.P., Baker, A.J.M., 2001. Hyperaccumulation of Zn by *Thlaspi caerulescens* can ameliorate Zn toxicity in the rhizosphere of co-cropped *Thlaspi arvense*. Environ. Sci. Technol. 35, 3237–3241.

Wu, Q.T., Wei, Z.B., Ouyang, Y., 2007. Phytoextraction of metal-contaminated soil by *Sedum alfredi* H: effects of chelator and co-planting. Water Air Soil Pollut. 180, 131–139.

Wuana, R.A., Okieimen, F.E., 2011. Heavy metals in contaminated soils: a review of sources, chemistry, risks and best available strategies for remediation. ISRN Ecol. Article ID 402647.

Xiao, X., Luo, S., Zeng, G., Wei, W., Wan, Y., Chen, L., Guo, H., Cao, Z., Yang, L., Chen, J., Xi, Q., 2010. Biosorption of cadmium by endophytic fungus (EF) *Microsphaeropsis* sp. LSE10 isolated from cadmium hyperaccumulator *Solanum nigrum L*. Bioresour. Technol. 101 (6), 1668–1674.

Xu, J., Yin, H., Li, X., 2009. Protective effects of proline against cadmium toxicity in micro propagated hyperaccumulator, *Solanum nigrum* L. Plant Cell Rep. 28 (2), 325–333.

Xu, J., Sun, J., Du, L., Liu, X., 2012. Comparative transcriptome analysis of cadmium responses in *Solanum nigrum* and Solanum torvum. New Phytol. 196, 110–124.

Xu, W., Lu, G., Wang, R., Guo, C., Liao, C., Yi, X., Dang, Z., 2015. The effect of pollination on Cd phytoextraction from soil by maize (*Zea mays* L.). Int. J. Phytoremediation 17, 945–950.

Yadav, S.K., 2010. Heavy metals toxicity in plants: an overview on the role of glutathione and phyto-chelatins in heavy metal stress tolerance of plants. South Afr. J. Bot. 76, 167–179.

Yan, Z., Zhang, W., Chen, J., Li, X., 2015. Methyl jasmonate alleviates cadmium toxicity in *Solanum nigrum* by regulating metal uptake and antioxidative capacity. Biol. Plant. 59, 373–381.

Yang, J., Li, K., Zheng, W., Zhang, H., Cao, X., Lan, Y., Yang, C., Li, C., 2015. Characterization of early transcriptional responses to cadmium in the root and leaf of Cd-resistant *Salix matsudana Koidz*. BMC Genomics 16 (1), 1.

Yoneyama, T., Gosho, T., Kato, M., Goto, S., Hayashi, H., 2010. Xylem and phloem transport of Cd, Zn and Fe into the grains of rice plants (*Oryza sativa* L.) grown in continuously flooded Cd-contaminated soil. Soil Sci. Plant Nutr. 56, 445–453.

Younis, U., Malik, S.A., Rizwan, M., Qayyum, M.F., Ok, Y.S., Shah, M.H.R., Rehman, R.A., Ahmad, N., 2016. Biochar enhances the cadmium tolerance in spinach (*Spinacia oleracea*) through modification of Cd uptake and physiological and biochemical attributes. Environ. Sci. Pollut. Res. 23, 21385–21394.

Yu, C., Peng, X., Yan, H., Li, X., Zhou, Z., Yan, T., 2015. Phytoremediation ability of *Solanum nigrum* L. to Cd-contaminated soils with high levels of Cu, Zn, and Pb. Water Air Soil Pollut. 226, 1–10.

Zhang, F.S., Li, L., 2003. Using competitive and facilitative interactions in intercropping systems enhances crop productivity and nutrient-use efficiency. Plant Soil 248, 305–312.

Zhou, Q.X., Song, Y.F., 2004. Principles and Methods of Contaminated Soil Remediation. Science Press, Beijing.

Zhou, J., Deckard, E.L., Ahrens, W.H., 2005. Factors affecting germination of hairy nightshade (*Solanum sarrachoides*) seeds. Weed Sci. 53, 41–45.

Zorrig, W., Rouached, A., Shahzad, Z., Abdelly, C., Davidian, J.C., Berthomieu, P., 2010. Identification of three relationships linking cadmium accumulation to cadmium tolerance and zinc and citrate accumulation in lettuce. J. Plant Physiol. 167, 1239–1247.

Zou, X.L., 2015. Phytoextraction of heavy metals from contaminated soil by co-cropping *Solanum nigrum* L. with ryegrass associated with endophytic bacterium. Sep. Sci. Technol. 50, 1806–1813.

## FURTHER READING

Koopmans, G.F., Römkens, P.F.A.M., Fokkema, M.J., Song, J., Luo, Y.M., Japenga, J., Zhao, F.J., 2008. Feasibility of phytoremediation to remediate cadmium and zinc contaminated soils. Environ. Pollut. 156, 905–914.

# CHAPTER 19

# Phytoremediation of Cadmium-Contaminated Soils by Using Legumes Inoculated by Efficient and Cadmium-Resistant Plant Growth-Promoting Bacteria

**Salwa Harzalli Jebara, Souhir Abdelkrim Ayed, Manel Chiboub, Imen Challougui Fatnassi, Omar Saadani, Ghassen Abid, Moez Jebara**
Laboratory of Legumes, Centre Biotechnology Borj Cedria, University Tunis El Manar, Hammam Lif, Tunisia

## Abstract

Legumes coinoculated by cadmium-resistant, plant growth promoting bacteria (PGPB) contribute to a type of phytoremediation for alleviating Cd contamination in the soil. PGPB are involved in plant growth via production of phytohormones and siderophores, in addition to symbiotic nitrogen fixation and phosphorus solubility capacity. The legumes-PGPB symbioses are reported to influence Cd bioavailability in legumes by various mechanisms, such as bioaccumulation, precipitation, complexation, and chelation. Cd toxicity has induced a wide range of physiological and biochemical tolerance mechanisms in legumes, such as the gene expression of metal binding proteins involved in chelation, and the transport of Cd, such as phytochelatin synthase (PCS) and metallothionin (MTs), and the activation of the antioxidant defense system. In this chapter, we demonstrate that legumes coinoculated by Cd-resistant PGPB could be useful in phytoremediation of Cd-contaminated soil and biofertilization.

**Keywords:** Cadmium, Gene expression, Legume, PGPB, Phytoremediation

## 1 INTRODUCTION

Industrial activities and intensive agricultural and waste water irrigation have caused the pollution of soils by heavy metals (HMs) (Chibuike and Obiora, 2014). HMs are nonbiodegradable, and accumulate in the environment, causing the contamination of the food chain, and negatively impacting human health (Clemens et al., 2013). Among HMs, cadmium is one of the most toxic pollutants (Nazar et al., 2012). It is a nonessential, highly phytotoxic metal that

*Cadmium Toxicity and Tolerance in Plants*
https://doi.org/10.1016/B978-0-12-814864-8.00019-X

damages general cell metabolism (Clemens et al., 2013). In plants, an excessive amount of Cd can impair important physiological and biological processes. It alters mineral nutrients' uptake; inhibits the stomatal opening by interacting with the water balance of the plant (Hossain et al., 2010); disturbs the Calvin cycle enzymes, photosynthesis, and carbohydrate metabolism; and lowers crop productivity (Shi et al., 2010).

Exposure to cadmium leads to the generation of the reactive oxygen species (ROS), which can cause oxidative stress when produced in large amounts (Cuypers et al., 2010). To repair the inhibitory effects of ROS, plants use antioxidative enzymes, including superoxide dismutase (SOD), ascorbate peroxidase (APX), peroxidase (POX), and catalase (CAT) (Wu et al., 2015). On the other hand, plants that have over-expressed or inhibited certain genes involved in chelation and transport of HM, such as phytochelatins synthase (PCS) and metallothioneins (MTs), contribute to HMs-tolerance, and could be candidates for the future engineering of plants for use in phytoremediation (Chiang et al., 2006).

Remediation of Cd through bioremediation is an environmentally friendly method that uses biological processes to remove HMs from the environment. One form of bioremediation is phytoremediation, which includes phytoextraction and phytostabilization (Ghosh and Singh, 2005; Dary et al., 2010). It is an effective solution for remediation of HMs-contaminated soils, as it aims to use metal accumulating plants and microorganisms to remove, transfer, or stabilize these contaminants from soil (Khan, 2005).

It has been reported that suitable legumes coinoculated by HMs-resistant plant growth promoting bacteria (PGPB) were part of a successful method for phytoremediation of HMs-contaminated soils (Fatnassi et al., 2015; Harzalli Jebara et al., 2015b; Saadani et al., 2016). Generally, legumes are nodulated by a microbial population other than rhizobia (Pilar and Ann, 2017) and rhizobia genera (Gyaneshwar et al., 2011), which improve the metal remediation capacity of symbiotic legumes (Gómez-Sagasti and Marino, 2015). PGPB, essentially rhizobia, provide legumes with nitrogen through symbiotic nitrogen fixation, in addition to their capacity to solubilize phosphorus, synthesize siderophores that can solubilize and sequester iron from the soil, and produce several phytohormones, such as indole acetic acids (IAA), which regulate plant growth and decrease the toxicity of metals by changing their bioavailability in plants (Ma et al., 2009). Legumes provide the microorganisms with root exudates and flavonoids that are important for their nutrition and nodulation, which provides nitrogen to legumes in N-limited environments and contaminated soils (Zaidi et al., 2012). For this purpose, we focused on the use of legumes coinoculated by a specific inoculum formed by a consortium of

PGPB in the phytoremediation of Cd-contaminated soils; in fact, the use of rhizobium-legumes' symbioses for efficient phytoremediation of Cd has been studied previously (Gómez-Sagasti and Marino, 2015; Chiboub et al., 2018; Osama et al., 2018).

## 2 HEAVY METALS' RESISTANCE OF PLANT-GROWTH-PROMOTING BACTERIA

The contamination of soil by HMs affects the survival of PGPB in the rhizosphere (Broos et al., 2005); however, certain effective and HMs-tolerant PGPB has the capacity to survive under HMs toxicity, and form an interesting consortium for phytoremediation use. Different soil bacteria demonstrated HMs resistance, such as *Rhizobium leguminosarum* bv. *Trifolii* that tolerate 7.1 mg Cd kg$^{-1}$ (Chaudri et al., 1992). Other genera of PGPB were obtained by the cultivation of four local legumes, *Vicia faba*, *Lens culinaris*, *Cicer arietinum*, and *Sulla coronaria* in HMs contaminated soils situated near mining and industrial zones. This study showed that *S. coronaria* was nodulated by a number of bacteria that have variable HMs tolerance; among them, *Pseudomonas* resisted up to (4.1 mM Cd). In *V. faba,* certain isolates, such as *Enterobacter cloacae,* were able to grow with 4.5 mM Cd and 2 mM Cu; whereas in *C. arietinum,* Cd tolerance did not exceed 3 mM. *L. culinaris* was nodulated by Pb-resistant strains such as *Agrobacterium tumefaciens* and *Rahnella aquatilis*, which tolerate up to 3.35 mM Pb; but Cd resistance didn't exceed 2.2 mM (Fatnassi et al., 2014). Other PGPB that exhibited HMs resistance are shown in Table 1. HMs resistance was confirmed by HMs resistance genes; in fact, PGPR-nodulating *S. coronaria* revealed Cd-resistant genes that encoded enzymes involved in Cd tolerance (Chiboub et al., 2016). Other heavy-metals-resistant PGPB revealed HMs-resistant genes, such as those nodulating *L. culinaris,* which revealed Pb-resistant genes (Harzalli Jebara et al., 2015b); whereas those nodulating *Lathyrus sativus* revealed a Cd-resistant gene (Fig. 1). The presence of Cu-resistant genes was revealed in *Sinorhizobium meliloti* (Carrasco et al., 2005), and in *S. meliloti* isolated from *Medicago lupulina* (Fan et al., 2011).

## 3 ROLE OF PLANT-GROWTH-PROMOTING TRAITS IN PLANT GROWTH OF LEGUMES UNDER CADMIUM CONTAMINATION

Cadmium contamination causes deleterious effects on plant growth and development, because its accumulation in plants inhibits cell functions such as photosynthesis, respiration, and water and nutrient uptake

**Table 1** Heavy metal tolerance range (CMI mM) of PGPR isolated from nodules of legumes grown in metal contaminated soils

| Isolat | Cd | Pb | Cu | Zn | References |
|---|---|---|---|---|---|
| *R. sullae* | 0.25 | 1.2 | 0.25 | 1.75 | Chiboub et al. (2016) and Saadani et al. (2016) |
| *R. sullae* | 1 | 1.1 | 0.05 | 1.75 | |
| *Pseudomonas fluorescens* | 1 | 1.2 | 0.35 | 1.6 | |
| *Pseudomonas* sp. | 4.1 | 1.2 | 0.05 | 1.5 | |
| *E. cloacae* | 3.2 | 4.05 | 2 | 1.75 | Fatnassi et al. (2014, 2015) |
| *R. leguminosarum* | 1.2 | 3.94 | 0.5 | 1.35 | |
| *Pseudomonas* sp. | 2.2 | 3.24 | 1 | 1.7 | |
| *Rhizobium* sp. CCNWSX0481 | 1.8 | 3.37 | 0.5 | 1.2 | |
| *Agrobacterium tumefaciens* | 0.25 | 3.35 | 0.1 | 1.2 | Harzalli Jebara et al. (2015a,b) |
| *Rahnella aquatilis* | 0.1 | 3.35 | 0.25 | 1.2 | |
| *Pseudomonas* sp. | 1.5 | 3.24 | 0.075 | 1.2 | |
| *Pseudomonas* sp. | 1.5 | 3.24 | 0.075 | 1.2 | |
| *Rhizobium* sp. | 0.1–1 | 3–6 | 0.5–2 | 0.5–2 | Pajuelo et al. (2008) |
| *Sinorhizobium* | N.A | 2 | 1.4 | N.A | |
| *Mesorhizobium* | 2.2 | N.A | 2.2 | 16–32 | Hao et al. (2012) |
| *Bradyrhizobium* | 1.5 | 2 | 1.5 | 5.1 | Dary et al. (2010) |

N.A, not available data.

**Fig. 1** Amplification of cadmium-resistant genes in different strains of *Rhizobium* sp. nodulating *Lathyrus sativus*. M: molecular marker (100pb). Primers used were described by Chiboub et al. (2016).

(DalCorso et al., 2010). The decrease in plant growth could be due to the antagonism between Cd and other mineral nutrients. In soybeans cultivated under Cd stress caused by the addition of 16 and 20 mg Cd kg$^{-1}$ soil, Cd caused deleterious effects on plant growth and nodulation (Sheirdil et al., 2012). In addition, Cd affects the structure of nodules and in the nitrogen fixation process in white lupin (Carpena et al., 2003).

However, the inoculation with rhizobium and/or PGPR induces an ameloration of plant growth. These results were observed in different legumes, such as *S. coronaria* cultivated in HMs and Cd contaminated soils, which showed amelioration of plant growth and nodulation after inoculation with HMs-resistant PGPB (Chiboub et al., 2018). Elsewhere, *Medicago sativa* inoculated by *S. meliloti* originated from a mining site cultivated for 60 days on soils containing 50 and/or 100 mg Cd kg$^{-1}$ soil showed that the inoculation hindered the occurrence of Cd-induced toxicity symptoms, increases biomass production, and improved nutrient acquisition comparatively with noninoculated plants (Ghnaya et al., 2015). Alfalfa cultivated in 1 mM Cd demonstrated deleterious impacts due to Cd uptake, and the alteration of photosynthetic efficiency; however, inoculation with *Rhizobium meliloti* ameliorated the plant's growth (Osama et al., 2018). Generally, HMs-resistant PGPB had beneficial effects on legume growth, essentially under HMs (Dary et al., 2010; Harzalli Jebara et al., 2015b; Saadani et al., 2016). These results suggested that HMs-resistant PGPB could be used as a biofertilizer, because they can improve soil fertility and crop productivity under stressful conditions (Katiyar et al., 2016).

It has been demonstrated that PGPB are characterized by multiple traits that promote plant growth, such as production of phytohormones, including indole-3-acetic acid (IAA); and solubilization of insoluble phosphate fixation of atmospheric nitrogen (Jebara et al., 2017), which additionally, under HM contamination, produced 1-aminocyclopropane-1-carboxylate (ACC) deaminase, leading to regulation of an ethylene concentration (Mishra et al., 2017). These PGP traits were demonstrated in PGPB-nodulating *S. coronaria* (Chiboub et al., 2016) and other PGPBs nodulating different other legumes, such as *L. culinaris* (Harzalli Jebara et al., 2015b), *V. faba* (Fatnassi et al., 2015), and *L. sativus* (Abdelkrim et al., 2018). These PGP traits are plant growth regulators, and phytostimulators of plant hormones such as auxins, cytokinins, and giberillins (Mehnaz et al., 2001).

## 4 ROLE OF PGPB ON Cd UPTAKE AND PHYTOREMEDIATION CAPACITY

Many reports demonstrated that legumes are clearly root-accumulators, and can tolerate high concentrations of some HMs (Pajuelo et al., 2008; Dary et al., 2010). Certain legume genotypes are tolerant to cadmium, such as *Lotus*, *Lupinus*, *Pisum*, and *Vicia* (Pajuelo et al., 2007). Generally, HMs tolerance of legumes depends on different factors, including the metal availability and mobility, its uptake, and translocation to the shoots (Sessitsch et al., 2011). On the other hand, legume-PGPB tolerance depends on their characteristics, including PGP substance production, nitrogen fixation, phosphorus solubility, and HMs bioaccumulation capacity (Jebara et al., 2017).

When legumes are exposed to high Cd concentrations, it results in high mobility of Cd, its translocation to shoots, and photosynthesis impairment (Gómez-Sagasti and Marino, 2015). However, inoculation with PGPB limits the deleterious effects of Cd exposure, as demonstrated in *S. coronaria* cultivated in hydroponic conditions, or in Cd-contaminated soils, which significantly increased Cd accumulation, especially in the roots (Chiboub et al., 2018). Therefore, the concentration of Cd in the roots and the shoots of *S. coronaria* are under the permissible level permitted ($10\,mg\,kg^{-1}$) for animal feeding (Barac et al., 2004), so, this symbiosis could be used in phytostabilization of moderately Cd contaminated soils. *M. sativa* inoculated by *S. meliloti* cultivated in the presence of $50\,mg\,Cd\,kg^{-1}$ in the soil enhanced Cd absorption by the roots and Cd translocation to the shoots; this association may be an efficient biological system to extract Cd from contaminated soils, and be useful for Cd phytoextraction (Ghnaya et al., 2015). Alfalfa coinoculated with *R. meliloti* and *Arbuscular Mycorrhizal Fungi* reduced Cd uptake and promoted plant growth (Osama et al., 2018). On the other hand, in the rhizosphere, there were root exudates and other free-living soil microorganisms that could influence the bioavailability to the plant.

PGPB developed many resistance mechanisms to immobilize, mobilize, or transform metals to reduce toxicity and metal uptake through the bioaccumulation process, including bioexclusion and biosorption, sequestration, exclusion, and complexation with metal-binding proteins (Gopalakrishnan et al., 2015; Emamverdian et al., 2015). Further, root exudates contained different components, such as organic acids, amino acids, which play an important role in phytoremediation by changing metal mobility, solubility, and bioavailability in soil by a range of mechanisms, such as precipitation, and chelation with ligands such as MTs and phytochelatins (PCs) (Cetin et al., 2011; Sofia et al., 2012; Chiang et al., 2011).

In the rhizosphere, the legume can communicate with the other soil bacteria of the rhizosphere by the root exudates containing different components that influence the HMs bioavailability, nitrogen fixation, and PGP production (Jebara et al., 2017), and act as chemoattractant signals and sources of carbon and nitrogen for microbes (Compant et al., 2010).

## 5 EFFECT OF Cd CONTAMINATION ON ANTIOXIDATIVE RESPONSES

Cd exposure of plants induces oxidative stress (Ahemad and Anis, 2014); it exerts its toxic effect on cells through formation of ROS (Souguir et al., 2008). Generally, plants contain different metal resistance systems to maintain metal homeostasis in cells by increasing levels of antioxidants in plants, as they may be used as early indicators of environmental stress (Wu et al., 2015). HMs, in particular Cd, reduce the GSH/GSSG ratio, and activate antioxidant enzymes such as SOD and GR (Romero-Puertas et al., 2007). The antioxidant response to cadmium exposure was described in the work of Lin et al. (2007), and showed that in *V. faba* exposed to Cd at 5 $\mu g\,mL^{-1}$, the activities of POD, CAT, and SOD were significantly decreased, thus leading to an accumulation of ROS. Sandalio et al. (2001) noticed a decrease in POX activity in pea plants under the influence of Cd. The modulation of antioxidant levels constitutes an important adaptive response in withstanding adverse conditions, and maintenance of a high antioxidant capacity in cells is linked to increased tolerance against heavy metal contamination (Baycu et al., 2006). Different responses were obtained in shoots and roots of *S. coronaria* cultivated hydroponically and coinoculated by *Pseudomonas fluorescens*, *Pseudomonas*, and *Rhizobium sullae*; the application of 50 $\mu M$ $CdCl_2$ stimulated all enzymes in shoots and decreased SOD and CAT activities in roots; whereas, 100 $\mu M$ of $CdCl_2$ increased SOD, APX, CAT, and GPOX activities in shoots and significantly increased CAT activity in roots (Chiboub et al., 2018). The response of antioxidant enzymes to HMs in inoculated legumes depends on the metal concentration, period of treatment and inoculation, and plant organs (Harzalli Jebara et al., 2015a; Jebara et al., 2017). Other responses were reported in other inoculated legumes exposed to different HMs, *V. faba* treated by 1 mM Cu and inoculated with *Rhizobium* sp. *CCNWSX0481*, *R. leguminosarum* bv *viciae, E. cloacae* and *Pseudomonas* sp. showed enhancement in SOD, CAT, and APX activities in roots, and only SOD and POX were activated in shoots (Fatnassi et al., 2015). *L. culinaris* grown hydroponically with 2 mM Pb treatment and inoculated by a consortium formed by

*A. tumefaciens*, *R. aquatilis*, and *Pseudomonas* sp. demonstrated an activation of SOD and GPOX; whereas CAT was inhibited (Harzalli Jebara et al., 2015a). These results demonstrated that PGPRs control antioxidative responses, and control ROS produced. Nodular antioxidant enzymes take part in heavy metal plant tolerance; thus, nodules are rich in enzymatic antioxidants that protect the plant from toxicity (Becana et al., 2010; Jebara et al., 2010). In nodules, leghemoglobine controls low $O_2$ levels. This process is essential to protect nitrogenase after Cd exposure contributing to the tolerance of legumes during metal phytostabilization, or phytoextraction strategies (Shilev et al., 2012).

## 6 EFFECT OF CADMIUM TREATMENT AND COINOCULATION ON PCS AND MTs GENE EXPRESSIONS

The exposure of plants to cadmium contamination induces the expression or inhibition of genes involved in tolerance mechanisms such as PCs and MTs.

### 6.1 Phytochelatins (PCs)

PCs are synthesized from the enzyme PCS. The expression PCS genes in transgenic plants is a promising method of increasing plants' resistance against Cd and phytoremediation use; as shown in transgenic tobacco plants, the expression of PCS enhanced their resistance to cadmium (Postrigan et al., 2012).

In the roots of *S. coronaria*, Cd induced the gene expression PCS in roots, though only at much higher exposure levels of Cd (200 μM Cd); whereas coinoculation inhibited it significantly (Fig. 2). These results suggested that PCS genes could be a Cd responsive gene that contributes to the heavy metal tolerance in *S. coronaria*, but coinoculation inhibited them significantly. It appears that PGPR-enhanced Cd accumulation in roots inhibits the PCS gene, but other PCS genes coding for other isoforms of PCs must be studied to elucidate the exact mechanism, because it has been demonstrated that PCs are the best-characterized heavy metal chelators in plants, especially in the context of Cd tolerance (Cobbett, 2000). After synthesis, PCs bind the heavy metal ions and facilitate their transport as complexes into the vacuole by several classes of proteins called heavy metal-transporting ATPases, such as Nramp and Zip (Singh et al., 2011; Clemens, 2006).

### 6.2 Metallothioneins

Like PCs, MTs are cysteine-rich metal-binding peptides. The expression of MTs genes varies with plants and HM. Generally, PCs in plants are produced

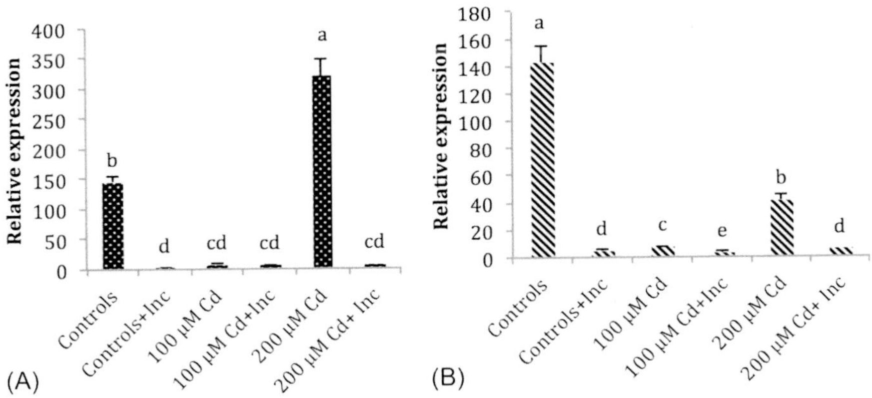


**Fig. 2** Relative expression ratios of the transcripts of (A) PCS and MT (B) genes in *Sulla coronaria* treated by 100 and 200 µM Cd, and inoculated by the inoculum (two *Rhizobium sullae*, *Pseudomonas* sp., and *Pseudomonas fluorescens*). The relative gene expression level as expressed by △CT was determined for each group and *vertical bars* represented the mean ± SD. Significant differences ($P < .05$) detected by Tukey's multiple comparison test are shown by different letters above the bars.

by Cd detoxification, but MTs appear to be capable of showing affinity with different metals, such as Cu, Zn, Cd, and As (Yang and Chu, 2011).

In *Arabidopsis thaliana*, MT gene expression is activated in response to Cu and Cd treatment, but not by Zn. In *V. faba*, which overexpresses AtMT2a and AtMT3 from *A. thaliana*, the plant can tolerate higher levels of Cd than untransformed cells (Lee et al., 2004). In soybeans, three isoforms of MTs were expressed under Cd exposure; MT1, MT2, and MT3 (Pagani et al., 2012).

In *S. coronaria* cultivated with 200 µM Cd, we marked inhibition of MT expression compared with the control group (Fig. 2). These results indicate that there is no correlation between MT2 expression and Cd accumulation in roots; for example, other isoforms must be studied to evaluate the role of MTs in plants exposed to Cd stress. Nevertheless, it appears that other characteristics were involved in MTs gene expression, such as genetic structure of plants and metal binding regions of plant MTs (Grennan, 2011), the sequence of gene coding for the isoforms (Dajbrowska et al., 2013).

## 7 CONCLUSIONS

HMs soil contamination is a serious threat to the environment and food safety. Cadmium is the worst contaminant of agricultural soils situated near mining and industrial zones. Phytoremediation is an eco-friendly way to

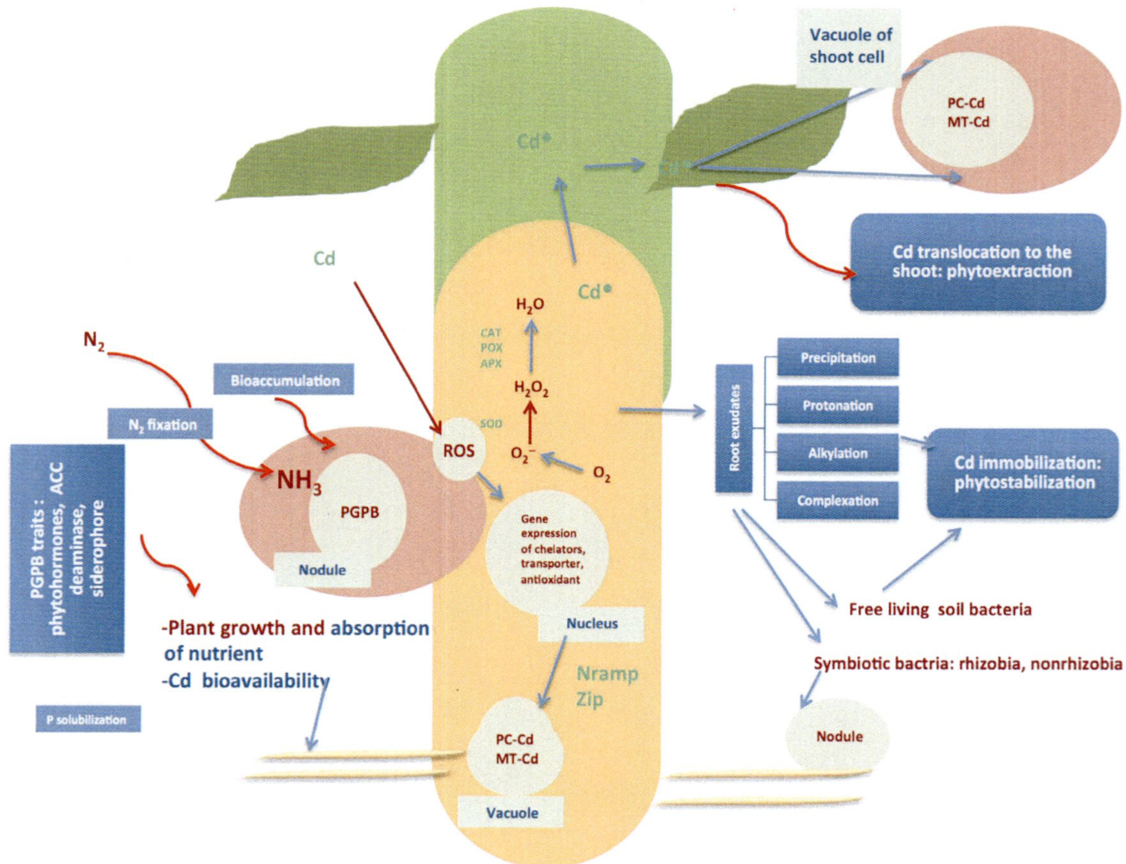


**Fig. 3** Different mechanisms used by legume-PGPB in phytoremediation of cadmium contaminated soils. PC, phytochelatins; MT, metallothionins.

cope with Cd soil contamination. In this chapter, we focused on the use of legume-PGPB in remediation of Cd-contaminated soil. Plant-growth-promoting bacteria possesses multiple traits that regulate plant growth and have the potential to detoxify pollutants. Different results demonstrated that these symbioses had the ability to enhance Cd uptake in plant organs, thus phytoextracting Cd from the soil, or phytostabilizing it in the soil, and enhancing its accumulation in the roots. The efficiency of a specific PGPB depends on variable parameters, such as soil characteristics, Cd concentration, and legume accumulation capacity. Theses symbioses respond to Cd toxicity by various mechanisms, as demonstrated in Fig. 3, including bioaccumulation, precipitation, complexation, and chelation, in addition to the expression of genes involved in chelation and transport. On the other hand, Cd contamination induces oxidative stress, and plants respond by the activation of antioxidant systems such as SOD, POX, APX, and CAT, which act in an integrated manner at excessive levels of Cd, and raise plant tolerance to Cd stress.

Therefore, others' investigations will further increase our understanding of the fundamental mechanisms involved in Cd tolerance processes of legumes-PGPB suited for phytoremediation of Cd-contaminated soils. The relationship between plant antioxidative defense mechanisms and HM chelators needs to be well defined and established to achieve sustainable agro-ecological conditions.

## REFERENCES

Abdelkrim, S., Jebara, S.H., Saadani, O., Chiboub, M., Abid, G., Jebara, M., 2018. Effect of Pb-resistant plant growth-promoting rhizobacteria inoculation on growth and lead uptake by *Lathyrus sativus*. J. Basic Microbiol. 1–11. https://doi.org/10.1002/jobm.201700626.

Ahemad, M., Anis, M., 2014. Changes in activity of antioxidant enzymes and photosynthetic machinery during acclimatization of micropropagated Cassia alata L. plantlets. In Vitro Cell. Dev. Biol. Plant 50, 601–609.

Barac, T., Taghavi, S., Borremans, B., Provoost, A., Oeyen, L., Colpaert, J.V., Vangronsveld, J., Van Der Lelie, D., 2004. Engineered endophytic bacteria improve phyto-remediation of water-soluble, volatile, organic pollutants. Nat. Biotechnol. 22, 583–588.

Baycu, G., Tolunay, D., Ozden, H., Guenenebakan, S., 2006. Ecophysiological and seasonal variations in Cd, Pb, Zn and Ni concentrations in the leaves of urban deciduous trees in Istanbul. Environ. Pollut. 143, 545–554.

Becana, M., Matamoros, M.A., Udvardi, M., Dalton, D.A., 2010. Recent insight into antioxidant defenses of legume root nodules. New Phytol. 188, 960–976.

Broos, K., Beyens, H., Smolders, E., 2005. Survival of rhizobia in soil is sensitive to elevated zinc in the absence of the host plant. Soil Biol. Biochem. 37, 573–579.

Carpena, R.O., Vázquez, S., Esteban, E., Fernández-Pascual, M., de Felipe, M.R., Zornoza, P., 2003. Cadmium-stress in white lupin: effects on nodule structure and functioning. Plant Physiol. Biochem. 41, 911–919.

Carrasco, J.A., Armario, P., Pajuelo, E., Burgos, A., et al., 2005. Isolation and characterisation of symbiotically effective Rhizobium resistant to arsenic and heavy metals after the toxic spill at the Aznalcollar pyrite mine. Soil Biol. Biochem. 37, 1131–1140.

Cetin, S.C., Karaca, A., Kizilkaya, R., Turgay, Q.C., 2011. Role of plant growth promoting bacteria and fungi in heavy metal detoxification. In: Sherameti, I., Verma, A. (Eds.), Detoxification of Heavy Metals. In: Soil Biology, vol. 30. Springer, Berlin, pp. 369–388.

Chaudri, A.M., McGrath, S.P., Giller, K.E., 1992. Survival of the indigenous population of *Rhizobium leguminosarum biovar trifolii* in soil spiked with Cd, Zn, Cu and Ni salts. Soil Biol. Biochem. 24, 625–632.

Chiang, H.C., Lo, J.C., Yeh, K.C., 2006. Genes associated with heavy metal tolerance and accumulation in Zn/Cd hyperaccumulator *Arabidopsis halleri*: a genomic survey with cDNA microarray. Environ. Sci. Technol. 40, 6792–6798.

Chiang, P.N., Chiu, C.Y., Wang, M.K., Chen, B.T., 2011. Low-molecular weight organic acids exuded by Millet (*Setaria italica* (L.) *Beauv.*) roots and their effect on the remediation of cadmium-contaminated soil. Soil Sci. 176, 33–38.

Chiboub, M., Saadani, O., Fatnassi, C.I., Souhir, A., Jebara, M., Harzalli Jebara, S., 2016. Characterization of plant growth promoting rhizobacteria efficient and resistant to cadmium isolated from *Sulla coronaria*. C.R. Biol. 339 (9–10), 391–398.

Chiboub, M., Harzalli Jebara, S., Saadani, O., Challougui Fatnassi, I., Abdelkerim, S., Jebara, M., 2018. Physiological responses and antioxidant enzyme changes in *Sulla coronaria* inoculated by cadmium resistant bacteria. J. Plant Res. 131, 99–110.

Chibuike, G.U., Obiora, S.C., 2014. Heavy metal polluted soils: effect on plants and bioremediation methods. Appl. Environ. Soil Sci.. https://doi.org/10.1155/2014/752708.

Clemens, S., 2006. Evolution and function of phytochelatin synthases. J. Plant Physiol. 163, 319–332.

Clemens, S., Aarts, M.G., Thomine, S., Verbruggen, N., 2013. Plant science: the key to preventing slow cadmium poisoning. Trends Plant Sci. 18, 92–99.

Cobbett, C.S., 2000. Phytochelatins and their roles in heavy metal detoxification. Plant Physiol. 123, 825–832.

Compant, S., Clément, C., Sessitsch, A., 2010. Plant growth-promoting bacteria in the rhizo- and endosphere of plants: their role, colonization, mechanisms involved and prospects for utilization. Soil Biol. Biochem. 42, 669–678.

Cuypers, A., Plusquin, M., Remans, T., Jozefczak, M., Keunen, E., Gielen, H., Opdenakker, K., Nair, A.R., Munters, E., Artois, T.J., Nawrot, T., Vangronsveld, J., Smeets, K., 2010. Cadmium stress: an oxidative challenge. Biometals 23, 927–940.

Dajbrowska, G., Mierek-Adamska, A., Goc, A., 2013. Characterisation of *Brassica napus* L. metallothionein genes (BnMTs) expression in organs and during seed germination. Aust. J. Crop. Sci. 7, 1324–1332.

DalCorso, G., Silvia, F., Antonella, F., 2010. Regulatory networks of cadmium stress in plants. Plant Signal. Behav. 5, 663–667.

Dary, M., Chamber-Perez, M.A., Palomares, A.J., Pajuelo, E., 2010. *In situ* phytostabilisation of heavy metal polluted soils using *Lupinus luteus* inoculated with metal resistant plant-growth promoting rhizobacteria. J. Hazard. Mater. 177, 323–330.

Emamverdian, A., Ding, Y., Mokhberdoran, F., Xie, Y., 2015. Heavy metal stress and some mechanisms of plant defense response. Sci. World J. https://doi.org/10.1155/2015/756120.

Fan, L.M., Zhan, Q.M., Liang, J.Q., HuiFen, L., Wang, E.T., Wei, G.H., 2011. Characterization of a copper-resistant symbiotic bacterium isolated from *Medicago lupulina* growing in mine tailings. Bioresour. Technol. 102, 703–709.

Fatnassi, C.I., Chiboub, M., Jebara, M., Jebara, H.S., 2014. Bacteria associated with different legume species grown in heavy-metal contaminated soils. Int. J. Agric. Policy Res 2, 460–467.

Fatnassi, C.I., Chiboub, M., Saadani, O., Jebara, M., Jebara, H.S., 2015. Phytostabilization of moderate copper contaminated soils using co-inoculation of *Vicia faba* with plant growth promoting bacteria. J. Basic Microbiol. 55, 303–311.

Ghnaya, T., Mnassri, M., Ghabriche, R., et al., 2015. Nodulation by *Sinorhizobium meliloti* originated from a mining soil alleviates Cd toxicity and increases Cd-phytoextraction in *Medicago sativa* L. Front. Plant Sci. 6, 863. https://doi.org/10.3389/fpls.2015.00863.

Ghosh, M., Singh, S.P., 2005. A review on phytoremediation of heavy metals and utilization of its byproducts. Appl. Ecol. Environ. 3, 1–18.

Gómez-Sagasti, M.T., Marino, D., 2015. PGPRs and nitrogen-fixing legumes: a perfect team for efficient Cd phytoremediation. Front. Plant Sci. 6, 81. https://doi.org/10.3389/fpls.2015.00081.

Gopalakrishnan, S., Sathya, A., Vijayabharathi, R., Kumar, R., Varshney, C.L., Gowda, L., Krishnamurthy, L., 2015. Plant growth promoting rhizobia: challenges and opportunities. 3 Biotech 5, 355–377.

Grennan, A.K., 2011. Metallothioneins, a diverse protein family. Plant Physiol. 155, 1750–1751.

Gyaneshwar, P., Hirsch, A.M., Moulin, L., Chen, W.M., Elliott, G.N., Bontemps, C., et al., 2011. Legume-nodulating betaproteobacteria: diversity, host range, and future prospects. Mol. Plant-Microbe Interact. 24, 1276–1288.

Hao, X., Lin, Y., Johnstone, L., Baltrus, D.A., Miller, S.J., Wei, G., Rensing, C., 2012. Draft genome sequence of plant growth-promoting rhizobium *Mesorhizobium amorphae*, isolated from zinc-lead mine tailings. J. Bacteriol. 194, 736–737.

Harzalli Jebara, S., Abdelkerim, S., Challougui Fatnassi, I., Chiboub, M., Saadani, O., Jebara, M., 2015a. Identification of effective Pb resistant bacteria isolated from *Lens culinaris* growing in lead contaminated soils. J. Basic Microbiol. 55, 346–353.

Harzalli Jebara, S., Saadani, O., Challougui Fatnassi, I., Chiboub, M., Abdelkerim, S., Jebara, M., 2015b. Inoculation of *Lens culinaris* with Pb-resistant bacteria shows potential for phytostabilization. Environ. Sci. Pollut. Res. 22, 2537–2545.

Hossain, M.A., Hasanuzzaman, M., Fujita, M., 2010. Up-regulation of antioxidant and glyoxalase systems by exogenous glycine betaine and proline in mung bean confer tolerance to cadmium stress. Physiol. Mol. Biol. Plants 16, 259–272.

Jebara, S., Drevon, J.J., Jebara, M., 2010. Modulation of symbiotic efficiency and nodular antioxidant enzyme activities in two *Phaseolus vulgaris* genotypes under salinity. Acta Physiol. Plant. 32, 925–932.

Jebara, S.H., Fatnassi, I.C., Ayed, A.S., Saadani, O., Chiboub, M., et al., 2017. Potentialities and limit of legume-plant growth promoting bacteria symbioses use in phytoremediation of heavy metal contaminated soils. Int. J. Plant Biol. Res. 5 (4), 1077.

Katiyar, D., Hemantaranjan, A., Singh, B., 2016. Plant growth promoting rhizobacteria-an efficient tool for agriculture promotion. Adv. Plants Agric. Res. 4 (6), 00163. https://doi.org/10.15406/apar.2016.04.00163.

Khan, A.G., 2005. Role of soil microbes in the rhizospheres of plants growing on trace metal contaminated soils in phytoremediation. J. Trace Elem. Med. Biol. 18, 355–364.

Lee, J., Shim, D., Song, W.Y., Hwang, I., Lee, Y., 2004. Arabidopsis metallothioneins 2a and 3 enhance resistance to cadmium when expressed in *Vicia faba* guard cells. Plant Mol. Biol. 54, 805–815.

Lin, R., Wang, X., Luo, Y., Du, W., Guo, H., Yin, D., 2007. Effects of soil cadmium on growth, oxidative stress and antioxidant system in wheat seedlings (*Triticum aestivum* L.). Chemosphere 69, 89–98.

Ma, Y., Rajkumar, M., Freitas, H., 2009. Isolation and characterization of Ni mobilizing PGPB from serpentine soils and their potential in promoting plant growth and Ni accumulation by *Brassica* spp. Chemosphere 75, 719–725.

Mehnaz, S., Mirza, M.S., Haurat, J., Bally, R., Normand, P., et al., 2001. Isolation and 16S rRNA sequence analysis of the beneficial bacteria from the rhizosphere of rice. Can. J. Microbiol. 47 (2), 110–117.

Mishra, J., Singh, R., Arora, N.K., 2017. Alleviation of heavy metal stress in plants and remediation of soil by rhizosphere microorganisms. Front. Microbiol. 8, 1706. https://doi.org/10.3389/fmicb.2017.01706.

Nazar, R., Iqbal, N., Masood, A., Khan, M.I.R., Syeed, S., Nafees, A.K., 2012. Cadmium toxicity in plants and role of mineral nutrients in its alleviation. Am. J. Plant Sci. 3, 1476–1489.

Osama, N.M., Ibrahim, M.Z., Ghazi, S.M., Shedeed, Z.A., Doaa, M.N., 2018. Remediation of cadmium toxicity on alfalfa (*Medicago sativa* L.) using biochar as a bioadsorbent, Rhizobium meliloti and arbuscular mycorrhizal fungi as biofertilizers. J. Bioremediat. Biodegrad. 9, 429. https://doi.org/10.4172/2155-6199.1000429.

Pagani, M.A., Tomas, M., Carrillo, J., et al., 2012. The response of the different soybean metallothionein isoforms to cadmium intoxication. J. Inorg. Biochem. 117, 306–315.

Pajuelo, E., Carrasco, J.A., Romero, L.C., Chamber, M.A., Gotor, C., 2007. Evaluation of the metal phytoextraction potential of crop legumes. Regulation of the expression of O-acetylserine (thiol) lyase under metal stress. Plant Biol. 9, 672–681.

Pajuelo, E., Rodriguez-Llorente, D., Dary, M., Palomares, A.J., 2008. Toxic effects of arsenic on *Sinorhizobium-Medicago sativa* symbiotic interaction. Environ. Pollut. 154, 203–211.

Pilar, M.H., Ann, M.H., 2017. The nodule microbiome: $N_2$-fixing rhizobia do not live alone. Phytobiomes J. 1 (2), 70–82.

Postrigan, B.N., Knyazev, A.B., Kuluev, B.R., Yakhin, O.I., Chemeris, A.V., 2012. Expression of the synthetic phytochelatin gene in tobacco. Russ. J. Plant Physiol. 59, 275–280.

Romero-Puertas, M.C., Corpas, F.J., Rodriguez-Serrano, M., Gomez, M., del Río, L.A., Sandalio, L.M., 2007. Differential expression and regulation of antioxidative enzymes by cadmium in pea plants. J. Plant Physiol. 164, 1346–1357.

Saadani, O., Challougui Fatnassi, I., Chiboub, M., Abdelkrim, S., Barhoumi, F., Jebara, M., Harzalli, J.S., 2016. In situ phytostabilisation capacity of three legumes and their associated plant growth promoting bacteria (PGPBs) in mine tailings of northern Tunisia. Ecotoxicol. Environ. Saf. 130, 263–269.

Sandalio, L.M., Dalurzo, H.C., Gomez, M., Romero-Puertas, M., del Rio, L.A., 2001. Cadmium-induced changes in the growth and oxidative metabolism of pea plants. J. Exp. Bot. 52, 2115–2126.

Sessitsch, A., Kuffner, M., Kidd, P., Vangronsveld, J., Wenzel, W.W., Fallmann, K., et al., 2011. The role of plant-associated bacteria in the mobilization and phytoextraction of trace elements in contaminated soils. Soil Biol. Biochem. 60, 182–194.

Sheirdil, M., Bashir, R.A., Hayat, K., Akhtar, M.S., 2012. Effect of cadmium on soybean (*Glycine max* L.) growth and nitrogen fixation. Afr. J. Biotechnol. 11, 1886–1891.

Shi, G., Liu, C., Cai, Q., Liu, Q., Hou, C., 2010. Cadmium accumulation and tolerance of two safflower cultivars in relation to photosynthesis and antioxidative enzymes. Bull. Environ. Contam. Toxicol. 85, 256–263.

Shilev, S., Naydenov, M., Prieto, M.S., Vassilev, N., Sancho, E.D., 2012. PGPR as inoculants in management of lands contaminated with trace elements. In: Maheshwari, D.K. (Ed.), Bacteria in Agrobiology: Stress Management. Springer, Berlin, pp. 259–277.

Singh, R.K., Anandhan, S., Singh, S., Patade, V.Y., Ahmed, Z., Pande, V., 2011. Metallothionein-like gene from *Cicer microphyllum* is regulated by multiple abiotic stresses. Protoplasma 248, 839–847.

Sofia, C., Sofia, P., Ana, L., Etelvina, F., 2012. The influence of glutathione on the tolerance of Rhizobium leguminosarum to cadmium. In: Zaidi, A., Wani, P.A., Khan, M.S. (Eds.), Toxicity of Heavy Metals to Legumes and Bioremediation. Springer-Verlag, Wien, pp. 89–100.

Souguir, D., Ferjani, E., Ledoigt, G., Goupil, P., 2008. Exposure of *Vicia faba* and *Pisum sativum* to copper-induced genotoxicity. Protoplasma 233, 203–207.

Wu, Z., Zhao, X., Scu, X., Tang, Q., Nie, Z., Qu, C., Chen, Z., Hu, C., 2015. Antioxidant enzyme systems and the ascorbate–glutathione cycle as contributing factors to cadmium accumulation and tolerance in two oilseed rape cultivars (*Brassica napus* L.) under moderate cadmium stress. Chemosphere 138, 526–536.

Yang, Z., Chu, C., 2011. Towards understanding plant response to heavy metal stress. In: Abiotic Stress in Plants—Mechanisms and Adaptations. InTech, Shanghai, China, pp. 59–78.

Zaidi, A., Wani, P.A., Khan, M.S. (Eds.), 2012. Toxicity of Heavy Metals to Legumes and Bioremediation. Springer, Dordrechthttps://doi.org/10.1007/978-3-7091-0730-0.

## FURTHER READING

Chen, W.M., Yu, C.H., James, E.K., Chang, J.S., 2008. Metal biosorption capability of *Cupriavidus taiwaninsis* and its effects on heavy metals removal by nodulated *Mimosa pudica*. J. Hazard. Mater. 153, 364–372.

Compant, S., Saikkonen, K., Mitter, B., Campisano, A., Mercado-Blanco, J., 2016. Editorial special issue: soil, plants and endophytes. Plant Soil 405, 1–11.

Corticeiro, S.C., Lima, A.I.G., Figueira, E., 2006. The importance of glutathione in oxidative status of *Rhizobium leguminosarum* biovar *viciae* under Cd exposure. Enzym. Microb. Technol. 40, 132–137.

DalCorso, G., Farinati, S., Maistri, S., Furini, A., 2008. How plants cope with cadmium: staking all on metabolism and gene expression. J. Integr. Plant Biol. 50, 1268–1280.

Glick, B.R., 2010. Using soil bacteria to facilitate phytoremediation. Biotechnol. Adv. 28, 367–374.

Lyzenga, W.J., Stone, S.L., 2012. Abiotic stress tolerance mediated by protein ubiquitination. J. Exp. Bot. 63, 599–616.

Manier, N., Deram, A., Broos, K., Denayer, F.O., Van Haluwyn, C., 2009. White clover nodulation index in heavy metal contaminated soils—a potential bioindicator. J. Environ. Qual. 38, 685–692.

Yurekli, F., Kucukbay, Z., 2003. Synthesis of phytochelatins in *Helianthus annus* is enhanced by cadmium nitrate. Acta Bot. Croat. 62, 21–25.

CHAPTER 20

# Phytoremediation of Cadmium-Polluted Water/Sediment by Aquatic Macrophytes: Role of Plant-Induced pH Changes

**Muhammad Tariq Javed*, Kashif Tanwir*, Muhammad Sohail Akram*, Muhammad Shahid†, Nabeel Khan Niazi‡, Sylvia Lindberg§**

*Department of Botany, Faculty of Life Sciences, Government College University, Faisalabad, Pakistan
†Department of Bioinformatics and Biotechnology, Government College University, Faisalabad, Pakistan
‡Institute of Soil and Environmental Sciences, University of Agriculture Faisalabad, Faisalabad, Pakistan
§Department of Ecology, Environment and Plant Sciences, Stockholm University, Stockholm, Sweden

## Abstract

Cadmium (Cd) is one of the most readily absorbed and rapidly translocated heavy metal in soil-plants systems. Aquatic plant species play an important role in the removal of Cd from waste waters passing through wetlands. Wetlands' macrophytes directly take up the metals, or indirectly affect the removal efficiencies, by increasing the residence time and reducing water velocities, which in turn enhances sedimentation. Different categories of wetland plants, i.e., emergent, surface floating, free floating, rooted leaves, submerged macrophytes, and trees have different ways of taking up the contaminants. In aquatic plants, Cd evokes a number of parallel and/or consecutive events at molecular, physiological, and morphological levels. The appropriate understanding of the mechanisms employed by macrophytes may optimize their use for remedial purposes. This chapter, therefore, focuses on summarizing the responses of aquatic macrophytes to Cd, with and their possible implementation in phytoremediation techniques. The principle components reviewed include Cd-induced growth inhibition, changes in water relations and ion metabolism, effects on photosynthesis, changes in enzyme activities, and formation of free radicals. Efforts have also been made to highlight the defense mechanisms such as cell wall immobilization, phytochelatin-based sequestration and compartmentalization processes, plasma membrane exclusion, stress proteins, peroxidases, metallothioneins, and so forth. In addition, understanding the role of Cd in modifying the rhizospheric pH also has great importance in understanding remediation, and hence, will be the subject of this chapter.

**Keywords:** Aquatic plants, Cadmium, Defense mechanisms, pH changes, Phytofiltration, Phytostabilization, Tolerance

*Cadmium Toxicity and Tolerance in Plants*
https://doi.org/10.1016/B978-0-12-814864-8.00020-6

## 1 INTRODUCTION

The term "phytoremediation" is explained as "the use of green plants and their associated microorganisms, soil amendments, and agronomic techniques to remove, contain, or render environmental contaminants harmless." Phytoremediation is attractive, as it is sustainable and relatively inexpensive as compared with other remediation schemes (Muehe et al., 2015), and it is considered a modified form of green technology to eradicate pollutants from atmospheric components. Various techniques that are used for the remediation of toxic metals from polluted soils are: phytostabilization, phytoextraction, rhizofiltration, and phytovolatilization. Countless factors affect the choice of the appropriate phytoremediation approach for soil refinement. They may include heavy metal concentration, soil properties and texture, climatic conditions, and plant species (Laghlimi et al., 2015).

Heavy metals are hazardous because they have high intensity of bioaccumulation in the living ecosystems. Bioaccumulation is the proliferation of toxic chemicals in a living organism with time, in contrast to the concentration in the atmosphere. Compounds recurrently accumulate in living organisms, are metabolized inside, and become a part of the biomass, or are released (Verma and Suthar, 2015).

Metal contamination of farm soils tend to create diverse ecological and environmental problems, including microbial plant growth, yield reduction and community shift, entry of metal into the food chain, and deterioration of soil (Sharma and Archana, 2016).

Cadmium that is abundant in zinc ores and during zinc production is produced as a by-product. It is also used in electroplating, batteries, and electronic appliances. Cadmium is a nonessential, poisonous heavy metal that can be easily transported from roots to shoots in plants (Sterckeman et al., 2015). Elevated Cd levels result in numerous adverse physiological, morphological, structural, and biochemical modifications in plants, i.e., water imbalance, growth inhibition, suppression of seed germination, and reduction in yield (Tran and Popova, 2013). Cadmium pollution of food and soil is a universal environmental issue that was generated by rigorous agricultural practices, urbanization, and industrialization (Khan et al., 2017).

## 2 HEAVY METALS AND WATER POLLUTION

The oldest scientific term for heavy metal can be located in the third Danish edition of Bjerrum's Inorganic Chemistry. Heavy metals have a specific density of $5\,g\,cm^{-3}$, and 53 out of 92 natural elements are heavy metals. Only a few of these metals are biologically important for plants, due to their high solubility (Hassan and Aarts, 2011). Trace elements are a complex series of metalloids and metals with an atomic mass above 20 that are harmful for animals and plants (Liu et al., 2015). During the past decade, a vast increase in metal pollution was recorded, due to manufacturing, agriculture, mining, smelting, and waste disposal practices, and the use of municipal water for soil amendments (Liu et al., 2013). Few heavy metals are necessary for physiological reactions in plants, e.g., zinc (Zn), manganese (Mn), nickel (Ni), and copper (Cu); while various others have no specific role in plant metabolism, such as lead (Pb), cadmium (Cd), and mercury (Hg) (Jia et al., 2018). The most dangerous heavy metal is Cd, due to its high mobility in soil plant systems and the fact that is has no role in plant nutrition (Tran and Popova, 2013).

## 3 CADMIUM AS A HEAVY METAL

Cadmium was discovered by Karl Hermann and Friedrich Stromeyer in 1817 from zinc oxide produced during the process of roasting zinc carbonate. Cadmium was found as cadmiate anions, and was insoluble in bases (Tolcin, 2015). Cadmium is a rare earth element, and it is always present in the form of salt. It is readily oxidized into cadmium oxide in the presence of air. This metal is highly reactive with water vapor; hydrogen chloride, carbon dioxide, sulfur dioxide, and sulfur trioxide produce cadmium sulfide, carbonate, chloride, or hydroxide. Often associated with ores of zinc, lead, and copper; the concentration of Cd in the earth's crust is fluctuating from 0.1 to 0.5 $mg\,kg^{-1}$. At 25°C, Cd has a specific density of 8.65 $g\,cm^{-3}$, 765 °C boiling point, and 321°C melting point (ATSDR, 2012). The aquatic wetland plants are capable of taking up higher concentrations of Cd as compared with terrestrial land plants, and are helpful in phytoremediation, i.e., *Eichhornia crassipes* (9.13 $mg\,kg^{-1}$), *Panicum antidotale* (7.99 $mg\,kg^{-1}$), *Lemna minor* (8.35 $mg\,kg^{-1}$), and *Potamogeton crispus* (7.15 $mg\,kg^{-1}$). The bioaccumulation factor for Cd was varied in terrestrial wetland plants as: 35.19 (*Sida acuta*), 826.72 (*Achyranthus aspera*), 619.80 (*Cannabis sativa*), and 11.29 (*Anagalis arvensis*) (Singh et al., 2017).

### 3.1 Natural and Anthropogenic Sources of Cadmium Toxicity

Cadmium has two major sources; one is natural, and other is anthropogenic (Fig. 1). Anthropogenic sources account for >90% of the total Cd present in surface water, with atmospheric particulate deposition from fossil fuel combustion (including coal) contributing to approximately 40% of the total Cd

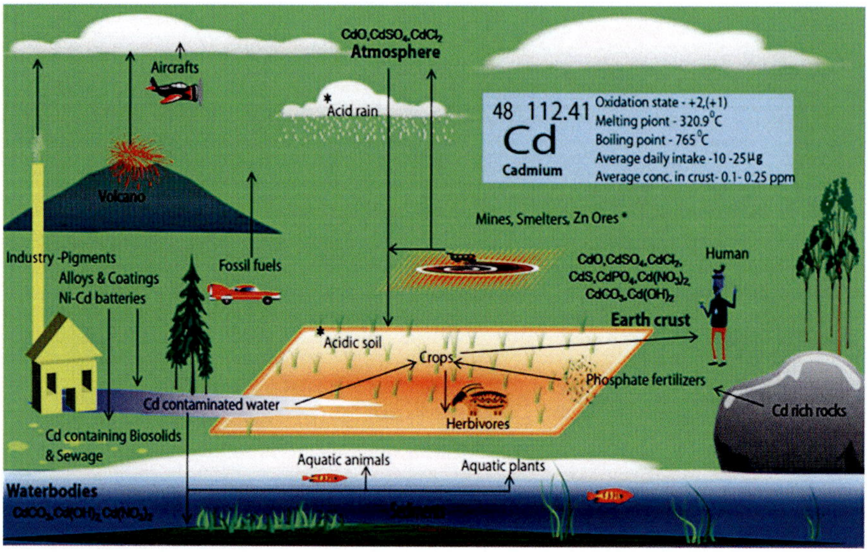


**Fig. 1** The anthropogenic and natural sources of cadmium (Cd) together with biogeochemical cycle showing the transportation of Cd in the atmosphere (Sebastian et al., 2014). *Acid rain and acidic soils have strong effect on Cd uptake in plants by lowering pH of soil which increases Cd uptake by plants.

present in surface water (US EPA, 2016). Generally, 25,000 tons of Cd is released into the atmosphere per year, and half of this is produced due to weathering of rocks, and the remaining half through volcanoes and forest fires. Cd is also released into the environment through human activities. The 90% of anthropogenic Cd is the outcome of industrial units and agricultural practices (Wood et al., 2012), including industrial waste from factories, i.e., the manufacturing of plastics, batteries that contain Cd, mining, electroplating, alloy preparation, paint pigments, and composts or fertilizers. Even municipal wastewater and sewage waste are sources of Cd pollution (Tran and Popova, 2013). The municipal wastewater-containing Cd is thrown into freshwater canals, streams, and rivers, which deteriorates the quality of freshwater (ATSDR, 2012). Cadmium concentrations in the world's oceans are estimated to range from $<0.005$ to $0.110\,\mu g\,L^{-1}$, with higher concentrations reported near some coastal areas. Cadmium concentrations in surface waters of impacted environments are frequently $2–3\,\mu g\,L^{-1}$ or greater (US EPA, 2016). Ugya (2015) investigated the ability of *L. minor* for the process of phytoremediation of petrochemical and refinery wastewater. After 21 days, the water was analyzed. The results of his study show that the maximum reduction in heavy metal was recorded as 95.4%, 100%,

94.3%, 99.3%, 93.3%, and 99.6% of Ag, Pb, Zn, Mn, Hg, and Cd, respectively. These investigations suggest that aquatic macrophytes, e.g., duckweed, are suitable candidates for effective treatment of water.

## 4 WETLANDS AND HYDROPHYTES

Wetlands, also referred to as "Earth's kidney," incorporate a wide series of wet environments, fluctuating from salt marshes to submerged coastal grass beds, boggy meadows, and swamp forests. The wetlands are important for the environment due to their natural cleansing ability for numerous pollutants, including pesticides and heavy metals (Joyce, 2012). The following four conditions are necessary for the formation of wetlands (Matache et al., 2013):

- Land cover should be shallow with a high groundwater table.
- Flooding-tolerant macrophytes are abundantly present in the area
- The area is rich in undrained hydric soils. The soil is adequately wet for an extensive period of time, resulting in anaerobic conditions that limit the specific plants that grow in it.
- The area should have nonsoil particles, i.e., rock or gravel, that are saturated with water at specific times throughout the fruiting season of plants.

Aquatic plants are a good option to use as an indicator for wetland pollution (Souza et al., 2013). Weis and Weis (2004) reported that the marsh plant species are used for the removal of metals from water. Wetlands are a relatively low cost and efficient choice, because these are easy to install and maintain, and produce limited quantities of by-products (Souza et al., 2013). Wetlands cover multifaceted processes, including biological, chemical, and physical mechanisms to eradicate several pollutants for improving water quality. The man-made, or artificial wetlands constructed for the purpose of elimination of heavy metals from wastewater are termed "constructed wetlands" (Heike et al., 2011). The constructed wetlands with aquatic microbial communities, hydrophytes plant roots, and subsidiary mineral matter are necessary for removal of pollutants, i.e., particulate organic matter and suspended solid metals, photogenic organisms, and dissolved nitrogen and phosphorus from municipal wastewater (Heike et al., 2011).

Anaerobic wetlands usually generate alkalinity and raise pH, while aerobic ones can treat net alkaline drainage, or may also serve as polishing agents. Sheoran and Sheoran (2006) suggested that for the handling of acidic water wastes, anaerobic bacteria may be especially effective.

## 4.1 Vegetation in the Wetlands

On the basis of their growth form, wetland plants can be classified as submerged, floating, floating-leaved, and emergent (Materac and Sobiecka, 2017).

### 4.1.1 Submerged

Submerged plants usually grow under water and are rooted in mud.

### 4.1.2 Floating Leave Plants

Their leaves are floating on the water surface, and they contain anchoring roots on the surface of soil.

### 4.1.3 Floating

The stem and leaves of these plants float submerged in water without any attachment to the sediment.

### 4.1.4 Emergent

Emergent plants spread their shoots and leaves above the water, while keeping their roots beneath the surface.

Some examples of wetland plant species are shown in Table 1 (Ikram et al., 2013; Materac and Sobiecka, 2017).

Some ferns, such as *Pistia* spp. and *Salvinia* spp. also grow in wetlands. Perhaps the most abundant are the royal fern (*Osmunda regalis* L.), and marsh fern (*Thelypteris palustris* L.) (Moore and Garratt, 2006).

Plants are often vital to wastewater handling in wetlands. A healthy stand of macrophytic vegetation is key for efficient performance of wetland treatment systems.

For use in such systems, Ikram et al. (2013) suggested that the following requirements of plants should be considered:

- Ecological adaptability (no weed or disease risk to the nearby natural or artificial ecosystems).
- Tolerance of native environments in terms of diseases, pests, and climate.
- Tolerance of hypertrophic waterlogged and contaminants' conditions.
- Rapid establishment, propagation, growth, and spread.
- High contaminant elimination ability, through indirect enhancement or uninterrupted assimilation of denitrification, nitrification, and other microbial methods.

**Table 1** Some examples of wetland submerged, floating leaved, floating, and emergent plant species

| Sr. no. | Common name | Scientific name |
|---|---|---|
| *Submerged plants* | | |
| 1 | Hornwort | *Ceratophyllum demersum* L. |
| 2 | Water weed | *Elodea canadensis* L. |
| 3 | Water milfoil | *Myriophyllum oliganthum* L. |
| 4 | Water celery | *Vallisneria americana* L. |
| | Celery leaf buttercup | *Ranunculus scleratus* L. |
| *Floating leaved* | | |
| 1 | American water lotus | *Nelumbo lutea* L. |
| 2 | White water lily | *Nymphaea odorata* L. |
| *Floating* | | |
| 1 | Duckweed | *Lemna* spp. |
| 2 | Greater duckweed | *Spirodela* spp. |
| 3 | Water meal | *Wolffia* spp. |
| 4 | Water hyacinth | *Eichhornia crassipes* L. |
| 5 | Water lettuce | *Pistia spathulata* Michaux |
| *Emergent* | | |
| 1 | Cattails | *Typha* spp. |
| 2 | Wooly bulrush | *Scirpus cyperinus* L. |
| 3 | Reed canary grass | *Phalaris arundinacea* L. |
| 4 | Bald cypress | *Taxodium distichum* L. |
| 5 | New England aster | *Aster novae-angliae* L. |
| 6 | Sea lavender | *Limonium carolinianum* L. |
| 7 | Water tupelo | *Nyssa aquatic* L. |
| 8 | Spearmint | *Mentha viridis* L. |
| 9 | Water cress | *Sisymbrium nasturtium-aquaticum* L. |
| 10 | Red maple | *Acer rubrum* L. |
| 11 | Frog-fruit | *Verbena nodiflora* L. |
| 12 | Knotweed | *Polygonum barbatum* L. |
| 13 | Foxtail flatsedge | *Cyperus alopecuroides* |
| 14 | Finger flatsedge | *Cyperus digitatus* |

## 4.2 Phytostabilization

Phytostabilization is a method of reducing the bioavailability or mobility of heavy metals, which are then dumped in the substrate and taken up by plant roots. The concentration of metals in sediments is decreased due to its higher uptake by roots. Uptake of Cd by roots depends on both plant and soil features, i.e., pH, CEC, chemical forms, source of elements in the soil, organic material, plant species, tissues, and age of the plant (Pavel et al., 2012). For the handling of contaminated soil, an important method of phytoextraction is also used in which plants with higher metal accumulation capacity are used for the removal of metals from wetland sediments. It is an economically practicable method of treating contaminated land (Aravind and Prasad, 2005). *Elodea canadensis* is an important submerged hydrophyte plant that is broadly scattered in the aquatic atmosphere, and capable of removing large concentrations of heavy metals from water reservoirs. *E. canadensis* triggers antioxidant enzymes to reduce the impacts of reactive oxygen species (ROS) on plants (Zouainia et al., 2016).

## 5 METAL REMOVAL MECHANISMS IN WETLANDS

In wetlands, heavy metal removal occurs in three main components, i.e., (a) vegetation, (b) hydrology, and (c) soil or substrate. The total removal of heavy metal ions from wetland water is estimated by combining processes of sedimentation, settling, photodegradation, cation exchange, microbial activity, phytoaccumulation, sorption, sedimentation, biodegradation, coprecipitation, and plant uptake. These processes directly depend upon the configuration of the sediment pH, substrate, plant species, and nature of the wastewater. Following are the various physio-chemical and biological processes used for this purpose.

### 5.1 Physical Processes

Sedimentation and settling are efficient physical methods for the elimination of cadmium from wetland water. In standing water, the elements that are thicker as compare with water will settle down. The transportation of metal takes place from water to sediments or plants, or vice versa.

In natural and constructed wetlands, sedimentation has been accepted as the main metal removal method. Sedimentation occurs when some other chemical procedures such as coprecipitation and precipitation aggregate the Cd into particles that are able to sink (Wani et al., 2017). For elements that are not as dense as water, the sedimentation takes place only after floc creation. High pH, concentration of suspended matter, high algal concentration, and ionic strength have enhancing effects on floc formation (Matagi et al., 1998). The resistance offered by rooted and floating plants in wetlands serves as a sediment ploy, as shallow water travels very slowly from wetlands.

Johnston (1993) shows that the length of wetlands and particle settling velocity determine the efficiency of suspended particles.

## 5.2 Chemical Processes

The chemical processes involved in metal removal from wetlands can be grouped as follows.

### 5.2.1 Sorption

Sorption is the shifting of ions from water to the sediment. Processes such as adsorption and precipitation ultimately cause the sorption of contaminants. In wetland soils, the sorption leads to short-term holding of numerous contaminants.

### 5.2.2 Adsorption

Cation exchange and chemio-sorption mostly cause the adsorption of heavy metals into the wetland sediment. The adsorption of metals in organic matter or clay is due to the electrostatic magnetism. After adsorption into clay or humic colloid, the heavy metals did not decompose, but their speciation might be altered with time, due to changes in sediment conditions (Batty et al., 2002; Wiebner et al., 2005). The *Iris sibirica* plant was planted in a constructed wetland during the low-temperature season. Microcosmic subsurface vertical flow constructed wetlands (MVFCWs) were fed through replicated Cd-contaminated river water for 3 months under a different Cd supply, and the seasonal growth changes and Cd uptake ability of *I. sibirica* were investigated. *I. sibirica* proves to be an excellent accumulator of Cd with the highest accretion of Cd in below-ground parts of 443.3 $mg\,kg^{-1}$ when exposed to Cd (8 $mg\,L^{-1}$) for 3 months. Under 1, 2, 4, and 8 $mg\,L^{-1}$ Cd supply in low-temperature seasons, the MVFCWs exhibit average Cd removals of 94.9%, 97.4%, 95.6%, and 96.2%, respectively. The predominant removal process of Cd in MVFCWs is substrate adsorption, which accounts for about 50% of Cd removal (Ma et al., 2017).

### 5.2.3 Precipitation

Precipitation is an adsorptive mechanism that limits the bioavailability of metals to many aquatic systems through the formation of insoluble heavy metal precipitates. Solubility product ($K_{sp}$) of metallic ions, the concentration of metal ions, the pH of the wetland, and relative cations are the key factors affecting the precipitation process. Precipitation only occurs when the product of cation and anion exceeds the solubility product ($K_{sp}$).

### 5.2.4 Coprecipitation

Coprecipitation is the amalgamation of trace ions into inorganic structures through recrystallization of minerals and solid solution development. Coprecipitation is also an adsorptive mechanism in which Cd usually coprecipitates with trace minerals in swamps. For example, cobalt and iron are coprecipitated in Mn-oxides, while copper, zinc, nickel, manganese, and so forth, are coprecipitated in iron oxides (Jadia and Fulekar, 2009). The pH condition of the mineral surfaces is the principle factor determining the coprecipitation of heavy metals.

### 5.2.5 Oxidation and Hydrolysis of Metals

Oxidation and hydrolysis are also important processes for removal of metals in wetlands. Iron, aluminum, and manganese mostly form insoluble compounds (oxides, hydroxide, and oxyhydroxide) as a result of hydrolysis and/or oxidation processes in wetlands. Iron is most susceptible to oxidation in wetlands. Iron removal depends upon pH, oxidation-reduction potential, and the presence of various anions (ITRC, 2003). $Fe^{2+}$ is highly soluble in water that has low dissolved oxygen with a pH close to 8. First, $Fe^{2+}$ needs to be oxidized to $Fe^{3+}$ at a pH of $<4$ or 5, because at low pH, the oxidation of ferrous to ferric will be catalyzed by acidophilic bacteria (*Thiobacillus* and *Ferrobacillus ferrooxidans* L.) (Robbins and Norden, 1994).

### 5.2.6 Metal Sulfides

Under anaerobic conditions, wetlands with appropriate substrate promote the growth of sulfate-reducing bacteria (SRB). In acid mine water rich in sulfate, these bacteria generate hydrogen sulfide. Most of the heavy metals react with hydrogen sulfide to form highly insoluble metal sulfides (Jha et al., 2016).

$$2CH_2O + SO_4^{2-} \rightarrow H_2S + 2HCO_3 \quad (1)$$

$$M^{2+} + H_2S + 2HCO_3 \rightarrow MS + 2H_2 + 2CO_3 \quad (2)$$

In Eq. (1), $CH_2O$ represents the organic matter. In Eq. (2), M represents the metal.

Thus, bacterial sulfate decreases due to the precipitation of metal sulfide solids. But for Fe, pyrite production is also conceivable.

$$Fe^{2+} + H_2S + S^0 \rightarrow FeS_2 + 2H^+ \quad (3)$$

Suspended substrate (SS) in natural waters controls the interaction, transportation, and biological effects of heavy metals in water bodies. The large amount of Cd carried by SS is an important source of Cd pollution in irrigation water. A novel procedure to remove trace Cd coexisting with SS from irrigation water by fertilizer flocculation was investigated. Four common fertilizers ($K_2HPO_4$, $(NH_4)_2HPO_4$, $KH_2PO_4$, and $K_2SO_4$) were used as flocculants. The aggregation of SS and its sedimentation by $K_2HPO_4$ were due to the combined effect of the oxidation of $Fe^{2+}$ to $Fe^{3+}$ and cation binding behavior toward SS, as a result of charge neutralization and electric double layer compression (Li et al., 2017).

### 5.2.7 *Metal Carbonates*

Heavy metals also produce carbonates when they react with water having high bicarbonates. Carbonate formation takes place only when limestone is present in the flow path of acid mine drainage (AMD), or when bacteria produce alkaline bicarbonate in wetland sediment that is sufficient (ITRC, 2003).

$$M^+ \left\{ \begin{matrix} SO_4 \\ \\ Cl_2 \end{matrix} \right. + Na_2CO_3 \longrightarrow MCO_3 + Na_2 \left\{ \begin{matrix} SO_4 \\ \\ Cl_2 \end{matrix} \right. \quad (4)$$

Lin et al. (2016) reported that carbonate precipitation is operative for elimination of nickel and lead.

## 5.3 Biological Removal Processes

The most vital pathway for heavy metal elimination in wetlands is the biological exclusion method.

### 5.3.1 *Role of Wetland Plants*

Plants of wetlands are primarily used in phytoremediation due to their active role in heavy metal elimination from waters of wetlands. Wetlands plants directly take up the metals (Fritioff and Greger, 2006), or indirectly affect the removal efficiencies, by increasing the residence time and reducing water velocities, and ultimately enhancing sedimentation of heavy metals. Sriyaraj and Shutes (2001) reported that submerged-rooted plants have some ability to extract metals, both from sediment and water, while rootless plants can take up metals from water only. The total bioaccumulation depends on the root uptake, but plants can also absorb relative amounts of metals by foliar absorption. Collins et al. (2005) showed that the plants of deep and shallow wetlands differed in terms of their shoot elemental concentration when subjected to metal-laden acid mine run-off from a coal-storage pile. Deep wetland plants had higher concentrations of sodium, manganese, and phosphate; and shallow wetland plants had higher concentrations of zinc, aluminum, and iron. According to Sheoran and Sheoran (2006), macrophytes such as *Typha angustata* L., *Typha elephantiana* L., *Desmostachya bipinnata* L., and *Saccharum bengalense* L. can tolerate high concentrations of several metals without showing negative effects on their growth. Barley et al. (2005) also reported higher metal uptake in roots. It means that wetland plants immobilize metals and store them in roots, and/or soil (phytostabilization).

During photosynthesis, plants absorb $CO_2$ from wetland waters, which ultimately increase the water's pH. When the plant community of the water column is highly productive, the pH increase can be three to four times per day. During the night, the pH will decrease again due to the release of $CO_2$ from plants (Cronk and Fennessy, 2001). The pH changes, as well as water and sediment chemistry, organic matter (substrate), and temperature, play an important role in metal uptake (Sheoran and Sheoran, 2006).

### 5.3.2 Role of Microorganisms

Microorganisms competently restrain toxic metals with their precipitation. Bacteria, algae, and fungi produce sulfide, hydroxide, phosphate, and carbonate, which form highly insoluble metal precipitates in reactions with metals (Li et al., 2017). Sobolewski (1999) reported that the metals were reduced to an immobile form by microbial action in wetlands. Both endophytic and rhizospheric bacteria may associate with plants (Cherian et al., 2012). Plant-bacteria connection causes many benefits for the host plant such as promoted growth and development. Both types of bacteria were shown to increase the plant tolerance to trace elements, e.g., by induction of plant defense mechanisms.

Two metal-resistant strains, *Neorhizobium huautlense* T1-17 and *Bacillus megaterium* H3, were examined for their nonmobilization of Cd in water and tissues of rice (*Oryza sativa* wuyun-23) growing in Cd-rich land. Strains T1-17 and H3 reduced 79%–96% of dissolved Cd in water, and enlarged grain biomass (Li et al., 2017).

Three Cd-tolerant bacterial strains (*Stenotrophomonas acidaminiphila*, *Pseudomonas aeruginosa*, and *Delftia tsuruhatensis*) were isolated and assessed for possible use in Cd bioremediation. *P. aeruginosa* was found to be highly tolerant to Cd ($2200\,mg\,L^{-1}$).

# 6 METAL UPTAKE, LOCALIZATION, TRANSLOCATION, AND DISTRIBUTION

## 6.1 Metal Uptake and Cellular Localization

The important known species of plants that are capable of extruding metals include Pontederiaceae, Typhaceae, Gramineae, Haloragaceae, and Ceratophyllaceae. The capability of a specific plant species to uptake a specified metal was closely connected with its capability to captivate the other metals. However, the absorption capabilities vary with the trends in roots > stems > leaves. The life cycle of aquatic plants and pH of the water also disturb the plant's capacity to intake elements. A higher surface area exposed to the plants strongly correlated with a higher uptake of metal ions from water (Li et al., 2015). Plants are capable of taking up metals from water, air, and soil. However, there is also a revert of heavy metals back into the water, soil, and air from the plant tissues. It means that the metal accumulation depends on the uptake, as well as on leakage of heavy metals into the surroundings (Greger et al., 2016).

Metal uptake by roots is the main method of metal accumulation in aquatic plants, except for fully submerged species. Metals are first taken up into the apoplast of the roots. Then some of the accumulated metals are further transported into the cells, but some become attached to the cell wall constituents, depending on the type of metal and plant genotype. There are two pathways by which the metals move from the roots to the stele for further transport to other plant parts.

**(a)** *Symplastic pathway:* In this pathway, the movement of metal ions takes place from cell to cell through channels in the cell wall, called plasmodesmata.

**(b)** *Apoplastic pathway:* In this pathway, the metal ions pass through intercellular spaces and cell walls, called apoplasm (Choppala et al., 2014).

## 6.2 Translocation to the Above-Ground Tissues

After uptake into the root, metals are transported further into the xylem vessels and are translocated to the stem and leaves. The mark of ascending translocation is reliant on plant species, environmental conditions, and the particular metal. Metal uptake mostly takes place in younger roots due to the absence of endodermal casparian strips in the walls of the roots' endodermal cells, which prevents the passage of water and other molecules. In the xylem vessels, the metals are either transported in complex form, in free form, or by interaction with nondiffusible anions of the cell wall of xylem

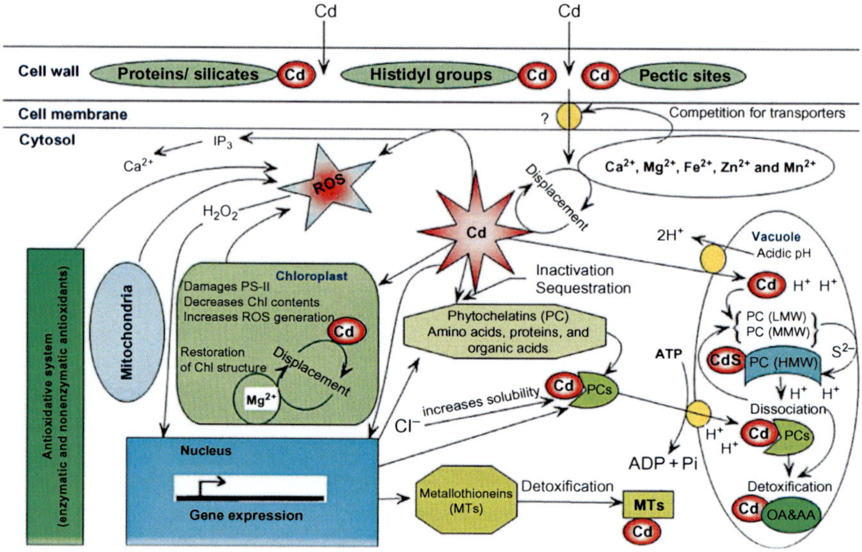


**Fig. 2** Schematic diagram of different cellular processes involved in Cd sequestration in plants (Choppala et al., 2014).

vessels. The translocation of Cd toward shoots takes place through the xylem, and is controlled by transpiration pull (Ueno et al., 2008). The Cd is also sequestered in the vacuole, or excluded into the apoplasts of roots by facilitated membrane pumps controlling the root-to-shoot Cd transport (Mohamed et al., 2012) (Fig. 2). For example, xylem loading producing adjustments in the frequency of root uptake was the essential method responsible for a critically improved rate of root-to-shoot transport in the Cd hyperaccumulator *Sedum alfredii* (Lu et al., 2008). Normal resistance-associated macrophage protein (NRAMP) family and cation/$H^+$ antiporters transfer vacuolar Cd to cytosol from where heavy metal transporting ATPases (HMAs) and other unfamiliar transporters load Cd to xylem. Consequently, a high percentage of Cd in the soluble portion, as well as its chemical form, could cause root-to-shoot transport of Cd, causing low Cd intensities in the shoots (Su et al., 2014).

## 7 PLANT RELATIONSHIPS WITH WETLAND SEDIMENTS

Sediments of wetlands are normally called a "sink" for heavy metals, and maximum concentrations of metals are in reduced form in the anoxic zone. In wetland sediments, the availability of metals is low as compared with the terrestrial land. Different physicochemical forms of metals have different bioavailability, e.g., exchangeable and water-soluble metals have maximum availability. Precipitation of metals as inorganic complexes and metals in humus have greater molecular-weights and are less available, as well as metals absorbed to hydrous oxides. Metals are also precipitated as insoluble sulfides, and metals bound within the crystalline lattice of minerals are actually not available (Gambrell, 1994). The penetration of plant roots is also inhibited due to reducing conditions that alter the uptake of pollutants and rhizosphere actions to lower levels (Williams, 2002).

In inlets, much of the metal pollutants bind with sulfides to become anoxic sediments., However, the plants are capable of oxidation of root zone sediments due to the presence of oxygen in aerenchyma tissue, and this oxidation can reactivate the contaminants, thus enhancing the low concentration of metals contaminants in wetland sediments. *Avicennia* is a type of mangrove that is able to oxidize the rhizosphere, thus decreasing sulfides and increasing metal availability in the transferrable form. A decrease in rhizospheric pH by plant organic acid exudates also results in an increase in metal availability (Doyle and Otte, 1997).

## 8 HEAVY METALS TOXICITY IN PLANTS

Heavy metals have several toxic effects on plants, particularly when the metals are transferred to the aerial tissues. Although toxic effects can be measured at the cellular, as well as biochemical level, most of the studies have focused on growth effects in response to contaminants. For example, MacFarlane et al. (2003) reported that gray mangroves' seedlings (*Avicennia marina* L.) showed significant reduction in seedling height, leaf number, and leaf area when exposed to 100 $\mu g g^{-1}$ sediment copper. At a higher level (400 $\mu g g^{-1}$), the total biomass and root growth were reduced, and at 800 $\mu g g^{-1}$ the emergence was totally inhibited. On the other hand, very little lead was transported to the above-ground tissues, and minimum negative effects were observed. By adding 0.05 mM cadmium to the nutrient medium, Abdel-Latif (2008) obtained a decrease in nutrient ions, chlorophyll content, and shoot and root growth of wheat seedling (*Triticum aestivum* L.).

### 8.1 Photosynthesis Inhibition

Cadmium retarded the electron transfer photosystem II (PSII) during photoactivation. Thus, the generation of stress resulted in the production of ROS in chloroplast. The Cd is also involved in ROS generation during electron transfer in the electron transfer chain (Heyno et al., 2008). Cadmium decreases the photosynthetic capability of plants by disturbing the enzymes of carbohydrate metabolism and the Calvin cycle (Shi et al., 2010).

The effects of $ZnCl_2$ (25 or 50 $\mu mol L^{-1}$), $CuCl_2$ (2.5 or 5 $\mu mol L^{-1}$), $CdCl_2$ (5 $\mu mol L^{-1}$), and combinations of $ZnCl_2$, or $CuCl_2$ with $CdCl_2$, were investigated on the growth, photosynthetic pigments, and PSII efficiency in *L. minor* L. (duckweed). All of the treatments caused inhibition of growth and significant metal accumulation in plant tissues after 4–7 days. The chlorophyll was decreased after 4 days, together with the quantum yield (Fv/Fm), and actual quantum yield ($\Delta F/F'm$) of PSII (Vidaković-Cifrek et al., 2015). The quick $Cd^{2+}$ uptake in PCC 6803 *Synechocystis* allows a direct contact of Cd with the photosynthetic apparatus, and shows a difference between primary and secondary effects. It was revealed that 40 mM $Cd^{2+}$ concentration in cells retarded $CO_2$-mediated electron transport. However, the effect of cadmium was first observed in photosystem I, in which the acceptor side for electrons was blocked, and an excitation pressure was built on the reaction centers, as shown by light-induced P700 redox transients, and changes in chlorophyll *a* and $O_2$ polarography. After a day in the light, the degradation of D1 protein and electron transport of PSII was observed. This data strongly suggested a cascade mechanism for the toxic effect of Cd, with a main response during darkness (Toth et al., 2012).

Cadmium convincingly resulted in a marked distortion of chloroplast ultrastructure, leading to a disturbed shape and inflated thylakoids (Najeeb et al., 2011). Chloroplast structure disturbance has been partly manifested by a notable decrease in chloroplast number and size, grana stacking, starch grain content, and accumulation of plastoglobuli observed in various plants such as *Picris divarticata* (75 μM, 14 days after treatment (DAT), *Hordeum vulgare* (5 μM, 15 DAT), and *Brassica* (Ying et al., 2010; Wang et al., 2011; Elhiti et al., 2012). Cadmium also decreases uptake of nutrients, i.e., Mn, Fe, and Mg; hence, a somewhat higher quantity of cellular Cd interferes with $Mg^{2+}$ supplement with proto-porphyrinogen, or it may be a source of chlorophyll devastation as a consequence of $Mg^{2+}$ substitution in both Chl *a* and *b* (Gillet et al., 2006). In conclusion, Cd affects photosynthesis either directly or indirectly, thus decreasing the crop yield.

## 8.2 Mineral Nutrition and Cadmium

It is well known that transport, uptake, and subsequent circulation of nutrients by the plants can be exaggerated by the occurrence of ionic Cd. In general, Cd has been shown to restrict Ca, P, Mg, K, and water in plants. The decrease of K and Ca accumulation in roots under Cd stress was reported in Cd hyperaccumulator *Atriplex halimus* (Nedjimi and Daoud, 2009) and in roots of sugar beet (Lindberg and Wingstrand, 1985). The $H^+$ATPase causes an electrochemical gradient across the plasma membrane, which is used for ion transport. The $H^+$ATPase present in the cortical and epidermal cell layers of plant roots is inhibited by cadmium. It was found that adsorption of important nutrients was decreased in maize due to the inactivity of $H^+$ATPase under elevated Cd (Astolfi et al., 2005). Kinetic studies also showed that the plasma membrane ATPase of sugar beet roots was inhibited by Cd in a competitive way, probably by binding of Cd to ATP, thus decreasing the substrate MgATP (Lindberg and Wingstrand, 1985) and caused a decreased uptake of potassium into the roots. In addition, it was also observed in *Populus jaquemontiana* that Cd toxicity decreased the uptake of essential nutrients due to its competition with the uptake of these nutrients (Solti et al., 2011). The accumulation of mineral nutrients in plant tissues is changed by Cd stress, which in response, spoiled the biochemical structures. While some other nutrients, i.e., Ca, Zn, Si, Fe, can initiate a counter-response against the toxic effects of Cd toxicity in plants by an increase in the production of phytochelatins (Sarwar et al., 2010; Greger et al., 2016). Cadmium is accumulated in roots by NRAMP and ZIP transporters across the plasma membrane by competing with nutrients (Lindberg and Wingstrand, 1985). Cadmium contact escalates the transcription factors' response (Shim et al., 2009), which would be a Cd tolerance mechanism (Gratao et al., 2012).

## 8.3 ROS Generation and Cadmium

Heavy metals are an important oxidative stress-causing toxin via the generation of ROS. Some toxic heavy metals, i.e., Cu and Fe, can cause damage in plants via involvement in oxidation reduction cycles, such as Fenton and Haber-Weiss responses. In contrast, cadmium is not capable of taking part in a single-electron transferring reaction, and is unable to generate ROS, i.e., singlet oxygen ($1O_2$), superoxide anion ($O_2^{\bullet -}$), hydroxyl radical ($OH^{\bullet}$), and hydrogen peroxide ($H_2O_2$); but it destroys the oxidative defense system of plants, and in response, generate ROS (Gratao et al., 2012). Pea and rice plants growing under Cd stress start the production of plasma–membrane-bounded NADPH oxidase in peroxisomes, and in response, produce ROS. The initiation of ROS production is very fast. For example, in seedlings of

*Pinus sylvestris* growing in Cd, toxicity starts the generation of ROS within 6 h. In *Medicago sativa,* contact with Cd for 6–24 h triggers the production of homoglutathione (hGSH), peroxides, and reduction of glutathione (GSH). A high Cd concentration causes tissue death in BY-2 tobacco due to an NADPH-oxidase-dependent increase of $H_2O_2$ together with the cellular fatty acid hydroperoxide, and $O_2$ accumulation (Gill and Tuteja, 2010). Cadmium was also involved in obliteration of the cell membrane, nucleic acids, proteins, and lipids, and the decrease in protein production and reduction in photosynthesis, which imparts a severe reduction in the growth and development of the plant. DNA damage was also reported due to Cd toxicity defined by the frequency of anomalies such as precocious separation, fragments, single and double bridges, laggards, and stickiness (Gill and Tuteja, 2010; Kranner and Colville, 2011). The isolation and production of GST genes is from *Trichoderma virens*, a fungus used for biocontrol and transplanting genes into tobacco plants by using the plasmid *Agrobacterium*. The results revealed that the plant with the *TvGST* gene expressed it, and it was found to be more tolerant against Cd toxicity as compared with the wild-type. Under Cd stress, the production of POD in *Raphanus sativus* L. was increased (El-Beltagi et al., 2010), but did not result in any specific alteration in leaves of pea plant. Similarly, Cd stress increased the production of ASH in barley, and decreased it in soybeans, the leaves of *Arabidopsis*, peas, and cucumber chloroplasts, and continued into the roots of *Populus canescens* (Gill and Tuteja, 2010). The increase in organic acids production during Cd stress increases the generation of antioxidants such as SOD and POD. It is likely that an increase in organic acid release increases nutrient accumulation and accomplishments of antioxidants, which allows the tolerant maize variety to adapt in Cd-polluted soil (Javed et al., 2017).

## 9 DEFENSE MECHANISMS AGAINST METAL PHYTOTOXICITY

Plants have established different processes to overcome the stress imposed by high metal concentration (Fig. 3). But not much is known about the control of metallic ion toxicity in plants. The metal defense response can be classified into two main categories; avoidance and tolerance.

### 9.1 Avoidance

Stress avoidance relies on preventing metal uptake by plants, or its quick exclusion from the cytosol. It is achieved in plants by the following ways.

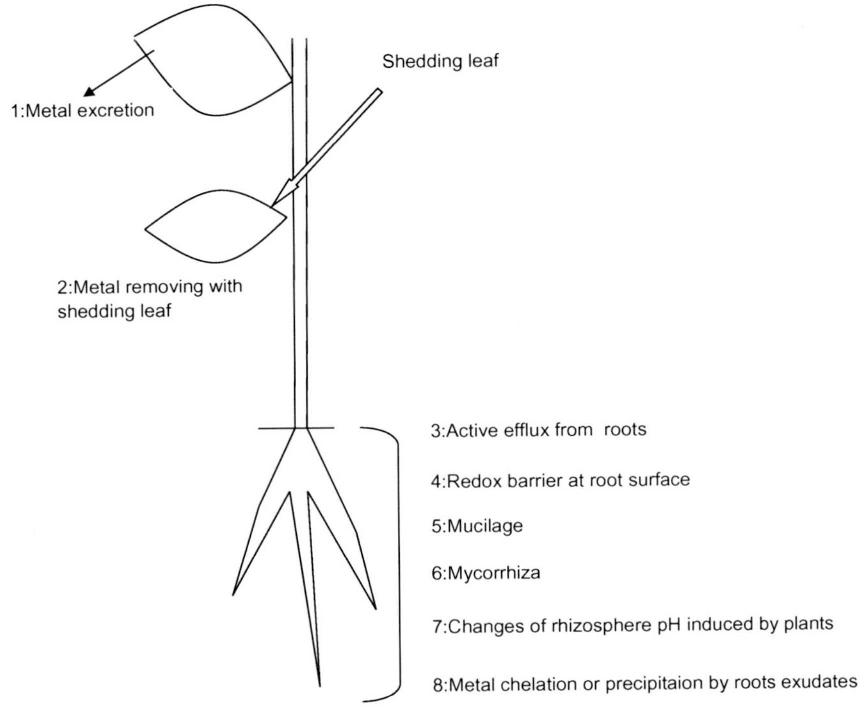


**Fig. 3** Summary of stress avoidance mechanisms in higher plants. 1: Metal removal by excretion. 2: Metal removal with shedding leaf. 3: Active efflux of metals from roots. 4: Redox barrier at root surface. 5: Resistance to metal influx by mucilage. 6: Fungal association with plant roots preventing metal influx. 7: pH changes induced by plants in the rhizosphere. 8: Metal chelation by root exudates.

### *9.1.1 Reduction of Active Metal Uptake*

The cell wall is the first mechanical and chemical fence against the influx of ions into the cytosol. The cell wall can retain the metal from a small amount, up to 90% of the total metal taken up. The metal can bind to pectin or cell wall proteins. Reduction of toxic ion uptake by plants is mainly due to rhizosphere modifications, such as:

**(a)** *pH changes induced by plants.*

**(b)** *redox barrier relocation into the plasma membrane of roots and associated microorganisms* (Prasad, 2001).

#### pH Changes Induced by Plants

Plants are capable of absorbing, accumulating, and translocating heavy metals from organic compounds present in the environment. In order to maintain their charge equilibrium, when they gain more cations than anions, they can gain protons in contrast (Hinsinger et al., 2003). Plant roots are organs that are responsible for increasing or decreasing rhizosphere pH. Roots alter the pH of the rhizosphere directly, which not only affects the uptake of nutrients from soil, but also affects the bioavailability of Cr, Cd, and Hg. The acidification of rhizosphere enhances the solubility of Zn- or Al-containing chemical compounds, enhancing accumulation of Zn or Al in plants (Calba et al., 2004). In alkaline pH of the rhizospheric soil, the Cu solubility in the soil was increased (Chaignon et al., 2004). *Aspalathus linearis* L. is a legume plant growing in high acidic soil and capable of $N_2$-fixination, and used as tea. This plant increases its pH in the rhizosphere, as compared with nonrhizospheric soil. The experimental study revealed that *A. linearis* L. increases its pH 2.8 units when roots are dipped in a nutrient solution, in comparison with uninoculated control soil having pH 1.5. Titrimetric approach showed that $HCO^{3-}$ and $OH^{-}$were the reasons for basicity in the nutrient solution in which *A. linearis* roots were dipped, and were accountable for the acidification of the rhizosphere pH (Muofhe and Dakora, 2000).

The same was also found in wetland plants; they are capable of changing the rhizosphere water pH, which prevented metal toxicity. The roots of rice plants alter the rhizosphere pH when growing in Cr toxicity, and especially when Cr concentrations in the nutrient solution was 50 μM or more (Zeng et al., 2007). *E. canadensis* L. is a submerged macrophyte, and is able to increase water pH, especially when growing in Cd toxicity.

According to Song et al. (2004), the rhizosphere pH basically formed from:

**I.** The absorption of anion and cation imbalance
**II.** Production of $CO_2$ by rhizosphere breath
**III.** The release of $H^+$, organic acids, and other chemical compounds from roots

This may be due to the excretion of buffering agents ($OH^{-1}$, $CO_3{}^{-2}$) by the plant roots. This release of buffering agents is due to cation/anion imbalance caused by the source of nitrogen, i.e., $NH_4{}^+$-uptake decreases pH, while $NO_3{}^-$-uptake increases pH, but the delay of escalation was lesser than that of reduction in the $NH_4{}^+$ treatment. Soil pH values near the root decreased by 1.1 pH units when ammonium was applied, and increased only by 0.25 units with nitrate application. Moreover, organic acids produced by the root can also change pH, i.e., citric acid is normally produced as anions, which consume protons, causing soil basicity. One possibility is that heavy metal stress inhibits the activity of ATPase, and thus increases the apoplast pH. $Cd^{2+}$ stress inhibits the ATPase activity of the plasma membrane (Lindberg and Wingstrand, 1985). This reduction in ATPase action resulted in decreased proton efflux, causing the change in rhizosphere pH (Astolfi et al., 2005).

VanLoon and Duffy (2005) reported that hydroxide and oxyhydroxide precipitates of metals increased with increasing pH. The pH also influences the binding of toxic metals with organic matter. At lower pH, the metal complexes become weaker (Clemensson-Lindell et al., 1992), and metal solubility increases, but it reduces metal uptake by plants due to enhanced interaction between $H^+$ and metal cations (Marschner, 2011).

*Eriophorum angustifolium* is a macrophytic plant and, when growing under Cd, it alters the rhizospheric pH toward basification (Javed et al., 2013). Terrestrial land plants can also alter the rhizospheric pH, like maize roots when growing in hydroponics carrying different Cd levels, they produced significant fluctuations in rhizospheric pH, which was different at lower and higher concentrations of Cd (Tanwir et al., 2015). The increase in pH of the mucilage plant at a lower Cd is perhaps increased, which can hinder the action of the proton pump in the plasma membrane (Janicka-Russak et al., 2008) and induce the exudation of organic acids. A similar kind of organic acid release was observed in hot peppers under low and high concentrations of Cd toxicity (Xin et al., 2015). However, a contrasting result of rhizospheric acidification of *Zea mays L.* growing in a mixture of heavy metals was earlier reported (Blossfeld et al., 2010). The release of organic acids is the result of anions, and anion exchange across the plasma membrane (Zhu et al. (2011). The anion release was stable by cations/proton exclusion, as observed in poplar roots under heavy metal stress (Qin et al., 2007).

#### Formation of Redox Barriers

The redox barrier is formed as an outcome of oxidizing action of plant roots and related microorganisms. Wetland plants such as *Spartina alterniflora* L., *Phragmites communis* L., *Typha latifolia* L., and *O. sativa* L. oxidize more soluble Fe (II) to less soluble, and less toxic forms of Fe (I II), and as a result, form iron plaque in their roots.

### *9.1.2 Mycorrhizas*

Symbiotic mycorrhizal fungi, such as AM, form a mutualistic symbiosis with the roots of most vascular plant species (Upadhyaya et al., 2010). Plant fungal mutualistic relationships may act as an antecedent in which it indicates the proclaimer of stress to symbiotic plants so that they can make their defensive mechanisms active to mitigate lethal effects of stress earlier than nonsymbiotic plants. The principal mechanisms adopted by mycorrhizal fungi to cancel out the impacts of HM stress on plants include acting as a barrier by dumping metals inside the cortical cells, deposition of metals in the cell wall or mycelium, and confiscating them in their vacuole or other organelles, releasing glutathione and heat-shock protein (Hildebrandt et al., 2007), chelating or precipitating metals in the soil medium via manufacturing glycoprotein, or forming phosphate metal multiplexes inside the hypha (Aloui et al., 2011), and dropping the concentration of metals by discriminating root and shoot growth (Mohammadi et al., 2011). The same result was obtained by Shivakumar et al. (2011) when working with green gram (*Vigna radiata*) grown in soil containing excessive Zn. It is reported that the changes in soil pH due to the activities of AM fungi are a major contributing factor to the immobilization of metals in the mycorrhizosphere region (Bano and Ashfaq, 2013). AM association can reduce or prevent the induction of ROS species, and also give a boost to the activity of detoxifying enzymes within plants. Abad and Khara (2007) showed that wheat plants colonized by different AM fungi species subjected to toxic levels of Cd had conspicuously more functionality of protective antioxidants, such as APX and GPX, in their roots and shoots compared with non-AM wheat. They also observed that, among fungal species, *Glomus veruciforme* and *Glomus etunicatum* were the most effective activators of the protective proteins. The decrease in lipid peroxidation and electrolyte leakage, and increased activities of peroxidase (POX), catalase (CAT), and superoxide dismutase (SOD) were observed in mycorrhizal *Cajanus cajan* L. under Cd and Pb-contaminated soils (Garg and Aggarwal, 2012).

### 9.1.3 Metal Release by Leaves

Excretion of metals (Cd, Fe, Cu, and Zn) from plants can occur via salt glands, hydathodes, or ectodesmata. Burke et al. (2000) studied the leaf elimination by *S. alterniflora* L. and *Phragmites australis* L. in a contaminated area in New Jersey, and found that leaves of *S. alterniflora* discharge two to four times extra copper, lead, zinc, and chromium than leaves of *P. australis*. The reason for this alteration in metal discharge from two species was the existence of salt glands is *S. alterniflora,* which are missing in *P. australis*.

## 9.2 Metal Tolerance

### 9.2.1 Translocation to Cell Compartments

The vacuole comprises >80%–90% of the cell volume in most of the cells, and acts as a storage pole for ions. Vacuolar compartmentalization of toxic metals is an effective decontamination mechanism in plants, as it prevents the direct interaction of heavy metals with metabolic processes. Some amino acids may also chelate metals in cytosol. Saifullah et al. (2014) described the vacuolar deposition of Pb, Zn, Cd, and Cu in both metal sensitive and resistant plants. Organic acids are also very effective chelators for toxic metal ions in cytosol and vacuoles.

### 9.2.2 Sequestration in the Cytoplasm

In many plant species, amplified concentrations of metals seem to encourage the making of cysteine-rich metal-binding polypeptides (Fig. 4). The most important of these are (a) *phytochelatins* and (b) *metallothioneins.*

#### Phytochelatins

Phytochelatins (PCs) are a group of metal complexing peptides with a universal structure (γ-Glu Cys) *n*-Gly, where $n=2$–11, and they are prompted in plants in reaction to heavy metals management. The formation of PCs is encouraged by several metals, i.e., Hg, Cd, Ag, Ni, Cu, Pb, Au, Zn, and As (Gill and Tuteja, 2010). Phytochelatins play a vital role in heavy metal control and decontamination through metal chelation, and their pronounced effect is in Cd tolerance (Yadav, 2010) (Fig. 5). Phytochelatins synthesized in cytosol bind Cd. and the Cd-PC complex is transported across tonoplasts by an ATP-binding transporter. Amalgamation of isoPCs generally relies on the accessibility of Gly, or an important enzyme, glutathione synthetase, in the cells, and may change to manufactured isoPCs when the plant is in a stress condition (Rea, 2012).

Lindberg et al. (2007) suggested a lower accumulation of $Cd^{2+}$, or even removal of $Cd^{2+}$ for protoplasts pretreated with $CdCl_2$. For the efflux of metal from cytosol, two main mechanisms exist; the first one is the transport of the Cd-PC complex, which is less important, and the second one is the $Cd^{2+}/H^+$ antiport, which is more important. For the transport of Cd-PC, complex pretreatment of seeds with Cd is necessary. When the transgenic *Arabidopsis* plant was cultivated in a high concentration of $Cd^{2+}$, it produced PCs rapidly, while in the case of low metal availability, it was reduced (Liu et al., 2013). In *Nitzschia palea*, with an increase of Cd metal, the binding to the PCs complex

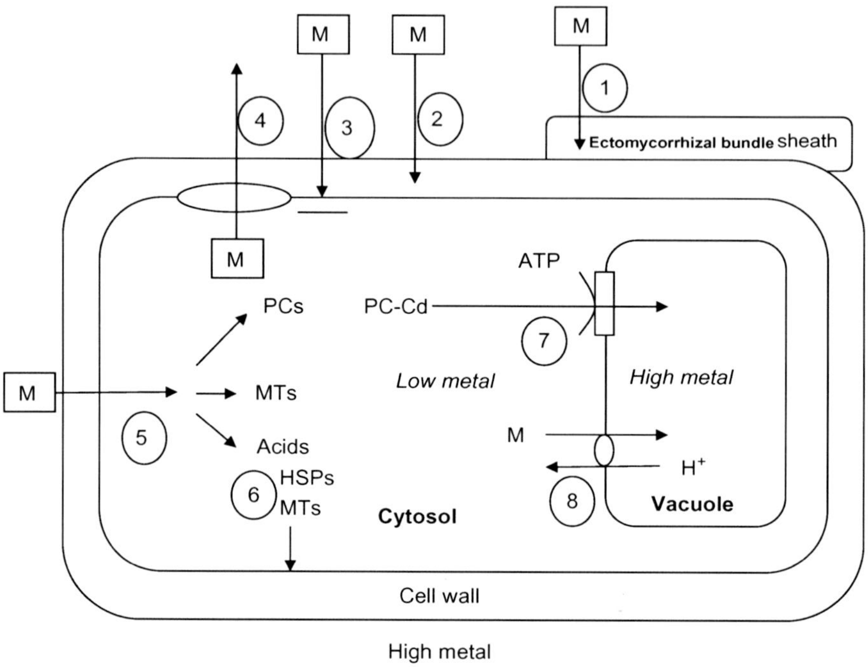


**Fig. 4** Summary of cellular detoxification mechanisms in higher plants. 1: Control of metal transport to roots by mycorrhizas. 2: Binding to cell wall and root exudates. 3: Reduced influx across plasma membrane. 4: Active efflux into apoplast. 5: Chelation in cytosol by various ligands. 6: Restoration and fortification of plasma membrane under stress conditions. 7: Transportation of PC-Cd complex into the vacuole. 8: Transportation and addition of metals in vacuole (Hall, 2002).

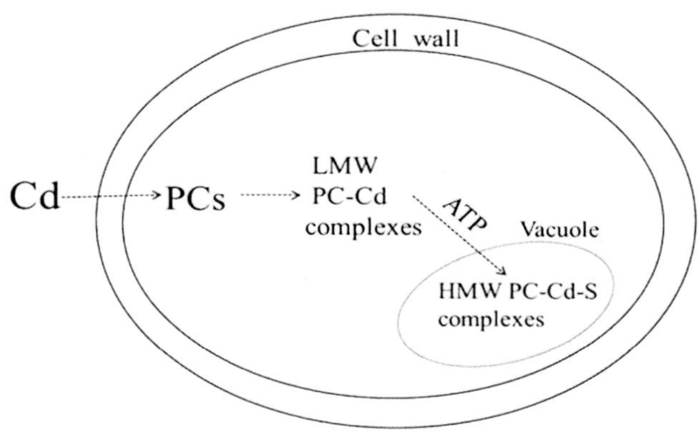


**Fig. 5** Mechanisms of phytochelatins' decontamination of Cadmium heavy metal (Liu et al., 2015).

was also increased (Figueira et al., 2014). The PCs temporarily bind with Cd and reduce its toxicity (Sylwia et al., 2010). Attachment of Cd with PCs, and their successive transportation to the vacuole, are essential for Cd tolerance (Sylwia et al., 2010). When Cd hyperaccumulating *S. alfredii* was treated with Cd, a large amount of Cd (47% in stems and 70% in leaves) was found to bind with malic acid (Tian et al., 2011). Their outcomes show that a major amount of Cd dumped in the parenchymatous tissue was present in the stem and was probably connected with the Ca channel and attached to malate. However, PCs are not a main component of cellular detoxification of heavy metals in plants (Liu et al., 2015).

#### Metallothioneins (MTs)

In 1957, MTs were first extracted from an equine kidney. These are small, low-molecular-weight metal binding proteins, rich with cysteine residues, and they are present in extensive variations of eukaryotic organisms, including invertebrates, fungi, plants, and mammals; and also in some prokaryotes (Du and Yang, 2014; Macovei et al., 2010; Verkleij et al., 2003). The PCs only have a role in Cd toxicity in plants. They are also involved in the decontamination of other metals, such as MTs, Zn, Cu, As, and Fe (Yang and Chu, 2011). MTs display dissimilar features and functions, based on their existence as a diverse organism. MTs are exceptionally diverse in their structural and molecular characteristics (Leszczyszyn et al., 2013). They play an active and diverse role in plants, as compared with other organisms. MTs are involved in detoxification and removal of Cd via intracellular metal ions, homeostasis, metal transport adjustment, and cellular sequestration (Guo et al., 2013). Along with metal detoxification, MTs are also involved in different cellular reactions, and known ROS scavenging (Wong et al., 2004), maintenance of the redox level (Macovei et al., 2010), repair of the plasma membrane (Mishra and Dubey, 2006), cell growth, and proliferation and repair of spoiled DNA (Grennan, 2011). Zhou et al. (2014) validated the role of biologically engineered *TaMT3* (a metallothionein T-3 isolated from *Tamarix androssowii*) in tobacco plants. They stated that MTs are involved in Cd stress tolerance by producing SOD, which elevated the capability of ROS production, which in turn leads toward reduced POD activity in tobacco plants.

## 10 PLANT-INDUCED pH CHANGES AND ITS ROLE IN PHYTOREMEDIATION

The pH of the surrounding medium is one of the important factors affecting the availability of metals to plants, both in terrestrial and aquatic systems. But this effect can be different for terrestrial and aquatic plant species, depending upon the surrounding medium, and the type of metal.

In hydroponics, metal accumulation increases as the cellular pH increases (Marschner, 2011), and this can be due to the competition between metal and hydrogen ions at the accumulation site. Nyquist and Greger (2008) indicated the ability of a wetland plant cv *E. canadensis.* to increase the nearby water pH by shoots, particularly when the primary surrounding water pH was low. This rise in pH was more pronounced if the surrounding medium carried free $Cd^{2+}$ ions. The mechanism by which plant shoots increase surrounding water pH is poorly understood, as only few studies have highlighted it so far. The role of free metal ions which trigger shoot-induced pH changes also needs further investigation. With an increasing pH, the metal uptake increased in *E. canadensis* and *Carex rostrata*, which can be due to weakened contact between $Cd^{2+}$ and $H^+$. The cell-bound fraction of Cd increased more at a higher pH, prevented communication between $Cd^{2+}$ and $H^+$.

Bravin et al. (2009) elaborated how root generated alkalization controlled the formation of copper gradients in the rhizospheric zone of wheat cropped in highly acidic, copper-contaminated soil. They observed that the rhizosphere pH improved over 6 mm in soil approaching up to +2.8 units, and the total concentration of Cu decreased by threefold in the rhizosphere solution. The plant-induced pH rise is important for metal uptake from polluted waters/sediments (Fig. 6). As compared with terrestrial plants, the pH-based remediation is better for aquatic submerged plants, whose leaves take up metals 10 times higher than terrestrial plants (Greger et al., 2016). Therefore, the establishment of aquatic-plant communities in waters polluted with metals, and characterized by low pH, will optimize the remedial potential.

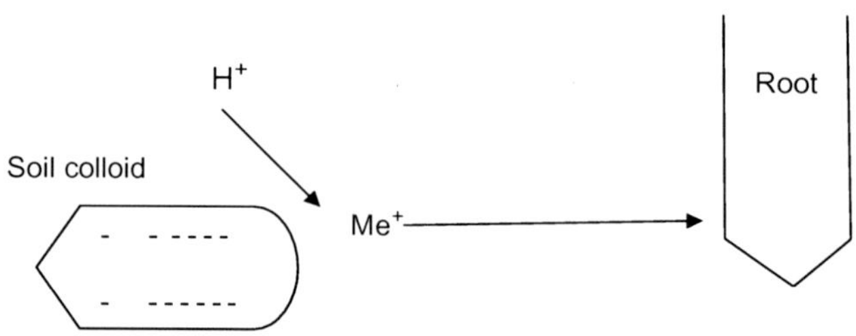


**Fig. 6** Showing the release of metal ions from soil colloid by $H^+$ ions at low pH and their uptake by plant roots (Greger et al., 2016). Metal uptake can be high (due to more available metal) or low due to interaction effect between metal and hydrogen ions.

Mine tailings (waste from mining activities) are stored in big reservoirs. As mentioned before, such mine tailings carry different dissolved metals, as well as a characteristically low pH. Therefore, their dispersal in the surrounding areas by percolation, drainage, or accidental breakdown of dams is highly unwanted. Although low pH only has temporary effects (as natural processes increase the pH again), the residual effects of metals may be long term, preventing vegetation establishment in the area. Plants have the ability to stabilize the toxic metals in the tailing reservoir, and that may depend on plant-induced pH-rise via roots. Stoltz and Greger (2002) studied the influence of two cotton grass species (*E. angustifolium* Honckeny and *E. scheuchzeri* Hoppe) on metal dispersal from limed and unlimed submersed mine tailings. They concluded that plants stabilize the Cd, Cu, Pb, and Zn of the drainage water in unlimed mine tailing by keeping pH high, a mechanism likely to be triggered by high free metal concentration.

However, root-induced changes in rhizosphere pH can be dramatic for barley and soybeans, as shown by Youssef and Chino (1989). These authors showed that alkalization occurred in the rhizosphere at an initially low pH (pH=4.8), while acidification occurred at higher initial pH values (pH=7.1). Likewise, in the rhizosphere of oilseed rape and tomato plants, acidification takes place in a calcarious soil, and basification in an acidic soil (Chaignon et al., 2004).

The optimal condition for phytoremediation by using duckweed was a medium containing Cd dose of 2 mg $L^{-1}$ at pH 7 (Verma and Suthar, 2015). The metal elimination from water is due to the outcome of an amalgamation of adsorption and absorption phenomena (Duman et al., 2009). Uysal and Taner (2010) also demonstrated the effect of pH, temperature, and initial load of metal on removal efficiency of *L. minor*.

## 10.1 Production of Siderophores

Under aerobic condition, the availability of iron for plants and microorganism in alkaline pH is lower than the requirement because Fe(III) is insoluble in water under these conditions. Siderophores are compounds secreted by plants and microorganisms, and are meant for the scavenging of $Fe^{3+}$. The plants of the Gramineae family secrete phytosiderophores, which are able to bind with $Fe^{3+}$, and take up the iron from soil. On the other hand, plants not belonging to Gramineae family convert insoluble $Fe^{3+}$ to water soluble $Fe^{2+}$ by the action of ferric chelate reductase enzyme. These soluble $Fe^{2+}$ ions are taken up by root epidermal cells by the high affinity transporter

IRT1 (Kobayashi and Nishizawa, 2012). The Cd transport across the membrane of root cells competes with iron transporters, due to its high affinity. Cd contests with iron, and causes iron starvation (Yoshihara et al., 2006). The Zn response to the IRT transporters due to deficiency of iron has been detected in Cd hyperaccumulator plants (*Thlaspi caerulescens*) (Lombi et al., 2002), as well as in nonhyper accumulators, i.e., *Oryza sativa* and *Arabidopsis thaliana* (Kobayashi and Nishizawa, 2012). These results suggest that the increase in iron concentration in soil facilitates countering the Cd toxicity, which indirectly improves the growth and yield of plants (Wu et al., 2012), indicating that higher iron concentration might be a result in captivity of Cd in the root zone of plants.

The addition of siderophore-producing bacteria into the soils that are facing a Cd toxicity problem would result in the reduction of the accumulation of metal. However, it depends upon the type of plant, metal, and bacterium (Ma et al., 2011).

## REFERENCES

Abad, A.K.J., Khara, J., 2007. Effect of cadmium toxicity on the level of lipid peroxidation and antioxidative enzymes activity in wheat plants colonized by arbuscular mycorrhizal fungi. Pak. J. Biol. Sci. 10 (14), 2413–2417.

Abdel-Latif, A., 2008. Phosphoenolpyruvate carboxylase activity of wheat and maize seedlings subjected to salt stress. Aust. J. Basic Appl. Sci. 2, 37–41.

Aloui, A., Recorbet, G., Robert, F., 2011. Arbuscular mycorrhizal symbiosis elicits shoot proteome changes that are modified during cadmium stress alleviation in *Medicago truncatula*. BMC Plant Biol. 11, 75.

Aravind, P., Prasad, M.N.V., 2005. Cadmium–zinc interactions in a hydroponic system using *Ceratophyllum demersum* L.: adaptive ecophysiology, biochemistry and molecular toxicology. Braz. J. Plant Physiol. 17, 3–20.

Astolfi, S., Zuchi, S., Passera, C., 2005. Effect of cadmium on $H^+$ ATPase activity of plasma membrane vesicles isolated from roots of different S-supplied maize (*Zea mays* L.) plants. Plant Sci. 169, 361–368.

ATSDR (Agency for Toxic Substances and Disease Registry), 2012. Toxicological Profile for Cadmium: Update. United States Department of Health and Human Services, Public Health Service, Atlanta, GA.

Bano, S.A., Ashfaq, D., 2013. Role of mycorrhiza to reduce heavy metal stress. Nat. Sci. 5 (12), 16–20.

Barley, R.W., Hutton, C., Brown, M.M.E., Cusworth, J.E., Hamilton, T.J., 2005. Trends in biomass and metal sequestration associated with reeds and algae at wheel jane biorem pilot passive treatment. Sci. Total Environ. 345 (1–3), 279–286.

Batty, L.C., Baker, A.J., Wheeler, B.D., 2002. Aluminium and phosphate uptake by *Phragmites australis* L.: the role of Fe, Mn and Al root plaques. Ann. Bot. 89, 443–449.

Blossfeld, S., Perriguey, P., Sterckeman, T., Morel, J., Losch, R., 2010. Rhizosphere pH dynamics in trace-metal-contaminated soils monitored with planar pH optodes. Plant Soil 330, 173–184.

Bravin, M.N., Tentscher, P., Rose, J., Hinsinger, P., 2009. Rhizosphere pH gradient controls copper availability in a strongly acidic soil. Environ. Sci. Technol. 43, 5686–5691.

Burke, D.J., Weis, J.S., Weis, P., 2000. Release of metals by the leaves of the salt marsh grasses *Spartina alterniflora* and *Phragmites australis*. Estuar. Coast. Shelf Sci. 51, 153–159.

Calba, H., Cazevieille, P., Thée, C., Poss, R., Jaillard, B., 2004. The dynamics of protons, aluminium and calcium in the rhizosphere of maize cultivated in tropical acid soils, experimental study and modelling. Plant Soil 260, 33–46.

Chaignon, V., Bedin, F., Hinsinger, P., 2004. Copper bioavailability and rhizosphere pH changes as affected by nitrogen supply for tomato and oilseed rape cropped on an acidic and a calcareous soil. Plant Soil 243, 635–639.

Cherian, S., Weyens, N., Lindberg, S., Vangronsveld, J., 2012. Phytoremediation of trace element contaminated environments and the potential of endophytic bacteria for improving this process. Crit. Rev. Environ. Sci. Technol. 42, 2215–2260.

Choppala, G., Bolan, S.N., Bibi, S., Iqbal, M., Rengel, Z., Kunhikrishnan, A., Ashwath, N., Ok, Y.S., 2014. Cellular mechanisms in higher plants governing tolerance to cadmium toxicity. Crit. Rev. Plant Sci. 33 (5), 374–391.

Clemensson-Lindell, A., Borgegard, S.O., Persson, H., 1992. Reclamation of Mine Waste and Its Effect on Plant Growth and Root Development: A Literature Review. Report 47Swedish University of Agricultural Sciences and Environmental Research, Uppsala, Sweden.

Collins, B.S., Sharitz, R.R., Coughlin, D.P., 2005. Elemental composition of native wetland plants in constructed mesocosm treatment wetlands. Bioresour. Technol. 96 (8), 937–948.

Cronk, J.K., Fennessy, M.S., 2001. Wetland Plants, Biology and Ecology. CRC Press.

Doyle, M.O., Otte, M.L., 1997. Organism-induced accumulation of iron, zinc and arsenic in wetland soils. Environ. Pollut. 96, 1–11.

Du, J.L., Yang, C.H.L., 2014. Advances in metallotionein studies in forest trees. Plant OMICS 5 (1), 46–51.

Duman, F., Leblebici, Z., Aksoy, A., 2009. Bioaccumulation of nickel, copper, and cadmium by *Spirodela polyrhiza* and *Lemna gibba*. J. Fresh Water Ecol. 24, 177–179.

El-Beltagi, H.S., Mohamed, A.A., Rashed, M.M., 2010. Response of antioxidative enzymes to cadmium stress in leaves and roots of radish (*Raphanus sativus* L.). Not. Sci. Biol. 2, 76–82.

Elhiti, M., Yang, C., Chan, A., Durnin, D.C., Belmonte, M., Ayele, B.T., Tahir, M., Stasolla, C., 2012. Altered seed oil and glucosinolate levels in transgenic plants overexpressing the *Brassica napus* shoot meristem less gene. J. Exp. Bot. 63, 4447–4461.

Figueira, E., Freitas, R., Guasch, H., Almeida, S.P., 2014. Efficiency of cadmium chelation by phytochelatins in *Nitzschia palea* (Kützing) W. Smith. Ecotoxicology 23, 285–292.

Fritioff, A., Greger, M., 2006. Aquatic and terrestrial plant species with potential to remove heavy metals from storm water. Int. J. Phytoremediation 5 (3), 211–224.

Gambrell, R., 1994. Trace and toxic metals in wetlands—a review. J. Environ. Qual. 23, 883–891.

Garg, N., Aggarwal, N., 2012. Effect of mycorrhizal inoculations on heavy metal uptake and stress alleviation of *Cajanus cajan* (L.) Millsp. genotypes grown in cadmium and lead contaminated soils. Plant Growth Regul. 66 (1), 9–26.

Gill, S.S., Tuteja, N., 2010. Reactive oxygen species and antioxidant machinery in abiotic stress tolerance in crop plants. Plant Physiol. Biochem. 48, 909–930.

Gillet, S., Decottignies, P., Chardonnet, S., Maréchal, P., 2006. Cadmium response and redoxin targets in *Chlamydomonas reinhardtii*: a proteomic approach. Photosynth. Res. 89, 201–211.

Gratao, P.L., Monteiro, C.C., Carvalho, R.F., Tezotto, T., Piotto, F.A., 2012. Biochemical dissection of diageotropica and never ripe tomato mutants to Cd stressful conditions. Plant Physiol. Biochem. 56, 79–96.

Greger, M., Kabir, A.H., Landberg, T., Maity, P.J., Lindberg, S., 2016. Silicate reduces cadmium uptake into cells of wheat. Environ. Pollut. 211, 90–97.

Grennan, A.K., 2011. Metallothioneins, a diverse protein family. Plant Physiol. 155 (4), 1750–1751.

Guo, J.L., Xu, L.P., Su, Y.C., 2013. *ScMT2-1-3*, a metallothionein gene of sugarcane, plays an important role in the regulation of heavy metal tolerance/accumulation. Biomed. Res. Int. 904769.

Hall, J.L., 2002. Cellular mechanism for heavy metal detoxification and tolerance. J. Exp. Bot. 53, 1–11.

Hassan, Z., Aarts, M.G.M., 2011. Opportunities and feasibilities for biotechnological improvement of Zn, Cd or Ni tolerance and accumulation in plants. Environ. Exp. Bot. 72, 53–63.

Heike, H., Christopher, P., Martina, W., Elizabeth, V.M., 2011. Technology Review of Constructed Wetlands. Gesellscaft fur Internationalle Zusammenarbeit/GIZ, Germany, pp. 9–17.

Heyno, E., Klose, C., Krieger-Liszkay, A., 2008. Origin of cadmium induced reactive oxygen species production: mitochondrial electron transfers versus plasma membrane NADPH oxidase. New Phytol. 179, 687–699.

Hildebrandt, U., Regvar, M., Bothe, H., 2007. Arbuscular mycorrhiza and heavy metal tolerance. Phytochemistry 68 (1), 139–146.

Hinsinger, P., Plassard, C., Tang, C., Jaillard, B., 2003. Origins of root-induced pH changes in the rhizosphere and their responses to environmental constraints: a review. Plant Soil 248, 43–59.

Ikram, S., Parvaiz, M., Ahmed, M., Ilyas, U., Khalid, A., Munir, M., Anwar, S., Afzal, H., 2013. Aquatic plants of Sialkot district, Pakistan. World Appl. Sci. J. 28 (6), 884–896.

ITRC, 2003. Technical and Regulatory Guidance Document for Constructed Treatment Wetlands. The Interstate Technology & Regulatory Council Wetlands Team, p. 128.

Jadia, D.C., Fulekar, M.H., 2009. Phytoremediation of heavy metals: recent techniques. Afr. J. Biotechnol. 8, 921–928.

Janicka-Russak, M., Kabala, K., Burzynski, M., Klobus, G., 2008. Response of plasma membrane H+-ATPase to heavy metal stress in *Cucumis sativus* root. J. Exp. Bot. 59, 3721–3728.

Javed, M.T., Stoltz, E., Lindberg, S., Greger, M., 2013. Changes in pH and organic acids in mucilage of *Eriophorum angustifolium* roots after exposure to elevated concentrations of toxic elements. Environ. Sci. Pollut. Res. 20, 1876–1880.

Javed, M.T., Akram, M.S., Tanwir, K., Chaudhary, H.J., Ali, Q., Stoltz, E., Lindberg, S., 2017. Cadmium spiked soil modulates root organic acids exudation and ionic contents of two differentially Cd tolerant maize (*Zea mays* L.) cultivars. Ecotoxicol. Environ. Saf. 141, 216–225.

Jha, P., Samal, A.C., Santra, S.C., Dewanji, A., 2016. Heavy metal accumulation potential of some wetland plants growing naturally in the city of Kolkata, India. Am. J. Plant Sci. 7, 2112–2137.

Jia, X., Otte, M.L., Liu, Y., Qin, L., Tian, X., Lu, X., Jiang, M., Zou, Y., 2018. Performance of iron plaque of wetland plants for regulating iron, manganese, and phosphorus from agricultural drainage water. Water 10, 42. https://doi.org/10.3390/w10010042.

Johnston, C.A., 1993. Mechanisms of water wetland water quality interaction. In: - Moshiri, G.A. (Ed.), Constructed Wetlands for Water Quality Improvement. Lewis Publishers, Boca Raton, FL, pp. 293–300.

Joyce, C., 2012. Preface: wetland services and management. Hydrobiologia 692, 1–3.

Khan, M.A., Khan, S., Khan, A., Alam, M., 2017. Soil contamination with cadmium, consequences and remediation using organic amendments. Sci. Total Environ. 601, 1591–1605.

Kobayashi, T., Nishizawa, N.K., 2012. Iron uptake translocation, and regulation in higher plants. Annu. Rev. Plant Biol. 63, 131–152.

Kranner, I., Colville, L., 2011. Metals and seeds: biochemical and molecular implications and their significance for seed germination. Environ. Exp. Bot. 72, 93–105.

Laghlimi, M., Baghdad, B., Hadi, H.E., Bouabdli, A., 2015. Phytoremediation mechanisms of heavy metal contaminated soils. Open J. Ecol. 5, 375–388.

Leszczyszyn, O.I., Imam, H.T., Blindauer, C.A., 2013. Diversity and distribution of plant metallothioneins: a review of structure, properties and functions. Metallomics 5 (9), 1146–1169.

Li, J., Yu, H., Luan, Y., 2015. Meta-analysis of the copper, zinc, and cadmium absorption capacities of aquatic plants in heavy metal-polluted water. Int. J. Environ. Res. Public Health 12, 14958–14973.

Li, B., Pang, L., Wai, D., Lei, M., Liu, B., Lin, Y., Li, Z., Gu, D., 2017. Enhanced flocculation and sedimentation of trace cadmium from irrigation water using phosphoric fertilizer. Sci. Total Environ. 601, 485–492.

Lin, X., Mou, R., Cao, Z., Xu, P., Wu, X., Zhu, Z., Chen, M., 2016. Characterization of cadmium-resistant bacteria and their potential for reducing accumulation of cadmium in rice grains. Sci. Total Environ. 569, 97–104.

Lindberg, S., Landberg, T., Greger, M., 2007. Cadmium uptake and induction of phytochelatins in wheat protoplasts. Plant Physiol. Biochem. 45, 47–53.

Lindberg, S., Wingstrand, G., 1985. Mechanism for $Cd^{2+}$ inhibition of ($K^{++}$ $Mg^{2+}$) ATPase activity and $K^{+}$ ($86Rb^{+}$) uptake in roots of sugar beet (*Beta vulgaris*). Physiol. Plant. 63, 181–185.

Liu, W.T., Ni, J.C., Zhou, Q.X., 2013. Uptake of heavy metals by trees: prospects for phytoremediation. Mater. Sci. Forum 743–744, 768–781.

Liu, W., Zhang, X., Liang, L., Chen, C., Wei, S., Zhou, O., 2015. Phytochelatin and oxidative stress under heavy metal stress tolerance in plants. In: Gupta, D., Palma, J., Corpas, F. (Eds.), Reactive Oxygen Species and Oxidative Damage in Plants Under Stress. Springer, Cham.

Lombi, E., Tearall, K.L., Howarth, J.R., 2002. Influence of iron status on cadmium and zinc uptake by different ecotypes of the hyperaccumulator *Thlaspi caerulescens*. Plant Physiol. 128, 1359–1367.

Lu, L.L., Tian, S.K., Yang, X.E., Wang, X.C., Brown, P., Li, T.Q., He, Z.L., 2008. Enhanced root-to-shoot translocation of cadmium in the hyperaccumulating ecotype of *Sedum alfredii*. J. Exp. Bot. 59, 3203–3213.

Ma, Y., Prasad, M.N.V., Rajkumar, M., 2011. Plant growth promoting rhizobacteria and endophytes accelerate phytoremediation of metalliferous soils. Biotechnol. Adv. 29, 248–258.

Ma, N., Wang, W., Gao, J., Chen, J., 2017. Removal of cadmium in subsurface vertical flow constructed wetlands planted with *Iris sibirica* in the low-temperature season. Ecol. Eng. 109, 48–56.

MacFarlane, G.R., Pulkownik, A., Burchett, M.D., 2003. Accumulation and distribution of heavy metals in the grey mangrove, *Avicennia marina* (Forsk.) Vierh.: biological indication potential. Environ. Pollut. 51, 123–139.

Macovei, A., Ventura, L., Dona, M., Fae, M., Balestrazzi, A., Carbonera, D., 2010. Effects of heavy metal treatments on metallothionein expression profiles in white poplar (*Populus alba* L.) cell suspension cultures. An. Univ. Oradea Fasc. Biol. 18, 274–279.

Marschner, P. (Ed.), 2011. Marschner's Mineral Nutrition of Higher Plants. third ed. Academic, London.

Matache, M.L., Marin, C., Rozylowicz, L., Tudorache, A., 2013. Plants accumulating heavy metals in the Danube river wetlands. J. Environ. Health Sci. Eng. 11, 3–9.

Matagi, S.V., Swai, D., Mugabe, R., 1998. A review of heavy metal removal mechanisms in wetlands. Afr. J. Trop. Hydrobiol. Fish. 8, 23–35.

Materac, M., Sobiecka, E., 2017. The efficiency of macrophytes for heavy metals removal from water. Biotechnol. Food Sci. 81 (1), 35–40.

Mishra, S., Dubey, R.S., 2006. Heavy metal uptake and detoxification mechanisms in plants. Int. J. Agric. Res. 1 (2), 122–141.

Mohamed, A.A., Castagna, A., Ranieri, A., Sanita di Toppi, L., 2012. Cadmium tolerance in *Brassica juncea* roots and shoots is affected by antioxidant status and phytochelatin biosynthesis. Plant Physiol. Biochem. 57, 15–22.

Mohammadi, K., Khalesro, S., Sohrabi, S., Heidari, G., 2011. A review: beneficial effects of the mycorrhizal fungi for plant growth. J. Appl. Environ. Biol. Sci. 1 (9), 310–319.

Moore, P.D., Garratt, R., 2006. Wetland Biodiversity. Wetlands. Chelsea House Publishers, New York, pp. 98–105.

Muehe, E.M., Weigold, P., Adaktylou, I.J., Friedrich, B.P., Kraemer, U., Kappler, A., Behrens, S., 2015. Rhizosphere microbial community composition affects cadmium and zinc uptake of the metal-hyperaccumulating plant *Arabidopsis halleri*. Appl. Environ. Microbiol.. https://doi.org/10.1128/AEM.03359-14.

Muofhe, M.L., Dakora, F.D., 2000. Modification of rhizosphere pH by the symbiotic legume Aspalathus linearis growing in a sandy acidic soil. Aust. J. Plant Physiol. 27, 1169–1173.

Najeeb, U., Jilanic, G., Alia, S., Sarward, M., Xua, L., Zhoua, W., 2011. Insights into cadmium induced physiological and ultra-structural disorders in *Juncus effusus* L. and its remediation through exogenous citric acid. J. Hazard. Mater. 186, 565–574.

Nedjimi, B., Daoud, Y., 2009. Cadmium accumulation in *Atriplex halimus* subsp. *schweinfurthii* and its influence on growth, proline, root hydraulic conductivity and nutrient uptake. Flora Morphol. Distrib. Funct. Plant Ecol. 204, 316–324.

Nyquist, J., Greger, M., 2008. Response of two wetland species to Cd exposure at low and high pH. Environ. Exp. Bot. 65 (2–3), 417–424.

Pavel, V.L., Diaconu, M., Bulgariu, D., Statescu, F., Gavrilescu, G., 2012. Evaluation of heavy metals toxicity on two microbial strains isolated from soil: *Azotobacter sp.* and *Pichia sp.* Environ. Eng. Manag. J. 11, 165–168.

Prasad, M.N.V., 2001. Metals in the Environment. CRC Press.

Qin, R., Hirano, Y., Brunner, I., 2007. Exudation of organic acid anions from poplar roots after exposure to Al, Cu and Zn. Tree Physiol. 27, 313–320.

Rea, P.A., 2012. Phytochelatin synthase: of a protease a peptide polymerase made. Plant Physiol. 145, 154–164.

Robbins, E.I., Norden, A.W., 1994. Microbial oxidation of iron and manganese in wetlands and creeks of Maryland, Virginia, Delaware, and Washington. In: Chiang, S.H. (Ed.), Proceedings of the XI Annual International Pittsburgh Coal Conference. In: vol. 2. pp. 1154–1159.

Saifullah, U., Sarwar, N., Bibi, S., Ahmad, M., Ok, Y.S., 2014. Effectiveness of zinc application to minimize cadmium toxicity and accumulation in wheat (*Triticum aestivum* L.). Environ. Earth Sci. 71, 1663–1672.

Sarwar, N., Saifullah, S., Malhi, S., Zia, M.H., Naeem, A., Bibi, S., Farida, G., 2010. Role of mineral nutrition in minimizing cadmium accumulation by plants. J. Sci. Food Agric. 90, 925–937.

Sebastian, J.R., Sebastian, D.J., Beck, K.G., 2014. Feral Rye Control in Colorado. Western Society of Weed Science Progress Report, Colorado Springs, CO.

Sharma, R.K., Archana, G., 2016. Cadmium minimization in food crops by cadmium resistant plant growth promoting rhizobacteria. Appl. Soil Ecol. 107, 66–78.

Sheoran, A.S., Sheoran, V., 2006. Heavy metal removal mechanism of acid mine drainage in wetlands: a critical review. Miner. Eng. 19, 105–116.

Shi, G., Liu, C., Cai, Q., Liu, Q., Hou, C., 2010. Cadmium accumulation and tolerance of two safflower cultivars in relation to photosynthesis and antioxidative enzymes. Bull. Environ. Contam. Toxicol. 85, 256–263.

Shim, M.S., Kim, J.Y., Jung, H.K., Lee, K.H., Xu, X.M., Carlson, B.A., Kim, K.W., Kim, I.Y., Hatfield, D.L., Lee, B.J., 2009. Elevation of glutamine level by selenophosphate synthetase 1 knockdown induces megamitochondrial formation in Drosophila cells. J. Biol. Chem. 284 (47), 32881–32894.

Shivakumar, C.K., Hemavani, C., Thippeswamy, B., Krishnappa, M., 2011. Effect of inoculation with arbuscular mycorrhizal fungi on green gram grown in soil containing heavy metal zinc. J. Exp. Sci. 2 (10), 17–21.

Singh, N., Kaur, M., Katnoria, J.K., 2017. Analysis on bioaccumulation of metals in aquatic environment of Beas River Basin: a case study from Kanjli wetland. GeoHealth 1, 93–105.

Sobolewski, A., 1999. A review of processes responsible for metal removal in wetlands treating contaminated mine drainage. Int. J. Phytoremediation 1 (1), 19–51.

Solti, A., Sarvari, E., Toth, B., Basa, B., Levai, L., Fodor, F., 2011. Cd affects the translocation of some metals either Fe-like or Ca-like way in poplar. Plant Physiol. Biochem. 49, 494–498.

Song, F.Q., Yang, G.T., Meng, F.R., Tian, X.J., Dong, A.R., 2004. The rhizospheric niche of seedlings of Populus ussruiensis colonized by arbuscular mycorrhizal (AM) fungi. Ecol. Environ. 13, 211–216.

Souza, F.A., Dziedzic, A.S., Cubas, S.A., Maranho, L.T., 2013. Restoration of polluted waters by phytoremediation using *Myriophyllum aquaticum* (Vell.) Verdc. Haloragaceae. J. Environ. Manag. 120, 5–9.

Sriyaraj, K., Shutes, R.B.E., 2001. An assessment of the impact of motorway runoff on a pond, wetland and stream. Environ. Int. 26 (5–6), 433–439.

Sterckeman, T., Goderniaux, M., Sirguey, C., 2015. Do roots or shoots control cadmium accumulation in the hyperaccumulator *Noccaea caerulescens*. Plant Soil 392, 87–99.

Stoltz, E., Greger, M., 2002. Cotton grass effects of trace elements in submersed mine tailings. J. Environ. Qual. 31, 1477–1483.

Su, Y., Liu, J., Lu, Z., Wang, X., Zhang, Z., Shi, G., 2014. Effects of iron deficiency on subcellular distribution and chemical forms of cadmium in peanut roots in relation to its translocation. Environ. Exp. Bot. 97, 40–48.

Sylwia, W., Anna, R., Ewa, B., Stephan, C., Danuta Maria, A., 2010. The role of subcellular distribution of cadmium and phytochelatins in the generation of distinct phenotypes of AtPCS1- and CePCS3-expressing tobacco. J. Plant Physiol. 167, 981–988.

Tanwir, K., Akram, S.A., Masood, S., Chaudhary, H.J., Lindberg, S., Javed, M.T., 2015. Cadmium-induced rhizospheric pH dynamics modulated nutrient acquisition and physiological attributes of maize (Zea mays L.). Environ. Sci. Pollut. Res. 22 (12), 9193–9203.

Tian, S., Lu, L., Labavitch, J., Yang, X., He, Z., Hu, H., Sarangi, R., Newville, M., Commisso, J., Brown, P., 2011. Cellular sequestration of cadmium in the hyperaccumulator plant species Sedum alfredii. Plant Physiol. 157, 1914–1925.

Tolcin, A.C., 2015. Cadmium (Advance Release). Minerals Yearbook. U.S. Geological Survey.

Toth, T., Zsiross, Z., Kis, M., Garab, G., Kovacs, L., 2012. Cadmium exerts its toxic effects on photosynthesis via a cascade mechanism in the cyanobacterium, *Synechocystis* PCC 6803. Plant Cell Environ. 35, 2075–2086.

Tran, T.A., Popova, L.P., 2013. Functions and toxicity of cadmium in plants recent advances and future prospects. Turk. J. Bot. 37, 1–13.

Ueno, D., Iwashita, T., Zhao, F.J., Ma, J.F., 2008. Characterization of Cd translocation and identification of Cd form in xylem sap of the Cd-hyperaccumulator *Arabidopsis halleri*. Plant Cell Physiol. 49, 540–548.

Ugya, A., 2015. The efficiency of *Lemna minor* L. in the phytoremediation of rani stream: a case study of Kaduna refinery and petrochemical company polluted stream. J. Appl. Biol. Biotechnol. 1, 011–014.

Upadhyaya, H., Panda, S.K., Bhattacharjee, M.K., Dutta, S., 2010. Role of arbuscular mycorrhiza in heavy metal tolerance in plants: prospects for phytoremediation. J. Phytology 2 (7), 16–27.

US EPA, 2016. Aquatic Life Ambient Water Quality Criteria for Cadmium-2016. Office of Science and Technology, Washington, DC (CAS Registry (333-41)).

Uysal, Y., Taner, F., 2010. Bioremoval of cadmium by *Lemna minor* in different aquatic conditions. Clean: Soil, Air, Water 38, 370–377.

VanLoon, G.W., Duffy, J.S., 2005. Environmental Chemistry: A Global Perspective, second ed. Oxford University Press, New York.

Verkleij, J.A.C., Sneller, F.E.C., Schat, H., 2003. Metallothioneins and phytochelatins: ecophysiological aspects. In: Abrol, Y.P., Ahmad, A. (Eds.), Sulphur in Plants. Springer, Dordrecht, Netherlands, pp. 163–176.

Verma, R., Suthar, S., 2015. Lead and cadmium removal from water using duckweed-*Lemna gibba* L.: impact of pH and initial metal load. Alex. Eng. J. 54, 1297–1304.

Vidaković-Cifrek, Z., Tkalec, M., Šikić, S., Tolić, S., Lepeduš, H., Pevalek-Kozlina, B., 2015. Growth and photosynthetic responses of *Lemna minor* L. exposed to Cd in combination with Zn or Cu. Arh. Hig. Rada Toksikol. 66, 141–152.

Wang, F., Chen, F., Cai, Y., Zhang, G., Wu, F., 2011. Modulation of exogenous glutathione in ultrastructure and photosynthetic performance against Cd stress in the two barley genotypes differing in Cd tolerance. Biol. Trace Elem. Res. 144 (1–3), 1275–1288.

Wani, R.A., Ganai, B.A., Shah, M.A., Uqab, B., 2017. Heavy metal uptake potential of aquatic plants through phytoremediation technique—a review. J. Bioremed. Biodegr. 8, 404. https://doi.org/10.4172/2155-6199.1000404.

Weis, J.S., Weis, P., 2004. Metal uptake, transport and release by wetland plants: implications for phytoremediation and restoration. Environ. Int. 30 (5), 685–700.

Wiebner, A., Kappelmeyer, U., Kuschk, P., Kastner, M., 2005. Influence of the redox condition dynamics on the removal efficiency of a laboratory-scale constructed wetland. Water Res. 29, 248–256.

Williams, J.B., 2002. Phytoremediation in wetland ecosystems: progress, problems and potential. Crit. Rev. Plant Sci. 21, 607–635.

Wong, H.L., Sakamoto, T., Kawasaki, T., Umemura, K., Shimamoto, K., 2004. Down-regulation of metallothionein, a reactive oxygen scavenger, by the small GTPase OsRac1 in rice. Plant Physiol. 135 (3), 1447–1456.

Wood, C.M., Farrell, A.P., Brauner, C.J., 2012. Homeostatis and Toxicology of Non-essential Metals: Volume 31B. Fish Physiology. Academic Press/Elsevier, Waltham, MA.

Wu, H., Chen, C., Du, J., 2012. Co-overexpression FIT with AtbHLH38 or AtbHLH39 in Arabidopsis-enhanced cadmium tolerance via increased cadmium sequestration in roots and improved iron homeostasis of shoots. Plant Physiol. 158, 790–800.

Xin, J., Huang, B., Dai, H., Zhou, W., Yi, Y., Peng, L., 2015. Roles of rhizosphere and root derived organic acids in Cd accumulation by two hot pepper cultivars. Environ. Sci. Pollut. Res. 22 (8), 6254–6261.

Yadav, S.K., 2010. Heavy metals toxicity in plants: an overview on the role of glutathione and phytochelatins in heavy metal stress tolerance of plants. S. Afr. J. Bot. 76, 167–179.

Yang, Z., Chu, C., 2011. Towards understanding plant response to heavy metal stress. In: Abiotic Stress in Plants-Mechanisms and Adaptations. In Tech, Shanghai, China, pp. 59–78.

Ying, R.R., Qiu, R.L., Tang, Y.T., Hu, P.J., Qiu, H., Chen, H.R., Shi, T.H., Morel, J.L., 2010. Cadmium tolerance of carbon assimilation enzymes and chloroplast in Zn/Cd hyperaccumulator *Picris divaricata*. J. Plant Physiol. 167, 81–87.

Yoshihara, T., Hodoshima, H., Miyano, Y., 2006. Cadmium inducible Fe deficiency responses observed from macro and molecular views in tobacco plants. Plant Cell Rep. 25, 365–373.

Youssef, R.A., Chino, M., 1989. Root-induced changes in the rhizosphere of plants. II. Distribution of heavy metals across the rhizosphere in soils. Soil Sci. Plant Nutr. 35, 609–621.

Zeng, F., Chen, S., Miao, Y., Wu, F., Zhang, G., 2007. Changes of organic acid exudation and rhizosphere pH in rice plant under chromium stress. Environ. Pollut. 155 (2), 284–289.

Zhou, B., Yao, W., Wang, S., Wang, X., Jiang, T., 2014. The metallothionein gene, *TaMT3*, from *Tamarix androssowii* confers $Cd^{2+}$ tolerance in tobacco. Int. J. Mol. Sci. 15 (6), 10398–10409.

Zhu, X.F., Zheng, C., Hu, Y.T., Jiang, T., Liu, Y., Dong, N.Y., Yang, J.L., Zheng, S.J., 2011. Cadmium-induced oxalate secretion from root apex is associated with cadmium exclusion and resistance in *Lycopersicon esulentum*. Plant Cell Environ. 34, 1055–1064.

Zouainia, S., Djebar, M.R., Sbartai, H., Cherait, A., Khene, L., Kahli, H., Berrebbah, H., 2016. Toxicological impact assessment of cadmium on aquatic macrophyte *Elodea canadensis*. *Studia Universitatis "Vasile Goldis" Arad. Seria Stiintele Vietii (Life Sciences Series)* 26 (3), 375–380.

## FURTHER READING

Dixit, P., Mukherjee, P.K., Ramachandran, V., Eapen, S., 2011. Glutathione transferase from *Trichoderma virens* enhances cadmium tolerance without enhancing its accumulation in transgenic *Nicotiana tabacum*. PLoS ONE 6(1)e16360.

# CHAPTER 21

# Phytoremediation of Cd-Contaminated Soil and Water

**Nadeem Iqbal*, Malik Tahir Hayat*, Bibi Saima Zeb*, Zaigham Abbas†, Toqeer Ahmed‡**

*Department of Environmental Sciences, COMSATS, Abbottabad, Pakistan
†Ministry of Climate Change, Islamabad, Pakistan
‡Centre for Climate Research and Development (CCRD), COMSATS, Islamabad, Pakistan

## Abstract

Phytoremediation is the best technique among other conventional methods to treat contaminated sites. It is an inexpensive, environmentally safe technique that can remove cadmium (Cd) from contaminated soil and water through different processes, such as phytoextraction, phytosequestration, and phytostimulation. The use of hyperaccumulator plants can enhance the removal of Cd from the polluted site and result in productive land. These plants accumulate heavy metals such as Cd into vacuoles of root cells and in the aerial parts of the plants, i.e., the shoot and leaf. Different mechanisms have been studied by which Cd is accumulated into the plants. There are different factors that can increase the removal of Cd through plants, such as microbial-assisted phytoremediation, the addition of chelating agents, and decapitation. The efficiency of phytoremediation can be enhanced by lowering the pH of the contaminated site, which ultimately helps in more removal of Cd from contaminated water and soil. There are several advantages of adopting this efficient and more productive technique. It can be the best option, or be used as an alternative to other traditional methods. This innovative and promising technique enhances the removal of Cd from contaminated sites, and converts these barren sites into productive lands.

**Keywords:** Phytoremediation, Cadmium, Phytoextraction, Phytosequestration, Microbe-assisted phytoremediation, Soil and water contamination

## 1 INTRODUCTION

Contamination of soil and water is a major problem causing drastic effects on all living organisms, and ultimately on humans. Soil and water pollution is caused due to many natural and anthropogenic sources. Natural sources include forest fires, volcanic eruptions, radioactivity, and weathering of rocks. Anthropogenic sources include consumption of fossil fuels, incineration of solid waste, and industrial processes such as smelting of ore, mining,

*Cadmium Toxicity and Tolerance in Plants*
https://doi.org/10.1016/B978-0-12-814864-8.00021-8

and production of fuel and energy. The world population is growing rapidly, and the availability of nonrenewable resources is gradually decreasing. It is time to turn our attention toward renewable or regenerative resources to meet the needs of the growing population. Both natural and anthropogenic activities have badly affected the balanced ecosystem. Therefore, it is necessary to recover contaminated soil and water (Pavaloaia et al., 2015).

Many toxic organic and inorganic compounds from various sources are being released into the environment, such as polycyclic aromatic hydrocarbons (PAHs), polychlorinated biphenyls (PCBs), and heavy metals, including cadmium (Cd), zinc (Zn), lead (Pb), chromium (Cr), and nickel (Ni). These heavy metals can cause disease in animals and humans. Direct and indirect sources of these heavy metals result in air, water, and soil pollution. The transformations of these heavy metals result in many economic, public, and environmental consequences. Heavy metals are not degraded by microbes, and they are persistent in the environment. Heavy metals are mostly carcinogenic in nature, and accumulate in the organs of organisms (plants and animals). These metals have various applications in consumer products, pulp and paper engineering works, the leather industry, materials used in photography, plastic stabilizers, batteries, and fertilizers (Ali et al., 2013). Cd has toxicological effects on humans and other living organisms. If a small amount of Cd is exposed for a short time, it can be harmful, and result in adverse health effects (Abu-El-Halawa and Zabin, 2017). Cd contamination in cultivated soils comes from several sources, including phosphate fertilizers, compost, manure, and sewage sludge. Cd causes serious threats to animals and humans by being taken up by plants, accumulating in them, and exceeding harmful levels (Cheng et al., 2017). Cd binds at active sites of enzymes, and is selected by competing with other essential nutrients, and then causes toxicity to the host cells of exposed organisms.

Therefore, different, effective ways are needed to treat Cd-contaminated soils to mitigate its risk. Different techniques have been used to treat contaminated soil and water. Phytoremediation is a promising technique that is cost effective and environmentally safe. This technique has the potential to accumulate heavy metals, such as Cd, in plants from soil (Cheng et al., 2017). The cost analysis of phytoremediation was carried out, and it was found that this technique is more cost effective than other treatment techniques, such as ion exchange, solvent extraction, adsorption, oxidation-reduction, and reverse osmosis. Additionally, phytoremediation improves the soil quality and soil biological processes (Wan et al., 2016).

## 2 PHYTOREMEDIATION

Phytoremediation is a type of bioremediation in which plants are used to degrade or immobilize contaminants in soil or water to mitigate the effects of toxic pollutants. This word is from the Greek language, and it refers to restoring contaminated sites through plants. Phytoremediation is an economical and environmentally safe technique in which contaminants are extracted from soil. Phytoremediation involves various processes, which are described as follows (Jiang et al., 2015).

Phytoremediation technology uses plants to remediate or treat contaminated soil and water to make them harmless. This technique has attracted the attention of stakeholders because it has useful applications in the remediation of soil and water, and improves the quality of land and water by managing waste. The contamination of these heavy metals has severe health risks and drastic impacts on human health. Heavy metals, including Cd, are mutagenic, teratogenic, and carcinogenic to human beings. These metals affect the endocrine and nervous systems, especially in children. Several physio-chemical methods have been used to treat Cd-contaminated soils, but traditional methods are too expensive to implement on a large scale. Such methods need more processing, and result in a change in the soil properties. Therefore, phytoremediation can be applied at a large scale with sustainable costs (Sharma and Pandey, 2014).

Phytoremediation exploits herbs, shrubs, and woody species because they have the potential to remove or accumulate different environmental pollutants, such as heavy metals, including Cd in soil and water. The technique can also make the pollutants harmless. Phytoremediation also mitigates the risk of dispersion of contaminants, and is helpful in decontaminating soil and water. Several factors affect the mechanism and efficacy of phytoremediation, such as bioavailability, soil characteristics, plant species, and the nature of the contaminants. Plants showing more efficient removal of heavy metal are called hyperaccumulators. These plants have the capacity to remove or accumulate large amounts of heavy metals, but have less biomass production. Biomass production of plants affects their extraction efficiency, because more biomass production will remove more heavy metal, but it will also need more harvests for the plants' removal. However, the quantity of total harvests will represent the total cost of the complete operation, including incineration and disposal (Sreelal and Jayanthi, 2017).

## 3 WHY DO WE NEED PHYTOREMEDIATION?

Phytoremediation is an environmentally safe approach that has the potential to use the capacity of plants to accumulate contaminants from the environment, and to degrade or accumulate these toxic pollutants. The accumulation in certain plants leverages their natural ability to render harmless pollutants in soil or water. With the knowledge of molecular and physiological changes, the technology of phytoremediation has been improved, and has emerged as an innovative technique. In the phytoremediation technique, more harmful heavy metals and toxic organic contaminants are targeted. Different types of engineering works and biological strategies have been developed to improve the process of phytoremediation. The feasibility of various plants has been confirmed using phytoremediation for environmental cleanup purposes (Ali et al., 2013).

## 4 PHYTOREMEDIATION AS AN ENVIRONMENTALLY FRIENDLY TECHNIQUE

Phytoremediation is a cost effective and environmentally friendly technology that is considered a green, or sustainable, remediation method. In this technique, plants (hyperaccumulators) are used to treat polluted sites (water or soil) to remove harmful contaminants. Phytoremediation is a promising technique that provides the best alternative to other conventional methods. It has been reported that more than 500 plant species are well recognized as accumulators of Cd and other heavy metals from polluted sites (Kramer, 2010; Hemen, 2011; Yu et al., 2013).

Plants are considered the primary acceptors of metals such as Cd. Therefore, plants remediate such contaminated sites to treat xenobiotic metals through phytoremediation, which is an effective and technique for removing pollutants from soil or water. It does not disrupt the environment and it is economically feasible. It has the ability to treat the polluted site to remove more than one contaminant. It is easy, and can be implemented in situ and ex situ. Due to the use of plants, it is pleasing aesthetically (Ma et al., 2005).

## 5 PROCESSES OF PHYTOREMEDIATION

The following sections describe the different remediation processes, including how contaminants are accumulated and removed.

## 5.1 Phytosequestration

Phytosequestration is also known as phytostabilization. In this process, chemicals or pollutants are absorbed or adsorbed on the surface of roots. Toxic pollutants are immobilized or sequestrated in the vacuoles of roots in contaminated soil or water. Contaminants are absorbed by the roots of plants and immobilized in the rhizosphere. It decreases the mobility of pollutants and their bioavailability in the food chain. This technique also prevents the movement of contaminants into the ground water. Contaminated sites are being treated using metal-tolerant species of plants. The heavy metals can include Cd, Cr, As, Cd, or Zn (Zhao et al., 2016). This technique initiates changes in the composition of the soil chemistry, providing the surface of the roots with the facility to absorb the metals (Yang et al., 2016). However, this method cannot remove pollutants from treated soil until the plants are eradicated.

## 5.2 Phytoextraction

This process is also called phytoaccumulation. Contaminants are taken up by the plants and stored in the tissues of the stems or leaves. In this process, pollutants are degraded and removed from the environment. It is a useful technique for removing heavy metals from the contaminated soil or water. Through the process of incineration of these plants, these heavy metals can be recovered for reuse, which is referred to as phytomining (Eissa, 2017).

Roots absorb pollutants from the soil and transport them to the shoots and leaves, where they are accumulated. Plants that have a high capacity to remove pollutants and produce a large biomass are ideal for this technique. The hyperaccumulator species has the ability to store pollutants, but exhibits less biomass production. Those species that produce more biomass, but perform less removal of heavy metals can be selected and used for this purpose. This process ends with harvesting of the plants, followed by their incineration, and landfilling (Mahar et al., 2016).

This technique is environmentally friendly, among other advantages. It poses no damage on the landscape. It protects the ecosystem and is a primary phytoremediation technique. It is inexpensive, and the most commercially acceptable technique (Shiri et al., 2015). Additionally, the rhizosphere has a low bioavailability of metal, and a low absorption rate of heavy metal from the roots. Metals are retarded within the roots (Fiorentino et al., 2013).

Research has been conducted to elaborate on the capability of plants to remove or accumulate the heavy metals from contaminated soil and water.

Specific research has also been conducted to explain the performance of the herbaceous plant *Aruddox donax* in Cd contaminated soil. A 20% concentration of Cd in leaves and 30% in the rhizome was accumulated, which was more efficient than the control group (Van Oosten and Maggio, 2014).

## 5.3 Phytostimulation

The activity of rhizobacteria is enhanced by the addition of mycorrhizal-helping bacteria. The removal of toxic heavy metals is increased by the addition of microbes in the roots of plants. This process is also known as rhizosphere remediation. It is a technique in which different compounds are released from the roots of plants, and they enhance microbial activity. It is a low-cost technology for removing Cd, heavy metals, and other organic compounds (Jia et al., 2016).

There are several reasons for phytostimulation to be used for rhizoremediation. A particular niche is provided by rhizosphere-to-soil microorganisms. Exudates released by roots provide nutrients and enhance microbial growth. A habitat is provided by the rhizosphere for the soil microbes. These microorganisms, in return, cause the degradation, or biotransformation, of many organic and inorganic compounds, such as heavy metals. Another method is the addition of microorganisms in the soil. These resistant microbial strains can cause the breakdown of organic compounds and the accumulation of toxic heavy metals such as Cd (Yanai et al., 2006).

# 6 PHYTOREMEDIATION MECHANISMS FOR Cd-CONTAMINATED SOIL/WATER

Contaminated soils can be treated in a number of ways to reduce Cd toxicity. The common methods are use of landfills, incineration, washing, leaching, vitrification, electroreclamation, and excavation. These methods are expensive and need capital investment. These techniques are nonbiological processes, and cause disruption of the surrounding environment (Gao et al., 2012). Usually, plants absorb Cd metal through the roots from the soil, which serves as a bioavailable pool of Cd metal, and nutrients required by plants as well. Different factors affect the bioavailability of Cd metal in soil. These factors can be organic matter, root exudates, pH, competitive cations, and microbial biomass (Sarwar et al., 2010).

Once a Cd metal is taken up by plants, it can either be stored or accumulated in root tissues, or translocated to upper parts of plants through apoplastic or symplastic routes. Hyperaccumulators have vital role in Cd metabolism: to keep it away from the other metabolic activities in order to hinder its deleterious effects on plants. Vacuoles can accumulate Cd metal in the shoots. There are five main steps involved in phytoextraction; Cd mobility in the rhizospheric zone, plant roots' uptake of Cd metal, the transfer or translocation of Cd ions to upper plant roots, plant tissues sequestration of Cd, and tolerance of Cd metal (Ali et al., 2013). If a plant is more tolerant to a specific metal, it will accumulate a huge portion of that metal inside its tissues, and undergo minimum adverse impacts. The potential of a plant to tolerate Cd metal depends upon various mechanisms. These mechanisms can be Cd metal binding to the cell wall, movement of Cd ions into vacuoles through active transport, or binding of proteins with metal chelates (Memon and Schröder, 2009).

## 7 USE OF PLANTS AND THEIR BIOCHEMICAL PROCESSES IN PHYTOREMEDIATION

Different essential nutrients are required by the plants, which are absorbed by the roots. These nutrients are N, P, K, Ca, S, Mg, Cl, Fe, Zn, Cu, B, Mn, Mo, and B. These nutrients are transported to the upper parts of plants, or the cells of roots through active transport or passive transport. Transport proteins are also associated with cell membranes. These nutrients are transported within the plants through two ways; apoplast and symplast (Lu et al., 2011).

Certain plant species (*Thlaspi caerulescens*) showed tolerance against Cd and Zn due to their metal chilate formation. Such plants reduce the cytosolic concentration of the metals, and store them in the vacuoles. Willows exhibit the potential to uptake different metals such as Cd, Cu, Zn, Fe, and Pb. As compared with other herbaceous plants, willows possess deeper root systems, and can decontaminate a large zone of contaminated soil (Xiao et al., 2015).

## 8 MOBILITY/AVAILABILITY OF Cd

Total concentrations of metals in contaminated soil cannot be predicted on the basis of their behavior. The toxicity and uptake of most of the metals is dependent on specific metals and free ions in the soil. Except for this general observation, free metal can also be available to plants, and is more active than the ionic form. Mostly, the concentrations of metals soluble in water are considered available forms to plants (Abe et al., 2008). Equilibrium partitioning is used to describe the metal distribution between pore water and the solid phase of soil. The total distributed or dissolved concentration of metal does not necessarily correspond to the amount that is available to plants. Complex ions, ion pairs, microparticulates, or polymers can decrease free ionic species in a solution. Speciation is the process in which the identification and quantification of various metals is carried out in a sample. Different models are present in order to calculate the metal speciation in water, soil, or sediments, i.e., the free ion activity (FIAM) model and wandermere humic aqueous model (WHAM) (Fang et al., 2012). Cd has better ability to become available for plant uptake as compared with other heavy metals, such as Cu, Ni, Pb, Zn, and Mn. Therefore, high Cd uptake by plants results from low concentrations of Cd in the soil. It has already been established that Cd is a nonessential trace element for the growth of plants (Peng et al., 2009).

## 9 FACTORS AFFECTING THE SOLUBILITY/BIOAVAILABILITY OF Cd IN SOIL

There are a number of factors that affect Cd availability in plants. One of them is pH, which affects the concentrations of soluble Cd in plants. At low pH, the solubility of Cd metal increases, and decreases at higher pH. In accordance with some research, at pH 7, all metal ions are removed from the solution as hydroxo species, which are being adsorbed on the soil's surface. The buffering capacity of the soil for an acidic surrounding can also affect the metal mobility in soil. Moreover, the leaching of these metals can increase, due to high pH. However, Cd adsorption was found at 3.7 and 4 pH values. At pH 6–7, precipitation of the insoluble solids becomes essential (Zhang et al., 2010).

It has been reported that the decrease in redox potential increased the solubility of Cd in soil. The mechanism that is behind that fact is the dissolution of Fe-Mn oxyhydroxides, which release the absorbed metals under reducing conditions. According to some scientists, the availability of Cd to plants is decreased in water logging conditions because of CdS formation. This will also result in Zn solubility in water. Many studies have shown that soil's organic matter also affects the Cd uptake by plants in a negative association (Arnold and McDonald, 1999).

## 10 FACTORS INCREASING Cd UPTAKE BY PLANTS

There are several ways to increase the uptake of Cd in the plants from soil.

## 10.1 Microbe-Assisted Phytoremediation

Several bacteria (rhizobacteria) have been successfully applied to plants, resulting in plant growth. These bacteria assisted in decreasing Cd toxicity and increased plant growth in Cd contaminated soils. These plant-growth-promoting bacteria transform Cd into a soluble and available form to plants by changing their chemical properties. These microbes use different binding compounds, such as siderphores, organic acids, biosurfactants, and exopolymers (Ullah et al., 2015). Another promising technique is microbe-assisted phytoremediation. The deficiency of nutrients in the soil can inhibit the growth of bacteria, and then inhabiting bacteria are activated by adding nutrients (bioaugmentation). Therefore, microbes are biostimulated with nutrients to increase microbial growth to remove organic and heavy metal pollutants (Rojjanateeranaj et al., 2017). Different types of crops (maize and soybean) have been used to treat Cd contaminated soil, which was assisted by the bacterial community. These crops were selected due to rapid growth, large biomass, and good energy production (Chen et al., 2017).

## 10.2 Decapitation Increased Phytoremediation

The decapitation method is practiced in agronomy, in which the number of branches is increased by enhancing apical dominance. It has been reported that removing or cutting of apical buds increased shoot growth in various plants. It is well known that Cd accumulates in the shoots of hyperaccumulators because it is a predominant part of the plants. Decapitation increased the process of transpiration, biomass production, and accumulation of Cd (Liu et al., 2017).

## 10.3 Effect of pH on Cd Availability in Plants

Soil properties are also affected by the availability of Cd in soil. Cd mobilization is caused by texture, nutrient levels, organic matter, and pH. It has been reported that the bioavailability of Cd to plants is greatly influenced by pH. An inverse relationship was found between pH and Cd uptake in general. The uptake of Cd in plants increases with a decrease in pH. When the pH of soil decreases, the sesquioxides and variable charged clays release Cd in the soil. The solubility of Cd compounds ($CdCO_3$ and $Cd_3(PO_4)_2$ also increases with decreasing pH. Therefore, the bioavailability of Cd to plants can be enhanced by increasing rhizospheric acidification, and it can be an efficient technique to increase the efficacy of phytoremediation, and lessens the remediation time period (Yanai et al., 2006).

### 10.4 Enhancement of Phytoremediation Using Mobilized Chelating Agents

To obtain more efficient removal of Cd, different chelating agents are used. Ethylenediaminetetraacetic acid (EDTA) has application in detergents and other cleaning products, the paper and pulp industries, and water treatment. EDTA can be used as a chelating agent because it has a high affinity to combine with heavy metals such as Cd, and forms metal and EDTA complex compounds. It is also being applied to remove different metals from contaminated soil and water. This agent can enhance the availability of Cd-bound EDTA to plants and increase the translocation of these metals from the roots to the shoots of plants. Studies have shown that the addition of EDTA in soil ($12\,mmol\,kg^{-1}$) enhanced the availability of Cd up to 233%. Therefore, the addition of chelate ligands to contaminated soils can increase the uptake of Cd to plants, and the removal of Cd from the environment (Eissa, 2017).

## 11 ADVANTAGES OF PHYTOREMEDIATION

Phytoremediation is a cost-effective technique. It is inexpensive due to that fact that very little labor is required to monitor plants. It does not interfere with the ecosystem, and adds an aesthetic value to the treated land through plant cover. Different heavy metals can be recovered and reused for several purposes in industries. It is an environmentally safe process, because it poses no potentially harmful impacts on the environment. This technique uses naturally occurring living organisms (Tahir et al., 2016).

## 12 DISADVANTAGES OF PHYTOREMEDIATION

Phytoremediation is a slow process, compared with other engineering techniques. It can take years to treat slightly polluted soil or water (Han et al., 2018). It is confined to the surface area and the depth of roots of plants. The contaminated area in soil is next to the roots of plants. It requires more time for plants' biomass production, and exhibits slow growth. It is impossible to prevent the complete leaching of pollutants into ground water. The toxicity caused by contaminants also affects the survival of plants. Phytoremediation also requires the safe disposal of contaminated plant material. The accumulation of toxic metals from soil in plants is passed through the food chain, and ultimately affects animals and humans.

## 13 CONCLUSION AND FUTURE PERSPECTIVES

Cd contamination in soil and water is a major concern to health and the environment because of its adverse impacts on food chain contamination, and other connected health risks. In this case, the technique of phytoremediation can be a helpful tool to tackle environmental pollution. This is an emerging and innovative method. This technique is economically feasible and more effective than other physical and chemical techniques. Hyperaccumulators can be efficiently used to remove large concentrations of Cd and other heavy metals from contaminated soil and water. Other conventional methods are expensive and a significant amount of labor is required. Phytoremediation is a slow process, yielding less biomass production above ground. Other approaches, such as use of microbes, decapitation, and low pH maintenance are also useful in combination with phytoremediation to decontaminate Cd-polluted soil and water. The phytoremediation technique requires further research to improve the abilities of plants in order to increase their effectiveness and efficiency. Phytoremediation can serve as a reliable alternative to traditional methods because it is time-saving and more cost-effective. Additionally, its harvests can be utilized for biofuel production. The incineration of metal-tolerant plants can yield valuable metals, which can be reused in industry. Further research should be done on transgenic plants to determine all of the environmental and economic benefits. It provides a better option compared with other conventional techniques. More knowledge of this technique is required in order to increase its efficiency and potential in remediating Cd-contaminated sites.

## REFERENCES


Abe, T., Fukami, M., Ogasawara, M., 2008. Cadmium accumulation in the shoots and roots of 93 weed species. Soil Sci. Plant Nutr. 54, 566–573.

Abu-El-Halawa, R., Zabin, S.A., 2017. Removal efficiency of Pb, Cd, Cu and Zn from polluted water using dithiocarbamate ligands. J. Taibah Univ. Sci. 11 (1), 57–65.

Ali, H., Khan, E., Sajad, M.A., 2013. Phytoremediation of heavy metals—concepts and applications. Chemosphere 91, 869–881.

Arnold, M.A., McDonald, G.V., 1999. Accelerator containers alter plant growth and the root zone environment. J. Environ. Hortic. 17, 168–173.

Chen, Z., Zheng, Y., Ding, C., Ren, X., Yuan, J., Sun, F., Li, Y., 2017. Integrated metagenomics and molecular ecological network analysis of bacterial community composition during the phytoremediation of cadmium-contaminated soils by bioenergy crops. Ecotoxicol. Environ. Saf. 145 (Suppl. C), 111–118.

Cheng, M., Wang, A., Tang, C., 2017. Ammonium-based fertilizers enhance Cd accumulation in *Carpobrotus rossii* grown in two soils differing in pH. Chemosphere 188 (Suppl. C), 689–696.

Eissa, M.A., 2017. Phytoextraction mechanism of Cd by *Atriplex lentiformis* using some mobilizing agents. Ecol. Eng. 108 (Part A), 220–226.

Fang, Y.Y., Cao, X.D., Zhao, L., 2012. Effects of phosphorus amendments and plant growth on the mobility of Pb, Cu, and Zn in a multi-metal-contaminated soil. Environ. Sci. Pollut. Res. 19 (5), 1659–1667.

Fiorentino, N., Fagnano, M., Adamo, P., Impagliazzo, A., Mori, M., Pepe, O., Zoina, A., 2013. Assisted phytoextraction of heavy metals: compost and Trichoderma effects on giant reed (*Arundo donax* L.) uptake and soil N-cycle microflora. Ital. J. Agron. 8 (4), 29.

Gao, Y., Miao, C.H.Y., Xia, J., Luo, C.H.Y., Mao, L., Zhou, P., Shi, W.J., 2012. Effect of citric acid on phytoextraction and antioxidative defense in *Solanum nigrum* L. as a hyperaccumulator under Cd and Pb combined pollution. Environ. Earth Sci. 65 (7), 1923–1932.

Han, Z., Guo, Z., Zhang, Y., Xiao, X., Xu, Z., Sun, Y., 2018. Adsorption-pyrolysis technology for recovering heavy metals in solution using contaminated biomass phytoremediation. Resour. Conserv. Recycl. 129 (Suppl. C), 20–26.

Hemen, S., 2011. Metal hyperaccumulation in plants: a review focusing on phytoremediation technology. J. Environ. Sci. Technol. 4 (2), 118–138.

Jia, H., Wang, H., Lua, H., Jiang, S., Dai, M., Liu, J., Yan, C., 2016. Rhizodegradation potential and tolerance of *Avicennia marina* (Forsk.) vierh in phenanthrene and pyrene contaminated sediments. Mar. Pollut. Bull. 110, 112–118.

Jiang, Y., Lei, M., Duan, L., Longhurst, P., 2015. Integrating phytoremediation with biomass valorisation and critical element recovery: a UK contaminated land perspective. Biomass Bioenergy 83, 328–339.

Kramer, U., 2010. Metal hyperaccumulation in plants. Annu. Rev. Plant Biol. 61, 517–534.

Liu, J., Zhang, X., Mo, L., Yao, S., Wang, Y., 2017. Decapitation improves the efficiency of Cd phytoextraction by *Celosia argentea* Linn. Chemosphere 181 (Suppl. C), 382–389.

Lu, H., Zhang, Y., Liu, B., Liu, J., Ye, J., Yan, C., 2011. Rhizodegradation gradients of phenanthrene and pyrene in sediment of mangrove (*Kandelia candel* (L.) Druce). J. Hazard. Mater. 196, 263–269.

Ma, J.F., Ueno, D., Zhao, F.J., McGrath, S.P., 2005. Subcellular localisation of Cd and Zn in the leaves of a Cd–hyperaccumulating ecotype of *Thlaspi caerulescens*. Planta 220, 731–736.

Mahar, A., Wang, P., Ali, A., Awasthi, M.K., Lahori, A.H., Wang, Q., Li, R., Zhang, Z., 2016. Challenges and opportunities in the phytoremediation of heavy metals contaminated soils: a review. Ecotoxicol. Environ. Saf. 126, 111–121.

Memon, A.R., Schröder, P., 2009. Implications of metal accumulation mechanisms to phytoremediation. Environ. Sci. Pollut. Res. 16, 162–175.

Pavaloaia, L., Georgescu, I., Georgescu, M., 2015. The system of green certificates—promoter of energy from renewable resources. Procedia Soc. Behav. Sci. 188 (Suppl. C), 206–213.

Peng, K.J., Luo, C.L., Chen, Y.H., Wang, G.P., Li, X.D., Shen, Z.G., 2009. Cadmium and other metal uptake by *Lobelia chinensis* and *Solanum nigrum* from contaminated soils. Bull. Environ. Contam. Toxicol. 83 (2), 260–264.

Rojjanateeranaj, P., Sangthong, C., Prapagdee, B., 2017. Enhanced cadmium phytoremediation of *Glycine max* L. through bioaugmentation of cadmium-resistant bacteria assisted by biostimulation. Chemosphere 185 (Suppl. C), 764–771.

Sarwar, N., Saifullah, Malhi, S.S., Zia, M.H., Naeem, A., Bibi, S., Farid, G., 2010. Role of plant nutrients in minimizing cadmium accumulation by plant. J. Sci. Food Agric. 90, 925–937.

Sharma, P., Pandey, S., 2014. Status of phytoremediation in world scenario. Int. J. Environ. Bioremediat. Biodegrad. 2 (4), 178–191.

Shiri, M., Rabhi, M., Abdelly, C., El Amrani, A., 2015. The halophytic model plant *Thellungiella salsuginea* exhibited increased tolerance to phenanthrene-induced stress in comparison with the glycophitic one *Arabidopsis thaliana*: application for phytoremediation. Ecol. Eng. 74, 125–134.

Sreelal, G., Jayanthi, R., 2017. Review on phytoremediation technology for removal of soil contaminant. Indian J. Sci. Res. 14 (1), 127–130.

Tahir, U., Yasmin, A., Khan, U.H., 2016. Phytoremediation: potential flora for synthetic dyestuff metabolism. J. King Saud Univ. 28 (2), 119–130.

Ullah, A., Heng, S., Munis, M.F.H., Fahad, S., Yang, X., 2015. Phytoremediation of heavy metals assisted by plant growth promoting (PGP) bacteria: a review. Environ. Exp. Bot. 117 (Suppl. C), 28–40.

Van Oosten, M.J., Maggio, A., 2014. Functional biology of halophytes in the phytoremediation of heavy metal contaminated soils. Environ. Exp. Bot. 111, 135–146.

Wan, X., Lei, M., Chen, T., 2016. Cost–benefit calculation of phytoremediation technology for heavy-metal-contaminated soil. Sci. Total Environ. 563–564 (Suppl. C), 796–802.

Xiao, N., Liu, R., Jin, C., Dai, Y., 2015. Efficiency of five ornamental plant species in the phytoremediation of polycyclic aromatic hydrocarbon (PAH)-contaminated soil. Ecol. Eng. 75, 384–391.

Yanai, J., Zhao, F.-J., McGrath, S.P., Kosaki, T., 2006. Effect of soil characteristics on Cd uptake by the hyperaccumulator *Thlaspi caerulescens*. Environ. Pollut. 139 (1), 167–175.

Yang, S.X., Liao, B., Yang, Z.H., Chai, L.Y., Li, J.T., 2016. Revegetation of extremely acid mine soils based on aided phytostabilization: a case study from southern China. Sci. Total Environ. 562 (2016), 427–434.

Yu, C.L., Liu, B., Yang, Y., Qin, Y.C., 2013. Effects of chemical control on *Solanum nigrum* L. remediation to cadmium (Cd) from seriously-polluted soil. J. Beijing For. Univ. 35 (5), 133–138.

Zhang, X., Xia, H., Li, Z., Zhuang, P., Gao, B., 2010. Potential of four grasses in remediation of Cd and Zn contaminated soils. Bioresour. Technol. 101, 2063–2066.

Zhao, L., Li, T., Yu, H., Zhang, X., Zheng, Z., 2016. Effects of [S,S]-ethylenediaminedisuccinic acid and nitrilotriacetic acid on the efficiency of Pb phytostabilization by *Athyrium wardii* (Hook.) grown in Pb-contaminated soils. J. Environ. Manag. 182, 94–100.

## FURTHER READING

Kavamura, V.N., Esposito, E., 2010. Biotechnological strategies applied to the decontamination of soils polluted with heavy metals. Biotechnol. Adv. 28, 61–69.

Zhang, H., Zheng, L.C., Yi, X.Y., 2009. Remediation of soil co-contaminated with pyrene and cadmium by growing maize (*Zea mays* L.). Int. J. Environ. Sci. Technol. 6, 249–258.

# CHAPTER 22

# The Role of Aquatic Macrophytes in Cadmium Phytoremediation of Contaminated Estuarine Environments

**Carmen A. Pedro*, Sílvia C. Gonçalves*,†**

*MARE—Marine and Environmental Sciences Centre, ESTM, Polytechnic Institute of Leiria, Peniche, Portugal

†MARE—Marine and Environmental Sciences Centre, Department of Life Sciences, Faculty of Sciences and Technology, University of Coimbra, Coimbra, Portugal

## Abstract

The considerable increase in the discharge of pollutants into the environment, chiefly into the soil and water of coastal environments, has created an alarming situation for these ecosystems' sustainability. Among the hazardous pollutants, trace metals, such as cadmium, constitute a serious threat to biodiversity, due to their high degree of toxicity and capacity to accumulate in the abiotic (sediments and water) and biotic (in the tissues of living organisms) compartments. Consequently, human health is also vulnerable to possible contamination through the food chain. Coastal environments, namely estuaries, are highly productive and sensitive areas of strong ecological and economic importance. However, those systems are also very vulnerable to metal contamination because they are located at river mouths, especially near industrial areas, and are considered sinks for metal pollutants. In this scenario, the macrophytes, namely halophytes, common in estuarine environments, play an important role, because they may control bioavailability and toxicity, as well as phytoremediate toxic metals. In this chapter, the importance of phytoremediation in environmental preservation will be discussed under the scope of estuarine environments, considering the ecological concerns. Also, the physiological mechanisms involved in plants' tolerance to trace metals, as well as the technological advances that improve their phytoremediation capacity, will be discussed. The phytoremediation potential of aquatic macrophytes is well-documented, and, in view of this, this chapter highlights several examples of potential candidates in restoration projects for cadmium phytoremediation in metal contaminated estuarine environments.

**Keywords:** Aquatic macrophytes, Cadmium, Estuarine environments, Halophytes, Phytoremediation, Trace metals contamination


https://doi.org/10.1016/B978-0-12-814864-8.00022-X

## 1 INTRODUCTION

The continuous increase of pollutants, namely trace metals, into the ecosystems is a major worldwide concern, because contamination of environmental matrices (air, water, soil, and biota) can result in drastic consequences for the ecosystems' health and function, as well as several adverse effects on human health (toxicity and carcinogenicity) (Amari et al., 2017; Cristaldi et al., 2017; Dixit et al., 2015). However, an adequate balance of the interactions between biotic and abiotic factors is not always easy to achieve, and often the perturbations in abiotic components create such a stressful condition for the survival of living organisms that the recovery could be impossible. Therefore, research efforts have focused on maintenance or restoration of the ecosystems' sustainability, namely in regard to estuarine environments (see, e.g., Curado et al., 2014b; Idaszkin et al., 2017; Pedro et al., 2015; Teixeira et al., 2014; Teuchies et al., 2013).

Conventional remediation methods (physical and chemical methods) have been used to remediate contaminated systems, namely the soil compartment, and included isolation and containment, mechanical separation, pyrometallurgical separation, chemical treatment (by reductive, as well as oxidative, mechanisms), and electrokinetics (Mulligan et al., 2001). However, the efficiency of those methods is far from satisfactory, because they are environmentally destructive, with negative effects on biodiversity, and they are highly expensive and difficult to implement (Lasat, 2000; Thakur et al., 2016). Additionally, irreversible ecological consequences may result from their use. An ecologically responsible alternative to these environmentally destructive methods was necessary, and phytoremediation (from the Greek *phyto*: plant; and from the Latin *remedium*: balance or remediation), a newly evolving field of science and technology (Meagher, 2000), is considered the most effective (see, e.g., Amari et al., 2017; Cheraghi et al., 2011; Manousaki and Kalogerakis, 2011; Salt et al., 1998; Thakur et al., 2016). According to Meagher (2000), phytoremediation techniques may even be considered the only effective means of recovering the hundreds of thousands of square miles of land and water contaminated due to anthropogenic activities.

Although it is a relatively recent technology that emerged in the 1990s, the concept of phytoremediation has been known for some time, at least 300 years, and cannot be traced to any particular source (Gomes et al., 2016; Lasat, 2000; Salt et al., 1998). Advances in scientific and technology knowledge allowed a series of fascinating scientific discoveries, which, combined with an interdisciplinary research approach, resulted in the development of the phytoremediation idea into a promising, cost-effective, and environmentally friendly technology (Salt et al., 1998). Moreover, it is a solar-driven technology with widespread public acceptance (Amari et al., 2017). Hinchman et al. (1995) defined phytoremediation "as the engineered use of green plants, including grasses, forbs, and woody species, to remove, contain, or render harmless environmental contaminants as heavy metals, trace elements, organic compounds, and radioactive compounds in soil or water." Later, Greipsson (2011) complemented this definition, and described phytoremediation as the use of plants and associated soil microbes to reduce the concentrations or the toxic effects of contaminants in the environment.

Among the most hazardous contaminants that hinder the balance of ecosystems, nonessential trace metals, such as cadmium (Cd), stand out due to their ubiquitous presence, high toxicity, nondegradable nature (they produce long-lasting effects upon the environment even after major sources have been removed), and accumulative properties (Ali et al., 2013; Dhanakumar et al., 2015; Pedro et al., 2016). The capacity to accumulate, not only in the abiotic compartments (sediments and water), but also in the biotic (in the tissues of several living organisms), results in trace metals' biomagnification along food webs, threatening human health through ingestion (Lee et al., 2000; Wang et al., 2017).

Estuarine environments are critical transition zones that establish the link between land and freshwater habitats, and the ocean, which is tidally-influenced. Those zones are ecologically relevant, providing essential ecosystem services, including nutrient cycling and production; regulation of fluxes of nutrients, water, particles, and organisms to and from land, rivers, and the ocean; serving as a habitat for several species (birds, mammals, fish, thousands of invertebrates); providing nursery areas (including for several commercially valuable fish species); providing flood control (porous, resilient salt marsh soils and vegetation absorb floodwaters and dissipate storm surges) and erosion protection of the coastline; and offering water and sediment remediation (Correia et al., 2012; Levin et al., 2001; MacFarlane et al., 2007; Sousa et al., 2008).

Paradoxically, estuaries also represent habitats at risk from anthropogenic contamination, often being considered sinks for pollutants, especially for trace metals, receiving important inputs from agricultural runoff and urban and industrial sewage (Doyle and Otte, 1997; Geffard et al., 2005; Pedro et al., 2016). Anthropogenic pressures subject estuaries to several changes, namely due to contaminants' retention in bottom sediments, which can impair the quality and normal function of these ecosystems (Cheng et al., 2000), because trace metals can become bioavailable to benthic organisms, and can eventually be transferred to upper trophic levels, affecting the marine food chains (Warwick et al., 1998).

Salt marshes, an extraordinary component of the estuarine ecosystem, are normally settled on intertidal areas of mud and sand flats, which are stabilized by vegetation and highly influenced by tides. The accumulation of several contaminants, namely trace metals, in salt marshes sediments, as in other coastal environments, is a fact, as proved by the immense literature produced on this subject (see, e.g., Aderinola et al., 2009; Almeida et al., 2013 and literature therein; Du Laing et al., 2009; Monterroso et al., 2003; Pedro et al., 2016; Uluturhan et al., 2011). Salt marshes are recognized by intergovernmental agreements (e.g., Ramsar Convention on Wetlands, http://www.ramsar.org) and/or directives (e.g., EU Habitats Directive) that provide the framework for national action and international cooperation for the conservation and wise use of wetlands, including salt marshes, and their resources.

Among the several factors (such as sediment type, tides, redox, pH, and organic matter) that influence the distribution and accumulation of trace metals in salt marsh sediments, the presence of vegetation assumes a main role. According to Pedro et al. (2015), the marsh capacity as a sink for trace metals, namely Cd, is closely related with the marsh maturation process, which in turn is related with the abundance and development of the vegetation.

The presence of aquatic macrophytes, namely halophytes, is of uppermost importance, because they contribute to the structure, as well as the function, of the salt marsh. On the other hand, these plants also have the capability to remove and store both nutrients and pollutants, such as the toxic trace metals (see, e.g., Levin et al., 2001; Reboreda and Caçador, 2007; Weis and Weis, 2004), playing an important role in estuarine ecosystem remediation.

Many studies have been developed and published recently under the scope of the phytoremediation capability of halophyte species, highlighting the relevance of this subject in the ecosystems' sustainability and, consequently, the well-being of mankind. Taking into account the toxicity and consequences of trace metals, this chapter aims to provide a brief overview of the recent achievements and progress, and to explain the importance of halophyte species in environmental remediation, namely as regarding

Cd-contaminated estuarine environments. Concerning the advancement in knowledge and technology in this field, and some limitations presented by traditional phytoremediation strategies, some ecological recommendations are also reported. In this chapter, a survey of the published literature, as well as of the website Halophytes database eHALOPH (version 3.15), using different combinations of keywords (Cd phytoremediation, halophytes, estuarine environments contamination, phytoremediation technology), was performed. This allowed a reasonable output of potential publications that was sorted manually. Nevertheless, this survey was not exclusive, and other works developed under the scope of cadmium phytoremediation by halophytes were also analyzed.

## 2 PHYTOREMEDIATION, A NATURAL STRATEGY THAT CAN ASSIST IN THE PRESERVATION OF ESTUARINE ENVIRONMENTS

Nutrients and some metals, such as Cu (Copper), Fe (Iron), Mn (Manganese), and Ni (Nickel), are needed to support life, because they are essential for vital physiological and biochemical functions. However, when nutrients and essential metals' concentrations are unnaturally elevated in an aquatic system, it can lead, respectively, to eutrophication or toxic effects, with all of the implicit tragic consequences to the environment. In the case of nonessential metals, such as Cd, Pb (Lead), Hg (Mercury), and As (Arsenic), no functions in biological systems are known, and they are extremely toxic, even at very low concentrations.

There has been a growing concern, since the last decades of the 20th century, regarding the effects of industrial discharges on trace metal levels in nearshore waters and estuaries (Baeyens et al., 1998), with research efforts focused primarily in these high productivity environments, because they are often directly and exposed to contamination from upstream catchments, as well as from surrounding areas (Andreotti et al., 2015; Cohen et al., 2001).

According to Singh et al. (2003), over recent decades, the annual worldwide release of trace metals into the environment reached 22,000 t (metric ton) for Cd, 939,000 t for Cu, 783,000 t for Pb, and 1,350,000 t for Zn. By the middle of the 20th century, 50% of all Atlantic Coast estuaries had been altered in some way by human activities and, according to a recent study conducted by the European Commission, $1.17 \times 10^6$ identified contaminated sites were reported in 33 countries in Europe, and only 58 have been identified as remediated (Cicero-Fernández et al., 2015).

Although positive measures have been taken to reduce pollution impacts on estuarine environments (such as Directive 2008/105/CE), xenobiotic substances, such as trace metals, are continuously being introduced into these ecosystems, and the adverse effects in marine and estuarine biota still manifest. Remediation of trace metals contaminated systems, namely from estuarine sediments, is mostly necessary in order to minimize their impact on ecosystems and ensure the ecological balance. Conventional remediation methods, including physical and chemical techniques, suffer from several limitations, namely high costs and ecosystems disturbance, sometimes with irreversible consequences. Therefore, phytoremediation and its development comes up as the "green" solution to the problem of trace metals pollution, as demonstrated by the numerous studies that have been conducted in recent years. This solar-driven remediation strategy can be used for removal not only of metals and radionuclides, but also organic pollutants, such as polynuclear aromatic hydrocarbons, polychlorinated biphenyls, and pesticides (Ali et al., 2013). As regarding estuarine environments' phytoremediation, salt marsh halophytes assume the main role, due to their capacity to absorb and accumulate trace metals from the sediments, with some species showing a great potential to phytoremediate metal-contaminated sediments (see, e.g., Almeida et al., 2004; Weis and Weis, 2004). To a certain extent, salt marsh vegetation prevents more widespread contamination of coastal waters (Valiela et al., 2004), as well as helping prevent erosion and metal leaching (Chaudhry et al., 1998).

In addition to the ability to uptake and accumulate contaminants in their tissues, plants can also promote changes in the surrounding sediments, such as the redox potential (by pumping of atmospheric oxygen by the root system) (Koop-Jakobsen and Wenzhöfer, 2015), and metal speciation, and therefore can affect the bioavailability and toxicity of trace metals. For example, Cd is typically very mobile in the estuarine environment (Förstner et al., 1986); however, this trace metal tends to associate with Fe/Mn oxides, organic matter, and the residual fraction between plants roots, becoming immobilized and relatively inert over long periods of time (Pedro et al., 2015).

Plants can be used for phytoremediation by distinct physiological processes that allow them to uptake contaminants from the environment and accomplish their detoxification (Fig. 1). The main phytoremediation strategies can be categorized as: phytoextraction (same as phytoaccumulation, phytoabsorption, or phytosequestration), phytostabilization (or phytoimmobilization), phytofiltration, phytovolatilization, and phytodegradation (or phytotransformation) (Ali et al., 2013; Garbisu and Alkorta, 2001; Gomes et al., 2016; Lutts and Lefèvre, 2015). Because the main objective of

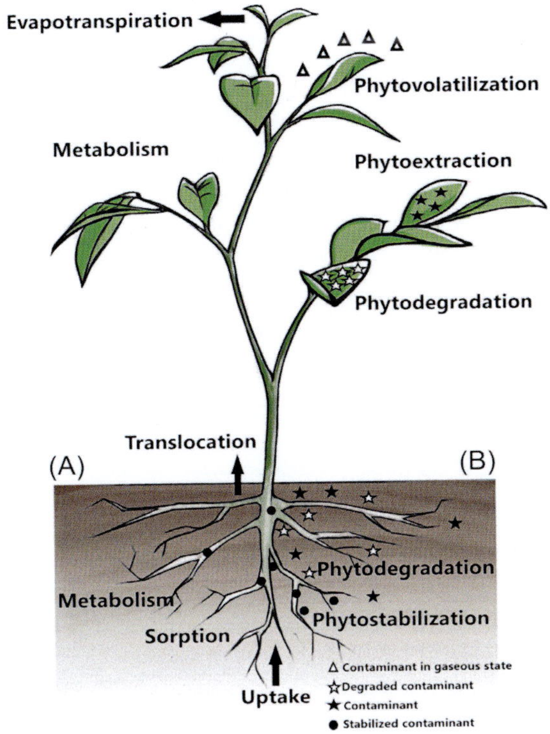


**Fig. 1** Plants' physiological processes implicit in phytoremediation (A) and different phytoremediation strategies (B). *(Based on Gomes, M.A., Da, C., Hauser-Davis, R., de Souza, A.N., Vitória, A.P., 2016. Metal phytoremediation: general strategies, genetically modified plants and applications in metal nanoparticle contamination. Ecotoxicol. Environ. Saf. 134, 133–147. doi:10.1016/j.ecoenv.2016.08.024.)*

the present work is Cd phytoremediation, only the useful and adequate remediation strategies for this trace metal will be described herein, i.e., phytoextraction, phytostabilization, and phytofiltration. In addition to being the most appropriate for Cd phytoremediation, those are also, among all strategies, commercially important (Thakur et al., 2016).

## 2.1 Phytoextraction

Phytoextraction involves the uptake of inorganic compounds, such as trace metals, from soil or water by a plant's underground organs, and their consequent translocation and accumulation in the above-ground organs, which means that trace metals will be concentrated in the most easily harvested parts (shoots and leaves). To complete the remediation and remove those pollutants from the environment, this technique comprises harvesting the plant, although generally only the above-ground organs are harvested (Ali et al., 2013; Chaudhry et al., 1998). These harvestable plants' biomass disposal, when contaminated with trace metals, should posteriorly suffer an adequate treatment, which involves a pretreatment (to reduce the volume) and incineration, washing, or liquid extraction as a final disposal. The final hazardous waste has to be deposited in specialized dumps or, if economically feasible, processed for phytomining (recovery of precious or semiprecious metals) (Ali et al., 2013; Lutts and Lefèvre, 2015; Sas-Nowosiemska et al., 2004) (Fig. 2). There is evidence that the trace metals phytoextraction market is expanding, in fact, according to Glass (2000), this expansion corresponds to a growth from 15 to 25 million USD in the year 2000 to 70–100 million USD in 2005. For the phytoextraction process to be economically viable and efficient, the plant must hyperaccumulate the contaminants, produce a large biomass (Chaudhry et al., 1998), and have a high capacity to translocate the metal to the shoots. In addition to plant species, phytoremediation efficiency is also influenced by metal bioavailability and speciation, as well as sediment characteristics (Ali et al., 2013; Li et al., 2010). However, the application of this phytoremediation strategy in any environmental monitoring project must consider the risk of food chain contamination, and that's why plants suitable for phytoextraction should be repulsive to herbivores (Sarma, 2011).

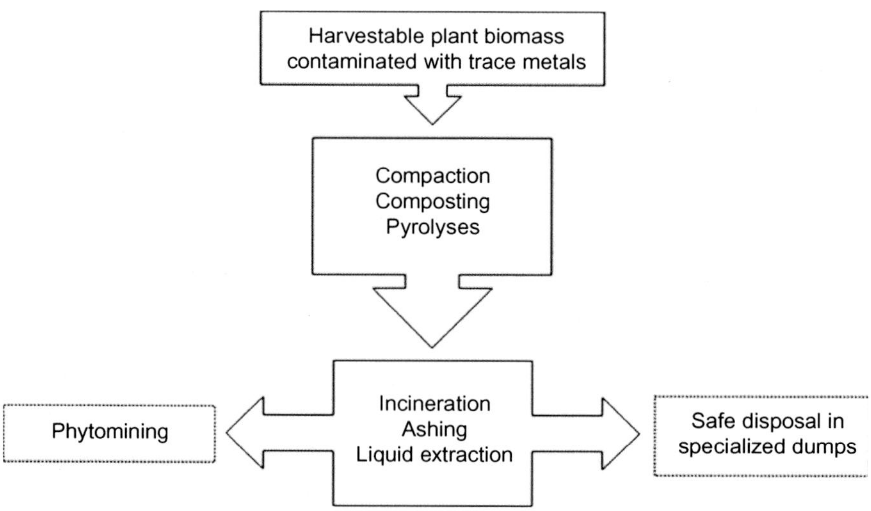


**Fig. 2** The most common treatment processes of postharvested plants used in phytoextraction.

## 2.2 Phytostabilization

The purpose of this phytoremediation strategy is not actually to remove contaminants, such as trace metals, but rather to stabilize them by accumulation in roots, or precipitation within the rhizosphere (Padmavathiamma and Li, 2007; Vangronsveld et al., 2009). Plants are used to immobilize the contaminants in the interface of roots and soil, or roots and water, and consequently, reduce their bioavailability (Ali et al., 2013; Chaudhry et al., 1998). This immobilization occurs through sorption by roots, precipitation, complexation, or metal valence reduction in the rhizosphere (Ghosh and Singh, 2005; Gomes et al., 2016), due to some plants' root exudates (Chaudhry et al., 1998) (Fig. 3). By excreting special redox enzymes, plants

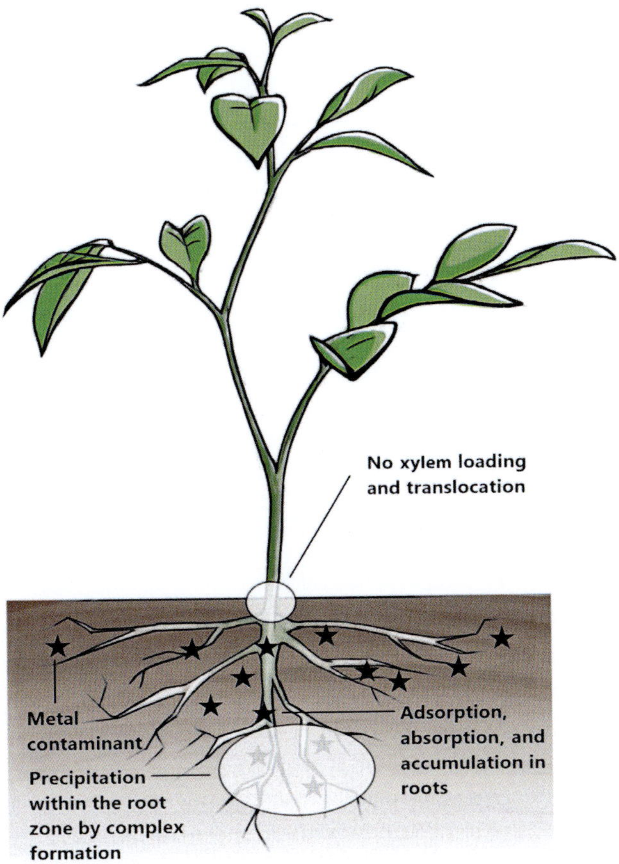


**Fig. 3** Phytostabilization mechanisms. *(Based on Padmavathiamma, P.K., Li, L.Y., 2007. Phytoremediation technology: hyper-accumulation metals in plants. Water Air Soil Pollut. 184, 105–126. doi:10.1007/s11270-007-9401-5.)*

skillfully convert hazardous metals to a relatively less toxic state, decreasing possible metal stress and damage (Ali et al., 2013). Although it's a temporary solution, because metals are only inactived and immobilized, phytostabilization prevents the contaminants' migration to groundwater, and reduces the risk of food chain contamination (Erakhrumen, 2007).

Among other characteristics, plants suitable for phytostabilization should have an extensive root system, and low translocation capacity (Krämer, 2005).

## 2.3 Phytofiltration

Phytofiltration may present some variant techniques, such as rhizofiltration (uses the plant root systems), blastofiltration (uses the seedlings), and caulofiltration (uses the excised plant shoots) (Gomes et al., 2016; Sarma, 2011). In any of these types of processes, pollutants are adsorbed from contaminated waters (surface waters, groundwater, or wastewaters) and are concentrated in the roots system, then precipitated as phosphates and carbonates, reducing their mobility (Salt et al., 1995; Chaudhry et al., 1998; Ali et al., 2013; Gomes et al., 2016). In particular, rhizofiltration has found innumerous applications, mainly using aquatic macrophytes (Gomes et al., 2016).

# 3 ECOLOGICAL CONSIDERATIONS

When it is intended to implement remediation strategies, namely in estuarine environments, it is important and essential to evaluate ecological issues, in order to avoid any unintended harm to the ecosystem, presently or in the future. In particular, one has to consider how the phytoremediation efforts might affect inhabiting fauna and flora, and the local ecological relationships (Padmavathiamma and Li, 2007). The basis used for evaluating and classifying the effectiveness of phytoremediation strategies is not standardized, and has to be evaluated on a case-by-case basis.

Remediation of estuarine environments, namely salt marsh waters and sediments, is crucial, not only for their conservation, but also to prevent the propagation of pollutants through the marine environments, as well as into food webs. Although phytoremediation is a promising approach for trace metals remediation, it suffers from some limitations, which should be carefully considered, because of the risk of food chain contamination in case of mismanagement and lack of proper care (Ali et al., 2013).

As regards trace metals, and as previously mentioned, there are several abiotic factors, as well as plants themselves, that influence their mobility, speciation, and consequently, bioavailability (Almeida et al., 2011;

Pedro et al., 2015; Weis and Weis, 2004). Consequently, phytoremediation efficiency is influenced by uneven distribution of contaminants, and by other factors, such as temperature, nutrients, precipitation and moisture, plant pathogens and herbivory, and rooting zone depth (Vangronsveld et al., 2009; Ali et al., 2013). Moreover, generally, ecosystems are polluted by a mixture of pollutants, and not just by a single one, which means that antagonistic and synergetic effects between pollutants should be taken into consideration when studying the remediation potential of plants (Almeida et al., 2013).

It is important to understand that technique selection has crucial ecological implications, because each phytoremediation technique leads to differences in the fate of the trace metals (retained in the sediments or accumulated in plants above- or below-ground organs). Also, it is essential to understand plants' capacity to tolerate contaminated environments and the mechanisms by which they take up trace metals, namely Cd, from the sediments and water, and accumulate or distribute it and detox it, because different plant species could present different patterns of trace metals' allocation (Briat, 2010; Idaszkin et al., 2017). For example, plants that accumulate metals in their aboveground tissues can act as phytoextractors; however, if these tissues are not harvested, metals may return into the environment, increasing their bioavailability (Reboreda and Caçador, 2007). Moreover, the risk of metals getting into food webs increases with the consumption of contaminated detritus by consumers (Burke et al., 2000). On the other hand, some halophytes, as *Spartina alterniflora* Brongn., have salt glands on their leaf surfaces that excrete salt, an important mechanism to maintain ionic balance in estuarine environments. These glands, however, can also excrete metals (Hg, Pb, and Cr). Burke et al. (2000) reported that *S. alterniflora* liberates larger quantities of metals into salt marsh ecosystems than does *Phragmites australis* (Cav.) Trin. Ex Steud (who generally concentrates metals in below-ground tissues), through both excretion and leaf deposition, an important fact to be taken into account in restoration efforts of contaminated salt marshes.

The knowledge of the different halophytes roles in estuaries is a critical requirement for the development of appropriate management strategies; however, it is still poorly described for Eurosiberian salt marshes (Almeida et al., 2011). According to Sarma (2011), the plant species selection is possibly the single most important factor for the success of the phytoremediation strategy. In fact, this author also states that the application of indigenous plant species for phytoremediation is often favored as it requires less management and acclimatizes successfully in native climate conditions and seasonal cycles. Moreover, the use of naturally present plants on the application of phytoremediation to recover impacted areas is important to preserve the natural biodiversity (da Silva et al., 2014a), as is the case of halophyte plants in contaminated estuarine environments' phytoremediation.

## 4 SALT MARSH HALOPHYTES POTENTIAL IN TRACE METALS PHYTOREMEDIATION

Salt marshes are complex coastal environments, usually located within estuarine systems, where they assume important structural and functional roles, as well as in regard to phytoremediation. The plants inhabiting these systems regulate many aspects of nutrient, particle, and organism dynamics above and below ground (Levin et al., 2001). Salt marshes are colonized by halophytic vegetation species, i.e., macrophytes adapted to complete their lifecycle in salty environments (Silvestri et al., 2005), where nonsalt tolerating plants (glycophytes) cannot grow or reproduce.

According to the United Nations Food and Agriculture Organization and the United Nations Environment Programme, there are currently about 4 million square kilometers of salinized land, and approximately 20% of agricultural land and 50% of crop land in the world is salt stressed (Manousaki and Kalogerakis, 2011). Therefore, halophytic plants represent a tremendous resource in biology to study the mechanisms of adaptation and evolution to/in hyperosmotic environments (Oosten and Maggio, 2015). Moreover, those plants show tolerance to high salinity levels, as well as to other toxic ions, including trace metals, offering greater potential for phytoremediation research (Manousaki and Kalogerakis, 2011). Some studies demonstrated that chloride salinity, mainly NaCl, increased Cd concentration in harvestable parts of cultivated plants (Lefèvre et al., 2009 and literature cited therein). Such a fact proves to be of extreme importance as regards phytoremediation purposes, namely phytoextraction, in estuarine environments. Furthermore, halophytes may have an additional advantage because they can be cultivated with saline irrigation water, which is a desirable feature because the quality of irrigation waters is decreasing (Manousaki and Kalogerakis, 2011).

Halophyte plants not only exhibit tolerance, but also influence the concentration and bioavailability of trace metals in sediments, namely Cd, being able to immobilize and store metals in sediments surrounding their roots (Almeida et al., 2011). Pedro et al. (2015) investigated the influence of vegetation (and it is absence) on metal mobility and availability in salt marsh sediments, and showed that there was an increase of Cd and Zn concentrations between roots in the sediment when compared with bulk sediments.

In the past few decades, the hyperaccumulation of trace metals in some plants has been recorded by many studies, which reveals the importance of further advanced research on a molecular basis of phytoremediation technology (Sarma, 2011). Hyperaccumulator plants are known to be able to accumulate (on dry weight basis) more than: 1% of zinc or manganese; 0.1% of nickel, copper, cobalt, arsenium, selenium and lead; and 0.01% of cadmium (Baker et al., 2000; Jiang et al., 2015; Lutts and Lefèvre, 2015; Redondo-Goméz et al., 2010; Rejeb et al., 2013), without exhibiting any detrimental symptoms in terms of growth or survival. Metal hyperaccumulation occurs in approximately 0.2% of all angiosperms (Krämer, 2010). As regards Cd hyperaccumulators, the majority of species belong to the family Brassicaceae, being a member of Crassulaceae (*Sedum alfredii* Hance), the only species outside (Sarma, 2011). Unfortunately, most, if not all of the known hyperaccumulators have a low annual biomass production, which renders these plants as poor candidates for phytoremediation. Moreover, those plants are salt-sensitive glycophyte plant species, and are therefore unable to survive in salt areas (Rejeb et al., 2013). On the other hand, the halophyte's protection mechanisms against salt stress is usually correlated with a more efficient antioxidant system (Kumari et al., 2015; Rangani et al., 2016), and therefore, these plants have been suggested to be naturally better adapted to cope with environmental stresses, including toxic trace metals (Manousaki and Kalogerakis, 2011).

The growing interest in metal tolerance of halophytes is mainly focused on their potential in phytoremediation of metal polluted, saline soils (Anjum et al., 2014; Oosten and Maggio, 2015; Wali et al., 2017). Moreover, in the past few years, some halophytes, as the hydrohalophyte *Arthrocnemum macrostachyum* (Moric.) K. koch, have proved to have the potential to apply for hyperaccumulators, which represents greater opportunities for phytoremediation research (Redondo-Goméz et al., 2010).

## 4.1 Fundamental Physiological Mechanisms of Tolerance to Trace Metals Accumulation

Trace metals can enter to root cells by the same uptake processes that move essential micronutrient metal ions (Ross and Kaye, 1994). For example, a competitive transport for Cd via voltage-gated cation (like $Ca^{2+}$) channels has been pointed out as a way of Cd absorption by roots (Raskin et al., 1997). The adaptive responses of plants to trace-metals-contaminated environments are efficient mechanisms that include many physiological, molecular, genetic, and ecological traits (Sarma, 2011). These mechanisms include the subcellular compartmentalization of the metal, and the sequestration of the metal by specially produced organic compounds, such as phytochelatins, concentrating metal in the plants roots (Ross and Kaye, 1994; Thakur et al., 2016). The major storage sites for Cd in plants are the cellular walls of roots cells, vacuoles of the epidermis, and the bundle sheath of leaves (Hu et al., 2010). For example, some halophyte species are succulents (e.g., *Salicornia*, *Sarcocornia*), which could sequester the metals in vacuoles in the aerial tissue, and in this way minimize the toxic effect of the accumulation of metals (Lokhande and Suprasanna, 2012).

Tolerance of halophytes to ionic and osmotic components of salt stress are linked to their ability to synthesize osmoprotectants in order to maintain a favorable water-potential gradient, and to protect cellular structures (Lefèvre et al., 2009; Lutts and Lefèvre, 2015; Wali et al., 2016). According to Przymusiński et al. (2004), tolerance to salt and to trace metals, such as Cd, may be related to common physiological mechanisms. Mechanisms underlying the salt-induced improvement of metal tolerance have especially been considered in the case of Cd (Wali et al., 2017). For example, studies by Lefèvre et al. (2009) on the halophyte *Atriplex halimus* L. suggest that chloride salinity improved the resistance to Cd toxicity through an increase in the synthesis of glycinebetaine, the most efficient osmoprotectant synthesized by Chenopodiaceae. In a study with *Sesuvium portulacastrum* (L.) L., Wali et al. (2016) concluded that in this halophyte, alleviation of Cd toxicity by NaCl is related to the modification of GSH (reduced glutathione) and proline contents, as well as stress hormone levels, thus protecting redox balance and photosynthesis. Other common tolerance mechanisms, besides the synthesis of antioxidants, is salt excretion (with the excretion of some metals through specialized salt glands onto the leaf surface) (Idaszkin et al., 2017; Manousaki and Kalogerakis, 2011, and literature cited therein). Fig. 4 summarizes the plant responses and mechanism of tolerance to several abiotic stresses. However, eventually, some considerations must be taken. According to Wali et al. (2017), who studied Cd effects in salt tolerance mechanisms in the halophyte *Sesuvium portulacastrum*, while the tolerance to high salinity allows the amelioration of Cd toxicity, Cd hampers salt tolerance mechanisms, leading to higher cytoplasmic $Na^+$ concentrations. Those results may reveal some limitations in the usefulness of halophytes in remediation strategies of saline environments, and suggest the need for further studies.

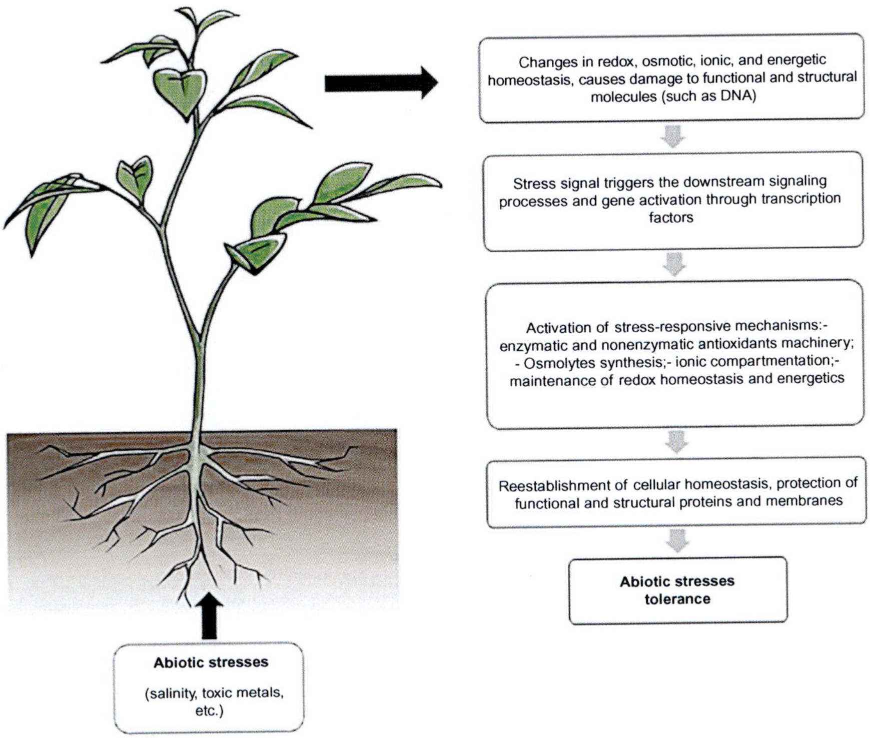


**Fig. 4** Generalized scheme for plant responses to abiotic stresses and mechanism of stress tolerance. *(Based on Lokhande, V.H., Suprasanna, P., 2012. Prospects of halophytes in understanding and managing abiotic stress tolerance. In: Ahmad, P., Prasad, M., (Eds.) Environmental Adaptations and Stress Tolerance of Plants in the Era of Climate Change. Springer, New York. doi:10.1007/978-1-4614-0815-4_2.)*

The reason why some plants have evolved the hyperaccumulation and metal tolerance phenotype is still not well understood. However, according to some authors (see, e.g., Boyd, 2007 and Poschendier et al., 2006), it could be related to some kind of metal defense function, such as poisoning plant tissues for herbivores and pathogens, known as the Elemental Defense Hypothesis.

## 4.2 Halophytes and Technology Advancement

Trace metals phytoremediation technology is still at the development stage. The most important challenge that lies ahead is how to improve the efficiency of phytoremediation by increasing the capacity of accumulated trace metals, or by improving key plant biological traits that should enhance those contaminants' uptake (Wu et al., 2009). Knowledge of the basic processes of metal transport, accumulation, and tolerance in both plants and other organisms is leading to ingenious strategies for the manipulation of these traits in transgenic plants, in order to improve their uptake, translocation, and tolerance, and therefore maximize the phytoremediation capacity (Cobbet, 2003; Kotrba et al., 2009). Genetically modified plants can be created to increase phytoremediation capabilities, showing advantages against both abiotic stress and the presence of trace metals in the environment (Gomes et al., 2016). In the perspective of phytoremediation, engineering plants able to take up more trace metals, namely Cd, from the sediments would require the use of plants that are more tolerant to trace metals, as well as being high-biomass producing and fast-growing (Briat, 2010; Pilon-Smiths and Freeman, 2006; Yang et al., 2005). To be suitable for phytoremediation, a plant should present a set of characteristics, which are illustrated in Fig. 5. While phytochelatins functioned as the first step in trace metals detoxification, plant vacuoles act as final detoxification storage; thus the identification and

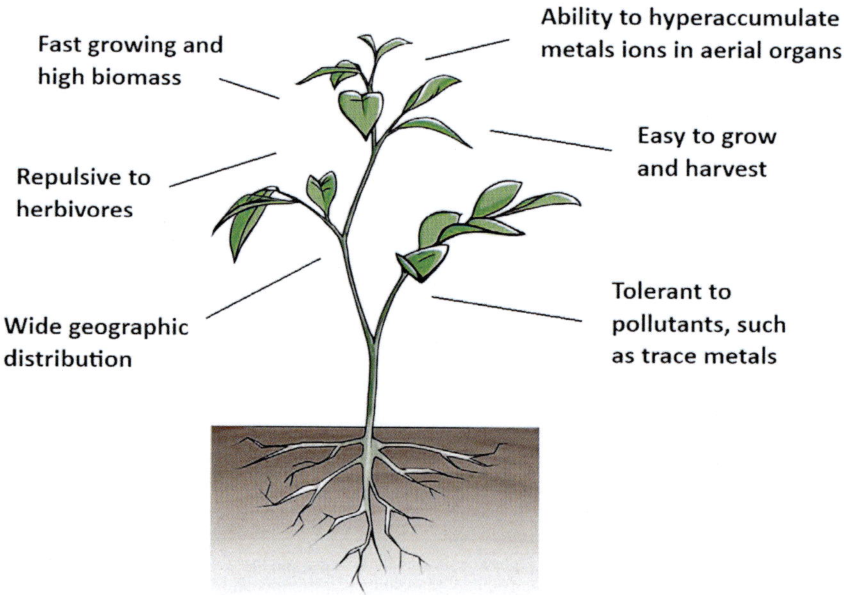


**Fig. 5** Characteristics exhibited by an ideal plant suitable for phytoremediation purposes. *(Based on Thakur, S., Singh, L., Wahid, Z.A., Siddiqui, M.F., Atnaw, S.M., Din, M.F.M., 2016. Plant-driven removal of heavy metals from soil: uptake, translocation, tolerance mechanism, challenges and future perspectives. Environ. Monit. Assess. 188–206. doi:10.1007/s10661-016-5211-9.)*

modulation of vacuolar phytochelatins transporters is of primary importance (Song et al., 2010).

Distinct scientific and technical challenges are implicit in trace metals phytoremediation advancement, because those contaminants cannot be degraded into nonharmful products, but still can be removed or sequestered from the sediments or water (Jiang et al., 2015). The intense research developed recently concerning phytoremediation allowed a significant improvement, namely as regards hyperaccumulator species responses, at the molecular and physiological level, to trace metal toxicity, tolerance, and accumulation. Those mechanisms, as well as underlying molecular and genetic determinants for these traits, are being well established in many organisms, and provide a solid basis for the selection of suitable genes to be (over)expressed in plants to improve their trace metal remediation capacity (Kotrba et al., 2009). Recent advances in the development of transgenic plants for trace metal remediation, namely as regards Cd, have been reported in a review by Kotrba et al. (2009). However, the potential ecological impact inherent in application of phytoremediation with transgenic plants should be carefully evaluated.

Another way to increase plants' tolerance and trace metals uptake, and thus improve metal phytoremediation, is by inoculation of plants with rhizospheric microorganisms resistant to trace metals (Khan, 2005). In a study developed by da Silva et al. (2014a), the rhizosphere of two salt marsh plants (*Juncus maritimus* Lam. and *P. australis*) was inoculated with a microbial consortium containing microorganisms resistant to Cd, and an amazing increase in this trace metal accumulation (up to seven times in *P. australis* stems and 48% in *J. maritimus* below ground structures) was observed. These results indicate that autochthonous bioaugmentation could be a valuable strategy to potentiate Cd phytoremediation by salt marsh halophytes.

Trace metals ions mobility in soil, and thus availability to plants, mainly depends on the pH and the presence of chelating agents, such as ethylene-diamine tetraacetic acid (EDTA) (Thakur et al., 2016). For this purpose, chelators such as EDTA, hydroxyethyl-ethylenediamine triacetic acid, ethylene-bis (oxyethylenetrinitrilo) tetraacetic acid, diethylene-triamine pentaacetic acid, ethylene-diamine disuccinic acid (EDDS), and soil acidification using organic acids have been investigated at the laboratory scale (Vangronsveld et al., 2009). In this context, Duarte et al. (2007) reported a gradual increase in Cd uptake by *Halimione portulacoides* (L.) Aellen in Verh. roots, with the increase in citric acid concentration added.

Once more, the ecological impact has to be considered, because those chelators, and the soil acidification, could be toxic to plants, and can cause groundwater pollution. In fact, the use of chelating agents such as EDTA is not allowed for field applications in many European Community countries (Vangronsveld et al., 2009).

## 5 CADMIUM TOXICITY

Cadmium (Cd) is a silver white metal, that is slightly bluish, belonging to the family of transition metals, and has eight stable natural isotopes, and an atomic weight of 112.4 (Amari et al., 2017). This trace metal enters the environment by natural and human processes, however, human sources, such as mining and urban processes, are responsible for contributing approximately 90% of the cadmium found in surface waters (US EPA, 2016). The main cadmium sources are nonferrous metal mining and refining, manufacture and application of phosphate fertilizers, fossil fuel combustion, and waste incineration and disposal (ATSDR, 2012). Recently, cadmium has been used in manufacturing nanoparticles used as a semiconductor in photovoltaic devices (Balmuri et al., 2017; US EPA, 2016). Due to its toxicity to humans and terrestrial and aquatic organisms, as well as the constant increase in the environment and bioaccumulation through the food chain, Cd has become the focus of intense research and regulation globally, being a high-priority contaminant (Amari et al., 2017; Malley, 1996). Cadmium is classified as a priority hazardous substance by the European Parliament and by the Council (Directive 2008/105/EC), US Environmental Protection Agency (US EPA, 2001), being also included in the OSPAR List of Chemicals for Priority Action (OSPAR, 2002).

Upon entering the freshwater or estuarine/marine aquatic environments, cadmium becomes strongly adsorbed into clays, muds, organic matter, and Fe-Mn hydrous oxides, which could be a determining factor in the retention of Cd in sediments. Cadmium adsorption-desorption reaction characteristics may control the amount of this trace metal in the soil solution and, consequently, affect Cd mobility, reactivity, and potential toxicity (Tian et al., 2017). However, the adsorption of Cd onto sediment is strongly influenced, not just by organic matter, but also by other abiotic factors, such as redox potential, pH, and the presence of attendant anions (Ololade and Ologundudu, 2007). For example, the presence of $Cl^-$ reduces Cd sorption in sediments, due to the formation of Cd-Cl complexes, as demonstrated in several studies (Tian et al., 2017). Also, the presence of some plant species, as mentioned before, may affect the bioavailability and toxicity of this contaminant.

The toxic effects of Cd on biological systems have been reported by several authors (see, e.g., Ali et al., 2013; Balmuri et al., 2017). As mentioned before, Cd is a nonessential metal with no biological function, and is recognized as an extremely significant pollutant due to its high toxicity to most organisms, with toxicities 2–20 times higher than many other trace metals (Vassilev et al., 1998), being the fourth most toxic to vascular plants (Ghosh and Singh, 2005).

Cadmium toxicity is attributed to the ion release as $Cd^{2+}$, which is absorbed by biological tissues (McGeer et al., 2011; Pedro et al., 2016). Apart from other toxicity mechanisms, such as increased oxidative stress, changes in the activity of antioxidant enzymes, and modifications to gene expression, $Cd^{2+}$ is known to interfere with the absorption of nutrients, as the essential metal $Ca^{2+}$, potentially affecting other regulated processes in cells and tissues of aquatic organisms (Balmuri et al., 2017; Lu et al., 2018). In aquatic animals, besides acute effects such as mortality, chronic exposure to cadmium can lead to adverse effects on growth, reproduction, immune and endocrine systems, development, and behavior (US EPA, 2016). The excess of Cd accumulation in plants can profoundly interfere with metabolism, affecting respiration, photosynthesis, and stomata opening and growth (He et al., 2017 and literature cited therein), which are a few of the reasons for decreased agriculture yield worldwide (Sharma et al., 2010). Similar to what happens with other toxic trace metals, Cd may disrupt the cellular membrane structure and permeability and affect the electron transport chain (Panda et al., 2017). Regarding human health, the contamination of food chains deserves special attention, because this is one of the main sources of Cd exposure to humans. According to Filipič (2012), this trace metal is a human carcinogen, teratogen, and mutagen, and it also can cause irreversible renal injury, osteoporosis, nonhypertrophic emphysema, anemia, eosinophilia, insomnia, and rhinitis.

## 6 USING HYDROHALOPHYTES AS FEASIBLE TOOLS IN CADMIUM PHYTOREMEDIATION

Inhabitants of estuarine environments have to deal with harsh conditions, every day. Salt marsh halophytes' ability to survive under those stress conditions allowed the development of several adaptation mechanisms to saline environments, as well as to flooding and pollution. As mentioned before, these environments are intensely subjected to anthropogenic pressures, especially as regards contamination, sometimes continuously, by trace metals.

Halophytes provide several ecosystem services, as already mentioned, highlighting the capacity to remediate trace metals, such as Cd and, thus reduce this toxic metal bioavailability into the other biota, as well as the dispersion to other aquatic/semiaquatic/terrestrial environments. While the ability of hyperaccumulator plants to phytoremediate polluted soils has been receiving increased attention, the natural ability of native halophytes' communities presents an important role in the dynamics of metals, dictating their mobility within the salt marsh environment, and to the surrounding ecosystems (Castro et al., 2009). Some salt marsh halophytes are able to stabilize and accumulate Cd from the water and the sediment because of their location, physiology, and defense mechanisms against trace metals' toxicity. Phytoextraction and phytostabilization are the two most referred to phytoremediation strategies performed by these plants (Alkorta et al., 2010; Garbisu and Alkorta, 2001; Manousaki and Kalogerakis, 2011; Nanda-Kumar et al., 1995).

Despite the ability of hyperaccumulator plants to phytoremediate trace metals, the natural potential of native salt marsh plant communities to remediate and stabilize these toxic contaminants from the surrounding water and sediments is an added value, especially given the adverse conditions existing in estuarine environments (Castro et al., 2009). Cadmium phytoremediation capacity of halophytes inhabiting estuarine environments has been triggering the attention of the scientific community and the general public, as demonstrated by the several studies developed and published. Considering the main objective of the present work, a selection of some halophytes, specifically hydrohalophytes, widespread in the salt marshes of temperate climates, was performed and presented in Table 1, evidencing their potential for Cd phytoremediation in salt marshes, some of which have already been mentioned throughout this work.

## 7 CONCLUSIONS

As an environmentally friendly technology, phytoremediation is a promising solution, or at least a strategy able to attenuate the impact of trace metals contamination, namely by cadmium in estuarine environments. Although considerable knowledge has already been recorded in recent decades, more fundamental research is still necessary to better understand and exploit the

**Table 1** Hydrohalophytes species with Cd phytoremediation potential

| Family | Scientific name | Common name | References |
|---|---|---|---|
| Amaranthaceae | *Arthrocnemum macrostachyumn* (Moric.) K. Koch | Glaucous glasswort | Conesa et al. (2011) and Redondo-Goméz et al. (2010) |
| | *Atriplex portulacoides* L. (syn. *Halimione portulacoides* (L.) Aellen in Verh.) | Sea purslane | Caçador et al. (2009), Duarte et al. (2010), Pedro et al. (2015), Reboreda and Caçador (2007), and Sousa et al. (2008) |
| | *Halosarcia indica* (Willd.) Paul G. Wilson (syn. *Salicornia brachiata* Roxb.) | nf | Sharma et al. (2010) |
| | *Salicornia europaea* L. | Common glasswort | Ozawa et al. (2009) |
| | *Salicornia ramosissima* J. Woods | Purple glasswort | Pedro et al. (2013) and Pérez-Romero et al. (2016) |
| | *Sarcocornia fructicosa* (L.) A.J. Scott | Shrubby swampfire | Caçador et al. (2009), Conesa et al. (2011), Duarte et al. (2010), and Pedro et al. (2015) |
| | *Sarcocornia perennis* (Mill.) A.J. Scott | Perennial glasswort | Caçador et al. (2009), Curado et al. (2014b), and Duarte et al. (2010) |
| | *Suaeda maritima* (L.) Dumort. | Herbaceous seepweed | El-Amier et al. (2017) |
| | *Suaeda monoica* Forssk. Ex J.F. Gmel | nf | Ayyappan and Ravindran (2014) |
| Cyperaceae | *Bolboschoenus maritimus* (L.) Palla (syn. *Scirpus maritimus* L.) | Sea clubrush | Madejón et al. (2006), Santos et al. (2015), and Teuchies et al. (2013) |

*Continued*

**Table 1** Hydrohalophytes species with Cd phytoremediation potential—cont'd

| Family | Scientific name | Common name | References |
|---|---|---|---|
| Juncaceae | *Juncus maritimus* Lam. | Sea rush | Almeida et al. (2004), Almeida et al. (2011), Conesa et al. (2011), da Silva et al. (2014a, b), Teixeira et al. (2014), and Teuchies et al. (2013) |
| Juncaginaceae | *Triglochin striata* Ruiz *et* Pav. | Streaked arrow grass | Almeida et al. (2011) |
| Poaceae | *Phragmites australis* (Cav.) Trin. Ex Steud | Common reed | Almeida et al. (2011), Cicero-Fernández et al. (2015), Cicero-Fernandés et al. (2017), Conesa et al. (2011), da Silva et al. (2014a,b), Laing et al. (2009), Southichak et al. (2006), Teixeira et al. (2014), Teuchies et al. (2013), and Ye et al. (1997) |
| | *Spartina densiflora* Brongn | Denseflower glasswort | Domínguez et al. (2010) |
| | *Spartina maritima* (Curtis) Fernald | Small cordgrass | Caçador et al. (2009), Curado et al. (2014a), Duarte et al. (2010), Pedro et al. (2015), and Reboreda and Caçador (2007) |
| | *Spartina patens* (Aiton) Muhl | Saltmeadow cordgrass | Almeida et al. (2011) |

nf, not found.

metabolic diversity of the halophytes themselves. Also, more data are still needed to quantify the underlying economics as support for public acceptance and to convince policy makers and stakeholders.

To improve phytoremediation efficiency, it is also important to understand the complex interactions between plants, sediments, contaminants, and microorganisms in the rhizosphere. The properties and capacities described for the halophytes confirm the fact that these plants will receive increasing attention by phytoremediation researchers in the future. It is important to emphasize the need for further studies, namely in genomic, proteomic, and metabolomics approaches for metal or salinity tolerance mechanisms in salt marsh halophytes. The technology advancement increases the phytoremediation strategies' efficiency for metal contaminated ecosystems, namely in estuarine environments.

## REFERENCES

Aderinola, O.J., Clarke, E.O., Olarinmoye, O.M., Kusemiju, V., Anatekhai, M.A., 2009. Heavy metals in surface water, sediments, fish and periwinkles of Lagos Lagoon. Am. Eurasian J. Agric. Environ. Sci. 5, 609–617.

Ali, H., Khan, E., Sajad, M.A., 2013. Phytoremediation of heavy metals: concepts and applications. Chemosphere 91, 869–881. https://doi.org/10.1016/j.chemosphere.2013.01.075.

Alkorta, I., Becerril, J.M., Garbisu, C., 2010. Phytostabilization of metal contaminated soils. Rev. Environ. Health 25 (2), 135–146. https://doi.org/10.1515/REVEH.2010.25.2.135.

Almeida, C.M.R., Mucha, A.P., Vasconcelos, M.T.S.D., 2004. Influence of the sea rush *Juncus maritimus* on metal concentration and speciation in estuarine sediment colonized by the plant. Environ. Sci. Technol. 38 (11), 3112–3118. https://doi.org/10.1021/es049932j.

Almeida, C.M.R., Mucha, A.P., Vasconcelos, M.T., 2011. Role of different salt marsh plants on metal retention in an urban estuary (Lima estuary, NW Portugal). Estuar. Coast. Shelf Sci. 91, 243–249. https://doi.org/10.1016/j.ecss.2010.10.037.

Almeida, C.M.R., Rocha, A.C., Mucha, A.P., Vasconcelos, M.T.S., 2013. Evaluation of the potential of salt marsh plants for metal phytoremediation in estuarine environment. In: -Gupta, D., Corpas, F., Palma, J. (Eds.), Heavy Metal Stress in Plants. Springer, Berlin, Heidelberg, pp. 225–239. https://doi.org/10.1007/978-3-642-38469-1_12.

Amari, T., Ghnaya, T., Abdelly, C., 2017. Nickel, cadmium and lead phytotoxicity and potential of halophytic plants in heavy metal extraction. S. Afr. J. Bot. 111, 99–110. https://doi.org/10.1016/j.sajb.2017.03.011.

Andreotti, F., Mucha, A.P., Caetano, C., Rodrigues, P., Gomes, C.R., Almeida, C.M.R., 2015. Interactions between salt marsh plants and Cu nanoparticles—effects on metal uptake and phytoremediation processes. Ecotoxicol. Environ. Saf. 120, 303–309. https://doi.org/10.1016/j.ecoenv.2015.06.017.

Anjum, N.A., Ahmad, I., Válega, M., Mohmood, I., Gill, S.S., Tuteja, N., Duarte, A.C., Pereira, E., 2014. Salt marsh halophyte services to metal-metalloide remediation: assessment of the processes and underlying mechanisms. Crit. Rev. Environ. Sci. Technol. 44, 2038–2106. https://doi.org/10.1080/10643389.2013.828271.

ATSDR, Agency for Toxic Substances and Disease Registry, U.S. Department of Health and Human Services, 2012. Toxicological Profile for Cadmium. Agency for Toxic Substances and Disease Registry, Atlanta, GA.

Ayyappan, D., Ravindran, K.C., 2014. Potentiality of *Suaeda monoica* (Forsk.). A salt marsh halophyte on bioaccumulation of heavy metals from tannery effluent. Int. J. Modn. Res. Revs. 2 (9), 267–274.

Baeyens, W., Elskens, M., Ryssen, R.V., Leermakers, M., 1998. The impact of the Scheldt input on the trace metal distribution in the Belgian coastal area (results of 1981–1983 and 1995–1996). Hydrobiologia 366, 91–108. https://doi.org/10.1007/978-94-017-3573-5_6.

Baker, A.J.M., McGrath, S.P., Reeves, R.D., Smith, J.A.C., 2000. Metal hyperaccumulator plants: a review of the ecology and physiology of a biological resources for phytoremediation of metal polluted soils. In: Terry, N., Banuelos, G.S. (Eds.), Phytoremediation of Contaminated Soil and Water. CRC Press, Boca Raton, pp. 85–107.

Balmuri, S.R., Selvaraj, U., Kumar, V.V., Anthony, S.P., Tsatsakis, A.M., Golokhvast, K.S., Raman, T., 2017. Effect of surfactant in mitigating cadmium oxide nanoparticle toxicity: implications for mitigating cadmium toxicity in environment. Environ. Res. 152, 141–149. https://doi.org/10.1016/j.envres.2016.10.005.

Boyd, R.S., 2007. The defense hypothesis of elemental hyperaccumulation: status, challenges and new directions. Plant Soil 293, 153–176. https://doi.org/10.1007/s11104-007-9240-6.

Briat, J.-F., 2010. Arsenic tolerance in plants: pas de deux between phytochelatin synthesis and ABCC vacuolar transporters. Proc. Natl. Acad. Sci. U. S. A. 107, 20853–20854. https://doi.org/10.1073/pnas.1016286107.

Burke, D.J., Weis, J.S., Weis, P., 2000. Release of metals by the leaves of the salt marsh grasses *Spartina alterniflora* and *Phragmites australis*. Estuar. Coast. Shelf Sci. 51, 153–159. https://doi.org/10.1006/ecss.2000.0673.

Caçador, I., Caetano, M., Duarte, B., Vale, C., 2009. Stock and losses of trace metals from salt marsh plants. Mar. Environ. Res. 67, 75–82. https://doi.org/10.1016/j.marenvres.2008.11.004.

Castro, R., Pereira, S., Lima, A., Corticeiro, S., Válega, M., Pereira, E., Duarte, A., Figueira, E., 2009. Accumulation, distribution and cellular partitioning of mercury in several halophytes of a contaminated salt marsh. Chemosphere 76, 1348–1355. https://doi.org/10.1016/j.chemosphere.2009.06.033.

Chaudhry, T.M., Hayes, W.J., Khan, A.G., Khoo, C.S., 1998. Phytoremediation—focusing on accumulator plants that remediate metal-contaminated soils. Australas. J. Ecotoxicol. 4, 37–51.

Cheng, C.Y., Stemberger, R.S., Klaue, B., Blum, J.D., Pickhardt, P.C., Folt, C.L., 2000. Accumulation of heavy metals in food web components across a gradient of lakes. Limnol. Oceanogr. 45 (7), 1525–1536. https://doi.org/10.4319/lo.2000.45.7.1525.

Cheraghi, M., Lorestani, B., Yousefi, N., 2011. Introduction of hyperaccumulator plants with phytoremediation potential of a lead–zinc mine in Iran. World Acad. Sci. Eng. Technol. 77https://doi.org/10.1999/1307-6892/1833.

Cicero-Fernandés, D., Peña-Fernández, M., Expósito-Camargo, J.A., Antizar-Ladislao, B., 2017. Long-term (two annual cycles) phytoremediation of heavy metal-contaminated estuarine sediments by *Phragmites australis*. New Biotechnol. 38, 56–64. https://doi.org/10.1016/j.nbt.2016.07.011.

Cicero-Fernández, D., Peña-Fernández, M., Expósito-Camargo, J.A., Antizar-Ladislao, B., 2015. Role of *Phragmites australis* (common reed) for heavy metals phytoremediation of estuarine sediments. Int. J. Phytoremediation. https://doi.org/10.1080/15226514.2015.1086306.

Cobbet, C., 2003. Heavy metals and plants—model systems and hyperaccumulators. New Phytol. 159 (2), 289–293. https://doi.org/10.1046/j.1469-8137.2003.00832.x.

Cohen, T., Hee, S.S.Q., Ambrose, R.F., 2001. Trace metals in fish and invertebrates of three California coastal wetlands. Mar. Pollut. Bull. 42 (3), 224–232. https://doi.org/10.1016/S0025-326X(00)00146-6.

Conesa, H.M., María-Cervantes, A., Álvarez-Rogel, J., Gonzáles-Alcaraz, M.N., 2011. Influence of soil properties on trace element availability and plant accumulation in a Mediterranean salt marsh polluted by mining wastes: implications for phytomanagement. Sci. Total Environ. 409, 4470–4479. https://doi.org/10.1016/j.scitotenv.2011.07.049.

Correia, M.J., Costa, J.L., Cainho, P., Félix, P.M., Chaves, M.L., Medeiros, J.P., Silva, G., Azeda, C., Tavares, P., Costa, A., Costa, A.M., Bernardo, J., Cabral, H.N., Costa, M.J., Fonseca, L.C., 2012. Inter-annual variations of macrobenthic communities over three decades in a land-locked coastal lagoon (Santo André, SW Portugal). Estuar. Coast. Shelf Sci. 110, 168–175. https://doi.org/10.1016/j.ecss.2012.04.028.

Cristaldi, A., Conti, G.O., Jho, E.H., Zuccarello, P., Grasso, A., Copat, C., Ferrante, M., 2017. Phytoremediation of contaminated soils by heavy metals and PAHs. A brief review. Environ. Technol. Innov. 8, 309–326. https://doi.org/10.1016/j.eti.2017.08.002.

Curado, G., Rubio-Casal, A.E., Figueroa, E., Castillo, J.M., 2014a. Potential of *Spartina marítima* in restored salt marshes for phytoremediation of metals in a highly polluted estuary. Int. J. Phytoremediation 16, 1209–1220. https://doi.org/10.1080/15226514.2013.821451.

Curado, G., Grewell, B.J., Figueroa, E., Castillo, J.M., 2014b. Effectiveness of the aquatic halophyte *Sarcocornia perennis* spp. *Perennis* as a biotool for ecological restoration of salt marsh. Water Air Soil Pollut. 225 (9), 1–14. https://doi.org/10.1007/s11270-014-2108-5.

da Silva, M.N., Mucha, A.P., Rocha, A.C., Teixeira, C., Gomes, C.R., Almeida, C.M.R., 2014a. A strategy to potentiate Cd phytoremediation by saltmarsh plants—autochthonous bioaugmentation. J. Environ. Manag. 134, 136–144. https://doi.org/10.1016/j.jenvman.2014.01.004.

da Silva, M.N., Mucha, A.P., Rocha, A.C., Silva, C., Carli, C., Gomes, C.R., Almeida, C.M.R., 2014b. Evaluation of the ability of two plants for the phytoremediation of Cd in salt marshes. Estuar. Coast. Shelf Sci. 141, 78–84. https://doi.org/10.1016/j.ecss.2014.03.004.

Dhanakumar, S., Solaraj, G., Mohanraj, R., 2015. Heavy metal partitioning in sediments and bioaccumulation in commercial fish species of three major reservoirs of river Cauvery delta region, India. Ecotoxicol. Environ. Saf. 113, 145–151. https://doi.org/10.1016/j.ecoenv.2014.11.032.

Directive 2008/105/EC of the European Parliament and of the Council of 16 December, 2008. On environmental quality standards in the field of water policy, amending and subsequently repealing Council Directives 82/176/EEC, 83/513/EEC, 84/513/EEC, 84/491/EEC, 88/280/EEC and amending Directive 2000/60/EC of the European Parliament and of the Council. O. J. L. 348 (24.12.2008), 84–97.

Dixit, R., Wasiullah, Malaviya, D., Pandiyan, K., Singh, U.B., Sahu, A., Shukla, R., Singh, B.P., Rai, J.P., Sharma, P.K., Lade, H., Paul, D., 2015. Bioremediation of heavy metals from soil and aquatic environment: an overview of principles and criteria of fundamental processes. Sustainability 7, 2189–2212. https://doi.org/10.3390/su7022189.

Domínguez, D.M., García, F.C., Raya, A.C., Santiago, R.T., 2010. Cadmium-induced oxidative stress and the response of the antioxidative defense system in *Spartina densiflora*. Physiol. Plant. 139, 289–302. https://doi.org/10.1111/j.1399-3054.2010.01.368.x.

Doyle, M.O., Otte, M.L., 1997. Organism-induced accumulation of iron, zinc and arsenic in wetlands soils. Environ. Pollut. 96 (1), 1–11. https://doi.org/10.1016/S0269-7491(97)00014-6.

Du Laing, G., Rinklebe, J., Vandecasteele, B., Meers, E., Tack, F.M.G., 2009. Trace metal behavior in estuarine and riverine floodplain soils and sediments: a review. Sci. Total Environ. 407 (13), 3972–3985. https://doi.org/10.1016/j.scitotenv.2008.07.025.

Duarte, B., Delgado, M., Caçador, I., 2007. The role of citric acid in cadmium and nickel uptake and translocation, in *Halimione portulacoides*. Chemosphere 69, 836–840. https://doi.org/10.1016/j.chemosphere.2007.05.007.

Duarte, B., Caetano, M., Almeida, P.R., Vale, C., Caçador, I., 2010. Accumulation and biological cycling of heavy metal in four salt marsh species, from Tagus estuary (Portugal). Environ. Pollut. 158, 1661–1668. https://doi.org/10.1016/j.envpol.2009.12.004.

El-Amier, Y.A., Alghanem, S.M., Alzuaibr, F.M., 2017. Bioaccumulation and translocation of heavy metals from coastal soil by wild halophytes. Am. J. Environ. Prot. 5 (2), 52–60. https://doi.org/10.12691/env-5-2-4.

Erakhrumen, A.A., 2007. Phytoremediation: an environmentally sound technology for pollution prevention, control and remediation in developing countries. Educ. Res. Rev. 2 (7), 151–156.

Filipič, M., 2012. Mechanisms of cadmium induced genomic instability. Mutat. Res. 733, 69–77. https://doi.org/10.1016/j.mrfmmm.2011.09.002.

Förstner, U., Ahlf, W., Calmano, W., Kersten, M., Salomons, W., 1986. Mobility of heavy metals in dredged harbor sediments. In: Sly, P.G. (Ed.), Sediments and Water Interactions. Springer, New York.

Garbisu, C., Alkorta, I., 2001. Phytoextraction: a cost-effective plant-based technology for the removal of metals from the environment. Bioresour. Technol. 77 (3), 229–236. https://doi.org/10.1016/S0960-8524(00)00108-5.

Geffard, A., Smith, B.D., Amiard-Triquet, C., Jeantet, A.Y., Rainbow, P.S., 2005. Kinetics of trace metal accumulation and excretion in the polychaete *Nereis diversicolor*. Mar. Biol. 147, 1291–1304. https://doi.org/10.1007/s00227-005-0044-z.

Ghosh, M., Singh, S.P., 2005. A review on phytoremediation of heavy metals and utilization of its products. Appl. Ecol. Environ. Res. 3 (1), 1–18.

Glass, D.J., 2000. Economic potential of phytoremediation. In: Raskin, I., Ensley, B.D. (Eds.), Phytoremediation of Toxic Metals: Using Plants to Clean up the Environment. John Wiley & Sons, New York, pp. 15–31.

Gomes, M.A., Da, C., Hauser-Davis, R., de Souza, A.N., Vitória, A.P., 2016. Metal phytoremediation: general strategies, genetically modified plants and applications in metal nanoparticle contamination. Ecotoxicol. Environ. Saf. 134, 133–147. https://doi.org/10.1016/j.ecoenv.2016.08.024.

Greipsson, S., 2011. Phytoremediation. Nat. Educ. Knowl. 3 (10), 7.

He, S.Y., Yang, X.E., He, Z., Baligar, V.C., 2017. Morphological and physiological responses of plants to cadmium toxicity: a review. Pedosphere 27 (3), 421–438. https://doi.org/10.1016/S1002-0160(17)60339-4.

Hinchman, R.R., Negri, M.C., Gatliff, E.G., 1995. Phytoremediation: Using Green Plants to Clean up Contaminated Soil, Groundwater and Wastewater. Argonne National Laboratory Hinchman, Applied Natural Sciences, Inc.

Hu, J.Z., Zheng, A.Z., Pei, D.L., Shi, G.X., 2010. Bioaccumulation and chemical forms of cadmium, copper and lead in aquatic plants. Braz. Arch. Biol. Technol. 53 (1), 235–240. https://doi.org/10.1590/S1516-89132010000100029.

Idaszkin, Y.L., Lancelotti, J.L., Pollicelli, M.P., Marcovecchio, J.E., Bouza, P.J., 2017. Comparison of phytoremediation potential capacity of *Spartina densiflora* and *Sarcocornia perennis* for metal polluted soils. Mar. Pollut. Bull. 118, 297–306. https://doi.org/10.1016/j.marpolbul.2017.03.007.

Jiang, Y., Lei, M., Duan, L., Longhurst, P., 2015. Integrating phytoremediation with biomass valorization and critical element recovery: a UK contaminated land perspective. Biomass Bioenergy 83, 328–339. https://doi.org/10.1016/j.biombioe.2015.10.013.

Khan, A.G., 2005. Role of soil microbes in the rhizospheres of plants growing on trace metal contaminated soil in phytoremediation. J. Trace Elem. Med. Biol. 18, 355–364. https://doi.org/10.1016/j.jtemb.2005.02.006.

Koop-Jakobsen, K., Wenzhöfer, F., 2015. The dynamics of plant-mediated sediment oxygenation in *Spartina anglica* rhizosphere—a planar optode study. Estuar. Coasts 38, 951–963. https://doi.org/10.1007/s12237-014-9861-y.

Kotrba, P., Najmanova, J., Macek, T., Ruml, T., Mackova, M., 2009. Genetically modified plants in phytoremediation of heavy metal and metalloid soil and sediment pollution. Biotechnol. Adv. 22, 799–810. https://doi.org/10.1016/j.biotechadv.2009.06.003.

Krämer, U., 2005. Phytoremediation: novel approaches to cleaning up polluted soils. Curr. Opin. Biotechnol. 16, 133–141. https://doi.org/10.1016/j.copbio.2005.02.006.

Krämer, U., 2010. Metal hyperaccumulation in plants. Annu. Rev. Plant Biol. 61, 517–534. https://doi.org/10.1146/aanurev-arplant-042809-112156.

Kumari, A., Das, P., Parida, A.K., Agarwal, P.K., 2015. Proteomics, metabolomics, and ionomics perspectives of salinity tolerance in halophytes. Front. Plant Sci. 6(537) https://doi.org/10.3389/fpls.2015.00537.

Laing, G.D., Van de Moortel, A.M.K., Moors, W., De Grauwe, P., Meers, E., Tack, F.M.G., Verloo, M.G., 2009. Factors affecting metal concentrations in reed plants (*Phragmites australis*) of intertidal marshes in the Scheldt estuary. Ecol. Eng. 35, 310–318. https://doi.org/10.1016/j.ecoleng.2008.01.002.

Lasat, M.M., 2000. Phytoremediation of metals from contaminated soil: a review of plant/soil/metal interaction and assessment of pertinent agronomic issues. J. Hazard. Subst. Res. 2, 1–25.

Lee, B.-G., Lee, J.-S., Luoma, S.N., Choi, H.J., Koh, C.-H., 2000. Influence of acid volatile sulphide and metal concentrations on metal bioavailability to marine invertebrates in contaminated sediments. Environ. Sci. Technol. 34 (21), 4517–4523. https://doi.org/10.1021/es001033h.

Lefèvre, I., Marchal, G., Meerts, P., Corréal, E., Lutts, S., 2009. Chloride salinity reduces cadmium accumulation by the Mediterranean halophyte species *Atriplex halimus* L. Environ. Exp. Bot. 65, 142–152. https://doi.org/10.1016/j.envexpbot.2008.07.005.

Levin, L.A., Boesch, D.F., Covich, A., Dahm, C., Erséus, C., Ewel, K.C., Kneib, R.T., Moldenke, A., Palmer, M.A., Snelgrove, P., Strayer, D., Weslawski, J.M., 2001. The function of marine critical transition zones and the importance of sediment biodiversity. Ecosystems 4, 430–451. https://doi.org/10.1007/s10021-001-0021-4.

Li, J.T., Liao, B., Lan, C.Y., Ye, Z.H., Baker, A.J.M., Shu, W.S., 2010. Cadmium tolerance and accumulation in cultivars of a high-biomass tropical tree (*Averrhoa carambola*) and its potential for phytoextraction. J. Environ. Qual. 39 (4), 1262–1268. https://doi.org/10.2134/jeq2009.0195.

Lokhande, V.H., Suprasanna, P., 2012. Prospects of halophytes in understanding and managing abiotic stress tolerance. In: Ahmad, P., Prasad, M. (Eds.), Environmental Adaptations and Stress Tolerance of Plants in the Era of Climate Change. Springer, New Yorkhttps://doi.org/10.1007/978-1-4614-0815-4_2.

Lu, Q., Zhang, T., Zhang, W., Su, C., Yang, Y., Hu, D., Xu, Q., 2018. Alleviation of cadmium toxicity in *Lemna minor* by exogenous salicylic acid. Ecotoxicol. Environ. Saf. 147, 500–508. https://doi.org/10.1016/j.ecoenv.2017.09.015.

Lutts, S., Lefèvre, I., 2015. How can we take advantage of halophyte properties to cope with heavy metal toxicity in salt-affected areas? Ann. Bot. 115, 509–538. https://doi.org/10.1093/aob/mcu264.

MacFarlane, G.R., Koller, C.E., Blomberg, S.P., 2007. Accumulation and partitioning of heavy metals in mangroves: a synthesis of field-based studies. Chemosphere 69, 1454–1464. https://doi.org/10.1016/j.chemosphere.2007.04.059.

Madejón, P., Murillo, J.M., Marañóin, T., Espinar, J.L., Cabrera, F., 2006. Accumulation of As, Cd and selected trace elements in tubers of *Scirpus maritimus* L. from Doñana marshes (South Spain). Chemosphere 65 (5), 742–748. https://doi.org/10.1016/j.chemosphere.2005.11.032.

Malley, D.F., 1996. Cadmium whole-lake experiment at the experimental lakes area: an anachronism? Can. J. Fish. Aquat. Sci. 53 (8), 1862–1870. https://doi.org/10.1139/f96-116.

Manousaki, E., Kalogerakis, N., 2011. Halophytes present new opportunities in phytoremediation of heavy metals and saline soils. Ind. Eng. Chem. Res. 50, 656–660. https://doi.org/10.1021/ie100270x.

McGeer, J.C., Niyogi, S., Smith, D.S., 2011. 3-Cadmium. In: Wood, C.M., Farrel, A.P., Brauner, C.J. (Eds.), Fish Physiology. In: vol. 31. Academic Press, pp. 125–184. https://doi.org/10.1016/S1546-5098(11)31025-4.

Meagher, R.B., 2000. Phytoremediation of toxic elemental and organic pollutants. Curr. Opin. Plant Biol. 3, 153–162.

Monterroso, P., Pato, P., Pereira, E., Vale, C., Duarte, A.C., 2003. Distribution and accumulation of metals (Cu, Cd, Zn and Pb) in sediments of a lagoon on the northwestern coast of Portugal. Mar. Pollut. Bull. 46 (9), 1200–1205. https://doi.org/10.1016/S0025-326X(03)00243-1.

Mulligan, C.N., Yong, R.N., Gibbs, B.F., 2001. Remediation technologies for metal-contaminated soils and groundwater: an evaluation. Eng. Geol. 60, 193–207 PII: S0013-7952(00)00101-0.

Nanda-Kumar, P.B.A., Dushenkov, V., Motto, H., Raskin, I., 1995. Phytoextraction: the use of plants to remove heavy metals from soils. Environ. Sci. Technol. 29, 1232–1238.

Ololade, L.A., Ologundudu, A., 2007. Concentration and bioavailability of cadmium by some plants. Afr. J. Biotechnol. 6 (16), 1916–1921. https://doi.org/10.5897/AJB2007.000-2290.

Oosten, M.J.V., Maggio, A., 2015. Functional biology of halophytes in the phytoremediation of heavy metal contaminated soils. Environ. Exp. Bot. 111, 135–146. https://doi.org/10.1016/j.envexpbot.2014.11.010.

OSPAR, 2002. Cadmium. Hazardous Substances Series. OSPAR Commission 2002. OSPAR publication number 151/2002.

Ozawa, T., Miura, M., Fukuda, M., Kakuta, S., 2009. Cadmium Tolerance and Accumulation in a Halophyte *Salicornia europaea* as a New Candidate for Phytoremediation of Saline Soils. vol. 60. Scientific Report of the Graduate School of Life and Environmental Sciences, Osaka Prefecture University, pp. 1–8.

Padmavathiamma, P.K., Li, L.Y., 2007. Phytoremediation technology: hyper-accumulation metals in plants. Water Air Soil Pollut. 184, 105–126. https://doi.org/10.1007/s11270-007-9401-5.

Panda, A., Rangani, J., Kumari, A., Parida, A.K., 2017. Efficient regulation of arsenic translocation to shot tissue and modulation of phytochelatin levels and antioxidative defense system confers salinity and arsenic tolerance in the halophyte *Suaeda maritima*. Environ. Exp. Bot. 143, 149–171. https://doi.org/10.1016/j.envexpbot.2017.09.007.

Pedro, C.A., Santos, M.S.S., Ferreira, S.M.F., Gonçalves, S.C., 2013. The influence of cadmium contamination and salinity on the survival, growth and phytoremediation capacity of the saltmarsh plant *Salicornia ramosissima*. Mar. Environ. Res. 92, 197–205. https://doi.org/10.1016/j.marenvres.2013.09.018.

Pedro, S., Duarte, B., de Almeida, P.R., Caçador, I., 2015. Metal speciation in salt marsh sediments: influence of halophyte vegetation in salt marshes with different morphology. Estuar. Coast. Shelf Sci. 167, 248–255. https://doi.org/10.1016/j.ecss.2015.05.034.

Pedro, C.A., Santos, M.S.S., Ferreira, S.M.F., Gonçalves, S.C., 2016. The presence of cadmium in the intertidal environments of a moderately impacted coastal lagoon in western Portugal (Óbidos Lagoon)—spatial and seasonal evaluations. Environ. Sci. Pollut. Res. 23, 1960–1969. https://doi.org/10.1007/s11356-015-5847-y.

Pérez-Romero, J.A., Redondo-Gómez, S., Mateos-Naranjo, E., 2016. Growth and photosynthetic limitation analysis of the cd-accumulator *Salicornia ramosissima* under excessive cadmium concentrations and optimum salinity conditions. Plant Physiol. Biochem. 109, 103–113. https://doi.org/10.1016/j.plaphy.2016.09.011.

Pilon-Smiths, E.A.H., Freeman, J.L., 2006. Environmental cleanup using plants: biotechnological advances and ecological considerations. Front. Ecol. Environ. 4 (4), 203–210. https://doi.org/10.1890/1540-9295(2006)004.

Poschendier, C., Tolrà, R., Barceló, J., 2006. Can metals defend plants against biotic stress? Trends Plant Sci. 11(6)https://doi.org/10.1016/j.tplants.2006.04.007.

Przymusiński, R., Rucińska, R., Gwóźdź, E.A., 2004. Increased accumulation of pathogenesis-related proteins in response of lupine roots to various abiotic stress. Environ. Exp. Bot. 52, 53–61. https://doi.org/10.1016/j.envexpbot.2004.01.006.

Rangani, J., Parida, A.K., Panda, A., Kumari, A., 2016. Coordinated changes in antioxidative enzymes protect the photosynthetic machinery from salinity induced oxidative damage and confer salt tolerance in an extreme halophyte *Salvadora persica* L. Front. Plant Sci. 7 (50)https://doi.org/10.3389/fpls.2016.00050.

Raskin, I., Smith, R.D., Salt, D.E., 1997. Phytoremediation of metals: using plants to remove pollutants from the environment. Plant Biotechnol. 8, 221–226. https://doi.org/10.1016/S0958-1669(97)80106-1.

Reboreda, R., Caçador, I., 2007. Halophyte vegetation influences in salt marsh retention capacity for heavy metals. Environ. Pollut. 146, 147–154. https://doi.org/10.1016/j.envpol.2006.05.035.

Redondo-Goméz, S., Mateos-Naranjo, E., Andrades-Moreno, L., 2010. Accumulation and tolerance characteristics of cadmium in a halophytic Cd-hyperaccumulator, *Arthrocnemum macrostachyum*. J. Hazard. Mater. 184, 299–307. https://doi.org/10.1016/j.hazmat.2010.08.036.

Rejeb, K.B., Ghnaya, T., Zaier, H., Benzarti, M., Baioui, R., Ghabriche, R., Wali, M., Lutts, S., Abdelly, C., 2013. Evaluation of the $Cd^{2+}$ phytoextraction potential in the xerohalophyte *Salsola kali* L. and the impact of EDTA on this process. Ecol. Eng. 60, 309–315. https://doi.org/10.1016/j.ecoleng.2013.07.026.

Ross, S.M., Kaye, K.J., 1994. The meaning of metal toxicity in soil-plant system. In: Ross, S.M. (Ed.), Toxic Metals in Soil-Plant System. John Wiley & Sons, New York, pp. 27–61.

Salt, D.E., Blaylock, M., Kumar, N.P.B.A., Dushenkov, V., Ensley, B.D., Chet, I., Raskin, I., 1995. Phytoremediation: a novel strategy for the removal of toxic metals from the environment using plants. Biotechnology 13 (5), 468–474. https://doi.org/10.1038/nbt0595-468.

Salt, D., Smith, R.D., Raskin, I., 1998. Phytoremediation. Annu. Rev. Plant Physiol. Plant Mol. Biol. 49, 643–668.

Santos, M.S.S., Pedro, C.A., Gonçalves, S.C., Ferreira, S.M.F., 2015. Phytoremediation of cadmium by the facultative halophyte plant Bolboschoenus maritimus (L.) Palla, at different salinities. Environ. Sci. Pollut. Res. 22 (20), 15598–15609. https://doi.org/10.1007/s11356-015-4750-x.

Sarma, H., 2011. Metal hyperaccumulation in plants: a review focusing on phytoremediation technology. J. Environ. Sci. Technol. 4 (2), 118–138. https://doi.org/10.3923/jest.2011.118.138.

Sas-Nowosiemska, A., Kucharski, R., Małkowski, E., Pogrzeba, M., Kuperberg, J.M., Kryński, K., 2004. Phytoextraction crop disposal—an unsolved problem. Environ. Pollut. 128, 373–379. https://doi.org/10.1016/j.envpol.2003.09.012.

Sharma, A., Gontia, I., Agarwal, P.K., Jha, B., 2010. Accumulation of heavy metals and its biochemical responses in *Salicornia brachiate*, an extreme halophyte. Mar. Biol. Res. 6 (5), 511–518. https://doi.org/10.1080/17451000903434064.

Silvestri, S., Defina, A., Marani, M., 2005. Tidal regime, salinity and salt marsh plant zonation. Estuar. Coast. Shelf Sci. 62, 119–130. https://doi.org/10.1016/j.ecss.2004.08.010.

Singh, O.V., Labana, S., Pandey, G., Budhiraja, R., Jain, R.K., 2003. Phytoremediation: an overview of metallic ion decontamination from soil. Appl. Microbiol. Biotechnol. 61, 405–412. https://doi.org/10.1007/s00253-003-1244-4.

Song, W.-Y., Park, J., Mendoza-Cózatl, D.G., Suter-Grotemeyer, M., Shim, D., Hörtensteiner, S., Geisler, M., Weder, B., Rea, P.A., Rentsch, D., Schroeder, J.I., Lee, Y., Martinoia, E., 2010. Arsenic tolerance in *Arabidopsis* is mediated by two ABCC-type phytochelatin transporters. Proc. Natl. Acad. Sci. U. S. A. 107 (49), 21187–21192. https://doi.org/10.1073/pnas.1013964107.

Sousa, A.I., Caçador, I., Lillebø, A.I., Pardal, M.A., 2008. Heavy metal accumulation in *Halimione portulacoides*: intra- and extra-cellular metal binding sites. Chemosphere 70, 850–857. https://doi.org/10.1016/j.chemosphere.2007.07.012.

Southichak, B., Nakano, K., Nomura, M., Chiba, N., 2006. *Phragmites australis*: a novel biosorbent for the removal of heavy metals from aqueous solution. Water Res. 40, 2295–2302. https://doi.org/10.1016/j.watres.2006.04.027.

Teixeira, C., Almeida, C.M.R., da Silva, M.N., Bordalo, A.A., Mucha, A.P., 2014. Development of autochthonous microbial consortia for enhanced phytoremediation of salt-marsh sediments contaminated with cadmium. Sci. Total Environ. 493, 757–765. https://doi.org/10.1016/j.scitotenv.2014.06.040.

Teuchies, J., Jacobs, S., Oosterlee, L., Bervoets, L., Meire, P., 2013. Role of plants in metal cycling in a tidal wetland: implications for phytoremediation. Sci. Total Environ. 445-446, 146–154. https://doi.org/10.1016/j.scitotenv.2012.11.088.

Thakur, S., Singh, L., Wahid, Z.A., Siddiqui, M.F., Atnaw, S.M., Din, M.F.M., 2016. Plant-driven removal of heavy metals from soil: uptake, translocation, tolerance mechanism, challenges and future perspectives. Environ. Monit. Assess. 188-206https://doi.org/10.1007/s10661-016-5211-9.

Tian, H., Kong, L., Megharaj, M., He, W., 2017. Contribution of attendant anions on cadmium toxicity to soil enzymes. Chemosphere 187, 19–26. https://doi.org/10.1016/j.chemosphere.2017.08.073.

Uluturhan, E., Kontas, A., Can, E., 2011. Sediment concentrations of heavy metals in the Homa Lagoon (Eastern Aegean Sea): assessment of contamination and ecological risks. Mar. Pollut. Bull. 62 (9), 1989–1997. https://doi.org/10.1016/j.marpolbul.2011.06.019.

US EPA (US Environmental Protection Agency), 2001. Update of Ambient Water Quality Criteria for Cadmium, EPA-822-R-01-001.

US EPA (US Environmental Protection Agency), 2016. Aquatic Life Ambient Water Quality Criteria Update for Cadmium, EPA-822-F-16-003.

Valiela, I., Rutecki, D., Fox, S., 2004. Salt marshes: biological controls of food webs in a diminishing environment. J. Exp. Mar. Biol. Ecol. 300, 131–159. https://doi.org/10.1016/j.jembe.2003.12.023.

Vangronsveld, J., Herzig, R., Weyens, N., Boulet, J., Adriaensen, K., Ruttens, A., Thewys, T., Vassilev, A., Meers, E., Nehnevajova, E., van der Lelie, D., Mench, M., 2009. Phytoremediation of contaminated soils and groundwater: lessons from the field. Environ. Sci. Pollut. Res. 16, 765–794. https://doi.org/10.1007/s11356-009-0213-6.

Vassilev, A., Tsonev, T., Yordanov, I., 1998. Physiological response of barley plants (*Hordeum vulgare*) to cadmium contamination in soil during ontogenesis. Environ. Pollut. 103, 287–293 PII: S0269-7491(98)00110-9.

Wali, M., Gunsè, B., Llugany, M., Corrales, I., Abdelly, C., Poschenrieder, C., Ghnaya, T., 2016. High salinity helps the halophyte *Sesuvium portulacastrum* in defense against Cd toxicity by maintaining redox balance and photosynthesis. Planta 244, 333–346. https://doi.org/10.1007/s00425-016-2515-5.

Wali, M., Martos, S., Pérez-Martín, L., Abdelly, C., Ghnaya, T., Poschenrieder, C., Gunsé, B., 2017. Cadmium hampers salt tolerance of *Sesuvium portulacastrum*. Plant Physiol. Biochem. 115, 390–399. https://doi.org/10.1016/j.plaphy.2017.04.014.

Wang, Y., Huang, Q., Lemckert, C., Ma, Y., 2017. Laboratory and field magnetic evaluation of the heavy metal contamination on Shilaoren Beach, China. Mar. Pollut. Bull. 117, 291–301. https://doi.org/10.1016/j.marpolbul.2017.01.080.

Warwick, P., Hall, A., Pashley, V., Lee, J.V., Maes, A., 1998. Zinc and cadmium mobility in sand: effects of pH, speciation cation exchange capacity (CEC), humic acid and metal ions. Chemosphere 36, 2283–2290. https://doi.org/10.1016/S0045-6535(97)10197-7.

Weis, J.S., Weis, P., 2004. Metal uptake, transport and release by wetland plants: implications for phytoremediation and restoration. Environ. Int. 30, 685–700. https://doi.org/10.1016/j.envint.2003.11.002.

Wu, H., Tang, S., Zhang, X., Guo, J., Song, Z., Tian, S., Smith, D.L., 2009. Using elevated $CO_2$ to increase the biomass of a *Sorghum vulgare* x *Sorghum vulgare var. sudanense* hybrid and *Trifolium pretense* L. and to trigger hyperaccumulation of cesium. J. Hazard. Mater. 170, 861–870. https://doi.org/10.1016/j.jhazmat.2009.05.069.

Yang, X., Feng, Y., He, Z., Stoffella, P.J., 2005. Molecular mechanisms of heavy metal hyperaccumulation and phytoremediation. J. Trace Elem. Med. Biol. 18, 339–353. https://doi.org/10.1016/j.jtemb.2005.02.007.

Ye, Z.H., Baker, A.J.M., Wong, M.H., Willis, A.J., 1997. Zinc, lead and cadmium tolerance, uptake and accumulation by the common reed, *Phragmites australis* (Cav.) Trin. Ex Steudel. Ann. Bot.-London 80, 363–370. https://doi.org/10.1006/anbo.1997.0456.

# INDEX

Note: Page numbers followed by *f* indicate figures and *t* indicate tables.

## D



## E

## F

## G



## H

## N

Agriculture



## T